Fins into Limbs

Fins into Limbs

Evolution, Development, and Transformation

Edited by Brian K. Hall

The University of Chicago Press
Chicago and London

Brian K. Hall is the George S. Campbell Professor of Biology at Dalhousie University. He is the author of many books, including *Evolutionary Developmental Biology, The Neural Crest in Development and Evolution,* and *Bones and Cartilage: Developmental and Evolutionary Skeletal Biology;* he is editor of *Homology: The Hierarchical Basis of Comparative Biology,* and coeditor of the three-volume *The Skull* and *Variation: A Central Concept in Biology.*

The University of Chicago Press, Chicago 60637
The University of Chicago Press, Ltd., London
© 2007 by The University of Chicago
All rights reserved. Published 2007
Printed in the United States of America

16 15 14 13 12 11 10 09 08 07   1 2 3 4 5

ISBN-13: 978-0-226-31336-8 (cloth)
ISNB-13: 987-0-226-31337-5 (paper)
ISBN-10: 0-226-31336-0 (cloth)
ISBN-10: 0-226-31337-9 (paper)

Library of Congress Cataloging-in-Publication Data

Fins into limbs : evolution, development, and transformation / edited by
    Brian K. Hall.
        p.   cm.
    Includes bibliographical references and index.
    ISBN-13: 978-0-226-31336-8 (cloth : alk. paper)
    ISBN-10: 0-226-31337-9 (pbk. : alk. paper)
    ISBN-10: 0-226-31336-0 (cloth : alk. paper)
    1. Extremities (Anatomy)—Evolution. I. Hall, Brian Keith, 1941–
QL950.7F56 2007
573.9′9833—dc22
                                                2006011177

This book is printed on acid-free paper.

Contents

Introduction

Brian K. Hall

Birds in a way resemble fishes. For birds have their wings in the upper part of their bodies and fishes have two fins in the front part of their bodies. Birds have feet on their under part and most fishes have a second pair of fins in their under-part and near their front fins.
—Aristotle, *De Incessu Animalium*

Recognition of the homology between fish fins and tetrapod limbs was known to philosopher-naturalists such as Aristotle over 2,700 years ago. "Modern" studies can be traced back to morphological studies that predate the publication of Darwin's *On the Origin of Species* in 1859. A classic study is the 1849 monograph *The Nature of Limbs*, by Richard Owen, which is to be reprinted by the University of Chicago Press (Owen 1849 [2007]).

In placing his study into the context of the anatomical sciences, Owen wrote, "I should define the present lecture as being: 'On the general and Serial Homologies of the Locomotive Extremities'" (Owen 1849, 2). Owen was concerned with the essential nature of fins and limbs as homologous elements. In recognizing homologies and in seeking unity of type, Owen was following a philosophical approach whose origins are Aristotelian. Monographs and popular accounts continue to explore the consequences of this homology (Hinchliffe and Johnson 1980; Hinchliffe et al. 1991; Zimmer and Buell 1998; Clack 2002b).

Owen used the word "Nature" in the title of his talk "in the sense of the German '*Bedeutung*' [signification] as signifying that essential character of a part which belongs to it in its relation to a predetermined pattern, answering to the 'idea' of the Archetypal World in the Platonic cosmogony, which archetype or primal pattern is the basis supporting all the modifications of such part" (2–3). Despite this affirmation of transformation only within the type, the last paragraph of Owen's text has been taken as indicating a glimmer of transformation between type, for which see discussions by Amundson (2007) and Hall (2007).

Fins and limbs (where limbs are defined as paired appendages with digits) are homologous as paired appendages. I should say *paired* fins—the median unpaired fins of amphibian larvae and fish larvae and adults are only discussed in passing. Owen (1849) recognized this homology: "The 'limbs' . . . are the parts called the 'arms' and 'legs' in Man; the 'fore-' and 'hind-legs' of beasts; the 'wings' and 'legs' of Bats and Birds; the 'pectoral fins' and 'ventral [pelvic] fins' of Fishes" (3), and he took for granted that "the arm of the Man is the fore-leg of the Beast, the wing of the Bird, and the pectoral fin of the Fish" (3) and that these are homologous parts.

At a second level in the biological hierarchy, the cartilaginous elements of fish fins are homologous with the most proximal (humerus/femur) and next most proximal (tibia-fibula/radius/ulna) elements of limbs. At a third level, the epithelial-mesenchymal interactions that initiate fin and limb buds, and at a fourth, the cellular condensations from which these cartilages arise in fins and limb, are homologous. Finally, the gene networks and cascades that underlie fin and limb development share a remarkable homology.

Although these five levels of homology justify discussing fins and limbs within a single volume, fins are not limbs. The most striking structural difference between the two types of appendages is that fins possess bony fin rays (lepidotrichia) that limbs lack, while limbs possess digits (and wrist/ankle elements, although this is more controversial) that fins lack.

As fins and limbs are homologous, and as tetrapods (vertebrates with limbs) arose from fish, the most likely scenario is that limbs arose from fins (although other scenarios have been proposed). As I argued elsewhere (Hall 2005), a shorthand way of viewing this transformation is that "fins minus fin rays plus digits equal limbs."

All the skeletal elements of tetrapod limbs are derived from embryonic mesoderm, as are the cartilaginous elements of fish fins. Fin rays are derived from cells of another germ layer, the neural crest. Transformation of fins to limbs therefore involved (again in shorthand) "suppression of the neural crest (fin-ray) component and elaboration of a distal mesodermal component from which digits arose."

Presentation, analysis, evaluation, and discussion of the wealth of fascinating detail underlying and supporting these two shorthand comments is a major aim of this book, which elaborates five major themes concerning fins and limbs:

- their development, growth, structure, maintenance, function, regeneration, and evolution;
- the transformation of fins to limbs at the origin of the tetrapods;
- transformation of limbs to flippers in those reptiles and mammals that became secondarily aquatic and of limbs to wings in flying tetrapods;
- adaptations associated with other specialized modes of life such as digging and burrowing; and
- reduction in digit number or loss of limbs in some taxa.

Reflecting major themes, the book is organized into three parts—evolution, development, and transformation. Throughout, the emphasis is on the skeletons of fins and limbs. Other organ systems—muscular, nervous, vascular, ligamentous, and tendinous—either are not considered or are treated only in passing. This is a book about the appendicular skeleton—the development, evolution, and transformation of fins and limbs.

The first chapter, by Peter Bowler, places fins and limbs into the context of studies spanning the 100 years between 1840 and 1940 and lays out the major themes and issues that concerned past works and continue to concern us today. These themes and issues include transformation of characters and of taxa; how fins and limbs arose; identification of the group from which amphibians arose; and functional, adaptive, and ecological explanations of transformation/evolution, all of which remain as alive today as they were 150 years ago, and all of which are addressed in this book. Bowler ends his analysis with the comment that this "short history of how biologists tackled the question of how the vertebrates emerged onto land illustrates the depth of the questions, and, despite over 150 years of concentrated effort, the comparative shallowness of our understanding of the causes of this remarkable transition," leaving the other authors to show how our understanding has advanced in the last decades.

Chapter 2 outlines our understanding of the first major transformation, which was from fins to limbs. The major structural changes are set out and illustrated beautifully. Chapter 3 examines the functions of fins and limbs as locomotory appendages and considers how approaches to that functional role have changed over the years. It provides the necessary historical perspective on limb function against which readers can evaluate the anatomical approaches summarized in chapter 1 with chapter 4, the final chapter in part 1 (Evolution), which examines fins and limbs in the context

of evolutionary novelty and innovation. If fins minus fin rays plus digits equal limbs, then digits are evolutionary novelties. Wrists and ankles may also be novelties. Formation of an additional digit (polyphalangy) may also constitute a novelty, depending on how the extra digit(s) arises. A duplicated digit V is not a novelty. Origination of a digit VI or transformation of a carpal bone or sesamoid to a digit are novelties.

Because chapter 4 is as much an analysis of limb development as it is a perspective on limb evolution, it forms a logical link to part 2 (Development). The eight chapters in part 2 deal with the development of fins and limbs, mostly during embryonic life but with discussion of postnatal growth and regeneration. Current understanding of the molecular underpinnings of fin and limb development is discussed in chapter 5. Neither the older literature on cell and tissue interactions nor the extensive experimental studies on normal and mutant embryos are discussed. For these topics see DeHaan and Ursprung (1965), Milaire (1974), Hall (1978, 2005a), Hinchliffe and Johnson (1980), Kelley et al. (1982), and Fallon and Caplan (1983).

Because skeletogenesis varies across taxa, chapters 6 and 7 treat chondro- and osteogenesis of fins and limbs in some detail. Chapter 8 provides a brief evaluation of the important role played by cell death (apoptosis) in fin and limb development. How joints arise and how endochondral ossification modulates postnatal growth are discussed in chapters 9 and 10. Regeneration of fins and limbs is the topic of chapters 11 and 12. Alert readers will see that the perspective in these two chapters is developmental and mechanistic rather than evolutionary. This was not an oversight by the authors but a response to the request to provide syntheses of our understanding of regeneration in the two classes of vertebrate paired appendages.

The seven chapters in part 3 (Transformation) introduce examples of transformation of fins and/or of limbs in evolutionary, adaptive, functional, and developmental contexts. Because the transformation of fins into limbs was associated with the origin of the first tetrapods—of amphibians—and because multiple lineages developed limbs, the evolution of amphibian limb skeletons is discussed in depth in chapter 13. Indeed, as the most detailed and thoughtful analysis available on this topic, this chapter provides an exemplary introduction to part 3. It may be usefully read in conjunction with chapter 2, which analyzes the evolutionary origin of limbs, and with the description in the journal *Nature* (2006, 440, 750–63) by Edward Daeschler and colleagues of the discovery in the Canadian Arctic of *Tiktaalik roseae*, a Devonian fishlike member of the tetrapod stem-group, with a mosaic of features intermediate between a fish with fins and a tetrapod with limbs. This animal—not quite a fish and not a full-limbed tetrapod—has the potential for a great deal of information regarding changes in fin-limb structure during the fish-to-tetrapod transition.

Chapters 14 and 18 are the two chapters that deal with aspects of limblessness and limb reduction. These fascinating topics could have an entire volume to themselves. I elected to present what are essentially case study approaches by confining the discussion to reptiles and mammals. Reduction of entire limbs (as in snakes and legless lizards) or of digits (as in representatives of all the tetrapod classes), a recurrent theme in limb evolution, is discussed in chapter 14 in the context of the diversity of limbs and the types of digit reduction seen in reptiles. The next three chapters explore the diversity of adaptive structural changes seen in terrestrial mammals (chapter 15), associated with flight (chapter 16), and displayed in tetrapods with digging and burrowing modes of life (chapter 17), some of which are associated with limb reduction, although this aspect is not addressed explicitly in chapter 17. Transformations and adaptations in the limbs of those reptiles and mammals that became secondarily aquatic are discussed in chapter 18. Chapter 19 treats what one could call extraskeletal elements associated with limbs—ossicles, sesamoids, and lanulae—that arise apart from the primary skeleton but are then incorporated into the appendicular skeleton. Because of its comparative analysis and perspectives on cell, tissue, and genetic aspects of transformation, this chapter illustrates nicely the problems confronting us when we attempt to understand and explain aspects of limb development, evolution, and transformation.

All the chapters are written by leading experts in their topics. It is a pleasure to thank these busy researchers for taking time from their laboratory or field studies to provide us with the benefit of their analyses. My thanks to Patricia (Paty) Avendaño for her assistance in copy editing the chapters, and to Mike Coates, Bob Carroll, and Marcello Ruta for most helpful comments on the index.

Part I
Evolution

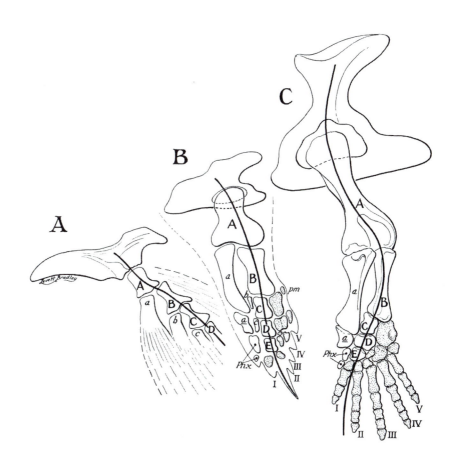

Chapter 1 Fins and Limbs and Fins into Limbs: The Historical Context, 1840–1940

Peter J. Bowler

THE HISTORY OF how biologists in the 100 years between 1840 and 1940 tackled the question of how the vertebrates emerged onto land provides insights into the ways in which evolutionary thinking itself has evolved. Whereas in the mid-19th century this and other major transformations were seen as episodes in the progress of life toward humankind, and evaluated by purely morphological evidence, we see in the late 19th century the growing importance of morphological study of the fossil record. By the early 20th century the emergence of new ways of looking at the earth's physical history, coupled with growing doubts about non-Darwinian mechanisms of evolution, encouraged biologists to attempt explanation of past transformations in terms of what we now call adaptive scenarios. We now know that limbs were developed first in completely aquatic creatures, which were thus preadapted to walking on land. How they developed these structures and eventually began to use them in a new way remains murky. Hence the present volume, which evaluates fins, limbs, and the transition from fins to limbs.

Transformation: An Evolutionary and Taxonomic Question

At first sight it may seem obvious that the question of how fins were transformed into limbs could only be asked after the theory of evolution had been accepted. In 1849, the doyen of British morphology, Richard Owen, published *On the Nature of Limbs*, an influential summary of a lecture delivered before the Royal Institution of Great Britain. Owen evaluated and discussed the *Bedeutung*—the signification or essential essence—of limbs as archetypes, using homology, "the relations of the parts of animal bodies understood by the German word 'Bedeutung'" (Owen 1849, 2). Indeed, Owen provided as a title for his lecture—in what he termed "the technical language of anatomical sciences"—"On the General and Serial Homologies of the Locomotory Extremities" (2). Limbs for Owen meant the arms and legs in man, fore- and hindlegs of beasts, wings and legs of bats and birds, and the pectoral and pelvic fins of fishes, taking for granted the general knowledge and acceptance of these appendages as "homologous parts." A dozen printings in Britain and the United States attest to the importance of this monograph. It is being reprinted again in 2007 (Owen 1849 [2007]).

Although he made an extended argument for the archetype as Platonic ideal, Owen was searching for laws that could explain the transformation of one type to another, as revealed in his concluding paragraph: "To what natural laws of secondary causes the orderly succession and progression of such organic phænomena may have been committed we are as yet ignorant.... [W]e learn from the past history of our globe that she [Nature] has advanced with slow and stately steps, guided by the archetypal light" (Owen 1849, 86). The Cambridge geologist Adam Sedgwick saw the significance of this search for "secondary causes," namely that Owen might have "meant to indicate some theoretical law of generative development from one animal type to another along the whole ascending scale of Nature" (Sedgwick 1850, ccxiv). This volume is timely, in part, because of the ongoing search for these elusive "theoretical law[s] of generative development."

Early studies of lungfish explored their relationship to fish on the one hand and amphibians on the other. Only in the 1860s, however, were serious efforts made to trace a plausible line of descent from fish to tetrapods when a group of "biologists," inspired by what we now call the "Darwinian revolution," began the attempt to reconstruct the history of life on earth from anatomical, embryological, and paleontological evidence. Darwin himself was reluctant to engage in this project; he feared that not enough evidence was available. So inspired were his followers by the idea of evolution that they felt it necessary to attempt the reconstruction. This was the task Gegenbaur, Haeckel, and others undertook as a means of adapting the science of morphology to the demands of the evolutionary perspective (Bowler 1996; B. K. Hall 2005b). This meant not only trying to understand how tetrapods had evolved from fish, but also identifying which kind (not kinds) of fish was the most plausible ancestor, and explaining how that ancestor had evolved as a fish. By the end of the 19th century it was accepted that, with hindsight, one could identify the critical phases in the evolution of life. The "conquest of the land" by the first amphibians was one such step.

Popular modern accounts of the history of life on earth tend to regard questions such as the origin of the amphibians as lying purely within the province of paleontology. Yet, in the late 19th century, comparative anatomy and embryology were thought to have an equal right to speak on these topics. (Those who studied the morphology of extant and extinct organisms traditionally are referred to as morphologists and paleontologists, respectively [Bowler 1996]. It is important to realize that both used morphological approaches.) In part because the fossil record of the 1870s was deficient in clues concerning all of the major steps in vertebrate history, morphologists took it upon themselves to identify the key transitions and the most likely ancestral forms. A number of problems plagued the reconstruction, whatever the source of evidence: (1) major disputes erupted over the determination of the most primitive members of each class, and (2) the status of crucial fossils was particularly open to challenge when, as with *Archaeopteryx,* they were clearly too late to be the actual missing link between the two groups whose characters they seemed to share.

Both morphologists and paleontologists—as defined above—had to confront the problem of parallel evolution. All the vertebrate classes were at one time or another alleged to be paraphyletic—that is, to be grades of organization reached independently by more than one lineage arising from the previous class. As two examples of conflicts: Paleontologists eventually dismissed lungfish such as *Ceratodus* as ancestors of the amphibians, claiming they independently evolved the ability to breathe air; a few zoologists argued that the Amphibia were diphyletic, some having their origin in the lungfish, others in the crossopterygian fishes favored by the paleontologists (see Thomson 1968, 1991).

A number of problems plagued the use of fossils to resolve issues of origins. Many paleontologists were anti-Darwinians and so were predisposed to accept evidence favoring the idea of predictable trends in evolution, horn size in titanotheres or increasingly elaborate sutures of the shell of ammonites being two examples. No matter what approach individuals brought to their studies, the fossil record rarely provided enough information to determine trends and/or to eliminate the possibility of convergence or parallelism. Early paleontologists often failed to consider functional interpretations of their finds. Later paleontologists such as William King Gregory, Robert Broom, and D. M. S. Watson were, however, far more willing to look for functional causes of change. They wanted to know how exactly the fin of a fish had been transformed into the limb of a tetrapod: what were the mechanical problems involved, and how had they been overcome? Functional changes in the limbs could be studied in considerable detail, however, without asking about the environmental conditions (stresses, some thought) that might have forced the animals to adopt a new means of locomotion. Functional morphology was still morphology, and it did not necessarily trigger an interest in the role played by external factors in determining an organism's behavior.

Those paleontologists who worked closely with geologists were more aware of the evidence for past climates and environments. The late 19th century saw a growing interest in the possibility that crucial breakthroughs in evolution might have been triggered by climatic stress. American paleontologists were especially active in this area, perhaps because they worked more closely with the geologists who were providing evidence of past climatic changes. Attempts were made to explain the sudden appearance of new classes as a response to the climatic stress induced by such events. Even so, few efforts were made to depict what would now be called an adaptive scenario to explain the precise circumstances that forced the modification of a species' structure in a particular direction. Alfred S. Romer's suggestion that the amphibians might have developed legs as a means of crawling to other pools in a world subject to increasing drought was one of the earliest suggestions of such a scenario, and it was not proposed until the 1930s (Romer 1933; see also Bowler 1996).

The Fin Problem

One of the most controversial issues that emerged from the study of fish evolution was the origin of the paired fins. It was natural to turn to those living vertebrates deemed to be the most primitive. Most turned to jawless and finless fish such as lampreys (cyclostomes). If cyclostomes were to be relied on, the most primitive vertebrates lacked paired fins. Consequently, and unless fins had arisen *de novo,* a preexisting structure that could have been transformed to produce fins had to be identified.

This topic was an important one, not least because the paired fins in one or more groups would subsequently be transformed into the limbs of tetrapods, an essential prelude to one of the most far-reaching revolutions in the history of the vertebrate phylum. Before tackling the problem, morphologists had to decide which was the most primitive form of the paired fins, since this would to some extent determine which form was the more likely source for these peculiar structures. Then they had to determine which line of limb evolution—and therefore which taxonomic group—made a plausible candidate for the transition to the legs of amphibians. Other limb forms would then have to be identified as specialized developments from the primitive original.

Two rival theories emerged rapidly in the post-Darwinian era and were debated fiercely into the 20th century (for a summary and relevant literature, see Bowler 1996, 219–229).

Carl Gegenbaur's work in comparative anatomy led him immediately to the idea of defining the most primitive form of the paired limbs, from which he sought to identify the most likely origin of these structures. In 1865, he showed how the shoulder girdle from which the forelimbs are suspended could be traced through the evolution of the higher vertebrates. He also dealt with the pectoral fins of fish, taking the elasmobranch (shark) form as the most primitive. By 1870 Gegenbaur had changed his views significantly; he now held that the forelimb of the African lungfish *Protopterus*—a whiplike rod with traces of rays on one side—illustrated the most primitive form. Later he identified the limb of the Australian lungfish *Ceratodus* (later known as *Neoceratodus*) as the primitive "archipterygium," the most basic form of the paired limbs. Gegenbaur argued that this limb had evolved from the gill arches of the early, limbless vertebrates. Significantly for the present volume, Gegenbaur held that the *Ceratodus* limb had evolved both into the various other forms of paired fins in fish, and also directly into the limbs of the first tetrapods.

Gegenbaur believed that the shark fin, which has strongly developed rays on one side, had been formed from the original archipterygium. Note that in his eyes, as in those of most of his contemporaries, the lungfish or Dipnoi were the most likely ancestors of the amphibians. It was thus possible to trace a direct line from the archipterygium of the Dipnoi to the amphibians, with the sharks and other fishes representing side branches leading to a purely finlike specialization.

Gegenbaur's theory was almost immediately challenged by the American James K. Thatcher and by the British evolutionary anatomist (and strong opponent of Darwinian selectionism) St. George Jackson Mivart. Thatcher and Mivart (and, independently, Francis Balfour) proposed that the paired fins had evolved from a continuous lateral fin that had once run down either side of the body in the earliest vertebrates. This interpretation, supported by embryological evidence and by evidence from adult anatomy, is now known as the Thatcher-Mivart-Balfour fin-fold theory of the origin of the paired fins. Its implication were twofold, important, and far-reaching: (1) the various complex fin structures all were specializations; and (2) there was no reason why a straight line of evolution should lead from lungfish to the first tetrapods.

By the end of the 19th century, the debate seemed to be going in favor of the fin-fold theory, although Gegenbaur's disciples continued to defend their master's interpretation. This issue exploded into the *Competenzkonflikt* between Gegenbaur and Anton Dohrn, a vicious debate over the relative standing of anatomical and embryological evidence that did much to discredit evolutionary morphology in Germany. Meanwhile, paleontologists were accumulating an ever-expanding wealth of fossil evidence, which seemed to offer some hope at last of determining the structure of the most primitive paired fins. Henry Fairfield Osborn reveled in the conclusion that paleontology had resolved a debate that could not be settled on purely morphological grounds (Osborn 1917).

The Origin of the Amphibians

The debate over the origin of the paired fins gained some significance because it served as a foundation for the equally controversial topic of the transformation of those fins into limbs (see Schaeffer 1965; Bowler 1996). But the question of the origin of the amphibians raised even wider issues. Morphologists and paleontologists alike had a field day arguing about the precise relationship between the paired fins and the tetrapod limbs. Much of the early discussion was of a purely morphological character, based on an analysis of the mechanical transformations required by the conversion of a fin into a limb. Only in the 20th century was there any serious discussion of the adaptive pressures involved.

There was also a major debate about which group of fish would have been ancestral to the amphibians. Haeckel made the natural assumption that the Dipnoi, the lungfishes, were the most likely candidates, and, as noted above, Gegenbaur developed this view. By the end of the 19th century, however, paleontologists' attention had increasingly switched to a group of fossil fish that seemed to provide a more plausible ancestry. A group of Paleozoic fishes, the Crossopterygians, had swim bladders thought to be homologous with lungs, and so were regarded as related to amphibians. Crossopterygians also had bony fins that might serve as the starting point for legs. The Dipnoi had specializations that suggested they were a side branch that had independently acquired characters resembling those of amphibians.

By the early decades of the 20th century, the crossopterygian ancestry of the tetrapods was taken for granted by most paleontologists. Those who studied living species were not so sure, however, and there were occasional warnings that the lungfish might turn out to be the closest living relative of the amphibians after all.

Lungfish as Ancestral Tetrapods

When specimens of the South American and African species of lungfish (*Lepidosiren, Polypterus*) were brought to Europe in the late 1830s, they immediately posed a problem for naturalists accustomed to making a clear distinction between fishes and amphibians (Kerr 1932; Bowler 1996; B. K. Hall 2001). Since their air bladders functioned as lungs enabling survival out of water, lungs seemed to serve as a bridge between the two classes. Von Bischoff described *Lepidosiren* as an amphibian. He thought that the lungs, internal nostrils, and structure of the heart were amphibian features and outweighed the scales and other fishlike characters (details in Patterson 1980). Specimens of the African lungfish *Protopterus* were brought to London by John Samuel Budgett (B. K. Hall 2001). Initially, Richard Owen described *Protopterus* as a teleost. Although Owen later admitted that he was mistaken on this point, he never wavered from his belief that lungfish were true fishes that happened to resemble amphibians in a few characters. Owen was supported by Louis Agassiz and other experts, so that by the middle of the 19th century it was taken for granted that the lungfish were indeed an order of fish, the Dipneusta or Dipnoi.

When he came to the origin of the tetrapods in his *History of Creation* (1876, 2:213), Ernst Haeckel proposed the Dipneusta as a transitional class between true fish and amphibians. Surviving lungfish were relics of a once numerous group, fossil evidence of which was provided by the teeth of *Ceratodus* in the Triassic rocks. The early Dipneusta were, in fact, the primary form from which the Amphibia had sprung.

Haeckel argued that the possession of a pentadactyle or five-digit limb by all tetrapods confirmed that they were a monophyletic group arising from the primitive amphibians. This latter point was taken for granted by all morphologists into the early 20th century.

The belief that the Dipneusta or Dipnoi were the ancestral form of the Amphibia became widely accepted in the 1870s and 1880s. As noted above, Dipnoi's status as the closest fish to the amphibians was built into Gegenbaur's theory of the origin of the vertebrate limbs. The discovery of the Australian lungfish, *Neoceratodus,* in 1870 suggested that the fossil Dipnoi, including *Ceratodus* itself, had well-developed bony fins. F. M. Balfour's *Treatise on Comparative Embryology* also placed the Dipnoi immediately preceding the hypothetical Proto-pentadactyloidei from which the Amphibia and the higher vertebrate classes had sprung (Balfour 1885, 3:327) When Richard Semon, a disciple of Haeckel, went to Australia in the 1890s, one of his chief objects was to study the embryology of *Neoceratodus* because it served as a link between fish and amphibians (see Semon 1899 and 1893–1915, vol. 1).

The Crossopterygians

By the end of the 19th century a powerful opposing movement had grown up based on the assumption that the dipnoans' resemblance to amphibians was superficial, a product of convergent evolution, and was not an indication of true genealogical relationship. The Dipnoi could not be ancestral to the amphibians; they had already developed specialized characters such as the crushing plates of the jaw by which the fossil *Ceratodus* was known. This structure was unlike anything possessed by amphibians, and indicated that the dipnoans must lie on a side branch that did not lead toward the "higher" class. The alternative hypothetical ancestor of the amphibians was another group of fish prominent in the Paleozoic, the crossopterygian or lobe-finned fishes. These also had well-developed bony fins, which, Gegenbaur's opponents claimed, offered a better starting point for the evolution of the tetrapod limb. In the most extreme version of this theory, the Dipnoi were derived from crossopterygians (Bowler 1996; B. K. Hall 2001).

The only living fishes included in the suborder Crossopterygidae created by Huxley in 1861—and which therefore became by definition "living fossils"—were the bichir *Polypterus* of the river Nile and its more specialized relative, the ropefish, *Calamoichthys calabaricus.* The presumed existence of living representatives of the crossopterygians became particularly significant later in the century when earlier members of the suborder were postulated as ancestors of the amphibians; morphologists expended a great deal of effort on *Polypterus* in the hope that it would throw light on this crucial transition. But even when he established the suborder, Huxley (1861) admitted that *Polypterus* exhibited significant differences from the other crossopterygians; its inclusion in the suborder was further questioned in the 20th century.

The claim that the crossopterygians offered a more plausible ancestry than the dipnoans for the amphibians, first suggested by H. B. Pollard (1891) and J. S. Kingsley (1892), soon gained wide support from influential figures such as Cope (1892b). Pollard argued that the skull structure of the Dipnoi differed from that of the Amphibia and that there was no evidence of a phase resembling the Dipnoi in the ontogeny of living Amphibia or in the fossil members of the group. His phylogenetic tree showed the Dipnoi as descendants of the crossopterygians, branching off in a direction different from that taken by the amphibians (Pollard 1891, 344).

Cope (1892b), originally a supporter of the lungfish-amphibian link, took note of Pollard and Kingsley's work and opted for the new theory, thereby extending it into the realm of paleontology. Cope argued that the structure of the paired fins in Dipnoi did not anticipate that of the tetrapod limb, but that fossil rhipidistians offered a better model on which the derivation of the limb could be based. In particular the fins of *Eusthenopteron* from the Devonian of New Brunswick almost realized Gegenbaur's ambition of demonstrating the derivation of the tetrapod limb from the fin of a fish (Cope 1892b, 279–280). Cope repeated these views in his influential book, *Primary Factors of Organic Evolution* (1896, 88–89). By throwing his weight behind the new theory Cope ensured that other paleontologists also took it seriously.

Perhaps the most decisive intervention in the debate came from the respected Belgian paleontologist Louis Dollo. His 1895 reappraisal of lungfish phylogeny transformed ideas about the group's evolution in a way that seemed to confirm their status as a specialized offshoot from the stem leading to the amphibians. Dollo interpreted lungfish evolution in ecological terms, as a specialization for living in impure water. Devonian lungfish such as *Dipterus* had moved into this environment, and the living members of the group illustrated stages of further specialization. The Australian *Ceratodus* still had working fins and could not live out of water, while *Protopterus* and *Lepidosiren* had better-developed lungs and almost totally degenerate paired fins. These later forms were adapted to living in the mud, and other fish that were adapted to the same environment shared a similar eel-like structure, acquired by convergent evolution (Dollo 1895, 9–100). Dollo then went on to look for the most likely ancestry of the earliest dipnoans and found it in the crossopterygians. The latter were already adapting in the same direction: they

were bottom dwellers rather than swimmers in the open water and their lobed fins had been developed to enable them to "walk" over the bottom surface (107). In effect, then, lungfish were the end product of a specializing trend started by Devonian crossopterygians.

Morphologists continued to make some input into the theory of crossopterygian ancestry, but attention was increasingly switching to the fossil record as the preferred source of information on the relationship between fish and amphibians. In 1896 a study of the early armored amphibians, the Stegocephalia, by Georg Baur lent support to the new theory:

1. The structure of the earliest amphibians could best be explained by supposing that they had evolved from crossopterygians.
2. Lungfish were specialized descendants of the earliest crossopterygians, from which the first amphibians also had evolved.

The same point was taken up in the early decades of the 20th century by D. M. S. Watson, who spent much of his career trying to identify trends in the evolution of the fossil amphibia (e.g., Watson 1919). Watson, Gregory, and others tried to explain the actual transformations that gave rise to the amphibians from a starting point in the osteolepid crossopterygians. Popular studies by paleontologists like Osborn (*Origin and Evolution of Life*, 1917) and Gregory (*Our Face from Fish to Man*, 1929) took the same position. A few years later, Alfred Sherwood Romer's textbook of vertebrate paleontology dismissed the lungfish as "not the parents but the uncles of the tetrapods" and sought the origins of the tetrapod limb in the crossopterygian fin (1933, 92 and 104). Fifty years later, D. E. Rosen et al. (1981) marshaled the evidence for lungfish as the sister group to the tetrapods.

More Than One Origin of the Amphibians?

Disagreement had thus emerged between the paleontologists, almost all of whom had adopted the crossopterygian theory, and those who dealt with living lungfish and amphibians, many of whom saw the similarities as being too close to be explained away by convergence.

It is a sign of paleontology's increasing dominance that we seldom hear of the rival theory, especially in popular accounts of the history of life on earth. One of the strangest products of this tension between the professionals was the suggestion developed by several Scandinavian biologists that the amphibia might be diphyletic, having two separate origins within different groups of fish. In 1933 Nils Holmgren published a study of amphibian limbs that stressed the differences between urodeles (salamanders) and anurans (frogs).

He seized upon this difference as a means of arguing that the Amphibia are an artificial group composed of two separate taxa. Existing theories of limb evolution were unsatisfactory because no one had admitted the possibility of the "amphibian" limb having been formed by two different routes. Holmgren (1933) argued that the stegocephalians had evolved from crossopterygian fish and had in turn given rise to the anurans and the reptiles. The urodeles had evolved separately, either from another crossopterygian source or, more likely, from the dipnoans (288; for a critical discussion see Schmalhausen 1968, chap. 19). Credibility of the morphological evidence for a relationship between the lungfish and at least one type of amphibian was thus salvaged at the price of splitting the old class Amphibia into two fundamentally different types. Jarvik (1942) stressed the possibility that several different groups of crossopterygians might have been preadapted for terrestrial life, so that the amphibians might have diverse origins within the crossopterygians themselves.

From Water to Land: The Habitat Transformation

The problem of explaining the transition to a new habitat on the land was a complex one. The physiological transformation was obvious enough: lungs had to replace gills as a means of respiration. Darwin himself had argued that this was not as great a problem as it might seem. Darwin pointed out that most fish have swim bladders that contain air and are used to regulate buoyancy, and so he could easily imagine how the bladder could be transformed into a lung in an animal that needed to breathe air (Darwin 1859, 190–191), although he had fallen into the trap of assuming that the structure typical of the fish must be more primitive than that of the higher vertebrates.

Most evolutionists agreed that lungfish—whether or not they are directly related to the amphibians—show the transitional phase in which the bladder has become modified to absorb air in circumstances where the fish has to exist at certain times out of water. The American biologist Charles Morris (1892) seems to have been the first to suggest the modern view that the original function of the swim bladder was respiratory—only in the later bony fish had it degenerated into a mere regulator of buoyancy as the gills took over the whole function of respiration. He pointed out that sharks did very well without a swim bladder, which certainly suggested that it was not a necessary fish structure. Most early 20th-century evolutionists rejected Morris's claim that fish with lungs had invaded the land, but there was certainly a strong presumption that the crossopterygians had bladders preadapted to breathing air, which would have prepared them to move into the new environment.

Swimming to Walking: The Functional Transformation

Transformations from water to land involved far more than the fins' acquiring the ability to move the body over the ground. As Dollo argued, the crossopterygian fin was preadapted to pushing the fish along the bottom in shallow water. It was relatively easy to suppose that the same structure could be used to propel a primitive amphibian over a muddy surface. But to move efficiently on the land the limbs had to become far more powerful and had to be anchored into the body in a way that would transmit the force efficiently. To function out of water the whole body had to be supported in such a way as to allow breathing to take place against the pressure created by gravity. A complex series of morphological transformations had to take place to give rise to the first amphibians.

Despite the lack of fossils illustrating the actual transformation, paleontologists became increasingly willing to use their studies of crossopterygians and primitive amphibians to explore the details of how the transformation might have taken place. In part, the problem would be solved by identifying homologies; which bones in the ancestral fin-support have been transformed into the bones of the tetrapod limb? This was not as straightforward a question as it might seem.

The fish fin is an essentially rigid structure articulating with the body only at the "shoulder." The tetrapod limb articulates at the "elbow" and "wrist" as well, and the upper and lower parts of the limb have evidently been twisted with respect to the body in a way that confused many early morphologists who tried to work out the homologies involved. Transformation of the shoulder and pelvic girdles also presented problems. The fish shoulder girdle is attached to the rear of the skull; to avoid transmitting the shock of each step to the head it must have been moved caudally (tailward) and become connected more closely with the spine. The pelvic girdle of the fish, which floats freely in the muscles, had to be enlarged and also become connected to the vertebral column.

Even when they came to an agreement over the basic transformations by which the lobe fin of a crossopterygian had been transformed into an amphibian leg, paleontologists were no longer satisfied. Increasingly, they saw themselves as functional morphologists, trying to understand the pattern of stresses and strains that would have shaped the transformation as the ancestral fish began to move out of the water. How had transitional forms coped with a way of life that was partly aquatic and partly terrestrial (see Coates and Ruta, chap. 2, and Akimenko and Smith, chap. 11 in this volume), and—perhaps more important—why would a fish have taken the risk of first venturing out into a new and hostile environment? (Also, why did terrestrial tetrapods make the secondary transition back to the water? See Thewissen and Taylor, chap. 18 in this volume.) Evolutionists of this school were no longer satisfied with the construction of phylogenetic trees based on morphological relationships. They were now beginning to construct adaptive scenarios to explain particular transformations, exploiting information about changing environments derived from geology.

In the final version of Gegenbaur's theory (mentioned above) the archipterygium modeled on the fin of *Ceratodus* was seen as the most primitive form that had been converted both into the fins of other fishes and into the amphibian limb. But few, apart from Gegenbaur's own disciples, were entirely happy with the theory. The archipterygium consisted of a central rod of bones with rays branching out symmetrically on either side. Yet the tetrapod lower limb consists of two bones, the radius and ulna in the anterior limb or arm, the tibia and fibula in the posterior limb or leg. These in turn must articulate in a particular way. In the human arm, the wrist is a simple hinge, while the elbow also permits the lower arm to rotate as a unit with respect to the upper. In the leg it is the opposite way around: the lower joint, the ankle, permits both bending and rotation, while the upper, the knee, is a simple hinge. Gegenbaur (1874) tried to identify the bones of the leg and arm with elements of the symmetrical archipterygium. He believed that the homologues of the main axis in the archipterygium were (for the forelimb) the humerus, the radius, and the first digit (497). The pentadactyle limb was thus derived from *only one side* of the archipterygium.

In 1876 T. H. Huxley published a study of *Ceratodus* in which he evaluated Gegenbaur's theory. Huxley noted the problem that in fish and tetrapods the limbs rotate in different directions with respect to the trunk (1876, 109–110). While accepting that the archipterygium of *Ceratodus* was the fundamental form of the limb, Huxley was forced to dissent from the rest of Gegenbaur's theory. As Huxley understood the homologies of the limb bones in fish and tetrapods, the rotations required by the theory would create torsion of the humerus, which he found quite implausible (Huxley 1876, 118).

Gegenbaur thought that the tetrapod limb was produced by a continuation of the same process as that which generated the asymmetrical fins of other fish. Huxley argued that abandoning this assumption made a simpler explanation possible. The tetrapod limb, or *cheiropterygium,* and the fish fin were developed by different kinds of specialization starting from the archipterygium. Huxley provided a diagram to illustrate the comparable bones in a shark fin and an amphibian limb (Huxley 1876, 20). Gegenbaur accepted Huxley's criticism; in later editions of his work Gegenbaur showed the main axis running through to the fifth digit (Gegenbaur 1878, 480)

Little further progress was made while the majority of biologists continued to believe that the lungfish were the starting point for amphibian origins. But when it was recognized in the 1890s that the crossopterygians offered a more plausible ancestral form, new developments became possible. It was immediately obvious that the fins of crossopterygians could much more easily have been transformed into tetrapod limbs than could the archipterygium of *Ceratodus.*

The pace of progress was slow over the following decades. Goodrich (1930), who saw his work as an attempt to understand evolutionary relationships, claimed that none of the efforts made to reconstruct the evolution of the tetrapod limb were convincing, concluding that "as yet nothing for certain is known about the origin of the cheiropterygium" (159–160). In the detailed study of amphibian limb anatomy that led him to propose that the class was diphyletic, Holmgren noted that "it is fairly clear that the problem of the origin of the tetrapod limb is today nearly as far from solution as it was in Gegenbaur's time" (1933, 208).

Inferring Function from Fossils

The early 20th century saw a rush of work by paleontologists seeking to exploit the new theory that the amphibians had evolved from crossopterygians.

William King Gregory (1915) recorded a remarkable coincidence of scientists independently moving toward the hypothesis that the fins of certain fossil crossopterygians could be used as a model for the origin of the early amphibian limb. Both Watson, an expert of fossil amphibians, and Robert Broom, better known for his work on the mammal-like reptiles, independently identified *Eusthenopteron* or the late Devonian *Sauripterus* as the best models from which to derive the tetrapod limb (Watson 1913; Broom 1913). Gregory records that he became aware of these publications while he was himself investigating the fin of *Sauripterus,* having been alerted to its amphibian-like structure by the publication of a photograph in a museum catalog (1915, 358). This fin has a single proximal element equivalent to the humerus, two distal elements equivalent to the radius and ulna, and a number of radials from which the digits might be derived. R. S. Lull (1917) reported these studies in his textbook on evolution and added an illustration of a fossil footprint from the upper Devonian, which seemed to indicate that the earliest amphibian foot had not yet developed the full complement of five digits (488–489).

Over the next couple of decades, a number of paleontologists tried to reconstruct the details of a process by which the crossopterygian fin could be transformed into the tetrapod limb. The best available fossil amphibians were studied in an attempt to understand the structure of the early amphibian limb and the way in which it was used. As Watson (1926) noted, the mere search for homologies was no longer satisfying: "the centre of interest has passed from structure to function, and it is in the attempt to realise the conditions under which the transformation took place, and to understand the process by which the animals' mechanism was so profoundly modified whilst remaining a working whole throughout, that the attraction of the problem lies" (189).

Gregory and his students, including Alfred Sherwood Romer (1933) and Roy Waldo Miner (1925), were most active in carrying forward the program sketched out in Watson's words (see Rainger 1991, chap. 9). They created a paleontology based on functional morphology (for which see Hall 2002), using living examples to reconstruct not only the skeleton but also the musculature of fossil species. Both fish and amphibian fossils were studied in an effort to bridge the gap.

In 1941 Gregory and Henry C. Raven published an extensive study of the evolution of the limbs using *Eusthenopteron* as a starting point. They used a large flexible model to demonstrate the different positions taken up by the limb as it became bent and twisted to form a functioning leg. They were particularly insistent that the transformation should be explained as far as possible by seeking transitions between forms already known from the fossil record. Even when the known fossils occurred too late in the record to be the actual ancestor (this was certainly the case with *Eusthenopteron*) the later form could be used as a model on the assumption that close relatives of the true ancestor might have survived unchanged into later epochs.

Adaptive Scenarios

In the late 19th century all morphologists, and most paleontologists, took it for granted that lungfish or crossopterygians had acquired the habit of moving around outside the water and investigated the morphological and functional changes that made this possible. They were not interested in postulating what a modern evolutionist would call an "adaptive scenario" to explain the transition.

The first steps toward what we might call a more Darwinian (i.e., adaptive) approach were prompted by the interaction between paleontologists and geologists, especially in America. Here new theoretical developments in geology encouraged the search for evidence of past climatic changes and were linked to an active use of vertebrate paleontology in stratigraphy. By the end of the 19th century geology was no longer dominated by a philosophy of complete, steady-state uniformitarianism. Geologists such as Thomas C. Chamberlin were now convinced that there were episodes of intense (but not actually catastrophic) change in the earth's physical conditions. From this source came the inspiration to inquire whether some of the more dramatic steps in the history of life might have been triggered by environmental stresses flowing from these catastrophic changes.

The American geologist Joseph Barrell was the first to apply the new philosophy of earth history to the question of the origin of land vertebrates. In 1906 he began a series of studies on sedimentation that provided information on the climates of the successive geological periods.

Over the following 10 years Barrell became convinced that climatic stress was the trigger for major evolutionary changes, and in 1916 he published a paper titled "Influence of Silurian-Devonian Climates on the Rise of Air-Breathing Vertebrates." The main driving force of evolution, Barrell maintained, was pressure of the environment on the organism. Periods of climatic stress imposed a more intense struggle for existence that eliminated the less hardy and adaptable types and favored the survival of advanced mutations (1916, 414). Barrell was not a convinced Darwinist. Like many of his contemporaries he thought that natural selection was not the sole driving force of evolution. He did insist, however, that it is "nevertheless a broad controlling force which compels development within certain limits of efficiency" (1916, 390) and thought that it coordinated changes in different parts of the organism.

Within the context of this rather vague sense of an environmental pressure upon the organism Barrell began to ask exactly what kind of incentive would have been enough to drive the ancestors of the amphibians out of the water. The physical environment was the trigger for change, although Barrell's theory did not explain why some fish eventually became so modified that they could live permanently on the land, an issue with which students of the transformation of fins into limbs, including those with chapters in this volume, continue to struggle today.

This short history of how biologists tackled the question of how the vertebrates emerged onto land illustrates the depth of the questions, and, despite over 150 years of concentrated effort, the comparative shallowness of our understanding of the causes of this remarkable transition.

Chapter 2 Skeletal Changes in the Transition from Fins to Limbs

Michael I. Coates and Marcello Ruta

FOR THE PURPOSES of this chapter, tetrapods are considered a sarcopterygian subset. The chapter is necessarily data-heavy, focusing primarily on a broad-based review of girdle, fin, and limb skeletons. The aim, as conceived by the editor, was to describe skeletal transformations spanning the transition from fin to limbs. However, to embed such changes in a meaningful context, it was rapidly apparent that a broader phylogenetic bracket was required. Therefore, lungfish, coelacanth, and a reasonably comprehensive summary of fossil nontetrapod sarcopterygian fins are also included. In fact, unless these data are placed side by side with basal tetrapod limbs, fins, and girdles, it is not at all clear how a minimum assessment of primitive conditions can be established.

Throughout the text the term "Tetrapoda" is used to mean the tetrapod total group (Patterson 1993). Crown or stem group memberships are specified as needed. Crown, stem, and total group terminology is far from universally accepted; we acknowledge that total group tetrapods include many taxa that would commonly be described as fish (e.g., the tristichopterid *Eusthenopteron*). Unfortunately, "fish" as a taxonomic term is imprecise, and the entire issue can be muddied with debates about the presence or absence of key characteristics and the minutiae thereof. For alternative and more elaborate hierarchies of names, see Ahlberg (1991), Ahlberg and Johanson (1998), and Johanson et al. (2003). Irrespective of whichever Tetrapoda definition is used (cf. Gaffney 1979; Lebedev and Coates 1995; Coates 1996; Ahlberg and Clack 1998; Anderson 2001; Laurin 1998a; Coates et al. 2002; Ruta et al. 2003), the transition from fins to limbs

concerns changes implied by the full array of paired appendage patterns in taxa branching from the entire tetrapod stem. Stem taxa provide the only direct morphological information on primitive fishlike conditions unique to the tetrapod lineage; there are no living finned tetrapods.

The chapter is divided into four sections. The first reviews the phylogenetic context of tetrapods within living and fossil sarcopterygians. The basis of the framework used for the present work is specified, and sources of recent, alternative hypotheses are included. The second part reviews appendicular skeletons throughout the Sarcopterygii excluding tetrapods (in the total group sense). The third part reviews tetrapod paired fins, limbs, and girdles. Each subsection of these two parts includes brief details of geological and stratigraphic range, primary recent data sources in the literature (much of which is unlikely ever to be online), and a description in the sequence of dermal skeletal, then endoskeletal pectoral and endoskeletal pelvic morphologies. Where appropriate, notes on variation within the group in question are added. The fourth part summarizes the implied transformational trends, examples of convergent events in other sarcopterygian lineages, the emerging pattern of characters, and thus implied transformational, distribution through phylogeny, and notes on functional implications.

Phylogenetic Context

Any discussion of evolutionary change requires a phylogenetic context. The fin-to-limb transition spans three areas of phylogenetic debate: the interrelationships of sarcopterygians as a whole, the composition of the tetrapod stem group, and the phylogenetic location and basal branching pattern of the tetrapod crown group. The crown group hypothesis defines, either implicitly or explicitly, those characteristics that might be used to construct a *Bauplan* of modern tetrapod limbs. Stem group hypotheses provide clues about the evolutionary direction and sequence of *Bauplan* assembly. And basal sarcopterygian interrelationships deliver a hypothesis of primitive conditions: the inferred set of characteristics present in the last common ancestor of tetrapods and their living sister group.

Predictably, the identity of the living sister group of tetrapods is disputed. Molecular data are equivocal about the candidacy of lungfishes (Dipnoi), the coelacanth (Actinistia), and lungfishes plus coelacanth (Zardoya and Meyer 2001); a third option presents Pisces as a whole—a crown group subtending *all* modern jawed fishes—as the tetrapod sister taxon (Arnason et al. 2001, and references therein). In fact, analyses of molecular sequences have delivered an unexpectedly wide range of hypotheses about sarcopterygian relationships among gnathostomes as a whole (earlier attempts reviewed in Forey 1998). The most widely discussed explanation of this failure of molecular data to deliver a consistent result is that modern osteichthyan lineages result from a rapid sequence of chronologically ancient (~400+ mya) branching events. In comparison, results of morphology-based analyses including fossils are conservative. Most computer-assisted analyses favor a lungfish-tetrapod grouping (Cloutier and Ahlberg 1996; Forey 1998; Zhu et al. 2001), although the coelacanth-tetrapod arrangement (Zhu and Schultze 2001) remains actively debated.

For present purposes, the most recent version of the lungfish-tetrapod hypothesis is used (Zhu et al. 2001). The branching pattern is shown in figure 2.1 with primitive exemplars of each major clade. Each of these early representatives (all are Devonian) of the major sarcopterygian fish groups differs significantly from their more recent relatives. The coelacanth, *Miguashaia*, lacks the muscular, lobate, anal, and second dorsal fins present in the extant *Latimeria*. The lungfish *Dipterus* retains the primitive complement of median fins instead of the continuous caudal-dorsal fin fold of all recent genera. Both stem tetrapods (*Gooloogongia* and *Osteolepis*) are conventionally fishlike.

The inclusion of fossil taxa in analyses of sarcopterygian phylogeny has generated several tetrapod stem group hypotheses. Significantly, the branching patterns of these tetrapod-like fish groups are in broad agreement (Cloutier and Ahlberg 1996; Ahlberg and Johanson 1998; Jeffery 2001; Zhu and Schultze 2001; Zhu et al. 2001) even though the tetrapod-lungfish-coelacanth issue remains unsettled.

These results represent a real advance on textbook summaries (e.g., R. L. Carroll 1987; Janvier 1996), and have moved far beyond conditions 25 years ago, when cladistic methods were first used to test accepted evolutionary scenarios of fish-tetrapod transformations (Patterson 1980; D. E. Rosen et al. 1981). The furor generated by this challenge to ancestor-descendant scenarios—which themselves were more or less direct descendants of works by Huxley (1861) and Cope (1871)—did much to force the debate about the relevance and utility of fossil data (Panchen and Smithson 1987). Primitive conditions and the potential to reveal instances of homoplasy (convergence) emerged as key attributes of fossils in phylogenies. The discovery of polydactylous tetrapod limbs underscored further the importance of fossils for revealing morphologies absent in the extant biota.

The most comprehensive analyses of the tetrapod stem (Johanson and Ahlberg 2001; Jeffery 2001) place *Kenichthys*, a sarcopterygian fish from the Middle Devonian (~380 mya) of China, as the most basal tetrapod in the broadest sense of the term (i.e., as member of the tetrapod total group; fig. 2.1B). The divergence date from shared ancestry with lungfishes (or coelacanths) is likely to have been Lower Devonian, ~400+ mya. However, *Kenichthys* is poorly preserved, and the median and paired fins are unknown (M. M. Chang and Zhu 1993). Thus, paired fin conditions at the very base of the tetrapod stem are better indicated by fossil outgroups, such as the porolepiforms.

Branching patterns at the apex of the stem group (fig. 2.2) are more intensely disputed, with widely differing theories about the position of the tetrapod crown-group node, and thus the basal divergence of lissamphibians from amniotes. Most of the changes usually associated with the fin-to-limb transition are completed within taxa branching from nodes below most of the hypothesized positions of the crown-group radiation. However, the most taxon-inclusive crown hypotheses incorporate the hexadactylous Late Devonian genus *Tulerpeton* as a basal stem amniote (Lebedev and Coates 1995; Coates 1996), and thus posit the lissamphibian-amniote split at a locus preceding the inferred origin of a five-digit manus and pes. The lissamphibian-amniote divergence is thereby pegged to a minimum date of around 360 mya. In contrast, the least inclusive hypothesis excludes a series of taxa from the crown group, so that several putative stem amphibians and stem amniotes are repositioned as stem taxa (Laurin 1998a; Laurin et al. 2000). Pentadactylous limbs thus evolve below the crown-group node, and the minimum age of the crown group is reduced to about 340 million years (Lower Carboniferous). Neither extreme is used directly in the present work. The simplified tree apex used here (fig. 2.2B) is abstracted from a combined reanalysis, which places

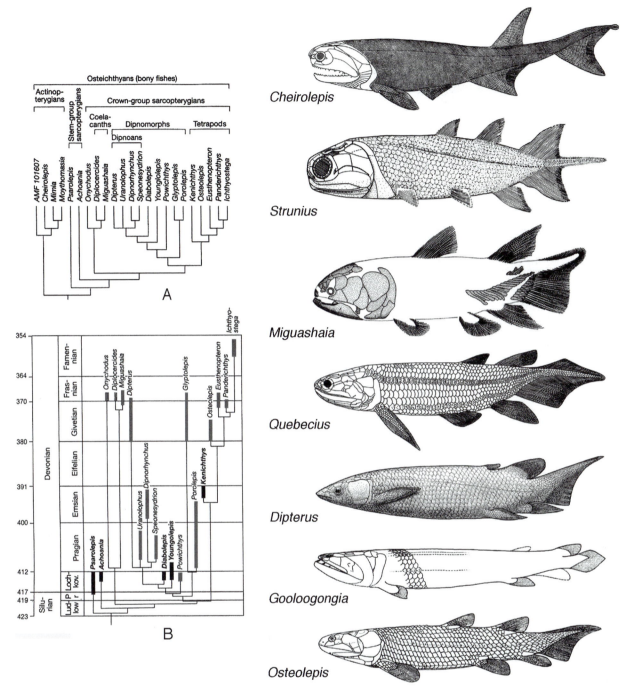

Figure 2-1 Sarcopterygian fish diversity and interrelationships. (A) A simplified cladogram for sarcopterygians, including stem tetrapods. (B) Plot of cladogram in A on a time scale (A, B, from Zhu et al. 2001, fig. 3a, b). Osteichthyan reconstructions (not drawn to the same scale) include the basal actinopterygian, *Cheirolepis canadensis* (after Pearson and Westoll 1979, fig. 16a); the onychodontiform, *Strunius walteri* (after Jessen 1966, fig. 7); the basal actinistian, *Miguashaia bureaui* (after Cloutier 1996, fig. 1b, and Forey 1998, fig. 11.13); the porolepiform, *Quebecius quebecensis* (after Cloutier and Schultze 1996, fig. 2b); the dipnoan, *Dipterus valenciennesi* (after Ahlberg and Trewin 1995, fig. 9a); the rhizodont, *Gooloogongia loomesi* (reversed from Johanson and Ahlberg 2001, fig. 18a); and the osteolepiform, *Osteolepis macrolepidota* (after Moy-Thomas and Miles 1971, fig. 6.1).

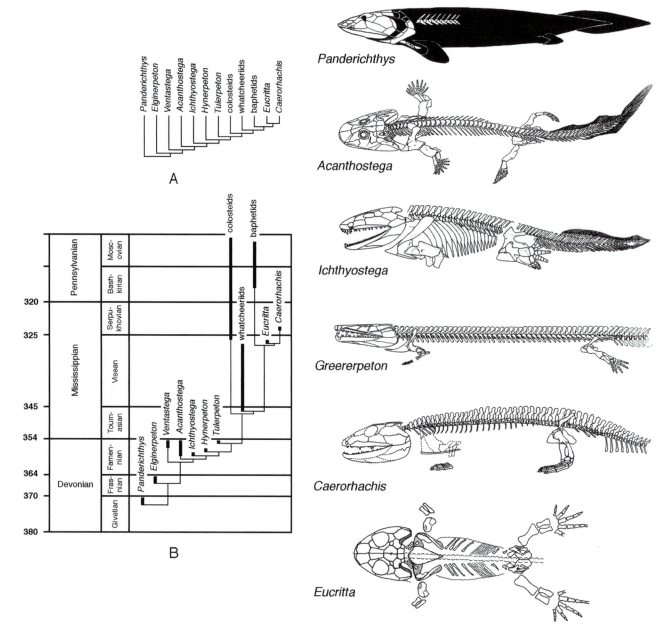

Figure 2-2 Diversity and interrelationships of postpanderichthyid stem tetrapods. (A) Simplified cladogram (combined from Coates 1996 and Ruta et al. 2003). (B) Plot of cladogram in A on a time scale (new). Reconstructions (not drawn to the same scale) include the most derived sarcopterygian showing fins, *Panderichthys rhombolepis* (after Coates 2001, fig. 1.3.7.1b); the two best-known Devonian limbed tetrapods, *Acanthostega gunnari* and *Ichthyostega stensioi* (after Coates 1996, fig. 31; and Coates and Clack 1995, fig. 1c); the best-known colosteid, *Greererpeton burkemorani* (after Godfrey 1989, fig. 1a); the putative basal stem amniote, *Caerorhachis bairdi* (after Ruta et al. 2001, fig. 1a); and the basal baphetid, *Eucritta melanolimnetes* (modified after Clack 2001, fig. 8a). In Ruta et al. (2003), *Eucritta* and *Caerorhachis* bracket the crown group node.

Tulerpeton plus several Lower Carboniferous taxa on the tetrapod stem, but the majority of early limbed tetrapods remain within the crown group (Ruta et al. 2003). The minimum date of 340 mya is robust to these changes because of the diversity of tetrapods first known from similarly aged strata.

Sarcopterygian Paired Fins and Girdles, Excluding Tetrapods

Out-group conditions (i.e., primitive actinopterygian patterns) show that pectoral and pelvic fins differ in shape and size, and that pectorals are generally bigger than pelvics

(Coates 1994, 2003; Bemis and Grande 1999). Pectoral fins supported by stout muscular lobes are present in primitive actinopterygians (Pearson and Westoll 1979), and it seems likely that the lobate pectorals of the living cladistia (*Polypterus* and *Erpetoichthys*) are plesiomorphic in this respect. Pelvic fins are primitively long based, with no muscular lobe extending from the body wall.

These differences in shape reflect differences in endoskeletal pattern. Sets of pectoral radials primitively include morphologically distinct pro- and metapterygial units, whereas pelvic radials are primitively small and often lack distinct morphological identities. Given the repeated conclusion that the sarcopterygian endoskeleton is homologous with the metapterygial unit of nonsarcopterygian fins (Rosen et al. 1981, among others), it is noteworthy that in nonsarcopterygians the metapterygium is mostly smooth and bears none of the pre- and postaxial processes that collectively represent a further, and in this respect largely overlooked, sarcopterygian synapomorphy (cf. character sets in Cloutier and Ahlberg 1996; Forey 1998; Zhu and Schultze 2001).

Psarolepis

Psarolepis romeri (Yu 1998), from the Lower Devonian of Yunnan, China, is among the earliest and most primitive known sarcopterygians. Because it is known from only fragmentary remains, the coherence of this taxon is open to question. If current interpretations are correct, then *Psarolepis* shows that the dermal skeletal component of the pectoral girdle primitively includes large spines preceding the fins (Zhu et al. 1999). Given the presence of paired fin spines in assorted out-groups of bony fishes, the implication is that such structures were lost independently in ray-finned and lobe-finned lineages. Further details of *Psarolepis* fins are unknown.

Coelacanths

The earliest coelacanths are known from the lowermost Upper Devonian (Frasnian), in excess of 370 mya (Schultze 1993). This marks the Actinistia (coelacanths) as among the youngest of the major sarcopterygian divisions. Coelacanths are conservative in terms of their gross morphology. Paired fins and girdles of the sole living example, *Latimeria chalumnae* (Millot and Anthony 1958; Forey 1998), are thus reasonable exemplars of conditions throughout most of the group's phylogeny (fig. 2.3).

The pectoral girdle (fig. 2.3A), like those of all osteichthyes excluding the crown-group tetrapods, is for the most part composed of dermal bones, the largest of which are the cleithrum and clavicle. An anocleithrum extends an-

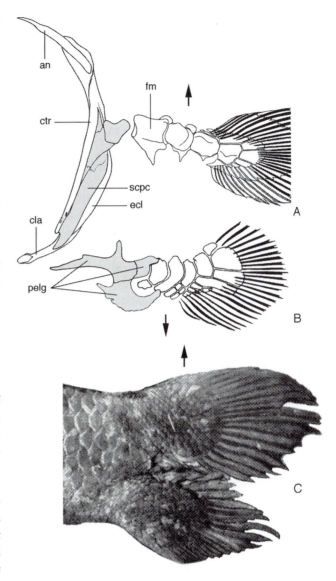

Figure 2-3 Appendicular and fin skeleton of the extant coelacanth, *Latimeria chalumnae*. (A) Right pectoral girdle and fin in mesial view (reversed from Millot and Anthony 1958, fig. 1840). (B) Left pelvic girdle and fin in dorsal view (after Forey 1998, fig. 8.4b). (C) Close-up view of pelvic fins in ventral view (after Millot and Anthony 1958, fig. 1832). Arrows indicate leading edge of fins. See appendix for abbreviations.

terodorsally from the apex of the cleithrum toward the rear of the skull roof; coelacanths also have an extracleithrum (fig. 2.3A, ecl), which is a specialization of the group. As in tetrapods, the dermal pectoral girdle is separate from the skull. There is no supracleithrum or posttemporal, as in other primitive or generalized bony fishes such as basal stem tetrapods. An interclavicle, ventormedially uniting both halves of the dermal girdle, seems also to be generally and primitively absent in coelacanths (*contra* Forey 1998).

The scapulocoracoid is a single endoskeletal unit and articulates over a broad area of cleithrum and clavicle (fig.

2.3A, scpc; Millot and Anthony 1958). Like those of the (fossil) porolepiforms and extant lungfishes, the scapulocoracoid is not perforated by nerve and vascular foramina, and the articular surface is strongly convex. This is the reverse of the tetrapod pattern, in which the scapulocoracoid bears a concave glenoid surface and the humeral head is convex. This pattern of scapulocoracoid morphology seems to have evolved early in actinistian phylogeny (Ahlberg 1989; Forey 1998).

The pectoral fin endoskeleton consists of four axial "mesomeres." In this context, the term "mesomere" is not associated with the synonymous mesodermal structures of vertebrate embryos. Here we follow the convention of using "mesomere" as the name for the subcylindrical radial segments of the principal axis in sarcopterygian fins (cf. Jarvik 1980; Ahlberg 1989; Janvier 1996; Forey 1998). The most proximal, first axial mesomere articulates with the girdle via a concave articular surface. The pectoral fin is slightly larger (fig. 2.3A, fm) than the pelvic; each mesomere bears prominent ridges for the attachment of segmentally arranged muscles. The ventral process, corresponding to the postaxial process of other sarcopterygian fins, is better developed than the dorsal processes. The radials are mostly preaxial in position, and not related in any clearly segmental pattern relative to the mesomeres.

The fin web, like that of the pelvic, is supported by lepidotrichia (fig. 2.3C). These consist of two parallel and symmetrical bony rods or strips termed "hemilepidotrichia" (Géraudie and Meunier 1984). Each is composed of successive segments separated by short gaps representing joints. Distally, these overlap the actinotrichia: unmineralized long tapering rods of elastoidin. Proximally, the hemilepidotrichia are separated where they overlap the extremities of the endoskeletal radials. The structure and distribution of these hard and soft fin rays, supporting the outermost parts of the adult fins, are remarkably similar to those of Actinopterygii (Géraudie and Meunier 1980, 1982; Géraudie and Landis 1982). In *Latimeria* the rays are distributed more or less evenly around the fins, although they extend more proximally along the leading (preaxial) edge of the pelvic than pectoral.

The pelvic girdle (fig. 2.3B) is embedded in an abdominal position. Endoskeletal, with no dermal component, the pelvic bones diverge and expand posteriorly, each with a convex articular surface. Adjacent to the articular area, a medial expanded (ischial) process attaches to the corresponding process of the opposite girdle half. A lateral (iliac) process is also present. Forey (1998) notes that the pelvic fin construction in *Latimeria* is the structural reverse of the pectoral (note orientations in fig. 2.3A, B). The four axial mesomeres are shorter than the pectoral mesomeres, and the entire fin is

slightly shorter and broader. Radials are more obviously arranged in a segmental manner and associated with particular mesomeres. Fin rays are distributed in a more asymmetric pattern around the fin perimeter.

Within the coelacanths (Actinistia), pectoral-pelvic similarity is phylogenetically deep (Forey 1998). No taxa show significant differences between fore and hind fin pairs, although few fossils preserve detailed fin morphologies. The most striking deviation from the pattern preserved in *Latimeria* is the advanced teleosts-like anterior translocation of the pelvic skeleton in a few fossil taxa, most notably *Laugia* (Forey 1998).

Onychodonts

Onyochodontiformes (struniiforms) are a poorly known group of sarcopterygians of which very few are known from body fossils. Occurrences are restricted, and extend from the Early to Late Devonian. As a possible subgroup arising from the coelacanth stem (Zhu et al. 2001; Zhu and Schultze 2001), or from some earlier node in sarcopterygian phylogeny (Cloutier and Ahlberg 1996), onychodontiform conditions could have a major bearing on hypotheses of primitive patterns of lobate paired appendages. In fact, few postcranial data have been described.

Detailed description of exceptionally well-preserved material of at least the forequarters of a species of *Onychodus* are currently in preparation (J. A. Long 2001). From this it is clear that the dermal pectoral girdle includes a large cleithrum and clavicle, and, like most osteichthyans, a supracleithrum connecting the girdle to the rear of the skull table (J. A. Long 2001). A median interclavicle is also considered present (Cloutier and Ahlberg 1996). Pectoral fin insertion is high, as in *Latimeria*, and a short projecting glenoid portion of the scapulocoracoid is visible in lateral aspect. Again, this resembles *Latimeria*, but the glenoid is now known to be concave (J. A. Long, Western Australian Museum, Perth, Australia, pers. comm.). It is also now known that the most proximal axial mesomere of the pectoral fin is large, with a perforated postaxial process (J. A. Long, Western Australian Museum, Perth, Australia, pers. comm.).

Jessen (1966) reconstructed the paired fins of a second genus, *Strunius* (fig. 2.1), as small and narrow based at pectoral and pelvic levels (endoskeletal patterns are unknown; the dermal pectoral girdle is incomplete). Lepidotrichia are present, but no unusual features are noted (Jessen 1966).

Lungfishes

Lungfishes (Dipnoi) have a fossil record extending back to the late Early Devonian (Schultze and Marshall 1993; Janvier

1996). The extant genera, *Neoceratodus* (Australian), *Lepidosiren* (South American), and *Protopterus* (African), are specialized in many respects relative to early fossil forms. Suborder Ceratodontoidei, effectively coextensive with the dipnoan crown group, encompasses all three extant taxa and dates from the Lower Triassic, some 245 mya. Undisputed members of the dipnoan stem group date from a minimum of 390 mya, leaving a 145 million year lineage of fossil taxa in which paired fins are known mostly from external morphology (summarized in Ahlberg 1989). In most regards these resemble the paired fins of *Neoceratodus*.

Neoceratodus forsteri has a dermal pectoral girdle consisting of three bones: the anocleithrum, cleithrum, and clavicle (fig. 2.4A). The long axis of the anocleithrum is directed forward and attached to the rear of the skull roof by a stout ligament. There is no direct osseous connection. The largest dermal bone of the girdle is the cleithrum, the ventral margin of which articulates with the clavicle. In early lungfish a supracleithrum is also present; an interclavicle seems to be primitively absent (Jarvik 1980; Janvier 1996).

The scapulocoracoid of *Neoceratodus*, like that of the extant actinistian *Latimeria*, is applied across a large area to the medial surface of the cleithrum, and extends ventrally onto the clavicle. There are no large canals or foramina, and the articular glenoid area consists of a distinctly convex knob. A ventral median cartilage (fig. 2.4A, vmc) bridges the dorsomedial surfaces of the clavicles; the origin of this specialized feature is uncertain (Goodrich 1930; Jarvik 1980). Like *Neoceratodus*, the scapulocoracoid in Devonian lungfishes is also large, but attached to the inner surface of the cleithrum via distinct buttresses that straddle supracoracoid and supraglenoid canals (fig. 2.4E). In this respect it resembles the scapulocoracoid of stem-group tetrapods (cf. fig. 2.7A), indicating some convergence between extant Dipnoi and Actinistia.

Lungfish paired fins are usually described as having a biserial endoskeletal pattern. A well-developed central axis of 18 or more mesomeres extends almost to the distal tip (fig. 2.4B, D). In *Neoceratodus* radials articulate with preaxial and postaxial surfaces to produce an elongate, leaf-shaped, distally tapered outline. In the pectoral fin the preaxial radials are related in a more clearly segmental pattern relative to the mesomeres than the postaxial radials. In *Lepidosiren* and *Protopterus* the distribution of these radials is much reduced and restricted to the trailing, postaxial, side of the elongate, whiplike, pectoral fins (30+ mesomeres). As in coelacanths, the most proximal mesomere has a concave surface articulating with the condyle-bearing pectoral girdle. The anteroposteriorly broad second mesomere is the ontogenetic product of incompletely separated axial and preaxial cartilages (Joss and Longhurst 2001). Unlike tetrapods, the

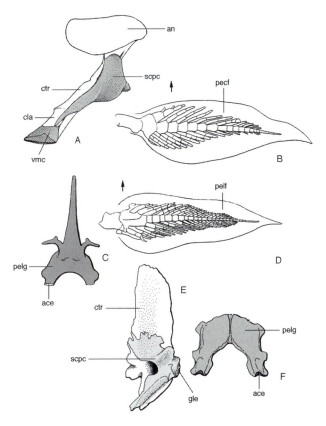

Figure 2-4 Appendicular and fin skeleton of dipnoans. (A) Right half of pectoral girdle of *Neoceratodus forsteri* in mesial view (after Jarvik 1980, fig 334b). (B) Right pectoral and (D) pelvic fins of *Neoceratodus forsteri* in extensor view (after Haswell 1882, pl. 1, figs. 1, 5). (C) Pelvic girdle of *Neoceratodus forsteri* in dorsal view (after Goodrich 1930, fig. 152). (E) Right half of pectoral girdle of *Chirodipterus australis* in mesial view (after Janvier 1996, fig. 4.82a1). (F) Pelvic girdle of *Chirodipterus australis* in ventral view (after Janvier 1996, fig. 4.82b). Arrows indicate leading edge of fins.

preaxial (presumed radial homologue) is smaller than the axial cartilage (ulnar homologue). Each mesomere bears ridges for the attachment of segmentally arranged muscles, although these are noticeably less pronounced than those of actinistians and stem-group tetrapods.

In *Neoceratodus*, fin rays are distributed more or less equally along pre- and postaxial edges of pectoral and pelvic fins. This contrasts with *Lepidosiren* and *Protopterus*, in which fin rays are much reduced and confined to the postaxial pectoral fin margin. Lungfish fin rays have a peculiar structure unlike the lepidotrichia and actinotrichia of actinopterygians and *Latimeria*. Named "camptotrichia" (Goodrich 1904), these consist partly of bone, either cellular or acellular, and partly of a "permanently pre-osseous tissue" permitting an exceptional degree of flexibility (Géraudie and Meunier 1984). Often cylindrical in cross section, each camptotrichium extends to the fin perimeter with no intervening actinotrichia. Camptotrichia show no symmetrical arrangement across opposing surfaces of the fin web; thus, each

camptotrichium might correspond to a single hemilepi-
dotrichium. Géraudie and Meunier (1984) comment that the
origin of camptotrichia lies phylogenetically deep within the
Dipnoi.

Lungfish pelvic girdles generally consist of two large
plates fused across the ventral midline by means of an extensive symphysial cartilage (fig. 2.4C, F). Each acetabular surface faces posteriorly and, in recent taxa at least, like the glenoid is convex (but see fossil examples in fig. 2.4). Goodrich
(1930) noted that, except for the absence of an ilium-like process, the (extant) dipnoan pelvis resembles that of the
urodele *Necturus*. A long, tapering epipubic process projects
anteriorly, and on either side a slender prepubic process em-
beds into an intermuscular septum. Figure 2.4 contrasts the
pelvis of *Neoceratodus* with the even broader and more
tetrapod-like pelvis of the Devonian taxon *Chirodipterus*.

Pelvic fin patterns resemble very closely those of pectoral
fins, although in *Protopterus* and *Lepidosiren* pelvic fin rays
are absent (Goodrich 1930).

Porolepiforms

Porolepiformes, an extinct clade of sarcopterygians, have a
fossil record extending from the Early Devonian to the base
of the Carboniferous (Schultze 1993; Ahlberg 1989). Few
fossil sarcopterygian fin skeletons are known in any detail
beyond those of taxa associated directly with the tetrapod
stem group; thus the fin skeletons of the porolepiform *Glyptolepis* are of particular significance (Ahlberg 1989).

The dermal skeletal pectoral girdle (fig. 2.5) includes the
standard set of bones: clavicle, large cleithrum, anocei-
thrum, and supracleithrum (Jarvik 1980; Ahlberg 1989). The
primitive presence or absence of an interclavicle among
porolepiforms is uncertain (Cloutier and Ahlberg 1996). The
scapulocoracoid consists of a broad, flat basal plate applied
closely to the mesial face of the cleithrum. From this projects
a wide but dorsoventrally thin mesial flange, the posterior
edge of which forms a rounded, strap-shaped glenoid. The
entire scapulocoracoid is a single ossification without any
sutures, foramina, or large canals (cf. *Chirodipterus* and
stem group tetrapods; figs. 2.4 and 2.7).

The pectoral fin endoskeleton (fig. 2.5B) is reminiscent
of conditions in the extant *Neoceratodus*. A well-formed
central axis consists of mesomeres with dorsal and ventral
processes. In proximal elements the ventral processes are
considerably larger than dorsal processes. From the third
mesomere onward, each bears pre- and postaxial radials.
Preaxial radials are slightly longer than postaxials. The long
tapering shape of the pectoral fin seems to include space for
18 or more mesomeres in total. Fin rays are well developed
and consist of conventional lepidotrichia. Basal parts of each

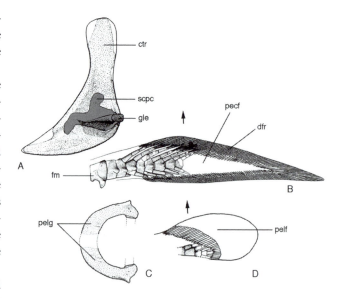

Figure 2-5 Appendicular and fin skeleton of porolepiforms. (A) Right half of
pectoral girdle of *Glyptolepis* sp. in mesial view (after Ahlberg 1989, fig. 3a).
(B) Left pectoral fin of *Glyptolepis ?leptopterus* (reversed from Ahlberg 1989,
fig. 5b). (C) Pelvic girdle and (D) pelvic left fin of *Glyptolepis ?leptopterus* (after
Ahlberg 1989, fig. 11). Arrows indicate leading edge of fins.

lepidotrichium are unjointed; postaxial lepidotrichia start at
the level of mesomere 4, and preaxial lepidotrichia at the
level of mesomere 5.

The pelvic girdle (fig. 2.5C) consists of a simple, curved
bar meeting its antimere at the ventral midline. The acetabulum is situated at the posterolateral extremity; the convexity
or concavity of the articular surface is unclear.

Pelvic fin morphology (fig. 2.5D) differs from that of the
long, tapered pectoral. Pelvic fins are shorter, proximodistally, and broader anteroposteriorly. The precise numbers of
mesomeres per fin is uncertain; radials articulate only with
the leading edge. Lepidotrichia, forming the broad, rounded
gross outline of the fin, articulate with only preaxial/anterior
and distal extremities of the endoskeleton.

Further variation in porolepiform pelvic fins is manifest
in basal porolepiforms such as *Quebecius* (fig. 2.1; Cloutier
and Schultze 1996, and references therein). The fin is remarkably broad-based and lacks any lobe whatsoever, thus
resembling pelvic fins in primitive ray-finned fishes.

Tetrapod Fins, Limbs, and Girdles

Rhizodontids

The Rhizodontida (Andrews and Westoll 1970a) is the most
basal group of stem tetrapods with evidence of paired fin
structures. Rhizodontid fossils are known from the Upper

Devonian to the Upper Carboniferous, and their paired fins have been long been noted for their particular resemblances to primitive tetrapod limbs (J. A. Long 1989; Daeschler and Shubin 1998; Vorobyeva 2000; M. C. Davis et al. 2001; Jeffery 2001; Johanson and Ahlberg 2001).

The dermal skeletal pectoral girdle (fig. 2.6A, E) includes a clavicle, large cleithrum, anocleithrum and supracleithrum. A slender interclavicle appears to be present (Andrews 1985; Davis et al. 2001; see Jeffery 2001 for alternative opinion). Clavicles and cleithra are particularly well ossified; Johanson and Ahlberg (2001) describe a large field of sensory canal pores on the cleithrum of the Upper Devonian genus *Gooloogongia*. The scapulocoracoid, which has been described in a couple of genera, *Rhizodus* and *Strepsodus* (Jeffery 2001), consists of a massive, well-ossified, medially projecting shelf bearing a posteriorly and slightly laterally directed glenoid. This shelf is supported by stout supra- and infraglenoid buttresses. The entire scapulocoracoid is a single ossification without any sutures, but the bone is penetrated by several foramina including a large glenoid foramen and canal. The entire scapulocoracoid is applied to the medial face of the cleithrum. The glenoid articular surface is concave.

Rhizodont pectoral fins are large (fig. 2.6B, D): broad distally, with well-developed lobes, endoskeletons, extensive scale cover, and elongate fin rays. Jeffery (2001) and M. C. Davis et al. (2001) document amply the historical significance of these fins in the search for ancestral tetrapod limbs. The classic exemplar rhizodont fin skeleton, that of *Sauripterus,* is now known from quite exceptional specimens from the Upper Devonian of Pennsylvania, including subadult as well as adult material (Daeschler and Shubin 1998; M. C. Davis et al. 2001).

The major axis of the *Sauripterus* fin (fig. 2.6B) consists of three mesomeres: presumed homologues of the humerus, ulna, and ulnare. These follow a 1:2 proximal to distal ratio of humerus articulating with ulna and radius, and ulna with ulnare and intermedium. Four radials articulate distal to the ulnare in *Sauripterus;* three in *Gooloogongia* (Johanson and Ahlberg 2001) and *Barameda* (J. A. Long 1989; fig. 2.6C), and two in *Rhizodus* (Jeffery 2001). Each distal radial articulates in a further 1:2 sequence, but the continuation of the mesomeric axis beyond the ulnare is obscure.

The humeral shaft is short and subcylindrical, with a strongly convex head and prominent ventral/postaxial and dorsal processes. The postaxial process (fig. 2.6B, C, ent) is far larger than any example noted thus far (i.e., any non-tetrapod sarcopterygian), and projects in the same direction as the entepicondyle of limb-supporting humeri. The dorsal processes resemble the ectepicondyle and supinator process of limb humeri (Jeffery 2001). However, as opposed to fins, in early tetrapod limb humeri the entepicondyle is broad proximodistally, whereas in rhizodonts the entepicondyle is broad dorsoventrally, the dorsal crest of which is more or less continuous with the aforementioned dorsal processes (Jeffery 2001). Rhizodont entepicondyle shape varies considerably; the likely functional significance of this is as yet unexplored.

Ulna shape is conservative throughout rhizodonts: short, broad, and subequal in length to the radius. The radius, in contrast, varies strongly between genera. In *Sauripterus* it bears a broad anteriorly projecting blade, in *Barameda* it is narrow and rodlike, and in *Rhizodus* squat, resembling an ulna. In all cases, further radials articulate with the distal end

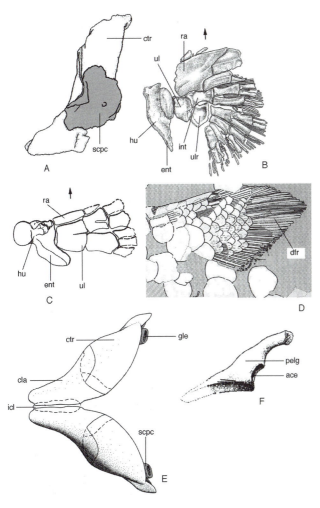

Figure 2-6 Appendicular and fin skeleton of rhizodonts. (A) Right scapulocoracoid of *Strepsodus sauroides* in mesial view (after Jeffery 2001, fig. 5f). (B) Left pectoral fin of cf. *Sauripterus* in extensor view (reversed from Davis et al. 2001, fig. 7b). (C) Right pectoral fin of *Barameda decipiens* in extensor view (reversed from Long 1989, fig. 11b). (D) Right pectoral fin of *?Strepsodus anculonamensis* (reversed from Andrews 1985, fig. 3). (E) Pectoral girdle of large Faulden rhizodont in ventral view (after Andrews 1985, fig. 13c). (F) Left half of pelvic girdle of *Gooloogongia loomesi* in lateral view (after Johanson and Ahlberg 2001, fig. 12b). Arrows indicate leading edge of fins.

of the radius. In fact, the profusion of distal radials in rhizodont pectoral fins is noteworthy in its own right, totaling in excess of 20 in *Sauripterus* (M. C. Davis et al. 2001).

Fin rays consist of specialized lepidotrichia (Andrews 1985; Jeffery 2001). Each is segmented only distally, and thus consists mostly of an elongate basal hemilepidotrichial segment. Like camptotrichia these may be circular in cross section and rarely if ever branch, and there appears to be a comprehensive loss of register between dorsal and ventral counterparts. Lepidotrichial overlap with the endoskeleton is extensive; rays and endoskeleton may, in turn, be overlapped by scale cover extending almost to the fin perimeter (?*Strepsodus*, Andrews 1985; fig. 2.6D).

Rhizodont pelvic fins are lobate, positioned toward the rear of the body, and known mostly from external morphology. Unlike the vast majority of sarcopterygian pelvics, they are much smaller than the pectorals. The pelvic girdle (fig. 2.6F) is barlike, with an unossified symphysis and a concave acetabulum flanked by pubic and iliac processes (Johanson and Ahlberg 2001). It is remarkably similar to that of the tristichopterid *Eusthenopteron* (fig. 2.7G).

Osteolepiforms

The group Osteolepiformes is now understood to be paraphyletic, and thus no longer constitute a formal taxon (Cloutier and Ahlberg 1996; Ahlberg and Johanson 1998). Instead, the osteolepiform fishes represent a grade or series of monophyletic groups branching from the tetrapod stem. The largest, best characterized, and most derived of these (in terms of proximity to limbed tetrapods) is the Tristichopteridae (otherwise referred to as the Eusthenopteridae; e.g., Schultze 1993). Others include the Osteolepididae, Canowindridae, and Rhizodopsidae (Schultze 1993). The stratigraphic range of osteolepiform fishes extends from the upper Middle Devonian to the Lower Permian. Relative to flanking groups on the tetrapod stem, although numerically diverse, osteolepiforms are morphologically conservative (Johanson et al. 2003). The paired fins of the exemplar taxon used here, the basal tristichopterid *Eusthenopteron foordi*, are thus plausible stand-ins for conditions in primitive members of other osteolepiform clades.

In comparison with rhizodontids, *Eusthenopteron* is by far the best-known fishlike stem tetrapod (see Jarvik 1980, 1996, and references therein). The appendicular skeleton has long been central to discussions about the fin-limb transition, of which Andrews and Westoll's (1970a) study of the *Eusthenopteron* postcranium remains an unparalleled source of detailed basic data.

On either side of the slender median interclavicle, each half of the dermal pectoral girdle includes a clavicle, clei-

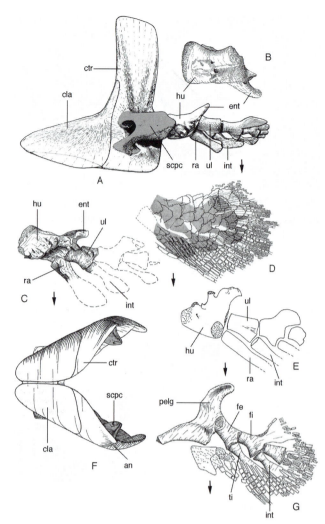

Figure 2-7 Appendicular and fin skeleton of osteolepiforms, based upon *Eusthenopteron foordi* (A–D, F–G) and *Sterropterygion brandei* (E). (A) Right pectoral girdle and fin in mesial view (after Jarvik 1980, fig. 100). (B) Right humerus in extensor view (reversed from Andrews and Westoll 1970, fig. 10a). (C) Right pectoral fin in flexor view (reversed from Andrews and Westoll 1970, fig. 6b). (D) Lepidotrichia and scales associated with right pectoral fin in extensor view (after Andrews and Westoll 1970, fig. 7a). (E) Right pectoral fin of *Sterropterygion* in extensor view (after Rackoff 1980, fig. 9). (F) Pectoral girdle of *Eusthenopteron* in ventral view (after Andrews and Westoll 1970, fig. 2e). (G) Left pelvic girdle, fin, and lepidotrichia in extensor view (reversed from Andrews and Westoll 1970, fig. 17). Arrows indicate leading edge of fins.

thrum, anocleithrum, and supracleithrum (fig. 2.7A, F). The latter contacts the posttemporal and thus articulates with the rear of the skull roof. The scapulocoracoid is attached to the medial wall of the ventrolateral angle of the cleithrum. The scapulocoracoid consists of a small, tripodal, endochondral ossification. The anterior pair of "feet" are the supraglenoid and infraglenoid buttresses, and the posterior "foot" is formed by the attachment of the main mass of bone to the cleithral wall. These attachments define a three-way canal or space communicating between supraglenoid and supracoracoid foramina, and a large subscapular fossa. The glenoid is concave and subelliptical, with the longest axis oriented horizontally.

The major axis of the pectoral fin (fig. 2.7B–D) consists of four mesomeres, and the characteristic 1:2 proximal to distal ratio is repeated throughout the fin endoskeleton. Only preaxial radials are present, as well as a pair of small terminal radials beyond the fourth mesomere. First, third and, occasionally, fourth mesomeres bear prominent postaxial processes. Thus the humerus homologue has a large, well-developed entepicondyle (fig. 2.7B, C). This narrows and curves to a slight hook distally; proximally, the process is continuous with a prominent ventral ridge sweeping obliquely across the underside of the humeral shaft. As in rhizodontids, dorsal processes resemble the ectepicondyle and supinator process of tetrapod limb humeri. It is noteworthy that while some variation is apparent, the same general features are identified easily in the pectoral fin of a Devonian osteolepidid, *Sterropterygion* (Rackoff 1980; fig. 2.7E).

The pelvic girdle (fig. 2.7G) is small and barlike, with posterodorsal (iliac) and anterior, ventromedial (pubic), processes. Note that this orientation follows Andrews and Westoll's (1970a) interpretation rather than Jarvik's (1980). As noted earlier, the pelvis of *Eusthenopteron* is remarkably similar to that of the primitive rhizodontid *Gooloogongia* (Johanson and Ahlberg 2001). The concave acetabulum faces posterolaterally and receives the convex head of the short cylindrical femur. The pelvic fin is slightly smaller than the pectoral, with an axis of only three mesomeres, and with a postaxial process extending from only the second of these (the fibular homologue). The significance of this "out-of-step" registration between pectoral and pelvic endoskeletons has been the source of much discussion, including speculation that the pelvic girdle originated as the most proximal fin segment embedded in hypaxial musculature (cf. Rackoff 1980; Rosen et al. 1981, and references therein).

Fin rays in osteolepiforms consist of conventional lepidotrichia (fig. 2.7D), segmented at even intervals throughout their length. Andrews and Westoll (1970a) note some elongation of proximal segments at the level of fin ray insertion, where they overlap postaxial processes and the spatulate ends of endoskeletal radials.

Panderichthyids

The Panderichthyida (Vorobyeva and Schultze 1991) includes three genera, *Panderichthys*, *Elpistostege*, and *Obruchevichthys*, all of which are confined to the Late Devonian (Frasnian). The monophyletic status of this clade is, however, questionable, and it may be that these genera constitute nothing more than another grade on the tetrapod stem (Ahl-

berg et al. 2000). *Panderichthys* is the only member known from complete specimens (Vorobyeva 1992).

The dermal pectoral girdle includes the full complement of bones described in osteolepiforms. Vorobyeva and Schultze (1991) comment on the narrow external exposure of these; otherwise they display the conventional, plesiomorphic, articulation with the rear of the skull table by means of posttemporal contact. However, unlike rhizodontid and osteolepiform clavicles, in *Panderichthys* the posterior rim of the clavicle (fig. 2.8D) is not bounded by any ventromedial extension of the cleithrum (Vorobyeva 1992). The interclavicle is small, and situated posteromedial to the anterior articulation of the clavicles (Vorobyeva and Schultze 1991).

The scapulocoracoid (fig. 2.8A, C, D) differs significantly from that of *Eusthenopteron* (Vorobyeva and Schultze 1991; Vorobyeva 1992). The scapulocoracoid is attached to the cleithrum across a single broad surface formed by fusion of the three buttresses also present in osteolepiforms. The coracoid plate is much enlarged, and in mesial view the ventral expansion of the endoskeletal girdle obstructs any view of the ventral rim of the cleithrum. Major canals perforating the scapulocoracoid are separated from the inner surface of the cleithrum by cartilage bone. The coracoid plate, perforated

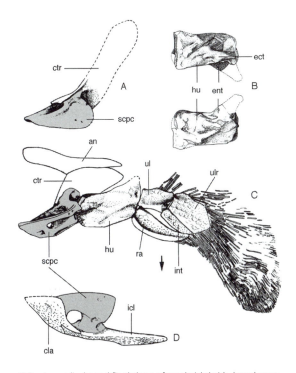

Figure 2-8 Appendicular and fin skeleton of panderichthyids, based upon *Panderichthys rhombolepis*. (A) Right scapulocoracoid in mesial view (after Vorobyeva 1992, fig. 60a). (B) Right humerus in extensor (*upper*) and flexor (*lower*) views (after Vorobyeva 2000, fig. 3b, d). (C) Right pectoral fin and girdle in ventral view (after Vorobyeva 1992, fig. 57b). (D) Right half of pectoral girdle in ventral view (after Vorobyeva 2000, fig. 9d). Arrows indicate leading edge of fins.

by a pair of small foramina, makes extensive contact ventrally with the clavicle (removed in fig. 2.8A, C; Vorobyeva and Kuznestov 1992). The glenoid is incompletely known, although described as a "shallow groove" (Vorobyeva and Schultze 1991), suggesting some similarity to the strap-shaped glenoids of early limb-supporting scapulocoracoids.

Pectoral (and pelvic) fins are located in unusually ventral positions relative to all examples described thus far. The pectoral endoskeleton (Vorobyeva 1992, 2000) is divided into only three segments proximodistally (fig. 2.8C). Each of these is morphologically distinct, and, unlike *Eusthenopteron*, *Sterropterygion*, and other sarcopterygian fins (with the notable exception of rhizodontids), there is no clear proximodistal iterative pattern. Humerus and ulna homologues could be described as first and second axial mesomeres, but there is no obviously axial characteristic to the plate distal to the ulna, which bears an uncertain relationship to the ulnare of tetrapod limbs. As noted by Vorobyeva and Kusnetsov (1992), this fin skeleton contains fewer elements than any other sarcopterygian example described thus far.

The humerus (fig. 2.8B) is uniquely—for a fin—like those of early limbs (Vorobyeva 2000). Perhaps most significantly, it is dorsoventrally compressed, and thus has separate extensor and flexor surfaces instead of the cylindrical shaft present in rhizodonts and osteolepiforms. The only prominent postaxial process of the entire fin endoskeleton is the entepicondyle of the humerus. The dorsal process is more similar to the ectepicondyles of early limb humeri: directed proximodistally, it terminates just above the ulnal condyle. The ulna is also dorsoventrally flattened, but like osteolepiform and other finned examples, subequal in length to the rodlike radius. The "ulnare" plate fits closely to the neighboring, slender, intermedium; there appear to be no further distal radials.

Fin rays consist of apparently conventional lepidotrichia that segment and branch only distally (Vorobyeva and Schultze 1991). Fin ray bases overlie the distal portions of the radius, intermedium, and "ulnare" plate.

Pelvic fins are much smaller than pectoral fins. Nothing has been described of the pelvic girdle and fin endoskeleton (Vorobyeva 1992, 2000). Absence of this information is one of the most outstanding gaps in the current data set.

Elginerpeton

Elginerpeton pancheni consists of an assortment of stem tetrapod jaw fragments from the Frasnian (Upper Devonian) of Morayshire, Scotland (Ahlberg 1995). A series of postcranial remains are known from the same locality and have been attributed to the same species (Ahlberg 1998). If interpreted correctly, then they occupy a crucial phylogenetic position, branching from the stem group below *Acanthostega* and above *Panderichthys* (Ahlberg 1998; Ahlberg et al. 2000).

In summary, the postcranial fragments include portions of scapulocoracoid and cleithrum; pelvic ilia; a complete right humerus; an incomplete right tibia; and part of a right femur. Frustratingly, nothing is known of radials, digits, or wrist and ankle bones. However, it is noteworthy that with the exception of the humerus, each of these (incomplete) bones is reasonably consistent with an *Acanthostega*- or *Ichthyostega*-like interpretation (see the summaries that follow). The humerus, however, is unique. The postaxial entepicondyle is extraordinarily broad proximodistally; the articular head unusually narrow; the radial condyle is directed ventrally; and, uniquely, there is no trace of an oblique ventral ridge.

Acanthostega

Acanthostega, from the Upper Devonian (Frasnian) of East Greenland, is the most basal tetrapod with digits known in any detail (Coates 1996; Clack 2002a). Despite its array of fishlike characters, the anatomy of its girdles and limbs departs significantly from that of osteolepiforms.

The pectoral girdle (fig. 2.9A, C) is detached from the back of the skull, and all dermal bones situated dorsal to the anocleithrum are lost. The cleithrum is reduced, posteroventrally, thereby exposing more of the scapulocoracoid in lateral view. In anterior view, a distinct postbranchial lamina forms an anteromedially directed flange. The clavicles are somewhat expanded anteriorly, but the most striking novelty is the enormous expansion of the interclavicle into a broad, lozenge- or kite-shaped plate, the posteromedial process of which extends beyond the posterior level of the scapulocoracoids (fig. 2.9C).

The endochondral pectoral girdle (fig. 2.9A, C) is in several respects similar to that of *Panderichthys*, except for the absence of large canals passing through the scapulocoracoid, and the presence of a broad fossa on the mesial surface. The coracoid region is broader than in *Panderichthys*, but the dorsal extent of the scapular region is similarly limited. The endochondral scapular process is not distinct from the large, dermal cleithrum, and a subvertical infraglenoid buttress is not strongly developed. The glenoid is much less twisted than in more derived tetrapods. The forelimb (fig. 2.9B) is paddle-shaped, and, as in *Panderichthys*, the complete endoskeleton can be divided into three segments: stylopodium, zeugopodium, and autopodium.

The humerus has the characteristic L-shaped form present in many early tetrapod limbs. Several canals open on its surface, including large entepicondylar foramina and an ectepicondylar foramen. A ventral humeral ridge is present,

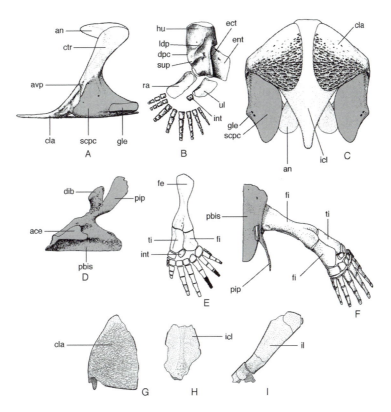

Figure 2-9 Appendicular and limb skeleton of *Acanthostega gunnari* (A–F) and *Ventastega curonica* (G-I). Pectoral girdle of *Acanthostega* in (A) left lateral and (C) ventral views (after Coates 1996, fig. 14a, b). (B) Reconstruction of left pectoral limb in extensor view (combined and partly reversed from Coates 1996, fig. 16d, 17b, h, 18a). (D) Pelvic girdle in left lateral view (reversed from Coates 1996, fig. 20c). (E) Reconstruction of left pectoral limb in extensor view (reversed from Coates 1996, fig. 24b). (F) Right pectoral limb and right half of pelvis in dorsal view (after Coates 1996, fig. 23c). (G) Right clavicle of *Ventastega* in ventral view (after Ahlberg et al. 1994, fig. 12a). (H) Interclavicle in ventral view (after Ahlberg et al. 1994, fig. 12c). (I) Left ilium in lateral view (after Ahlberg et al. 1994, fig. 13a).

although low and rounded relative to examples in osteolepiforms. In anterior aspect, a slanting, short deltopectoral crest lies considerably below the level of the radial facet. The posterior edge of the humerus bears a small acuminate process (Coates's [1996] "process 2"), and the proximal edge of the subrectangular entepicondyle forms a slightly obtuse angle with the long axis of the humeral shaft (such as it is). A distinct trough finished with periosteal bone separates radial and ulnar facets. The ectepicondyle, again like that of *Panderichthys*, consists of a proximodistally elongate ridge, but in *Acanthostega* this projects above and beyond the dorsal rim of the ulnar facet. Unlike more derived humeri, in this the ectepicondyle is not in line with the process for the m. latissimus dorsi (fig. 2.9B, ldp).

The radius is long and spatulate: the proximal subcylindrical shaft is dorsoventrally flattened distally. The ulna is subequal in length, like many preceding examples. The wrist is unossified with the exception of a short cylindrical intermedium, articulating with the anterodistal extremity of the ulna.

Eight digits are present. In common with those of all other tetrapod limbs, these digits consist of two or more spool-shaped bones/cartilages exhibiting a one-to-one pattern of proximodistal articulation; they form an anteroposteriorly arranged set or radiating series; they bear no simple ratio of unit-to-unit correspondence with more proximal limb parts.

There are no dermal fin rays.

Relative to all pelvises mentioned thus far, that of *Acanthostega* is much larger and morphologically far more complex (fig. 2.9D, F). Like all fin-bearing examples, the two halves are undivided plates of endochondral bone. These are united ventromedially via an elongate, presumably fibrous, puboischiadic symphysis. Opposing sides of the pubic region flare anteriorly and laterally to define an almost bowl-shaped forward-facing space; the posterior interischiadic volume is much narrower. The anterior edges of the pubic region are incompletely ossified, and laterally, this unfinished surface is continuous with the anterior extremity of the acetabular facet. The acetabular areas are bounded by a strongly ossified ventral shelf, prominent posterior buttress, and less pronounced superior buttress. More than one obturator foramen is present. The ischiadic region is expanded as a broad posteriorly extended plate, and the iliac region is extended dorsally into a complex forked process. The medial surface of each iliac process articulated with the bladed extremity of a simple sacral rib. The stout iliac neck is placed distinctly behind the level of the supraacetabular buttress.

The hindlimb (fig. 2.9E, F), like the pelvis, is also profoundly different from all pelvic appendages described thus far. The hindlimb is larger than the forelimb, and the bones of the hindlimb zeugopod, the tibia and fibula, are significantly shorter than the stylopod, the femur. Like the humerus, the femur bears a series of large processes indicating much greater elaboration of the appendicular muscles. Unlike the humerus, the femur has a distinct shaft region. Proximally, this is broadened to produce a femoral head, the concave underside of which forms a shallow trochanteric fossa. Distally, the femur expands to produce articular surfaces for the tibia and fibula. Ventrally, the femoral shaft bears a large and thickly ossified adductor blade, the central portion of which bears a rugose surface (fourth trochanter), and, proximally, a small acuminate internal trochanter.

The tibia and slightly smaller fibula are broad, flat, and subrectangular, with no semblance of a shaft region in either (fig. 2.9E). In dorsal and ventral aspects there is no trace of an interepipodial space; this is only visible and anterior and posterior aspects (see Coates 1996). More than any other parts of the hindlimb, these flattened epipodials contribute most to its paddlelike shape. Unlike the wrist, the ankle is ossified. Tibiale, intermedium, and fibulare are squarish, similarly sized bones that interarticulate anteroposteriorly as well as proximodistally.

An arc of distal elements, the identity of which as distal tarsals or metatarsals is uncertain, lies proximal to the eight digits. The two most anterior digits (including a diminutive digit I and rather squat digit II) are distal to the tibiale. Digits III–IV occupy the central part of the foot. More posteriorly, digits VI–VIII articulate with the fibulare.

Ventastega

Ventastega, from the Late Devonian of Latvia (Ahlberg et al. 1994), co-occurs with *Panderichthys*. Remains (fig. 2.9G–I) include fragmentary girdle elements with an array of generalized plesiomorphic features. The clavicle is broad and triangular, with smoothly curved margins, and much more expanded anteriorly than posteriorly (more so than in *Acanthostega*). An interclavicle is poorly preserved but shows the broad dimensions characteristic of early, limbed tetrapods. Shallow areas overlapped in life by the clavicular plates indicate that the clavicles were separated only narrowly at the ventral midline. The sole fragment of the pelvic girdle is a posterodorsal process of an isolated ilium. More elongate and strap-shaped than in *Acanthostega*, it has a gently arcuate mesial surface and a spatulate and slightly flared distal end. A small notch at the base of its posterior margin sug-

gests the presence of a stout iliac neck. Like the expanded interclavicle, this isolated piece is, again, indicative of a limb-bearing girdle.

Ichthyostega

Ichthyostega is the classic "earliest tetrapod," from the Upper Devonian (Frasnian) of East Greenland (extensive literature on this genus summarized in Jarvik 1980, 1996). The appendicular skeleton is known in some detail, although the forelimb autopod, the manus, remains incompletely known.

Each half of the dermal shoulder girdle (fig. 2.10A, B) includes a substantial cleithrum and clavicle; left and right clavicles articulate via a broad area of overlap with the median interclavicle. As in *Acanthostega*, the anteroventral rim of each cleithrum extends as a medially directed flange and forms a narrow postbranchial lamina. The scapulocoracoid is large and well ossified, and like those of *Panderichthys* and *Acanthostega* consists predominantly of a ventromedially broad coracoid plate. Medially, the buttress configuration is quite unlike those of more derived tetrapods: the infraglenoid buttress is extraordinarily broad, so that the subscapular fossa is narrow and restricted to the anterior third of the medial surface. The glenoid has the characteristic strap shape of early tetrapod examples.

The humerus (fig. 2.10C, F) resembles a robust version of that in *Acanthostega*: many of the same process and crests are present although more highly sculpted. Most unusually, the radial articular surface faces anteroventrally. The radius is short and rather squat, and the ulna is similarly proportioned but with a well-ossified olecranon process extending around the "elbow." As restored by Jarvik (1996), it appears that the short forearm (zeugopod) was held in an almost fixed, permanently flexed, posture.

The pelvic girdle (fig. 2.10D) consists of a pair of large, well-ossified plates in which, once again, there are no traces of sutures marking the limits of pubis, ishium, and illium. The ilium is produced dorsally into a thick neck supporting dorsal and posteriorly directed processes; the long axis of the acetabulum is oriented from anteroventral to posterodorsal; the unfinished acetabular surface is continuous with the anterior pelvic rim. The ischiadic region is broad, and a substantial symphysial surface indicates that pelvic halves were united throughout most of their length. An articular area for a sacral rib is present at the iliac apex.

The hindlimb (fig. 2.10E, G) shares many features with that of its contemporary, *Acanthostega* (Jarvik 1980, 1996; Coates and Clack 1990; Coates 1996). In both taxa the limbs are paddlelike. The femur has a substantial adductor blade, the tibia and fibula are broad and flat, and both have a well-ossified ankle. However, in these early ankles the skeletal

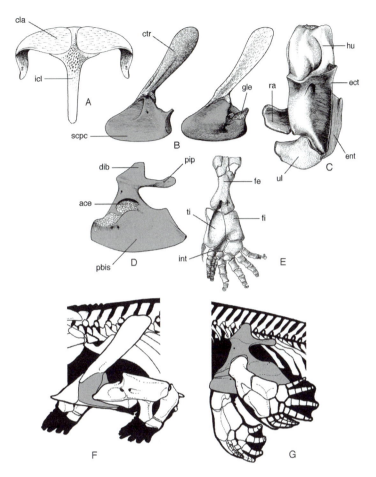

Figure 2-10 Appendicular and limb skeleton of *Ichthyostega stensioei*. (A) Dermal pectoral girdle in ventral view (combined from Jarvik 1996, fig. 41a, b). (B) Right cleithrum and scapulocoracoid in mesial view (*left*) and left cleithrum and scapulocoracoid in lateral view (*right*; after Jarvik 1996, fig. 42a, b). (C) Left pectoral limb in extensor view (combined from Jarvik 1996, figs. 45a, d, 46d). (D) Pelvic girdle in left lateral view (after Jarvik 1996, fig. 48a). (E) Left pelvic limb in extensor view (after Coates 1991, fig. 2a). (F) Pectoral and (G) pelvic girdles and limbs and their position relative to the axial skeleton (after Coates 2001, fig. 1.3.7.1d).

pattern is simple. Distal tarsals are absent, and at least two digits articulate directly with a massive fibulare. In *Ichthyostega* the toes include seven members arranged in two sets: four large posteriorly and three small anteriorly. Unusually, for material as rare as Devonian tetrapod limbs, it is possible to show that this strange pattern was a conserved feature, because it is preserved in three or more specimens. In the most complete specimen an apparently weakly ossified spur extends distally from the leading edge of the tibia, preceding the clustered, anterior, three small digits.

Tulerpeton

Tulerpeton, from the Upper Devonian of Central Russia, was the first polydactylous early tetrapod to be discovered and recognized as such (Lebedev 1984; Lebedev and Coates 1995). Its limb morphology departs in several important ways from that of *Acanthostega* and *Ichthyostega*.

In the dermal pectoral girdle (fig. 2.11A, B), the anteriorly expanded clavicles meet anteriorly and resemble a less strongly sculptured version of those of *Greererpeton* (see below; fig. 2.12C), including a stout triangular ascending process. The rhomboidal interclavicle is drawn posteriorly into a robust stem—again as in more derived early tetrapods. The cleithrum, separated from the scapulocoracoid, is a robust rod without clear evidence of a postbranchial lamina, but retains primitively a robust, expanded dorsal end. An anocleithrum is present. The scapulocoracoid is incompletely preserved, but shows a series of derived features including, dorsally, the earliest example of an enlarged scapular region, and, ventrally, a distinct infraglenoid buttress immediately below the glenoid facet. A single, laterally opening, supraglenoid foramen lies lateral to the supraglenoid triangular depression.

In the forelimb (fig. 2.11C), the humerus has slightly more elegant proportions than those of *Acanthostega* and *Ichthy-*

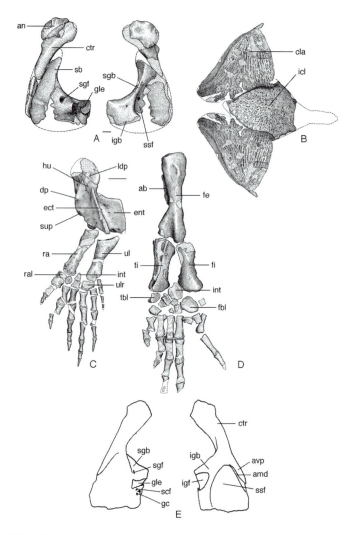

Figure 2-11 Appendicular and limb skeleton of *Tulerpeton curtum* (A–D) and *Hynerpeton bassetti* (E). (A) Left scapulocoracoid in lateral (*left*) and mesial (*right*) views (after Lebedev and Coates 1995, fig. 3). (B) Dermal pectoral girdle in ventral view (combined from Lebedev and Coates 1995, figs. 4a, 17a). (C) Left pectoral limb (combined and reversed from Lebedev and Coates 1995, figs. 5a, 8a, c, e). (D) Left pelvic limb (combined and reversed from Lebedev and Coates 1995, figs. 10b, 11a, h, 12a). (E) Left scapulocoracoid of *Hynerpeton* in lateral (*left*) and mesial (*right*) views (after Daeschler et al. 1994, fig. 1a, b).

ostega, with moderate torsion between proximal and distal extremities, and the beginnings of a shaft region are apparent. The ectepicondyle is aligned proximally with the latissimus dorsi process, and projects distally slightly anterior to the ulnar facet. The deltopectoral crest lies subcentrally along the anterior margin, and a distinct notch separates the supinator process from the bulbous radial condyle. Radius and ulna are slender and elongate, slightly shorter than the humerus, and subcylindrical in cross section. The ulna has a robust olecranon process. There is a simple ossified wrist including a cross-articulating intermedium, but there are no distal carpals. Six digits are present, the metacarpals of which for digits 1, 4, and 5 are markedly asymmetrical. Furthermore, the individual phalanges are elongate, and very dissimilar to the stout phalanges of *Acanthostega*.

A poorly preserved ilium, with a stout fanlike dorsal blade and slightly flattened proximal part of a posterodorsal process, is the only fragment of the pelvic girdle. The hindlimb (fig. 2.11D) is in much better condition, with a particularly well-formed femur. This includes a broad intertrochanteric fossa, a robust internal trochanter separated by a distinct notch from the shaft, a robust adductor blade, a broad, elongate intercondylar fossa, a small but distinct fibular fossa, and a short interpopliteal space. The tibia and fibula each have subcylindrical shafts and, although still primitively robust, they delimit a broad interepipodial space.

The ankle includes at least 12 bones, with a distinct distal tarsal series separating metatarsals from the fibulare and broad intermedium. The intermedium has a proximal notch, as in several early tetrapods associated with the base of the amniote stem group. The phalanges and metatarsals are elongate and more robust than those of the manus; once again there appear to have been six digits.

Hynerpeton

Hynerpeton (fig. 2.11E), originally described from an isolated left scapulocoracoid and cleithrum (Daeschler et al. 1994), originates from the Upper Devonian of Pennsylvania, USA. Although most comparable to conditions in *Tulerpeton,* the smooth cleithrum remains fused to the scapulocoracoid. The coracoid region is relatively thin. The glenoid is directed posterolaterally and strongly buttressed dorsally, and the articular surface shows a degree of helical twisting comparable with that of *Tulerpeton.* The supraglenoid foramen lies inside the triangular supraglenoid area delimited anteriorly by the thick ridge of the supraglenoid buttress. The infraglenoid buttress is well developed. A broad subscapular fossa with prominent dorsal scarring along its dorsal edge is visible medially.

Colosteids

Colosteids range from the late Viséan to the late Moscovian (330–300 mya) of the Carboniferous, and are recorded in both Great Britain and North America. The best-known representative is the superficially crocodile-like genus *Greererpeton,* with an elongate skull, long trunk, and small limbs (Smithson 1982; S. J. Godfrey 1989). These tetrapods lie close to the crown-tetrapod split and have at times been proposed as members of the latter (Panchen and Smithson 1988), although more recent studies place them as derived members of the tetrapod stem.

The massive, rhomboidal interclavicle (fig. 2.12B, C) bears a triangular posterior process and a broad, rectangular anterior. The clavicles have anteriorly expanded ventral plates and short, robust ascending processes. Each rodlike cleithrum retains a narrow postbranchial lamina, but they are otherwise rather advanced with an anteroposteriorly expanded, slightly fore-turned dorsal extremity. No anocleithra are known. The stout scapulocoracoids have a short scapular process and a thick, subelliptical coracoid plate. These two portions are fused; as in other early examples, no sutures are observed within the endoskeletal girdles. The poorly preserved glenoids seem to have been narrow and elongate and not helical along their lengths (S. J. Godfrey 1989), unlike those of most other early tetrapods.

In the forelimb skeleton (fig. 2.12A), the humerus is distinctly L-shaped, with only limited torsion along its proximodistal axis. The robust ectepicondyle is aligned proximally with the latissimus dorsi process and projects distally

immediately posterior to the radial facet. Compared with humeral length, the humeral head is narrower than in *Acanthostega.* The supinator process is little more than a poorly pronounced bulge, the deltopectoral crest is slightly more robust than in *Acanthostega,* and the rectangular entepicondyle (subtriangular in juveniles) extends at a distinct right angle to the humeral posterior margin (hence its description as "L-shaped"). The slender and elongate radius and ulna delimit a well-formed interepipodial space. As in *Tulerpton* (fig. 2.11C), the ulna has a distinct olecranon process. Radius and ulna are less than two-thirds as long as the humerus. No wrist elements are known. The manus is pentadactyl (Coates 1996), with a provisional phalangeal formula of 2-3-3-4-3.

The pelvic girdle (fig. 2.12D) is more conventionally tetrapod-like than all examples described thus far. Pubis, ilium, and ischium are suture-separated entities. The ilium has a single posterodorsal blade, and a large obturator foramen perforates the pubic plate. Immediately anterior to the acetabulum, the ilium shows a narrow strip of finished bone, separating socket from the anterior margin of the girdle. The femur (fig. 2.12E) is similarly conventional, with a distally reduced adductor blade (Coates 1996). The fibula displays some torsion along its proximodistal axis, and the distal end is flared and neatly delimited from the shaft. The tibia displays a prominent cnemial crest and, in the largest individuals, is less than half femoral length. The foot has stout digits with asymmetrycal phalanges, reconstructed with a formula of 2-2-3-4-2+. Except for the absence of four distal tarsals, the ankle is well known. The massive intermedium contacts the fibulare, a large proximal centrale, and the most posterior of the three distal centralia. The tibiale contacts the most anterior of the three distal centralia as well as the proximal centrale.

Whatcheeriids

Whatcheeria deltae (Lombard and Bolt 1995) from the Chesterian/Viséan of Iowa, USA, is known from abundant material and awaits full description. The most unusual feature of the dermal pectoral girdle is the long stem of the interclavicle (fig. 2.13A), which parallels conditions in *Ichthyostega* and several stem amniotes.

In the forelimb, the humerus (fig. 2.13B) is massive and displays considerably more torsion than in *Greererpeton* or *Acanthostega.* The angular form resembles conditions in *Tulerpeton* (fig. 2.11C), although the ectepicondyle is more robust distally, and the massive entepicondyle is proximodistally extended. Radius and ulna are described as robust, the latter with a well-developed olecranon process.

The pelvis (fig. 2.13C) includes an ilium with a short,

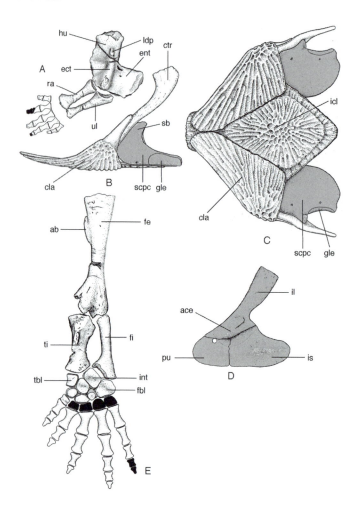

Figure 2-12 Appendicular and limb skeleton of colosteids, based on *Greererpeton burkemorani.* (A) Left pectoral limb (combined and reversed from Godfrey 1989, figs. 18a, 20c, i; and after Coates 1996, fig. 37g). Pectoral girdle in (B) left and (C) ventral views (after Godfrey 1989, fig. 14a, b). (D) Pelvic girdle in left lateral view (reversed from Godfrey 1989, fig. 22a). (E) Left pelvic limb with reconstructed pes (combined from Godfrey 1989, figs. 24o, 25b, n, 26b).

stout neck and two processes. As in *Acanthostega,* several obturator foramina are present in the anterior half of the puboischiadic plate. The acetabulum, however, has a completely finished rim, as in colosteids and higher taxa. The hindlimb has femur of comparable size to humerus (fig. 2.13D). The ventral surface has a prominent adductor crest or ridge, but the extensive blade described in *Acanthostega* (Coates 1996) and colosteids (S. J. Godfrey 1989; Clack and Carroll 2000, their fig. 5) is absent, as are distinct internal and fourth trochanters. Tibia and fibula are similarly robust and only slightly shorter than the femur. They bracket a small, subcircular interepipodial space, quite unlike the spindle-shaped space of other early tetrapods. No articulated hand or foot is known; absence of wrist and ankle bones indicates that these elements were probably unossified (Lombard and Bolt 1995; Bolt and Lombard 2000). *Whatcheeria* material includes numerous phalanges and some articulated digits. Digit counts and phalangeal formulae are unknown; short and flat phalangeal shapes indicate that manus

and pes were paddlelike (cf. forelimb phalanges in *Pederpes,* Clack 2002b) and recall those of Devonian taxa, with the possible exception of *Tulerpeton.*

A further *Whatcheeria*-like form, the Scottish *Pederpes finneyae,* has the distinction of being the only articulated tetrapod from the Tournaisian (Ivorian, 348–344 mya; Clack 2002b). The humerus has a comparatively shorter, trapezoidal entepicondyle. A spikelike latissimus dorsi process is reminiscent of the condition in *Baphetes* (Milner and Lindsay 1998; fig. 2.13D). A very small digit on the manus is thought to suggest the presence of polydactyly. The only other preserved digit is short, broad, and tapered.

The ilium has a stout and poorly developed neck and two flat processes, as in *Whatcheeria.* Likewise, femur, tibia, and fibula match closely the morphology of their *Whatcheeria* homologues. The pes has five robust digits, three of which are complete. Importantly, the metatarsals are bilaterally and proximodistally asymmetrical, as in *Greererpeton,* and many more recent examples. This feature provides a valuable clue

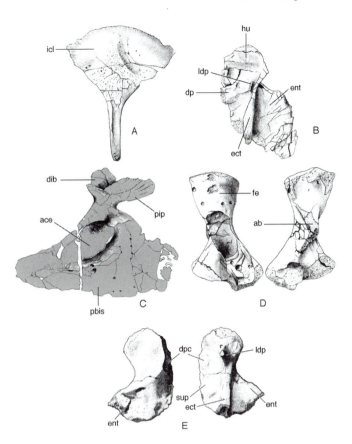

Figure 2-13 Appendicular and limb skeleton of whatcheeriids (A–D, *Whatcheeria deltae*) and baphetids (E, *Baphetes* cf. *kirkbyi*). (A) Interclavicle in ventral view (after Lombard and Bolt 1995, fig. 7a). (B) Left humerus in extensor view (after Lombard and Bolt 1995, fig. 7B). (C) Left half of pelvic girdle (reversed from Lombard and Bolt 1995, fig. 8). (D) Left femur in extensor (*left*) and flexor (*right*) views (after Lombard and Bolt 1995, fig. 9a, b). (E) Left humerus of *Baphetes* in flexor (*left*) and extensor (*right*) views (after Milner and Lindsay 1998, fig. 9a, b).

about foot realignment, from the laterally oriented paddles of Devonian forms to an anteriorly directed stance, more suited for walking (Clack 2002b).

Baphetids

Baphetids, formerly known as loxommatids, are known mostly from cranial material, dating from the Viséan to Westphalian. Assumed to be crocodile-like piscivores, the only attributable postcranial material, of *Baphetes* cf. *kirkbyi*, includes a humerus, radius, tibia, and fibula (Milner and Lindsay 1998). Manus and pes material are unknown. On the humerus (fig. 2.13E), the spike-shaped latissimus dorsi process resembles that of whatcheeriids. The distal end of the humerus appears incomplete, but articular surfaces for radius and ulna are identified, with the radial area in line with the ectepicondyle. The radius is about half as long as the humerus. Of the hindlimb and girdle, an iliac neck is preserved, extending into dorsal and posterior processes. Tibia and fibula show some resemblance to those of *Tulerpeton* (Lebedev and Coates 1995), quite unlike the broad examples of *Whatcheeria*.

Horton Bluff Material

Isolated humeri and femora indicate the presence of limbed tetrapods at the Tournaisian Horton Bluff locality, of much the same age as *Pederpes* (Clack and Carroll 2000). Two kinds of L-shaped humerus are known. One resembles that of *Greererpeton* (fig. 2.14E), with squat proportions; another resembles that of *Tulerpeton* (fig. 2.14F). The femora are stout (fig. 2.14G), with an expanded, elongate adductor blade and robust, distally confined adductor crest. Isolated scapulocoracoids (fig. 2.14A, B) show multiple foramina, as in examples from Devonian tetrapods. Subtriangular clavicular plates (fig. 2.14C) and kite-shaped interclavicles (fig. 2.14D) are also recorded.

Eucritta and *Caerorhachis*

Eucritta and *Caerorhachis* (fig. 2.2) are included because in at least one recent large-scale systematic analysis they straddle the base of the tetrapod crown-group radiation (Ruta et al. 2003). Limb conditions in these taxa, although not particularly well preserved, thus bracket an arbitrary

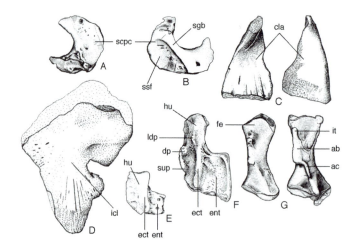

Figure 2-14 Isolated tetrapod girdle and limb remains from Horton Bluff. Two different right scapulocoracoids in (A) lateral and (B) mesial views (after Clack and Carroll 2000, fig. 3c, d). (C) Dorsal (*left*) and ventral (*right*) views of right clavicle (after Clack and Carroll 2000, fig. 4d, e). (D) Interclavicle in ventral view (after Clack and Carroll 2000, fig. 4a). (E, F) Two different left humeri in extensor view (after Clack and Carroll 2000, fig. 5b, e). (G) Two different right femora in flexor view (after Clack and Carroll 2000, fig. 5f, g).

marker for the near-end of the fish-tetrapod and fin-to-limb transition.

In *Eucritta* (Clack 1998b, 2001), the dermal component of the pectoral girdle is less extensive than in all examples described thus far. The kitelike interclavicle is similar to those of other primitive taxa, but the clavicular plates are narrowly triangular and poorly expanded anterorly. The humerus is crescentic, with a subtriangular entepicondyle, and, as in *Baphetes* (Milner and Lindsay 1998), this peculiar form may result from incomplete ossification. The epipodials are unremarkable; the manus unknown. The pelvic girdle includes ilia with blades and postiliac processes, but, most notably, the pubic region is unossified. The hindlimb is approximately one-third larger than the forelimb, but little detail is preserved of the skeletal morphology. Tibia and fibula are almost two-thirds the length of the femur; the ankle is unossified, and the pes is pentadactylous.

Remains of the forelimb and girdle in *Caerorhachis* are negligible (Holmes and Carroll 1977; Ruta et al. 2001), but the pelvic apparatus is informative. The platelike girdle includes only the barest suggestion of division between pubis, ilium, and ischium: traces of sutures are just visible in the acetabulum. The socket itself has a robust supra-acetabular buttress but retains a primitively unfinished anterior continuity with the pubic leading edge, as seen in Devonian examples. The ilium has a pronounced neck, a dorsal blade, and a distinct, elongate posterior process. The hindlimb includes an elongate femur exposed in extensor view, showing that the adductor blade was confined to the proximal half of the shaft. A trough separates the internal trochanter from the femur head. The contrast between femoral proportions and those of the stout tibia and fibula is striking: the latter are only 40% of femoral length. Parts of the ossified ankle are

preserved, including a proximally notched intermedium, much as in *Tulerpeton* (fig. 2.11D, int). Unlike the whatcheeriid examples, the metatarsals in this Carboniferous tetrapod are bilaterally symmetrical.

Trends: From Comparative Hierarchy to Transformational Scenarios

The following section deals with morphology, phylogenetic pattern, and implied directions of change. It is not the intention here to digress into discussions of developmental-transformational scenarios of tetrapod limb origin. It should be noted, however, that the general pattern sketched out below is incompatible with many, if not all, of the standard transformational hypotheses (for a succinct and informed summary, see Janvier 1996, 266–270).

Dermal bones are a major part of the pectoral girdle in most osteichthyans but are reduced to the point of near or complete absence in (limbed) tetrapods. However, the pattern of loss is not a simple sequence of accumulating absences and proportional size reductions. Initially, bone reduction is restricted to the dorsal part of the girdle. Loss of the supracleithrum and/or anocleithrum achieves separation of the pectoral girdle from the skull, although it is noteworthy that anocleithra persist in an increasing variety of taxa close to the crown node. Loss of surface ornament in these dorsal parts of the girdle indicates that dermal bones had become increasingly subdermal. Ventrally, the trend is quite different. The ventral clavicular plate is expanded, especially along the posterior margin, and the interclavicle is transformed from a slender midline plate to a huge kite-shaped shield. Thus, the (fore) fin-to-limb transition might be better characterized as one of dermal bone redistribution rather than bone reduction, at least in terms of gross area covered. It may also be significant that all of these changes are clustered along the *Panderichthys-Acanthostega* internodal stretch of the tetrapod stem (fig. 2.2B). Post-*Acanthostega* and *Ichthyostega*, the most obvious changes are of cleithral separation from the scapulocoracoid and progressive reduction posteroventrally.

Dermal fin-ray (lepidotrichial) loss occurs, as far as we can tell, simultaneously at pectoral and pelvic levels. Dermal rays of paired fins closest to the origin of digits display no obvious specializations suggesting developmental or functional change. This is in marked contrast to those of rhizodontid fins, which resemble lungfish camptotrichia.

Scapulocoracoid enlargement occurs prior to fin-ray loss and digit origination. It involves change from a tripodal to a platelike attachment to the dermal girdle, and enlargement, primarily, of the coracoid region. Posteroventral expansion of the girdle also results in increased lateral exposure, and reorientation of the glenoid from a more or less posterior to posterolateral orientation. The glenoid is elongated, and assumes the characteristic strap shape of early tetrapods. Helical curvature of the articular surface is a later novelty acquired within the limbed taxa. The most striking change coincident with fin-ray loss and digit origination is the absence/closure of large canals penetrating the scapulocoracoid unit. Origin of a scapular blade, and thus dorsal expansion of the endoskeletal girdle, is a phylogenetically later event, the beginnings of which are only discernible with detachment of the cleithrum. Likewise, the internal buttressing of the glenoid seems to occur in a variety of ways among late Devonian tetrapods; once again after, rather than coincident with, the origin of digits.

Throughout most of the tetrapod stem group, the pelvic girdle seems to have been morphologically conservative. The form in rhizodontids is very similar to that in tristichopterids. In both, the girdle consists of a simple pubo-iliac bar, with a modestly expanded body surrounding the acetabulum. Conditions in *Panderichthys* are unknown, but the pelvis in *Acanthostega* is considerably different. An ischiadic plate is present; the ilium has gained a neck with bladelike processes (associated with a sacral attachment) extending dorsally and posteriorly; and the acetabulum is enlarged with buttresses posteriorly and dorsally. The postacetabular buttress is largest initially, whereas the supraacetabular becomes more prominent in later groups. Other post-*Acanthostega* changes include full ossification of the anterior rim of the pelvic plate and acetabular rim, and an anterior shift in the level of the iliac neck, until positioned directly above the supraacetabular buttress.

Changes affecting fin and limb endoskeletons are not easily summarized, in part because of the apparent independence of changes at pectoral and pelvic levels. In general there appears to be trend of proximodistal truncation, extending from the tetrapod stem-group base through to the kind of pectoral fin present in *Panderichthys*. This is accompanied by an increase in the morphological individuality of the ever fewer "segments" of the fin skeleton. Rhizodontid pectorals show a quite separate radiation in fin form to that which occurred at the advent of digits. It seems unlikely that any rhizodont fin provides much evidence of conditions at the base of the tetrapod stem. Fishlike tetrapod conditions between rhizodontids and *Panderichthys*, if represented adequately by *Sterropterygion* and *Eusthenopteron*, are marked by further anatomical conservatism.

From *Panderichthys* onward, something like a recognizable tetrapod humerus is present: dorsoventrally flattened with an easily identified entepicondyle and ectepicondyle. In these respects it differs strongly from rhizodontid examples. In contrast, the midsection of the fin skeleton is the least changed: the zeugopod with subequal ulnar and radial homologues is little more than a dorsoventrally compressed version of the pattern in *Eusthenopteron*; likewise for *Acanthostega*. Distally, the nascent pectoral autopod of *Panderichthys* includes no direct clue of digit origin: simply an anteroposteriorly broad ulnaric plate and a slender, but primitively elongate, intermedium.

From *Acanthostega* onward, limbs with an autopod including digits are present, and these digits occur first at pectoral *and* pelvic levels. As noted elsewhere (Coates et al. 2002), absence of pelvic fin data in *Panderichthys* leaves a morphological gulf between conditions in *Eusthenopteron* and *Acanthostega*, not least because hindlimbs in the earliest limbed tetrapods are generally more derived than their pectoral counterparts (Coates et al. 2002). Comparison of *Acanthostega* and *Eusthenopteron* femora reveals few similarities; in particular, the massive adductor blade of the former has no clear homologue in the fin skeleton of the latter. Similarly, the broad postaxial process of the *Eusthenopteron* fibula has no equivalent among early limb fibulae.

Polydactylous patterns among the late Devonian *Acanthostega*, *Ichthyostega*, and *Tulerpeton* are varied in terms of digit and phalanx numbers. The most generalized, or least individualized, of the digit sets is that of the *Acanthostega* forelimb. Elaboration of the mesopod (wrist/ankle bones) seems to have been an incremental process, once again occurring in the hindlimb prior to the forelimb. Evidence for this is provided not only by the subcylindrical intermedium in the *Acanthostega* wrist, but also by the more completely ossified and cross-articulating ankle. Only in *Tulerpeton* and higher taxa are distal tarsals present. Furthermore, the altogether more conventionally limblike appendages of *Tulerpeton* hint at the hidden diversity of the late Devonian tetrapod radiation.

Most phylogenies now indicate that pentadactylous limbs evolved once (Ruta et al. 2003; *contra* Coates 1996), and the most primitive examples of these are probably those of colosteids. From this point onward, limb skeletons exhibit a reduction in the number of humeral (and other) foramina, loss of primitively paddlelike forms, proximalization of the major points of muscle insertion, increased size difference between fore- and hindlimbs, and increasing asymmetry of the phalanges, especially the metacarpals and metatarsals. The latter point is related to the forward turning of the manus and pes when in "neutral" posture (Clack 2002b).

Homoplasy

Only a narrow view of fin diversity and change in taxa more or less associated with the tetrapod stem lineage is provided by the preceding summary. Homoplasy is edited out of the picture. There is a danger here of throwing the proverbial baby out with the bathwater, because if homology is distinguished from homoplasy with some degree of accuracy, "we not only arrive at more valid phylogenies, but also reveal very interesting convergences in evolution" (Raff 1996, 50). A less restricted perspective shows that several changes characteristic of the fin-to-limb transition have indeed occurred repeatedly within sarcopterygians.

In the dermal pectoral girdle, separation from the skull via loss of the supracleithrum has occurred on at least two further occasions: in lungfishes and in coelacanths. Unlike tetrapods, in neither of these examples is dermal bone reduction above glenoid level accompanied by dermal bone expansion below. However, all three examples show evidence of the remaining dorsal bones acquiring an increasingly subdermal location.

Dermal fin-ray loss is not unique to tetrapod sarcopterygians. Fin rays are also reduced and lost in the paired fins of lineages leading to two out of the three living lungfishes. *Lepidosiren* and *Protopterus* both show an extraordinary degree of paired fin-axis elongation, in each it is accompanied by fin-ray reduction to the point of apparent absence at pelvic levels, and only limited retention at pectoral levels (along the proximal trailing edge in *Protopterus*). This is noteworthy even if only for the fact that dermal fin-ray absence can occur without any semblance of digit evolution.

The scapulocoracoid has enlarged from an assumed primitive, small, and tripodal condition on three occasions independently of near-crown tetrapods: in coelacanths, lungfishes, and rhizodontids. In each of these examples, the endoskeletal girdle acquired a platelike area of adhesion to the dermal bones, and this was associated with reduction and, sometimes, complete closure of scapulocoracoid canals. However, none of these three transformations from small to large is associated with the kind of coracoid plate expansion found in early tetrapods.

Pelvic girdle enlargement has occurred more rarely among sarcopterygians, and seems to be restricted to lungfish. There is some evidence of pelvic enlargement in primitive examples, as exemplified by *Chirodipterus*, but the greatest similarities are present among extant taxa. In modern lungfish the pelvic girdle is united ventromedially along an extended symphysis, and projects anteriorly with a significant area of prepubic cartilage with dorsal processes. Furthermore, in *Protopterus*, the degree of resemblance to tetrapods led Goodrich (1930, 194) to note: "Except for the absence of an ilium the girdle resembles that of the Urodela, in particular of *Necturus*." The unfinished anterior pelvic rim in early, limbed tetrapods implies the presence of prepubic cartilage, but the extent of such growth is unknown.

Finally, at present there is only one convergent example of the anteroposterior expansion of the distal zone of a paired fin into something resembling a hand plate or foot plate. The occurrence of this feature in rhizodontids' pectoral fins, and thus basal stem tetrapods, may be significant. However, this expansion is not identical to the autopod of a digited limb. In addition to the persistence of dermal fin rays, the multiple endoskeletal radials occupying the pseudo-autopod are indistinguishable from more proximal radials, with which they articulate in a straightforward manner.

The Fin-to-Limb Transition: Character Distribution

A previous discussion grappling with this topic concluded that changes associated with the transition were distributed throughout the tetrapod stem (Coates et al. 2002). However, increased phylogenetic resolution shows significant conservatism in the osteolepiform-tristichopterid part of the stem, and a concentration of changes from the panderichthyid node onward. The problem is compounded by the uncertainty surrounding *Elginerpeton*, and especially by the lack of information about pelvic fin and girdle conditions in *Panderichthys*. The net result is a block of character-state changes that may be located at, or very close to, the node subtending *Acanthostega* and all other limbed tetrapods. It should also be noted that even within the Sarcopterygii, a group characterized by an extraordinary degree of pectoral-pelvic appendage similarity, similar features originate in pectoral and pelvic limbs at different points in phylogeny (usually in the pelvic before pectoral; Coates et al. 2002). Concerted change at both levels appears to be the exception rather than the rule. This has two important implications: first, in early tetrapods, at least, pectoral conditions are not reliable indicators of pelvic conditions; second, any bauplan or archetype of *the* tetrapod limb is an artifice both between and within individuals.

Functional Implications

There is compelling evidence that the vast majority of changes involved in the fin-to-limb transition were initiated and largely completed in animals that were mostly, and primitively, aquatic (Coates 1996; Clack 2002a). At a future date, and in the context of a broader consideration of stem tetrapod characteristics and paleoenvironmental settings, these transformations will probably be compiled into a scenario of "correlated progression" (T. S. Kemp 1982). But, as yet, insufficient evidence is apparent for the desired "lock-step" (Sumida 2003) of directed adaptive change. For the present, the following notes are included.

Loss of the dermal bone connection between pectoral girdle and skull was undoubtedly permissive to the evolution of a neck, like those of crown group tetrapods. However, similar absences in coelacanths (and even chondrichthyans) have no association with necklike functions. Alternatively, Johanson et al. (2003) postulated the presence of a necklike arrangement in a large tristichopterid, even though the dermal bone connection is clearly retained.

Cleithrum reduction is linked intimately to presence or absence of a functioning internal gill skeleton and chamber, because the cleithrum provides the postbranchial lamina. The expansion of the clavicle-interclavicle complex, coordinated with digit acquisition and fin-ray loss, probably served a mixture of protective and supportive roles. It must have confered some rigidity to the forequarters, consolidated the articular base for the forelimb, and provided a more effective cradle for carrying the trunk forward. Clack (2002a) has speculated that the interclavicle posterior prolongation acted like an amniote sternum, and provided a surface for anterior trunk muscle insertion.

Along with coracoid enlargement, the dorsoventral flattening of the humerus and corresponding change from subcircular to strap-shaped glenoid, prior to the origin of digits, provides an important hint about the specialized habits of at least some of the apical stem tetrapods. Vorobyeva and Kuznetsov (1992) recognized that this reduced the shoulder joint in *Panderichthys* to a hinge and that subequal radius and ulna lengths effectively blocked rotation at wrist level. The pectoral fin was interpreted as an anchoring point, pressed into the substrate via well-developed depressor muscles inserted on the expanded coracoid. Forward movement was achieved by means of axial (vertebral column) flexion and rotation about the anchorage, the stubby, robust fin skeleton being suited for resisting the resultant compressive phase. All of these points are equally applicable to *Acanthostega*, but in this taxon digits, enlarged hindlimbs, and a sacral pelvis are also present. Limb rotation below knee level was either extremely limited or nonexistent, but, unlike the shoulder, the hip joint seems to have maintained a degree of rotational freedom. In many respects, hindlimb conditions in *Ichthyostega* and *Acanthostega* resemble those of cetacean pectoral flippers (Coates and Clack 1995). It has therefore been argued that these functioned as relatively inflexible paddles, providing forward thrust, but unlikely to leave the kinds of trackways attributed to tetrapods in Devonian deposits (Clack 1997). These tetrapods might be attributed more convincingly to more derived, but conceivably coexistent, tetrapods with the kinds of limbs and girdles present in Tournaisian forms.

Functional differences between pectoral and pelvic appendages are entrenched deep in phylogeny, and there is an ever-present danger of generating over prescriptive interpretations of unusual fin morphologies. For example, the slender paired fins of an African lungfish would hardly seem appropriate tools for quadrupedal and frequently bipedal (pelvic-level) aquatic walking, and yet they are (as is confirmed by the authors' personal observations). As a concluding remark, and pointer to quite independent and parallel transformations from vertebrate fins to limblike function and organization, we recommend a glance at the station-holding and walking appendages of modern sharks and rays (Goto et al. 1999; Wilga and Lauder 2001; Lucifora and Vassallo 2002).

Appendix: List of Abbreviations

ab	adductor blade	int	intermedium
ac	adductor crest	is	ischium
ace	acetabulum	it	internal trochanter
amd	anteromedial depression	ldp	latissimus dorsi process
an	anocleithrum	pbis	puboischiadic plate
avp	anteroventral process	pecf	pectoral fin
cla	clavicle	pecg	pectoral girdle
ctr	cleithrum	pelf	pelvic fin
dfr	dermal fin rays	pelg	pelvic girdle
dib	dorsal iliac blade	pip	postiliac process
dp	deltoid process	pu	pubis
dpc	deltopectoral crest	ra	radius
ecl	extracleithrum	ral	radiale
ect	ectepicondyle	sb	scapular blade
ent	entepicondyle	scf	supracoracoid foramen
fe	femur	scpc	scapulocoracoid
fbl	fibulare	sgb	supraglenoid buttress
fi	fibula	sgf	supraglenoid foramen
fm	first proximal mesomere	ssf	subscapular fossa
gc	glenoid canal	sup	supinator process
gle	glenoid	tbl	tibiale
hu	humerus	ti	tibia
icl	interclavicle	ul	ulna
igb	infraglenoid buttress	ulr	ulnare
igf	infraglenoid fossa	vmc	ventral median cartilage
il	ilium		

Acknowledgments

We thank the editor, Professor Brian Hall, for the invitation to contribute to this volume.

Chapter 3 A Historical Perspective on the Study of Animal Locomotion with Fins and Limbs

Eliot G. Drucker and Adam P. Summers

IN THIS CHAPTER, we present a historical overview of the study of aquatic and terrestrial locomotion by means of the paired appendages. As with nearly any historical perspective on animal biology, the foundation of this discussion rests on the original and often insightful empiricism of Aristotle. In several surviving fourth-century BC texts (*History of Animals, Movement of Animals, Progression of Animals, Parts of Animals*), this keen observer of the natural world made the first analytic contributions to the study of animal locomotion, providing descriptions of the various gaits and of motion of the center of gravity of the body, as well as making the distinction between upright and sprawling posture during forward progression. So central is locomotion to our perception of animals, however, that it is not surprising that even earlier observations have been recorded. Images of moving animals are preserved in the earliest human communications, Neolithic representational art at least 17,000 years old (Bataille 1955). In the Lascaux caves of southern France there are anatomically accurate depictions of quadrupedal animals at rest and in motion that reflect an artist's clear understanding of terrestrial gaits. The kinematics of gait was used to demonstrate movement: galloping animals are shown with legs extended fore and aft, a position that cannot be maintained at rest (fig 3.1).

Although the homology of paired fins and limbs was recognized quite early (see Wagner and Larsson, chap. 4 in this volume, for a review; see also Clack 2002b), study of the use of these appendages in locomotion at first was not evenly focused on land and in water. The dominant mode of loco-

motion in aquatic animals was seen initially as whole-body (axial) undulation, while on land the limbs were the obvious prime movers. Only in the last 50 years has there been a realization that many, perhaps most, fishes use multiple fins, including the paired appendages, as propulsors. A nearly simultaneous interest in the several independent evolutions of axial undulation as a terrestrial mode of locomotion (see Kley and Kearney, chap. 17 in this volume; see also Jayne 1988 and Gillis 1998) has resulted in a complementary development: paired fins have received much recent attention as important propulsors, and axial undulation is now coming under increasing scrutiny in terrestrial vertebrates.

Our objective is to provide a historical account of advances in the field of aquatic vertebrate locomotion as related to the paired fins of fishes, followed by a synoptic review of innovations and breakthroughs in research into terrestrial limb-based locomotion. The central position of terrestrial locomotion in biomechanical inquiry through the centuries has led to several major historical perspectives on the field (for disparate examples see Marey 1874; Nigg et al. 2000; S. Vogel 2001). In contrast, the relatively recent rise of interest in aquatic flapping flight has yielded much primary literature but, until now, no synthetic historical review.

Paired Fins as Propulsors

Before the advent of modern imaging technology and other tools of the experimental zoologist, insight into how fishes

Figure 3-1 Neolithic drawings of a running horse (*right*) and walking bovids (*left*) from the walls of the Lascaux Caves in southern France. The posture and limb position show an understanding of gait in prehistoric times.

swim was gained principally through behavioral observations. To the naked eye, large-scale motions of the body and tail of freely swimming fishes are the most readily apprehended, while relatively subtle excursions of the smaller fins can be easily overlooked. Perhaps for this reason, the earliest students of fish locomotion restricted their greatest attention to axial rather than appendicular components of the propulsive anatomy. Subsequent workers, following this ancient precedent, have contributed an extensive body of literature on the locomotor biology of axially propelled fishes. As measured by the volume of published work, knowledge of the locomotor function of the paired appendages of fishes (pectoral fins anteriorly and pelvic fins posteriorly) is scant by comparison. Relative neglect of the paired fins by experimentalists, even after the means to slow motion on film became available, reflects an impression that only in recent decades has begun to be dispelled: namely, that action of the paired fins makes an insignificant contribution to progression underwater compared to "conventional" (i.e., axial) swimming (Lindsey 1978). A review of research conducted on the pectoral and pelvic fins from before modern times to the current day, however, reveals the importance of these paired fins both as primary propulsors for forward propulsion, and as critical ancillary propulsors for maneuvering.

Early Observations

The earliest discussion of fish locomotion is from Aristotle in his works *History of Animals, Movement of Animals, Progression of Animals,* and *Parts of Animals.* He noted the number of fins that different fishes possess and argued that those with four fins (i.e., the paired pectoral and pelvic fins) use them as the primary means of propulsion, while those with fewer fins swim by undulating the body. Although an idealized distinction, Aristotle's observation was remarkably acute and set the stage for future mechanical analyses of paired-fin function. Further progress, however, would come only after a long hiatus. During the Middle Ages, advances in any area of zoology were few; the works of ancient authors became widely available only in the Renaissance. Guillaume Rondelet (1554), referring frequently to Aristotle, repeated the idea that some fishes swim with the fins and others with the body (Alexander 1983b).

In the 17th century, interest in fish locomotion was renewed by Giovanni Borelli, a prominent professor of mathematics and physics in Pisa. In 1680, Borelli published *On the Movement of Animals,* apparently the first treatise on biomechanics. He considered Aristotle's opinion that fish swim principally with the paired fins to be erroneous. Noting that considerably more muscle is found in the body than in the fins, and that the fins themselves are flexible and have short lever arms, Borelli proposed that the side fins are "unable to beat and push water with the velocity and power necessary to move the body of the fish forward." In support of this contention, he cited his observation that fish swimming in a pool keep the paired fins furled against the body, rather than oscillating them in an extended position. Further, he conducted an experiment in which the pectoral fins of living fishes were removed. The observation that animals thus modified could still swim was taken as proof that the paired fins are not involved in forward swimming (such deduction of function through anatomical reduction is rarely a sound approach, as noted by J. E. Harris 1938, 32; Videler 1993, 94).

Borelli regarded the pectoral fins instead as primarily playing a role in maneuvers such as braking. Results of an additional experiment in which the pelvic fins of fishes were resected were used to support the proposition that these fins are used to maintain body posture and stability. The pioneering experimental approach of Borelli yielded hypotheses about swimming that have, in large part, been supported by modern experimental analyses. Some of these proposals, however, particularly the notion that the pectoral fins are not involved in rectilinear progression, have been refuted by later investigators.

The Rise of Quantitative Analysis

In the two centuries following Borelli, fish swimming research again fell largely into abeyance. The end of the 19th century brought a revolutionary technological advance that for the first time allowed precise and detailed observation of rapid animal movement. Cinematography was first applied to the study of aquatic locomotion by Marey (1890), who recorded sequential motion pictures on strips of celluloid film at the Naples Aquarium. These films included sequences of swimming by a tethered skate (*Raja*) in which posteriorly directed traveling waves of bending are visible along the pectoral fin (fig. 3.2A). It was "only a matter of time and patience," Marey acknowledged, before freely swimming fishes could be similarly imaged. Marey's work inspired the subsequent application of cine-photography by Magnan (1930),

who filmed a variety of cartilaginous and bony fishes swimming unrestrained in the aquarium at Nice. In addition to recording both lateral and ventral views of fishes swimming steadily, Magnan provided the first images of fishes using the paired fins to turn, hover, and execute other unsteady maneuvers (fig. 3.2B).

The earliest use of film enabled researchers to document qualitative patterns of paired-fin motion in greater detail than was possible before. However, these workers (Marey in particular) made no attempt to analyze the forces involved in the swimming movements recorded. An important milestone in the history of fish locomotion science was the initiation of a quantitative, engineering approach to the study of moving underwater. Houssay (1912) is credited with inaugurating the use of mechanical analysis of fish swimming forces (Alexander 1983b), although Regnard (1893) made similar experimental measurements of thrust using his "dynamometer" apparatus, in which fish swam while tethered to suspended weights. Sir James Gray (1968), perhaps the most influential researcher on fish locomotion of the early 20th century, continued in this vein by making detailed measurements of propulsor motion from films of swimming fishes and by calculating the resulting fluid forces exerted. Gray,

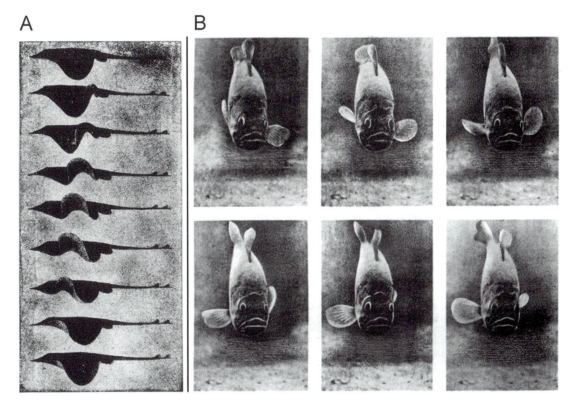

Figure 3-2 Early cinematographic records of paired-fin locomotion in fishes. (A) Sequential lateral images of a tethered skate (*Raja*) exhibiting anterior-to-posterior waves of bending along the pectoral fin (read from bottom to top; from Marey 1894). (B) A bass (*Micropterus*) filmed in a public aquarium at 8 frames per second while hovering by means of alternating left- and right-side pectoral fin oscillation (from Magnan 1930).

however, like most other mechanists dating back to Regnard, focused exclusively on the motions and forces of axial undulation.

The first English-language analysis of the mechanics of paired-fin locomotion is given by Breder (1926) in his valuable synthetic review, *The Locomotion of Fishes* (see Tsvetkov 1992 for a review of earlier work by Russian authors). Breder drew inspiration from the fact that much of the existing work on fish locomotion was "in the nature of abstract philosophical discussion" rather than "based on actual experiment and study of living fishes." Breder's research, conducted at the New York Aquarium, involved direct observation of swimming fishes whose rapid fin motions were perceived with the aid of a stroboscopic device (i.e., an intermittently shuttered light source). Breder's 1926 paper provides both a mechanical analysis of various types of swimming movement and a systematic review of the locomotor behaviors exhibited by different groups of fishes. Perhaps most influential was his categorization of the modes of aquatic progression based on propulsor kinematics. Three swimming modes involving the pectoral fins were identified, each named after a taxonomic exemplar. Rajiform locomotion, exhibited by *Raja* in the photographs of Marey (1890; fig. 3.2A), involves longitudinal undulation of broad, horizontally aligned pectoral fins in a fashion analogous to the whole-body undulation observed in axial swimmers. Diodontiform swimming, employed by the eponymous order of fishes, is characterized by undulation of short-based, vertically oriented pectoral fins. Labriform locomotion involves synchronous oscillatory flapping of the left- and right-side pectoral fins in fishes such as *Scarus* and *Tautoga* (Breder 1926), although this mode is used by many nonlabroid fishes as well.

Breder's contributions included consideration of the paired fins as both primary locomotor organs for straight-ahead swimming and as vanes for swimming maneuvers involving change in heading and speed. He illustrated that yawing turns of the body may be effected by abruptly extending one pectoral fin during rectilinear swimming (fig. 3.3A). He also considered how fishes slow their forward progress. Breder understood that, for elongate fishes with ventrally positioned paired fins, extension of the pectoral fins during braking will cause a downward pitching moment of the head, or somersaulting tendency, which must be balanced by action of the pelvic fins posteriorly to avoid an uncontrolled maneuver (fig. 3.3B). Among his other insights, Breder pointed out that the paired fins can be used intermittently for propulsion even in species relying on axial undulation as the principle locomotor method. In addition, he noted that fishes may exhibit a combination of undulatory and oscillatory fin motions (e.g., Diodontiform + Labri-

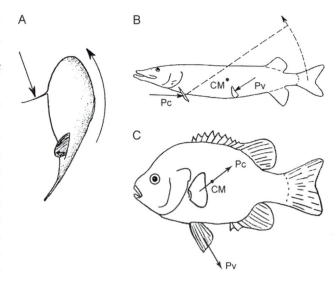

Figure 3-3 Early mechanistic analyses of paired-fin maneuvering locomotion by fishes. (A) Dorsal view of a pufferfish (*Lagocephalus*) executing a turning maneuver by means of unilateral pectoral fin extension. Arrow at left indicates the direction of the force applied to the fin by the surrounding water; arrow at right indicates the direction of body rotation (from Breder 1926). (B) Illustration of the forces acting on the paired fins of a long-bodied teleost (*Esox*) during a braking maneuver. Protraction of the pectoral fins anterior to the center of mass of the body (CM) results in a reaction force (Pc) that both decelerates the body and exerts a pitching moment tending to "trip up" the fish (i.e., cause the caudal portion of the body to rise, as indicated by the dashed-line arrow). Simultaneous action of the pelvic fins posterior to the center of mass exerts a complementary force whose reaction (Pv) balances this pitching moment and preserves a stable, horizontal body posture during the maneuver (from Breder 1926). (C) Forces acting on the paired fins of a short-bodied percoid teleost (*Lepomis*) during braking. By virtue of their dorsolateral position on the body, the pectoral fins may exert a force whose reaction (Pc) passes through the center of mass, thereby eliminating the destabilizing pitching moment experienced by elongate fishes. The pelvic fins act simultaneously to create a downward force (Pv) that neutralizes the lifting force on the pectorals and prevents the body from rising in the water column during deceleration (modified from Harris 1938).

form), resulting in kinematics of an intermediate type. This recognition of the complexity and subtlety of swimming movements in fishes served to focus the research efforts of many subsequent investigators.

Working in France and apparently unaware of Breder's work, Magnan argued similarly that the paired fins play an active role in both forward locomotion and maneuvering. Magnan (1929) undertook a broadly comparative study of pectoral and pelvic fin morphology, measuring fin surface area, shape, and position relative to the center of gravity of the body for more than 150 species of cartilaginous and bony fishes. He noted that interspecific variation in these anatomical parameters should directly influence the ability to generate propulsive force and to maintain postural stability.

In a continuation of his mechanical study, Magnan (1930) explicitly examined the functional role of individual fins in steady and unsteady swimming. For fishes swimming straight ahead at low speed, he examined the action of both undulatory and oscillatory pectoral fin motions. Through fin-removal studies he inferred the importance of the pectoral and pelvic fins in resisting rotational instabilities of the body. Magnan (1930) also performed a careful analysis of hovering behavior in bass (*Micropterus*), which uses alternating strokes of the left and right pectoral fins to maintain a fixed position in the water column. This analysis included a plot of the change in angular orientation of the pectoral fin blade over time, the first published kinematic profile in a study on paired-fin locomotion.

Another leading biomechanist of this time was John Harris, who collaborated with Gray as an undergraduate at the University of Cambridge (J. Gray 1933) and later worked with Breder in New York. Harris conducted a series of experiments at the Guggenheim School of Aeronautics (New York University) in which he placed a life-sized model of a dogfish shark (*Mustelus*) in a wind tunnel whose wind speed was set to give a biologically relevant Reynolds number, and examined the effects of the median and paired fins on static stability of the body (J. E. Harris 1936). He measured a large lifting force generated by the pectoral fins, which, because of their position anterior to the center of gravity of the body, also exerted an upward and backward pitching moment that tended to raise the head. He surmised that during steady swimming this rotational tendency is balanced by an opposite-sign torque generated by the heterocercal tail to maintain a level course of progression. By removing the pelvic fins both from his aerodynamic model and from living dogfish, Harris concluded that these posteriorly situated paired fins make a negligible contribution to the balance of forces and moments on steadily swimming sharks such as *Mustelus*.

In later work at the Tortugas Laboratory of the Carnegie Institution of Washington, Harris (1937) explored the functional role of the paired fins of bony fishes through anatomical study and direct observation of swimming behavior in the field. He noted an increased mobility of the paired fins in teleost fishes, as compared to those of selachians; speculated about activity patterns of the controlling musculature; and inferred swimming forces from cinematographic records of fin motions. Harris (1938) concluded that many teleosts use the paired fins not just as lifting planes but as sophisticated control devices to maintain body stability (the same conclusion was reached by Schmalhausen [1916]). Through an analysis of forces arising in reaction to paired-fin motion, Harris argued that the derived position of the pectoral and pelvic fins on the body of teleosts like sunfish (*Lepomis*) greatly reduces destabilizing pitching moments during braking (fig. 3.3C).

Despite its explicit treatment of the forces arising from paired-fin motion, the works of Breder, Magnan, and Harris

can be described as qualitative mechanical modeling, as compared to the more computationally intensive modeling that followed. Beginning in the 1950s, mathematicians and engineers began to formulate precise hydrodynamic theories to describe the mechanics of aquatic animal locomotion (reviewed by P. W. Webb 1978; Alexander 1983b). A burst of activity in this area in the 1960s and 1970s included the groundbreaking work of Lighthill (1960, 1969, 1970, 1971) and T. Y.-T. Wu (1971a, 1971b, 1971c, 1971d). A key contribution of this hydromechanical modeling was to provide a means of estimating the average thrust and power generated by swimming movements solely on the basis of trailing-edge propulsor kinematics. Such modeling, however, was restricted to analysis of swimming modes based on axial undulation (although it could also be applied, by analogy, to rajiform pectoral fin undulation, as pointed out by P. W. Webb [1975]).

The first formal application of hydrodynamic theory to study paired-fin swimming was developed by Blake (1979a, 1979b) who modeled the undulatory pectoral fins of mandarin fish (*Synchiropus*) as "actuator-discs" that increase the pressure of and impart momentum to the surrounding fluid. The mechanics of oscillatory pectoral fin motions have been modeled through applications of blade-element theory, which treats spanwise sections of the propulsor as individual two-dimensional hydrofoils, and sums total locomotor forces over the span (Blake 1979c, 1980, 1981, 1983a). Despite the limiting assumptions of these theories (reviewed by Blake 1983b), current applications, especially of blade-element theory, continue to provide new insight into the hydromechanics of paired-fin propulsion in fishes (e.g., Daniel 1988; J. A. Walker and Westneat 2000).

Modern Analytic Approaches

Recent advances in paired-fin swimming research have been supported by a platform of new experimental technologies. The advent of compact laboratory swim tunnels in the mid-1960s provided a controlled environment for empirical study of the swimming abilities of fishes. In the past 20 years, flow tanks of varying designs have been used by physiologists to measure both the aerobic metabolism and locomotor performance of labriform swimmers. Respirometry flumes have yielded information about oxygen consumption rates and the cost of transport for labriform swimmers (Gordon et al. 1989; G. R. Parsons and Sylvester 1992). Through increasing velocity trials, maximum sustained pectoral-fin swimming speed has been determined for a variety of teleosts (e.g., Davison 1988; S. D. Archer and Johnston 1989; Stobutzki and Bellwood 1994; Drucker and Jensen 1996a). Although generally slower than axial swimmers of the same size, labriform swimmers can attain impressively high sustained speeds with the pectoral fins alone (up to 4–6 body lengths per second; Whoriskey and Wootton 1987; J. A. Walker and Westneat 1997).

A second key technological advance was in the area of film and video imaging. Unlike the first sequential photographs of fish locomotion, which were taken at 8 to 20 images per second (Marey 1890; Magnan 1930), more contemporary cinematographic efforts are characterized by frame rates ranging from 50 to 1000 Hz. This improved temporal resolution facilitated detailed study of the rapid locomotor motions of the paired fins. The first high-speed motion picture (300 Hz) of an undulating pectoral fin was taken by Breder and Edgerton (1942). Kinematic parameters describing the oscillatory pectoral fin stroke, including beat frequency and amplitude, were first reported by P. W. Webb (1973) for shiner surfperch (*Cymatogaster*) and since then have been measured for numerous other species swimming over a range of speeds (e.g., S. D. Archer and Johnston 1989; Rosenberger 2001; Mussi et al. 2002; J. A. Walker and Westneat 2002). General changes in the shape and orientation of the pectoral fin blade during the stroke cycle have been described by P. W. Webb (1973) and Blake (1977); more recent studies have tracked the movement of anatomical landmarks in two perpendicular cine views simultaneously to allow quantitative analysis of three-dimensional fin excursions (Geerlink 1983; Gibb et al. 1994; Drucker and Jensen 1997; Walker and Westneat 1997).

Modern imaging tools have been employed to study labriform swimming kinematics in fishes varying in body size (Drucker and Jensen 1996a, 1996b; Mussi et al. 2002) and ecological association (Drucker 1996), as well as in taxa that recruit the pectoral fins in combination with multiple median fins for forward propulsion (Arreola and Westneat 1996; B. Wright 2000; Hove et al. 2001; Korsmeyer et al. 2002). Film and video have, in addition, provided insight into patterns of paired-fin use during swimming maneuvers, such as turning (Gerstner 1999; Drucker and Lauder 2001b; Webb and Fairchild 2001), braking (Videler 1981; Geerlink 1987; Drucker and Lauder 2003), and negotiating obstacles (P. W. Webb et al. 1996; Schrank and Webb 1998). Increasingly detailed study of pectoral fin kinematics in recent years has also led to refinement of the swimming-style categorization developed by Breder (1926). To describe the locomotion of fishes that flap the pectoral fins specifically along a dorsoventral axis, P. W. Webb (1994) proposed the embiotociform and mobuliform modes (named after the surfperches and the manta and devil rays, respectively).

Modern techniques for studying locomotor muscle physiology have provided further information about paired-fin function. Electromyography, first used with fishes in the 1950s and 1960s to study respiration and feeding (reviewed in Alexander 1983b), has relatively recently been applied to examine patterns of activity in the muscles of the fins. Lauder (1995) provided the first report of electromyographic activity in sunfish (*Lepomis*) during pectoral fin swimming. Subsequently, *in vivo* pectoral-fin motor patterns have been measured in the labrid fish *Gomphosus* (Westneat 1996; Westneat and Walker 1997), two species of surfperches (Drucker and Jensen 1997), and the stingray *Taeniura* (Rosenberger and Westneat 1999). Through the synchronous recording of electromyographic activity and fin kinematics, these studies have shed light on how the pectoral musculature controls propulsor motion. Previously, functional inferences about these muscles were based solely on myological architecture and muscle mass measurements (e.g., Williamson 1893; Danforth 1913; Ganguly and Nag 1964; Jessen 1972; Mansuri 1976; Geerlink 1989; Tsvetkov 1992).

Study of paired-fin function has also been facilitated by the development of techniques to investigate muscle physiology *in vitro*. Histochemical assays have been used to determine the oxidative and glycolytic capacities of pectoral fin muscle and thereby interpret labriform swimming performance in the field (Johnston and Camm 1987; Davison 1988). Luiker and Stevens (1993) examined the performance of isolated pectoral adductor muscle of *Lepomis* using the work-loop technique (Josephson 1985). Direct measurement of force and length change in this muscle undergoing oscillatory contractions revealed the contraction frequencies at which maximal net work and power are produced. These frequencies are within the range of fin beat frequencies used by sunfish during labriform locomotion.

Recent advances in theoretical analysis of paired-fin mechanics have been supported by the availability to zoologists of improved computing power for evaluating complex mathematical models. The two-dimensional blade-element theory of labriform swimming developed by Blake (1979c, 1980, 1981, 1983a) involves quasi-steady-state analysis, in which steady forces based on fixed lift and drag coefficients are calculated at multiple fin positions during the stroke and then summed to give the total force generated over the entire beat cycle. As noted by Dickinson (1996), actual paired appendages moving in fluid undergo complex three-dimensional excursions and rotational accelerations that result in highly time-dependent force coefficients. Quasi-steady-state "lift-based" and "drag-based" mechanisms of pectoral fin propulsion (e.g., P. W. Webb and Blake 1985) obscure the fact that both lift and pressure drag are arbitrary components of a single, unsteady circulatory force arising from vorticity bound to the fin.

Daniel (1988) used a combination of blade-element analysis and unsteady airfoil theory to investigate the mechanics of rajiform locomotion. He specifically assessed the effects of pectoral fin shape and deformation over time on locomotor forces. Combes and Daniel (2001) employed unsteady potential flow analysis to predict the forces generated by the flapping pectoral wings of ratfish (*Hydrolagus*). This analysis explicitly accounted for unsteady effects of fluid motion arising from wing acceleration and circulation-based phenomena.

Current modeling efforts have also included computationally intensive numerical studies involving solution of the Navier-Stokes equations (e.g., H. Liu et al. 1997; Carling et al. 1998; H. Liu et al. 1998). This approach allows the investigator to simulate unsteady three-dimensional flows around an oscillating virtual propulsor of realistic, time-dependent geometry. Propulsive forces are calculated by integrating the surface pressure over the body or fin at each time step throughout the simulation. Unlike earlier mathematical models, these computational fluid dynamic (CFD) analyses take into account the viscous effects in water, allowing more accurate estimates of body drag (Carling et al. 1998). Ramamurti et al. (2002) provided the first CFD study of pectoral fin propulsion in an investigation of force production by the labriform swimmer *Gomphosus* (plate 3.1A). Incorporating empirically determined fin kinematics from J. A. Walker and Westneat (1997), these authors found that more accurate estimates of lift and thrust magnitude were given by their unsteady model than by steady-state computations.

A complementary, experimental approach to the study of paired-fin mechanics is the use of wake visualization. Blake (1976) conducted a vector analysis of the forces produced by the pectoral fins of the seahorse *Hippocampus* on the basis of near-body particle motion recorded by a single cine camera. A similar technique was used to study pectoral fin forces in *Synchiropus* during forward swimming at a variety of speeds (Blake 1979b). Flow induced by the paired fins has also been examined by injecting dye streams into the water near the body. Using this technique, Arnold et al. (1991) showed that salmon parr holding stationary in a current reorient the incident flow with their pectoral fins to create an upwash. From this flow, a negative lift force was inferred to aid the maintenance of body position in fast streams.

In the past 10 years, biologists interested in the mechanics of aquatic propulsion have increasingly adopted more quantitative methods for measuring wake flow. Digital particle image velocimetry (DPIV) is one such technique, originally developed for examining man-made flows in engineering applications (Willert and Gharib 1991). DPIV involves laser illumination of densely seeded, reflective particles in the water surrounding a freely swimming animal, high-speed imaging of the particles in one or more video views, and computational resolution of particle displacement in consecutive video frames (Stamhuis and Videler 1995; U. K. Müller et al. 1997; Lauder 2000; Gharib et al. 2002). The result of each DPIV analysis is a matrix of uniformly distributed velocity vectors describing the average magnitude and orientation of wake flow over the course of the video framing period. The advantage of this flow visualization approach over the manual tracking of individual particles or dye fronts is that it can be used to make reliable estimates of momentum injected into the wake and, from the rate of change in momentum, propulsive fluid force (Drucker and Lauder 2002a, 2002b). In contrast to CFD models, which yield instantaneous forces acting on the propulsor, DPIV provides data on stroke-averaged forces. This technique has been used to document the force balance on sunfish during slow, steady labriform locomotion (Drucker and Lauder 1999) and to examine the forces produced by surfperch during high-speed swimming (Drucker and Lauder 2000; plate 3.1B). In addition, DPIV has clarified the function of the paired fins during unsteady maneuvering behavior (e.g., Wilga and Lauder 1999, 2000; Drucker and Lauder 2001b, 2003) and has shed light on the mechanistic basis of interspecific variation in top labriform swimming speed (Drucker and Lauder 2000).

Limbs as Propulsors

The technological innovations that spurred advances in our understanding of the paired fins of fishes made similar and simultaneous impressions on the study of terrestrial animal locomotion. Here we review the impact of both ancient observations and modern biomechanical analyses on the field of limb-based locomotion. These analyses involve the measurement and modeling of propulsive forces (both muscular and mechanical), high-speed photography and video, and exercise respirometry. Two techniques, ground-reaction force measurement and energetic analysis, have profoundly expanded our understanding of limbed locomotion, but for a variety of technical reasons they have not been as extensively employed to study fin-based locomotion.

In many ways, terrestrial limbed locomotion is more amenable to the application of experimental technologies than aquatic locomotion. For example, the work done by the leg muscles of running turkeys has been measured by the simultaneous application of sonomicrometry (to determine changes in muscle length), tendon buckles (muscle forces) and force plates (ground reaction forces; see Roberts 2002 for a review). The small size and broad insertions of the fin adductors and abductors make tendon buckles and sonomicrometry impractical for examining fin-based swimming, although similar techniques have been applied in studies of axial undulation (Shadwick et al. 2002; Donley and Shadwick 2003). Furthermore, water, with its greater density than air, poses insurmountable challenges to the use of force plates in the aquatic medium, thus rendering one of the most powerful tools for studying terrestrial locomotion inapplicable for the study of propulsion underwater.

Classic Observations

Aristotle lays claim to a great many firsts in the study of limbed locomotion. He defined one of the principal problems in limbed locomotion by clearly describing gaits, distinguishing a walk from a run, and positing mechanical differences between them. Although some of his locomotor observations have not stood the test of time, including the proposals that motion is always initiated by the right side of an animal, and that four contact points are needed for vertebrate locomotion (*Parts of Animals*), some questions he raised are still under active investigation. For example, Aristotle realized that in a walking gait the center of mass rises to its highest point at mid-stance and descends to its lowest at mid-swing. He also pointed out the distinction between a parasagittal gait and a sprawling gait, later understood in terms of fundamentally different mechanics. That he attributed the existence of the latter posture to a need on the part of sprawlers to get into holes in the ground has perhaps served as a goad to researchers seeking an explanation more in keeping with the phylogenetic and functional distribution of this gait (e.g., Jenkins et al. 1997; Reilly 1998; Parchman et al. 2003).

As in the study of fin-based locomotion, mechanical models of limbed locomotion were first proposed by Borelli (1680) in his compendium of theory and anatomical experimentation *De Motu Animalium*. The propositions in this surprisingly wide-ranging volume include analyses of the forces at joints, forces exerted by muscles, and ground reaction forces. However, the analysis is largely static and primarily attempts to explain intrinsic mechanisms rather than the extrinsic effects of locomotion.

Kinematics

Stop-action photography was employed from its infancy in the examination of limbed locomotion. In direct contrast to Borelli's fixation on the internals of movement, the work of Marey (1874) and Muybridge (1887) was completely focused on the externals. The particulars of gait, especially in horses, were of intense interest to sportsmen, bettors, the military, and scientists (Marey 1874; Pettigrew 1874; W. P. Wainwright 1880). The footfall patterns for walking, trotting, cantering, and galloping had been diagrammed for a wide variety of two- and four-legged creatures (using notation still

employed today), but the exact timing of the footfalls was a matter of some dispute. Leyland Stanford, robber baron and university founder, provided Eadweard Muybridge with the opportunity to use stop-action photography, the precursor to high-speed cinematography, to settle the question of whether all four limbs are ever off the ground at once when a horse is at the gallop. Muybridge produced a now-famous sequence of photos clearly demonstrating that Stanford's horse had an airborne phase while galloping (Muybridge 1972). These widely publicized results ushered in an era of examining the timing and patterns of footfalls during limbed locomotion.

Swinging and Bouncing

D'Arcy Thompson (1917) realized that when running there really is a "spring" in one's step, and Sir Thomas Gray (1968) made clear the importance of understanding the forces of locomotion in addition to movement patterns. A walking gait is fundamentally pendular: that is, the center of mass swings back and forth over a leg of very nearly fixed length. In contrast, running gaits like the trot, canter, and gallop are bouncing gaits, in which, during the early stance phase, the body mass transfers kinetic energy to a spring, represented by the muscles and tendons of the leg, reclaiming it later in stance to launch the next stride (see Alexander 1984 for an extensive treatment of this phenomenon).

Since terrestrial walking is a pendular gait, leg length determines stride length and maximum walking speed (reviewed in Alexander 2002). The ability to calculate these parameters has given paleontologists a window into the behavior and lifestyle of extinct animals (R. W. Frey 1975; Padian 1997). There is also an extensive literature on the transition from walking to running, which occurs when the inertial forces of locomotion overcome gravitational constraints, leading to an aerial phase (for a review of the mechanics see S. Vogel 2003). Observation of animals spanning a wide range of sizes, and experimental manipulation of inertia and gravity has confirmed that the transition from walking to running is determined by Froude number, the ratio of inertial to gravitational forces (Kram et al. 1997; Y. H. Chang et al. 2000; Hutchinson et al. 2003).

To explain gait transitions within the bouncing running gaits, physiologists turned to the respirometer. Oxygen consumption is an explicit measure of the energetic cost of locomotion from which cost of transport and energetic efficiency can be computed (e.g., C. R. Taylor et al. 1982). One interesting result of this research program is support for the hypothesis that cost of transport is a determining factor in gait transitions. While Froude number dictates the walk-trot transition, it is a loose constraint and does not apply to the trot-gallop transition. There is evidence that as speed increases trotting becomes energetically expensive, and a transition to galloping decreases oxygen consumption (Hoyt and Taylor 1981; Minetti 1998). Many fishes switch from paired-fin to axial locomotion at high swimming speeds, a gait transition proposed to represent "physiologically equivalent" levels of exercise in animals of different size (Drucker 1996). It would be interesting to know whether this transition is also determined energetically.

In relating forces to kinematics, the study of terrestrial locomotion is strides beyond aquatic and aerial flapping flight. Force plates work well on land but are impractical for measuring forces in swimming fishes and flying birds. Digital particle image velocimetry (DPIV) has emerged as the only practical method for empirically determining the forces developed during swimming. In free-flying birds, forces transmitted to the environment have been measured with DPIV (Spedding et al. 2003), and internal forces have been determined by the clever use of strain gages attached to the humeral crest (M. R. Williamson et al. 2001; Tobalske et al. 2003). Mathematical modeling of flight in fluid is an alternative method of determining forces and has been extensively used both by the avian research community (e.g., Lockwood et al. 1998; Rayner 2001a) and to study the locomotion of fishes (reviewed in the section "Modern Analytic Approaches"). Instrumented mechanical models, while extensively used to analyze arthropod flight (e.g., Ellington et al. 1996; Dickinson et al. 1999), have been less commonly applied to fishes and birds, perhaps because the "wing" deforms considerably during flapping in the latter two groups.

Areas for Future Research

In spite of the substantial gains made to date in our understanding of how fishes use the paired fins during swimming, a number of important questions about their function remain. Two promising areas for future research are refinement of hydromechanical analysis, and synthesis of comparative biomechanical data.

Extending our understanding of paired-fin mechanics will be facilitated by interdisciplinary investigation of locomotor force production. We need to integrate computational fluid dynamic analysis (CFD) with quantitative wake visualization to test assumptions inherent in both approaches. Data from CFD and DPIV studies on the same propulsor system will validate the strengths of each technique and identify areas where theory and reality do not match well. New flow visualization methods that provide information about wake velocity in three dimensions are being developed (Herrmann et al. 2000; Kähler and Kompenhans 2000; Pereira and Gharib 2002) and should allow improved estimates of wake

momentum. In addition, microelectricalmechanical systems (e.g., Abeysinghe et al. 2002) that will measure the pressure distribution over flapping fins *in vivo* may soon be available. Such pressure data are critical for direct determinations of fluid force. It is also vital that we bridge the gap between hydrodynamic and musculoskeletal performance studies. At present, the mechanisms by which internal muscular forces are transmitted, via the fins, to the external fluid environment remain obscure. Combining the techniques of *in vitro* muscle physiology with methods for calculating wake momentum flux will enable estimation of the overall mechanical efficiency of paired-fin propulsion.

Synthetic analysis of comparative biomechanical data will become a fruitful tool for understanding fin function. Fishes are characterized by pronounced morphological diversification of the paired fins. Historical transformations in the shape, orientation, and position of pectoral and pelvic fins are well documented (Drucker and Lauder 2002b), yet the hydrodynamic consequences of this evolutionary variation are poorly understood (theoretical and empirical work on this topic includes Combes and Daniel 2001; P. W. Webb and Fairchild 2001). Research on taxonomically diverse clades will allow a more broadly comparative analysis of the relationship between propulsor anatomy and force production.

In addition, new insights into paired-fin swimming performance may be gained by combining laboratory and field studies of locomotor behavior. The touchstone of a well-engineered flume is the minimization of flow heterogeneity. Yet fishes in the real world seldom if ever see such constrained flows. Recent studies indicate that trout exhibit different patterns of pectoral fin recruitment when swimming in a flow tank as opposed to their natal streams (McLaughlin and Noakes 1998; Drucker and Lauder 2003). Data from the field on paired-fin use (Fricke and Hissmann 1992) and swimming speed (P. C. Wainwright et al. 2002) complement observations made in a controlled laboratory setting by expanding both behavioral repertoire and performance range.

The existing body of research on paired-fin locomotion effectively counters early impressions that axial undulation is the only swimming mode of import to fishes. Up to 20% of living fishes rely on the pectoral fins as the primary means of propulsion (Westneat 1996), and most of the remainder undoubtedly employ both sets of paired fins on an intermittent basis to control maneuvers and maintain body stability. Future technological and analytic advances will allow further strides in understanding the functional role of the appendicular propulsors in aquatic animal locomotion.

The challenges for the study of terrestrial locomotion are primarily in translating techniques and generalizations learned in the lab to the real world of animal movement. Two principal areas are of particular interest: intermittent loco-

48 **Eliot G. Drucker and Adam P. Summers**

motion (see, e.g., Weinstein and Full 1999; Girard et al. 2001; Gleeson and Hancock 2001) and the integration of movement with ecology (Jayne and Irschick 2000; Irschick and Garland 2001).

Outside the laboratory most terrestrial animals do not move in a stereotyped, invariant fashion for very long (migrations are an exception); rather, they turn, start, slow down, stop, and start again. The repetitive nature of continuous locomotion allows relatively easy access to kinematics and forces through the use of treadmills and trackways; however, intermittent locomotion, with its accelerations, decelerations, and sudden turns, presents more challenges. It seems likely that the solution to these difficulties is going to involve telemetered force and kinematic data from free-ranging animals, though eliciting a defined maneuver (turn, jump, etc.) from an animal on an instrumented trackway also has promise.

When examining limbed locomotion in the laboratory, researchers generally try to elicit particular gaits and speeds from an animal. While these data have a great deal to say about the performance, kinematics, and force production in the limb during locomotion, they have little bearing on the use of the limbs in day-to-day life. This field performance is of paramount importance to the organism, as it plays a role in ecological niche and often has a very direct bearing on fitness. There are several approaches to understanding locomo-

tor performance in the wild including telemetry, focal animal studies, and analysis of trace evidence. Telemetry is useful for understanding integrated performance over longer durations, although the limited temporal resolution makes it difficult to determine moment-to-moment activity. Focal animal studies can work well for assessing locomotor performance in large animals that move in open areas. For most terrestrial animals, however, this sort of observation is so intrusive that it will alter behavior.

Recent ichnology, the study of footprints and other traces of extant animals, shows promise for understanding daily locomotor performance in narrowly defined habitats. For example, in desert sand-dune communities trackways of lizards, mammals, and birds are an exact record of movement. The stride length and print size can be used to estimate speed and size of the animal, and in communities with few species it is possible to know who made the track. This technique has been used to estimate the locomotor habits of sand-dwelling lizards to a surprising level of detail, including a comparison of speed at different inclines, histograms of preferred speeds, and maximal realized sprint speeds (Jayne and Irschick 2000). In this case moving locomotor research from the lab into the field provides an example of the overestimation of maximal sprint speed under the controlled conditions of a treadmill versus the unpredictable and variable outdoors.

Chapter 4 Fins and Limbs in the Study of Evolutionary Novelties

Gunter P. Wagner and Hans C. E. Larsson

FROM THE VIEWPOINT of an advanced primate on the phylogenetic tree a handful of evolutionary transitions stand out as particularly momentous. Naturally those are the points in phylogeny where a major bifurcation occurs along our lineage. The origin of tetrapods and the transition that separates us from the evolutionary fate of fishes are among the most intriguing points in our history. During this transition from an aquatic ancestor to terrestrial descendents a number of important anatomical and functional changes took place in the skull, the girdles, the axial skeleton, and much more (Clack 2002b). But few of those changes have attracted as much attention as the origin of tetrapod limbs from sarcopterygian fins. That interest has many roots, including the comparatively simple anatomy of limbs, the intuitive relationship with terrestriality, and the large amount of experimental data available about the development of limbs (Hinchliffe and Johnson 1980; A. S. Wilkins 2002). The origin of tetrapod limbs has long been a paradigm of a major evolutionary transformation of a morphological character, and recent progress in the developmental genetics of limbs makes the study of the fin-limb transition a particularly promising subject.

In this chapter we summarize facts and ideas about the fin-limb transition from the standpoint of the study of evolutionary novelties. Before we consider the evolution and development of fins and limbs we reflect on the notion of evolutionary novelties to the extent necessary to motivate a developmental approach to their study.

How Should We Study Evolutionary Novelties?

Is the study of evolutionary innovations and novelties a research program distinct from the study of adaptation? (Love [2003] proposed to distinguish between the origin of new body parts, to be called novelties, and new functions, to be called innovations. We think it is useful to distinguish between morphological and functional changes and thus respect this distinction in this paper.) Clearly, if the answer to this question is negative—that is, if the study of novelties is no different from the study of adaptations—then it would be useless to even speak of "novelties," since the term would not introduce a scientifically meaningful distinction. Among biologists, the jury is still out on this question. On the one hand there is Ernst Mayr, who already in 1960 had argued that innovations are a much neglected problem of evolutionary biology. On the other hand there is Brian Charlesworth, whose contribution to the Field Museum Conference "Evolutionary Innovations" (Nitecki 1990) is titled "The Evolutionary Genetics of Adaptations." Obviously Charlesworth rejected the assumption underlying the whole conference, namely that evolutionary innovations require special attention.

In this chapter we assert that tetrapod limbs are an evolutionary novelty, in fact that limbs are a paradigm of an evolutionary novelty. Thus, we have to explain what this assertion implies. To explain the unique agenda that characterizes research into the origin of novelties we start with a short summary of the adaptationist program. The aim is to make clear the different assumptions and approaches that go into the study of adaptations as opposed to the study of novelties.

It is widely accepted that adaptations are traits or features of an organism that owe their existence to the action of natural selection (Futuyma 1998). The agenda of the adaptationist program is to explain the origin of features that enhance the survival and reproductive success of individuals (Mayr 1983; Stearns 1986). It is well established that natural selection is the cause and thus the explanation of adaptive traits. Putting aside all the complications that result from historical contingencies and the many problems in demonstrating the role of natural selection, there is a common denominator to all attempts to study adaptations. That common denominator is the assumption that natural selection is causally sufficient to explain the outcome of adaptive evolution (e.g., R. N. Brandon 1990). If natural selection is a sufficient cause (given certain boundary conditions) of the evolution of adaptive traits then this outcome has to be reproducible. Hence, repeated evolution of the same or similar traits under similar ecological or functional conditions is seen as evidence in favor of adaptation. (The fact that some apparent morphological novelties can originate multiple times in a clade has been used to challenge the very notion of evolutionary novelty [Eberhard 2001], but the evidence in this case is far from conclusive [G. P. Wagner and Müller 2002].) Similarly, the ability to evolve a certain trait in the laboratory is a good argument that natural selection is in fact the cause for that trait. Finally, the ability to predict the outcome of evolution from functional optimality considerations is also a strong argument in favor of adaptation (Orzack and Sober 2001). Hence, explaining a trait as an adaptation implies a research program with the aim of showing that natural selection was a sufficient cause for the evolution of the trait. The goal is to identify the functional and ecological factors causing the fitness differences that led to the selection of this trait. This is the heart of the adaptationist research program (Mayr 1983; Orzack and Sober 2001).

Natural selection requires the availability of heritable variation, and demonstrating the heritability of the trait under consideration is a standard part of the adaptationist research program (Stearns 1992). But, on the other hand, the dependency of natural selection on heritable variation adds nothing to the explanatory force of adaptationist research programs, because the heritable variation does not determine the outcome of the evolutionary process (Amundson 1989). The crucial causal factor is still natural selection. The availability of heritable variation can be thought of as a boundary condition rather than a cause of adaptation (Sterelny 2000). But there are certain traits for which the presence of heritable variation is not guaranteed. For instance, there is no heritable variation in *Drosophila* for directional asymmetry in the location of ocellae (Maynard-Smith and Sondhi 1960), and it is unlikely that one can demonstrate the presence of selectable variation, or any variation, for flight feathers in an alligator population. Nevertheless, flight feathers evolved from archosaur epidermal scales (Prum and Brush 2002). This leads to the question of how feathers arose with their radically different morphology compared to scales. This is the novelty problem.

The agenda for the study of novelties is to explain the origin of characters that open up new functional and morphological possibilities to the lineage possessing the character. There is no question that such traits exist. The evolution of feathers just mentioned is not only of utility for various new functions such as heat insulation, flight, and communication: feathers also increased the range of possible morphologies an archosaurian epidermal appendage could assume. One may select on the shape of a crocodile epidermal scale for a long time and chances are that there would be no branched pennate appendages among the results of the selection experiment. The reason is that the range of morphologies characteristic of feathers requires a quite specific set of developmental features that feathers have but scales do not (M. P. Harris et al. 2002; Prum 1999; M. Yu et al. 2002). Most likely, feathers first evolved for heat regulation and assumed more sophisticated functions later on (Prum and Brush 2002).

Another epidermal appendage also most likely evolved first as insulation but never gained the ability to support flight: namely, hair in mammals. Mammals, of course, did evolve flying species. In fact, the second largest group of mammals is bats, with 925 described species (Koopman 1993). The reason for the distinct morphological and functional versatility of hairs and feathers is the developmental differences between them and the implications of these differences for the variational properties of feathers (Prum 1999). Even though the selection pressure for the origin of hairs and feathers was presumable the same—heat insulation—the outcome is different in potential. Hence, natural selection does not explain the difference between feathers and hair. And the reasoning is the same for the difference between the ancestral character of feathers—scales—and feathers themselves, the developmental pathway creating a divergent range of opportunities for further change in one character (feathers), but not in others (scales and hair).

From this sketchy example it is clear that developmental pathways can promote or limit the ability to realize certain morphologies. This idea is also known as developmental constraint (Maynard-Smith et al. 1985). Without going into complicated conceptual issues one can say that novelties are characterized by new morphogenetic possibilities that do not exist in the ancestral character or body plan (G. P. Wagner and Müller 2002). The question then is whether these novelties are just some special adaptations or whether their study requires a different research program. Certainly novelties can be seen as adaptations since they most likely were of some utility to the organisms in which they first arose and were thus subject to natural selection (although that is not necessarily the case; see Gould and Lewontin 1978, for instance). But is that all? Does natural selection explain the most important consequence, namely the new opportunities that come with the novelty?

A number of features characteristic of novelties make it unlikely that the adaptationist program will give us a satisfactory explanation. For one, novelties are rare. One can only assume that the necessary genetic variation does not arise in high frequency in natural populations, or that it requires a specific genomic predisposition to happen. In contrast, genetic variation for other traits arises at a high degree of regularity—for instance, genetic variation in abdominal bristle number in *Drosophila* (Falconer and Mackay 1996) and in other traits (Roff 1997). This phenomenological fact is supported by the recent discovery that novelties are often realized by "genetic rewiring" of developmental genes rather than by tweaking of existing regulatory relations (E. Davidson 2001). It seems that new regulatory interactions arise at a much lower rate than quantitative variation, but this is still an open empirical question. Furthermore, the specific new potential of a novelty can hardly be "seen" by the natural selection that originally selected the new trait. Natural selection is short-sighted, looking only for the immediate benefit. Thus, it is unlikely that natural selection can provide a satisfactory account of the fact that feathers turned out to be able to support flight, but hair did not.

Table 4.1 summarizes the issues separating the study of evolutionary adaptations from evolutionary novelties, as we have outlined them. Adaptation is about the origin of features that increase the fitness of the individual possessing the trait. Novelties are about the new variational opportunities that a novel character allows. Adaptations are explained by natural selection. Novelties are explained by the derived developmental features that allow new phenotypic opportunities—that is, new variational opportunities (G. P. Wagner and Müller 2002). The adaptationist program implies research in functional, behavioral, and ecological factors determining the fitness of organisms (Stearns 1992). The study of novelties implies a developmental research program (S. B. Carroll et al. 2001; G. B. Müller and Wagner 1991; Raff 1996). The goal is to identify those developmental changes that created new body parts and their characteristic variational potential (Prum 1999).

While the goal of the study of evolutionary novelties may seem clear, the process of characterizing and defending supporting data is less clear. The starting point for such a research program must lie in the careful characterization of morphological differences. The discovery of the mechanisms necessary for the development of the derived morphological features must follow. Among those mechanisms are probably also those derived developmental processes that brought about the novelty and that account for the special properties of the new character. But of course, not every developmental mechanism necessary for the development of a derived character was instrumental for the origin of the character (G. P. Wagner et al. 2000). At least two common possibilities demonstrate this pitfall.

First, the function of a developmental gene could be phylogenetically older than the novel character. For instance, the zone of polarizing activity (ZPA) is a mesenchymal signaling center that secretes Shh protein at the posterior margin of the limb bud and is essential for the development of digits and for digit identity (Riddle et al. 1993). A similar signaling center, however, has been shown to exist in zebrafish paired fins (Riddle et al. 1993). Thus, the evolution of the ZPA was not instrumental for the origin of digits.

Second, a gene that is essential in a derived species could have acquired the new function after the character evolved. The best-documented example is the function of the gene *bicoid* (*bcd*) in *Drosophila* body axis determination. This gene exists only in a subgroup of flies; in basal insects and other arthropods, axis determination is caused by different mechanisms (S. Brown et al. 2001). Hence *bcd*, though essential in some flies, was not instrumental in the evolution of the bilaterian body axis.

There is currently no coherent methodology for the study of the evolution of novelties. Clearly the comparative developmental biology (evolutionary developmental biology) of derived and plesiomorphic characters has to be and is part of the research methodology employed. But the challenge is to find evidential criteria to distinguish between chance associations and truly causal relationships between developmental changes and morphological changes in evolution. We return to this problem after we consider whether the tetrapod limb

Table 4-1 Major features of evolutionary adaptations and novelties

	Evolutionary adaptation	Evolutionary novelty
Phenotypic change	Modification of characters that increase the fitness of an individual	Origin of new variational opportunities within a lineage
Causal explanation	Natural selection	Derived developmental features
Genetic change	Allele frequency	"Genetic rewiring"

can productively be considered an evolutionary novelty and what issues need to be addressed in a research program on the origin of the tetrapod limb.

The Phylogenetic Context of the Fin-Limb Transition

The anatomical and paleontological facts on the origin of tetrapods have been summarized recently in a monograph (Clack 2002b). In this section we only want to point out those facts that are most relevant to the investigation of the developmental basis of tetrapod limb origin.

There is a broad consensus that living tetrapods are a monophyletic group nested within the sarcopterygian fishes (Cloutier and Ahlberg 1996). The sister group of sarcopterygian fishes are the ray finned fishes, or actinopterygians, with teleosts as the crown group. Recent molecular data confirm that Polypteriformes (bichers) are the most basal ray finned fish lineage, followed by a group containing the Acipenseriformes (sturgeons), *Amia* and lepisosteiforms (gars; J. G. Inoue et al. 2003). For the present discussion we use a simplified phylogeny to place the anatomical data into a phylogenetic context (fig. 4.1). Data from comparative anatomy indicate that the proximal parts of the tetrapod limb, represented by the upper and lower limb (stylo- and zeugopod) are inherited from the sarcopterygian fins; that is, they are plesiomorphic for tetrapods. The distal part of the limbs, the hand or foot (autopod), is an evolutionary novelty; that is, is apomorphic for tetrapods. More detailed discussions of these anatomies are presented by Coates and Ruta (chap. 2 in this volume) and Carroll and Holmes (chap. 13 in this vol-

ume). For the purpose of the present chapter, we summarize the evolution of the appendicular endoskeleton.

As diverse as paired appendage skeletons are, some generalities can be made among the earliest forms. Eugnathostomata (the clade composed of Chondrichthyes and Teleostomi) have one or more columnar endochondral elements supporting their paired appendages. These elements bridge the appendicular girdles (pectoral and pelvic) to the more distal structures in the appendages. These basal elements are generally columnar in outline and lack complex branching patterns. Most non-eugnathostomatan gnathostome fossils do not have endochondral structures preserved in their paired appendages. An exception are ptyctodontid placoderms, which have three endochondral elements at the base of their pectoral appendages.

Chondrichthyans exhibit some diversity of the number of basal endochondral elements in their paired appendages. A number of taxa, such as Xenocanthida, express only a single element at the base of their pinnate-patterned pectoral fins. Others, such as Cladoselachida, have up to eight basal endochondral elements to form broad pectoral fins. Pelvic fins are often reduced and are difficult to compare with pectoral fins in many taxa. However, most forms have a number of simple, columnar elements articulating to the pelvis. For simplicity, only the pectoral fin will be discussed hereafter. Pelvic appendages are generally reduced or highly modified in numerous lineages, which reduces their value for understanding the origin of the tetrapod limb.

Acanthodian fishes, the probable sister taxa to Osteichthyes, ossify much of their endochondral skeleton. Their paired appendages consist of a large preaxially positioned fin spine with a set of ossified endochondral elements at the base of each fin. The arrangement of the elements is unclear; they appear to be nodular with no apparent pattern of arrangement.

Osteichthyes appear to have a tribasal pectoral fin, or variations of it (fig. 4.2). Paleoniscoid fishes are the most basal actinopterygians known only from fossils. These taxa are small and have generally poorly preserved appendicular skeletons. However, some taxa, such as *Moythomasia*, appear to have a preaxially located propterygium and a postaxial metapterygium that supports a number of radial elements. At least five basal radials lie between the pro- and metapterygium. These may be homologous to the mesopterygium.

The chondrification patterns of the pectoral fins in *Danio*, *Polyodon*, and *Acipenser* have been well described (Grandel and Schulte-Merker 1998; M. C. Davis et al. 2004; Mabee and Noordsy 2004). Differences in interpretation exist between these authors, however. Mabee and Noordsy (2004)

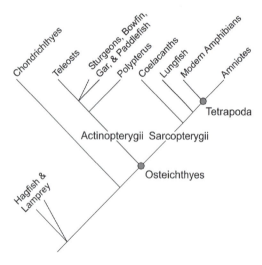

Figure 4-1 Simplified phylogeny of the living relatives of tetrapods. The grouping of sturgeons, bowfins, and gars in one clade is based on the results of J. G. Inoue and collaborators (2003).

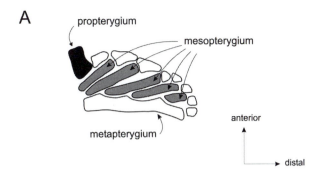

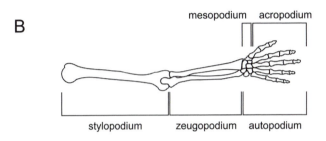

Figure 4-2 Schematic representation of the morphology and terminology of the appendicular endochondral skeleton of a tribasal fish and a tetrapod. (A) A *Polyodon* (paddlefish) pectoral fin (adapted from Grande and Bemis 1991). (B) A human arm. Each appendage is oriented with anterior (preaxial) upward and distal to the right. Figures are not to scale.

describe the early pectoral fin development of *Polyodon* as arising out of a single cartilaginous plate, termed the endochondral disc. The plate divides into three basal elements. The preaxial element further divides to form a propterygium and the first basal radial. The middle element forms the second basal radial, while the postaxial element forms the metapterygium and the radials it supports. M. C. Davis et al. (2004) describe a slightly different pattern in the same taxon. An endochondral disc initially chondrifies preaxial to the metapterygium. The endochondral disc subsequently divides into the propterygium and both basal radials. M. C. Davis et al. (2004) describe a similar pattern in *Acipenser*.

Danio has no metapterygium (see below). However, an endochondral disc does chondrify initially (Grandel and Schulte-Merker 1998). These authors interpret the disc as a functional requirement to support the pectoral fin during larval life. From comparisons with the early development of chondrichthyans, nonteleost actinopterygians, and teleosts, these authors suggest that the endochondral disc and its early function of larval fin support may be features of all teleost fishes. In *Danio*, the disc divides into four basal radials. These authors identify an element that lies preaxially to the basal radials as a distal radial. Mabee and Noordsy (2004) interpret this preaxial distal radial as the propterygium.

The nature of the basal endochondral elements of actinopterygians remains pluralistic. However, in light of the early patterns of chondrification and for the sake of discussion, we treat the elements as a postaxial metapterygium and a preaxial endochondral disc that divides into a preaxial propterygium and a set of basal radials. This interpretation conflicts least with the fossil record, the patterns seen in basal actinopterygians, such as *Polypterus*, which has a columnar pro- and metapterygium bounding a platelike mesopterygium. A tripartite arrangement of basal elements seems to best explain the variation of structures across living and fossil fishes. Further division of the mesopterygial portion of the endochondral disc may explain the multiple basal radials of Cladoselachida and paleoniscoid fishes.

Three basal elements are found in the majority of extant chondrichthyans, suggesting that a tribasal pectoral appendage is the ancestral condition of Eugnathostomata. This type of fin has been found to be a basal condition for sharks and rays in some systematic studies of chondrichthyans (Compagno 1973). This hypothesis is not entirely secure, as some studies reject this phylogenetic reconstruction (Maisey 1984). Further still, the phylogenetic position of many basal chondrichthyans is unresolved. However, herein we agree with Mabee (2000) and treat the plesiomorphic condition for chondrichthyans and osteichthyans as tribasal.

In all cases, basal endochondral elements of the paired appendages remain relatively simple columnar structures. In addition, the elements do not appear to be derived from a postaxial metapterygial chondrification and a preaxial endochondral disc that divides into a propterygium and mesopterygium. The mesopterygium may remain a single platelike element, as in many chondrichthyans and *Polypterus*, or divide into a number of basal radials. If the pattern described is as invariant (in both the evolution and the known development of the structures) as we present it, the tripartite pattern may represent a useful morphological unit to discuss the issue of the development and evolution of the appendicular skeleton.

A trend identified in Actinopterygii is the loss of the metapterygium in all Euteleostei. The only basal elements that remain in these taxa are a propterygium (if Mabee and Noorsky [2004] are correct in interpreting the preaxial "proximal radial" as a propterygium) and four basal radials (derived from the mesopterygium).

Sarcopterygian fishes present a different trend of fin evolution. The least inclusive group with preserved fin skeletons are Actinistia (coelacanths), which show no indication of a pro- or mesopterygium and retain only an ossified metapterygium. Knowledge of the embryology and detailed adult soft anatomy of the extant actinistian, *Latimeria* is insufficient to state whether or not a cartilaginous remnant of a pro- or mesopterygium exists. The metapterygium is retained and comprises the entire ossified endochondral fin skeleton. The metapterygium is segmented to form a string of repeating elements that often have extra endochondral elements on the preaxial side of each segment joint. These elements and those of other sarcopterygians have been homologized with radials of actinopterygians (Mabee 2000; Metscher and Ahlberg 1999). However, the absence of a mesopterygium suggests that these structures may not be homologues of radials but derivatives of the metapterygium. The endochondral skeleton terminates in a fan of unequally sized plates that support the dermal lepidotrichia.

A similar fin endochondral skeleton is present in extant dipnoans and extinct porolepiforms, although their fins are quite elongate in comparison to those of other sarcopterygians. The metapterygium is further divided into a string of segments. Additional segmented endochondral structures articulate on the pre- and postaxial sides of the metapterygial string at their joints. Some of the additional structures articulate directly to a metapterygial segment in extant dipnoans.

Rhizodontids appear to be the first group of sarcopterygians along the line toward tetrapods to have organized their fins, at least proximally, like those of tetrapods. These fish had a single proximal stylopodium (humerus/femur) that articulates with two elements in the position of a tetrapod zeugopodium (ulna and radius/fibula and tibia). Coelacanths and dipnoans have only a single element contacting their single stylopodia. Rhizodonts, such as *Sauripterus*, have as many as eight endochondral elements in positions similar to tetrapod digits (Daeschler and Shubin 1998).

Tristichopterids, such as *Eusthenopteron*, retain the proximal tetrapod-like skeleton of rhizodontids, but without the extended distal elaborations. Only the preaxial sides of the joints of the metapterygial segments support an unsegmented endochondral element. The distal most metapterygial segment (the fourth) supports two. *Panderichthys* fins are similar to those of *Eusthenopteron* but have only three metapterygial segments before ending in a fan of lepidotrichia.

The earliest sarcopterygians with a discrete autopodium are *Acanthostega, Ichthyostega,* and *Tulerpeton.* These forms retain the single-element stylopodium, a dual-element zeugopodium, and an autopodium composed proximally of a transverse series of mostly nodular elements and distally of a series of elongate digits. Although the digits are remarkably similar to each other, differences in digit size and in phalangeal number suggest some degree of unique identity to each digit.

Limb Development

Limbs, as well as fins, develop from ectodermal outgrowths filled with mesenchyme, the so-called fin/limb buds (Hinchliffe and Johnson 1980). Many aspects of early limb bud development are similar to the development of paired fins in zebrafish. In this section we only focus on those aspects of limb development that directly relate to the derived characters of the tetrapod limbs: that is, the autopodium.

Development of the Autopodium

The autopodium is the most distal of the three main limb segments (stylo-, zeugo- and autopodium). The mechanisms by which these three segments are determined in development are controversial—progress zone model (Saunders 1948) versus prespecification model (Dudley et al. 2002)—but it is clear that the development of the autopodium involves quite distinct developmental events from those of the proximal parts (reviewed in G. P. Wagner and Chiu 2001). The limit between the developing zeugopodium and the autopodium is first marked by the boundary between the expression domains of *Hoxa11* (proximal) and *Hoxa13* (distal; Haack and Gruss 1993). *Hoxa13* is thus a good candidate for an autopodial "organ identity gene." The evidence supporting this hypothesis is threefold. Ectopic expression of *Hoxa13* in the proximal parts of the chick wing leads to cartilage reminiscent of the wrist (Yokouchi et al. 1995). Shifting the *Hoxa11/Hoxa13* expression boundary distally leads to malformation of the wrist region (Mercanter et al. 1999). Knockout mutations of *Hoxa13* only affect the autopod in the limb and lead to loss of digit I (Fromental-Ramain et al. 1996b), while a mutation in the coding region of *Hoxa13, hypodactyly (Hd)* leads to more severe digit loss (Post and Innis 1999). *Hoxa13$^{-/-}$, Hoxd13$^{-/-}$* double knockout leads to a complete loss of the autopodium. Hence *Hoxa13* has partially overlapping functions with *Hoxd13* in determining the autopodium.

The locally exclusive expression of *Hoxa11* and *Hoxa13* in the limb bud is likely to be a derived characteristic, since these genes are expressed in overlapping domains in zebrafish and paddlefish (Sordino et al. 1995; Metscher et al. 2005). Little is known about the mechanisms regulating *Hoxa11* and *Hoxa13* expression. *Hoxa13* is induced by FGFs from the apical ectodermal ridge (AER; Vargesson et al. 2001). There is some evidence that suppression of *Hoxa11* in the distal limb bud is caused by *Hoxa13* (Post and Innis 1999). In terms of the evolutionary sequence of limb evolution this observation suggests that the typical tetrapod expression pattern may have evolved through the acquisition of mutually repressive regulatory links between *Hoxa13* and *Hoxa11* (Metscher et al. 2005).

Development of Digits

Both *Hoxa13* and *Hoxd13* are necessary for digit development, but it is noteworthy that *Hoxa13* knockouts affect mesenchymal digit condensations, while *Hoxd13* knockouts affect the growth of a normal complement of digits (Dollé et al. 1993; Fromental-Ramain et al. 1996b; Post and Innis 1999). Mortlock and collaborators (1996) previously reported that hypodactylous mice have a deletion within *Hoxa13*. Heterozygous mutants lack digit I, while homozygous mutants lack all but one digit (Robertson et al. 1997). This work was extended by Zákány and collaborators (1997) by investigating knockouts of multiple *AbdB* related *HoxD* genes and *Hoxa13*. These authors concluded that *HoxD* gene mutations lead to perturbations in digit development and to polydactyly, while *HoxA* mutations are necessary for digit loss. From these observations Zákány and collaborators suggest that *Hoxa13* acts upstream of *HoxD* genes and in evolution was the first gene recruited into the development of the autopodium (Zákány et al. 1997). In a second step, the 5' *HoxD* genes were recruited into autopodial development for limiting digit number to five and for the origin of distinct digit identities.

The regulation of *HoxD* gene expression has been studied extensively. In short, expression of 5' *HoxD* genes (*Hoxd13*, *-12*, *-11*, *-10*, and *-9*) in the limb has three phases, roughly corresponding to their role in the development of the three main limb segments (Nelson et al. 1996). The expression in these three phases is regulated by distinct cis-regulatory elements. For instance, the elements responsible for *Hoxd11* expression in the zeugopod is located in the intergenic region between *Hoxd12* and *Hoxd11* (element RXI [Hérault et al. 1998]), while autopodial expression is regulated by a global control element 250kb 5' of the *HoxD* cluster (Spitz et al. 2003).

Classical experiments on chick wing buds by Summerbell et al. (1973) led to the discovery of a signaling center at the posterior margin of the limb bud called the zone of polarizing activity or ZPA. The ZPA influences both digit number and digit identity. The primary signal has been shown to be sonic hedgehog (Shh; Riddle et al. 1993). The effect of the Shh signal is mediated through a homologue of the *Drosophila* gene *cubitus interruptus* (*ci*), *Gli3*. Knockout phenotypes of *Gli3* and *Shh* show that the ability to form digits is dependent upon *Shh*, in the presence of *Gli3*, but independent of *Shh* in the absence of *Gli3* (Litingtung et al. 2002; te Welscher et al. 2002b). *Shh^{-/-}* knockout mice yield only a

Table 4-2 Summary of the relevant developmental information outlined herein for each of the three major steps of autopodial development

Autopodial field	Digits	Digit identity
Hoxa-11 and *Hoxa-13* mutually exclude each other	Phase I of Shh expression establishes a *Gli3* activator:repressor gradient that, in turn, regulates at least *Hoxd-11* and *-13*	Phase II of *Shh* expression reestablishes (?) a *Gli3* activator:repressor gradient that, in turn, regulates *Hoxd-9* through *-13*. Later Hoxd~*Gli3* complexes maintain *Shh* target gene activation in the absence of *Shh*.

single-digit, ectopic expression of the repressor form of *Gli3*, and an absence of an AER (Litingtung et al. 2002). Digits do form in the absence of *Gli3* but are polydactylous and lack position-specific identities. Hence, the role of *Shh* in digit development is to modulate both the number and the morphology of the digits (table 4.2).

In this context an intriguing fact is the existence of a physical interaction between the AbdB-derived HoxD proteins and *Gli3* (Y. Chen et al. 2004). In late autopodial development, when Shh signaling is absent but digit identity is still labile, *Gli3*R forms a complex with Hoxd12 and other Hoxd proteins (e.g., *Gli3*~Hoxd12). This complex is a transcriptional activator of Shh target genes. The authors speculate that this interaction is a novel character of tetrapods. A preliminary sequence comparison reveals that the *Gli3*R amino acid sequence has two tetrapod-specific amino acid motifs, batches of amino acid residues conserved among tetrapods but variable or different in fishes. This suggests that *Gli3* underwent adaptive evolution in the stem lineage of tetrapods, possibly in connection with the evolution of digit development.

In mice, loss of *Shh* leads to complete loss of *HoxD* and *Hoxa13* expression. This is so because in the absence of *Shh*, *Gli3* is processed to produce the repressor variant of *Gli3* protein. Shh signaling suppresses this processing step, leading to an abundance of the activator form of *Gli3*, allowing Gremlin to activate FGF expression in the AER and allowing the expression of *Hox* genes (Litingtung et al. 2002). A loss of function mutation of *Shh* in zebrafish, however, *sonic you* (*syu*), does not affect *Hoxd11*, *Hoxd13* expression, at least initially (Neumann et al. 1999). Hence, it is possible that the evolution of digit initiation and identity is coupled with the two proposed phases of Shh function and the interaction of Shh with *Gli3* during each of those phases. Regulatory links between *Gli3* and *Hoxd13* and *Hoxd11* during the first phase of *Shh* expression and *HoxD* genes during the second phase of *Shh* expression may account for developmental changes that allowed for the evolution of digits and, subsequently or concurrently, digit identity (table 4.2).

Digit identity has been shown to be influenced by the activity of the interdigital mesenchyme, where the interdigital mesenchyme determines the identity of the digit anterior to it (Dahn and Fallon 2000). It is likely that this signal is mediated through *Bmp4* expression in these cells. *Bmp4* affects digit development indirectly through its influence on the activity of the AER (Selever et al. 2004).

Scenarios for the Origin of the Tetrapod Limb

The Metapterygial Axis

All discussions of fin-limb homology must begin with the metapterygial axis, which is the proximodistal trajectory through the metapterygium. It is clear from the phylogenetic summary above that this structure persisted in sarcopterygian fishes and was the ancestral structure from which the tetrapod limb evolved. This axis has long been used as a topologic framework to aid in identifying putative homologies. The proximal metapterygial mesomere has been widely accepted as homologous to the stylopodium (humerus/femur). The zeugopodium (radius and ulna/fibula and tibia) has been generally homologized with the second mesomere and the first anterior radial that is articulated to and derived from the metapterygium. The projection of the axis further into the tetrapod limb has been a source of debate. Gegenbaur (1878), D. M. S. Watson (1913), and Jarvik (1980) suggested that the metapterygial axis passed through the posterior elements of the autopodium. Watson and Jarvik went so far to suggest that the axis passed into digit IV of a pentadactylous autopodium. They claimed that the first preaxial radial was homologous with the radius, the second with the intermedium, and the third with the first centrale. Digits V and the pisiform were equated with the postaxial processes on the third and fourth mesomere of *Eusthenopteron*, in spite of the lack of obvious separation of this process from the mesomere in fossils (Andrews and Westoll 1970b). However, digits I through III were equated with the preaxial radials. In this scenario, the autopodium is completely derived from plesiomorphic metapterygial elements. Steiner (1934) provided some embryological support for this hypothesis. While describing a lizard in early development, he noted similarities between the elements forming in the arm and the elements found in adult *Eusthenopteron*. The metapterygial axis appeared to pass through the ulna and ulnare while the radius and intermedium appeared in positions similar to the first and second radials of *Eusthenopteron*.

Westoll (1943) suggested an alternative hypothesis when he described the axis as passing through the first two metapterygial mesomeres (as Watson and Jarvik) and then bending preaxially to pass through the intermedium. The axis was suggested to pass toward digits I and II. Since there are no postaxial radials known in *Eusthenopteron* or *Panderichthys*, all digits were suggested to be neomorphic structures. Holmgren (1933) had also suggested that digits were neomorphic structures. However, he suggested that the metapterygial axis extended into the ulna, ulnare, and fourth (lateral) centrale, a pattern preferred by D. M. S. Watson (1913) and Jarvik (1980). Holmgren hypothesized that the distal carpals and all digits were neomorphic structures attached to the end of a metapterygial axis and radials.

Developmentally the situation is different in urodeles, but there are good reasons to think that this situation is phylogenetically derived (G. P. Wagner et al. 1999; Shubin and Wake 2003). Salamanders with larval forms appear to develop digits I and II from a single commune basale that was derived from the metapterygial axis. A range of explanations have been suggested for this peculiarity, including a polyphyletic origin of tetrapods (Holmgren 1933; Jarvik 1980), a reversed digital arch (see below), an adaptation to pond larval lifestyle (D. B. Wake and Shubin 1994), and a loss of preaxial digits with a gain in postaxial digits (G. P. Wagner et al. 1999). This peculiarity appears to be unique to larval salamanders and will not be discussed further here.

Digital Arch Model

Shubin and Alberch (1986) presented an intriguing scenario to explain the origin of the autopodium. While documenting the chondrification pattern of a large number of tetrapods, they described a pattern in which modern frogs and amniotes appear to follow a stereotypical pattern of precocial chondrification along the posterior margin of the developing limb. This pattern results in the early chondrification of what has been termed the primary axis (Burke and Alberch 1985; G. B. Müller and Alberch 1990; Shubin and Alberch 1986). The primary axis extends from the ulna/fibula to the ulnare/fibulare and then into the autopodium through the lateral centrale and digit four of pentadactyl amniotes; that is, it is identical to the metapterygial axis *sensu* Watson and Jarvik (see above). The next digits to chondrify are, in order, digits III and II. Digits I and V follow but without any generally consistent pattern (Burke and Alberch 1985; G. B. Müller and Alberch 1990; Shubin and Alberch 1986).

The stereotypical pattern of chondrification led these workers to propose the so-called digital arch model. This model suggests that the metapterygial axis passes into the fourth distal carpal/tarsal and bends preaxially to continue through distal carpals/tarsals three, two, and one. The remaining carpals/tarsals are described as originating from preaxial radials while the digits are postaxial radials. This model suggests that the preaxial segmented radials and postaxial bifurcations of the proximal limb are reversed in the autopodium to produce postaxial segmented radials with preaxial bifurcations.

Some molecular evidence, summarized by Shubin and colleagues (1997), has been used to support the digital arch model. For instance, Nelson et al. (1996) described the expression patterns of the *AbdB*-related *HoxA, B, C,* and *D* genes. Their thorough study revealed three distinct phases of gene expression. In summary, phase I correlates with an initial proximodistal limb bud elongation; phase II is generally characterized by a posterior expression of the 5′ *HoxD* genes, generally following the metapterygial axis; and in phase III the *HoxD* genes experience an anterior expansion of their expression pattern. This anterior expansion of gene expression was interpreted as consistent with the digital arch being homologous to the metapterygial axis (Shubin et al. 1997). This observation gains its evidential quality for the homology of the digital arch through the assumption that *HoxD* expression is a molecular marker for the metapterygial axis. This is a highly questionable assumption, as *HoxD* expression in the pectoral fin of zebrafish is similar to phase II expression in tetrapods (Neumann et al. 1999; Sordino et al. 1995), even though paired fins of teleosts do not have a homologue of the metapterygium (Mabee 2000). Furthermore, there are considerable problems with the notion that the digital arch is a developmentally significant structure.

The theory of the digital arch is based on the perceived near-complete conservation of the stereotypical chondrification pattern. A strongly conserved pattern suggests that it is maintained because it either fulfills an important functional role, like the cartilage plate of larval teleost pectoral fins (Grandel and Schulte-Merker 1998), or has an essential developmental role (Riedl 1978). Close consideration of the data, however, does not support the notion that the digital arch is highly conserved and developmentally significant (Cohn et al. 2002; Galis et al. 2001). Digits I and V chondrify independently from the digital arch in an alligator and a turtle (Burke and Alberch 1985; G. B. Müller and Alberch 1990), and a digital arch cannot be seen in *Xenopus levis* (Blanco et al. 1998). Cell proliferation in the limb bud follows a general proximo-distal direction, without respecting the direction of digital arch differentiation (Tschumi 1957; Vargesson et al. 1997). Furthermore, mechanical barriers that split the anterior from the posterior part of the developing autopodium do not affect differentiation of the anterior digits (Hinchliffe and Gumpel-Pinot 1981; Stephens and McNulty 1981). Hence, the sequence of differentiation seen in normal chondrogenesis is not a causally relevant event for the development of the anterior digits.

At this point a few comments seem to be in order on a recent sweeping critique (Cohn et al. 2002) of Shubin and Alberch (1986). The criticism of Cohn and collaborators is aimed at three potentially independent issues: the existence of the metapterygial axis, the question of whether the digital arch represents a developmental axis, and the model of condensation and branching used to explain the developmental pattern of skeletogenesis. The authors cite convincing evidence that the model of condensation and bifurcation is causally irrelevant for the development of digits in chickens (Cohn et al. 2002). But the existence of a metapterygial axis as a morphological entity cannot be refuted only because a mechanistic model explaining its morphology turned out to be wrong. Regardless of which proximate mechanisms cause the specification of the elements in the metapterygial axis, there is ample evidence that what Shubin and Alberch called the primary axis is real. First there is the fact that the posterior elements are the first to differentiate in all amniotes and anurans examined. The digit derived from the primary axis is the most stable in the face of digit reduction (Alberch et al. 1979; Raynaud and Clergue-Gazeau 1986). Normally the absence of Shh signaling leads to the almost complete loss of autopodial elements. But even in the complete absence of Shh a distal element in the position of a digit is often formed in the posterior limb bud (Chiang et al. 2001). Note that the metapterygium is already present in sharks (Braus 1906a), which lack Shh expression in the fin buds (M. Tanaka et al. 2002). Hence, the independence of the primary axis from Shh signaling is consistent with its hypothetical derivation from the metapterygium. Therefore, we prefer the conclusion that the developmentally most stable part of the distal limb skeleton is related to the evolutionarily most stable part of autopodial patterning, namely the primary axis (the developmental structure), which is the same as the metapterygial axis *sensu* Watson and Jarvik (which is an anatomical structure). We thus think that the autopodium contains an ancestral morphological/developmental element, the metapterygial axis, which is phylogenetically older than the autopodium per se.

Neomorphic Autopodium

As discussed above, the autopodium has been considered a *de novo* structure. A number of more recent arguments have come forth in support of a neomorphic autopodium (Ahlberg and Milner 1994; Capdevila and Izpisúa-Belmonte 2001; Coates 1996; Sordino and Duboule 1996; G. P. Wagner and Chiu 2001). Morphological evidence for a neomorphic autopodium rests on the suite of autopodial elements that are notably absent in the closest sister taxa to tetrapods, *Eusthenopteron* and *Panderichthys*. Neither fish has a set of proximal nodular elements followed by a set of distal elongate elements that characterize the autopodium. In fact, *Panderichthys* does not even have endochondral bone distal to its putative homologues of the ulnare and intermedium. Molecular arguments for a neomorphic autopodium stem from the unique gene expression patterns and gene regulations found at the initiation of the autopodial determination phase of Hox gene expression, phase III (Nelson et al. 1996; Spitz et al. 2003).

G. P. Wagner and Chiu (2001) stressed the importance of HoxD gene regulation during the three phases. Phases I and II are regulated by a number of enhancer elements (Beckers et al. 1996; Hoeven et al. 1996) while phase III expression is regulated by a single "global" enhancer (Hérault et al. 1999; Spitz et al. 2003). Zákány et al. (1997) suggested an autopodial enhancer functions downstream of *Hoxa13*, a gene that is known to regulate the autopodial field (Mortlock et al. 1996). These unique regulations of *Hoxd* genes during phase III as well as the results of knockout mutations (Zákány et al. 1997) led G. P. Wagner and Chiu (2001) to propose that the first step in the origin of the autopodium was the evolution of an autopodial morphogenetic field. A morphogenetic field is a part of the embryo that has its own locally autonomous patterning mechanisms, which make it an independent development module (*sensu* S. F. Gilbert et al. 1996). The origin of the autopodial field is associated with the nonoverlapping proximodistal expression of *Hoxa11* and *Hoxa13* found in tetrapods during phase III of *Hox* expression (Nelson et al. 1996) but not found in zebrafish (Sordino et al. 1995) or in paddlefish (Metscher et al. 2005). However, *hypodactyly* (*Hd*) mutant mice (Post and Innis 1999) and *Shh* knockout mice (Litingtung et al. 2002) both yield phenotypes with a single digitlike structure—not a complete absence of digits. *Hd* is known to be the result of a functional deletion of *Hoxa13* (Mortlock et al. 1996). This digitlike structure appears to be in the position of digit IV, suggesting that it may be part of the primary axis discussed above. Combining the experimental results on limb development with the paleontological data summarized above suggests a series of genetic events that may have been involved in the origin of the autopodium.

Evolution of the Autopodial Field

The autopodial field is a morphogenetic field (in the sense of Gilbert et al. 1996) at the distal end of the limb bud from which the autopodial structures develop. At least two genetic events have been hypothesized to be responsible for the origin of the autopodial field (Zákány et al. 1997; G. P. Wagner and Chiu 2001). (1) The mutual exclusion of *Hoxa11* and

Hoxa13 expression domains. (2) The acquisition of new downstream target genes by *Hoxa13*. The latter events would have established *Hoxa13* as an autopodial organ identity gene (a role it shares with *Hoxd13* in modern tetrapods).

Note that the evolution of an autopodial field does not need to coincide with the evolution of the specific autopodial morphology. It need only have established a developmental module distinct from the proximal parts of the fin/limb bud; see Raff (1996) for the notion of a developmental module. Thus, these hypothesized events can have taken place anytime between the most recent common ancestor of sarcopterygian fishes and the most recent common ancestor of tetrapods or even earlier (fig. 4.3). Preliminary results on coding sequence evolution of *Hoxa13* show a significant increase in the number of nonsynonymous sequence changes on the stem lineage of tetrapods after the most recent common ancestor of lungfish and tetrapods and before the most recent common ancestor of recent tetrapods (Wagner and Takahashi, in preparation). These elaborations appear to be endochondral derivatives of the metapterygium that bifurcate and segment preaxially and postaxially. Perhaps these elaborations are some of the first morphological signatures of an autopodial field. In contrast, the gene *Hoxa11*, involved in the differentiation of the zeugopodium, which is a plesiomorphic character, does not show an increased rate of nonsynonymous substitutions in the stem lineage of tetrapods. This pattern supports the idea that *Hoxa13* acquired novel target genes coincidental with the origin of the autopodium.

The argument that patterns of coding sequence evolution can be indicative of changes in the functional role of transcription factor genes is supported by the recent finding that *Hoxa11* experienced strong directional selection in the stem lineage of placental mammals (Lynch et al. 2004). This observation is significant because this gene acquired a derived function in female reproductive tract development and function. *Hoxa11* knockout mice are sterile because the endometrial stroma cells degenerate and embryo nidation is impossible.

Evolution of Digits

While the mechanistic data summarized above suggest that the *HoxD* genes were recruited into autopodial function after *Hoxa13* (Zákány et al. 1997), the evolutionary context of this event is not clear. It could be either during the evolution of the digits or during the evolution of digit identity. The difficulty lies in the partial redundancy between *Hoxa13* and *Hoxd13* in mice (Fromental-Ramain et al. 1996b) and the fact that *HoxD* gene function is involved also in digit identity determination. On the other hand, paleontological evidence has been used to suggest that digit identity and the

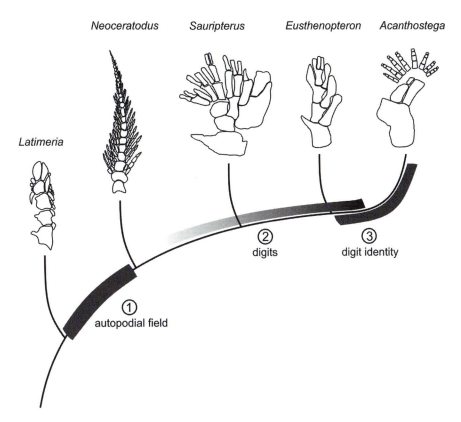

Neoceratodus   *Sauripterus*   *Eusthenopteron*   *Acanthostega*

Latimeria

② digits   ③ digit identity

① autopodial field

Figure 4-3 Phylogenetic representation of the evolutionary model of autopodial origin proposed herein. The model is composed of three distinct events. (1) The autopodial field is established along the branch between lungfish and *Sauripterus*. (2) Digits are established somewhere within Rhipidistia, perhaps at the level of rhizodontids, such as *Sauripterus*. (3) Digit identity may have been established simultaneous with the origin of digits or subsequently. (*Latimeria*, the modern coelacanth, is adapted from Forey 1998; *Neoceratodus*, the modern Australian lungfish, and *Eusthenopteron* are adapted from Jarvik 1980; *Sauripterus* is adapted from Daeschler and Shubin 1998; *Acanthostega* is adapted from Carroll and Holmes, chap. 13 in this volume.) All appendages are from the left side with distal toward the top and anterior (preaxial) toward the right. Figures are not to scale.

pentadactyl pattern evolved after the origin of digits (Coates and Clack 1990). However, as argued above, a slight degree of individuality is present in the earliest forms, such as *Acanthostega* and *Ichthyostega*. In any case, the recruitment of *HoxD* genes in the evolution of digits is currently equivocal. *HoxD* genes may have been recruited in the evolution of digit identity subsequent to the origin of digit development, or involved with both events. It remains entirely possible for downstream targets of *HoxD* genes to be incorporated later into the identification of each digit and thus blur the current view of its involvement in digit origin and identity.

The evolution of digit development must have occurred either before or synchronous with digit identity. Logically, digit identity cannot have occurred before digit development. The developmental mechanisms responsible for these separate, but clearly nested, mechanisms may center on the regulatory interactions of *Shh* and *Gli3* with each other and with *HoxD* genes (summarized above). The first phase of *Shh* expression may be tied with digit initiation, the second phase with digit identity (Drossopoulou et al. 2000). The first phase appears to regulate the development of digits

about a digitlike structure that lies within the primary axis. The dissociation of the primary axis digit from the others is shown by the maintenance of this digitlike structure in *Shh*$^{-/-}$ mice (Litingtung et al. 2002) and *Hd* mice (Robertson et al. 1997; Post and Innis 1999) while all other digits are lost. Whether or not the recruitment of the *HoxD* genes was in the context of digit origins or digit identity evolution, the recruitment should have happened after the most recent common ancestor of lungfishes and tetrapods, assuming that lungfishes are the closest living relatives of tetrapods. The latest time for this event is just before the most recent common ancestor of amniotes; we still cannot eliminate the possibility that regulation of digit identity evolved independently in the amniote and amphibian lineages (Stopper and Wagner 2005). A comparison of *Shh* function in amniotes and amphibians should address this question.

This scenario outlined above is certainly incomplete and will need to be expanded as new data are obtained. For instance, there is no genetic scenario known to us for the origin of the mesopodium (wrist and ankle) because the genetics of its development is not well understood.

A Model of a Mixed Plesiomorphic/Apomorphic Nature of the Autopodium

Taken together, the evidence for the reality of the primary axis and its possible homology to the metapterygial axis (see above), as well as all the evidence for the neomorphic character of the autopodium, calls for a reconsideration of these various scenarios, since these scenarios, as stated, are logically incompatible. If one claims that all of the autopodium is entirely neomorphic, then it cannot contain an ancestral element such as the primary axis, and its derivative, the anlage of digit IV in the pentadactyl limb. Likewise, the claim that the autopodium is entirely derived from the metapterygium is also flawed (see above).

To attempt to resolve some of the current contradictions, we propose a synthetic scenario in which the autopodium evolved through (1) an initial autopodial developmental field, (2) the modification of an ancestral structure, the metapterygial axis, into a digit within this field, (3) the reiteration of this derived structure anterior and posterior of the primary axis, giving rise to the other digits, and (4) the acquisition of specific identities to these digits. The evolution of an autopodial field may be associated with the exclusive expression of *Hoxa11* and *-13* (G. P. Wagner and Chiu 2001). The modification of the metapterygial axis within this field to form a digit and the initiation of a number of other digits may be related to the first phase of *Shh/Gli3* expression and their regulation of *HoxD* genes. The identity of these digits may then have evolved simultaneously or subsequently with the development of the second phase of *Shh/Gli3* expression. In this scenario, each digital ray is developmentally independent from other such rays and does not require the assumption that the digital arch is causally relevant for the development of anterior digits. The developmental innovations of the distal autopodium (acropodium: metapodials and digits) consists in (1) the developmental independence from proximal modules and (2) the ability to reiterate the development of a primary proximo-distal axis of growth and differentiation in an anterior-posterior arrangement of digital rays. The morphological pattern of the digital arch is the result of this reiterated differentiation rather than a causally necessary progression.

How Should We Test These Models?

The biggest intellectual danger of any evolutionary research program is the temptation to find satisfaction in telling ingenious "just so" stories (Gould and Lewontin 1978). We think that developmental evolution, as one of the youngest members of the evolutionary sciences, is in particular danger in falling into this trap, just as other branches of evolutionary biology were in the past (Chiu and Hamrick 2002; Fröhlich 2003; Gibson 1999; G. P. Wagner et al. 2000). We agree with Gibson (1999) and Fröhlich (2003) that the way out of this corner has to involve a combination of experimental and hypothesis-driven research programs. But what kinds of hypotheses are critical to an evaluation of these scenarios? And what is the inference method underlying the testing of these hypotheses? There is no generally accepted methodology for evolutionary developmental biology to answer these questions. In a recent manuscript (Larsson and Wagner, in preparation) we proposed an inference method that could guide hypothesis testing in developmental evolution.

The general problem we want to address can be presented with the following anecdote. When one finds a dead person the forensic scientist is charged with finding the "cause" of the person's death. In what sense can we say the forensic scientist investigates the cause of a death if the activity he or she usually engages in is not to kill more people in order to experimentally demonstrate a cause-effect relationship? Of course, if causality could only be proven experimentally by repeating a process in the lab, this is what we should expect from a forensic scientist. In reality, however, the forensic scientist uses a special application of what we call the "forensic inference principle." Different causes of death lead to different ways for a body to be dead. Or, in other words, the structure of the effect (i.e., the dead body) reflects the structure of the cause (a gunshot, drowning, strangling, etc.). Why is this a valid scientific inference? The reason is the symmetry principle (J. Rosen 1983). *The cause must have at least as much structure as the effect.* If the skull is pierced and lungs are dry then death occurred not because the person drowned.

Of course, the same inference method is also used in sciences other than forensic medicine, and thus is perhaps as old as the scientific enterprise itself. For instance, much of the power of population genetics rests on this method of inference. If a decline in fitness of a population is caused by inbreeding (either behaviorally or ecologically induced) then the degree of heterozygosity all over the genome has to be low. Hence, inbreeding as the cause of population decline is predicted to have a "footprint" in the genetic composition of the population. Lower than expected heterozygosity is legitimate evidence for past inbreeding. All this is well known and time tested, but *it is worth pointing out that this form of inference is as legitimate as the experimental method.* It is along those lines that we propose a solution to the developmental evolution inference problem (Larsson and Wagner, in preparation).

In analogy to this "forensic inference principle" we propose that different devo-evo scenarios lead to different predictions about the developmental pathway and mechanisms active in the lineage with the derived character state. As the victim and the crime scene contain the record of what happened during the crime, the derived lineages and their development contain the traces of the evolutionary modifications that led to the novel character. Hence, it should be possible to discriminate between different scenarios based on a detailed description of the mode and mechanisms of the development of the derived character. This is not an entirely new idea, as it was used in one of the first experimental devo-evo investigations in history, the paper by Hermann Braus (1906b) on the fact that shark skeletogenesis is independent of muscle bud development. These experiments were conducted to test a prediction of the lateral fin-fold theory for the origin of paired fins (see Nyhart 2002). But we think that a systematic application of this inference method to large amounts of morphological and mechanistic data has the potential to positively exclude some scenarios and uniquely support others. The heuristic and critical value of this approach may be demonstrated in our presentation of the origin of the autopodial field. Genetic signatures in *Hoxa13* sequences associated with the origin of the autopodial field

support the hypothesis of G. P. Wagner and Chiu (2001) that the autopodial field originated in the stem lineage of tetrapods. Additional functional studies are required to further test this conclusion, but the method of investigation is at least compatible with that of hypothesis-testing science. The other hypothesized evolutionary transitions proposed above will also have to be tested with comparative and functional studies. Taken together, these studies will provide the empirical give-and-take that will lead to a deeper understanding of the fin-limb transition. Given the rapid progress of developmental biology of limb development, and improved methods in the analysis of coding and noncoding sequences the chances for this approach to succeed are particularly high at this historical juncture. The next step in the study of the developmental evolution of the fin-limb transition must examine each theory for the origin of tetrapod limbs in order to determine what they predict about the development of the tetrapod limb as well as about the sequence evolution of genes involved in limb development, and then test these predictions.

Part II

Development

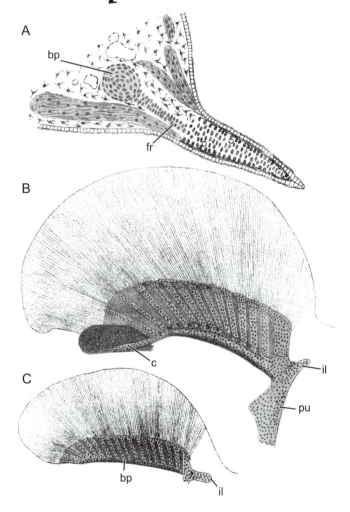

Chapter 5 The Development of Fins and Limbs

Mikiko Tanaka and Cheryll Tickle

FIN/LIMB DEVELOPMENT has been studied in the embryos of different vertebrates ranging from fish to mammals, with the zebrafish, chicken, and mouse being the most common models. Recent research into the genetic basis of limb development shows that at least some of the cellular and molecular mechanisms are conserved. Initiation of development of both fins and limbs is probably very similar and leads to the formation of small buds. Patterning of both fin buds in bony fish and limb buds in higher vertebrates involves establishment of an autonomous developmental system in which cell-cell interactions occur. In higher vertebrates, cells at the tip of the limb bud develop into digits, and this seems to involve the establishment of a signaling system for each individual digit primordium. The analysis of fin and limb development gives insights into vertebrate limb evolution, morphological diversity, and human congenital limb defects.

Outline of Fin/Limb Development in Different Vertebrates

Two sets of paired appendages, which protrude ventro-laterally from the body wall, are common to gnathostomes (jawed vertebrates) including cartilaginous fishes (chondrichthyans), actinopterygians (ray finned fishes), sarcopterygians (lobe finned fishes and tetrapods). The anterior paired appendages (pectoral fins/forelimbs) are usually located at the anterior end of the trunk, and the posterior paired appendages (pelvic fins/hindlimbs) tend to be located at the posterior end. In some modern teleosts—for example, some perciformes fishes, including cichlids—pelvic fins are located in front of the pectoral fins. The region between the paired appendages is known as the flank or interlimb. The skeleton of these paired appendages generally comprises—from proximal to distal—(1) girdle, (2) upper arm/thigh, (3) lower arm/shank, and (4) hand/foot. Figures 5.1B and C show chick wing and mouse forelimb respectively, with (1) scapula, (2) humerus, (3) radius and ulna, (4) metacarpals and digits. It should be noted that teleost paired fins do not possess the distal elements (fig. 5.1A). In addition to the skeleton, there is a well-developed musculature (e.g., 50 or more individual muscles in mammalian limbs), tendons and ligaments, and so forth. Analysis of chick/quail chimeras has shown that the limb connective tissues are derived from lateral plate mesoderm, while myogenic cells of the muscles originate in the somites and migrate into the limb-forming regions (Chevallier et al. 1977; Christ et al. 1977). Somitic cells also give rise to limb endothelial cells (Ambler et al. 2001). Recent lineage analysis of somite-derived cells in chick limb buds shows, rather surprisingly, that myogenic and endothelial cells can share a common somitic cell precursor (Kardon et al. 2002). The limb/fin bud is vascularized from an early stage, and nerves grow into the developing fin/limb once the tissue pattern has been established.

Fin/limb development has been mainly investigated in three model vertebrates—zebrafish, chick, and mouse. Zebrafish pectoral fin buds form from a very thin layer of somato-

A Zebrafish pectoral fin

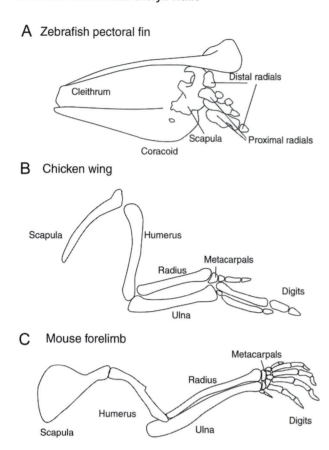

B Chicken wing

C Mouse forelimb

Figure 5-1 Skeleton of anterior paired appendages of three model vertebrates. (A) Ventrolateral view of zebrafish pectoral fin and girdle skeleton; dermal fin rays are omitted (modified from Coates 1995). (B) Skeletal pattern of chicken wing. (C) Skeletal pattern of mouse forelimb. Not drawn to scale.

pleural lateral plate mesenchyme, which thickens locally (Grandel and Schulte-Merker 1998). By 26 hours postfertilization, a small pectoral fin bud has formed with, initially, an apical ectodermal thickening along its entire distal margin (plate 5.1A). This thickening becomes transformed, at about 36 hours, into an apical fold. At 48 hours, mesenchyme cells, thought to be derived from the neural crest, invade the apical fold to give the fin fold, which gives rise to the adult fin. The pelvic fin buds develop much later at about three weeks postfertilization (Grandel and Schulte-Merker 1998).

Early stages of limb bud development are similar in both chick and mouse. The somatopleural mesenchyme of the lateral plate which gives rise to the limb bud is much thicker in chick and mouse embryos than in zebrafish embryos. In chick embryos, limb-forming regions become distinguishable by the thickening of the lateral plate mesoderm and formation of the apical ectodermal ridge in the overlying ectoderm (reviewed in Saunders 1977; plate 5.1B). The stages in early chick wing bud formation are shown in transverse section in plate 5.1D. Wing buds are first apparent at about 3 days of development, and the apical ectodermal ridge is seen

as a rim at the tip (plate 5.1B). Over the next 3 days, the buds grow out from the body and, by 6 days, the general shape of the wing is apparent. The thickening in the leg region occurs later than in the wing region, but wing and leg buds form at about the same time. In mouse embryos, the limb buds first appear at about 9.5 days of development, but the apical ectodermal ridge forms later than in the chick (P. Martin 1990). The buds elongate and undergo shape changes over the next 5 days (plate 5.1C). The shape of mouse limb buds, even at early outgrowth stages, differs from that of chick limb buds. There is another striking difference with respect to the vasculature: in chick embryos, the limb bud has an avascular rim, while in mouse limb buds, the marginal vein lies almost immediately beneath the apical ectodermal ridge.

Limb Initiation

Transplantation experiments in early chick embryos show that cells are determined to form limbs long before any buds are visible (reviewed in Saunders 1977). Furthermore, there are indications that interactions between segmental plate (the future somite-forming region) and lateral plate mesoderm may play a role in limb determination. In chick embryos, the interlimb region in addition to the presumptive limb-forming region initially becomes thickened. In fish embryos, the early pectoral fin rudiment is elongated anteroposteriorly and later becomes more localized.

There is evidence that suggests that the interlimb region of vertebrate embryos can form limbs. Thus, for example, Balinsky showed that grafting a nose rudiment to the flank of a newt embryo could induce a supernumerary limb, which resembles either a forelimb or hindlimb depending on whether it lies nearer the normal fore- or hindlimb (Balinsky 1933). More recently, it was shown that flank regions from chick embryos allowed to develop in isolation can also form limbs (Stephens et al. 1989).

Molecular Basis of Limb Positioning along the Antero-posterior Body Axis

Fin/limb development must be initiated at the correct position along the head-to-tail axis of the body. There is substantial evidence that *Hox* gene expression in somitic mesoderm specifies axial pattern (reviewed in Gruss and Kessel 1991), and it has been suggested that patterns of *Hox* gene expression in lateral plate mesoderm similarly may specify position along the antero-posterior axis to give forelimb, interlimb, hindlimb (Cohn et al. 1997). Limb position is shifted slightly in mice mutant for *Hoxb5*, and a rostral shift of the shoulder girdle is seen (Rancourt et al. 1995). Posterior displacement of the hindlimbs has also been reported in mice in which the gene encoding a member of the TGF beta family, Gdf11, has been functionally inactivated. This change in limb position is accompanied by changes in axial Hox gene expression (McPherron et al. 1999). *Tbx* genes, which are expressed in lateral plate mesoderm, may also play a role in limb bud positioning along the antero-posterior axis. Recent experiments have shown that manipulating *Tbx3* expression can produce shifts in limb bud position in chick embryos (Rallis et al. 2005), while overexpressing *Tbx18* can lead to a transient extension of the wing bud (Tanaka and Tickle 2004).

Initiation of Fin/Limb Buds

Several extracellular signaling molecules and transcription factors have been reported to be involved in initiating limb bud development in chick and mouse embryos. More recently, initiation of fin bud development in zebrafish embryos has also been studied, and the same molecules seem to be employed.

Fibroblast growth factors (Fgfs) were the first signaling molecules to be found that can experimentally initiate limb development (reviewed in G. R. Martin 1998). Application of a bead soaked in Fgf to the interlimb region of chick and mouse embryos can result in induction of an ectopic limb or limb bud; Fgf1, 2, 4, 7, 8, and 10 have all been shown to be effective (Cohn et al. 1995; Ohuchi et al. 1995; Yonei-Tamura et al. 1999a; Tanaka et al. 2000). Of these, Fgf8 and Fgf10 seem to play pivotal endogenous roles in limb bud initiation and outgrowth in the embryo by forming a regulatory loop (fig. 5.2). Once expression of *Fgf10* is initiated in presumptive limb mesenchyme, this leads to outgrowth of lateral plate mesoderm to form a bud and, later, to expression of *Fgf8* in the apical ectodermal ridge. The apical ridge secretes Fgf8 and other Fgfs (see later), and these Fgfs also promote limb outgrowth, at least in part, by maintaining the expression of *Fgf10* in the underlying mesenchyme (Ohuchi et al. 1997, see also later). In mouse embryos, it has been demonstrated that deficiency of *Fgf10* results in the loss of limbs (Sekine et al. 1999; Min et al. 1998). However, limb buds are initiated in *Fgf10* null mice (Sekine et al. 1999), suggesting that there are limb initiation signals that act prior to Fgf10. Transcripts of the gene encoding Fgfr2IIIb, a receptor for Fgf10 (Ornitz et al. 1996) is expressed in limb bud ectoderm (Noji et al. 1993). Mouse embryos in which the entire immunoglobin-like domain III of Fgfr2 is deleted lack limb buds (Xu et al. 1998), suggesting that Fgfr2 is required for limb bud induction via Fgf10.

Involvement of Fgf signaling in pectoral fin bud initiation has also been demonstrated in zebrafish embryos. *Fgf10* seems to play a role in fin bud initiation similar to that re-

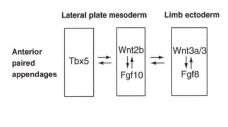

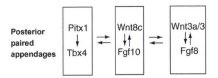

Figure 5-2 Some of the key genes implicated in initiation of fin/limb development. In lateral plate mesoderm, Tbx5 and Tbx4 activate Wnt2b/Fgf10 and Wnt8c/Fgf10 signals, respectively in anterior and posterior paired appendages. Wnt/Fgf signals feedback on *Tbx5* and *Tbx4* genes to maintain their expression. Fgf10 activates Wnt3a/3/Fgf8 signals in the limb ectoderm and induces apical ectodermal ridge formation. It should be noted that (1) Wnt2b and Wnt8c have not been shown to be involved in mouse, and (2) some work suggests that the *Tbx5* gene lies downstream of Wnt/Fgf signaling (see text for details).

ported in limb bud initiation in higher vertebrates (J. K. Ng et al. 2002), but in zebrafish, in contrast to chick and mouse limb buds, *Fgf8* only starts to be expressed after formation of pectoral fin buds. Recent experiments in zebrafish embryos showed that Fgf24, which belongs to the Fgf8/17/18 subfamily, activates *Fgf10* expression in the lateral plate mesoderm (Fischer et al. 2003).

Wnt signaling molecules can also trigger limb formation in chick embryos (Kawakami et al. 2001). Two different members of the Wnt family are expressed in the lateral plate mesoderm of limb-forming regions in chick embryos (fig. 5.2). *Wnt2b* is expressed in the presumptive wing region, while *Wnt8c* is expressed in the presumptive leg region. Wnt2b and Wnt8c have been shown to induce *Fgf10* expression via a beta-catenin-dependent pathway in wing- and leg-forming regions, respectively. In addition, grafts of Wnt-expressing cells to the interlimb region of chick embryos can induce ectopic limbs. Recent studies show that *Wnt2b* is involved in pectoral fin bud initiation in zebrafish embryos (J. K. Ng et al. 2002), but expression of neither *Wnt2b* nor *Wnt8c* has been detected in limb buds of early mouse embryos (Agarwal et al. 2003). Nevertheless, there is evidence that Wnt signaling is important in early mouse limb development because mice in which genes encoding Lef1 and Tcf1—two of the transcription factors implicated in transducing Wnt signals—are simultaneously functionally inactivated develop only rudimentary limb buds (Galceran et al. 1999). While the general program of limb initiation and identification seems to be conserved between zebrafish, chick, and mouse embryos, it is perhaps not so surprising that the functions of specific Wnts may have diverged in different vertebrates.

During limb development in chick embryos, *Wnt3a*, expressed in the apical ectodermal ridge, has been shown to be involved in initiating *Fgf8* expression in the apical ectodermal ridge (Kengaku et al. 1998). In mice, *Wnt3a* is not expressed in the limb ectoderm (Roelink and Nusse 1991; Parr et al. 1993; Takada et al. 1994), and mouse embryos lacking *Wnt3a* activity do not exhibit any limb defects (Takada et al. 1994). Again, however, the fact that the ectoderm over the rudimentary limb buds in *Lef1-/-Tcf1-/-* mutants does not express *Fgf8* (Galceran et al. 1999) suggests involvement of Wnt signaling in mice and recently it has been shown that Wnt3 seems to play this role (Barrow et al. 2003).

The T-box transcription factors, *Tbx5* and *Tbx4*, are expressed in forelimb/pectoral fin– and hindlimb/pelvic fin–forming regions, respectively (Gibson-Brown et al. 1996; Tamura et al. 1999), and have been shown to be involved in initiating limb development. The roles of *Tbx5* in fin/limb initiation have been studied using mouse, chick, and zebrafish embryos.

Deletion of *Tbx5* function in the developing forelimb of mouse embryos results in absence of all skeletal elements of the forelimb, including scapula and clavicle of the pectoral girdle, and this is accompanied by loss of *Fgf8* and *Fgf10* expression (Rallis et al. 2003). This phenotype is more severe than that observed in the forelimb of *Fgf10* null mutant mice, suggesting that *Tbx5* lies upstream of Fgf10 in limb initiation. Indeed, there is evidence that there is a Tbx5 binding site in the *Fgf10* promoter region. Thus, Tbx5 appears to act as primary initiator of forelimb bud formation in mice (Agarwal et al. 2003; Rallis et al. 2003). Consistent with this, analysis of *Fgf10* and *Tbx5* expression in *Lef1/Tcf1* double mutant mouse embryos suggests that *Tbx5* also acts upstream of Wnt signaling in apical ridge formation and that Wnt signaling is required via *Fgf8* to maintain *Fgf10* (Agarwal et al. 2003).

Tbx5 is also required for zebrafish pectoral fin development. In zebrafish lacking *Tbx5* activity, pectoral fin structures and shoulder girdle are absent (D. G. Ahn et al. 2002; J. K. Ng et al. 2002). Ng and colleagues postulated that the role of *Tbx5* in pectoral fin bud initiation and outgrowth is mediated by interactions with *Wnt2b* and *Fgf10* (J. K. Ng et al. 2002). These authors also made use of chick embryos, which allowed them to perform misexpression experiments in a temporally and spatially restricted manner. They injected a retrovirus expressing either *Tbx5* or a truncated form of *Tbx5* into the lateral plate mesoderm of chick embryos. *Tbx5* misexpression induced formation of additional

limb bud–like structures with ectopic expression of *Fgf10* and *Wnt2b*, while misexpression of truncated *Tbx5* resulted in the truncation of wing bud. Furthermore, they also injected a retrovirus expressing *Axin*, a negative regulator of Wnt/beta-catenin pathway, in the presumptive wing region and showed a requirement for Wnt/beta-catenin signaling for *Tbx5* expression in chick limb buds. From this and other results in zebrafish, Ng and colleagues concluded that Wnt2b functions upstream of Tbx5. Similar misexpression experiments in chick embryos with *Tbx5* and dominant negative forms of *Tbx5* have been carried out by Takeuchi et al. (2003), but from their results they concluded that *Tbx5* lies upstream of both Wnt and Fgf signaling in the limb induction process, as in mice (Agarwal et al. 2003; fig. 5.2).

Roles of *Tbx4* in hindlimb initiation have also been studied in both chick and mouse embryos. In chick embryos, misexpression of dominant-negative forms of the *Tbx4* gene in the prospective leg fields at early stages results in a complete legless phenotype with disruption of pelvic girdle formation (Takeuchi et al. 2003) and is associated with absence of expression of *Wnt8c*, *Fgf10*, and *Fgf8*. Furthermore, overexpression of *Tbx4* in the interlimb region induces an additional leg-type limb structure with induction of *Wnt8c*, *Fgf10*, and *Fgf8* expression, indicating that *Tbx4* as well as *Tbx5* may lie upstream of Wnts. In mouse embryos, although loss of *Tbx4* does not affect limb bud initiation, hindlimbs of *Tbx4* mutants fail to develop and *Fgf10* expression in the mesenchyme is not maintained (Naiche and Papaioannou 2003).

Recent analysis suggests that Pitx1 and Pitx2, members of the bicoid-related family of homeobox transcription factors that have been implicated in patterning of lateral plate mesoderm derivatives, play a role in regulating *Tbx4* expression in hindlimb regions. *Pitx1* is initially expressed throughout the posterior lateral plate mesoderm in mouse and chick embryos, but it then becomes localized to hindlimb buds (Lanctot et al. 1997; M. Logan and Tabin 1999), while *Pitx2* participates in the establishment of left-right asymmetry and is expressed in the left presumptive hindlimb region (M. Logan et al. 1998; Piedra et al. 1998; Ryan et al. 1998; Yoshioka et al. 1998). Analysis of mice deficient in both *Pitx1* and *Pitx2* showed that *Tbx4* expression in the limb buds is much reduced (Marcil et al. 2003).

Other genes are also expressed in limb-forming regions, including those encoding the transcription factors, snail and twist (Isaac et al. 2000). In early *Drosophila* embryos, these two transcription factors act together to specify mesoderm formation (Leptin 1991) and *snail*, and a related gene, *escargot*, are also expressed in Drosophila embryonic wing discs (Fuse et al. 1996). *Snail* expression is rapidly induced when an Fgf bead is applied to the interlimb region in chick embryos (Isaac et al. 2000), but its precise role in this process is not yet clear. It is striking that a zebrafish *snail* gene is also expressed in pectoral fin buds of zebrafish embryos (reviewed in Tickle 2002). Expression of *twist* also appears to be regulated by Fgf signaling (Tavares et al. 2001). In *twist*[-/-] mouse embryos, limb buds form but do not grow out and *Fgf10* expression is not maintained (Z. F. Chen and Behringer 1995; Zuniga et al. 2002; O'Rourke et al. 2002; see later).

Molecular Basis of Limb Positioning along the Dorso-ventral Axis

In vertebrates, two pairs of discrete buds that will give rise to the limbs not only arise at specific positions along the head to tail axis of the embryo but also are lined up along the sides of the body. The mechanisms that position limbs with respect to the dorso-ventral axis have been studied using both chick and mouse embryos. It has been shown that distinct ectodermal dorsal and ventral compartments exist in both presumptive limb and interlimb regions of early chick embryos. These compartments resemble those in *Drosophila* in that they are cell lineage restricted (Altabef et al. 1997), and the dorso-ventral compartment boundary in the ectoderm seems to be essential for the positioning of the apical ectodermal ridge (Tanaka et al. 1997; Michaud et al. 1997). Even ectopic limb buds induced by Fgf application to the interlimb region develop at the dorso-ventral boundary of the body (Cohn et al. 1995).

Engrailed-1, which encodes a transcription factor, is expressed in the ventral compartment in chick embryos (Altabef and Tickle 2002). In mice, *Engrailed-1* is initially expressed in a similar pattern but later extends throughout the apical ridge (Kimmel et al. 2000), while in zebrafish pectoral fin buds, *Engrailed-1* has been reported to be expressed in ventral ectoderm (Neumann et al. 1999; Grandel et al. 2000). However, misexpression of *Engrailed-1* in chick embryos does not alter the cell lineage restriction and compartmentalization of the ectoderm (Altabef et al. 2000). Instead, *Engrailed-1* seems to be a consequence of compartmentalization and to play an important role in positioning the apical ridge.

Fate mapping studies have shown that precursor cells of the apical ectodermal ridge are widely distributed throughout chicken limb ectoderm (Altabef et al. 1997). These ridge precursor cells are mixed with cells destined to be the non-ridge ectoderm and then seem to converge toward the dorso-ventral boundary of the limb field to form the apical ectodermal ridge. In mouse embryos, ridge precursor cells are distributed in ventral limb ectoderm, and these cells assemble along the dorsal border of the apical ridge; then the entire domain compresses symmetrically toward the dorso-

ventral border, which is maintained, at least in the early stages, by *Engrailed-1* (Kimmel et al. 2000). Either knockout or misexpression of *Engrailed-1* results in disruption of apical ridge formation in both mouse and chick embryos (Laufer et al. 1997; Rodriguez-Esteban et al. 1997; Tanaka et al. 1998; Loomis et al. 1996; C. Logan et al. 1997; Loomis et al. 1998), and in the chicken limbless mutant, in which the apical ridge fails to form, *Engrailed-1* is not expressed (Noramly et al. 1996). It has been shown that Bmps act upstream of *Engrailed-1* in apical ridge formation in chick limb buds (Pizette et al. 2001). In mice in which the gene encoding Bmp receptor-1 is functionally inactivated in ventral limb ectoderm, defects in ridge formation occur (K. Ahn et al. 2001).

There are also several lines of evidence that Notch signaling plays a role in establishing the apical ridge at the dorso-ventral boundary in the ectoderm. Experiments in chick embryos suggested that *Radical fringe* is involved (Laufer et al. 1997), and it is now known that fringe modulates Notch activity. Although mice in which the gene encoding *Radical fringe* has been functionally inactivated do not show limb defects, the apical ridge is enlarged in mice in which *Jagged2*, a gene encoding a Notch ligand, is functionally inactivated (Jiang et al. 1998). It has also been shown that a relative of the gene encoding the Cut transcription factor, an effector of Notch signaling in Drosophila, is expressed along the sides of the apical ridge in chick limb buds and that misexpression of this gene disrupts the apical ridge (Tavares et al. 2000).

Outgrowth of Fin/Limb Bud

The transition from a bulge in the presumptive limb-forming region of the body wall to a bud that grows out under its own steam is a critical step in limb development and involves signaling by the apical thickening/ridge rimming the bud. The apical ectodermal ridge is a thickening in the ectoderm and consists of tightly packed elongated cells with extensive gap junctions. The role of the ridge in promoting limb bud outgrowth was first demonstrated in chick embryos by Saunders (1948). When the apical ridge is cut away, limb bud outgrowth ceases and a truncated limb develops. It was further demonstrated that the extent of truncation depends on the time at which the ridge is removed (Saunders 1948; Summerbell 1974). When the ridge is removed at early limb stages, only proximal structures develop, while when the ridge is removed at very late stage, just the tips of the digits may be missing.

It is now known that signaling by the apical thickening/ridge is mediated by Fgfs. *Fgf8* is expressed throughout the apical ridge of both mammalian and chick limb buds (plate 5.2), while *Fgf4*, *Fgf9*, and *Fgf17* are expressed in the posterior parts of the ridge. *Fgf2* is expressed in both ectoderm and mesenchyme (M. P. Savage and Fallon 1995). *Fgf8* and *Fgf4* have also been reported to be expressed in the apical fold of zebrafish buds (Grandel et al. 2000). As outlined above, apical ridge removal from chick limb buds leads to limb truncations, but outgrowth (and patterning) can be rescued by stapling beads soaked in Fgf to the bud tip (Niswander et al. 1993; Fallon et al. 1994). Furthermore, when *Fgf8* and *Fgf4* together are conditionally inactivated in the apical ectodermal ridge of mouse limb buds leading to complete absence of Fgf signaling from the earliest stages of development, no limbs form and only a girdle develops (Sun et al. 2002; Boulet et al. 2004).

Maintenance of Fgf signaling from the apical ectodermal ridge at early limb bud stages is critical to sustain limb bud development, and this depends on reciprocal signaling by the mesenchyme. The products of a number of genes such as *twist* seem to be essential to maintain the apical ridge. As already mentioned, in *twist* [-/-] mouse knockouts, limb buds form but are not sustained (Z. F. Chen and Behringer 1995), and there is evidence that expression of Fgfs in both ectoderm and mesenchyme is disrupted (Zuniga et al. 2002; O'Rourke et al. 2002). In *Drosophila*, expression of *twist* is known to be activated by *dorsal*, a member of the NF-kappaB family. When NF-kappaB signaling is blocked in chick limb buds, *twist* expression in the mesenchyme is reduced and bud outgrowth is disrupted (Bushdid et al. 1998; Kanegae et al. 1998). More recently it has been found that mice lacking IKK alpha, which is needed for activation of NF-kappaB, have limb bud abnormalities and small limbs (K. Takeda et al. 1999; Hu et al. 1999). All this is consistent with a genetic cascade, similar to that involved in mesoderm patterning of *Drosophila*, playing a role in stabilizing limb bud formation.

As the limb bud grows out, a zone of undifferentiated mesenchyme cells is maintained at the tip, while, proximally, cells differentiate. This results in laying down the skeleton in a proximal to distal sequence. One of the actions of the apical ectodermal ridge (which can be replaced by Fgf) is the maintenance of this region of undifferentiated cells. A number of genes have been reported to be expressed beneath the apical ridge and controlled by apical ridge signaling including the homeobox gene, *Msx1* (Davidson et al. 1991), and *slug* (Buxton et al. 1997; Ros et al. 1997). The expression of *Msx1* is particularly interesting because there have been reports that *Msx1* can maintain cells in an undifferentiated state or even lead to dedifferentiation of mammalian myotubes (Odelberg et al. 2000). *Wnt5a* is also expressed at high levels in distal mesenchyme (Parr et al. 1993), and when *Wnt5a* is functionally inactivated in mice, the limbs are short and the phalanges of the digits are missing (Yamaguchi et al. 1999).

Molecular Basis of Limb Bud Patterning

The results of experimental analysis in chick embryos are consistent with the idea that cell position in the developing limb bud is specified with respect to a three-dimensional coordinate system related to the three main axes of the limb: proximo-distal (shoulder to finger tips), dorso-ventral (back of hand to palm), and antero-posterior (thumb to little finger). Cells would then interpret this positional information to form the appropriate structures (Wolpert 1969). As already outlined, the apical ectodermal ridge was shown to be required for limb bud outgrowth along the proximo-distal axis; the ectoderm covering the sides of the limb bud for dorso-ventral patterning and the polarizing region, a small group of mesenchyme cells at the posterior edge of the limb bud (little finger edge), for specifying position across the antero-posterior axis (reviewed in R. L. Johnson and Tabin 1997). At least one cell-cell signaling molecule associated with each of the three signaling regions in chick and mouse limb buds has been identified, and, in addition, some of the molecules involved in the response have also been discovered (fig. 5.3). There is evidence that there is a similar signaling network in zebrafish fin buds (see below). Nearly all the work in zebrafish has been carried out on pectoral fin buds, which develop in early embryos.

Proximo-distal Patterning of the Fin/Limb Bud

One important issue is how limb bud outgrowth and Fgf signaling by the apical ridge are related to specification of the pattern along the proximo-distal axis of the limb. In experiments in which ectoderm and mesenchyme of chick limb buds at different stages were separated and then recombined, the recombined limb buds were found to develop normally, showing that signaling by the apical ridge is permissive (Rubin and Saunders 1972). Furthermore, in other experiments in which the undifferentiated tip of a young chick limb bud was grafted to a stump of an old limb bud and vice versa, the tips behaved autonomously, showing that interactions with the stump do not play a role in patterning (Summerbell et al. 1973). Based on these and other observations, it was suggested that a timing mechanism operates in the zone of undifferentiated mesenchyme at the tip of the limb bud (Summerbell et al. 1973). According to this model, the function of Fgf signaling is to maintain the zone of undifferentiated mesenchyme at the limb bud tip in which the timing mechanism operates. This is reminiscent of the mode of action proposed for Fgf signaling in laying down the somites along the main body axis in which Fgf signaling specifies the timing of segmentation (Dubrulle et al. 2001). It may also be significant that Wnt3a, an extracellular signal that, in the chick limb,

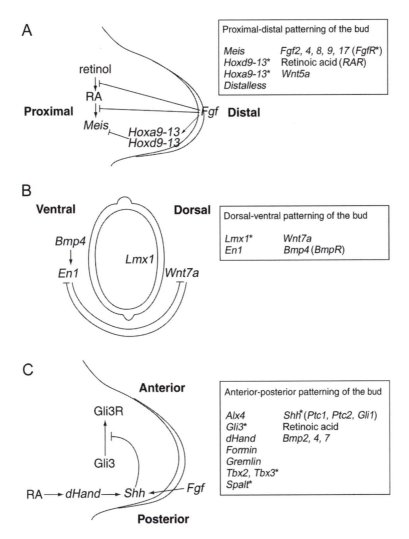

Figure 5-3 Some of the key genes for fin/limb buds implicated in patterning (see text for more details). (A) Signaling cascades for limb outgrowth and specification of position along the proximo-distal axis. (B) Signaling cascades for limb specification along the dorso-ventral axis. (C) Signaling cascades for establishment of antero-posterior pattern of limb bud. More details about some of the genes involved in patterning of fin/limb bud are indicated in the closed boxes. It should be noted that some of these genes have not yet been investigated in fish fin buds. Asterisks indicate genes known to be responsible for congenital limb malformations in human patients.

regulates apical ridge formation (a role taken on by Wnt3 in mouse—see fig. 5.2), plays a major role in the segmentation clock (Aulehla et al. 2003), as does Notch signaling, which is also involved in apical ridge formation in the limb (Palmeirim et al. 1997).

Recently, the progress zone model for proximo-distal limb patterning has been challenged. It has been proposed instead that proximo-distal pattern is established in the early bud and that prespecified pattern elements then expand and differentiate in a proximo-distal sequence during limb bud outgrowth under the influence of Fgfs (Dudley et al. 2002). It has also been suggested that Fgf signaling from the ridge may set up a gradient in the early bud that specifies proximo-distal pattern (Papageorgiou and Almirantis 1996), while an-

other model has proposed that there are opposing gradients of Fgf and retinoic acid signaling (Mercader et al. 2000). A review by Mariani and Martin provides a useful summary of how both the progress zone and the prespecification model can account for various experimental findings (Mariani and Martin 2003).

Over the last few years, it has emerged that the level of Fgf signaling in zebrafish, chick, and mouse fin/limb (and elsewhere in the embryo) is regulated by negative feedback loops in the MAP kinase signal transduction pathway (see, e.g., Eblaghie et al. 2003; Kawakami et al. 2003). These feedback loops operate at different levels in the pathway and control the amount of activated ERK entering the nucleus. One of these feedback loops involves a MAP kinase phosphatase, and expression of the gene encoding this MAP kinase phosphatase has been shown to be regulated by Fgf (Eblaghie et al. 2003; Kawakami et al. 2003). Other transcriptional targets of Fgf signaling have also been identified, including the genes encoding two transcription factors, Erm and Pea3 (Kawakami et al. 2003), and tracing their targets could give new insights into proximo-distal patterning mechanisms in the limb.

Fgf signaling from the ridge is involved in elaboration of position-dependent patterns of gene expression along the proximo-distal axis of the limb. The vertebrate homeobox genes, *Meis-1* and *-2*, orthologs of the *Drosophila homothorax* gene, are expressed in the body wall in mouse and chick embryos in both limb and interlimb regions and later in early limb buds (Mercader et al. 1999). Experimental analysis in developing chick limb buds shows that during bud outgrowth *Meis-1* and *-2* expression becomes localized to the proximal part of the limb bud by being repressed distally by Fgf signaling from the apical ridge and that expression is maintained proximally by retinoic acid signaling (Mercader et al. 2000). A similar dynamic pattern of expression is exhibited by a *Meis* gene in zebrafish, *Meis 3.1*. This is initially expressed throughout the developing pectoral fin bud, but it later becomes expressed proximally (Waskiewicz et al. 2001). When *Meis-1* is overexpressed in chicken limb buds, distal development is disrupted (Mercader et al. 1999; Capdevila et al. 1999). Ectopic expression of *homothorax* in *Drosophila* legs also alters proximo-distal pattern, suggesting a parallel in patterning mechanisms between insect and vertebrates and a conserved role for *Meis/Homothorax* (Mercader et al. 1999).

Detailed analysis of *Drosophila* leg development has uncovered a number of other transcription factors responsible for proximo-distal patterning (G. Campbell 2002; Galindo et al. 2002; see also Y. Inoue et al. 2002 for cricket leg). In this context, it is interesting that vertebrate orthologs of *distalless*, a gene that is necessary for formation of the tip of the insect leg, are expressed in the apical ectodermal ridge in both mouse and chick limb buds and that digit abnormalities result when two of these genes, *Dlx5* and *Dlx6*, are simultaneously inactivated in mice (Robledo et al. 2002). Furthermore, the gene encoding another transcription factor, dachshund, which is involved in proximo-distal pattern in *Drosophila*, is expressed in vertebrate fin/limb buds (Hammond et al. 2002; Ayres et al. 2001; Caubit et al. 1999). Recent work in chick limb buds suggests that Dach1 may act as an intracellular modulator of Bmp signaling and thus contribute to regulating proximo-distal pattern (Kida et al. 2004).

There is increasing evidence from mouse knockouts that different 5′ members of the *Hoxd* and *Hoxa* complexes are required for development of different limb "segments" along

the proximo-distal axis (Wellik and Capecchi 2003). The patterns of expression of 5′ *Hoxa* and *Hoxd* genes have been described in detail in mouse and chick embryos (Dollé et al. 1989; Izpisúa-Belmonte et al. 1991; Nelson et al. 1996), and *Hoxa* and *Hoxd* genes are expressed in zebrafish fin buds (Sordino et al. 1995). There is good evidence that elaboration of these domains of Hox gene expression are dependent on Fgf signaling (Hashimoto et al. 1999; Vargesson et al. 2001). When *Hoxd13* and *Hoxa13* are both functionally inactivated, digit development is severely impaired (Fromental-Ramain et al. 1996b), while when *Hoxd11* and *Hoxa11* are knocked out, it is the lower arm/shank that is affected (A. P. Davis et al. 1995). The limb phenotypes of knockouts of *Hox10* paralogs suggest that these pattern the upper leg (Wellik and Capecchi 2003), while those of double knockouts of *Hoxd9* and *-a9* suggest that these genes are important for upper arm development (Fromental-Ramain et al. 1996b).

Dorso-ventral Patterning of the Fin/Limb Bud

Experiments in chick embryos in which ectodermal hulls from right limb buds were recombined with mesodermal cores from left limb buds so that dorsal ectoderm overlies ventral mesoderm and vice versa showed that the ectoderm covering the sides of the limb buds controls dorso-ventral polarity of distal limb structures (MacCabe et al. 1974; Akita 1996). Furthermore, ectopic limb buds induced by grafting an apical ectodermal ridge to the dorsal or ventral surfaces of host limb buds have either double-dorsal or double-ventral patterns, respectively.

It has now emerged that Wnt7a signaling by dorsal ectoderm of mouse and chick limb buds plays a role in specifying dorsal limb structures. The dorsal ectoderm of both chick and mouse limb buds expresses *Wnt7a* (Dealy et al. 1993; Parr and McMahon 1995). Furthermore, mice in which *Wnt7a* is functionally inactivated have paws with a double ventral pattern, although proximal structures appear unaffected (Parr and McMahon 1995). One of the genes expressed in dorsal mesenchyme in response to Wnt7a signaling is *Lmx1*, which is an *LIM-homeodomain*-containing gene. Misexpression of *Lmx1* in ventral mesenchyme of chick limb buds leads to dorsalization, while, conversely, inhibiting *Lmx1* expression in dorsal mesenchyme leads to ventralization (Riddle et al. 1995; A. Vogel et al. 1995; Rodriguez-Esteban et al. 1998).

The development of ventral limb structures seems to depend on expression of *En1* in ventral limb bud ectoderm. This is in addition to the role of *En1* in apical ridge formation (see earlier). When *En1* is functionally inactivated in mice, *Wnt7a* is now expressed in ventral ectoderm, suggesting that *En1* normally represses expression of *Wnt7a* (fig. 5.3). Furthermore, the paws of $En1^{-/-}$ mouse embryos have a double dorsal pattern (Loomis et al. 1996). In chick limb buds, it has been shown that $En1$ expression in ventral ectoderm and apical ridge is controlled by Bmp signaling (Pizette et al. 2001), and in mice in which the gene encoding the Bmp receptor 1 A is functionally inactivated, there are not only defects in apical ridge formation (see earlier) but also defects in dorso-ventral patterning (K. Ahn et al. 2001).

Antero-posterior Patterning of the Fin/Limb Bud

Pattern across the antero-posterior axis of the vertebrate limb is controlled by signaling of the polarizing region at the posterior margin of the bud. The polarizing region was discovered by Saunders in chick limb buds (Saunders and Gasseling 1968). When tissue from the posterior margin of one chick wing bud is grafted to the anterior margin of another chick wing bud, six digits form instead of three, and an additional set of digits, 4 3 2, develops in mirror-image symmetry with the normal set, 2 3 4. The conclusion from the results of many embryological experiments is that signaling by the polarizing region is dose dependent and long range. This is consistent with a model in which the polarizing region produces a morphogen, which diffuses across the limb bud to establish a concentration gradient (Wolpert 1969, 1989). The polarizing region expresses sonic hedgehog, a cell-cell signaling molecule, and sonic hedgehog produced by the polarizing region is now known to be pivotal in controlling antero-posterior limb bud patterning (Riddle et al. 1993).

Shh is expressed at the posterior of both chick and mouse limb buds (Riddle et al. 1993; Echelard et al. 1993) and also in zebrafish fin buds (Krauss et al. 1993; also stickleback—see plate 5.2). When *Shh*-expressing cells or beads soaked in Shh are applied to the anterior margin of chick wing buds, concentration-dependent changes in digit pattern are induced (Riddle et al. 1993; Yang et al. 1997). Furthermore, ectopic *Shh* expression has been detected in many polydactylous mouse mutants (see, e.g., Masuya et al. 1997). There has been considerable debate about whether Shh can act at a distance. Two reports suggest that Shh diffuses at least some distance across the limb bud (Gritli-Linde et al. 2001; Zeng et al. 2001), and there is evidence that cholesterol modification of Shh is required for its long-range action (P. M. Lewis et al. 2001). Thus, like polarizing region signaling, Shh signaling is concentration dependent and long range. Recent work has raised the possibility that the length of time cells are exposed to Shh in addition to the concentration of Shh may be important (Harfe et al. 2004; S. Ahn and Joyner 2004).

Mesenchymal differences responsible for the anterior-posterior polarity of the proximal part of the limb and establishing *Shh* expression are set up very early prior to limb bud

formation (reviewed by Panman and Zeller 2003). Thus, the gene encoding the transcription factor, dHand, is initially expressed throughout limb-forming regions in both chick and mouse embryos. Later, expression becomes localized to the posterior region of the buds. When *dHand* is misexpressed at the anterior margin of either chick or mouse limb buds, ectopic *Shh* expression is induced, resulting in formation of additional digits (Fernandez-Teran et al. 2000; Charité et al. 2000). In mice and zebrafish deficient in *dHand*, antero-posterior patterning is defective (Charité et al. 2000; Yelon et al. 2000). In *dHand*$^{-/-}$ mouse embryos, not only is there absence of posterior *Shh* expression in the limb buds, but also expression of genes normally seen anteriorly is now extended posteriorly, suggesting that dHand may also act as a repressor (te Welscher et al. 2002a). Work in both chick and mouse shows that *Hoxb8* may be another gene that is important in establishing the polarizing region in the forelimb (Charité et al. 1994; Stratford et al. 1997; H. C. Lu et al. 1997). The involvement of *Hoxb8* provides a potentially important link with main body axial patterning.

The products of genes such as *dHand* and *Hoxb8* probably cooperate with the vitamin A derivative, retinoic acid. Retinoic acid was the first chemical to be identified with polarizing activity, being able to induce additional digits when applied to the anterior margin of chick wing buds (Tickle et al. 1982). It was subsequently shown that the basis of the polarizing activity of retinoic acid is the induction of ectopic *Shh* expression in anterior mesenchyme of the host limb bud (Riddle et al. 1993). Retinoic acid is also able to induce anterior expression of *Shh* in zebrafish pectoral fin buds (Akimenko and Ekker 1995). Application of retinoid antagonists to limb-forming regions in chick embryos (Stratford et al. 1996; H. C. Lu et al. 1997) and functional inactivation of a gene encoding an enzyme required for retinoic acid synthesis in mice and zebrafish show that retinoic acid is required for fin/limb bud development (Grandel et al. 2002; Niederreither et al. 2002).

Shh expression in posterior limb bud is maintained by Fgf signaling by the ridge (Niswander et al. 1994; Laufer et al. 1994), and Wnt7a from dorsal ectoderm also seems to play a role (Yang and Niswander 1995; Parr and McMahon 1995). In turn, Shh maintains *Fgf4* expression in the ridge (Niswander et al. 1994; Laufer et al. 1994). These interactions thus coordinate signaling with respect to all three axes of the developing limb. Furthermore, the role of *Shh* in maintaining Fgf signaling in the ridge explains why there are distal limb defects in *Shh*$^{-/-}$ mouse embryos. When *Shh* is functionally inactivated in mice, the limbs are very reduced distally, although a nail develops at the tip (Chiang et al. 2001; Kraus et al. 2001). In a chicken mutant, oligozeugodactyly, that has no detectable *Shh* expression in limb buds, distal limb structures, particularly posterior ones, are also lost (Ros et al. 2003). Similarly in the fin buds of the zebrafish *sonic you* mutant, in which *Shh* is disrupted, the apical ectodermal fold does not develop and outgrowth and patterning of the fin bud fails (Neumann et al. 1999).

A deeper understanding of the role of Shh signaling in the limb has depended on analysis of the Shh signal transduction pathway. The components of the Hedgehog (Hh) signal transduction pathway were first identified in *Drosophila*, and nearly all of the same components have since been found in vertebrates. Hh binds to a cell surface receptor, Ptc (in vertebrates, there are two *Ptc* genes), and this allows signaling through the transmembrane protein smoothened and then via an intracellular complex containing Gli transcription factors (there are three Gli's in vertebrates). Gli2 and 3 can act as either transcriptional activators (in presence of Hh signaling) or transcriptional repressors (in absence of Hh signaling), while Gli1 acts only as an activator (reviewed Koebernick and Pieler 2002).

Genes encoding components of the hedgehog signal transduction pathway outlined above have been shown to be expressed in chick and mouse limb buds. *Ptc1* and *Ptc2*, genes whose expression is activated in response to Shh, are expressed at high levels in posterior mesenchyme; *Ptc2* is expressed also in the apical ectodermal ridge (Pearse et al. 2001). *Ptc* is also expressed in zebrafish fin buds (Hoffman et al. 2002). The binding of Shh to Ptc restricts diffusion of Shh, and this is also aided by another vertebrate-specific membrane protein, Hip (Hedgehog-interacting protein), which is also expressed in response to Shh signaling (Chuang and McMahon 1999). *Gli1* is another direct target of Shh signaling, and is expressed in posterior limb mesenchyme in mouse and chick embryos (Platt et al. 1997; Marigo et al. 1996), while *Gli2* and *Gli3* are expressed more anteriorly (Schweitzer et al. 2000).

Analysis of Gli3 status in vertebrate limb buds (B. Wang et al. 2000; Litingung et al. 2002) together with the phenotype of *Gli3⁻/⁻Shh⁻/⁻* mouse mutants (Litintung et al. 2002; te Welscher et al. 2002b) has provided new insights into how Shh signaling operates during limb development. It now appears that the main function of Shh is to relieve repression of transcriptional targets by *Gli3* (reviewed Panman and Zeller 2003). In the absence of Shh signaling in the limb bud, the repressor form of *Gli3* predominates and shuts down nearly all distal development, while in the absence of *Gli3*, irrespective of whether *Shh* is expressed or not, the limb is polydactylous and all the digits have the same morphology. Two chicken polydactylous mutants called *talpid²* and *talpid³* have a series of morphologically uniform digits, although *Shh* expression is restricted posteriorly as normal (Francis-West et al. 1995; Caruccio et al. 1999). Furthermore, the repressor

form of *Gli3* is absent in the anterior of limb buds in *talpid²* mutants (B. Wang et al. 2000). In *talpid³* limb buds, it has been suggested that Shh diffuses more widely due to failure to induce high-level *Ptc* expression in response to Shh (K. E. Lewis et al. 1999).

Among target genes repressed by *Gli3* anteriorly in vertebrate limb buds are 5′ *Hoxd* genes, such as *Hoxd13*, which, along with *Hoxa13*, is required for digit formation (te Welscher et al. 2002b; Litintung et al. 2002). When Shh is applied to the anterior of chick wings—and in polydactylous mouse mutants in which *Shh* is expressed anteriorly—ectopic anterior domains of *Hox* gene expression are produced. This ectopic expression can now be interpreted as being due to the inhibition of processing of *Gli3* to the repressor form. This de-repression of *Gli3* also leads to changes in gene expression in the ridge, and cooperation by the ridge is known to be required for ectopic *Hox* gene expression (see, e.g., Izpisúa-Belmonte et al. 1992a). In addition, *Hoxd13* is expressed uniformly across the tip of *talpid³* limb buds (Izpisúa-Belmonte et al. 1992b).

The dual functions of Shh signaling in determining digit number and identity may not be direct but instead depend on further downstream signals. Digit number is related to the length of the apical ridge, and it was postulated that posterior limb mesenchyme produces an apical ridge maintenance factor (Zwilling and Hansborough 1956). This factor has been identified as the Bmp antagonist, gremlin, and Bmps are expressed at high levels in the apical ectodermal ridge. *Gremlin* expression in the posterior region of the bud is controlled in mouse and chick limbs by Shh signaling, and the action of Gremlin results in maintenance of *Fgf4* expression in posterior ridge (Zuniga et al. 1999; Khokha et al. 2003). In contrast, the apical ridge regresses in the anterior part of the limb bud, where *Bmp4* is also expressed at high levels in the mesenchyme (Pizette and Niswander 1999).

Experiments in chick embryos show that Shh can inhibit *Bmp4* expression in anterior limb mesenchyme (Tumpel et al. 2002), suggesting a parallel with the signaling system that patterns the dorso-ventral axis of the neural tube. Furthermore, in mouse embryos in which *Bmp7*, another Bmp expressed in mesenchyme and ridge, is functionally inactivated, additional digits can develop (Hofmann et al. 1996). Similar defects were also seen in *Bmp7* null mouse embryos produced by insertional mutagenesis (Jena et al. 1997). In doubly heterozygous *Bmp7* and *Bmp4* mice, again, additional digits form (Katagiri et al. 1998), suggesting that Bmp signaling anteriorly represses digit formation. Other genes expressed in the anterior region of the bud include *Alx4*, which is responsible for preaxial polydactyly in a mouse mutant (M. Takahashi et al. 1998).

Digit identity may be controlled by Bmps that act on cells

already primed by Shh (Drossopoulou et al. 2000). Another Bmp gene, *Bmp2*, together with *Bmp7*, is expressed predominantly in posterior mesenchyme and in the apical ridge in chick and mouse limb buds (P. H. Francis et al. 1994; Lyons et al. 1995; Lewis et al. 2001). In chick limb buds, anterior application of Shh leads to ectopic expression of *Bmp2* and *Bmp7* (Yang et al. 1997), and grafts of *Bmp2*-expressing cells at the anterior margin of chick wing buds can induce an additional anterior digit (Duprez et al. 1996c). When noggin (a Bmp antagonist) is applied sequentially after Shh application to the anterior margin of chick wing buds, the additional digits are not patterned (Drossopoulou et al. 2000). Furthermore, manipulations of Bmp receptor expression in both chicken and mouse limb buds lead to alterations in limb patterning (Kawakami et al. 1996; Z. Zhang et al. 2000). Thus, when gene constructs encoding dominant negative Bmp type I and II receptors are expressed throughout chick limb buds, posterior digits are lost (Kawakami et al. 1996), while, in mice, in which a constitutively active Bmpr-IB was expressed throughout hindlimb buds, additional digits develop anteriorly often associated with lack of an identifiable digit 1 (Z. Zhang et al. 2000). It is not clear in this latter case whether the status of the Gli3 repressor is affected.

The expression of *Bmp2* in response to Shh in vertebrate limb buds mirrors a signaling cascade in *Drosophila* wing development in which *dpp* is expressed in response to Hh. Vertebrate orthologs of two of the target genes of dpp signaling in Drosophila wing, *Tbx3* and *Tbx2* (*omb* orthologs) and *Sall1* (mouse), *cSal1* (chick) and *cSal2* (chick; *spalt* orthologs) are expressed in fin/limb buds. *Tbx3* and *Tbx2* are expressed in anterior and posterior stripes in both mouse and chick limb buds and in zebrafish fin buds (Gibson-Brown et al. 1996; Yonei-Tamura et al. 1999b). Experimental manipulation of chick wing buds and analysis of gene expression in mouse mutants shows that posterior expression of *Tbx3* is downstream of polarizing region signaling (Tumpel et al. 2002). Overexpression of Tbx3 in chick leg buds has been reported to lead to posteriorization of digits (T. Suzuki et al. 2004), while when *Tbx3* is functionally inactivated in mice, posterior forelimb skeletal elements are missing and distal elements are severely reduced in hindlimbs (T. G. Davenport et al. 2003). The various vertebrate *spalt* genes have dynamic patterns of expression during limb development (Farrell and Munsterberg 2000; Farrell et al. 2001). Expression of *csal1* in chick limb buds appears to depend on signals operating at the tip of the limb including Wnts, Fgfs, and Bmp2 (Farrell and Munsterberg 2000; Capdevila et al. 1999), but *spalt* mouse knockouts reported to date do not have limb defects (Sato et al. 2003). A *spalt* ortholog has also been shown to be expressed in the posterior region of pectoral fin buds of Medaka embryos (Koster et al. 1997).

Digit Patterning

By the time mouse and chick digit primordia appear, the limb bud has grown considerably. Furthermore, fate mapping experiments in chick limb buds show that the posterior part of the early limb bud expands to produce the entire hand plate (Bowen et al. 1989; Vargesson et al. 1997). During the last few years, it has become apparent that morphogenesis of the primordium of each individual digit of the chick leg and digit II in the chick wing, even at these late stages in limb development, is surprisingly plastic. It had been known for a long time that the apical ridge persists over the tip of developing digits, and its removal at late stages in chick limb development leads to truncations (Rubin and Saunders 1972). Furthermore, *Fgf8* has been shown to be expressed in the apical ridge over the tips of the digits at these late stages (Gañan et al. 1998; Salas-Vidal et al. 2001). More recently, it has been found that local grafts of interdigital tissue or application of either Shh or the Bmp antagonist noggin can lead to alterations in digit morphogenesis, resulting not only in truncated digits but also, more remarkably, in digits with additional phalanges (Dahn and Fallon 2000; Sanz-Ezquerro and Tickle 2000, 2003a). The endogeneous signaling network that determines phalange number during digit morphogenesis has yet to be elucidated, although it has been suggested that Bmp signaling is involved (Dahn and Fallon 2000). This may affect the duration of Fgf signaling, and it was suggested that a special program controls formation of the digit tip when Fgf signaling switches off (Sanz-Ezquerro and Tickle 2003b).

Hoxd10 through *-13* and *Hoxa13* are expressed at high levels in the digit-forming region of the limb in both mouse and chick embryos (Dollé et al. 1989; Nelson et al. 1996). Detailed genetic manipulations of the 5′ region of the *Hoxd* cluster have revealed its importance in digit formation (Kmita et al. 2002). Interestingly, this phase of *Hox* gene expression is not seen in teleost fin development. This is consistent with the idea that digits evolved as new structures in higher vertebrates (see below), although there are other interpretations.

Implications for Morphological Diversity, Evolution, and Human Congenital Limb Defects

Morphological Diversity

Present knowledge of fin/limb development relies heavily on analysis of zebrafish, chick, and mouse embryos. Nevertheless, the little information that is available from embryos of other higher vertebrates suggests that generally the same

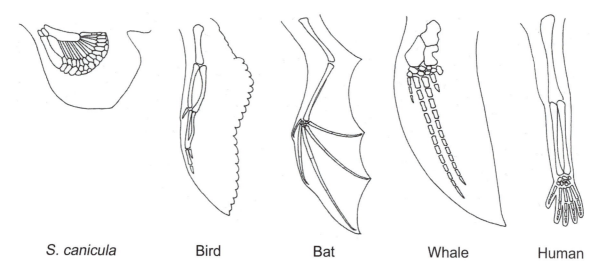

S. canicula   Bird   Bat   Whale   Human

Figure 5-4 Morphological variation in limb anatomy: Anterior appendages of vertebrates. All appendages are shown in dorsal view, anterior to the left. The outline of appendage and skeleton is indicated in each case.

principles are involved. One of the biggest challenges is to understand the basis for morphological diversity of fin/limb anatomy (fig. 5.4), especially given the striking conservation of the patterning mechanisms involved in early development of fins/limbs (fig. 5.3).

Recent research has made a start at understanding the basis for morphological differences between fore- and hindlimbs. Fore- and hindlimbs can vary considerably in their anatomy even though the same signaling networks are involved in their development. Some evidence suggests that these morphological differences may be due to differential expression of *Tbx* genes, with *Tbx5* being expressed in forelimbs but not hindlimbs, and *Tbx4* in hindlimbs only (Gibson-Brown et al. 1996). The same two genes are also expressed in the pectoral and pelvic fin buds, respectively, of zebrafish (Tamura et al. 1999). Another gene, Pitx1, is also specifically expressed in hindlimb regions. When either *Tbx4* or *Pitx1*, which regulates *Tbx4* expression, is misexpressed in forelimb regions or *Tbx5* is misexpressed in hindlimb regions in chick embryos, there are alterations in limb anatomy consistent with a change in limb type (Takeuchi et al. 1999; M. Logan and Tabin 1999). In these experiments, however, the endogenous limb-specific gene is still expressed, and this could account for the lack of complete limb type conversion. This problem has been addressed by sophisticated genetic experiments in mice (Minguillon et al. 2005). Conditional *Tbx5* mouse mutants were generated in which the forelimb was completely absent and then the ability of *Tbx4* to rescue the defects was tested. If *Tbx4* specifies hindlimb identity, the rescued limb would be expected to have a hindlimb phenotype, but instead the rescued limbs

have a forelimb phenotype. Only when *Pitx1* was expressed with *Tbx4* in the conditional mutant was a hindlimb phenotype obtained, suggesting that *Pitx1* plays a role in specifying hindlimb identity.

Animals that have undergone major secondary changes in limb anatomy during evolution might be expected to show embryonic differences in the basic limb signaling networks. Limb reductions have occurred in both fish and reptiles, and knowledge of the genetic and molecular basis of limb development in model vertebrates brings a new perspective. Three-spine stickleback, *Gasterosteus aculeatus*, show natural variation in anatomy including ranging degrees of pelvic reduction, affecting both the girdle and spines, which are modified fins. These changes in stickleback morphology have occurred in response to changes in the environment over a relatively short evolutionary period, in parallel and independently in different populations of fish. Recent mapping and gene expression studies suggest that *Pitx1* may be a key gene in pelvic reduction; furthermore, the same gene seems to be involved in pelvic reduction in different populations from Scotland, Canada, and Alaska (N. J. Cole et al. 2003; Shapiro et al. 2004; and Cresko et al. 2004). An intriguing similarity between pelvic reduction in three-spine stickleback and limb development in *Pitx1*$^{-/-}$ mouse embryos is the striking left/right asymmetry in fin/limb reduction. Even though *Pitx1* expression cannot be detected in the pelvic regions of pelvic-reduced stickleback, it is expressed elsewhere. Therefore, it has been suggested that it is changes in regulatory sequence(s) that normally drive *Pitx1* expression in the pelvic regions that are responsible for pelvic reduction (Shapiro et al. 2004). In another teleost fish, the puffer fish,

Takifugu rubripes, loss of pelvic structures is also associated with lack of *Pitx1*, expression but in addition one of the *Hox* genes normally expressed in lateral plate mesoderm in association with limb initiation is not expressed in pelvic regions (Tanaka et al. 2005). Therefore, it has been suggested that the genetic basis for reduction in puffer fish may be different from that in stickleback (Tanaka et al. 2005).

The involvement of *Hox* genes has also been suggested to underlie limblessness in a snake. Burmese pythons have no trace of forelimbs, and in python embryos the pattern of *Hox* gene expression along the main body axis appears homogeneous and could account for failure of forelimb initiation (Cohn and Tickle 1999). Burmese pythons have rudimentary hindlimbs, but *Shh* expression could not be detected in the hindlimb buds of python embryos (Cohn and Tickle 1999). Rather interestingly, the anatomy of python hindlimbs bears some resemblance to that of the limbs of *Shh⁻/⁻* mice, in that pelvic and proximal structures are present, and even though distal structures are absent, claws form at the tips. Loss of *Shh* expression must be secondary, and indeed posterior limb bud tissue from the python embryos can express *Shh* when grafted to the anterior margin of a chick wing bud (Cohn and Tickle 1999). Some lizards also show varying degrees of evolutionary limb reductions. One example that has been studied is the Australian skink genus *Hemiergis*, which contains species and populations differing with respect to digit number. It has been shown that the duration of Shh signaling seems to be correlated with digit number, with *Shh* expression persisting for longer in the limb buds of *Hemiergis* species with more digits (Shapiro et al. 2003).

Evolution of Limbs

Analysis of developmental mechanisms involved in fin/limb initiation in experimental vertebrates has shed light on the organization of the vertebrate body plan, which is relevant to the evolutionary origins of limbs. The finding that the interlimb region of higher vertebrates has the potential to form a limb (Cohn et al. 1995; Tanaka et al. 2000) has important implications for the lateral fin-fold theory. The fin-fold theory is a longstanding hypothesis about the evolution of vertebrate limbs and suggests that the two pairs of limbs in tetrapods evolved by subdivision of an elongated single fin (Thacher 1877; Mivart 1879; Balfour 1881). In support of this theory, classical studies on shark development described fin development from a continuous thickening of the ectoderm. Although a recent reexamination of *S. canicula* dogfish embryos using scanning electron microscopy did not detect lateral fin folds, the expression of *Engrailed-1* suggests

that the body of the dogfish embryo is compartmentalized dorso-ventrally as in higher vertebrates (Tanaka et al. 2002).

The fin-fold theory raises questions about how differences arose between paired fins and the specific patterns of *Tbx* gene expression came to be established. The cephalochordate Amphioxus has been reported to have a single Amphi*Tbx4/5* gene (Ruvinsky et al. 2000), and separate *Tbx5* and *Tbx4* genes appear to have arisen by duplication of the single ancestral *Tbx4/5* gene. *Tbx4/5* could have duplicated prior to the acquisition of two pairs of fins, or two pairs of fins could have been acquired prior to the duplication of *Tbx4/5*. The fact that dogfish have separate *Tbx4* and *Tbx5* genes suggests that *Tbx4/5* cluster duplication took place before the ancestor of cartilaginous fishes or in that lineage (Tanaka et al. 2002).

Recent analysis of fin development in embryos of cartilaginous fish has also revealed a departure from conserved fin bud signaling systems. In dogfish fin buds, no early *Shh* expression could be detected (Tanaka et al. 2002), suggesting that the fin/limb bud patterning system may not be the same as in teleosts. Interestingly, a limb bud–specific *Shh* regulatory element has been identified in preaxial polydactylous mutants in human and mouse (Lettice et al. 2002). The conserved specific enhancer region was identified by sequence comparison between mouse, human, chick, and Fugu, and is located within the *Lmbr1* gene. Even though this enhancer region is more than 1 Mb away from the *Shh* gene, it can control *Shh* expression in the polarizing region. A construct containing the *Fugu* regulatory element sequence can drive expression of a reporter gene in both fore- and hindlimbs in mouse embryos. This suggests that *Shh* expression is under the same regulatory influences in both *Fugu* fin buds and limb buds of higher vertebrates (Lettice et al. 2003). Furthermore, when this conserved enhancer region is eliminated in mice, there is loss of limb-specific *Shh* expression and truncated limbs develop, similar to those in *Shh⁻/⁻* mouse embryos (Sagai et al. 2005). Interestingly, this conserved sequence could not be amplified from DNA of a cartilaginous fish or four-limbless snakes or a limbless newt (Sagai et al. 2004). This again seems to point to changes in a region-specific regulatory sequence as being the mechanism for evolutionary change.

Late stages of limb development in which the digits are produced seem to be special to higher vertebrates. Again, in this context, it is interesting that a "digit enhancer" has been located in mouse that regulates transcription of the four *Hoxd* genes expressed in the digit-forming region (Kmita et al. 2002). It would be predicted that this function evolved after the ancestor of the teleosts or in the lineage leading to higher vertebrates.

Human Limb Defects

Identification of the molecular basis of vertebrate limb patterning has provided direct links to human clinical genetics. A number of recent reviews have documented human limb malformations for which the genes responsible are now known (e.g., Cohn and Bright 1999; Wilkie 2003; Ferretti and Tickle 2006). Many genes responsible for human conditions are familiar to developmental biologists; for example, genes encoding Shh and components of its transduction pathway, Fgf receptors, and *Tbx* and *Hox* genes (fig. 5.3; genes affected in human conditions are marked with asterisks). Further details are emerging about how different mutations in the same gene can lead to specific limb abnormalities for, for example, *Fgfr* (Wilkie et al. 2002) and *Hoxd13* (Caronia et al. 2003). The many different human conditions will provide new opportunities for exploring genotype/phenotype relationships and understanding how the precise anatomy of the limb develops.

Acknowledgments

Mikiko Tanaka is supported by an UHERA fellowship, Cheryll Tickle by The Royal Society and MRC. We thank Angie Blake and Andy Bain for help in preparing the manuscript.

Chapter 6 Mechanisms of Chondrogenesis and Osteogenesis in Fins

P. Eckhard Witten and Ann Huysseune

Recent "fish" (Chondrichthyes and primary aquatic Osteichthyes) possess both paired pectoral and pelvic and unpaired dorsal, anal, and caudal fins. We focus on the structure of the skeletal fin supports, and on the chondro- and osteogenic processes involved in the development of the fin skeleton. Skeletal development is commonly viewed from either an osteological (anatomical) or a developmental (molecular) perspective. Placing fin skeletal histogenesis in the center of our considerations, we aim to bridge the gaps between the many published studies on the osteological development of fish larvae, and an ever-growing number of papers dealing with the genetic and molecular control of fish skeletal tissue formation. Most data concerning fin development that reach beyond the anatomical level—that is, to the histological, cellular, and genetic levels—have been gathered on teleost genera such as *Cyprinus, Fundulus, Oryzias,* and *Tilapia.* A considerable amount of developmental, molecular, and genetic data now come from the zebrafish, *Danio rerio,* since this teleost was introduced as a "model organism" for developmental and genetic studies.

In the course of evolution, fins developed into an astounding diversity of forms and functions, processes that certainly had an impact on fin chondro- and osteogenesis (for examples see Rosa-Molinar et al. 1994; Moser 1996; J. F. Webb 1999). The variety of fin forms is most obvious in teleosts. Harder (1964) illustrates examples of some 20 different types of dorsal fins, solely for teleosts that still use their dorsal fins for swimming (see Drucker and Lauder 2001a for biomechanical considerations). In addition, many

fish species have transformed particular fins into organs with new functions. Among these are sucking, luring prey, gliding, walking, clasping, clinging, chemo-sensing, camouflage, defense against predators, and reproduction (N. B. Marshall 1976; Lagler et al. 1977; Kotrschal et al. 1985; Brandstätter et al. 1990; R. C. Peters et al. 1991; Rosa-Molinar et al. 1994; Kottelat and Lim 1999; Everly 2002; Lucifora and Vassallo 2002). Particular fins (often the pelvic) have been reduced in evolution, or are susceptible to loss (Villwock 1984; Adriaens et al. 2002). A description of the chondro- and osteogenic processes underlying the formation of these specialized fins would certainly go beyond the scope of this chapter. Thus, we concentrate on the processes that contribute to the development of the skeleton of those fins that are still used for propulsion and stabilization.

A Short Introduction to the Anatomy of Unpaired and Paired Fins

Fins are supported by a dermal skeleton and also by an endoskeleton (plate 6.1; see Starck 1979 and Stiassny 2000). Exceptions are the larval fin fold and the adipose fin. Adipose fins are present in salmoniform teleosts and seven other groups of basal euteleosts (Reimchen and Temple 2004). Exceptionally, the adipose fin may be supported by elongate neural spines, as in some clariid catfishes (Teugels 1983).

The endoskeleton in unpaired fins connects the free part of the fin directly to the axial skeleton. In paired fins, the endoskeleton is connected to the girdles, which are composed of endoskeletal elements and, in osteichthyans, also of dermal bones (Starck 1979; McGonnell 2001). The endoskeleton of the paired pelvic fins is supported by a pelvic girdle composed of endoskeletal elements only. The girdle of the pelvic fin originally has no connection to the axial skeleton but attaches to the pectoral girdle in derived teleosts. The pectoral girdle connects to the head. The endoskeleton of all fins in all fish groups is initially made of cartilage. Most elements of the primary cartilaginous endoskeleton ossify in bony fish.

The dermal skeleton, which supports the free part of all fins, consists of fin rays (plates 6.1, 6.2), which connect to the endoskeleton via cartilaginous joints. Fin rays stabilize the embryonic fin fold and the free part of the adult fin of chondrichthyans and osteichthyans. The different types of fin rays have been summarized under the term "dermotrichia" (Goodrich 1904; Francillon-Vieillot et al. 1990). They comprise both nonmineralizing elements made up of elastoidin (ceratotrichia in chondrichthyans, actinotrichia in osteichthyans), and mineralizing elements composed of bone (lepidotrichia in osteichthyans). Within osteichthyans, dipnoans are exceptional in possessing camptotrichia instead of lepidotrichia (Francillon-Vieillot et al. 1990). The classification of camptotrichia is a matter of discussion; they are regarded either as a type of lepidotrichia (Starck 1979) or as differing significantly from both actinotrichia and lepidotrichia (Géraudie and Meunier 1982). Below we briefly recapitulate fin anatomy as far as it is required to understand the histogenesis of fin skeletal elements.

Dorsal and Anal Fins

Median (dorsal and anal) fins are supported by primarily segmentally arranged endoskeletal supports, made up initially of cartilage: the radialia or pterygiophora (plate 6.1). In chondrichthyans and in many osteichthyans, each pterygiophore is composed of three parts: a proximal, a middle and a distal radial (Starck 1979; Stiassny 2000). Within the teleosts, there is an evolutionary tendency to decrease the number of radial elements through fusion or reduction (Bertin 1958; Stiassny 2000). The radials (pterygiophores) of the dorsal and anal fins start their development in the median muscle septum. In chondrichthyans, radials extend into the basal parts of the fin itself and do not ossify. In osteichthyans, radials do not extend into the free part of the fin and ossify by perichondral bone formation. Radialia that support the unpaired fins are thought to derive from hemal and neural spines (Starck 1979). In Neopterygii, one radial supports one lepidotrichium per muscle segment (Harder 1964; Mabee et al. 2002).

Caudal Fin

The genesis of the caudal fin differs from the anal and the dorsal fins; the vertebral column directly participates in the support of the caudal fin (plate 6.3; Nybelin 1963). The last part of the vertebral column in teleosts—the urostyle—represents phylogenetically the fusion product of a number of caudal vertebrae. Nevertheless, the urostyle develops as one piece in teleosts. In contrast, preural vertebrae develop as separate elements, but can display fusion phenomena. The endoskeletal fin support elements that insert on the urostyle and on the preural vertebrae derive from the neural (epurals) and hemal (hypurals) arches and spines. Some species have in addition uroneurals, vestiges of vertebral arches that are not incorporated into the urostyle (Bertin 1958; Stiassny 2000). However, Starck (1979) argues that because of their superficial position, and because uroneurals develop without cartilaginous precursor, they are evolutionary related to scales. In terms of processes of cartilage and bone formation, caudal fin development, despite its particular anatomy, displays several similarities with the development of other fins.

Paired Fins

Within the paired fins, the pectoral fins arise early in ontogeny and are supported in adult fish by an elaborate shoulder girdle apparatus, composed in osteichthyans of endoskeletal (scapula, coracoid, mesocoracoid) and dermal elements (cleithrum, postcleithrum, supracleithrum, posttemporale, thoracale, and interclavicula; plate 6.1; see also Starck 1979 and Stiassny 2000). Chondrichthyans lack dermal components, and the pectoral girdle is solely represented by a pair of cartilages, fused into a single scapulocoracoid bar in more advanced members of the group (McGonnell 2001). Distal to the girdle, the pectoral fin endoskeleton is composed of various, initially cartilaginous, elements: one to three basal elements in chondrichthyans, a serial row in dipnoans and crossopterygians, and three basal elements in polypterids, all carrying various numbers of smaller cartilaginous elements, the radials (Starck 1979). The origin of the various elements is still discussed, connected to the theories about the origin of paired limbs (Starck 1979; Hinchliffe 2002). Reviewing the fossil record (osteostracans) and pectoral fin development in actinopterygians, Janvier et al. (2004) suggest a monobasal condition (an undivided cartilaginous endoskeletal disk) as the basic type of gnathostome pectoral fin support. Subdivision of the endoskeletal disk, as it occurs during chondrichthyan and actinopterygian development, then gives rise to radials (Grandel and Schulte-Merker 1998; Cohn et al. 2002; Janvier et al. 2004; M. C. Davis et al. 2004; Mabee and Noordsy 2004; see also plates

6.17–6.20). A condition occurring in chondrichthyans and in basal osteichthyans is a tribasal pectoral fin support, consisting of three elements (anterior to posterior): a propterygium, a mesopterygium, and a metapterygium (Wilga and Lauder 2001; M. C. Davis et al. 2004). Whether all three elements derive from subdivision of an endoskeletal disk is, however, a matter of debate (Cohn et al. 2002). Studies of pectoral development of primitive actinopterygians (Acipenseriformes) led M. C. Davis et al. (2004) to conclude that the propterygium and the mesopterygium are homologous to the endoskeletal disk in teleosts, while the mesopterygium, which supports the pectoral fin in sarcopterygians, develops from a separate cartilage anlage.

The pelvic fins develop late in ontogeny and have a reduced and exclusively endoskeletal girdle. One piece of cartilage supports the pelvic fins in sharks (ischiopubic), two pieces in holocephalians. Osteichthyans have paired pelvic elements (the basipterygia), separated or variously fused (plates 6.4–6.6). As in the pectoral fin, fin rays in osteichthyans articulate with the pelvic girdle via a varying number of radials, which, however, can be absent in teleosts (plate 6.2). In this case the lepidotrichia directly articulate with the basipterygia (Lagler et al. 1977; Starck 1979; Stiassny 2000).

Whether or not the girdles should be considered part of the fin endoskeleton is a matter of debate. Some comparative anatomists view the girdle as a part of the extremity (e.g., because of the supposed homology of the girdle element with the basal metapterygium of chondrichthyans). Some developmental biologists view the girdle separate from the extremity, because its development is not affected in experiments where the apical ectodermal ridge (AER), necessary for limb formation, is ablated (Grandel and Schulte-Merker 1998). We include the girdles in our description, because of the intimate developmental relationship between the endoskeletal elements of the shoulder girdle (scapula, coracoid) and the fin endoskeleton proper (the radials), at least in osteichthyans.

Timing of Appearance of the Fins in Development and Evolution

As shown for *Polypterus senegalus*, the unpaired median fin fold precedes the development of the pectoral fin buds, which themselves precede the development of the pelvic fin buds (Bartsch et al. 1997). Although this sequence is basically maintained in teleosts, in some species ossification of the pelvic fin rays precedes ossification of the pectoral fin rays (Richards et al. 1974; Methven 1985). The timing of appearance of the fins in development coincides with the order in which fins appeared during evolution: unpaired preceded paired fins in evolution and evolved early in the craniate lineage (about 400 mya). Paired pectoral fins appeared before paired pelvic fins (Coates 1994; B. K. Hall 1998; Coates and Cohn 1998, 1999; Mabee et al. 2002).

Tissue Composition of Paired and Unpaired Fins

Actinotrichia

Actinotrichia are short, tapered rods, usually arranged parallel to the long axis of the fin, and sometimes branched distally (plates 6.2 and 6.7, and see Francillon-Vieillot et al. 1990). In polypterids, they can also be oriented perpendicularly to the long axis of the fin (Géraudie 1988).

Actinotrichia are found in the embryonic fin fold, along the outer edge of the fin in Actinopterygii, and in the adipose fin in teleosts (Starck 1979; Géraudie and Meunier 1982; Becerra et al. 1983). Only in polypterids has their presence been reported in the joints between adjacent lepidotrichial elements (Géraudie 1988). In the sarcopterygian lineage, actinotrichia are present in coelacanths (Géraudie and Meunier 1980) and in young dipnoans (Arratia et al. 2001).

Actinotrichia are nonmineralized elements that consist of a special type of collagen called elastoidin (Géraudie and Meunier 1982; Marí-Beffa et al. 1989; Francillon-Vieillot et al. 1990). The collagenous nature of elastoidin has since long been supported by histochemical and biochemical studies (N. E. Kemp 1977 and references therein; Becerra et al. 1983; Marí-Beffa et al. 1989); a second, noncollagenous, glycosylated, tyrosine- and cysteine-rich protein has been detected (Francillon-Vieillot et al. 1990). Actinotrichia exhibit a regular cross-banding, with a periodicity varying from 49 to 65 nm, according to the species studied (Géraudie and Landis 1982). The polarity of the fibrils is not necessarily the same within an actinotrichium (N. E. Kemp and Park 1970).

Ceratotrichia

Ceratotrichia, the fin rays supporting the fins of chondrichthyans (N. E. Kemp 1977; Starck 1979; Francillon-Vieillot et al. 1990), are flexible fibrous rods, unsegmented and only rarely branched (Starck 1979). They are longer and larger than actinotrichia (Francillon-Vieillot et al. 1990).

The prefix "cerato" is misleading, since ceratotrichia are not horny. Like actinotrichia, they are nonmineralized and made up of elastoidin (Francillon-Vieillot et al. 1990). The ceratotrichia of the fins in adult elasmobranch fins contain large amounts of tyrosine and cysteine (Géraudie 1977, and references therein). The periodicity of the banding pattern in shark elastoidin is like that of conventional collagen fibrils (i.e., 64 nm; N. E. Kemp 1977). Remarkably, whereas ordinary collagen fibrils can have opposite polarity (opposing anisometry), the polarity in ceratotrichia is always the same (N. E. Kemp 1977).

Lepidotrichia

Lepidotrichia (soft fin rays) occur in Crossopterygii and Actinopterygii (Starck 1979) but not in Dipnoi (Francillon-Vieillot et al. 1990). They are absent in the teleost adipose fin, which is solely supported by actinotrichia (Géraudie and Landis 1982).

Lepidotrichia are usually flexible rods, composed of successive segments, separated by gaps representing joints (plates 6.2, 6.8). Only the basal part that articulates with the endoskeleton remains unsegmented. Each segment consists of two demirays (N. E. Kemp and Park 1970) or hemisegments (Lanzing 1976). Advanced teleosts develop part of their fin rays into long, stiff elements devoid of joints and supposedly derived from the basal (unsegmented) part of the lepidotrichia. These constitute the so-called spiny fin rays that typify the fins of Acanthopterygii (plates 6.1, 6.7). Spiny fin rays are to be distinguished from yet another category of fin rays, the spines of cyprinids and silurids. These two families possess a small number of lepidotrichia that have lost their segmented appearance; their demirays can be ligamentously connected or fused (Bertin 1958; Drucker and Lauder 2001a).

True lepidotrichia are usually dichotomously branching toward the margin of the fin (Géraudie and Landis 1982; Becerra et al. 1983). Branching may differ in closely related species: in the brachiopterygian *Calamoichthys calabaricus*, the lepidotrichia are not dichotomized, whereas in its relative *Polypterus senegalus*, they are (Géraudie 1988). Also, within one species, branched and unbranched lepidotrichia can coexist within one fin, although some lepidotrichia never branch (e.g., in the caudal fin in zebrafish; Géraudie et al. 1995).

Lepidotrichia are composed of mineralized bone, and the matrix probably contains collagen type I, as in regular bone matrix (Géraudie and Landis 1982). The orientation of the collagen defines different zones in the lepidotrichia. In polypterids, the outer zone is made up of woven bone, whereas the core of the lepidotrichium is made up of lamellar bone (Géraudie 1988). Becerra et al. (1983) identified two layers with different collagen fibril orientation in several teleost species. In *Tilapia*, three zones can be recognized: an outer zone with parallel running fibers, a transition zone, and an inner woven-fibered zone (the oldest part; Lanzing 1976).

Like other bone in the osteichthyan skeleton, the bone making up lepidotrichia can be acellular—anosteocytic, without enclosed osteocytes—or cellular—osteocytic, with osteocytes in the matrix (Moss 1961a, 1961b, 1963; Weiss

and Watabe 1979; Ekanayake and Hall 1987, 1988; Meunier and Huysseune 1992; Huysseune 2000). The lepidotrichia of polypterids, for example, are cellular, although the cells are not numerous (Géraudie 1988). Becerra et al. (1983) suggest that the cellularity (the amount of enclosed osteocytes) depends on the thickness of the hemisegment. This is in line with observations (see below) that young cellular bone often is acellular (Huysseune 2000). Besides osteocytes, vascular canals can occur in the lepidotrichial matrix; they are numerous in the lepidotrichia of polypterids (Géraudie 1988).

In *Lepisosteus, Polypterus*, and *Calamoichthys* (Géraudie and Landis 1982; Géraudie 1988), patches of ganoine (an enamel substance, cf. Huysseune and Sire 1998) cover the lepidotrichia. This finding has been used as an argument in favor of the hypothesis that lepidotrichia derive from scales.

Camptotrichia

Dipnoans possess camptotrichia instead of lepidotrichia (Géraudie and Meunier 1982; Arratia et al. 2001). Camptotrichia are straight cylindrical rods, tapering and branching distally (Francillon-Vieillot et al. 1990). In the dorsal fin of adult *Protopterus*, the camptotrichia form two parallel rows, separated by mesenchymal cells (Géraudie and Meunier 1982). Two zones have been recognized in these camptotrichia: a superficial region of acellular fibrous bone, and a deep, mostly unmineralized zone (Géraudie and Meunier 1982).

Pterygiophora

Radialia of the dorsal, anal, and paired fins, and epurals and hypurals from the caudal fin, are primarily cartilaginous elements (plates 6.3, 6.9). Pterygiophora in Osteichthyes become surrounded by perichondral bone and may eventually be replaced by endochondral bone. In the unpaired fins, usually only the central parts of the basal and middle radial ossify by perichondral bone formation. The distal radial, which articulates with the basis of the lepidotrichia, often remains entirely cartilaginous (Stiassny 2000). In addition, in all fins, apolamellae without cartilaginous precursor (also called *zuwachsknochen* or secondary membrane bone) can develop as outgrowths of the perichondral bone (Patterson 1977; Starck 1979; Witten and Villwock 1997).

Many of the characteristics of chondrichthyan and osteichthyan skeletal tissues (matrix and cellular constituents) have been reviewed recently by Huysseune (2000). Hence, below we only briefly consider the most important facts.

The median fin pterygiophores and the caudal fin hypurals are initially made up of hyaline cartilage, usually cell-rich hyaline cartilage (CRHC; plate 6.10); this is the cartilage type that can undergo perichondral and subsequent endochondral ossification (Benjamin 1990; Benjamin et al. 1992). The hyaline cartilage of osteichthyans much resembles that of mammals (see, e.g., Muir 1995, for a review). Matrix components and skeletal cell enzymes largely correspond in teleosts and in tetrapods. Not all components have been explicitly identified in teleost fins, but basic skeletal components should also be present in the fin skeleton. Identified components of teleost cartilage matrix are chondroitin-6-sulphate, hyaluronic acid, collagen type II, and chondromodulin-1 (Huysseune 1989; Hinchliffe et al. 2000; Weiss Sachdev et al. 2001; Witten and Hall 2002); chondroitin-6-sulphate and collagen type II have been identified in cartilage of zebrafish pectoral fins (Hinchliffe et al. 2000; Weiss Sachdev et al. 2001). In sharks, in addition to collagen type II, the "bone" characteristic collagen type I is also a principal cartilage matrix component (Rama and Chandrakasan 1984; Sivakumar and Chandrakasan 1998). Likewise, pectoral fin cartilage in the skate *Raja kenojei* contains collagen type I and II as major, and type XI as a minor collagen (Mizuta et al. 2003).

In osteichthyans, including tetrapods, bone is in principle composed of three fundamental constituents: an organic matrix, a mineral phase, and cells (osteoblasts, osteocytes, and osteoclasts). The organic matrix is predominantly collagenous. The orientation of the collagen fibrils defines the major categories of bone in osteichthyans: woven-fibered bone, with random orientation of the fibrils; parallel-fibered bone, in which the fibrils show a predominant orientation, usually in the long axis of the bone; and finally, though much rarer, lamellar bone, which is characterized by lamellae in which all the collagen fibrils show the same orientation, while the fibril orientation changes from one lamella to the other (Meunier 1983; Francillon-Vieillot et al. 1990; Huysseune 2000; Witten and Hall 2002). The bone matrix of teleosts furthermore contains osteocalcin (bone Gla-protein), a vitamin K–dependent, calcium-binding protein required for adequate maturation of the hydroxyapatite crystal (Pinto et al. 2001). Also, SPP1 (osteopontin) was recently identified in teleosts and suggested to play a role in hydroxyapatite crystallization (Kawasaki et al. 2004).

The mineral phase of the fish endoskeleton (including calcifying cartilage of chondrichthyans) and the dermal skeleton, is mainly composed of calcium phosphate forming hydroxyapatite crystals (Hamada et al. 1995; Diekwisch et al. 2002; A. Kemp 2002; Takenaka et al. 2003). As in mammals, bone-forming cells (osteoblasts) in teleosts express high levels of alkaline phosphatase (Miyake et al. 1997; Witten 1997; Witten et al. 2001). Accordingly, alkaline phosphatase-positive osteoblasts have been observed to secrete the perichondral bone of the hypurals (plate 6.11, and see Witten and Villwock 1997). Bone resorbing cells (osteoclasts) are

also involved in teleost skeletal development and express naphthyl acetate esterase (as precursor cells), tartrate resistant acid phosphatase, and V-ATPase (Witten 1997; Witten et al. 1999, 2000), enzymes also present in mammalian osteoclasts (Connor et al. 1995; B. S. Lee et al. 1996; Ballanti et al. 1997; Halleen and Ranta 2001).

Incorporated bone cells (osteocytes) are found in cellular bone only and show variable levels of canaliculi number and distribution. Both cellular and acellular bone contain vascular canals of variable diameter (see Meunier 1983; Francillon-Vieillot et al. 1990; Meunier and Huysseune 1992; Huysseune 2000, for more details).

Chondrichthyans lack bone in the endoskeleton (but see Peignoux-Deville et al. 1982, 1985; Bordat 1987, for a different opinion) and thus have no bone component in the fin endoskeleton either. Bone, dentin, and enamel only occur in the dermal skeleton (placoid scales and teeth). In the fins, dermal bones are restricted to the placoid scales that cover the fins.

It is important to note that fish, particularly bony fish, display an astounding diversity of skeletal tissues, consisting of various types of cartilage and bone, as well as tissues with characteristics intermediate between cartilage and bone and between bone and dentin (plate 6.12, and see Stéphan 1900; Blanc 1953; Beresford 1981, 1993; Meunier 1989; Benjamin 1990; M. M. Smith and Hall 1990; Benjamin et al. 1992; Huysseune and Sire 1990, 1998; Huysseune 2000; Witten and Hall 2002; B. K. Hall and Witten, forthcoming).

The widest variety of skeletal tissues in teleosts is found in the head. Although the variety of skeletal tissues that contributes to the fin endoskeleton in bony fish is restricted, two types of cartilage (different from hyaline cartilage) are frequently found in the dorsal and caudal fin in representatives of several orders of teleosts (Benjamin et al. 1992). Fibro/cell-rich cartilage (FCRC), a highly cellular fibrocartilage, is found on the articulating surface between pterygiophores and fin rays (Benjamin et al. 1992) or is attached to the lepidotrichia (Ferreira et al. 1999). Hyaline-cell cartilage (HCC) forms expansions of the periosteum of the hypurals in some species and may develop as a secondary cartilage (Benjamin et al. 1992). In contrast to the hyaline cartilage that makes up the skeletal precursor of the radials (plates 6.9, 6.12), none of these cartilage types undergoes endochondral ossification (Benjamin et al. 1992). In addition to these cartilage types, chondroid bone has been reported by Benjamin et al. (1992) to occur beneath the articular surfaces of both dorsal fin pterygiophores and fin rays, especially in cypriniforms. In accordance with the existence of such a continuum of skeletal tissues, the matrix that composes the tissue is sometimes difficult to define as cartilaginous or bony, and its cells are difficult to assign to chondrocytic or osteoblastic/osteocytic phenotypes. Attempts, albeit only in the head region of teleosts, have used histochemical or immunohistochemical techniques (Huysseune 1989; Huysseune and Verraes 1990; Witten and Hall 2002).

Girdle Elements

The tissues that compose the girdle elements are in principle the same as those encountered in the pterygiophores. So cartilage (predominantly hyaline cartilage) initially makes up the elements of the endoskeleton (the scapula, coracoid, and mesocoracoid in the shoulder girdle, and the basipterygia in the pelvic girdle). In contrast, dermal bone (bone with no cartilaginous precursor) makes up the bones of the dermal skeleton (which is exclusively associated with the pectoral girdle). The cartilage frequently undergoes perichondral ossification (similar to the situation shown in plate 6.10), which can be followed by the deposition of bone trabeculae inside the bone tube (e.g., Ferreira et al. 1999). As in pterygiophores, ample membrane bone arises from the perichondral bone collar, especially in the pelvic fins (plates 6.5, 6.6).

The Pattern of Skeletal Tissue Formation in Fins

The sequence of development of fin skeletal elements in different primitive and derived teleost species is largely conserved (Faustino and Power 1999) and can start before (e.g., in *Oryzias latipes*) or after hatching (e.g., in *Dentex;* Iwamatsu 1994; Koumoundouros et al. 2001).

First, the fin fold forms, followed by the actinotrichia, which develop first in the fin bud apex and later reach to the base of the fin fold (Géraudie 1977; Géraudie and Landis 1982). In adult bony fish, actinotrichia persist at the distal margin of the fin (Starck 1979; Géraudie and Landis 1982; Francillon-Vieillot et al. 1990). The formation of actinotrichia is followed by the development of the cartilaginous endoskeletal elements (radials, epurals, and hypurals, later ossifying by perichondral bone formation) and usually precedes formation of the bony fin rays (lepidotrichia; Géraudie and Landis 1982; Cubbage and Mabee 1996; Borday et al. 2001; Mabee et al. 2002).

Mabee et al. (2002) summarized much of the literature available on patterns of skeletal differentiation in actinopterygian fins and found the following consistent patterns: (1) the rays (lepidotrichia) and distal radials of the pectoral fins develop from anterior to posterior (dorsal to ventral) in all actinopterygians; the rays of the pelvic fins also develop from anterior to posterior (distal to proximal), with a few exceptions; (2) the rays and radials (hypurals) of the caudal fin develop bidirectionally in all actinopterygians; and (3)

the rays and radials differentiate in a similar direction in the dorsal and anal fins in all actinopterygians except for engraulids. Most notably, the direction of development of fin rays (lepidotrichia) mirrors that of the endoskeletal supports (radials, hypurals) in all fins and for all species for which both were reported. Mabee et al. (2002) consider this shared direction of differentiation as a phylogenetically conserved module indicating a high level of integration between endoskeleton and dermal skeleton. It does not imply, however, that the development of the endoskeletal element necessarily precedes that of the lepidotrichia. In the pelvic fins of zebrafish, for example, lepidotrichia ossify before the radials develop (Cubbage and Mabee 1996; Grandel and Schulte-Merker 1998), while in rainbow trout pelvic fins, radials are not required for lepidotrichial morphogenesis (Géraudie 1981).

Despite some variation, the appearance of skeletal elements of the fin skeleton in ontogeny in general still coincides with the appearance of these elements in the course of evolution: (1) elastoidin fibers were the first elements that stabilized the fin fold in Cephalochordata; (2) cartilaginous pterygiophora (radialia) were the next elements to evolve in Craniata; and (3) lepidotrichia arose in Osteichthyes (Mabee et al. 2002).

In our further considerations, we deviate from the natural sequence in which fin skeletal elements appear and discuss the histogenesis of the elements of the dermal skeleton and the endoskeleton separately.

Development of the Fin Dermal Skeleton

Development of the Fin Fold and Colonization by Skeletogenic Cells

The unpaired dorsal, anal, and caudal fins arise in an initially continuous median larval fin fold, a phylogenetically conserved structure that evolved already in craniate ancestors (Mabee et al. 2002). Whether the fin fold that precedes the paired fins represents the rudiment of a lateral fin fold from which the paired fins have evolved is still discussed (Schaeffer 1977; B. K. Hall 1998; M. Tanaka et al. 2002). As the pectoral and pelvic fins develop, the intervening fold disappears (Schaeffer 1977).

The early fin bud of teleost paired fins first shows a thickened epidermis, which subsequently folds into a closely apposed epidermal bilayer, the fin fold, with a hairpin shape on transverse section (Hinchliffe 2002). There is evidence that the process that leads to the formation of the folded fin epithelium in the median fins and in the pectoral fin is similar (van Eeden et al. 1996). Interestingly, a proper differentiation of the larval fin fold does not seem to be a prerequisite for the formation of adult fin structures (van Eeden et al. 1996).

At the beginning of the formation of the dermal skeleton, but after initiation of actinotrichial development (Géraudie 1977), mesenchyme cells migrate into the fin fold, as early as 48 hours postfertilization in zebrafish (Grandel and Schulte-Merker 1998). The source of these mesenchymal cells is still a matter of speculation. According to Géraudie and François (1973) and Schaeffer (1977), they derive, in paired fins, from thickenings of the somatic layers of the lateral plate mesoderm and undergo interactions with the overlying epithelium. In contrast, trunk neural crest may induce the median fin fold and provide the mesenchyme occupying the center of the fold (Schaeffer 1977). Experimental evidence for the contribution of the trunk neural crest to the fin-fold mesenchyme is summarized by Balinsky (1975), who cites among others Du Shane (1935) and his experiments in salamander embryos. According to Du Shane (1935), removal of the neural crest tissue prevents the development of a fin fold, whereas translocation of the trunk neural crest tissue to another side of the body induces the development of a dislocated fin fold at the site of tissue implantation. The assumption that all scales are neural crest–derived, and the assumed homology between scales and lepidotrichia, indeed implies that lepidotrichia are neural crest–derived structures (B. K. Hall 1998). M. M. Smith et al. (1994) provided the first, and so far only, experimental evidence that cells of neural crest origin populate the caudal fin in the zebrafish and thus potentially contribute to lepidotrichial formation.

Formation of Elastoidin Fibers (Actinotrichia, Ceratotrichia)

In both median and paired fin folds, actinotrichia begin to develop in the distal part of the fold (van Eeden et al. 1996; Mabee et al. 2002). Bouvet (1974a) and Géraudie (1977, 1980, 1981) conclude from their studies that actinotrichia—unlike ceratotricha in sharks (N. E. Kemp 1977) and unlike the succeeding lepidotrichia—develop initially without the participation of mesenchymal cells and that their collagen is exclusively secreted by epidermal cells. Mesenchymal cells might nevertheless regulate the final growth in length and width of the actinotrichia (Géraudie 1977). Actinotrichia may be used as a substrate for migrating mesenchymal cells (A. Wood and Thorogood 1984; Marí-Beffa et al. 1989). In the brachiopterygians *Polypterus senegalus* and *Calamoichthys calabaricus*, separate actinotrichia likely fuse to give larger units; actinotrichia in these species also grow by accretion of collagen fibrils at their surface (Géraudie 1988).

Forming ceratotrichia are surrounded by a peritrichial layer composed of fibroblasts, not different from ordinary connective tissue fibroblasts (N. E. Kemp 1977). Ceratotrichia grow by apposition of collagen fibrils from the peritrichial matrix and perhaps also by direct intercalation of tropocollagen molecules. Unlike what could be expected from their composition, N. E. Kemp (1977) found no morphological evidence for incorporation of extrafibrillar matrix (such as noncollagenous proteins, or carbohydrates) into the ceratotrichium.

The process whereby forming fin rays are separated from the basal lamina by mesenchymal cells, to sink in, can be repeated, leading to more than one generation of fin rays existing in parallel. Consequently sharks possess both outer and inner layers of ceratotrichia (Starck 1979).

Formation of Lepidotrichia

Lepidotrichia differentiate within the extracellular collagenous basal lamella of the epidermal-dermal interface and subsequently become progressively separated from the epidermis by the invasion of mesenchymal cells (Starck 1979; Géraudie and Landis 1982; Sire and Huysseune 2003). According to Géraudie and Landis (1982) both epidermal and mesenchymal cells are involved in the process of lepidotrichial differentiation; Bouvet (1974b) believes that the collagenous lepidotrichial matrix is formed by mesenchymal cells and that the basal lamella is not involved. These apparently conflicting views might well be reconciled by assuming that fin rays develop from mesenchymal cells as a result of epithelial-mesenchymal interactions (Sire and Huysseune 2003).

Mineral is deposited in areas where a fine granular ground substance is present between the collagen fibrils, an observation that frequently has been made in teleost bone mineralization (e.g., Huysseune and Sire 1992). The mineral appears to be deposited in association with the collagen fibrils during initial and subsequent steps of mineralization (Géraudie and Landis 1982).

When lepidotrichia first appear, they are composed of acellular bone. Later, osteoblasts can become incorporated to form cellular bone around the initial acellular matrix (Meunier 1983). Lengthening of lepidotrichia is accomplished by the distal addition of segments (Iovine and Johnson 2000; Borday et al. 2001; Akimenko et al. 2003). Unusually long fins, as in the *long-fin* zebrafish mutant, have more, but not significantly longer, lepidotrichial segments (Géraudie et al. 1995). Collagen fibrils constituting the ligament between successive lepidotrichial segments can become mineralized and so participate in bone formation of the lepidotrichium (Géraudie 1988). The way joints form histologically has not been reported to our knowledge, although a gene that may be involved in controlling segment boundary formation has been identified, *evx1* (Borday et al. 2001). Furthermore, segment length and segment number are controlled through different genetic pathways (Iovine and Johnson 2000). There is also evidence to suggest that actinotrichia are capable of being incorporated into lepidotrichia in teleosts, as well as in brachiopterygians and in dipnoans (Géraudie and Landis 1982). An intriguing question that remains unanswered so far is what governs the remarkably synchronous morphogenesis in the two demirays.

Formation of Dermal Bones

The formation of dermal bones associated with the shoulder girdle proceeds according to a stereotypic pattern. First, a condensation of osteogenic cells forms. As these cells differentiate into osteoblasts, they begin to secrete unmineralized matrix in an initially polarized way, delimiting the first visible anlage of the bone. Mineralization of the matrix then takes place, leaving an osteoid seam on the surface. The osteoid rim can be narrow or wider, depending on the balance between speed of deposition of new matrix and speed of matrix mineralization. Often, the bone shows a preferential side of deposition, which is reflected in the characteristics of the osteoblasts: large, plump, and highly basophilic cells along the deposition side, very thin and elongated (resting cells) along the quiescent side. Notably, dermal bones, like the cleithrum, are among the first skeletal elements to ossify in larvae (plate 6.13). The reason for the early ossification of the cleithrum is likely related not to its function as pelvic fin support but to its involvement in the mechanism of mouth opening (Koumoundouros et al. 2001). In general, the bones that form first in the larva are those associated to movement and subjected to stresses, and the pattern of ossification follows the application of mechanical stress (Weisel 1967).

Are Lepidotrichia and Scales Homologous Structures?

Several arguments, including the fact that each segment of a lepidotrichium consists of two demirays or hemisegments, have led to the hypothesis that lepidotrichia evolved from scales (Jarvik 1959). The occurrence of ganoine on the lepidotrichia in lepisosteids and polypterids (Géraudie 1988; Francillon-Vieillot et al. 1990) has added further arguments, as have observations on secretion and mineralization of the lepidotrichial matrix (Lanzing 1976).

Schaeffer (1977) proposed that scales and lepidotrichia could be structurally similar because they might be manifestations of a single, modifiable, morphogenetic system. The idea of a "modifiable morphogenetic system" implies that lepidotrichia and scales display similarities because both structures derived from ancestral dermal skeletal elements

through a conserved developmental mechanism, involving epithelio-mesenchymal interactions. Although Schaeffer's ideas were meant to raise doubts about the homology between lepidotrichia and scales, he provides a possible mechanism for their development, rather than arguments against their homology (shared developmental mechanisms are often considered an argument in favor of homology). In our view the seemingly opposing views of Jarvik (1959) and Schaeffer (1977) could well be reconciled by addressing the question of whether lepidotrichia evolved from primitive or more derived dermal skeletal elements.

Based on developmental and structural data, Sire and Huysseune (2003) consider lepidotrichial anlagen to share properties with osteogenic dermal papillae such as those putatively involved in the formation of the deep part of primitive (e.g., ganoid) scales. Ancestral dermal skeletal elements, such as the scales found on the trunk of Ordovician vertebrates, were composed of a deep osteogenic component, consisting of bone, and a superficial odontogenic component, consisting of dentine and/or ganoine. During vertebrate evolution, the superficial, odontogenic component of scales was considerably modified or lost (Huysseune and Sire 1998). Sire and Huysseune (2003) propose that the lepidotrichia represent the deep osteogenic component of the ancestral scale, while the superficial, odontogenic component was lost.

Are Actinotrichia and Lepidotrichia Homologous Structures?

Most authors agree on the homology of actinotrichia and ceratotrichia, since elastoidin fibers evolved first in the median fin fold of cephalochordates and are retained in basal gnathostomes such as chondrichthyans (Mabee et al. 2002). Can elastoidin fibers and lepidotrichia also be considered as homologous structures? Starck (1979) and other authors (see also Goodrich 1904; Jarvik 1959) argue for homology and summarize lepidotrichia and elastoidin fibers (actinotrichia) under the term "dermotrichia." The assumption that scales and lepidotrichia derive from neural crest cells (M. M. Smith et al. 1994) raises, however, questions about the homology of elastoidin fibers and lepidotrichia; elastoidin fibers are already present in the fin fold of ancient cephalochordates (Mabee et al. 2002), a group that lacks a neural crest (B. K. Hall 2000). Furthermore, lepidotrichia (but not elastoidin fibers) are mesenchymal- (likely neural crest–) derived structures, whereas the first actinotrichia form without the participation of mesenchyme; only later do mesenchymal cells contribute (Géraudie 1977). Observations that zebrafish mutants that lack the embryonic fin fold (and actinotrichia) can develop regular fins (van Eeden et al. 1996) also suggest that elastoidin fibers and lepidotrichia derive from different sources and are under different genetic control. Finally, unlike lepidotrichia, elastoidin fibers are not segmented (Géraudie and Meunier 1980). These differences between elastoidin fibers and lepidotrichia suggest that bony fin rays (lepidotrichia) are a new character of Osteichthyes and not a continuation of preexisting elastoidin fibers.

Development of the Fin Endoskeleton

Origin of the Skeletogenic Cells

The origin of cells that contribute to the formation of the fin endoskeleton has not been elucidated. Géraudie and François (1973) traced the origin of the trout pelvic fin endoskeleton and pelvic girdle back to mesenchymal cells of somitic origin, but they did not exclude a possible contribution from the lateral plate mesoderm. Given the supposed homology of vertebrate girdles, one must assume that the cells that give rise to the endoskeletal girdle elements indeed derive from somitic mesoderm, as they do (at least partly) in tetrapods (Burke 1989; Huang et al. 2000). However, based on the homology between paired fins and limbs, the paired fin endoskeleton (the radials) supposedly derives from lateral plate mesoderm. Given the intimate developmental association between girdle and endoskeleton in the paired fins (see below), this supposed dual origin obviously raises a problem.

There is less ambiguity concerning the unpaired fins. Accepting that the radials of unpaired fins evolved from the processes of neural and hemal arches (Starck 1979), the scleroblasts that form the endoskeleton of median fins would derive from trunk somitic mesoderm.

Genetic and Molecular Control of Cartilage and Bone Formation

While most of what we know about the interplay of genes and gene products (transcription factors, growth factors, and their receptors) involved in skeletal development derives from studies on tetrapods, more and more data from teleosts are becoming available. Many data derive from fin regeneration experiments and relate in particular to fin ray (dermal bone) development (see Akimenko and Smith, chap. 11 in this volume; Géraudie et al. 1995; Laforest et al. 1998; Quint et al. 2002; Akimenko et al. 2003). Data concerning factors that regulate the development of the fin endoskeleton largely derive from studies on pectoral fins, since much interest is paid to the paired fins in connection with tetrapod limb evolution (Grandel and Schulte-Merker 1998; Jeffery 2001; Tanaka et al. 2002; Mabee et al. 2002).

Given that paired fins and limbs are homologous appendages (Hinchliffe 2002; Tanaka et al. 2002), most authors agree on the presence of an AER or an AER-like zone in teleosts, preceding the development of the paired fins. There is, however, some confusion in the literature whether the term "AER" refers to the thickened epithelium prior to the formation of the fin fold (Wood 1982) or to the epithelial bilayer (fin fold) proper (the apical fold of Grandel and Schulte-Merker 1998; Grandel et al. 2000). Géraudie (1980) calls the fin fold a "pseudo-apical ridge" to stress the differences with tetrapods. Grandel and Schulte-Merker (1998) and Grandel et al. (2000) reserve the term "fin fold" for the apical fold once invaded by mesenchymal cells.

Regardless of whether or not teleosts have a true AER or an AER-like structure, factors that regulate fin patterning—whether dorsal/ventral, proximal/distal, and anterior/posterior patterning—also affect the differentiation of skeletal elements. As in tetrapods, retinoic acid triggers the expression of genes that are essential for pectoral fin development (Akimenko and Ekker 1995; Suzuki et al. 2000), and, as in the early tetrapod limb, signaling centers are set up that involve genes like *dlx*s, *fgf*s, *shh*, and *en1* (Akimenko et al. 1994; Grandel et al. 2000). In zebrafish, a subset of cells in the posterior margin of pectoral fin buds expresses *shh*, a pattern consistent with the role of *shh* in mediating the zone of polarizing activity (ZPA) in tetrapods (Akimenko and Ekker 1995). Unlike in teleosts, *shh* expression has not been detected in the chondrichthyan fin bud (Tanaka et al. 2002). In line with these findings, Sagai et al. (2004) have recently identified a cis-acting regulatory sequence of *Shh*, highly conserved among teleosts and in tetrapods, but absent in cartilaginous fishes. Still, *dHand*, a gene involved in the regulation of *shh* expression, is expressed in the posterior part of the paired fin buds and the posterior margins of the dorsal and anal fin buds during dogfish (*Scyliorhinus canicula*) development (Tanaka et al. 2002). Noggin, another ZPA-related factor (a secreted protein functioning as BMP antagonist), is expressed in cartilage precursor cells of the zebrafish pectoral fin anlage (H. Bauer et al. 1998). As in tetrapods, early chondrogenic condensations appear to depend on expression of cadherins. Blocking of cadherin-2 (*cdh2*) via morpholino injection prevents chondrogenic condensation and endoskeleton formation in the developing pectoral fin buds in zebrafish (Q. Liu et al. 2003).

Despite similarities, evident structural differences between the limb and the fin skeleton exist and are reflected in divergent gene expression patterns. The early expression of *Hoxa* and *Hoxd* genes is characteristic for limb and fin buds, whereas in later developmental stages the distal-anterior expansion of *Hoxd* is restricted to tetrapod limb buds, a pattern that reflects the neomorphic character of the tetrapod autopod (Sordino et al. 1995; Coates and Cohn 1998; Gardiner et al. 1998). Furthermore, two members of the *Msx* gene family are expressed during tetrapod limb formation, but four members are expressed during zebrafish fin development (Akimenko et al. 1995).

With respect to the process of differentiation of cells into scleroblasts (e.g., chondroblasts, osteoblasts) and subsequent differentiation of skeletal tissues, we have reason to assume that the early developmental processes in amniotes and bony fish largely follow corresponding pathways (B. K. Hall and Miyake 1995; H. Bauer et al. 1998; Inohaya and Kudo 2000; B. K. Hall 2000; Olsen et al. 2000; Crotwell et al. 2001). B. K. Hall and Miyake (1995, 2000) reviewed the involvement of *Hox* genes and other transcription factors, growth factors, cell surface receptors, cell adhesion molecules, and extracellular matrix molecules in the four basic processes of skeletal cell and tissue differentiation: migration of skeletogenic cells to the site of future skeletogenesis, epithelial-mesenchymal interactions, cell condensation, and cell differentiation. Several factors reviewed by these authors have been diagnosed in teleosts, and some factors can be linked to the development of the fin endoskeleton.

TGF-β superfamily members expressed during zebrafish fin development are *contact, bmp2b, bmp4,* and *gdf5* (growth and differentiation factor 5; Bruneau et al. 1997; H. Bauer et al. 1998; Crotwell et al. 2001). Transcripts of *contact* are found in the precartilaginous condensation and in the differentiating scapulocoracoid cartilage of the pectoral girdle (Bruneau et al. 1997). *bmp2b* and *bmp4* are expressed in the precartilaginous plate of the pectoral fin in a complementary fashion to the BMP antagonist Noggin (H. Bauer et al. 1998). *Gdf5* is expressed between cartilage condensations of the anal, dorsal, and caudal fin and around the tips of developing cartilages, consistent with a role for this factor in cartilage growth and differentiation (Crotwell et al. 2001). *Gdf5*, but also *bmp2b* (Crotwell et al. 2004), is expressed in the segmenting regions of the radials, and therefore both may play a role in segmentation, as their orthologues do in tetrapod joints. Other genes or gene products identified as involved in skeletal development in teleosts include chondromudulin-1 (chm-1), *sox9* and *cbfa1/runx2*. ChM-I is a 25 kDa secreted glycoprotein that stimulates the growth and maturation of chondrocytes and inhibits vascular invasion into cartilage during endochondral bone formation in birds and mammals. The glycoprotein is present in the chondrogenic region of the pectoral fin in zebrafish (Weiss Sachdev et al. 2001).

Sox9 regulates the expression of collagen II in developing cartilage in mice, and the presence of *sox9* transcripts in the developing cartilage of the medaka (*Oryzias latipes*) pectoral fin (Yokoi et al. 2002) suggests a similar role in teleosts.

Cbfa1/Runx2, another key factor in the process of skeletal formation, is required for ossification. *Cbfa1* expression has been identified by *in situ* hybridization in immature osteoblasts and chondrocytes of developing skeletal elements in *Oryzias latipes* (Inohaya and Kudo 2000).

Most similarities between teleosts and tetrapods with respect to the genetic regulation of the skeleton, skeletal matrix composition, and enzymatic activity of skeletal cells are found in early developmental stages. As development of the skeleton progresses, teleosts and tetrapods differ, and not only because of the development of the dermal skeleton (fin rays) and of the autopodium, respectively. In teleosts, a wide variety of skeletal tissues can develop (Huysseune 2000). In derived teleosts, osteocytes are not enclosed in the bone matrix (acellular bone; cf. the section "Tissue Composition of Paired and Unpaired Fins"). Moreover, bone resorption can rely on mononucleated osteoclasts (instead of the multinucleated giant cells seen in mammals), although growth of the endoskeleton of unpaired fins can occur largely without bone resorption (Witten 1997; Witten and Villwock 1997).

Although apoptosis has been reported in the distal periphery of the median fin fold (L. K. Cole and Ross 2001), the lack of apoptosis during differentiation of the radials in the zebrafish pectoral fins is another distinctive feature (Grandel and Schulte-Merker 1998). For a while, the cartilage of the fin endoskeleton—enclosed by a bone tube—does not undergo endochondral ossification. Finally, fish do not develop a bone marrow that contains hematopoietic tissue (Huysseune 2000; Witten et al. 2001; Witten and Hall 2002). These differences between teleost and tetrapod skeletons that surface in the course of skeletal development certainly require special control mechanisms in fish, on both genetic and cellular levels (Witten et al. 2001). However, so far, most knowledge regarding the regulation of late skeletal differentiation in teleosts has been gathered at the level of hormonal control (for further readings see, e.g., C. W. Taylor 1985; Persson et al. 1998; Sasayama 1999; Trivett et al. 1999).

Cartilage and Bone Histogenesis

Although the formation of cartilage in the fin endoskeleton has been intensely studied on the anatomical level (Alcian blue staining of whole mounts), astoundingly little information is available on cartilage formation at the cellular level. However, it may be assumed that the process largely follows the scheme that has been observed in other instances of cartilage formation in chondrichthyans and osteichthyans. As has been illustrated for the development of a cranial cartilage (Huysseune and Sire 1992), the first step is a blastema (condensation) stage, characterized by closely packed prechondroblasts with no evidence of cartilage matrix components.

In the subsequent (primordium) stage, the cells, now chondroblasts, are usually stacked close together, but little matrix components have been deposited. Finally, during the differentiation phase, these chondroblasts begin to secrete matrix and come to lie further apart (plate 6.12, and see Huysseune and Sire 1992). During this process, the cartilage cells show an increasing state of differentiation, as indicated by a decrease of the nucleo-cytoplasmic ratio, and an increase in the number of organelles associated with secretory activity.

Irrespective of whether perichondral, endochondral, or membranous bone apolamellae form, the cells involved in the process of bone formation in the endoskeleton are osteoblasts (plate 6.14). They derive from osteoprogenitor cells, the mesenchymal source of which has not been unequivocally identified in fish. Osteoblasts display various morphologies, depending on their secretory activity, but also on their position on the bone: pear-shaped, spindle-shaped, or cuboidal with a pseudo-epithelial arrangement (Huysseune 2000). Secretory osteoblasts have a polarized appearance, with a highly basophilic cytoplasm indicative of intensive protein production. Unlike in cellular bone, osteoblasts in acellular bone formation show a polarized secretion of osteoid matrix, continuously withdraw from the surface, and thus are never incorporated into the matrix (Weiss and Watabe 1979; Meunier 1983; Ekanayake and Hall 1987, 1988; Huysseune 2000).

In Osteichthyes, perichondral bone formation is the basic process of ossification of the fin endoskeleton (plate 6.10). Different from mammals, this process is not often coupled to endochondral bone formation (Witten and Villwock 1997; B. K. Hall 1998; Huysseune 2000). Perichondral bone is laid down at the immediate contact of the cartilaginous template by cells that were formerly part of the perichondrium but have now taken up the characteristics of osteoblasts and secrete the bone matrix. Nevertheless, an admixture of cartilage and bone matrix is not excluded (Huysseune and Sire 1992; Huysseune 2000; Verreijdt et al. 2002).

Upon the beginning of perichondral bone formation, the former perichondrium has become a periosteum, and further thickening of the bone is carried out through deposition of bone by the osteogenic cells in the periosteum, in a process that should now properly be called "periosteal ossification." Thus, a typical pterygiophore consists of a persisting cartilage rod inside a bone tube with cartilage sticking out as a condyle (plate 6.5). In older individuals, however, cartilage inside the bone tube can be resorbed, while cartilage remains at both ends of the bone shaft.

Replacement of cartilage by spongiosa (endochondral bone formation) can be observed in larger teleost species (e.g., carp, salmon; plate 6.15). The way this happens is not completely understood, but Haines (1934) and Huysseune

(2000) propose a model according to which, after the initiation of resorption and prior to further cartilage resorption, thin seams of bone are deposited on the resorption front. When the cartilage further retreats, these seams are left behind as bone trabeculae. This process creates a ladderlike construction of trabeculae crossing the bone tube (e.g., Ferreira et al. 1999), different from the arrangement of endochondral bone of tetrapods. As fish have no hematopoietic tissue inside the bone marrow, the remaining spaces are usually filled with fat tissue, besides nerves and blood vessels and some connective tissue cells (plate 6.16, and see Huysseune 2000; Witten et al. 2001; Witten and Hall 2002). The perichondral bone collar together with the endochondral bone trabeculae can be either cellular or acellular. If cellular bone is formed, it is not uncommon, as with lepidotrichia, that osteoblasts only become entrapped at a later stage.

Further growth process of the radials, including a considerable reshaping, involves not only the lengthening of the bony shafts but also formation of secondary membrane bone (Witten and Villwock 1997); *zuwachsknochen, sensu* Starck 1979 (plates 6.5–6.6, 6.10). This type of bone forms as apolamellae, in a way similar to dermal bones.

Since part of the caudal fin endoskeleton involves the posterior part of the vertebral column, some attention should also be paid to the ossification processes of vertebrae. Most elements of the caudal fin complex are cartilage bones (hypurals and epurals) and remain cartilaginous, even if the vertebral centra undergo direct ossification without a chondrogenic phase, as in many teleosts (M. M. Smith and Hall 1990; Koumoundouros et al. 1999; Gavaia et al. 2002). In chondrichthyans, the cartilage of the vertebral centra, and of the neural and hemal arches, does not ossify but undergoes more or less extensive mineralization, forming discrete calcified areas. Interestingly, the chondrocytes in these calcified areas neither hypertrophy nor die (Moss 1977).

Subdivision of Cartilages

One aspect of fin skeletal development that deserves special attention is the way cartilages of the endoskeleton develop into distinct anatomical units.

The fate of these cartilages differs: cartilage can become either subdivided, or cartilage anlagen can fuse. Cartilage subdivision plays a role in median (dorsal and anal) fin and in pectoral fin differentiation. The cartilage anlagen of the osteichthyan pelvic fin basipterygia, on the other hand, display fusion rather than subdivision. Depending on the species, cartilage anlagen in the pelvic fin can remain separated or can fuse into a single complex in various ways (Lagler et al. 1977). As can be seen in the pupfish *Cyprinodon macularius*, primarily separated cartilages, which together form the anlage of one basipterygium, can fuse into a single piece of cartilage, a process that is further completed through perichondral ossification and membrane bone formation (plates 6.4–6.6).

In the pectoral fins of teleosts, parts of the shoulder girdle and the radials develop through subdivision of a single cartilage condensation (plates 6.17–6.19). A continuation of the anlage of the shoulder girdle and the fin endoskeleton can also be observed in some sharks, and in primitive osteichthyans such as *Acipenser* and *Amia* (Grandel and Schulte-Merker 1998). To generate the elements of the pectoral girdle and fin endoskeleton in zebrafish, a cartilaginous plate is subdivided into scapula, coracoid, mesocoracoid, and four radials. Only distal radials are added as separately originating cartilaginous elements (Grandel and Schulte-Merker 1998), in a way similar to the addition of distal radials in the median fins (see below).

As in the zebrafish, the scapulocoracoid cartilage and the cartilaginous fin plate (i.e., the precursor of the proximal radials, or actinosts) in *Sparus aurata* seem to develop from one piece of cartilage, or are at least jointed (Faustino and Power 1999). In other species, two separate cartilage anlagen are reported to develop: a scapulocoracoid cartilage, which ossifies into the coracoid and the scapula (both nevertheless remaining joined by cartilage), and a cartilaginous blade that first cleaves in the center, then dorsal and ventral, to form the (proximal) radials, which also ossify (Potthoff and Kelley 1982; Potthoff et al. 1988; Balart 1995; Mabee and Trendler 1996; Cubbage and Mabee 1996). It may well be that apparent differences concerning the subdivision of cartilage anlagen are due to the techniques used and the number of stages observed. Indeed, Grandel and Schulte-Merker (1998) observed that the scapulocoracoid cartilage and the preradial disc are initially continuous in the zebrafish, and split up thereafter. On the other hand, and based on cleared and Alcian blue–stained specimens, Cubbage and Mabee (1996) report them as separate entities. We suspect that other instances of apparent separate development may be secondary, and that one continuous condensation may well be more common than reported (plates 6.17, 6.18). This is of high significance, both in terms of evolutionary relationship between girdle and endoskeleton and in terms of morphogenetic processes (see below).

Distal radials can originate from the margin of a continuous cartilaginous plate by subdivision (e.g., Potthoff et al. 1988), as separate cartilage condensations (e.g., Faustino and Power 1999; M. C. Davis et al. 2004), or through both mechanisms (Balart 1995). These elements usually remain cartilaginous. Similarly, a distal radial, or propterygium, which supports the dorsalmost fin ray, can arise from the scapular

cartilage (Potthoff and Kelley 1982), from the bladelike cartilage (Kohno and Taki 1983; Balart 1995), or as a separate anlage (Faustino and Power 1999). Its origin and function suggest that the propterygium is indeed a rudiment of a fifth radial.

The way the cartilage subdivides is intriguing. In the pectoral fin endoskeleton, this apparently occurs without apoptosis (Grandel and Schulte-Merker 1998). This is similar to the noninvolvement of apoptosis in digit separation in amphibians but different from amniotes, where massive cell death occurs to transform the webbed regions of the footplate into interdigital spaces (Cameron and Fallon 1977; W. Wood et al. 2000; Zuzarte-Luís and Hurlé 2002). We should, however, point out that unlike in tetrapods, fully differentiated cartilage has to be "removed" during fin skeletogenesis. In fact, several papers report the formation of elongated "holes" (e.g., Potthoff and Kelley 1982), or "crevices" (e.g., Faustino and Power 1999; Balart 1995) that then enlarge to the border of the cartilage blade. Grandel and Schulte-Merker (1998) report "decomposition" of the matrix.

Whatever the mechanism of subdivision, cartilage separation is followed by the formation of perichondral bone, coinciding with the expression of alkaline phosphatase in peripheral cartilage-like cells prior to and during bone formation (plate 6.20). This process could suggest that the former cartilage cells, and not new osteoblasts, mediate the bone formation. Consequently, in the absence of apoptosis, separation of cartilage elements may involve metaplasia: differentiation and dedifferentiation of cartilage cells.

The precursor of the endoskeleton of the dorsal and the anal fins (i.e., the radials) is a cartilage rod (plate 6.9, and see Potthoff 1975; Potthoff and Kelley 1982; Crotwell et al. 2001). As shown by Potthoff (1975) in a developmental study on black fin tuna (*Thunnus atlanticus*) the separation of the primary cartilage rod is caused by the formation of two centers of ossification, forming two independent bone shafts along the cartilage rod, which subsequently leads to the separation into two parts. It was believed that all three parts of a radial derive from one cartilage anlage, but new observations suggest that the initial radial is only separated into two parts (the proximal and the middle radial), whereas the distal radial (which remains cartilaginous) arises independently (Koumoundouros et al. 2001). In the swordfish, where there are only two radials, both appear as one piece of cartilage; then the distal radial separates from the proximal (Potthoff and Kelley 1982). Distal and proximal radial are however separated *ab initio* in *Pagrus major* (Kohno and Taki 1983). Again, some variation may be explained by the techniques used for observations: cleared and stained specimens provide other data than histological sections, and, in double

A

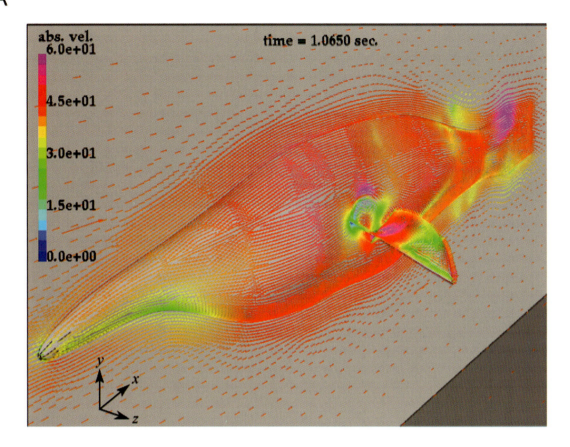

B

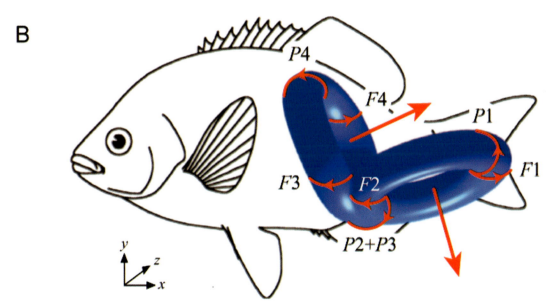

Plate 3.1. Two modern methods for estimating pectoral fin forces during swimming. (A) Computational fluid dynamic model of aquatic flapping flight by the bird wrasse (*Gomphosus;* after Ramamurti et al. 2002). The velocity of water flow over the body and pectoral fin at one instant in the fin beat cycle is represented by pseudocolors. Locomotor forces are determined in this theoretical model by integrating surface pressures over multiple time steps throughout the pectoral fin stroke. (B) Reconstruction of pectoral-fin wake morphology in freely swimming black surfperch (*Embiotoca*) by means of digital particle image velocimetry (from Drucker and Lauder 2000). Vortices shed by the left pectoral fin within the parasagittal and frontal planes are represented by curved arrows labeled P and F, respectively. Straight-line arrows signify fluid jets passing through toroidal vortex rings in the wake. In this empirical method, locomotor forces are calculated from the rate of change in wake momentum.

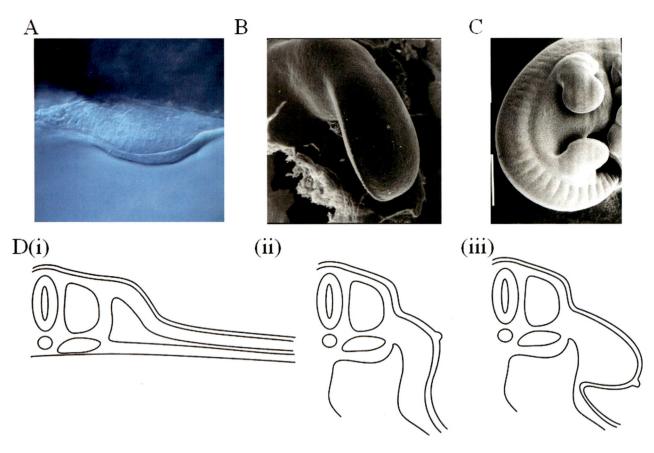

Plate 5.1. Fin/limb buds in three model vertebrates. (A) Zebrafish pectoral fin bud with apical ectodermal ridge (from Kimmel et al. 1995; photograph courtesy C. Kimmel). (B) Scanning EM of chick wing bud showing apical ectodermal ridge. (C) Scanning EM of mouse limb buds (photograph courtesy P. Martin). (D) Diagrams of the formation of the chick wing bud: transverse sections at wing level of chick embryos at stage (i) 16, (ii) 18, and (iii) 22.

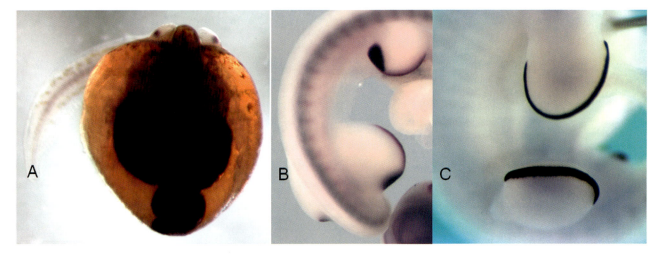

Plate 5.2. Expression of genes encoding signaling molecules in fin/limb bud development. (A) Expression pattern of *Shh* in stickleback fin buds (photograph courtesy N. Cole). (B) Expression pattern of *Shh* (posterior mescenchyme) and *Fgf8* (apical ectodermal ridge) in chick limb buds (photograph courtesy E. Tiecke). (C) Expression pattern of *Fgf8* (in apical ectodermal ridge) in mouse limb buds (photograph courtesy M. Eblaghie).

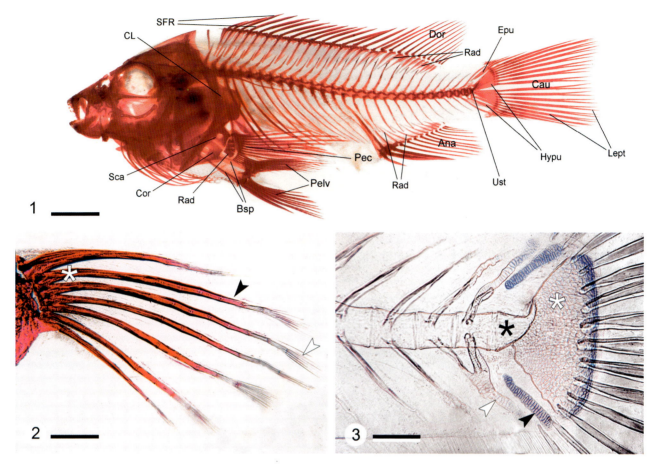

Plate 6.1. Skeleton of a juvenile Nile tilapia, *Oreochromis niloticus* (30 mm TL) showing extensively ossified fin dermal and endoskeleton elements in this advanced teleost. Ana, anal fin; Bsp, basipterygia/pelvic girdle; Cau, caudal fin; CL, cleithrum; Cor, coracoid; Dor, dorsal fin; Epu, epuralia; Hypu, hypuralia; Lept, lepidotrichia; Pec, pectoral fin; Pelv, pelvic fin; Rad, radialia/pterygiophora; Sca, scapula; SFR, spiny fin rays; Ust, urostyle. Whole mount staining with Alizarin red S, × 5, bar = 3 mm.

Plate 6.2. Pelvic fin of the Desert pupfish, *Cyprinodon macularius* (20 mm TL). The free part of the fin of osteichthyans is supported by segmented bony fin rays called lepidotrichia (black arrowhead), whereas from early ontogeny onward, the distal fin edge is stabilized by actinotrichia (white arrowhead). The basal part of each lepidotrichium is long and unsegmented (white asterisk). Note the absence of radials: the lepidotrichia articulate directly with the girdle (see also plate 6.6). Whole mount staining with Alizarin red S, × 15, bar = 1 mm.

Plate 6.3. Developing endoskeleton of the caudal fin of the Desert pupfish, *Cyprinodon macularius* (5 mm TL), stained with Alcian blue for cartilage. The cartilage of the caudal fin endoskeleton anlage (black arrowhead) becomes surrounded by perichondral bone, here indicated by areas of nonstaining cartilage matrix (white arrowhead and white asterisk). *C. macularius* shows in an exemplary way how fusion of skeletal elements contributes to the formation of the caudal fin of an advanced teleost. Formerly single hypurals are fused into a continuous hypural plate (white asterisk). The hypural plate itself inserts on the urostyle (black asterisk), which represents the fusion product of several caudal vertebrae. Whole mount staining with Alcian blue, × 150, bar = 100 μm (from Witten 1992).

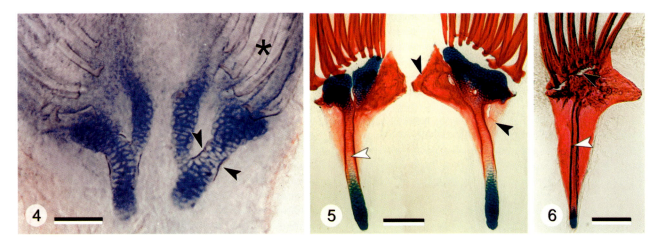

Plate 6.4. Anlage of the endoskeleton of the pelvic fins (basipterygia) in the Desert pupfish, *Cyprinodon macularius* (10 mm TL), composed of cartilage. The different cartilage elements are not easily homologized with pelvic girdle elements from other vertebrate groups. Black arrowheads indicate the onset of perichondral bone formation. Lepidotrichia are already present (black asterisk). Rostral to the bottom, caudal to the top. Whole mount staining with Alcian blue, × 250, bar = 60 μm.

Plate 6.5. The development of the endoskeleton of the pelvic fins (basipterygia) continues by lengthening of the cartilage core and by perichondral bone formation, as shown in *Cyprinodon macularius* (14 mm TL). Cartilage enclosed by perichondral bone (white arrowhead) persists. In addition to perichondral bone formation, membrane bone widens the basipterygia (black arrowheads). Rostral to the bottom, caudal to the top. Combined Alcian blue, Alizarin red S whole mount staining, × 85, bar = 150 μm.

Plate 6.6. Final stage of the development of the basipterygium in the Desert pupfish, *Cyprinodon macularius* (20 mm TL). Apart from the caudal tip, cartilage is now completely enclosed by perichondral bone (white arrowhead). Arising from the bone collar, intensive formation of (secondary) membrane bone has shaped the pelvic fin girdle. Different from other teleosts, the cartilage that represents the radials (which normally remain entirely cartilaginous), has been lost in the process of ossification (black arrowhead). Combined Alcian blue, Alizarin red S whole mount staining, × 40, bar = 300 μm.

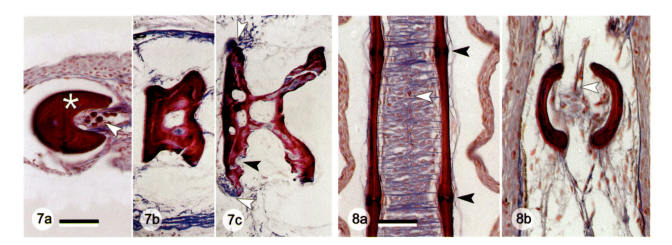

Plate 6.7. Cross section on three different levels—(a) distal to (c) proximal—through a spiny fin ray from the anterior part of the dorsal fin of a juvenile Nile tilapia, *Oreochromis niloticus* (18 mm TL), a representative of the acanthopterygian lineage. Unlike lepidotrichia, spiny fin rays of acanthopterygians are not segmented. Still, and similar to lepidotrichia, actinotrichia (white arrowhead in a) are present at the distal edge of the spiny fin ray (white asterisk in a). The presence of bone-forming cells (white arrowheads in c) and of resorption lacunae (black arrowhead in c) suggests that basal parts of the spiny fin ray are subject to remodeling. Azan staining, × 400, bar = 30 μm.

Plate 6.8. (a) Longitudinal and (b) cross section through a lepidotrichium of the caudal fin of a juvenile Nile tilapia, *Oreochromis niloticus* (18 mm TL). The two demirays that constitute a fin ray are visible. The demirays are connected by collagen fibers, and there is internal blood support (white arrowheads). Black arrowheads in (a) label the joint between successive hemisegments. Azan staining, × 400, bar = 30 μm.

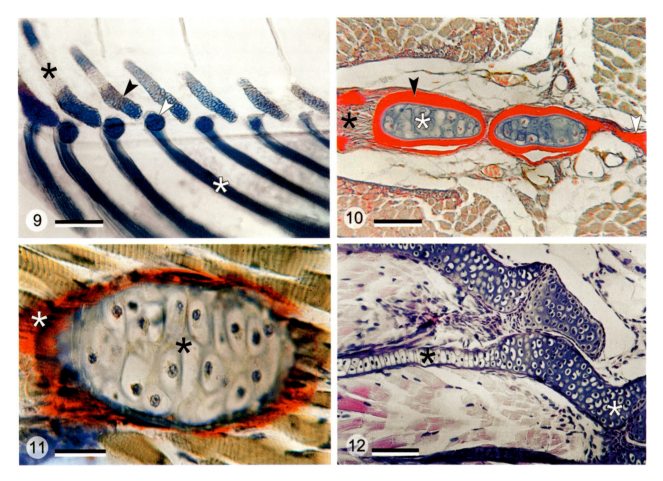

Plate 6.9. Developing endoskeleton (radials) of the anal fin of a juvenile Desert pupfish, *Cyprinodon macularius* (9 mm TL) stained for cartilage with Alcian blue. In the dorsal and anal fins of teleosts, originally three radials support the fin rays; *Cyprinodon* and other derived teleosts have only two radials. Proximal radials (black arrowhead), undergo perichondral ossification (black asterisk), whereas the distal radial that articulates with the fin ray remains cartilaginous (white arrowhead). At this early ontogenetic stage, the fin rays (white asterisk) also stain with Alcian blue, because their matrix contains acidic proteoglycans. Whole mount staining with Alcian blue, × 150, bar = 100 μm (from Witten 1992).

Plate 6.10. Perichondral bone (black arrowhead) surrounds the hypural cartilage (white asterisk) in a juvenile Nile tilapia, *Oreochromis niloticus* (18 mm TL). At this stage, the cartilage inside the bone collar still persists. The hypurals are widened by the formation of (secondary) membrane bone (white arrowhead), originating from the perichondral bone collar (see also plates 6.5 and 6.6). Base of the lepidotrichia (ventral) is to the left (black asterisk). Azan staining, × 300, bar = 50 μm (from Witten 1995).

Plate 6.11. The onset of perichondral bone formation around a hypural of the Desert pupfish, *Cyprinodon macularius* (14 mm TL). Osteoblasts surround the cartilage anlage (black asterisk) and express high levels of alkaline phosphatase activity (white asterisk). Demonstration of alkaline phosphatase with an azo-coupling procedure, × 800, bar = 20 μm (from Witten 1992).

Plate 6.12. Sagittal section through the radials of the dorsal fin of an adult zebrafish, *Danio rerio,* showing variation of cartilage morphology. The distal part of the radial consists of typical hyaline cartilage with considerable matrix—matrix-rich hyaline cartilage (white asterisk)—cf. Benjamin (1990). The proximal part of the radials consists of cell-rich hyaline cartilage, with little matrix surrounding the chondrocytes (black asterisk). Ventral to the left. Hematoxylin Eosin staining, × 250, bar = 60 μm.

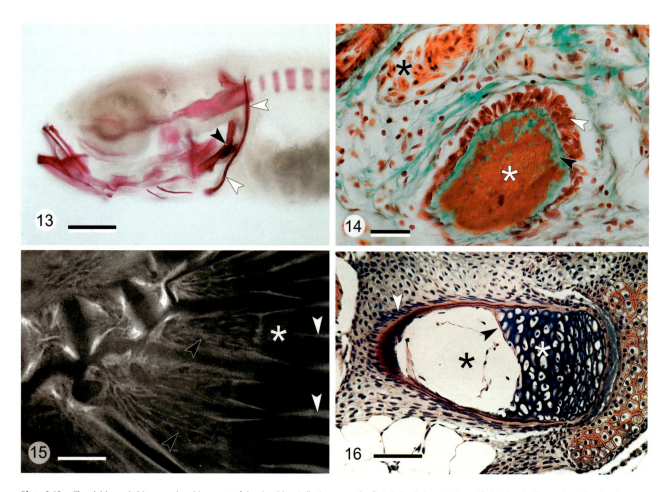

Plate 6.13. The cleithrum (white arrowheads), as part of the shoulder girdle, is among the first skeletal elements to ossify in teleosts. The early ossification (here shortly after hatching) is not connected to the development of the pelvic fin but related to the function of the jaws. The black arrowhead points to developed teeth already present on the pharyngeal jaws. Common carp, *Cyprinus carpio* (8 mm TL), whole mount staining with Alizarin red S, × 25, bar = 600 μm.

Plate 6.14. Typical membrane bone formation in an adult Atlantic salmon, *Salmo salar.* In the absence of any cartilaginous precursor, osteoblasts (white arrowhead) directly secrete the nonmineralized collagenous bone matrix (osteoid, black arrowhead) that subsequently mineralizes (white asterisk). The black asterisk indicates a blood vessel, located in the vicinity of the bone-forming site. Masson's Trichrome staining, × 400, bar = 30 μm.

Plate 6.15. Late skeletal development: radiograph of hypurals of the caudal fin in an adult Atlantic salmon, *Salmo salar,* showing that inside the perichondral bone collar, cartilage has been replaced by spongiosa-like bone structures (black arrowheads). Terminal cartilage persists (not visible on the X-ray; white asterisk) for the articulation with the lepidotrichia (white arrowheads). × 10, bar = 1.5 mm.

Plate 6.16. Hypural of an adult zebrafish, *Danio rerio.* Cartilage (white asterisk) inside the perichondral bone collar has been removed and replaced by fat (black asterisk). Black arrowhead points to the resorption front, where a very thin seam of bone has been deposited. White arrowhead indicates osteoblasts actively involved in periosteal ossification. Hematoxylin Eosin staining, × 250, bar = 60 μm.

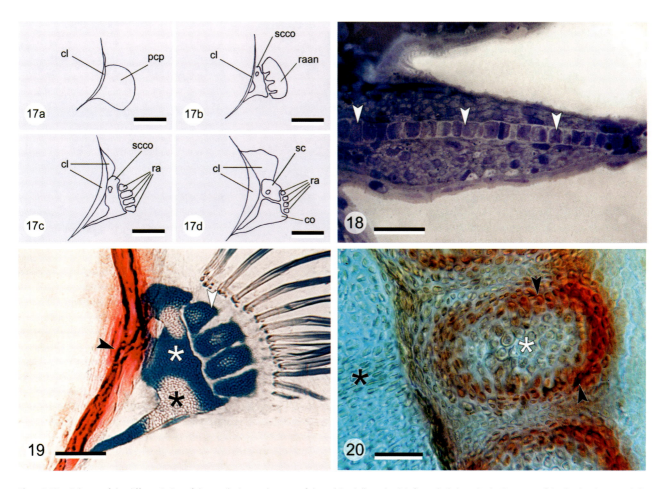

Plate 6.17. Scheme of the differentiation of the cartilaginous elements of the pelvic girdle and pelvic fin endoskeleton in the Desert pupfish, *Cyprinodon macularius* (4–20 mm TL). (a) The scapula, coracoid, and radials differentiate through cartilage subdivision from one precartilaginous plate. (b) Cartilage subdivision first separates a scapulocoracoid and a plate that gives rise to the radials. Next, (c) the radials differentiate before (d) the scapulocoracoid subdivides into the scapula and coracoid. Along with the differentiation of skeletal elements, the cartilage of all elements becomes surrounded by perichondral bone, which is extended by membrane bone. cl, cleithrum; co, coracoid; pcp, precartilaginous plate; ra, radials; raan, radial anlage; sc, scapula; scco, scapulocoracoid. a, b: × 50, bar = 200 μm; c: × 25, bar = 400 μm; d: × 12.5, bar = 800 μm (modified from Witten 1992).

Plate 6.18. The continuous precartilaginous plate in the pectoral fin anlage of the zebrafish, *Danio rerio* (4 days postfertilization; white arrowheads). The plate consists of a monolayer of large rounded cells with little matrix in between. Toluidine blue staining, × 800, bar = 20 μm.

Plate 6.19. Development of the shoulder girdle and radials in the Desert pupfish, *Cyprinodon macularius* (10 mm TL), at a developmental stage corresponding to the schematic drawing in plate 6.17c. The cartilage of the scapulocoracoid (white asterisk) is not yet subdivided, but has two ossification centers. Initiation of perichondral bone formation is indicated by areas of unstained cartilage matrix (black asterisk). Bone of the future scapula (upper half) and future coracoid (lower half) arises from the two ossification centers. Black arrowhead indicates the cleithrum; white arrowhead, a radial. Combined Alcian blue, Alizarin red S whole mount staining; × 80, bar = 200 μm.

Plate 6.20. Differentiation of the radials (white asterisk) in the Desert pupfish, *Cyprinodon macularius* (14 mm TL). In the process of perichondral bone formation a ring of alkaline phosphatase-positive cells (black arrowheads) is associated with each radial. A black asterisk indicates the scapulocoracoid cartilage. Azo-coupling method for alkaline phosphatase demonstration, × 300, bar = 50 μm.

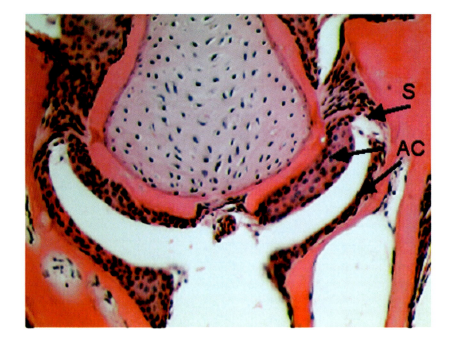

Plate 9.1. Section through the basal joint of the caudal fin of *Gnathonemus petersii* (elephant nosefish). The articular cartilage (AC) lies above the largely acellular bone. The synovium (S) encloses the joint cavity at the periphery although the bilayered structure is not easily distinguishable in this section. Hematoxylin and eosin.

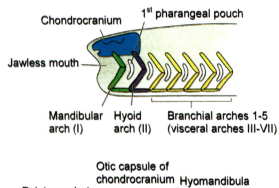

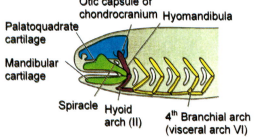

Plate 9.2. Diagram depicting the evolution of the craniofacial skeleton from the agnathous (hagfishes and lampreys) condition (*top*) to the jawed condition of lower (*middle*) and higher vertebrates (*bottom*). (Adapted from Liem et al. 2000.)

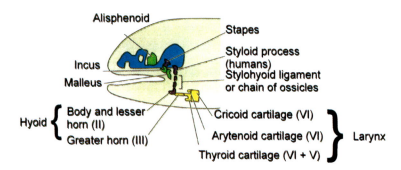

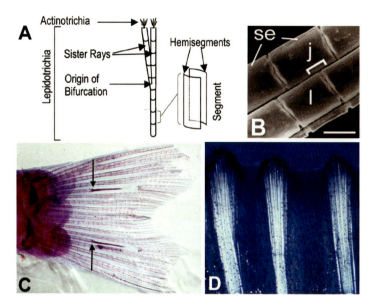

Plate 11.1. (A) Schematic representation of the dermal skeleton of a zebrafish fin ray (= lepidotrichia). A ray splits, at the origin of bifurcation (= fork), into two sister rays. (B) SEM of lepidotrichia of two paralllel segmented hemirays. (C) Whole mount of a caudal zebrafish fin stained with alizarin red showing the mineralized bony rays. Central portions of two individual rays were ablated, forming a gap within the rays. The rays are regenerating from the proximal part of the stump (arrows) only through a unidirectional process. (D) Actinotrichia, tufts of collagen-like fibers. se, segment; j, joint; l, lepidotrichia. Scale bars = B, 100 μm. (Adapted from Akimenko et al. 2003; Borday et al. 2001; Santamaría et al. 1996.)

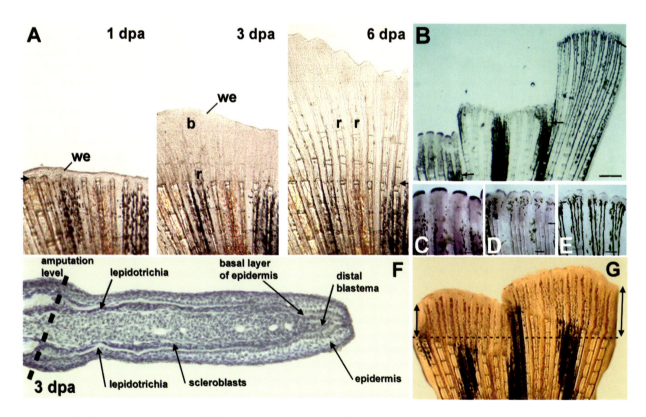

Plate 11.2. (A) Whole mounts of the same caudal fin regenerate at 1 day, 3 days, and 6 days postamputation (dpa). The arrows indicate the level of the cut. (B) Differential expression of the *msxB* gene following amputation of the zebrafish caudal fin at three different levels along the proximodistal axis at 4.5 days after amputation. (C–E) Higher-magnification photographs of the fin shown in (B); (C) proximal cut; (D) intermediate cut; (E) distal cut. Level of amputation indicated by arrows. (F) Longitudinal section of 4 dpa regenerate stained with Hematoxylin and Eosin. (G) Caudal fin of a zebrafish 5 days following an oblique cut. Note that lateral rays on both sides of the caudal fin (double-headed arrows) are cut at the same proximal-distal level but with a different slope (dotted line). Scale bars = (B) 500 μm; (C–E) 100 μm. b, blastema; r, ray; we, wound epidermis. (Adapted from Akimenko et al. 1995; Akimenko et al. 2003; Géraudie and Borday Birraux 2003; F–G: A. Smith and M.-A. Akimenko, unpublished data.)

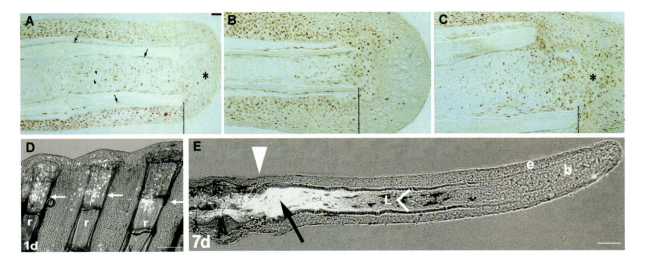

Plate 11.3. (A–C) Cell proliferation during blastema formation in the regenerating fin of the goldfish *Carassius auratus*. Longitudinal sections from regenerating fin rays of 4 dpa (A), 5 dpa (B), 6 dpa (C), immunostained for PCNA. Dashed lines indicate amputation planes. (D–E) Examination of cell migration in the connective tissue. DiI was injected within the intraray tissue of caudal fins of zebrafish 1 day before amputation. Fins were amputated one segment (D) and two segments (E) away from the site of injection. Cell movement was observed on whole-mount fins 1 dpa (D) or on longitudinal sections 7 dpa (E). Arrows in A indicate the site of injection. The white arrowhead in E indicates the level of amputation. The black arrow in E indicates the group of labeled cells. b, blastema; e, epidermis; L, lepidotrichia; r, ray. (Adapted from Poleo et al. 2001; Santos-Ruiz et al. 2002.)

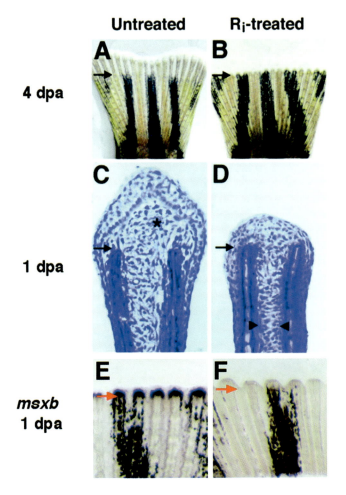

Untreated   **R$_i$-treated**

4 dpa

1 dpa

msxb
1 dpa

Plate 11.4. Fgf signaling is required for blastema formation. (A) Fin from untreated zebrafish at 4 dpa. (B) Fin from fish treated immediately after amputation for 4 days with the pharmacological inhibitor of Fgfr1, SU5402. These fins show no growth. (C–D) Hematoxylin stain of 1 dpa fins from untreated fin showing a normal blastema (*) (C) or fin treated with SU5402 for 24 hr leading to a lack of blastema (D). (E–F) Whole-mount *in situ* hybridization of *msxb* on 1 dpa fins showing strong blastema expression in the untreated fish (E) compared to the reduced or absent expression in fish treated with SU5402 for 24 hr (F). Arrowheads indicate amputation plane. (Adapted from Poss et al. 2000b.)

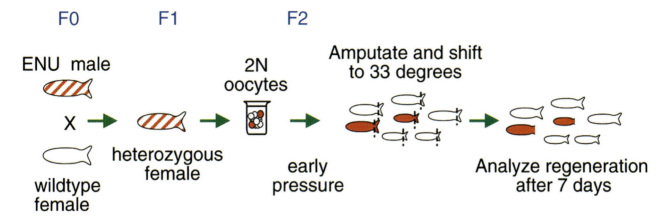

F0 F1 F2

ENU male

X →

wildtype
female

heterozygous
female

2N
oocytes

early
pressure

Amputate and shift
to 33 degrees

Analyze regeneration
after 7 days

Plate 11.5. Schematic of genetic screen for zebrafish temperature-sensitive mutants of fin regeneration. Zebrafish males are treated with ENU (*N*-ethyl-*N*-nitrosourea). Mutagenized families were generating by parthenogenesis: this involves fertilization of oocytes with ultraviolet (UV) irradiated sperm followed by treatment with hydrostatic pressure. Such a procedure generates diploid progeny, some homozygous for the ENU-generated mutations. The fish are raised for two months at 25°C; then their caudal fins are amputated and the animals shifted to 33°C for 7 days before assessing regeneration. Regeneration mutants can easily be detected after 4–7 days' incubation at 33°C. (Adapted from Poss et al. 2002.)

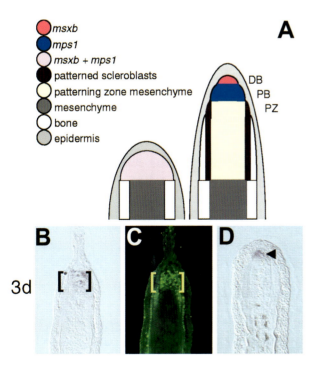

msxb
mps1
msxb + mps1
patterned scleroblasts
patterning zone mesenchyme
mesenchyme
bone
epidermis

A

DB
PB
PZ

B **C** **D**

3d

Plate 11.6. (A) Cellular and molecular model for fin regeneration. Blastema formation is characterized by the induction of *msxb* and *mps1* expression. During outgrowth the blastema can be subdivided into 3 domains. The distal blastema (DB) defined by msxb (orange), the proximal blastema (PB) by mps1 (blue), and the patterning zone (PZ) by newly patterned scleroblasts (brown) and differentiating mesenchyme (yellow). (B–C) Colocalisation of mps1 and PCNA in the PB of 3-day fin regenerate; (B) whole-mount *in situ* hybridization of *mps1*; (C) immunodetection of PCNA protein. (D) Whole-mount *in situ* hybridization of *msxb* in 3-day fin regenerate. *msxb* expression is restricted to the DB. (Adapted from Poss et al. 2002.)

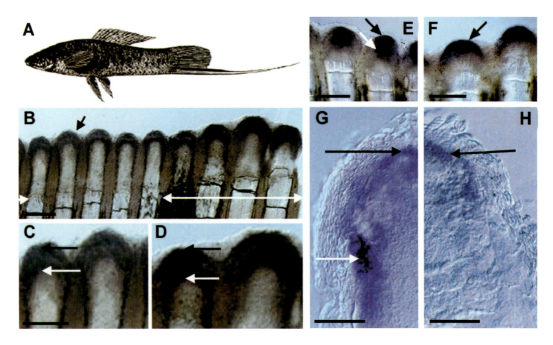

Plate 11.7. Expression of *msxc* and *msxE/1* genes during caudal fin regeneration of the *Xiphophorus helleri* (swordtail). (A) The adult male features a prominent sword, which is a sex-specific modification of ventral rays of the caudal fin. (B) At 6 dpa *msxC* expression is seen in both the distal blastema and lateral mesenchyme at higher levels in sword rays (white arrow) than in nonsword rays (black arrow). (C–D) Magnification of representative sword (D) and nonsword (C) rays, with distal (black arrow) and proximal (white arrow) expression domains. (E–F) At 5 dpa msxc is expressed in 2 domains, in the distal tip of the blastema (black arrow in F) and more proximally in the lateral mesenchyme (white arrow in F). (G–H) Expression of *msxE/1* is found exclusively in the distal tip of the blastema. Whole-mount (E, F) and longitudinal sections (G, H) of the same fin. White double arrows in (B) indicate sword rays. Scale bar = (B, C) 150 µm; (E, F) 50 µm; (G, H) 100 µm. (Adapted from Zauner et al. 2003.)

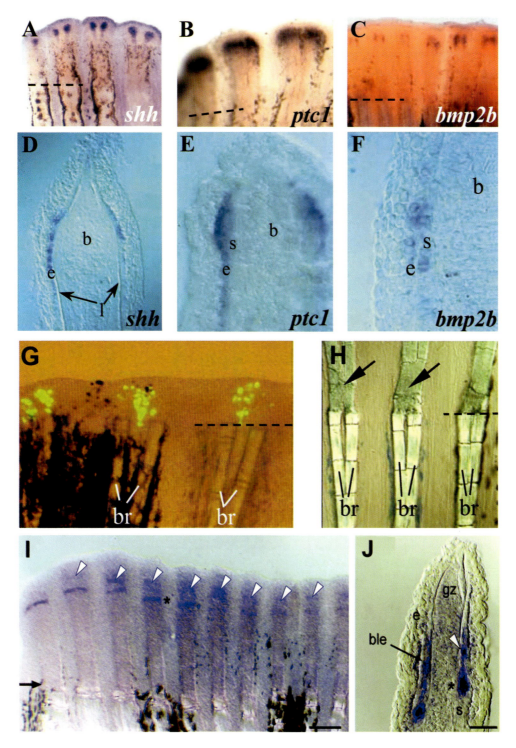

Plate 11.8. (A–F) Expression patterns of *sonic hedgehog* (*shh;* A, D), *patched 1* (*ptc1;* B, E) and *bone morphogenetic protein 2b* (*bmp2b;* C, F) in 4-day regenerated fins. Expression in the regenerate detected through whole-mount *in situ* hybridization (A–C) and longitudinal section (D–F). On sections, *shh* expression (D) is restricted to a limited number of cells of the basal epithelial layer. *ptc1* (E) and *bmp2b* (F) transcripts are found not only in the basal epithelial layer but also in adjacent scleroblasts. (G) Ectopic expression of EGFP 24 hr after injection of plasmid DNA containing the EGFP reporter gene placed under the control of the CMV promoter, in the tissue separating the branches of the lepidotrichia (br) of 3 adjacent fin rays. Injections were performed 2 days after amputation. (H) Whole-mount fin, 10 days after amputation, showing the fusions of ray branches induced by the ectopic expression of *shh* after microinjection of a construct expressing *shh* under the control of the CMV promoter using protocol described in (G). (I) *evx1* expression in a 4 dpa regenerate is observed at the level of the last formed segment boundary and in a more distal location (white arrowheads) that may prefigure the next segment boundary. (J) longitudinal section of a 4 dpa regenerate showing the distal *evx1* expression in scleroblasts located at the level of the putative joint (asterisks) and in distal differentiating scleroblasts (white triangle). The dotted lines in (A), (B), (C), (G), and (H) and the arrow in (I) indicate the level of amputation. b, blastema; br, bifurcating rays; e/ble, basal epithelial layer; l, regenerating lepidotrichia; s, scleroblasts. (Adapted from Akimenko et al. 2003; Borday et al. 2001.)

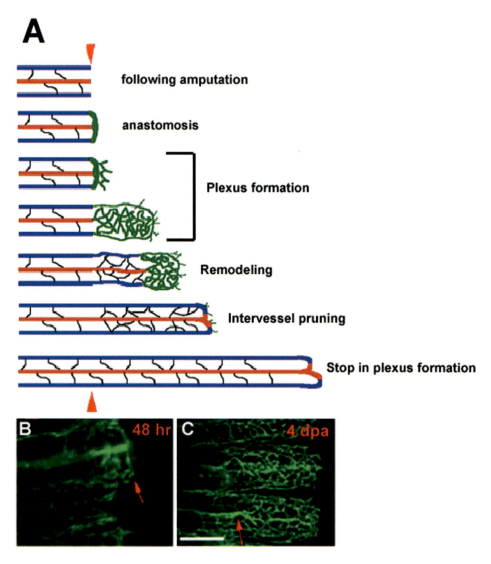

Plate 11.9. (A) Schematic of the stages of blood vessels regeneration. The red and blue lines depict arteries and veins, respectively. Black lines depict intervessel commissures. New vessels are in green. (B) By 48 hours postamputation, anastomoyic bridges are formed between artery and veins within the same ray. (C) At 3 dpa, a plexus with dense unstructured vessels is formed distal from the amputation plane. (Adapted from C. C. Huang et al. 2003.)

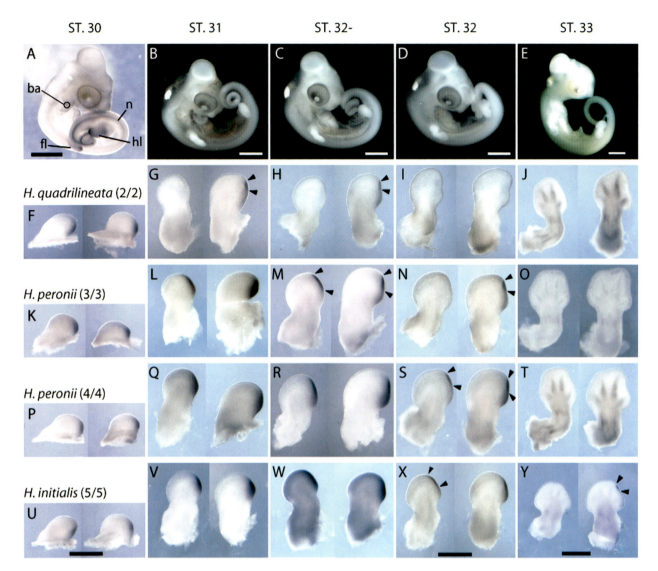

Plate 14.1. Sonic hedgehog (SHH) expression in the limbs of embryonic stages 30–33 *Hemiergis*. (A) Embryo of *H. quadrilineata* (2/2) showing SHH expression in the second branchial arch (ba), forelimb (fl), hindlimb (hl), and notochord plus neural tube (n). (B–E) Unstained embryos of *Hemiergis* spp. at stages 31–33. (F–Y) Limbs of (F–J) *H. quadrilineata* (2/2), (K–O) *H. peronii* (3/3), (P–T) *H. peronii* (4/4), (U–Y) *H. initialis* (5/5) embryos in dorsal view. In all panels, the forelimb is on the left and the hindlimb is on the right; anterior is to the left and distal is up. (G) In stage 31 *H. quadrilineata* (2/2), SHH is not expressed in the forelimb, and expression is restricted to the posterodistal part of the hindlimb (arrowhead); no such restriction is observed in the other three morphs (L, Q, V). (H) SHH is not expressed in the forelimb of stage 32 *H. quadrilineata* (2/2), and expression remains distally restricted in the hindlimb (arrowheads mark proximal and distal boundaries of intense expression). (M) Restricted expression is also seen in some *H. peronii* (3/3) embryos at this stage (arrowheads), but not in (R) *H. peronii* (4/4) or (W) *H. initialis* (5/5) limbs. (I) At stage 32, SHH expression is not detected in any forelimbs or in most hindlimbs of *H. quadrilineata* (2/2). (N) Expression is also absent from the forelimbs of *H. peronii* (3/3) and is distally restricted in hindlimbs. In contrast, *H. peronii* (4/4) and *H. initialis* (5/5) maintain SHH expression in both fore- and hindlimbs at this stage. (S) Expression is distally restricted in both sets of limbs in *H. peronii* (4/4). (X) *H. initialis* (5/5) forelimbs showed slight (as shown) or no (not shown) distal restriction of SHH expression at stage 32; no such restriction was observed in hindlimb expression. (J, O, T, Y) At stage 33, posterior mesenchymal SHH expression is not detected in the limbs of (J) *H. quadrilineata* (2/2) and (O, T) *H. peronii* (3/3 and 4/4). (Y) In *H. initialis* (5/5), however, distal expression foci persist in most hindlimbs at this stage. Scale bar = 1 mm for embryos (A–E) and for stage 33 limbs (as in Y). Scale bar = 0.5 mm in all other panels (as in U, X). (After Shapiro et al. 2003.)

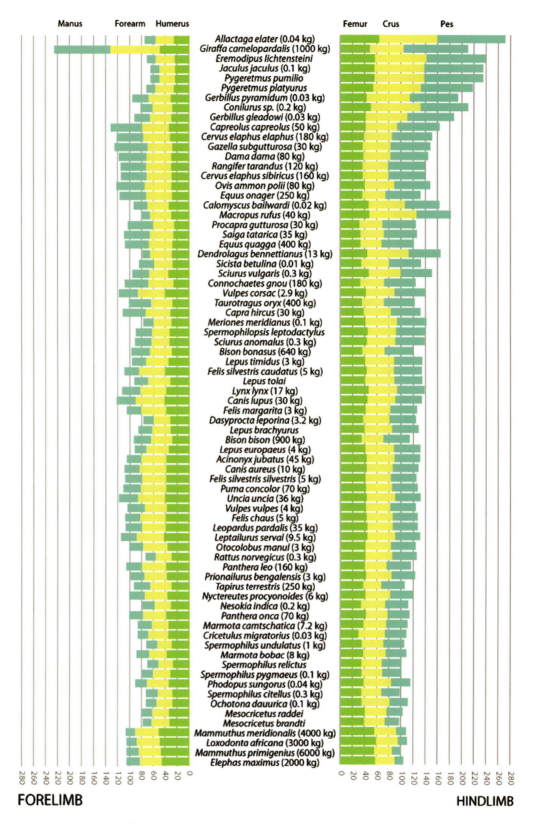

Manus **Forearm** **Humerus** | **Femur** **Crus** **Pes**

Allactaga elater (0.04 kg)
Giraffa camelopardalis (1000 kg)
Eremodipus lichtensteini
Jaculus jaculus (0.1 kg)
Pygeretmus pumilio
Pygeretmus platyurus
Gerbillus pyramidum (0.03 kg)
Conilurus sp. (0.2 kg)
Gerbillus gleadowi (0.03 kg)
Capreolus capreolus (50 kg)
Cervus elaphus elaphus (180 kg)
Gazella subgutturosa (30 kg)
Dama dama (80 kg)
Rangifer tarandus (120 kg)
Cervus elaphus sibiricus (160 kg)
Ovis ammon polii (80 kg)
Equus onager (250 kg)
Calomyscus bailwardi (0.02 kg)
Macropus rufus (40 kg)
Procapra gutturosa (30 kg)
Saiga tatarica (35 kg)
Equus quagga (400 kg)
Dendrolagus bennettianus (13 kg)
Sicista betulina (0.01 kg)
Sciurus vulgaris (0.3 kg)
Connochaetes gnou (180 kg)
Vulpes corsac (2.9 kg)
Taurotragus oryx (400 kg)
Capra hircus (30 kg)
Meriones meridianus (0.1 kg)
Spermophilopsis leptodactylus
Sciurus anomalus (0.3 kg)
Bison bonasus (640 kg)
Lepus timidus (3 kg)
Felis silvestris caudatus (5 kg)
Lepus tolai
Lynx lynx (17 kg)
Canis lupus (30 kg)
Felis margarita (3 kg)
Dasyprocta leporina (3.2 kg)
Lepus brachyurus
Bison bison (900 kg)
Lepus europaeus (4 kg)
Acinonyx jubatus (45 kg)
Canis aureus (10 kg)
Felis silvestris silvestris (5 kg)
Puma concolor (70 kg)
Uncia uncia (36 kg)
Vulpes vulpes (4 kg)
Felis chaus (5 kg)
Leopardus pardalis (35 kg)
Leptailurus serval (9.5 kg)
Otocolobus manul (3 kg)
Rattus norvegicus (0.3 kg)
Panthera leo (160 kg)
Prionailurus bengalensis (3 kg)
Tapirus terrestris (250 kg)
Nyctereutes procyonoides (6 kg)
Nesokia indica (0.2 kg)
Panthera onca (70 kg)
Marmota camtschatica (7.2 kg)
Cricetulus migratorius (0.03 kg)
Spermophilus undulatus (1 kg)
Marmota bobac (8 kg)
Spermophilus relictus
Spermophilus pygmaeus (0.1 kg)
Phodopus sungorus (0.04 kg)
Spermophilus citellus (0.3 kg)
Ochotona dauurica (0.1 kg)
Mesocricetus raddei
Mesocricetus brandti
Mammuthus meridionalis (4000 kg)
Loxodonta africana (3000 kg)
Mammuthus primigenius (6000 kg)
Elephas maximus (2000 kg)

280 260 240 220 200 180 160 140 120 100 80 60 40 20 0 | 0 20 40 60 80 100 120 140 160 180 200 220 240 260 280

FORELIMB **HINDLIMB**

Plate 15.1. Graph of standardized lengths of forelimb (left) and hindlimb (right) segments from 75 mammalian species. Species are sorted from those with the longest to the shortest pes. Segment lengths were standardized against the length of the thoracic and lumbar vertebrae. Most of the species represented are cursorial, saltatory, ambulatory, or graviportal and are drawn primarily from the orders Artiodactyla, Perissodactyla, Carnivora, Rodentia, and Proboscidea. Saltatory species (e.g., *Allactaga elater, Eremodipus lichtensteini, Pygeretmus pumilio, Jaculus jaculus,* and *Macropus rufus*) are easily picked out by their combination of long hindlimb and short forelimb segments. Note the proportionally long proximal relative to distal segments in the graviportal proboscideans (the last four species). These data demonstrate that there is only a loose correlation between the lengths of proximal and distal segments, between forelimb and hindlimb segments, and between segment proportions and body mass. (Original data from Gambaryan 1974.)

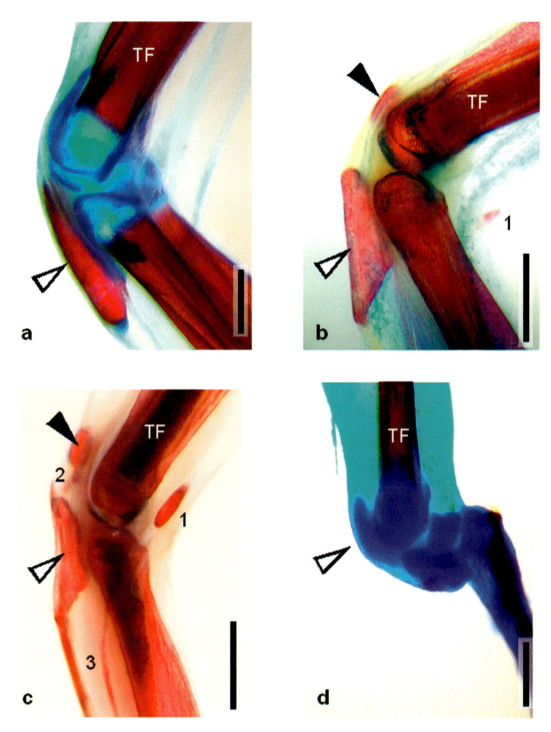

Plate 19.1. *Hymenochirus boettgeri* hindlimb whole mounts double-stained for cartilage and bone, detailing the position of the distal *os sesamoides tarsale* (OST; white arrowhead), and other neighboring ossicles adjacent to the tibiofibular-tarsometatarsal joint. (a) Metamorphosed subadult, Nieuwkoop and Faber 1967 (NF) stage 66+. Note that while the bulk of the sesamoid is mineralized (Alizarin-positive), the distal pole and articulating surfaces remain cartilaginous (Alcian-positive). (b) Adult, with fully mineralized (Alizarin-positive) distal OST, proximal OST (black arrowhead) and os tibialis anticus (1). (c) Aged adult, demonstrating hyper-mineralization of the connective tissue adjacent to tibiofibular-tarsometatarsal joint, including mineralized distal OST, proximal OST, *os tibialis anticus,* multiple "accessory" mineralizations (2) and mineralized tendons (3). (d) NF stage 56 regenerate, following amputation at the knee joint at NF stage 52. The regenerated distal OST is cartilaginous (Alcian-positive) has been displaced proximally. The presence of other ossicles remains unclear. Abbreviation: TF, tibiofibula. Scale bar = 1 mm.

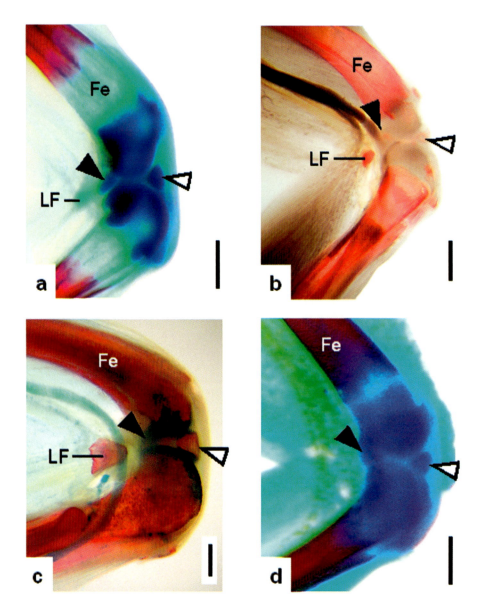

Plate 19.2. *Hymenochirus boettgeri* hindlimb whole mounts double-stained for cartilage and bone (a, c, and d) or bone and nerve (b), detailing the knee joint and the position of the anterolateral (white arrowhead) and posterolateral (black arrowhead) lunulae and the lateral fabella (LF). (a) Metamorphosed subadult, Nieuwkoop and Faber 1967 (NF) stage 66+. Both lunulae and the lateral fabella are composed of unmineralized cartilage (Alcian-positive). (b) Adult, with both lunulae and the lateral fabella mineralized (Alizarin-positive). The large femoral nerve (Sudan Black-positive) is seen coursing from the upper right hand corner of the image toward the knee joint. (c) Aged adult, demonstrating hyperossification of the anterolateral lunulae and lateral fabella (Alzarin-positive); the posterolateral lunulae is less pronounced. The large femoral artery is also seen passing anterior to the lateral fabella. (d) NF stage 56 regenerate, following amputation at the knee joint at NF stage 52. The regenerated anterolateral and posterolateral lunulae are vaguely evident as Alcian-positive nodules (arrowheads); the lateral fabella is not. Abbreviation: Fe, femur. Scale bar = 1 mm.

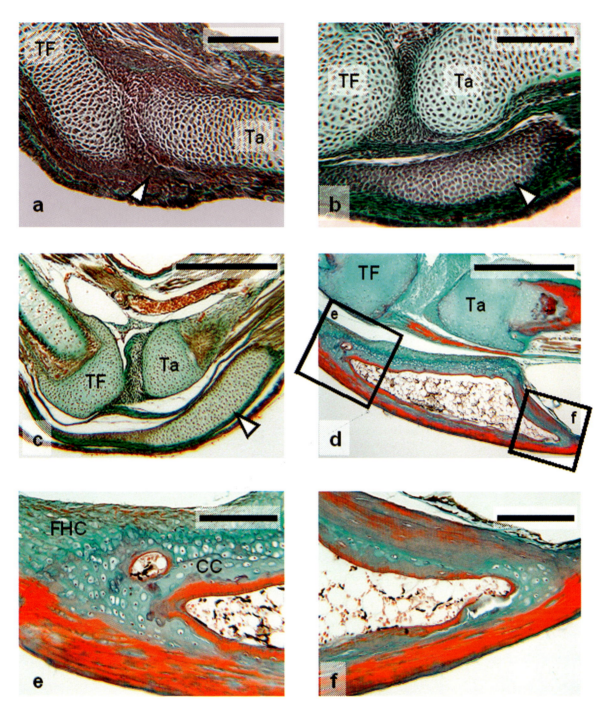

Plate 19.3. *Hymenochirus boettgeri* hindlimb serial sections stained with Masson's trichrome, detailing the histogenesis of the distal *os sesamoids tarsale* (OST; white arrowhead). (a) Tadpole, Nieuwkoop and Faber 1967 (NF) stage 56; the distal OST begins as a condensation of chondroblasts adjacent to the developing tibiofibular-tarsometatarsal joint. Scale bar = 100 μm. (b) Tadpole, NF stage 58; chondroblasts of the distal OST begin to hypertrophy. Note that differentiation of these cells trails that of the tibiofibula (TF) and tarsometatarsus (Ta). Scale bar = 100 μm. (c) Froglet, NF stage 62; the distal OST is composed of hyaline cartilage, with many chondrocytes embedded within matrix. Scale bar = 300 μm. (d) Adult; cartilage of the distal OST begins to mineralize and undergo endochondral ossification. Scale bar = 500 μm. (e) Close-up of the proximal pole of the distal OST depicted in (d), demonstrating the complex histology of this skeletally mature element. To the right, the aponeurosis plantaris grades into fibrocartilage. The fibrocartilage with hyaline cells (fibrohyaline-cell cartilage, FHC; Benjamin 1990) merges into a hyaline cartilage that is becoming calcified (CC). Deep to the calcified cartilage, endochondral bone is forming. Scale bar = 150 μm. (f) Close-up of the distal pole of the distal OST depicted in (d), demonstrating a similar histology to the proximal pole except with less fibrohyaline-cell cartilage. Scale bar = 150 μm.

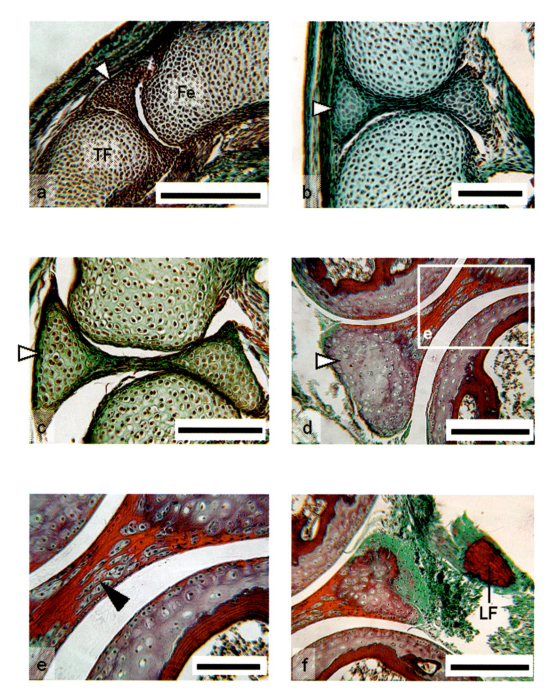

Plate 19.4. *Hymenochirus boettgeri* hindlimb serial sections, stained with Masson's trichrome, detailing the histogenesis of the anterolateral (white arrowhead) and posterolateral lunulae. The femur (Fe) is always toward the top of the image, the tibiofibula (TF) is always toward the bottom of the image. (a) Tadpole, Nieuwkoop and Faber 1967 (NF) stage 56; the lunulae begin as a condensations of chondroblasts adjacent to the developing femoror-tibiofibular joint. Scale bar = 100 μm. (b) Tadpole, NF stage 58; chondroblasts of each lunulae begin to hypertrophy. Note that differentiation of these cells trails that of the femur and tibiofibula. Scale bar = 100 μm. (c) Froglet, NF stage 62; the lunulae are composed of hyaline cartilage, with many chondrocytes embedded within matrix. Scale bar = 500 μm. (d) Adult; the anterolateral lunula is almost entirely calcified although no true bone is present. Scale bar = 1 mm. (e) Close-up of the band of tissue connecting the anterolateral and posterolateral lunulae, depicting the isogenous rows of chondrocytes (black arrowhead) and large collagen bundles. Scale bar = 300 μm. (f) Adult; the posterolateral lunulae is heavily calcified except for its posteriormost border, which remains as unmineralized connective tissue. The lateral fabella (LF) is also present as a nodule of calcified cartilage. Scale bar = 1 mm.

staining, bone may be partially decalcified, or persisting cartilage may be masked by bone.

Interestingly, a shift can exist between the development of the pterygiophore cartilage and the time that its distal tip segments away from the main part to form distal and proximal radial, respectively. For example, in the zebrafish, the radials in the dorsal and anal fins develop bidirectionally, but segmentation starts anterior and sweeps posteriorly (Crotwell et al. 2001).

During the development of the caudal fin, neither fusion nor branching of the cartilage occurs (although the development of the caudal fin in a flatfish species presents a clear exception; Gavaia et al. 2002). Instead, extra cartilages can appear, which may nevertheless represent rudiments of formerly fully developed epurals and hypurals.

Shubin and Alberch (1986), in an influential paper, proposed that three general mechanisms underlie the formation of cartilage condensations in limbs: *de novo* formation, branching, and segmentation. The type of separation observed, for example, in the pectoral fin, does not fit any of these categories. First, although the cartilage may initially subdivide in two elements (fitting the definition of segmentation), each segment then again subdivides. Second, segmentation occurs perpendicular to the axis of the limb. The segmentation process, as defined by Shubin and Alberch, only accounts for sequential segmentation along an axis. Both observations contradict the model proposed by Shubin and Alberch and lend support to the recent criticism of this model (Cohn et al. 2002). In any event, this phenomenon of subdivision of differentiated cartilage provides an intriguing field of study of morphogenetic processes occurring in vertebrate cartilages.

Late Events in Skeletogenesis

Developmental studies commonly consider early life stages, but fewer studies cover juvenile life stages, and little information is available about developmental processes in mature and aging fish (Reznick et al. 2002; Gerhard et al. 2002). Notable exceptions are studies on adult salmon bone (e.g., Kacem et al. 2000; Kvellestad et al. 2000; Witten and Hall 2002). Continuation of development and growth throughout life (Reznick et al. 2002) is certainly not without effect on the fin skeleton (plate 6.7c, and see Witten et al. 2001). Accordingly, the bone matrix of the fin skeleton can show signs of growth arrest or of resorption, in the form of so-called cement lines (Castanet et al. 1993). Indeed, bone is a remarkable recorder for the physiological status of the animal, as well as for the environmental conditions that the animal meets, and so are parts of the fin skeleton. Researchers, in particular in fisheries

and aquaculture, have taken advantage of the marks left inside the bone tissue of fin rays, to estimate age and growth rates (e.g., Ferreira et al. 1999; Ihde and Chittenden 2002).

Based on what is known about stress-induced bone transformation (Bertram and Swartz 1991; Bertram et al. 1997; Kostenuik et al. 1999), changes in mechanical forces should also alter the structure of the fin skeleton. The impact of mechanical forces on the fin skeleton of a juvenile fish is certainly quite different from mechanical forces that affect the fin skeleton of large adults. Still, we have little information whether the fin skeleton of a 50-year-old carp (*Cyprinus carpio*) or a 140-year-old perch (*Sebastes aleutianus*; cf. Reznick et al. 2002) has changed its structure.

Different from cartilage—which can grow through cell division and alterations in matrix composition—changes of the structure of bone can only occur through remodeling, a process that involves the removal of existing bone by resorbing cells (osteoclasts) and formation of new bone by osteoblasts (plate 6.7c). Extensive bone remodeling can be observed in the developing skeleton of cyprinids (*Cyprinus carpio* and *Danio rerio*; Witten et al. 2000, 2001), and in the skeleton of adult salmon (Kacem et al. 1998; Witten and Hall 2002). In contrast, studies on derived teleost species with acellular bone suggest a limited extent of bone remodeling (by mononucleated osteoclasts) or its absence (Moss 1962; Witten 1997; Witten et al. 1999). Accordingly, the fin endoskeleton of juvenile teleosts (e.g., the cichlid *Oreochromis niloticus*) does not display signs of bone remodeling (Witten and Villwock 1997).

Studies on zebrafish may provide a possible clue why various teleosts display widely different patterns of bone remodeling. In zebrafish, the appearance of multinucleated osteoclasts and bone remodeling is age dependent. Active resorbing mononucleated osteoclasts appear at 20 days postfertilization. Multinucleated osteoclasts appear only 10 days later, and bone remodeling does not start before day 60 days postfertilization (Witten et al. 2001). Thus, heterochronic shifts in the appearance of bone-resorbing cells could be one reason for dissimilarities in the occurrence of bone remodeling in various teleost groups.

Extensive and disproportionate enlargement of fin skeletal elements is another phenomenon of late skeletogenesis. In relatively large (mature) individuals, girdle elements and pterygiophores are liable to undergo hyperostosis, which is a localized bone expansion, based on extensive membrane bone formation. The intraspecific predictability and site-specificity of hyperostoses in a number of marine teleosts led Smith-Vaniz et al. (1995) to propose that hyperostosis is not a pathologic condition (osteoma), as suggested in earlier literature, but an expression of a genetically controlled process, occurring in relatively large (mature) individuals. Having all attributes of cellular bone, hyperostoses are interpreted by Smith-Vaniz et al. (1995) as extensive bone remodeling foci, even in marine fish with an apparently otherwise quiescent (acellular) skeleton.

Conclusions and Directions for Future Research

The fin skeleton emerges as a challenging subject to study developmental processes from the perspective of sequential initiation, symmetry control, regulation of cartilage and bone differentiation, and segmentation and fusion processes, to name but a few. At the end of the chapter it is, however, obvious that much more knowledge has to be gathered before the rope that we pulled across the water can be turned into a solid bridge. J. F. Webb (1999) outlined the problem by emphasizing that the development of the vast majority of fish species (more than 24,000 taxa in 482 families) has not been studied, and in some cases even data for the most conspicuous taxa are lacking.

An increasing amount of data suggests that the early histogenesis of the fin and the tetrapod limb skeleton is regulated by similar, if not the same, genetic and biochemical factors. Still, from a developmental perspective, intriguing questions remain to be answered regarding the source of the mesenchymal cells that populate the different fins, and the origin of the united cartilaginous plate that represents the precursor of both the shoulder girdle and the pectoral fin endoskeleton. An answer to this question, of course, also bears upon the evolutionary relationships between the appendicular skeletal elements in different vertebrate groups.

For the venture to understand histogenetic processes, fins are also challenging objects, especially for studies on cartilage condensation and differentiation. What mechanisms control whether radials, variably, emerge as independent condensations or bud off from an existing cartilage? Why does this budding occur in a precartilaginous condensation in some cases and in fully differentiated cartilage in others, and through which mechanisms? And what orchestrates the development of radials and lepidotrichia, such that, depending on the case, radials develop prior to or after lepidotrichial morphogenesis? In the course of such studies we may come across even more skeletal tissue types than the already vast array currently known to exist in the fin endoskeleton.

Given its severe reduction, the pelvic fin endoskeleton in teleosts has received little attention so far, compared to the pectoral fin endoskeleton. In view of the current interest in the evolutionary origin of the paired appendages and the contrasting opinions regarding nonequivalence (Coates and Cohn 1999) or equivalence (Tanaka et al. 2002) of pectoral and pelvic appendages, a challenge will be to further dissect the processes underlying pectoral and pelvic fin skeletal development and to assess whether pelvic fins develop through the same mechanisms as pectoral fins.

Acknowledgments

P. Eckhard Witten gratefully acknowledges funding granted by the Deutsche Forschungsgemeinschaft (DFG; Ro 381/9-1) and the German-Canadian Cooperation in Science and Technology (CAN 01/014). Ann Huysseune acknowledges funding from the Bijzonder Onderzoeksfonds of Ghent University (project no. 011B2396). Both authors acknowledge support from European COST ACTION B23: oral facial development and regeneration.

Chapter 7 Mechanisms of Chondrogenesis and Osteogenesis in Limbs

Scott D. Weatherbee and Lee A. Niswander

Chondrogenesis in Limbs

Generalized Steps during Limb Skeleton Development

One major limb morphogenetic event is the formation of the bony skeleton. The skeleton forms a framework for the overall shape of the limb and provides sites for muscle attachment for use in locomotion. Despite the wide range of tetrapod limb morphologies, they all have a conserved basic skeletal architecture (fig. 7.1), which comprises a proximal stylopod (e.g., humerus in the forelimb), a medial zeugopod (e.g., radius/ulna), and a distal autopod (e.g., wrist bones and fingers). Temporally, the limb skeletal elements form in a proximal-to-distal sequence from the stylopod to the autopod.

Limb skeleton formation occurs via endochondral ossification, whereby a cartilage template is laid down and later replaced by bone. The number of mesenchymal precursors that adopt a chondrocyte fate and their subsequent proliferation and deposition of extracellular matrix (ECM) determine the size and shape of each cartilage template. The ultimate size and shape of the limb elements are related in an unknown way to early limb patterning signals and late events that determine and refine the unique features of each element. This is a complex process beginning with aggregation of mesenchymal cells into prechondrogenic condensations (fig. 7.2). Within the condensation primordia, these cells initially express type I and type IIa collagen (Lui et al. 1995).

These chondroprogenitors then undergo changes in cell shape and gene expression as they differentiate into chondrocytes. These cells begin to express type IIb collagen, as well as type IX and type XI collagen, while turning off type I collagen expression (Ryan and Sandell 1990). During chondrocyte differentiation, cells on the periphery of the cartilage elements differentiate into a fibroblastic cell layer termed the perichondrium, which sheathes the cartilage rudiment. Initially, the chondrocyte cells are largely proliferative, but later, beginning in the middle of each cartilage element, chondrocytes undergo a multistep program of maturation and differentiation. This leads to the formation of prehypertrophic and finally terminal hypertrophic chondrocytes (see fig. 7.3a). In regions of the condensation where joints will form, condensed chondroprogenitors do not differentiate into chondrocytes and instead adopt a joint fate. This involves expression of new genes and a down-regulation of genes specific to the chondrogenic lineage. The details of joint initiation, formation, and maintenance are examined by Farnum (chap. 10 in this volume).

As chondrocytes mature, morphological changes also take place in the connective tissue or perichondrium surrounding the cartilage element. The perichondrium initially comprises chondroblasts with a similar morphology to proliferating chondrocytes. As the cartilage element begins to mature, chondroblasts surrounding the diaphysis, the central shaft of the bone, become flattened. However, cells in the articular region of the epiphysis, or ends of the bones, maintain their rounded shape (fig. 7.3a). As the chondrocytes begin to hypertrophy, the perichondrium of the diaphysis differentiates into the periosteum. The formation of a bone collar around

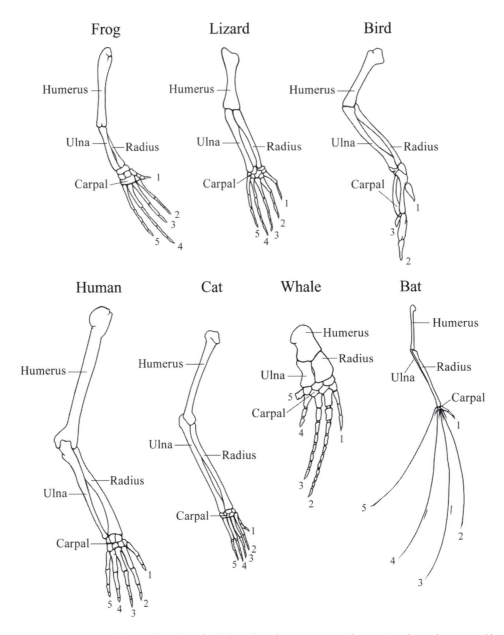

Figure 7-1 A sampling of tetrapod forelimb skeletons. All vertebrate forelimbs are homologous structures whose anatomy has undergone considerable diversification in the evolution and adaptation of these various vertebrate lineages. The divergence in skeletal structure reflects their unique lifestyles. (Adapted from Ridley 1996; artwork by Leanne Olds.)

the diaphysis results from the differentiation of the innermost periosteal cells into osteoblasts and their secretion of a bone matrix that becomes progressively calcified (fig. 7.3a).

Within the cartilage elements, the region of terminal hypertrophic chondrocytes is invaded by blood vessels, followed by subsequent invasions of osteoclasts and osteoblasts, which begin to replace cartilage with bone. During chondrocyte differentiation, cells from the perichondrium that flank the hypertrophic chondrocytes differentiate into bone-producing osteoblasts and form the periosteum. In the

region around the cellular hypertrophy, the surrounding matrix and vascularized tissue undergo calcification. Growth plates form at the ends of the long bones and separate the distal, cartilaginous epiphyses from the medial, bony diaphyses. These growth plates are made up of chondrocytes in different stages of hypertrophy. As terminally differentiated chondrocytes are continuously replaced by bone, the balance between chondrocyte proliferation and differentiation must be tightly controlled in order to allow proper longitudinal growth of the skeletal elements.

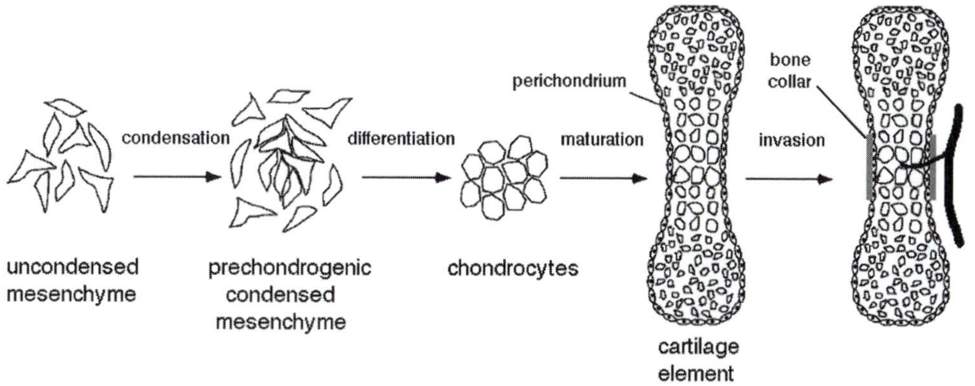

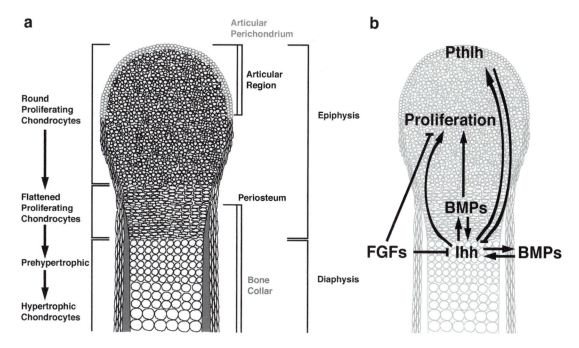

Figure 7-2 Endochondral bone formation in limb development. Mesenchymal cells in the limb condense and cells within the condensations differentiate into chondrocytes. Chondrocytes at the center of the condensation mature and become hypertrophic, while those at the end of the bone continue to proliferate, lengthening the bone. The bone collar forms adjacent to the hypertrophic chondrocytes, from cells in the perichondrium. In the mineralized region of apoptosing chondrocytes, invasion of the vasculature brings osteoblast and osteoclast precursors into the center of the bone.

Figure 7-3 Molecular regulation of chondrocyte development. (a) Schematic representation of the different zones of maturation within the cartilage element. Cartilage differentiation occurs progressively as chondrocytes mature from a population of highly proliferative cells (at the epiphysial end) to hypertrophic chondrocytes. (b) Integration of FGF, BMP, and IHH/PTHLH signaling pathways during chondrogenesis. (a adapted from Hartmann and Tabin 2000; b adapted from Minina et al. 2002.)

The coordination of chondrocyte differentiation and maturation is tightly regulated both temporally and spatially. A number of signaling pathways (FGF, BMP, PTHLH, IHH, WNT) are involved in orchestrating this multistep process leading from mesenchymal condensation to bone formation. Mutations in genes from these pathways adversely affect skeletal development and it is from studying these mutations in mice and through experimental manipulations in chicks that much has been learned about tetrapod limb development.

Fibroblast Growth Factor (FGF) Signaling

Signaling by FGFs plays a critical role in controlling chondrocyte differentiation. Mutations in human *FGF receptor 3* gene (*Fgfr3*) results in a number of dwarfism syndromes including hypochondroplasia, achondroplasia, and thanatophoric dysplasia due to aberrant activation of FGF signaling (Bellus et al. 1995; Rousseau et al. 1994; Shiang et al. 1994; Tavormina et al. 1995). Mouse models of the human achondroplasia phenotype have been established by expressing mutated forms of FGFR3 in the mouse cartilage anlagen (L. Chen et al. 1999; Iwata et al. 2000, 2001; Naski et al. 1998; Segev et al. 2000). These mice display shortened skeletal elements due to reduction in the number of proliferating and hypertrophic chondrocytes. The opposite result, an expanded zone of proliferating chondrocytes, can be achieved by a genetic deletion of *Fgfr3* (Colvin et al. 1996; Deng et al. 1996) or *Fgf18* (Z. Liu et al. 2002; Ohbayashi et al. 2002).

These studies have led to the conclusion that FGF signaling is a negative regulator of chondrocyte proliferation and that FGF18 may be the preferred ligand for FGFR3 in chondrocytes. FGF signaling is thought to be mediated by activation of the Janus Kinase/signal transducer and activator of transcription-1 (JAK-STAT1) pathway, which in turn leads to the upregulation of the cell cycle inhibitor p21/WAF1/CIP1, which inhibits chondrocyte proliferation (C. Li et al. 1999; Sahni et al. 1999, 2001). In addition to inhibiting proliferation, FGF signaling induces the onset of hypertrophic differentiation and accelerates the differentiation process (Minina et al. 2002).

Parathyroid Hormone-like Peptide (PTHLH) and Indian Hedgehog (IHH) Signaling

Pthlh is expressed in the periarticular region and acts via receptors expressed on chondrocytes (K. Lee et al. 1995). The receptor for PTHLH (PTHR) is most strongly expressed in the transition zone between chondrocyte proliferation and hypertrophy (Vortkamp et al. 1996). *Pthlh* (and *Pthr*) mutant mice are born with long bone defects including very short limbs (Karaplis et al. 1994). Looking more closely, one sees that *Pthlh-/-* chondrocytes are arrayed irregularly and *Pthlh* function seems to be necessary for normal proliferation of chondrocytes. In *Pthlh* mutants lacking this proliferation, there is premature differentiation of chondrocytes into hypertrophic chondrocytes resulting in early bone formation, which causes shorter limbs. Overexpression of *Pthlh* or constitutively active PTHR has the opposite effect, a delay in endochondral ossification due to persistent chondrocyte proliferation and abnormal chondrocyte hypertrophy (Weir et al. 1996; Lanske et al. 1996).

Pthlh expression and function are under the control of IHH signaling. IHH is synthesized and secreted by chondrocytes in the growth plate in the region of transition between proliferating and hypertrophic chondrocytes (Bitgood and McMahon 1995). The IHH signal appears to act directly on the chondrocytes, as well as by an unknown mechanism, perhaps through the perichondrium, on the periarticular region to control the levels of *Pthlh* expression. *Ihh* overexpression in chicks causes an increase in *Pthlh* expression, whereas *Ihh-/-* mice lack *Pthlh* expression (Vortkamp et al.

1996; St-Jacques et al. 1999). *Ihh* overexpression leads to broader and shorter cartilage elements (Vortkamp et al. 1996), a similar phenotype to that of overexpression of *Pthlh* or constitutive activation of PTHR (Schipani et al. 1997). The cause appears to be a delay in the onset of hypertrophic differentiation of chondrocytes. The skeletal elements persist as continuous cartilaginous cores lacking normal growth plates. Conversely, *Ihh* mutant mice show a significant decrease in chondrocyte proliferation and accelerated terminal differentiation (St-Jacques et al. 1999).

Additional studies have placed IHH upstream of *Pthlh*. *Pthlh* is required for the response to IHH signaling and the chondrocytes that express *Pthlh/Pthr* are the physiological target of the IHH pathway. Further, activation of IHH signaling in wild-type mice, but not *Pthlh* mutant mice, inhibits hypertrophic differentiation of chondrocytes (Lanske et al. 1996; Vortkamp et al. 1996). A model suggests that IHH serves as a measure for the number of chondrocytes undergoing hypertrophic differentiation, while PTHLH prevents chondrocytes from initiating the differentiation process. Thus, these factors maintain the balance between proliferating and hypertrophic chondrocytes.

Bone Morphogenetic Protein (BMP) Signaling

BMPs were first isolated as chondrogenic factors. Their potency in promoting chondrogenesis *in vitro* and *in vivo* likely reflects their ability to stimulate multiple steps of chondrogenesis. BMPs act at early steps to regulate the aggregation of mesenchymal cells into prechondrogenic condensations and the differentiation of chondroprogenitors into chondrocytes. Inhibition of BMP signaling by expression of the BMP antagonist Noggin results in absence of early condensations (Pizette and Niswander 2000; Capdevila and Johnson 1998). As the cartilage elements continue to develop, many BMP ligands (BMP 2, 3, 4, 5, 7) are expressed in the perichondrium flanking *Ihh* expression in the mouse (Pathi et al. 1999; Zou et al. 1997; Daluiski et al. 2001; Haaijman et al. 2000), and *Bmp7* is expressed in proliferating chondrocytes. BMPs signal through the type II receptor, BMPRII, and two type I receptors, BMPRIA and BMPRIB. *BmprIa* is expressed fairly ubiquitously and then becomes concentrated in the prehypertrophic chondrocytes, along with *BmprII*, while *BmprIb* is expressed in the mesenchymal prechondrogenic condensations and immature chondrocytes and later in the perichondrium (DeWulf et al. 1995; Zou et al. 1997; Kawakami et al. 1996; Yamaji et al. 1994; Merino et al. 1998; Yi et al. 2000; Haaijman et al. 2000). Overexpression of BMPs or constitutive activation of BMPRIA or BMPRIB results in greatly enlarged skeletal elements by stimulating mesenchyme cells to undergo chondrogenic aggregation and differentiation and by increasing chondrocyte proliferation within the cartilage element (Duprez et al. 1996a, 1996b; Zou et al. 1997). BMPs also negatively regulate the process of hypertrophic differentiation by delaying the maturation of terminally hypertrophic cells, thus further contributing to enlargement of the elements (Minina et al. 2001).

Bmp5^-/- mice have slightly shorter long bones, and some condensations are missing, which suggests some functional redundancy between the different family members (D. M. Kingsley et al. 1992; J. A. King et al. 1994). Indeed, downregulation of BMP signaling via expression of Noggin or dominant-negative BMP receptors results in greatly reduced bone growth (Capdevila and Johnson 1998; Pizette and Niswander 2000). Normally, *Noggin* is expressed in prechondrogenic condensations and chondrocytes. In *Noggin^-/-* mice, the limb skeletal elements are enlarged and joint formation is disrupted (Brunet et al. 1998). This highlights the critical balance that must be achieved *in vivo* between BMPs and their antagonist in the proper regulation of cartilage size.

The type I receptors BMPRIA and BMPRIB also appear to have partially redundant functions in skeletal development. The role of *BmprIa* in limb chondrogenesis has been difficult to analyze due to the early lethality of *BmprIa^-/-* mutants (Mishina et al. 1995); however, *BmprIb^-/-* mutants have been examined. Loss of BMPRIB activity does not affect prechondrogenic mesenchyme or the initial condensations, but there are defects in the differentiation and proliferation of chondrocytes in the phalanges (Yi et al. 2000; S. T. Baur et al. 2000). Constitutively active forms of either BMPRIA or BMPRIB promote chondrogenesis and osteogenic differentiation while dominant negative forms of BMPRIB (but not BMPRIA) block these events (Zou et al. 1997; Enomoto-Iwamoto et al. 1998; D. Chen et al. 1998). The importance of BMP signaling is further evidenced in human syndromes. Human brachydactyly, characterized by shortening of the digits, is associated with mutations in *BmprIb* (Lehmann et al. 2003) and *Gdf5*; a TGFβ ligand (Polinkovsky et al. 1997) and *Noggin* are required for joint formation (Y. Gong et al. 1999). Together these case studies in humans and experimental studies on loss and gain of BMP function in chick and mouse embryos provide evidence for the reiterative role of BMP signaling in the regulation of proliferation and differentiation throughout the process of chondrogenesis.

Interaction between the IHH, PTHLH, BMP, and FGF Signaling Pathways

The key signaling pathways highlighted above are critically required for the regulation of chondrocyte proliferation and differentiation. Interactions among these molecular signals serve to coordinate the size and state of differentiation of the cartilage elements. As described above, IHH modulates the levels of *Pthlh* and thus sets up a negative feedback loop that coordinates the number of chondrocytes undergoing hypertrophic differentiation. IHH and BMP signals interact, as well as intersecting, at several stages of chondrocyte development. IHH signaling induces the expression of *Bmp2* and *Bmp4* in the perichondrium (Pathi et al. 1999), and in turn BMP signaling positively regulates the expression of *Ihh* in cells outside the range of the PTHLH signal. BMPs act though IHH to regulate *Pthlh* expression and the onset of hypertrophic differentiation (Minina et al. 2001). However, IHH and BMPs act independently and in parallel to regulate chondrocyte proliferation (Minina et al. 2001).

FGF signaling acts antagonistically to IHH and BMP signaling. FGFs negatively regulate chondrocyte proliferation, while IHH and BMPs act in parallel to positively regulate proliferation, yet FGFs act independently of both IHH and BMPs (Minina et al. 2002). FGF also accelerates the onset and rate of hypertrophic differentiation, again in contrast to the activities of BMP and IHH. FGFs negatively regulate, and BMP positively regulates, the expression of Ihh in hypertrophic chondrocytes. Thus, the balance of FGF and BMP signals regulates the level of *Ihh* expression (fig. 7.3b). This balance modulates the IHH-PTHLH feedback loop. The complex interplay between these four signaling systems ultimately dictates the distance from the joint at which hypertrophic differentiation takes place (Minina et al. 2002).

WINGLESS/INT (WNT) Signaling

A number of components of the WNT signaling pathway are expressed during chondrogenesis and osteogenesis. *Wnt5a* is expressed at the border between proliferative and prehypertrophic chondrocytes (Y. Yang et al. 2003), and *Wnt5a*, *Wnt4*, and *Wnt14* are expressed in the joints (Hartmann and Tabin 2000, 2001) while *Wnt5b* and *Wnt11* are expressed in prehypertrophic chondrocytes (Church et al. 2002; Hartmann and Tabin 2000; Lako et al. 1998). *Wnt5b* and *Wnt5a* are also expressed in the perichondrium. Overexpression experiments with *Wnt* genes suggest that they act at multiple steps of chondrogenesis and those different WNT ligands may play different roles. WNT4 appears to block initiation of chondrogenesis and accelerates terminal chondrocyte differentiation and perichondrium maturation (Hartmann and Tabin 2000). Overexpression of *Wnt11*, *Wnt5a*, or *Wnt5b* results in truncated limbs, but the underlying causes are not the same (Kawakami et al. 1999; Church et al. 2002). WNT11 does not appear to delay chondrocyte differentiation, so the mechanism underlying limb truncations in these animals is unknown. WNT5a and WNT5b both act to delay hypertrophy of chondrocytes, but WNT5a inhibits proliferation of chondrocytes, whereas WNT5b promotes their proliferation. Moreover, these WNTs appear to differentially regulate the transition from resting to proliferating chondrocytes (WNT5a inhibits this transition, whereas WNT5b promotes it) and the transition from proliferating to prehypertrophic chondrocytes (promoted by WNT5a and inhibited by WNT5b). In this manner, WNT5a and WNT5b act to coordinate chondrocyte proliferation and differentiation to regulate longitudinal growth.

In loss-of-function studies, *Wnt5a*$^{-/-}$ limbs are shortened and the distal digits are missing (Yamaguchi et al. 1999). Studies of the digit phenotype suggest an early role for WNT5a at or before the stage of mesenchyme condensation. Moreover, WNT5a acts later to regulate both chondrocyte differentiation, by controlling the transition from proliferating to prehypertrophic chondrocytes, and osteoblast differentiation (Y. Yang et al. 2003). Interestingly, WNT5a appears to function in the distal limb to down-regulate canonical WNT signaling by promoting the degradation of β-CATENIN, a downstream mediator of canonical WNT signaling (Topol et al. 2003). The antagonism of canonical WNT signaling by WNT5a is required for the proper regulation of β-CATENIN levels during skeletal formation. This is consistent with the data that β-CATENIN acts as a negative regulator of chondrocyte differentiation (Ryu et al. 2002).

FRZB-1, a WNT antagonist, is expressed in the proliferative and prehypertrophic chondrocytes and in the articular cap (Hoang et al. 1996; Hoang et al. 1998; Ladher et al. 2000; Wada et al. 1999). Interestingly, *Frzb-1* overexpression causes phenotypes similar to those obtained by the overexpression of *Wnt* genes, namely shortening of skeletal elements, joint fusions, and delayed chondrocyte maturation, while constitutively active Lymphoid Enhancer Factor 1 (LEF-1; a downstream mediator of WNT signaling) promotes chondrocyte maturation, hypertrophy, and calcification (Kitagaki et al. 2003; Enomoto-Iwamoto et al. 2002). It is unclear how these contrasting results can be reconciled. It is likely that more detailed analysis of how these molecular manipulations affect the different zones of resting, proliferating, and differentiated chondrocytes, coupled with additional loss and gain of function experiments that are restricted in time and space will be revealing. It will also be of interest to determine the relationship between WNT and other signaling pathways. Studies to date indicate that WNT regulation of chondrocyte differentiation occurs independently of the IHH/PTHLH pathway (Y. Yang et al. 2003). The relationship of WNT signaling to FGFs and BMPs has yet to be established. Moreover, the transcription factors that directly mediate the transcriptional effects of these signals have not been identified genetically, although there are clear candidates from biochemical studies and molecular misexpression.

Sox Genes and Chondrogenesis

Some key transcription factors involved in chondrogenesis and osteogenesis have been identified. However, their relationship to the above signaling pathways is just beginning to be elucidated. Multiple members of the SOX family of transcription factors regulate skeletal development. *Sox5*, *Sox6*, and *Sox9* are expressed in mesenchymal condensations and in all chondroprogenitor cells and differentiated chondrocytes except the hypertrophic chondrocytes (E. Wright et al. 1995; L. J. Ng et al. 1997; Q. Zhao et al. 1997; LeFebvre et al. 1998). Mice that lack both *Sox5* and *Sox6* exhibit severe chondrodysplasia, with almost a complete lack of cartilage (Smits et al. 2001). Chondrogenic cells in these mutants arrest at the stage of mesenchymal condensations, demonstrating that SOX5 and SOX6 are needed for proper differentiation of chondrocytes.

Sox9 mutations were first identified in humans as the underlying cause of campomelic dysplasia, which results in hypoplasia of most endochondral bones (Foster et al. 1994; T. Wagner et al. 1994). Conditional mutations in mice that remove expression of *Sox9* from limb mesoderm have uncovered the essential role of SOX9 for chondrocyte differentiation and survival. Loss of *Sox9* expression results in a nearly complete loss of mesenchymal condensations in the developing limbs (Akiyama et al. 2002). The mesenchymal cells that would normally contribute to these condensations instead undergo apoptosis. SOX9 appears to lie upstream of the other *Sox* genes in prechondrocyte differentiation. While *Collagen 2aI*, an early cartilage matrix marker, is still expressed in *Sox5/Sox6* double mutants, it is lost in *Sox9* mutants. In addition, the expression of *Sox5* and *Sox6* is lost in *Sox9* mutants, indicating that SOX9 is the key upstream transcriptional regulator of mesenchyme condensation and differentiation (Akiyama et al. 2002). The complete loss of these genes results in defects in early steps of chondrogenesis. The role of SOX9 in later steps of chondrogenesis and osteogenesis has been examined by conditional removal of *Sox9* activity using a *Collagen 2a1*-promoter driving Cre recombinase in cartilage elements after mesenchyme condensation and initial differentiation into chondrocytes. This leads to decreased chondrocyte proliferation and disrupted chondrocyte differentiation including downregulation of IHH and PTHLH signaling components (Akiyama et al. 2002). Furthermore, it has been shown that SOX9 transcriptional activity can be enhanced by PTHLH signaling, and this may indicate that SOX9 mediates some of the effects of PTHLH on chondrocyte maturation (W. Huang et al. 2001). Indeed, loss of one copy of *Sox9* results in premature bone formation (Bi et al. 2001), indicating that the *Sox* genes are likely to play reiterative roles throughout skeletal development.

Hox Genes and Chondrogenesis

The HOX homeobox transcriptional regulators are also critical for proper chondrogenesis. In particular, the posterior group *Hox* genes (*Hox9–13*) from the *Hoxa* and *Hoxd* clusters are necessary for limb skeletal development (Dollé et al. 1993; Small and Potter 1993; A. P. Davis and Capecchi 1994, 1996; A. P. Davis et al. 1995; Fromental-Ramain et al. 1996a, 1996b; E. M. Carpenter et al. 1997; Wellik and Capecchi 2003). Paralogs of the *Hoxa* and *Hoxd* gene clusters (for instance *Hoxa9* and *Hoxd9*) are functionally redundant, and they serve in a dose-dependent manner to regulate the size of specific cartilage elements.

The expression of the *Hox* genes in the developing limbs is temporally and spatially dynamic. For the purposes of this review, it is relevant that the *Hox* members are expressed in largely complementary patterns with *Hoxa9/d9* within the stylopod (humerus/femur) precursor region, *Hoxa11/d11* in the zeugopod (ulna and radius/tibia and fibula) region, *Hoxa12/d12* in the wrist/ankle region, and *Hoxa13/d13* in the autopod region (Nelson et al. 1996; fig. 7.4). Complete loss of these pairs of genes results in severe reductions of the corresponding skeletal element. For instance, loss of all four alleles of *Hoxa11* and *Hoxd11* leads to greatly reduced zeugopod elements, whereas loss of all alleles of *Hoxa13* and *Hoxd13* result in loss of the autopod (A. P. Davis et al. 1995; Fromental-Ramain et al. 1996b; Wellik and Capecchi 2003). The retention of one copy of either *Hoxa* or *Hoxd* gene results in a partial rescue of the corresponding element, and retention of two copies of either gene can rescue more fully. This highlights the functional redundancy of these paralogous genes and their dose-dependent requirement during limb skeletal development.

A surprising finding was that these *Hox* genes are not required for limb patterning per se, as the early cartilage condensations are established relatively normally (A. P. Davis et al. 1995; Boulet and Capecchi 2004). Instead, the *Hox* genes act later to regulate longitudinal growth of the individual elements. Chondrocyte proliferation and differentiation continues, but the growth plates are not established properly in *Hox* compound mutants, leading to dramatic misshaping and shortening of specific skeletal elements (Boulet and Capecchi 2004).

Osteogenesis in Limbs

Generalized Steps in Osteogenesis

Following their differentiation, hypertrophic chondrocytes become surrounded by a calcified ECM and then die via

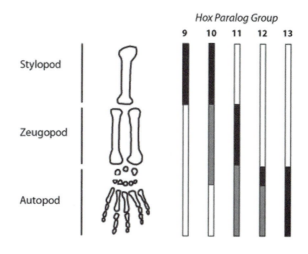

Hox Gene Patterning in the Forelimb

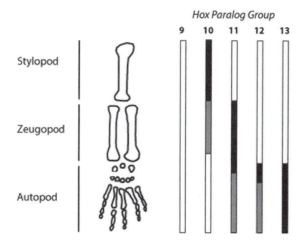

Hox Gene Patterning in the Hindlimb

Figure 7-4 *Hox* gene regulation of limb skeletal development. Schematic representation of the functional domains (black regions) of posterior group (5') *Hox* genes in limb patterning. *Hox9* and *Hox10* paralogs function together to pattern the forelimb stylopod. *Hox9* paralogs have no function in the hindlimb. *Hox10* paralogs also contribute somewhat to the zeugopod (gray region). *Hox11* paralogs function mainly in the developing zeugopod, with some contribution to the autopod (gray region). *Hox13* paralogs act mainly in the autopod (*Hox12* can substitute for *Hox13* in the autopod). (Adapted from Wellik and Cappechi 2003.)

apoptosis. This mineralized ECM is then invaded by the vasculature from the bone collar. This vascular invasion brings in progenitors of osteoblasts (bone forming cells) and osteoclasts (bone resorbing cells). Osteoblasts are of mesenchymal origin and are derived from the stromal cells of the periosteum. Osteoblasts replace the cartilaginous ECM with a bone ECM (reviewed in Karsenty and Wagner 2002). In contrast, osteoclasts are large multinucleated cells that are derived from monocytes in blood and marrow (Udagawa et al. 1990; Chambers 2000). Osteoclasts are crucial to proper skeletal formation as they are the principle cells involved in bone resorption and, together with osteoblasts, regulate bone remodeling.

Invasion of the Hypertrophic Cartilage by the Vasculature

As the blood vessels from the bone collar invade the ECM surrounding the hypertrophic chondrocytes, they bring osteoblast and osteoclast cells into the center of the bone. Several key factors have been implicated in the invasion process. These include Matrix Metalloproteinase 9 (MMP9) and Vascular Endothelial Growth Factor (VEGF). MMP9 is expressed in chondroclasts, cartilage-resorbing bones derived from bone marrow (Vu et al. 1998; Engsig et al. 2000). Mutations in *Mmp9* cause a transient shortening of the long bones and uncovered a role for MMP9 in cartilage removal and angiogenesis promotion (Vu et al. 1998). Further studies with *Mmp9*-deficient chondrocytes showed that they are defective in their ability to release VEGF, resulting in reduced proliferation and migration of vascular endothelial cells (Engsig et al. 2000). Loss of VEGF activity specifically in chondrocytes results in a similar phenotype to that observed in *Mmp9* mutant mice, namely a sharp reduction in vascular invasion, leading to an expansion of the hypertrophic chondrocyte zone and reduced bone length (Haigh et al. 2000). The transcription factor RUNX2/CBFA1/OSF2, which is expressed in prehypertrophic chondrocytes, is required for induction of *vegf* expression in hypertrophic chondrocytes. Thus in *Runx2/Cbfa1* mutants, there is no vascular invasion in any skeletal elements (Zelzer et al. 2001). This suggests that VEGF, synthesized by hypertrophic chondrocytes, is held in the cartilaginous ECM. VEGF is released when the ECM is degraded by MMP9 and then it can bind its receptor on vascular endothelial cells, thus inducing vascular invasion followed by cartilage replacement by bone (Karsenty and Wagner 2002).

The Genetic Mechanisms Regulating Osteoblast Differentiation

In addition to its expression in prehypertrophic chondrocytes, *Runx2/Cbfa1* is the earliest marker of osteoblasts and is expressed in mesenchymal cells (osteochondroprogenitors) that will become either osteoblasts or chondrocytes (Ducy et al. 1997). Later *Runx2/Cbfa1* is expressed primarily in perichondrial and osteoblast cells and at lower levels in hypertrophic chondrocytes (I. S. Kim et al. 1999). Loss of *Runx2/Cbfa1* in mice results in a normally patterned skeleton, yet it is composed only of chondrocytes that secrete a cartilage matrix, and no bone is formed, showing that RUNX2/CBFA1 is critically required for osteoblast formation (Komori et al. 1997; Otto et al. 1997). Ectopic expres-

sion of *Runx2/Cbfa1* causes cells to express other osteoblast-specific genes like *Osteocalcin* (*Osc*). Moreover, transgenic expression of *Runx2/Cbfa1* can rescue the defects of *Runx2/Cbfa1⁻/⁻* mice and can cause hypertrophy and mineralization in regions that normally remain cartilaginous (S. Takeda et al. 2001; Ueta et al. 2001). Human patients heterozygous for *Runx2/Cbfa1* loss of function alleles display one of the most frequent skeletal disorders: cleidocranial dysplasia (CCD; Mundlos et al. 1997; B. Lee et al. 1997). *Runx2/Cbfa1* expression is also lost in *Sox9* mutant limbs suggesting that *Sox9* is necessary for osteoblast differentiation, further implicating the SOX transcription factors in the regulation of multiple steps of chondrocyte and osteoblast differentiation.

Additional transcription factors involved in the regulation of osteoblast function include the *Dlx* and *Msx* genes. *Dlx5* and *Dlx6* have somewhat redundant functions and are expressed in the periosteum surrounding cartilage templates (Simeone et al. 1994). While loss of *Dlx5* results in a mild delay in chondrogenic differentiation and bone formation (Acampora et al. 1999; Bendall et al. 2003), deletion of both *Dlx5* and *Dlx6* results in a delay or absence of endochondral ossification in all bones of the appendicular and axial skeleton (Robledo et al. 2002). Although *Runx2/Cbfa1* is expressed at normal levels in the chondrium and perichondrium of *Dlx5/Dlx6* mutants, fewer cells in the perichondrium express *Runx2/Cbfa1*. *Msx2* is expressed in resting and proliferative chondrocytes in the growth plate as well as the perichondrium, the preosteoblasts, and osteoblasts of the femur and tibia. *Msx2* mutants show a decrease in *Runx2/Cbfa1*, a lack of osteoblast formation, and downregulation of a number of chondrogenic markers (Satokata et al. 2000). These data suggest that *Msx2* is necessary for osteoblast formation while *Dlx5/Dlx6* are required for osteoblast maturation.

A number of signaling molecules are also thought to play a role in osteoblast differentiation. *Fgf18* mutants display a delay in ossification, possibly due to defects in the activation of *Fgfr1* and *Fgfr2* in the perichondrium (Z. Liu et al. 2002; Ohbayashi et al. 2002). In addition to its role in chondrocyte differentiation, IHH has also been shown to affect osteoblast differentiation. *Ihh⁻/⁻* mutant mice lack bone in skeletal elements that normally form via endochondral ossification. *Runx2/Cbfa1* expression is decreased in these mutants, and chondrocyte apoptosis is not followed by osteoblast differentiation (St-Jacques et al. 1999). Overexpression of *Ihh* in all chondrocytes promotes an expansion of the bone collar toward the epiphyses. Likewise, genetic ablation of the transmembrane protein SMOOTHENED (SMO, which is essential for transducing all HEDGEHOG signals), from the perichondrial cells results in the loss of the normal bone collar. *Smo⁻/⁻* cells do not contribute to the osteoblast population but instead give rise to ectopic chondrocytes (F. Long et al. 2004). This highlights a critical role for IHH in generating the osteoblast lineage.

A role for the WNT signaling pathway in osteogenesis has also been uncovered. In contrast to the role of IHH signaling in osteoblast differentiation, WNT signaling seems to regulate osteoblast proliferation. Two proteins (LRP5, LRP6) related to the Low-Density Lipoprotein Receptor have been shown to act as coreceptors for the canonical WNT signaling pathway (Tamai et al. 2000). Mutations in the human *Lrp5* gene have been linked to defects in bone accrual. Gain of function mutations in *Lrp5* result in high bone density (Boyden et al. 2002; Little et al. 2002), while inactivating mutations in humans (Gong et al. 2001) and mice (Kato et al. 2002) cause a low bone mass phenotype. *Lrp5* is widely expressed during development, including expression in osteoblasts lining the endosteal and trabecular bone surfaces. LRP5 acts as a WNT receptor in osteoblasts, and other components of the Wnt signaling pathway—including *Wnt1*, *frizzled2*, *frizzled6*, *Lef-1*, and *Tcf-4*—are expressed in primary osteoblasts (Kato et al. 2002). *Wnt1* is absent from, and *Tcf-4* expression is markedly decreased in, *Lrp5−/−* osteoblasts. The underlying defect in the loss-of-function studies appears to be reduced proliferation of osteoblasts, which causes a decrease in bone formation (Kato et al. 2002). Interestingly, *Runx2/Cbfa1* expression is normal in these mice, indicating that osteoblast differentiation is normal in *Lrp5* mutants and the bone phenotype must occur in a RUNX2/CBFA1-independent manner.

The Genetic Mechanisms Regulating Osteoclast Differentiation

Osteoclasts, a cell type derived from monocytes, play a vital role in skeletal formation as they are the primary cell type that is responsible for bone resorption due to their unique capacity to degrade mineralized tissues. Together with osteoblasts, oosteoclasts play a pivotal role in bone remodeling. For example, the common human disorder osteoporosis is caused by a loss of bone, which can occur by a disruption in either osteoblast or osteoclast function. Loss of osteoclasts or their function results in osteopetrosis, a group of diseases characterized by exceedingly dense bone and loss of the bone marrow cavity. Multiple genes have been shown to regulate different steps of osteoclast biology. At least 24 genes have been shown to regulate osteoclastogenesis and osteoclast activation (Marks 1989; McLean and Olsen 2001; Boyle et al. 2003). Disruption of these genes can lead to blocked osteoclast function, resulting in osteopetrosis, or increased activation of osteoclasts, giving rise to osteopenia or reduced mineralization of the bone.

The transcription factor PU.1 is the earliest marker of osteoclast differentiation and is essential for macrophage differentiation (Tondravi et al. 1997). *PU.1* mutant mice lack both macrophages and osteoclasts. The maturation of macrophages into osteoclasts requires interactions between osteoclast precursors and the marrow stromal cells. Osteoclasts and their precursors also express c-FOS, the receptor for Macrophage Colony-Stimulating Factor (M-CSF) and Receptor for Activation of Nuclear Factor κB (RANK), which is activated by the NFκB-like ligand called RANKL (also called ODF/OPGL/TRANCE). These two stromal-derived hematopoietic factors, M-CSF and RANKL, are both necessary and sufficient for osteoclastogenesis (Yasuda et al. 1998; Lacey et al. 1998; Hofbauer et al. 1999). They activate signal transduction through c-FOS and RANK on the osteoclast precursor cells to promote osteoclastogenesis (Nakagawa et al. 1998; Hsu et al. 1999). Together, M-CSF and RANKL are required to induce expression of genes that typify the osteoclast lineage and are required for osteoclast maturation such as *tartrate-resistant acid phosphatase (TRAP)*, *cathepsin K (CATK)*, *calcitonin receptor,* and the *β3-integrin* (Lacey et al. 1998).

M-CSF causes osteoclast precursors to proliferate while RANKL stimulates the pool of M-CSF-expanded precursors to differentiate into osteoclasts (Burgess et al. 1999). This differentiation process involves a number of structural changes in the osteoclast precursors that prepare it to resorb bone. c-FOS also regulates the commitment of hematopoitic precursors to become osteoclasts instead of mature macrophages. Loss of *c-Fos* or *NFκB* components (p50 and p52 subunits) results in an arrest in osteoclast differentiation (Grigoriadis et al. 1994; Iotsova et al. 1997; Franzoso et al. 1997), while overexpression of c-Fos causes chondroblastic osteosarcomas (Ruther et al. 1989).

The interaction between RANK and RANKL can be disrupted by osteoprotegrin (OPG), which exerts an inhibitory effect on osteoclastogenesis by acting as a secreted decoy receptor for RANKL. OPG competes with RANK for RANKL and hence modulates osteoclast differentiation (Lacey et al. 1998). Overexpression of OPG blocks osteoclast production and leads to osteopetrosis, while mice lacking OPG develop severe osteoporosis and have accelerated osteoclastogenesis (Simonet et al. 1997). RANK associates with TNF receptor-associated factor 6 (TRAF6) intracellulary. TRAF6−/− mice have dysfunctional osteoclasts (Lomaga et al. 1999). Together with RANKL's role in osteoclastogenesis, this suggests a role for RANK in osteoclast activation. Thus, signaling by M-CSF/c-FOS and RANKL/RANK is required for osteoclast formation, and together they induce osteoclast differentiation and activate mature osteoclasts to resorb mineralized bone.

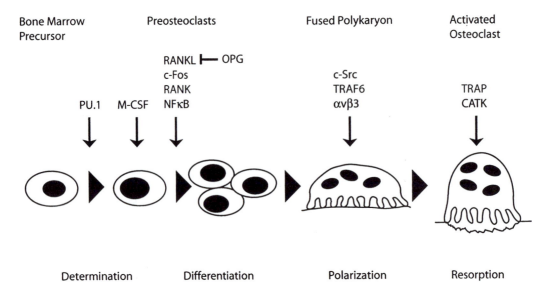

Figure 7-5 Regulation of osteoclast differentiation. Schematic representation of the development of osteoclasts from hematopoietic precursors. Osteoclasts are fused polykaryons arising from multiple individual cells. Several of the key factors required for osteoclast formation and function are illustrated. Mouse mutants defective for factors required early in osteoclast differentiation lack osteoclasts, while mutants lacking factors at later steps of oseoclast maturation have nonfunctional osteoclasts. Both classes of mutants are osteopetrotic. (Adapted from Teitelbaum 2000.)

Once osteoclasts are formed, they require the actions of a number of genes in order to properly resorb bone. The initial events in bone degradation are recognition of the bone by the osteoclasts and the attachment of osteoclasts to the bone matrix. This involves changes in the morphology of the osteoclast cell such that a ruffled membrane forms, which exposes the resorptive organelle to the bone and allows enzymes such as TRAP and CATK to begin degradation of the bone mineral and collagen matrices. Recognition of bone appears to be largely regulated by αvβ3 INTEGRIN, expressed on the osteoclast (Engleman et al. 1997). Loss of the *β3-integrin* subunit results in abnormal ruffled membranes and poor bone resorption (McHugh et al. 2000). Attachment via αvβ3 INTEGRIN causes activation of *c-Src* (Engleman et al. 1997; Chellaiah et al. 1998). *c-Src*-deficient mice produce large numbers of osteoclasts; however, they fail to form a ruffled membrane and thus do not resorb bone, which results in osteopetrosis (Soriano et al. 1991). Signaling from RANK culminates in changes in morphology and gene

expression patterns that characterize the active osteoclast (fig. 7.5). The overall balance between osteoblast and osteoclast function needs to be strictly controlled to maintain the healthy biology of the bone throughout the lifetime of the organism.

The formation of the limb skeleton is regulated by a network of gene interactions that serve to promote or to inhibit chondroctye proliferation and differentiation as well as bone formation and resorption. This intricate relay is necessary to balance and coordinate the number of cartilage cells and the rate in which they are replaced by bone. Fine-tuning of the limb skeletal program likely underlies the vast array of species-specific morphologies. The mystery remains as to how these concerted yet common pathways are influenced by the genes that specify the pattern of the limb, thus forming the skeletal elements with the correct shape and size and complex set of articulations to create a fully functional skeleton.

Chapter 8 Apoptosis in Fin and Limb Development

Vanessa Zuzarte-Luís and Juan M. Hurlé

POPTOSIS IS AN active cellular suicide program that is essential for the construction, maintenance, and repair of tissues during embryonic development and also in adult animals. All nucleated animal cells, including invertebrate cells, constitutively express all the proteins required to initiate the process of apoptosis (reviewed by Jacobson et al. 1997). This fact suggests that programmed cell death is a basic feature of all animal cells.

The idea of cell death as an important physiological mechanism is almost as old as the discovery that organisms are made of cells. It was first described in amphibian development by Vogt in 1842 (see P. G. Clarke and Clarke 1995) and later in other developing invertebrates and vertebrates. The term "apoptosis" was proposed by J. F. Kerr and colleagues (1972), who described a process of cell death that was commonly observed in many tissues and cell types. Judging from its frequency and the similar morphological features dying cells share, the authors proposed that this cell death results from an endogenous program.

We know now that apoptosis is strictly genetically controlled. Studies in the nematode *Caenorhabditis elegans* allowed the identification of the molecular activators of this process, signals later identified in vertebrates. The molecular machinery responsible for apoptosis exhibits a high degree of evolutionary conservation (fig. 8.1). Four functional groups of genes have been identified as regulating apoptosis in *C. elegans* (*Ced-3, Ced-4, Ced-9,* and *Egl-1*). In vertebrates, these functional groups are conserved, and each group includes

many different genes. The homologue of *Ced-3* in vertebrates is the large family of caspases, which are the direct effectors of the death program. The homologue of *Ced-4* is *Apaf-1* (*Apoptotic Protease-Activating Factor*), the prototype of a family of pro-apoptotic factors with the role of activating caspases. *Ced-9* in *C. elegans* inhibits *Ced-3* and *Ced-4*. In vertebrates this gene is represented by the large *Bcl-2* gene family, which includes inhibitors of cell death and pro-apoptotic factors. *Egl-1* (*Egg laying defective*) in *C. elegans* promotes apoptosis by inhibiting Ced-9. The homologues of *Egl-1* in vertebrates include several members that repress the anti-apoptotic activity of *Bcl-2* (see review by Hurlé and Merino 2002).

Interest in the process of apoptosis and its regulation has grown dramatically in the last few years, because many human viral and degenerative diseases have been found to be related to the misregulation of the apoptotic program in specific cell populations. In addition, advances in knowledge of apoptotic machinery have provided a conceptual framework for the development of new therapeutic strategies for a large number of diseases, including cancer.

Histogenetic and Morphogenetic Cell Death in Animal Development

During embryonic development cell death occurs as a physiological and genetically controlled process, the alteration of

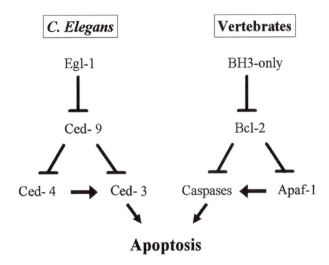

Figure 8-1 Schematic representation of the apoptotic molecular machinery in *C. elegans* and their corresponding homologues in vertebrates.

which leads to many malformations. The distinct patterns of apoptosis observed in animal species at different stages of development and in various organs have different significances. In some cases the apoptotic process sculpts the shape of developing organs. In other cases apoptosis implies removal of cells associated with tissue maturation or remodeling, but without changing organ morphology. These two types of apoptotic processes are usually termed, respectively, *morphogenetic* and *histogenetic apoptosis* (Glücksmann 1951).

Cell death is a constant feature associated with developing appendages (see Zuzarte-Luís and Hurlé 2002). Areas of cell death associated with morphogenesis have been reported in amniote embryos including birds, reptiles, and mammals (see below). The most remarkable areas of morphogenetic apoptosis appear in the undifferentiated mesoderm of the limb bud, where they regulate the amount of mesodermal cells available to form the skeleton. Elimination of interdigital mesoderm in amniote species with free digits constitutes a paradigmatic example of morphogenetic cell death. Apoptosis is also observed in the developing chondrogenic skeletal condensations marking the position of the future joints. Additionally, cell death is also detected in the ectoderm of the apical ectodermal ridge (AER). Since the AER directs outgrowth of the limb bud, the occurrence in this area of cell death indirectly exerts a central morphogenetic function modulating proliferation and survival of undifferentiating mesenchymal cells of the limb bud (Dudley et al. 2002).

Interestingly, areas of cell death similar to those found in vertebrate developing appendages are seen in one species of insect, the flesh fly, *Sarcophaga bullata* (Whitten 1969).

Patterns of Apoptosis in Vertebrate Developing Appendages

Fishes

In fishes, cell death occurs in the course of the development of many organs. In some cases, as in the developing sensory organs and brain, these apoptotic processes have a clear morphogenetic function. During fin development cell death occurs both in epidermal cells and in developing cartilage.

In zebrafish (*Danio rerio*) a single embryonic rudiment, the median fin fold, gives rise to the unpaired dorsal, anal, and caudal fins. At 20 hours postfertilization (hpf) the median fin fold appears. Apoptosis is detected at 22 hpf when apoptotic cells are present at the most distal part of the fin fold, and less frequently in proximally. Later, at 24 hpf, clusters of apoptotic cells are detected in the median fin fold at the distal tip of the fin, especially in the region of the future anal fin. The apoptotic cells are present until stage 72 hpf, always in the most distal part of the median fin fold. Although this cell death process is a constant feature during fin development there is no correlation between fin morphogenesis and cell death, suggesting that in this case apoptosis is not morphogenetic (Cole and Ross 2001). Evidence for a function of apoptosis in fin development was also obtained in regenerating fins treated with retinoic acid, but again there is not a clear relationship between apoptosis and fin morphogenesis (Géraudie and Ferretti 1997).

Amphibians

A remarkable feature in amphibians, in contrast to amniote animals, is that there are no areas of cell death associated with the formation of free digits. In fact, in several species of anurans and urodeles, the process of apoptosis seems to be absent from normal limb development and regeneration (Cameron and Fallon 1977; Vlaskalin et al. 2004). However, it is possible that this is not a general feature in amphibians, as Franssen et al. (2005) recently reported that in *Desmognathus aeneus*, a direct-developing salamander, there is a pattern of cell death resembling that present in amniota.

In *Xenopus laevis*, the formation of free digits in the forelimb has been correlated with different proliferation rates in prospective digital and interdigital areas. In contrast, the proliferation rate in the webbed hindlimb of *Xenopus* is comparable in digital and interdigital areas; the two areas elongate at essentially the same rate throughout development.

While areas of cell death appear not to be a general feature of amphibian limb morphogenesis, administration of retinoic acid to regenerating limbs causes cell death by apoptosis in blastema mesenchyme in a pattern that appears to correlate with abnormalities observed in the treated regenerating limbs (Géraudie and Ferretti 1997).

Reptiles

In 1977, Fallon and Cameron described the patterns of mesodermal cell death that accompany the formation of the digits in three species of reptiles, two of the most primitive extant species of reptiles—the snapping and the painted turtles *Chelydra serpentina* and *Chrysemys picta*—and a more phylogenetically advanced lizard species, *Eumeces fasciatis*.

The patterns of mesodermal cell death that occurs during formation of the digits in the three species correlates precisely with adult limb morphology. In the turtles, the areas of cell death are limited to the distal parts of the interdigital areas, with little degeneration in the proximal parts. Adults of these species have webbed digits for almost all the length of the digits. In contrast, the lizards that have completely free digits exhibit during limb development a high rate of cell death in the interdigital tissue.

According to Fallon and Cameron (1977) the fact that the turtles—the first species to branch off to the reptilian line—exhibit a correlation between cell death and limb morphology allows the conclusion that cell death became part of the formation of free digits with the emergence of the amniotes.

In addition to mesodermal cell death, apoptosis is also an important feature of the ectoderm of the apical ectodermal ridge (AER). The AER maintains cells in an undifferentiated and proliferating state, therefore inducing limb outgrowth. In serpentiform reptiles with rudimentary limbs the regression of the limb primordium is mediated by massive cell death in the AER (Raynaud 1990).

Another species that presents a particular limb morphology related to patterns of cell death is the chameleon, *Chamaeleo chamaeleo*. Chameleons have a very specialized prehensile autopodium whose morphology is unique among vertebrates. A prominent cleft divides the autopod into two sets of digits. As in other reptilian species, ectodermal cell death is observed in the AER; however, it is not related to ridge regression. The most intense area of mesodermal cell death occurs in the zone of the formation of the autopodial cleft and precedes the occurrence of interdigital cell death that results in the freeing of the digits. It has been suggested that the area of cell death of the autopodial cleft may be a specialized form of the normal interdigital area of cell death; it is precocious in its time of appearance and wider along the distal margin of the limb (Hurlé et al. 1987).

Avians

Chick embryos are the best-studied model system for cell death during limb development. A number of well-defined areas of cell death appear concomitant with formation of the skeletal primordia of the limbs.

In early chick limb buds, two areas of massive cell death appear in the undifferentiated mesenchyme in the anterior and posterior margins of the proximal segment of the limb. These areas, known as the anterior and posterior necrotic zones (ANZ and PNZ), relate to the reduction in the number of digits that occurs later in development. Later in development another area of cell death, the opaque patch (OP), appears in the central mesenchyme delimitated by the condensations of the two skeletal pieces of the zeugopod (tibia/fibula, ulna/radius). Other areas of cell death appear during digit formation. These arise in the undifferentiated mesenchyme located between the developing digital rays, and for that reason they were termed the interdigital necrotic zone (INZ; fig. 8.2).

The temporal and spatial pattern of distribution of these areas in the limb is constant between individuals within species. Between species the differences in these patterns are significant (as they are for reptiles). However, they always correlate with the morphology of the limb. In species with free digits, such as the chick (Saunders and Fallon 1967; Pautou 1975) and quail (Fallon and Cameron 1977), the areas of cell death extend throughout all the interdigital space. In species with webbed digits, such as the duck (Saunders and Fallon 1967; Hurlé and Colvee 1982), interdigital cell death is limited to the distal part of the interdigit. In species having free digits but a membranous fold along the margins of the digits, such as the moorhen (*Gallinula chloropus*) and the coot (*Fulika atra*), interdigital cell death is restricted to the central part of the interdigital tissue (Hurlé and Climent 1987).

Interdigital cell death is inhibited in mutants that exhibit syndactyly (Hinchliffe and Thorogood 1974; Hoeven et al. 1994; Zakeri and Ahuja 1994). When embryos are treated with drugs that inhibit cell death, the limbs exhibit soft tissue syndactyly (Tone et al. 1983).

As previously mentioned, cell death is also observed in the AER. The occurrence of AER apoptosis appears to regulate the spatial and temporal extension of that structure (Todt and Fallon 1986; Gañan et al. 1998).

Mammals

Analyses of programmed cell death in mammals have been performed mainly in embryos of rats and mice, but there is

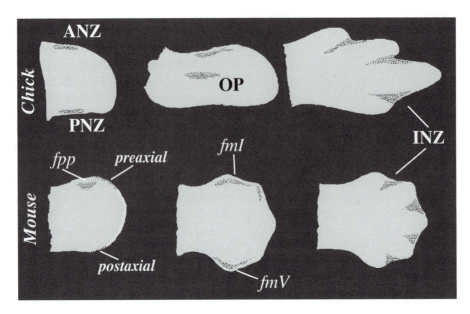

Figure 8-2 Schematic representation of the major areas of cell death of the developing limb in chick and mouse embryos. ANZ, anterior necrotic zone; PNZ, posterior necrotic zone; OP, opaque patch; INZ, interdigital necrotic zone; fpp, foyer preaxial primaire; fmI, foyer marginal I; fmV, foyer marginal V.

also evidence for its occurrence in moles (*Talpa*) and in humans (Milaire 1977; Kelley 1970).

The cell death patterns found in rat and mouse limb buds are practically identical and differ from those present in birds (Milaire 1967; 1976). In contrast to the results obtained in avian species, in which morphogenesis has been correlated with mesodermal cell death, in mammals the major morphogenetic role has been assigned to ectodermal cell death.

Very early in rat and mouse limb bud development two areas of ectodermal cell death are observed in the AER, a small area in the postaxial margin and a larger one in the preaxial margin. These areas, which are well defined and appear before the AER is completely formed, are responsible for regression of the extreme ends of the AER. Inhibition or delay of this ectodermal cell death causes an enlargement of the AER followed by an increase in the amount of subridge mesenchyme, resulting in polydactyly (Naruse and Kameyama 1982).

Later in development and preceding morphological changes associated with digit outgrowth, another wave of ectodermal cell death is observed. It starts in the postaxial portion of the ridge in association with the prospective digit V and adjacent to the interdigital area for digit IV. During a short period of time the AER opposite prospective digits III and IV remains free of apoptotic cells. Later the AER becomes apoptotic through all its length. The number of dead cells will later decrease in the flattening AER during digit outgrowth except in an area facing the prospective digit I that remains strongly apoptotic (Milaire and Roze 1983).

Three main areas of physiological cell death affect different areas of the footplate mesoderm in developing rat and mouse limb buds. The first to appear is a transitory preaxial area called the "foyer preaxial primaire" (fpp). The occurrence of this area of cell death has been related to the local decrease in the ectodermal influence exerted by the AER undergoing apoptosis or with a local drop in the concentration of the morphogen diffusing from the zone of polarizing activity, as suggested to explain the occurrence of the ANZ in the chick (Hinchliffe 1981).

In rats and mice no areas of cell death correspond to the posterior and anterior necrotic zones observed in chick limb development, although a parallel could be established between ANZ and fpp; it has been suggested that the fpp accounts for the reduction of the quantity of preaxial mesodermal cells. In chick embryos, occurrence of the areas PNZ and ANZ is directly correlated with the reduction of the number of digits (three in the wing and four in the leg). Since both rat and mouse have five digits the function of the fpp needs further clarification. However, in the forelimb of the mole, *Talpa*, the fpp is absent in correlation with the formation of the additional falciform digit in this species (Milaire 1977).

Later in development two other areas of cell death appear simultaneously in the subridge mesoderm: "foyer marginal I" (fmI), located in the preaxial margin, and "foyer marginal V" (fmV) located in the postaxial margin. The first, fmI, extends over digits I to III in forelimbs and from digits I to half digit II in hindlimbs. The postaxial death domain, fmV, covers digit V and extends to the border of digit IV, in both fore-

and hindlimb. The onset of this cell death is coincident with growth of the digital buds and corresponds topographically to the previously described AER death domain. It has been suggested that this mesodermal cell death is the result of the local decrease of the influence of the ectodermal layer.

The last cell death domains observed in mammalian limb morphogenesis are localized in the interdigital mesoderm and account for the separation of the digits. These INZs involve two successive waves of cell death, one occurring in the most superficial layer of subectodermal cells, the second affecting a group of deeper cells interposed between the precartilaginous rudiments of the phalanxes. Together these two domains of cell death contribute to the complete separation of digits (fig. 8.2).

Bmps Are the Apoptotic Inducing Signals

The first studies that permitted clarification of the molecular mechanisms that underlie the induction of apoptosis during limb development were performed mainly in chicken embryos. More recently, knockout studies have allowed mechanisms to be addressed in mice.

Bone morphogenetic proteins (Bmps) were identified as triggering apoptotic signals for both the ectoderm of the AER (Gañan et al. 1998; C. K. Wang et al. 2004) and mesodermal cells (Gañan et al. 1996; Zou and Niswander 1996; Yokouchi et al. 1996; Kawakami et al. 1996; Macias et al. 1997; Guha et al. 2002). These secreted factors belong to the transforming growth factor β superfamily and play important roles in other development mechanisms.

The identification of Bmps as the apoptotic triggering signals was first supported by their pattern of expression. Bmp-2, Bmp-4, Bmp-7 and the recently described Bmp-5 (Zuzarte-Luís et al. 2004) are expressed in undifferentiated limb mesoderm, in interdigital mesoderm, and in the AER, coinciding with areas of cell death. Gain-of-function experiments in chick embryos result in massive apoptosis in the undifferentiated mesoderm and ectoderm. As in the avian embryos, exogenous Bmps are apoptotic triggering signals for limb mesodermal tissue in the developing mouse (Tang et al. 2000).

However, clear evidence for this function has not been obtained from transgenic approaches. Bmp-2 and Bmp-4 homozygous mutant mice die before the onset of limb cell death (Winnier et al. 1995; Zhang and Bradley 1996). Bmp-7 knockout mice exhibit mild patterning defects in the limb in the form of preaxial polydactyly but do not exhibit syndactyly (G. Luo et al. 1995). Single Bmp-5 and Bmp-5/Bmp-7 double knockout mice also lack interdigital phenotype (J. A. King et al. 1994; Solloway and Robertson 1999). The fact that

at least four different members of the Bmp family (Bmp-2, Bmp-4, Bmp-5, and Bmp-7) are involved in limb development indicates that there may be a high redundancy between them, which would explain the knockout phenotypes.

In most embryonic models Bmp signaling is regulated by Bmp antagonists that modulate the intensity and/or the spatial distribution of the Bmp signal. Noggin, gremlin, DAN, and Drm are some antagonists expressed in the developing limb (Brunet et al. 1998; Capdevila and Johnson 1998; Capdevila et al. 1999; Francis-West et al. 1999; Merino et al. 1999a, 1999b; Pearce et al. 1999; Pizette and Niswander 1999). Gain-of-function experiments result in inhibition of cell death in both mesoderm and ectoderm (Pizzete and Niswander 1999; C. K. Wang et al. 2004).

One of these antagonists, Gremlin, is also expressed in the interdigital webs between the digits in the feet of ducks, while in the chick interdigital expression of this Bmp antagonist is down-regulated prior to the onset of INZ. Furthermore, implanting beads loaded with Gremlin into the interdigital mesenchyme of chick leg buds inhibits the INZ and induces a membranous syndactyly similar to that found in ducks (Merino et al. 1999b).

Bmps are also involved in the control of limb patterning (Pizette et al. 2001) and in regulating chondrogenic differentiation (Macias et al. 1997). Local treatments with any of the Bmps mentioned above induce intense growth and differentiation in the prechondrogenic mesenchyme. These proteins signal through a serine/threonine receptor kinase complex composed of type I and type II receptors. Ligand binding leads to an association of these two receptor types, phosphorylation of the type I receptor by the type II receptor, and subsequent propagation of the intracellular signal. The chondrogenic effect of Bmps appears to be mediated by the type Ib receptor (Yi et al. 2000). Inhibition of apoptosis has been obtained in overexpression experiments using dominant negative type IB and type IA Bmp receptors (Zou and Niswander 1996; Yokouchi et al. 1996; Zou et al. 1997). However, it is likely that the phenotype was secondary to depletion of Bmps, which bind to the overexpressed receptors. Recent studies using the Cre/loxP system support a specific function of BMP receptor type IA in the control of apoptosis (Rountere et al. 2004). Furthermore, interdigital induction of the type IB Bmp receptor gene by application of TGFβ1 beads is followed by inhibition of apoptosis and formation of an ectopic digit (Merino et al. 1998).

Intracellular Pathways

It is known that Bmps signal through a family of intracellular proteins called Smads. After binding of the Bmp molecule to the receiver, the Smad cascade that includes the BMP-responding smads (1, 5, and 8), the co-smad (4), and the inhibitory smads (6 and 7) is activated and these proteins are translocated into the nucleus were they activate gene transcription. Recently Bmps were shown also to be able to signal through the MapK pathway. Erk, Jnk, and p38 kinase are able to mediate Bmp signaling (Kawakami et al. 2003; Grotewold and Ruther 2002; Zuzarte-Luís et al. 2004).

As described above, the molecular machinery resulting in caspase activation has been observed in the course of the apoptotic process. There is experimental evidence for a role of caspase-3, caspase-9, and caspase-2 in the control of limb-programmed cell death (Milligan et al. 1995; Jacobson et al. 1996; Mirkes et al. 2001; Nakanishi et al. 2001; Zuzarte-Luís et al. 2006). Some members of the general apoptotic machinery present in the vertebrate cells, including the pro-apoptotic factors *DIO-1* (*Death Inducer-Obliterator-1*) and *Gas1* and *Gas2* (*Growth Arrest Specific*) were identified in the limb (Garcia-Domingo et al. 1999; K. K. Lee et al. 1999, 2001). Involvement of *Apaf-1* in limb-programmed cell death is supported by the occurrence of a reduced pattern of interdigital apoptosis and persistence of interdigital webs in mice mutant for this gene (Cecconi et al. 1998, but also see Chautan et al. 1999).

Bax, a proapototic member of the *Bcl-2* family, is expressed in the areas of cell death (Dupe et al. 1999), while several antiapoptotic members of this family, including *Bcl-2*, *Bcl-x*, and *A1*, are expressed in the digital rays but not in the interdigital spaces of the mice autopod (Novack and Korsmeyer 1994; Carrio et al. 1996). Moreover, the interdigital regions prior to the onset of apoptosis express *Bag-1*, which encodes an antiapoptotic protein that binds to Bcl-2 (Crocoll et al. 2002). Another antiapoptotic factor, *Dad-1* (*Defender Against apoptotic cell Death*) has been implicated in the control of limb-programmed cell death; heterozygous mutant mice for this gene display soft-tissue syndactyly (Nishii et al. 1999).

Other caspase-independent pathways may be simultaneously activated, and they can even sustain tissue degeneration when apoptosis is abolished (Chautan et al. 1999). In any case, whether the mechanism by which Bmps activate the latest step of the degenerative cascade is caspase dependent or independent is not known.

Regulation of Programmed Cell Death by Fgf and RA Signaling

It is now clear that regulation of programmed cell death by Bmps is closely integrated with other signaling pathways implicated in the control of outgrowth and differentiation of limb buds.

Fgf signaling is currently considered as responsible for outgrowth of limb buds. However, gain- and loss-of-function experiments have demonstrated that Fgfs cooperate with Bmps in the control of mesodermal apoptosis (Montero et al. 2001). When Fgf signaling is blocked by local application of Fgf inhibitors, Bmps are not sufficient to trigger apoptosis. Furthermore, we have provided evidence suggesting that the reduced pattern of interdigital apoptosis observed in the interdigital webs of the duck is due to a decrease in Fgf signaling rather than the absence of Bmps.

How Bmps and Fgfs cooperate in the control of apoptosis awaits clarification, but the expression of several genes potentially involved in apoptosis requires the integrity of both Fgf and Bmp signaling (Montero et al. 2001). Recently, Kawakami and colleagues (2003) described an apoptotic pathway in the limb controlled by Fgfs through the activation of ERK.

Retinoic acid signaling exerts major roles in limb patterning, including the control of apoptosis. In mice, inhibition of interdigital cell death and subsequent syndactyly has been reported in a variety of mutations of retinoic acid receptor genes (see Dupe et al. 1999). Furthermore, the phenotype of the hammer toe mutant caused by defective apoptosis can be partially rescued by administration of retinoic acid to the pregnant females (Ahuja et al. 1997). In the chick, it has been observed that retinoic acid acts in concert with Bmps to establish the interdigital regions (Lussier et al. 1993; Rodriguez-Leon et al. 1999). The function of retinoic acid signaling consists of promoting the apoptotic effect of Bmps and at the same time inhibiting the chondrogenic effect of these factors. This may be of considerable importance for normal morphogenesis since in the developing autopod Bmps not only induce apoptosis but also promote a dramatic growth of the cartilage.

Chapter 9 Joint Formation

Charles W. Archer, Gary P. Dowthwaite, and
Philippa Francis-West

A CENTRAL CHARACTERISTIC of the vertebrate form is the possession of an internal skeleton that first comprises a supporting structure and, together with articulations, provides a means of both movement and locomotion. Joints have been classified according to their structure, which, in turn, reflects the degree of movement that any one joint affords. Earlier investigators described four types of joint. First, a synarthrosis, in which the skeletal elements are joined to each other by cartilage, fibrocartilage, or other connective tissues. Second, a schizarthrosis, in which the space between the skeletal elements (referred to as an interzone) comprises connective tissue containing a single cavity or a (usually small) number of cavities. Third, a hemiarthrosis or periarthrosis, which has a single joint cavity but in which the skeletal elements remain united around the periphery. Fourth and finally, a eudiarthrosis (more recently referred to as a diarthrosis), which comprises separate articulating elements and joint cavity that is limited peripherally by synovial tissues (Bernays 1878; Lubosche 1910; Haines 1942b, 1942c; plate 9.1). More recently, this classification has been refined to include the degree of movement that the joint allows. Consequently, the synarthrosis is maintained and the bones are in almost direct contact but interspersed by connective tissue, and there is no appreciable movement as, for example, between cranial bones. Second is an amphiarthrosis, which has contiguous osseous surfaces connected by flattened discs of fibrocartilage as in the vertebrae and pubic symphysis, which allows for limited movement. Third and last is the diarthrosis, which is freely movable and possesses a cavity that is lined by a synovial membrane. This type of joint may or may not possess ligamentous and meniscal structures internally (Gray 1988).

A true skeleton is present in the subphylum vertebrata, and members show all types of articulations described above. However, in terms of evolution, it may be argued that it was the emergence of the diarthrodial joint that has allowed the greatest range of movement and, thus, facilitated adaptation from the aquatic to the terrestrial environment. In addition, it can also be argued that the evolution from the Agnatha (jawless fishes) to the Gnathostomata (jawed vertebrates) was driven by the acquisition of a hinged mandible that possessed a diarthorodial joint as the primitive condition and was subsequently adapted in other locations within the body. Thus, while the cyclostomes are confined to a suctorial mode of feeding, which effectively meant that they were also confined to prey who were of similar size or larger, the adaptation of the mandible allowed for a much greater range of prey of virtually any dimension. For these reasons, this chapter will focus mainly on this joint type, with relevant references to the other types where appropriate.

The Origin of the Mandible

In the agnathous condition, the jawless mouth is supported by the branchial arches, which supported the branchial pouches containing the gills. In addition, these formed skeletal support that allowed for breathing and the suction mode of feeding (plate 9.2). Thus, the mandibular arch in association with the chondrocranium formed the jaws comprising the upper palatoquadrate and lower mandibular cartilages. In modern mammals, this arrangement equates with the joint between the malleus and incus, which remains diarthrodial in its structure (Mallo 2001).

Chondrichthyes

In the Elasmobranchii (cartilaginous fishes), the fins are largely stiff and the joints are represented by syntharthroses, schizarthroses, and hemiarthroses (Bernays 1878; Lubosche 1910). The jaw in Scyliorhinidae (Carcharhiniformes—catsharks) and Squalidae (Squaliformes—dogfishes) is a hemiarthrosis, while, in contrast, in Chimaeridae (Holocephali) it appears more diarthrodial in structure. However, intriguingly, the synovial membrane is only apparent histologically on one side of the joint. It is clear that more specific analysis is required either immunologically or histochemically of what comprises a synovium in these lower vertebrates in order to confirm the true existence of this diarthrodial joint form. This particular facet will be addressed again when we consider tetrapods. Nonetheless, there is one other report that the Rajidae (Lubosche 1910) possessed true diarthrodial joints. It is questionable whether diarthrodial joints generally arose within the cartilaginous fish, especially as they enter the paleontological record later than early bony fish, with no fossil records represented within the Lower Devonian period. Subsequently, Haines (1942a) suggested that elasmobranches may have lost the diarthrodial condition, as seems to have occurred in some bony fish, such as the sturgeon, which will be discussed below.

Osteichthyes

Haines's assertion that the diarthrodial condition arose in the quadrate/mandibular apparatus is based on studies on the microscopic structure of primitive fish jaws (Haines 1937). These studies included a descendent of the paleoniscids *Polypterus* and the lungfish *Protopterus* and a later study on two surviving holosteans, *Lepidosteus osseus* (longnose gar) and the more advanced *Amia calva* (bowfish). As can be seen from a drawing taken after Haines (1942a), the jaw of *Lepidosteus* spp. Figure 9.1 shows classical diarthrodial structure even when the elements remain cartilaginous, as in *Protopterus*. At the microscopic level in the jaw of *Lepidosteus*, other distinctive features included a layer of calcified cartilage containing hypertrophic chondrocytes integrating into the underlying bone, hyaline cartilage, and a layer of articular fibrocartilage, which is also found in many birds and the secondary cartilaginous joints of mam-

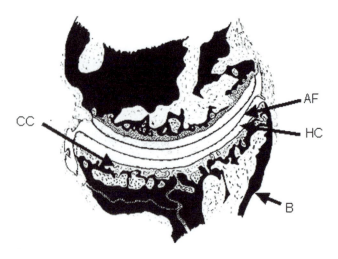

Figure 9-1 Diagram showing the structure of the mandibular/palatoquadrate joint in *Lepidosteus* spp. demonstrating the eudiarthrodial nature of the joint. AF, articular fibrocartilage; HC, hyaline cartilage; CC, calcified cartilage; B, bone. (Adapted from Haines 1942a.)

mals. Unlike mammalian joints, those of the early fish do not seem to have a fibrous capsule and instead are surrounded by loose connective tissue, but the synovial membrane is typically bilayered (Haines 1942a).

D. M. S. Watson and Gill (1923) have shown from the fossil record that earlier dipnoans from the Devonian period had articular elements that were bony in nature and supported overlying cartilage, suggesting that the living dipnoans exhibited secondary modifications. When the fin joints of these more primitive bony fishes are examined, diarthrodial joints are not observed, but the more simple arrangements are, with the syntharthrosis occupying the more distal smaller joints and the schizoarthrosis and hemiarthrosis found in the more proximal larger joints, respectively (Lubosche 1910). Lastly, in the case of *Amia*, both Lubosche (1910) and Haines (1942a) described the structure in the joint between the proximal radial and the girdle and basal cartilages. While Lubosche describes this joint as being hemidiarthrodial, Haines, by contrast, argues that it represents a true diarthordial joint based on the observations of a well-defined joint cavity that contains minimal connective tissue and a two-layered synovium.

It is evident that more modern bony fish have utilized the diarthrodial joint more widely; it is likely to be related to greater maneuverability during swimming and feeding and, also, to be related to greater size, as the larger joints at the base of the fins are often diarthrodial.

Tetrapods

In consideration of tetrapod limbs, the diarthrodial condition is the norm. However, the urodele and, as observed in fish fins, the larger more proximal joints may be diarthrodial, but the more distal joints may be a synarthrosis either fibrous or even cartilaginous. This condition may also be observed in the distal joints of the anura. Haines (1942b) argued that the simplified structure seen in urodeles probably represents a secondary modification associated with the evolutionary changes to a more flexible type of animal from armored predecessors. A major problem in interpreting structure is that secondary modifications can be mistaken for the primitive condition and vice versa. Thus, detailed knowledge of the developmental mechanisms is required to unequivocally resolve these potential misinterpretations. The primitive joint structure is considered to lack a joint capsule supporting the synovial lining (two-layered) and can be observed in both amphibia and reptilia (e.g., *Emys* and *Sphenodon;* Haines 1942b). In addition, as in birds, the primitive condition also possesses a fibrous/fibrocartilaginous articulating surface overlying the hyaline cartilage proper of the epiphysis. While the above observations are generalizations, a clearer but narrower picture can be achieved by the comparison of a single joint such as the knee. Again, this has been performed by Haines (1942c), who compared the knees of 24 vertebrate tetrapods and whose results are summarized below. In some respects the knee joint is not ideal, since it possesses so many secondary modifications such as ligaments and menisci, yet the presence or absence of these too can be used to identify secondary reversion to a simpler form. It is considered that *Crocodilus, Sphenodon,* and lizards represent the primitive form of knee yet possess cruciate ligaments, menisci, and a single joint cavity shared by the femur/tibia/fibula. Chelonians are specialized in that they have a "firm" articulation between the medial condyle of the femur and tibia with a consequent reduction in the meniscus on the medial side. As mentioned earlier, the urodeles have secondarily reduced or lost their cavity and associated structures such as the menisci and ligaments. In both the monotremes and marsupials, the femora-fibular articulation is retained as in the reptilian condition with some modifications such as subdividing the joint cavity by connective tissue septa, as in monotremes. Lastly, in eutherian mammals, the femora-fibular articulation is lost, the latter becoming closely bound to the tibia.

Histological Features of Joint Epiphyses: Evolutionary Trends

All of the above discourse relates to the general morphology of diarthrodial joints but pays little attention to the histological appearances of these joints and what this may tell us about their evolutionary relationships and growth mecha-

nisms. In particular, we will consider the histology of the epiphyses being the major structural component of the joint and also an integral part of the growth of the elements through endochondral ossification and, in particular, secondary centers of ossification. We will paraphrase a seminal article by Wheeler Haines (1942b), who described this aspect in great detail (fig. 9.2).

In bony fishes, the cartilaginous epiphyses lie at the ends of the diaphyses "like a cork into the neck of a bottle" (Haines 1942a). The articular surfaces tend to be fibrous, a feature seen in all vertebrate classes. Whether this fibrous layer constitutes a perichondrium is questionable, and perhaps only the presence of stem/progenitor cells within the structure and an immunological analysis of the collagen types are likely to resolve this question. Most of the epiphyses comprise a mass of rounded chondrocytes, which Haines describes as undifferentiated but which are in fact differentiated in that they are surrounded by abundant cartilage matrix. However, in the basal regions near the diaphysis, which we can term the metaphysis, the chondrocytes become flattened and then hypertrophied. In some species the cartilage matrix surrounding the hypertrophic chondrocytes may become calcified, and this calcified matrix is resorbed by elements of the bone marrow that ramify the basal cartilage to form marrow processes. Thus, the fundamental mechanism of longitudinal bone growth through what is essentially cartilage replacement is present in bony fishes and can be considered a true endochondral ossification.

Two other fundamental epiphyseal structures can be found in bony fish. First is calcified cartilage in the center of the epiphyses of the epibranchial bone of *Argyrosomus hololepidotus* and *Gadus morhua* (Haines 1934, 1938) that occupies the correct position to be a forerunner of a secondary center of ossification. Second, in *Salmo salar* and probably other genera, is a closing plate of endochondral bone (fig. 9.3). Thus, it can be seen that all of the essential features of a true epiphysis incorporating an endochondral growth mechanism are present in the bony fishes.

Tetrapod Modifications

Like fishes, the early tetrapods also seem to have possessed cartilaginous epiphyses that lacked secondary centers of ossification at least up to the end of the Permian age. However, within extant species, only the reptilia (*Chelonia* and *Crocodilia*) have retained the primitive condition, since other reptilia, amphibians, and birds show modifications of the epiphyses. A major modification of the tetrapod epiphyses that is likely to relate to land dwelling is a reduction in the zone of round cells, which thus brings the zone of flattened cells

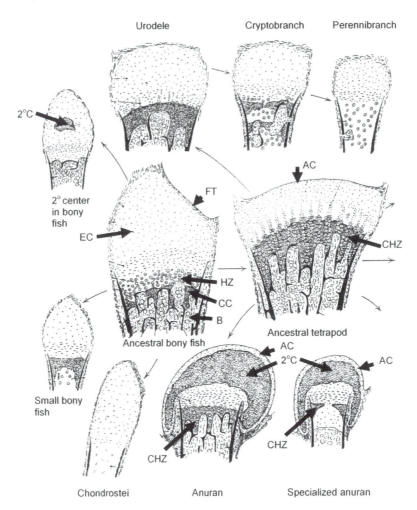

Figure 9-2   Diagrammatic representation of the evolution of epiphyses in fish and amphibian tetrapods (adapted from Haines 1942b). AC, articular cartilage; B, bone; CC, calcified cartilage; CHZ, calcified hypertrophic zone; EC, epiphyseal chondrocytes; FT, fibrous tissue; HZ, hypertrophic zone; 2°C, secondary center.

closer to the articular surface. It is argued that this modification makes for a firmer epiphysis, which would be advantageous in the land environment.

Undoubtedly, the most primitive extant tetrapod that possesses secondary centers of ossification is the reptilian tuatara *Sphenodon* spp. (Haines 1939). Similar centers of ossification have been described in the fossil record going back to the Jurassic age in *Sapheosaurus* (fig. 9.4). These centers are represented by large masses of calcified cartilage, which may occupy a greater part of the adult epiphysis, leaving a thin layer of articular cartilage.

A further modification is the partitioning of the flattened cell zone into columns. The likely mechanism of this partitioning reflects the mode of growth of this zone, in which its expansion results from the division of a founder population at the top of the column and, during cytokinesis, the progeny come to lie beneath the mother cell (Kember et al. 1990). While in the mammal this distinct column alignment occurs

early in the development of the cartilaginous anlagen, in other groups such as reptiles (*Emys orbicularis*) and birds (*Gallus domesticus*) the flattened cells are initially nonaligned, but a looser alignment appears later during postembryonic development (Haines 1942b; fig. 9.4). One of the consequences of the columnar arrangement is that it is maintained during hypertrophy and calcification. Thus, only the longitudinal septae between hypertrophic cells are calcified, and they then form templates for endochondral bone to be deposited. The noncalcified transverse septae are broken down by matrix metalloproteinases, and the hypertrophs undergo apoptosis or transdifferentiation into osteoblasts depending on the circumstances (Roach et al. 1995).

The Origin of Joint Cavities

It is clear that the incorporation of a cavity between two or more opposing skeletal elements allows for much greater

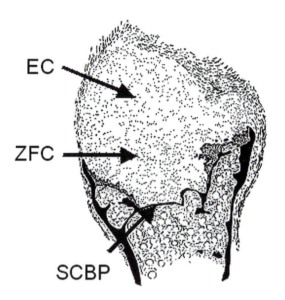

Figure 9-3 Diagram showing the near complete subchondral bony plate in *Salmo salar* (after Haines 1942a). EC, epiphyseal chondrocytes; SCBP, subchondral bony plate; ZFC, zone of flattened chondrocytes.

range of motion between elements compared with those that are joined by fibrous, fibrocartilaginous, or cartilaginous tissue. Hence, for larger joints in particular, selective advantages can be readily appreciated, especially in the case of tetrapods, in relation to the major joints of the limbs and girdles. A central question relates to the fundamental mechanism of cavitation and how this may have arisen. It has been discussed above that in the schizarthrosis, small cavities intersperse the connective tissues (interzone) between the elements. It is tempting to propose that this represents the primitive condition based upon the assumption that the cavities are rich in hyaluronan, although, again, this aspect has yet to be determined.

While a number of factors have been proposed that include the death of the cells that comprise the interzone to date, the strongest experimental data underlying the mechanism of joint cavitation concern differential hyaluronan (HA) synthesis under the influence of mechanical stimuli. The long chain glycosaminoglycan HA and its principal cell sur-

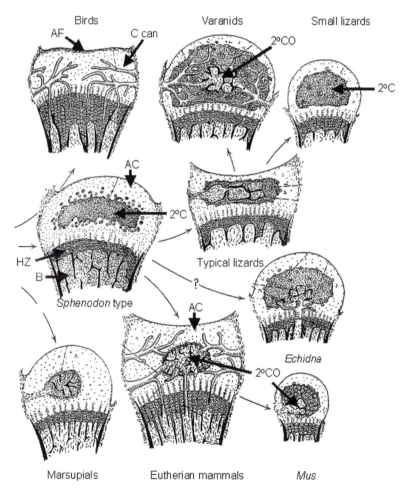

Figure 9-4 Diagrammatic depiction of the possible evolutionary pathways of higher vertebrate epiphyses (adapted from Haines 1942b). AC, articular cartilage; AF, articular fibrocartilage; B, bone; C can, cartilage canals; HZ, hypertrophic zone; 2°C, secondary center; 2°CO, secondary center of ossification.

face receptor, CD44, are differentially expressed at the joint interzone and developing articular surfaces (Craig et al. 1990; C. W. Archer et al. 1994; J. C. Edwards et al. 1994; Pitsillides et al. 1995; Dowthwaite et al. 1998). Using enzyme histochemistry, it was demonstrated that uridine diphosphoglucose dehydrogenase activity (UDPGD; an enzyme essential in the synthesis of UDP-glucuronate and hence HA) is significantly increased just prior to cavitation in the interzone and in the articular surfaces and synovium during cavitation (J. C. Edwards et al. 1994; Pitsillides et al. 1995). Thus, HA synthesis is increased at the time of tissue separation. The interaction between HA and CD44 can induce both cell adhesion and cell separation depending on the concentration of HA surrounding a cell population via receptor saturation, with increasing HA concentrations leading to cell separation (Toole 1991). Thus, during limb development the expression of CD44 in the interzone and at the developing articular surfaces, and the increased HA synthesis associated with cavitation, can facilitate tissue separation and create a functional joint cavity. Indeed, the addition of exogenous HA oligosaccarides (six dissacharide units) to the developing joint region inhibits joint cavitation by displacement of nascent HA from its receptor (Dowthwaite et al. 1998). Additionally, the results of these experiments resembled the fused joints originally described in immobilization experiments by Fell and Canti (1934), suggesting that differential HA synthesis during joint cavitation is modulated by mechanical influences.

In a reductionist approach, it was shown that the application of mechanical strain to isolated articular fibrocartilage cells from the chick did indeed modulate HA synthesis in these cells (Dowthwaite et al. 1999). The application of 3,600 $\mu\varepsilon$ (microstrain) for 10 minutes produced significant increases in HA release from isolated cells, which were accompanied by a significant increase in UDPGD activity (Dowthwaite et al. 1999) compared with static and flow controls. These increases in HA synthesis and synthetic ability were accompanied by increased CD44 expression and alterations in the expression of hyaluronan synthase (HAS) genes (Dowthwaite et al. 1999).

It therefore seems that joint cavitation is dependent upon differential HA synthesis and that this differential HA synthesis is regulated by mechanical stimuli. Thus, in mechanistic terms, it can be suggested that the ability to respond to the mechanical cues provided by joint motion resulting in the upregulation of HA synthesis and its accumulation between the opposed elements was a central event in joint evolution and was increasingly adopted by the higher vertebrates. These notions can be easily tested by the comparative culture of interzone cells from either teleost or urodele joints that are synarthroses or diarthrodial and subjecting them to mechan-

ical stimulation and testing their ability to regulate HA synthesis in response to that stimulation.

Hyperphalangy in Secondarily Aquatic Mammals

The molecular mechanisms that both specify and regulate joint formation are slowly being elucidated. These will be considered in detail below. In part, during terrestrial evolution there has been a tendency toward a reduction in the number of joints. At an experimental level this does not lend itself to the investigation of regulatory mechanisms. Furthermore, in recombination experiments of limb mesenchyme within ectodermal jackets, the resulting skeletal patterns seldom, if ever, produce an increase in the number of joint levels along the proximo-distal axis (Craig 1987; Hardy et al. 1995). However, an increase in the numbers of digits along this axis is seen in the secondarily aquatic taxa, namely the cetaceans (dolphins and whales) with flipper morphology, and shows many similarities to the fossil records of the ichthyosaurs, plesiosaurs, and mosasaurs. These aspects are reviewed in detail by Fedak and Hall (2004), and will be paraphrased here. In this context, hyperphalangy is defined as an increase in the number of phalanges within the autopodium (hand/foot) and consequently in the number of joints. This condition is distinct from polyphalangy, which consists of an extra row of digits of a branching nature, and polydactyly (Meteyer 2000; Cohn et al. 2002), which is represented by an entire extra phalanx on the lateral margin of the autopod.

Consequently, in hyperphalangy, there is an increase in the number of phalangeal elements above the normal mammalian number in each digit ranging from medial to lateral. Therefore, as shown in figure 9.5, the formula is 2-3-3-3-3. As pointed out by Fedak and Hall (2004), hyperphalangy is rare among terrestrial amniotes but does occur within the squamate taxa (lizards), but only by a single phalangeal element (maximum of six) over the normal plesiomorphic condition.

The fossil record shows abundant evidence of hyperphalangy among the diapsid ichthyosaurs (Triassic to Cretaceous periods) and shows many similarities to modern cetaceans. The more ancient ichthyosaurs possessed a modest hyperphalangy (2-4-4-4-1) where later specimens had up to 30 linearly aligned phalangeal elements (fig. 9.6). Unlike the ichthyosaurs, the early cetacean records show little evidence of hyperphalangy (*Rodhocetus balochistanensis*). This pattern clearly contrasts with extant cetaceans, which may have to 14 phalangeal elements and, may be asymmetrically patterned on either side of the body. The reason for the hyper-

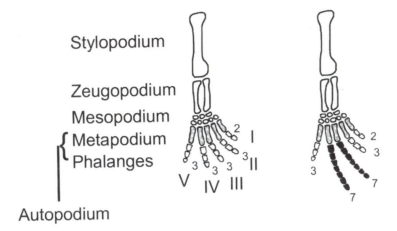

Figure 9-5 Diagram demonstrating digital hyperphalangy compared with the basic 2-3-3-3-3 pattern observed in many mammals. (After Fedak and Hall 2004.)

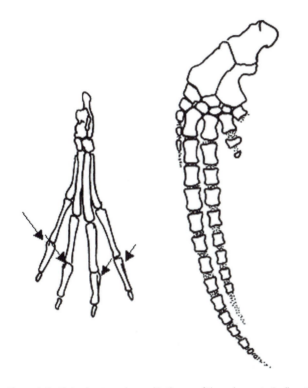

Figure 9-6 Skeletal pattern observed in the pes of the archaeocete *Rodhocetus balochistanensis* and the forelimb pattern of the long-finned pilot whale (*Globicephala melas*) showing hyperphalangy in two digits. (After Fedak and Hall 2004.)

practical level, this proposal is much more difficult. However, we do know that digit length, and hence phalange/joint number, is determined by the maintenance of the apical ectodermal ridge (AER) and fibroblast growth factor (FGF) signaling (Sanz-Ezquerro and Tickle 2003b). The timing of *Fgf8* expression in the AER in the chick wing overlying the different digits is differentially regulated, being first switched off over digit IV, then II and finally III, correlating with the ultimate length of the digit (Sanz-Ezquerro and Tickle 2003b). In the white-sided dolphin the AER is also preferentially maintained over digits II and III, which have 10 and 6 phalanges, respectively (Richardson and Oeschläger 2002). Ectopic FGF treatment in the chick has also been shown to increase the number of phalanges, which supports this correlation. Fgf8 expression is initially regulated by Shh in the ZPA and is thought to be later regulated by Ihh signaling in the condensing cartilage of the digits. Following ectopic application of FGFs, the increase in length of the phalange that is needed before segmentation occurs is intrinsically different for each digit (Sanz-Ezquerro and Tickle 2003b). This intrinsic difference in phalange length is also highlighted by analysis of other species. For example, the bat has extremely long phalanges. Studies in the chick have indicated that signals specifying joint position arise from the mesenchyme posterior to each digit (Dahn and Fallon 2000; Sanz-Ezquerro and Tickle 2003b).

Specification of Joint Pattern

In general terms there are three scenarios in joint formation. First, in the case of long bone elements, cartilage differentiates across the presumptive joint locations since the conden-

phalangic condition is not readily obvious, but it is thought to be related to better maneuverability and navigation during swimming.

Clearly, in terms of dissecting the molecular mechanisms of joint specification/phalange number, more detailed studies on cetacean embryos would be highly informative. At the

sations are continuous, and in the case of complex joints such as the elbow it appears as a branching continuum of cartilage matrix (Craig et al. 1987). At the presumptive joint sites, the chondrocytes then appear to flatten and the matrix becomes nonchondrogenic, largely comprising type I and III collagens with little protoeglycan. This structure is now termed the interzone, which is a signaling center acting on the opposing elements and is discussed below. Second, in the case of elements such as tarsals and carpals, chondrogenesis begins in the center of the condensations. As the condensations expand through matrix accumulation, the cells of the periphery stretch to form a boundary perichondrium that abuts the perichondrium of the neighboring element. Thus, an interzone is not clearly defined beyond resident cells between perichondria. Nevertheless, the expression of various signaling molecules suggests that an interzone is present (Settle et al. 2003). Third is the situation found in secondary cartilaginous joints. In this case, bone formation presages cartilage, and, in response to mechanical stimulation, progenitor cells within the periosteum that would normally adopt the osteogenic phenotype become chondrogenic and form a joint with the neighboring cartilage element. Little is known of the molecular mechanisms of this type of joint formation. However, we do know that signaling molecules associated with primary joint formation such as GDF-5 are also expressed during the formation of this joint type (Buxton et al. 2003).

In terms of molecular mechanisms we know most about long bone elements. In these cases the formation of a joint requires the reversal of the chondrogenic phenotype, and this is achieved by blocking prochondrogenic signaling. Thus, the knockout of the bone morphogenetic protein (BMP) antagonist *Noggin*, which is expressed throughout the condensations including the presumptive joint line, results in mice in the failure of joint formation. Early joint markers such as GDF-5 are also absent from the interzone, demonstrating that inhibition of BMP activity is central to the specification and early morphogenesis of joints (Brunet et al. 1998). The expression of other BMP antagonists is more specific; *chordin*, unlike *noggin*, is preferentially expressed in the interzone (Francis-West et al. 1999). The requirement for a reduction in chondrogenic potential during joint formation is also seen in the expression patterns of the powerfully prochondrogenic BMP-7. This factor is expressed along the perichondria of the cartilaginous primordia, but at the position of presumptive joints, its expression is absent. Furthermore, application of beads soaked in BMP-7 to the interdigital mesenchyme leads to failure or incomplete joint formation accompanied by whole or partial fusions of the joint (Macias et al. 1997). Other experiments using beads soaked in BMP2 and BMP 4 produce similar results in addi-

tion to the retroviral overexpression of these factors within the chick limb (Duprez et al. 1996a).

Initially, it was thought that because GDF-5 is expressed in the interzone region it had a role in joint specification. Given that GDF-5 belongs to the TGF superfamily and is a BMP that normally imbues chondrogenic potential, this seems somewhat paradoxical (Storm et al. 1994). However, we now know that GDF-5 has two roles in skeletogenesis: first, to promote condensation of the mesenchyme, and then to promote proliferation of chondrocytes within the developing epiphyses (Francis-West et al. 1999). Nevertheless, there remains evidence that the GDF family has a role in the maintenance/early development of some joints, but the mechanisms are still far from clear. The GDF-5 mutant *brachypod* mouse has a proximal interphalangeal joint missing from the digital region in addition to a variety of other anomalies. Other joints appear to be regulated by a combination of factors that include the related molecules GDF-6 and -7. These molecules have overlapping but distinct expression patterns differentially marking the joints. For example, GDF-6 is expressed in the carpals and tarsals, whereas GDF-5 is not. Conversely, GDF-5 marks the digit joints, whereas GDF-6 does not, with the exception of the one interphalangeal joint. Gene inactivation of Gdf-6 results in loss of joints in the wrists and ankles (Settle et al. 2003). Double knockout of these two genes affects additional joints (Settle et al. 2003). The mechanism/timing for the joint loss is slightly different, possibly reflecting different mechanisms of joint development in the various regions of the limb. Following gene inactivation of GDF-6, the joint loss is apparent early in the wrists and ankles. However, the additional joints lost in GDF-5/-6 double mutants initially form but are lost secondarily during fetal and early postnatal life, which argues for a role of GDF-5 and -6 in the maintenance of these joints. Dissection of the Gdf6 promoter has given insight into how the various joints may have evolved. Distinct elements that drive the expression of Gdf6 in the proximal joints (elbow/knee) versus the more distal ones (wrist/ankle) have been identified: in the former case a highly conserved 2.9kb element has been shown to be sufficient to drive expression (Mortlock et al. 2003).

There are few data on the lower vertebrate classes, but given the conserved nature of the BMPs, it is not surprising to find that *contact* (the Piscean homologue of GDF-5) is expressed between the dorsal fin and fin radials in *Danio rerio* (zebrafish) providing evidence that the BMPs have played a fundamental role in the evolution of joint morphogenesis. The precise role of these factors in joint specification, however, remains less clear.

In terms of specification, the growth factor Wnt9A (previously known as Wnt14) is expressed in the interzone and is capable of inducing other interzone-associated factors and hence lies upstream of GDF-5, CD44, and *chordin* and *autoaxin* (Hartmann and Tabin 2001). Overexpression of Wnt9A leads to inhibition of the chondrogenic phenotype during early differentiation and the adoption of the expression pattern described above. To date, these experiments pertain to the chick alone, but it may be expected that all the vertebrate classes are likely to regulate joint specification by Wnt9A, especially since it has been shown that fish express GDF5. The ortholog Wnt9 has been identified in hagfish (*Eptatretus stoutii*) and thresher shark (*Alopias vulpinus*), but evidence of Wnt9/9A involvement in lower vertebrate classes is awaited. Another potential intracellular target gene of Wnt9A is the transcription factor Cux1, which can also inhibit chondrogenic differentiation and which is expressed in the interzone, but again information across vertebrate groups is lacking.

The secondary loss of joints following gene inactivation of GDF-5 and -6 also argues that joint number may not be dependent only on the presence of joint specifying signals but may be modulated later by lack of joint maintenance signals. This is also illustrated by gain-of-function studies in mice in which FGF signaling has been increased in the epiphyses of the cartilage elements. This results in the loss of joints and is similar to the symphalangism seen in Apert's syndrome, which is due to increased FGF signaling (Ibrahimi et al. 2001; Q. Wang et al. 2001).

Clearly, in the case of hyperphalangy described above, obvious potential molecular mechanisms involve the possible duplication of expression of Wnt9A within existing elements, that is to say, additional segmentation in a model originally suggested by Caldwell (1997b) or its additional distal expression concomitant with the prolonged survival of the AER and continued expression of FGF-8, again suggested in an alternative model suggested by Howell (1970). Clearly, at the time of their proposal the molecular mechanisms had not been elucidated.

Another type of joint is seen in the zebrafish fin dermoskeleton, which consists of the unmineralized actinotrichia and the segmented bony rays, the lepidotrichia, which are joined by collageneous ligaments. The expression of Evx-1, which is related to the pair-rule segmentation gene, even

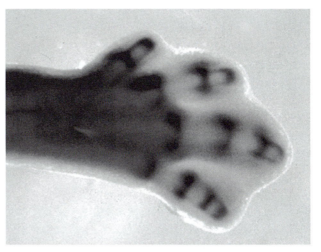
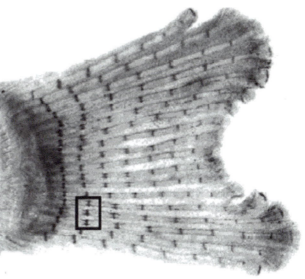

Figure 9-7 Whole-mount *in situ* hybridization of GDF-5 in the developing chick hindlimb (*top*) and of Evx-1 in the tail fin of *Danio rerio* (zebra fish; *bottom*).

skipped, in *Drosophila*, precedes joint formation and marks the developing joints (fig. 9.7; Borday et al. 2001). Although its precise role is unknown, given the role of the orthologous gene in *Drosophila* it is tempting to suggest that it is involved in joint specification.

Chapter 10 Postnatal Growth of Fins and Limbs through Endochondral Ossification

Cornelia E. Farnum

THE ARRAY OF DIVERSITY in the adult form of fins and limbs, be they fossil or extant, is a profound example of coevolution of form and function (fig. 10.1). Most limbs and many fins would at some point in early development be similar in external form to those seen in the stage 14 (6 mm) human embryo in figure 10.1, yet the specific developmental fate of limb tissues becomes stabilized and determined at an even earlier stage. Not only in prenatal development, but also during postnatal growth, timing differences in the same set of developmental processes can lead to structural differences in form, while the same genome can bring different phenotypic patterns into realization in response to the epigenetic environment (Sordino et al. 1995).

Issues associated with the interplay of genetic and epigenetic factors responsible for the postnatal growth of limbs and fins are explored in this chapter. Cohn and Tickle (1996) describe four main phases of limb development: (1) initiation of the limb bud; (2) specification of limb pattern; (3) differentiation of tissues and shaping of the limb; and (4) growth of the miniature limb to the adult size. The last of these is the subject of this chapter. The discussion is restricted to understanding mechanisms involved in growth in length of limbs and fins through endochondral ossification after birth/hatching. The most complete data come from analyses of postnatal growth in rats and mice, primarily because of the revolution in our understanding of basic regulatory mechanisms through the use of transgenic technology. Studies in the chick provide comparative data for birds. Less complete

data are available from studies of *Xenopus* and *Rana* spp. for amphibians, and the zebrafish (*Danio rerio*) for fish. Reptiles lack an indicator species as an agreed-upon prototype. The assumption is that the same interactive developmental processes are conserved in all tetrapod classes, but it is recognized that, when there is depth of understanding in only a few species, there is a risk in generalizing to the breadth of diversity of adult form that catches the imagination as seen in figure 10.1 (Enlow 1969; Tickle and Münsterberg 2001; Hinchliffe 2002).

An exciting aspect of the study of bone elongation through endochondral ossification is that it lends itself to a wide variety of approaches. Descriptive morphology of carefully chosen ontological sequences such as that of the extensive work of Haines starting in the 1930s (Haines 1934, 1938, 1939, 1941, 1969) has made fundamental contributions. Indeed, the lack of this kind of careful descriptive morphology at multiple stages of postnatal development in a broad range of species is a fundamental cause of the limitations of our comparative understanding, such as within the *Reptilia*. Stereology and morphometry added another dimension, with data on bone elongation presented in terms of three-dimensional kinetic parameters of cellular activity (Wilsman et al. 1996a, 1996b; Hunziker and Schenk 1989; Hunziker et al. 1999; Vanky et al. 1998). Reductionist approaches have investigated (1) the biochemical diversity of bone and cartilage matrices; (2) phenotypic characteristics of chondrocytes and osteoblasts manipulated in cell culture; and (3) cellular

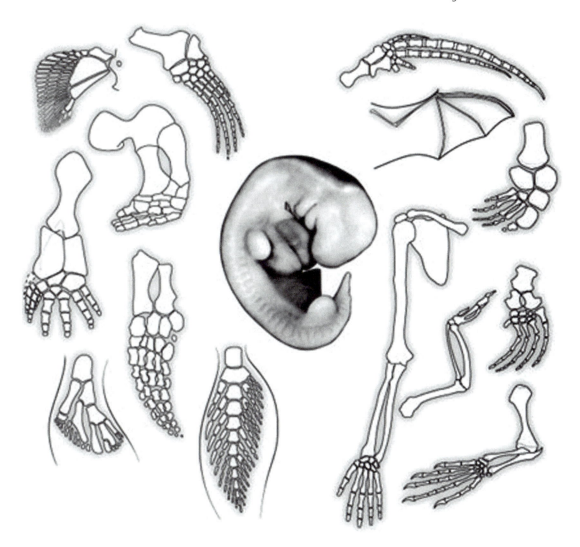

Figure 10-1 Examples of limb and fin diversity from multiple fossil and extant species. The question is, to what extent are the postnatal developmental processes similar, despite the apparent diversity and complexity of final form? (Human embryo picture courtesy of Dr. Drew Noden. Drawings by Michael A Simmons.)

regulatory pathways at multiple levels of interaction from cell-cell communication to the response of cartilage and bone cells to biomechanical signals. Analysis of phenotypes of transgenic animals with a wide array of specific gene alterations that affect skeletal formation pathways has proven to be a powerful way to bridge the insights of molecular genetics to the biology of the intact, functioning animal (Karsenty 1999). The study of naturally occurring disease in the multiple skeletal dysplasias also has yielded fundamental insights into the range of gene mutations that are compatible with life (Dreyer et al. 1998; International working group on the constitutional diseases of bone 1998; Superti-Furga et al. 2001; Hochberg 2002a; Unger 2002).

Comparative zoology and paleontology add yet another dimension with the study of such phenomena as the transi-

tion from limbs back to fins in cetaceans and fossil marine reptiles (Berta 1994; Tamura et al. 2001; Caldwell 2002; Thewissen et al. 1994); limb loss in snakes and mechanisms that inhibit limb formation (Cohn and Tickle 1999; Cohn 2001; Raynaud 1990); digit reduction in lizards (Gans 1975; Greer 1987; Shapiro 2002); miniaturization in multiple animal groups (Hanken 1982, 1985, 1993); the growth of antlers, a hard tissue that is created at a phenomenal rate (C. Li and Suttie 1994; Kierdorf et al. 1994; C. Li et al. 2002); and fracture healing, which recapitulates the sequence of endochondral ossification even after normal bone growth in length in the individual has ceased (D. R. Carter et al. 1998a; Ferguson et al. 1998, 1999; Gerstenfeld et al. 2003). Each of these approaches provides a perspective and insight into the growth of fins and limbs; by the very diversity of sources of insights, one begins to understand both the complexity of the processes, and the fundamental simplicity of the basic components involved.

Fins and Limbs: Starting Premises

There seem to be many points of current debate on the fin-limb transition. Did the tetrapod-like body appear before the limb itself (Ahlberg and Milner 1994)? Is limb morphological diversity based on a generic limb plan developed at the inception of the tetrapod lineage as a consequence of the self-organizing properties of mesenchyme (Newman 1993)? Did hindlimbs develop before forelimbs, or vice versa (Tamura et al. 2001), or did they develop simultaneously (Ahlberg and Milner 1994; Coates 1994; Coates and Cohn 1999)? Are digits an adaptation to the terrestrial environment, or did they first appear in the aquatic environment (Laurin et al. 2000)? Did tetrapods first walk on the bottom of water environments before they moved onto land (Edwards 1989; Shubin 1995)? Was polydactyly the primitive condition for the autopod (Coates and Clack 1990; Coates 1994; Laurin et al. 2000)? What were the particularly critical developments: the freeing of fins from the body axis and establishment of a separate limb axis (Tanaka et al. 2002), or the development of the autopodium and digit specification, with the ability of the distal part to evolve independently of the proximal part (Sordino et al. 1995; Cohn 2000; Cohn et al. 2002; G. P. Wagner and Chiu 2001)?

Despite debates about the details of the transition, there is consensus on multiple points that are fundamental for understanding growth of limbs and fins in modern animals:

1. Cartilage and bone are both very old tissues; cartilage did not necessarily precede bone in an evolutionary sense, but it does in the formation of endochondral bone (Fell 1925; B. K. Hall 1975; Moss and Moss-Salentijn 1983; De Ricqlès et al. 1991; M. M. Smith and Hall 1990; Maisey 1988).

2. The basic mechanisms involved in endochondral ossification are very old; epiphyseal structure and endochondral bone are relatively advanced in fish, and therefore tetrapods inherited a relatively advanced form of epiphyseal structure from their fish ancestors (Haines 1938; Maisey 1988; Hinchcliffe 2002; Shubin 2002). Skeletal tissues of the first vertebrates in the fossil record were as diverse and specialized as those of present-day vertebrates (Sledge 1966; B. K. Hall 1975; M. M. Smith and Hall 1990; Francillon-Viellot et al. 1991; Newman 1993; Sordino et al. 1995).

3. One reason that there could be rapid changes in limb morphology is the modular organization of limbs: changes occurring in the various modules can be disassociated from each other (Raff 1996). As changes evolve, there are modifications of the arrangements, but not of the fundamental units or properties (Ede 1971; B. K. Hall 1975; Newman 1993; Shapiro 2002; Shubin 2002).

4. Material properties developed early and are conserved; major changes leading to the array of diversity of form in modern species involve the size and shape of the elements, but not the material properties (Biewener 1982; Currey 1987; G. M. Erickson et al. 2002).

There are, in one sense, two basic kinds of bone formation—perichondral and endochondral (Moss and Moss-Salentijn 1983; Caldwell 1997c; Coates 1994; D. R. Carter et al. 1998b). In the first tetrapods, perichondral bone formation was the primary mechanism of growth; indeed, it is likely that such large animals as some of the dinosaurs achieved bone growth in length entirely through periosteal bone formation (Reid 1984). Conceptually, one can consider that in modern animals endochondral and periosteal bone go on together, in different proportions; one or the other proceeds faster and therefore advances growth in length for a given group of animals. Increase in the proportion of endochondral bone usually is accompanied by an increase in articular complexity (Caldwell 1997c). Although the focus of this chapter is on regulation of endochondral bone elongation, a major theme will be that endochondral bone elongation is intimately dependent upon regulatory interactions with the perichondrium, even in those mammals that have the most highly developed systems for rapid endochondral bone growth (F. Long and Linsenmeyer 1998). Osteogenesis in the perichondrium/periosteum is always coordinated with chondrogenesis/osteogenesis in the growth plate (Ornitz and Marie 2002). It is likely that this interdependence of periosteum and endochondral ossification evolved from the periosteal bone formation strategies already developed in early vertebrates.

An important concept when thinking about bone growth in length is that as length increases, the basic form of the bone is maintained (fig. 10.2). Postnatal growth in length, even for moderately sized animals, can be an order of magnitude or more, and thus, to maintain a constant shape or form to the bone, constant remodeling must occur (Enlow 1962; Hurov 1986; Bertram and Biewener 1990; Biewener and Bertram 1993). This is achieved primarily through periosteal resorption coupled to endosteal formation. Although this

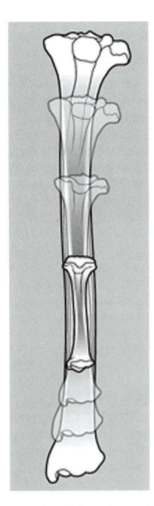

Figure 10-2 During postnatal growth, bones elongate while constantly remodeling so that their fundamental shape is maintained. As shown in this figure, there are different final contributions to overall length from the two ends, which is the result of both how fast growth occurs at each end and the length of the time growth is maintained. (Drawing by Michael A. Simmons.)

phenomenon is beyond the scope of the present chapter, it again emphasizes the significance of the periosteum in endochondral bone elongation (Horner et al. 1999, 2001; de-Crombrugghe et al. 2002). Periosteal ossification, always occurring in the outermost layer of the growth plate region, creates a ring or sheath of periosteal bone that constrains bone shape (Balmain et al. 1983; Caldwell 1997c).

Morphology of the Epiphysis

Two basic tenets of bone elongation are (1) increase in length during growth occurs only at the ends (the epiphyses); and (2) once diaphyseal bone has formed, there can be modulation of shape and width, but not increase in length. Haines, in his

thorough studies of epiphyseal structur, presented convincing evidence that the fundamental structure of the epiphysis among tetrapods is similar (Haines 1934, 1938, 1939, 1941). Figure 10.3 presents the comparative morphology of the epiphysis of a very young fish, amphibian, and reptile (10.3a, b, c, respectively), with the epiphysis being defined as the entire area from the articulating surface to the metaphyseal ossification front (Fell 1925). In these species at this age the epiphysis is entirely cartilaginous; the only distinguishing feature is a slight pattern of organization to the chondrocytes. A generalization is that the most superficial subset of these chondrocytes will contribute cells of the articulating surface (be it ultimately fibrocartilage or hyaline articular cartilage), while the remainder are either (1) responsible for bone growth in the diaphyseal direction, or (2) will contribute to the ultimate replacement of cartilage by bone in the epiphysis itself. Figure 10.3d is of the proximal tibial epiphysis of a four-day-old rat, which demonstrates a significantly greater extent of organization of epiphyseal chondrocytes than do the other three, even in the very early neonatal period.

For understanding bone elongation during endochondral ossification, the fundamental point is that chondrocytes of the epiphysis undergo a differentiation cascade, characterized by only two stages. There is proliferation of chondrocytes, potentially leading to an increase in cellular number; this is followed by chondrocytic terminal differentiation, which is characterized by an increase in cellular size during a stage that historically has been called hypertrophy. Interstitial growth is achieved by increase in cellular number, matrix synthesis, and chondrocytic size increase during hypertrophy. Interstitial growth during the chondrocytic differentiation cascade ultimately results in growth in length of the bone. When chondrocytes die, bone is laid down on the former cartilage matrix.

By examination of the three epiphyses of figure 10.3, areas of proliferating chondrocytes and areas of hypertrophic chondrocytes can be identified by their cellular shape and organization—relatively flat and small for proliferating chondrocytes viewed in this orientation, and relatively round for hypertrophic chondrocytes. It follows, then, that endochondral bone formation is achieved as a replacement phenomenon: cartilage grows and is replaced by bone. During active growth the number of cells contributing to growth can be modulated by changes in the rates of chondrocytic proliferation versus chondrocytic death at the cartilage/bone junction. But basically, and fundamentally, endochondral bone elongation occurs through some combination of chondrocytic proliferation, chondrocytic hypertrophy and matrix synthesis. This triad is fundamental to fin and limb elongation through endochondral ossification, and has been hy-

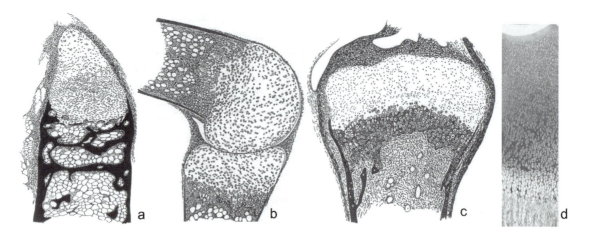

Figure 10-3 Bone elongation by endochondral ossification occurs through interstitial growth of chondrocytes in the cartilaginous epiphyses at the ends of the bone. In fish and amphibia (a and b) the epiphysis remains cartilaginous throughout the period of growth; in reptiles (c) the epiphysis may remain cartilaginous or may develop a secondary center of ossification. Although in the bones of most mammals secondary centers of ossification develop postnatally, in the early postnatal period the epiphysis is entirely cartilaginous (d, from the proximal tibia of a 4-day-old rat). Interstitial growth occurs as chondrocytes of the epiphysis undergo a differentiation cascade characterized by proliferation and cellular enlargement (hypertrophy). (a from Haines 1939; b from Haines 1941; c from Haines 1938; used with permission from Blackwell Publishing Ltd.)

pothesized to have been present even in the earliest tetrapods (Haines 1938; Francillon-Viellot et al. 1990).

Longitudinal growth is slow in epiphyses such as those shown in figures 10.3a, b, c, where only a small percentage of the total chondrocytes are aligned in the direction of growth. This is in contrast to the epiphysis from a two-week-old duckling shown in figure 10.4 (Barreto and Wilsman 1994). At low magnification (10.4a) the entire epiphysis is shown, including the articulating surface of fibrocartilage. Figure 10.4b shows the proliferative and hypertrophic regions, with a loose alignment of cellular columns parallel to the direction of longitudinal growth. This micrograph also shows important features of avian growth plates from precocial species. Similar to the epiphyses from fish, amphibians, and reptiles seen in figure 10.3, a major portion of the hypertrophic zone is surrounded by large metaphyseal vessels. Replacement of cartilage by bone occurs adjacent to these vessels, and bone formation also occurs along the periosteal interface surrounding the entire epiphysis (Hunt et al. 1979; Takechi and Itakura 1995). Therefore, the ability of hypertrophic chondrocytes to contribute to bone elongation is arrested at the tip of these metaphyseal vessels, meaning that endochondral bone formation lags behind bone elongation. This brings back the theme of the interplay of perichondral bone formation and endochondral bone formation. In all species there is probably a mix of the two, with differences in which actually drives the elongation process. When the rate of bone elongation is driven by the rate of periosteal bone formation, cartilage grows at the same pace; only the rate of replacement of cartilage by bone lags behind (D. R. Carter et al. 1998b).

Figure 10-4 In avian growth plates large metaphyseal vessels penetrate into the hypertrophic cell zone, indicating that the replacement of cartilage by bone through endochondral ossification lags behind periosteal bone formation. In panel (a), the extent of the articular cartilage (AC) and the growth plate (GP) are indicated. In the growth plate the boundaries of the proliferative zone (pz) and hypertrophic zone (hz) are indicated based upon morphological criteria of cellular size. Panel (b) shows how metaphyseal vessels penetrate well into the hypertrophic zone, leaving long tongues of cartilage penetrating into the metaphysis. Bar = 100 μm. (a reprinted from Barreto et al. 1993, with permission of the American Association for the Advancement of Science. b from Baretto and Wilsman 1994, used with permission from Elsevier.)

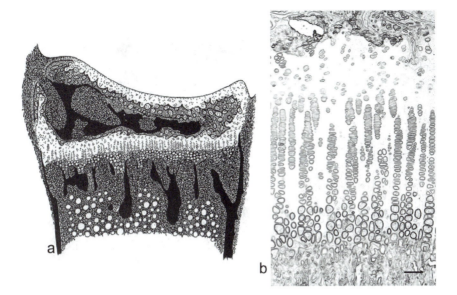

Figure 10-5   Panel (a) is the epiphysis of a reptile in which a secondary center of ossification forms. By comparison with figures 10.3a, b, c, the growth plate chondrocytes are more highly aligned into columns in the direction of growth, suggesting an efficiency in the translation of chondrocytic interstitial growth into bone elongation. Panel (b), from the distal radius of a four-week-old miniature Yucatan pig, demonstrates the spatial representation of the chondrocytic differentiation cascade. Proliferative cells form columns resembling stacked pancakes; rather abruptly cells change both size and shape during the hypertrophic stage; and the essentially straight contour of the cartilage to bone junction indicates that replacement of cartilage by bone occurs simultaneously with penetration of the metaphyseal vasculature. (b) Bar = 100 μm. (a from Haines 1941, used with permission from Blackwell Publishing Ltd.)

In most mammalian species, as in many reptiles and amphibians, a secondary center of ossification develops postnatally within the cartilaginous epiphysis, effectively separating the articular surface from the actively elongating region of cartilage, the growth plate or physis (fig. 10.5a). Haines speculated that the separation of the articular cartilage from the growth plate cartilage allowed each to become more efficient in one direction (Haines 1939, 1941). Articular cartilage provides a viscoelastic, load-bearing surface in articulating joints. In contrast, matrix structural properties and organization associated with the growth plate maximize the efficiency of promoting longitudinal growth and matrix mineralization (Haines 1938, 1969; McKibbin and Holdsworth 1967; Noonan et al. 1998). Indeed, current biochemical and molecular approaches confirm that very early in development articular and growth plate chondrocytes diverge in their phenotypic expression (Pacifici et al. 1995; Iwamoto et al. 2000).

The morphology of many mammalian growth plates conveys a sense of purpose for the rapid elongation of bone—that is, cells are in long columns highly aligned in the direction of growth, situated between the secondary center and the metaphyseal bone (fig. 10.5b). The highly organized nature of these growth plates allows definitions of zones, based on apparent shape. The cells that will be active in longitudinal growth are the proliferative cells in which the axial ratio of height to width is usually less than 0.5, and the hyper-

trophic cells where the axial ratio is usually 0.9 or greater. More rounded cells further proximal to the proliferative cells have traditionally been called *reserve* or *resting cells*. The columnar organization of chondrocytes in the growth plate is a spatial representation of any one cell temporally as it goes through its differentiation cascade. These growth plates have a dual blood supply, with the epiphyseal vasculature considered to be the primary nutrient vasculature for cells at all levels of the growth plate, and the metaphyseal vasculature providing signals for the replacement of cartilage by bone (Trueta and Amato 1960). The nature of the dual blood supply is particularly important for the understanding of multiple regulatory mechanisms discussed later in the chapter.

The epiphysis, with its function in bone elongation, is considered to be phylogenetically much older than secondary centers; perhaps the latter formed as the need for support in a terrestrial environment increased (Haines 1969; Moss and Moss-Salentijn 1983). It is likely that many avian species once had secondary centers and that they have been lost in essentially all modern birds (with the possible exception of one in the proximal tibia in many birds, and additional ones in very large birds such as the ostrich). The timing of the appearance of secondary centers is considered to be constant enough to use as a marker of biological age in individuals of a given species (Hare 1961; Hughes and Tanner 1970; B. T. Shea 1993), as is the timing of closure (Ogden et al. 1981). Indeed, the order of appearance within the skeleton is remarkably constant across species (Kohn et al. 1997). Carter and colleagues have hypothesized that the specific timing of appearance of secondary centers depends upon the biomechanical environment surrounding the epiphysis of a growing bone (D. R. Carter et al. 1987; D. R. Carter and Wong 1988; Wong and Carter 1990).

Many bony prominences associated with large muscles develop secondary centers of ossification that maintain endochondral ossification in a tensile environment. These so-called tension physes occur in multiple places where tendons of large muscle groups attach proximally on a bone. The tendon of the supraspinatus to the greater tubercle of the humerus, the tendon of the triceps brachii to the olecranon of the ulna, the tendon of the quadriceps femoris to the tibial tuberosity, and the tendon of the gastrocnemius to the calcaneus are examples. An interesting hypothesis is that, phylogenetically, these may have arisen from sesamoid bones, such as the ulnar patella in reptiles (Haines 1969; fig. 10.6). This emphasizes the point that secondary centers, like the primary centers of carpal and tarsal bones, and sesamoid bones such as the patella represent a manifestation of the ability of connective tissue to ossify under a variety of conditions, and that the evolution of diversified function is a sequel to this property.

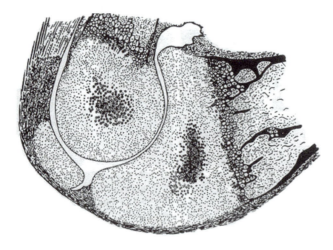

Figure 10-6 This representation of a reptilian elbow joint shows an ulnar-patella associated with the triceps brachii tendon of some reptiles, shown here intercalated within the tendon of the triceps brachii just proximal to its insertion on the olecranon process of the ulna. The olecranon in other animals has a growth plate associated with it that grows in a tensile environment. This "tension physis" contributes primarily to the shape of the proximal ulna, with very minor contribution to the length of the ulna. The hypothesis has been made that sesamoid bones associated with large tendons may have evolved into a permanent association with the bone, thus forming the so-called tension physes associated with bony prominences. (From Haines 1969, used with permission from Dr. Carl Gans.)

Overview of Endochondral Ossification

Bone, as an organ, not only provides the structural support system of limbs in terrestrial animals, it is also a key organ of the body for calcium and phosphorus homeostasis, and a major source of hematopoiesis for both red blood cells and the immune system (Moss and Moss-Salentijn 1983). Thus, when we analyze the process of chondrocytic activity in growth plate cartilage that results in interstitial growth and postnatal bone elongation, we may find that some of the activity is directed not to growth in length of the limb but to some future function of bone as an organ. The clearest example is mineralization of the cartilage in the hypertrophic zone, as demonstrated in figure 10.7. This mineralization is independent of the growth function and represents the initial scaffold on which osteoblasts will begin to deposit a bone matrix forming the primary trabeculae; later, secondary trabeculae are added that are independent of the cartilaginous framework (Maisey 1988; A. White and Wallis 2001). A second example is the recent finding demonstrating that a primary role for type X collagen in the hypertrophic cell zone of mammals is in the preparation of an appropriate matrix microenvironment for maturation of immune cells; a primary deficit in transgenic animals with aberrant type X collagen is a compromised immune system, as well as short stature (Jacenko et al. 2002). Therefore, the matrix made by hypertrophic chondrocytes of growth plate cartilage has continuing functions after the chondrocytes themselves have died and the matrix has been ossified.

The rate and duration of endochondral ossification, as well as variations in the final form of the limb/fin, are the result of a complex interplay of both genetic and epigenetic factors, several of which are shown in figure 10.8 for growth plates from a mammal; they are fundamentally the same for endochondral ossification in other tetrapods (B. K. Hall 1975). The final form and length of a bone are influenced by early patterning, local growth factors and transcription factors involved in autocrine and paracrine loops, multiple systemic hormones, and epigenetic influences ranging from disease states and nutritional quality to the biomechanical environment of a given joint. As a broad generalization, it has been suggested that cartilage grows primarily by intrinsic factors, while bone grows primarily by extrinsic factors, but this generalization does not do justice to the complex interplay of genetic and epigenetic factors during endochondral bone formation in postnatal animals (Sledge 1966; Ede 1971).

A significant question concerns the extent of the role of early embryonic patterning as a determinant of adult form. In very early development there are remarkable similarities in vertebrate limb/fin development at the cellular and molec-

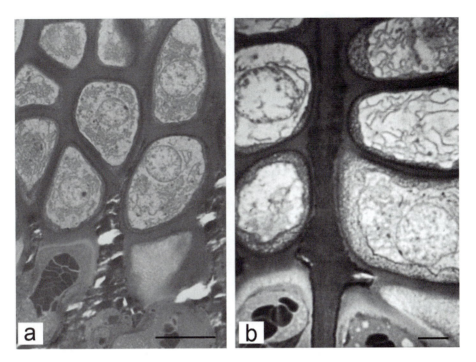

Figure 10-7　Mineralization of the matrix between columns of chondrocytes occurs in the distal hypertrophic cell zone, shown in the growth plate from the distal radius of a four-week-old miniature Yucatan pig (a), and at higher magnification (b). Mineralization is represented by the dark material that in panel (a) looks cracked or broken. This is because, in nondecalcified material, mineral will break out during the microtoming process. Vascular endothelial cells containing red blood cells can be seen adjacent to terminal hypertrophic chondrocytes in panel (b); in this figure it also is clear that mineralization is confined to the matrix between the hypertrophic cells. (a) bar = 25 µm; (b) bar = 10 µm.

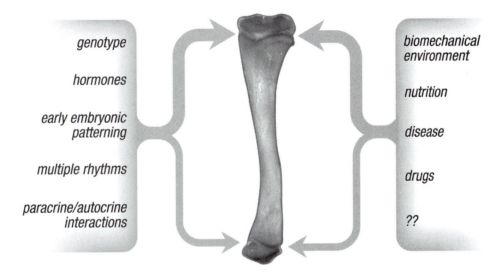

Figure 10-8　Multiple inputs at all levels of biological organization impact on bone growth and affect both the rate of elongation and the quality of the bone that is formed. Dissecting the significance of each of these requires both reductionist and whole-animal approaches. (Figure by Michael A. Simmons.)

ular levels (Coates 1995; Erlebacher et al. 1995; Cohn 2000; Hinchcliffe 2002; Shubin 2002). Much less is known about the extent to which there are similarities in their postnatal regulatory mechanisms (J. Lewis and Martin 1989). The pattern of expression of homeobox genes confers information to cells of the early mesenchymal lineages that specifies over-

all pattern, shape, and identity of individual elements (Newman 1993; Erlebacher et al. 1995; Mortlock et al. 1996; G. P. Wagner and Chiu 2001; S. Ahn et al. 2002; Shubin 2002). But does patterning extend to how many stem cells there will be per growth plate, and how many stem cell divisions there will be per stem cell (J. Lewis and Martin 1989; Tyrcha 2001)?

Does early patterning also determine directionality of growth (Rooney et al. 1984)? The almost limitless diversification of vertebrate limb and fin patterns, referring once again to figure 10.1, would seem to challenge the idea of any simple direct linkage between Hox gene expression and all the intricacies of final morphology (Coates 1994).

Analysis of shape has been one way of addressing the issue of patterning as well as some of the complexities of the genetic/epigenetic interplay during bone growth. In the *short-eared mouse* with mutations of the *Bmp-5* gene, changes in the shape of the sternum and ears, as well as altered curves on several long bones, are already seen in the mesenchymal condensations (Kingsley 1994). The analyses of multiple *Bmp-5* mutants as well as animals lacking the entire *Bmp-5* gene have demonstrated the stringency of early patterning for shape determination of multiple skeletal elements, and how vertebrates use this ancient family of signaling molecules as critical regulators in shape formation. In this model shape cannot be thought of entirely independently of elongation, since BMP-5 mutants also have subtle changes in rates of long bone growth postnatally (Bailòn-Plaza et al. 1999).

Some shape patterns emerge during postnatal development, with final shape clearly modulated by the biomechanical environment. An example is analysis of development of the complex contoured shape of the distal femoral growth plate in the rabbit. Deep contours of the distal femoral growth plate develop through localized positional differences in rates of growth in different regions of the growth plate in early postnatal development (Lerner and Kuhn 1997; Lerner et al. 1998). Shape of the growth plate contour develops over a six-week period, ultimately resulting in an elaborate interlocking pattern of oppositely positioned grooves and papillae (fig. 10.9). This shape cannot be patterned solely by gene expression in early development, since its final form depends upon appropriate biomechanical forces in the postnatal period.

Strategies of Growth

At the level of the growing cartilage, only three cellular activities define the chondrocytic differentiation cascade during endochondral ossification in all species that have been studied: cellular proliferation, cellular hypertrophy, and matrix synthesis. Interstitial growth occurs by different quantitative mixes of these in different species, as, for instance, a strategy for rapid growth based primarily on hypertrophy in the rat versus primarily on proliferation in the chick, and (in multiple species) a strategy for slow elongation based on matrix synthesis. Proliferation, matrix synthesis, and hypertro-

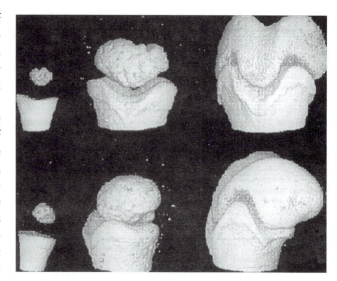

Figure 10-9 Complex contours are formed across the distal femoral growth plate of the rabbit postnatally, due to differential rates of growth across the growth plate. This figure shows the progression of these shape changes in two views at 1, 14, and 42 days. It is hypothesized that the biomechanical environment of a given growth plate is important for the development of these kinds of shape modulations to the growth plate. (From Lerner and Kuhn 1997; used with permission from the Orthopaedic Research Society.)

phy all are considered to have been present during endochondral ossification in the earliest tetrapods. The assumption is that the major components of the matrix of growing cartilage—multiple collagens, proteoglyans, metalloproteinases—are also all very ancient (Exposito et al. 2002; Moss and Moss-Salentijn 1983). Thus, the fundamental cellular activities that result in growth operate in all tetrapods, even if the associations and positions of skeletal elements constantly change.

At the level of the individual or the species, changes in final size during evolution can be achieved by changing either the duration of growth or the rate of growth, or a combination of these. As an example, currently it is thought that many dinosaurs reached their final gigantic sizes by growing very rapidly over a relatively short period of time, perhaps as little as seven years (Horner et al. 2000; Padian et al. 2001). This is in contrast to giant crocodiles, which are thought to have maintained a slow rate of growth, probably ever decreasing, throughout their lives (Erickson and Brochu 1999). In modern animals, even just within mammals, there is tremendous variability for what developmental stage animals are born in (marsupials vs. hyenas); how fast they mature (rodents vs. humans); and the form of the trajectory of growth during postnatal life (bears vs. other carnivores). Miniaturization has been achieved in some species through slowing of overall rate, in others through truncating the period of growth (Hanken 1982, 1985, 1993). In a classical experiment by Rutledge et al. (1974), selection for increase in length of the tail over several generations in inbred mice had two endpoints. In one, increased tail length was achieved by increase in the number of caudal vertebrae; in the other, increase in tail length was achieved by an increase in the size of individual vertebrae.

In domestic dogs, distinct periods of prenatal and postnatal growth have different rate trajectories that may be regulated by different systemic hormones (Wayne and Ruff 1993). A similar division of the postnatal growth curve into two periods has been identified in mice at the molecular level (Cheverud et al. 1996), with the suggestion that the same genes may be operating in both phases of growth, but under the influences of different hormonal environments. Therefore, the basic system, built upon responses within a rather simple differentiation cascade of one cell type, has phenomenal plasticity to modulate both rate of growth and length of time of growth, with individual elements of the modular construction being able to respond independently of the others (Raff 1996).

A distinction must be made between determinate growth—defined as growth that ends at a given time and cannot resume—and indeterminate growth, which has the potential to continue for the life of the animal. Indeterminate growth can occur either because the epiphysis remains cartilaginous or because, in the presence of a secondary center, no bony union ever occurs between the epiphysis and the diaphysis (Haines 1969; Moss and Moss-Salentijn 1983). Not all vertebrates that lack secondary centers can, a priori, be assumed to have indeterminate growth. However, in general, addition of bone at the chondro-osseous junction is possible throughout life in epiphyses where no secondary center develops and where the articular surface is not separated from bone by a bony plate (Moss and Moss-Salentijn 1983). This occurs in most fish, reptiles, and amphibians. Whether growth is determinate or indeterminate, the trajectory is usually that of an early period of the most rapid growth, with a declining rate that plateaus at what would be analogous to skeletal maturity in species with determinate growth, even if bone elongation never fully ceases.

A different strategy is effective for achieving either indeterminate growth or growth that occurs well into adulthood, as in the African elephant. Growth continues because growth plates remain open for most of adulthood, as shown in several growth plates from a 27-year-old female animal (fig. 10.10). Slow growth of the scapula, radius, ulna, and vertebra, clearly an ongoing potential in this elephant because of the open growth plates, would lead to continued increase in height throughout adult life (Smuts and Bezuidenhout 1993; Bezuidenhout and Seegers 1996). Hanken (1982) hypothesized that one purpose of secondary centers is to ensure that

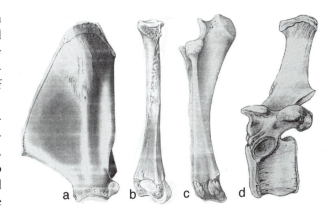

Figure 10-10 Four different bones of the skeleton of a 27-year-old elephant (not to scale). In each there is obvious evidence of open growth plate(s). In the scapula (a) there is an open growth plate along the entire proximal border. In the radius (b) and ulna (c) the distal growth plate is open. The lumbar vertebra (d) has open growth plates on both the body and the spinous process. These growth plates indicate the animal was still growing. (a-c from Smuts and Bezuidenhout 1993; d from Bezuidenhout and Seegers 1996; all are used with permission from *Onderstepoort Journal of Veterinary Research*.)

growth is determinate, and in most mammals epiphyseal closure is thought to be a very rapid event (A. B. Dawson 1929; Haines 1938). An exception is rodents, among which many growth plates remain open throughout life, even though growth effectively ceases, essentially decoupling cessation of growth from the formation of a bony union between the metaphysis and the secondary center (A. B. Dawson 1929).

The Concept of Differential Growth

Perhaps one of the most remarkable features of postnatal skeletal development is the concept of differential growth. This refers to the phenomenon that different ends of a given long bone will contribute differentially to the total increase in length of a bone, both because at any given time they are growing at different rates, and also because they grow for different lengths of time (reviewed by Farnum 1993). At any given point in time and until all growth has ceased, essentially every growth plate in the body is growing at its own unique rate. Differential growth is primarily a postnatal phenomenon: until birth, in most bones the two ends have grown at the same rate and have contributed equally to the length of the bone.

The magnitude of the inequality increases as postnatal growth continues, although the final total contributions can be quite equal (approximately 60% from the distal growth plate, 40% from the proximal growth plate in the radius of the dog), or very unequal. A good example of the latter is the ulna, where essentially 100% of growth is contributed from the distal end in the dog, with the proximal end contributing only to the formation of the olecranon (see before in discussion of secondary centers, and also Carrig and Wortman 1981). Table 10.1 gives data for four growth plates of the rabbit, over a 50-day period postnatally.

The relative distributions of growth potential between two growth plates of a given bone vary among species, but the pattern of which end contributes most is the same across a wide range of species. As an example, the proximal humerus and the distal femur contribute most in all species looked at, although the extent of their relative contributions have variations among species (Hughes and Tanner 1970). Metacarpals/tarsals and phalanges consistently have only one growth plate, the former at their distal ends and the latter at their proximal ends (Haines 1974).

Differential growth has a number of implications. Differential closure allows for the estimation of the biological age of a species by radiography (Ogden et al. 1981). In the leg, cellular activity in two growth plates, each growing at a different rate and each staying open for a different length of time, must ultimately result in two bones of equal length; the same is true for the forearm. If one of these growth plates closes prematurely, one of many documented angular limb deformities or leg length discrepancies will result (Carrig and Wortman 1981). The seriousness of premature closure obviously is related both to which growth plate is affected and to how close it is to the natural time of closure. The bilateral symmetry of skeletal form and the consistencies of patterns of closure across species suggest that very early patterning events play a large role in determining the broad pattern of differential growth, even though it is expressed primarily after hatching/birth. Regulatory controls of differential growth in the postnatal animal remain totally unstudied at the molecular level.

Conversion of the Chondrocytic Differentiation Cascade into Longitudinal Growth

Although it is recognized that there is a need to understand both the genetic basis of longitudinal growth and its regulation and control at the cellular level, almost all knowledge of comparative developmental genetics is about prenatal development; thorough studies of postnatal longitudinal growth at this level are lacking in all but a very few species (A. S. Wilkins 2002). This section will summarize current understanding of how the chondrocytic differentiation cascade is converted into longitudinal growth in the only species that has been studied in detail—the rat. However, given the nature of skeletal evolution, it can be expected that as future studies explore this question in a variety of species, knowledge gathered in detail in one species will serve as an excellent comparative benchmark.

Differential growth has served as an exceptional noninvasive experimental system, since growth plates growing at very different rates can be analyzed simultaneously without any experimental perturbation to the system. The histological appearance of the growth plate is the presentation of a spatial representation of a temporal differentiation process. Therefore, stereological analysis across the height of the growth plate of changes in such morphometric parameters as cellular or matrix volume fraction ($V_{v\,cells}$ and $V_{v\,matrix}$), cellular numbers in each zone ($N_{cells\,proliferative}$ and $N_{cells\,hypertrophic}$), cellular densities, cellular volume in each zone ($V_{cells\,proliferative}$ and $V_{cells\,hypertrophic}$), and cellular heights, widths, and axial ratios (as a measure of shape change), provides a way to analyze cells at multiple points during differentiation (Cruz-Orivé and Hunziker 1986). This approach can be used in conjunction with bromodeoxyuridine labeling to identify cells in the proliferative pool and, as necessary, as a way to estimate cell cycle times (fig. 10.11a; Farnum and Wilsman 1993). In addition, any of several fluorescent calcium chelators (oxytetracycline, calcein, alizarin red) that are deposited at sites of active mineralization, and thus are initially laid down in largest amounts at the chondro-osseous junction of the metaphysis, can be used to estimate 24- or 48-hour incremental growth (fig. 10.11b).

Stereology, bromodeoxyuridine localization, and tetracycline-based growth measurements can all be made on the same histological section, providing a powerful approach to analyze measurements as time-dependent rate parameters of chondrocytic kinetic activity during differentiation. The three-dimensional nature of the data allows estimations, on a 24-hour time scale, of volumes of matrix turned over, numbers of chondrocytes born/turned over, densities of chondrocytes, time spent in each stage of differentiation, and life span of terminal chondrocytes. Ultimately it provides a way

Table 10-1 Differential elongation rates (μm/day) in four lapine growth plates

Group age (days)	Proximal tibia (μm/day)	Distal tibia (μm/day)	Proximal radius (μm/day)	Distal radius (μm/day)
20	554±36	480±41	170±32	444±40
30	561±30	489±43	145±14	486±30
40	518±34	436±41	96±15	470±39
50	426±28	334±19	70±14	402±11
60	378±29	268±26	34±17	358±40
70	362±26	234±27	39±16	352±19

SOURCE: Data from Thorngren and Hansson 1981, 122.
NOTE: This table presents data for differential growth rates in four growth plates of the rabbit over a 50-day postnatal period. Rates were measured using 24-hour oxytetracycline labeling. The differential rate ratio of the proximal tibia compared to the proximal radius is only ~3.3 at day 20 but is ~12 at day 70.

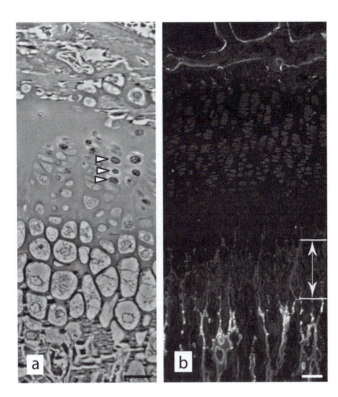

Figure 10-11 Incorporation of bromodeoxyuridine (BrdU), given a half-hour prior to euthanasia, into DNA in the S phase of mitosis is visualized as a dark reaction product in proliferative cell nuclei. Arrowheads in panel (a) point to three labeled nuclei in one column, and others can be seen in adjacent columns. The calcium chelator oxytetracycline (OTC), given 24 hours prior to euthanasia, is initially deposited at the chondro-osseous junction. After 24 hours the leading edge of the fluorescent OTC band is visualized in the metaphysis. The distance from this band to the chondro-osseous junction represents bone elongation in the 24-hour period (vertical double headed arrow). (a) Bar = 25 μm; (b) bar = 50 μm.

of calculating the percentage of growth attributed to activity during proliferation versus hypertrophy, and how much of total growth is the result of matrix production versus cellular number increase during proliferation or cellular volume increase during hypertrophy (Breur et al. 1991; Wilsman et al. 1996b).

This methodology, originally developed by Cruz-Orivé and Hunziker (1986) specifically for analysis of growth plate cartilage—a highly anisotropic tissue—has now been used in multiple investigations to analyze comparative aspects of growth plates growing at different rates (Wilsman et al. 1996b; Cruz-Orivé and Hunziker 1986; see Farnum and Wilsman 2002 for a recent review). Two studies in the past decade have used this methodology specifically to study differential growth. Hunziker and Schenk (1989) compared growth rates of the rat proximal tibia at three different ages: 21 days, 35, days and 80 days, representing growth rates of ~275, ~330, and ~85 μm/day, respectively; shown in figure 10.12. Wilsman et al. (1996b) studied four different growth

plates in the four-week-old rat representing an eightfold difference in rate of growth: ~50 μm/day in the proximal radius, ~140 μm/day in the distal tibia, ~220 μm/day in the distal radius, and ~400 μm/day in the proximal tibia.

Several consistent generalizations have emerged from these studies. First, one can consider all the variables that have been shown to have a strong positive correlation with rate of growth. These include—for the proliferative cell population—the height (volume) of the zone, the number of cells, and the rate of matrix production. These are identical for the hypertrophic cell population where there is a strong positive correlation of rate of growth with the height (volume) of the hypertrophic cell zone, the number of cells, and the rate of matrix production (Breur et al. 1991). Cell cycle time (rate of cellular production) in the proliferative cell zone also has a positive correlation with rate of growth, as does the number of cells turned over a day at the chondro-osseous junction in the hypertrophic cells zone (rate of cellular loss; Wilsman et al. 1996a, 1996b). In the steady state, production and loss are identical.

Two additional features of the hypertrophic cell zone also have a strong positive correlation with rate of growth. The first, indicated in figure 10.13, is the final size of hypertrophic cells. Hypertrophic cell final volume and height are both variables that have a strong positive correlation with rate of growth (Breur et al. 1991). Interestingly, the slope of the line plotting final hypertrophic cell size against rate of growth varies for different species, indicating that there is a different efficiency to increase in cellular volume as growth increases. The second is the differential height change of hypertrophic cells compared with proliferative cells. This is presented for the rat proximal tibia in figure 10.14, where the change in height$_{proliferative\ to\ hypertrophic}$ is considerably less for the cells represented in figure 10.14a, c (80-day-old rat) than for those presented in figure 10.14b, d (35-day-old animal).

Figure 10.15 summarizes data from the study of four growth plates with an eightfold difference in rate of growth in the four-week-old rat (Wilsman et al. 1996b). As expected, at all rates of growth there are contributions *both* during the proliferative cell phase and during the hypertrophic cell phase. Secondly, at all rates of growth there is a significant contribution from matrix production, but this contribution is *more* significant the slower the rate of growth. Third, at *all* rates of growth, more growth occurs during the hypertrophic phase than during the proliferative phase. And finally, and most important, the faster the rate of growth, the more significant is cellular volume increase and height increase to overall rate of growth. Therefore, in this rapidly growing mammal, cellular enlargement during hypertrophy is the most important variable in understanding not only how the chondrocytic differentiation cascade is converted into

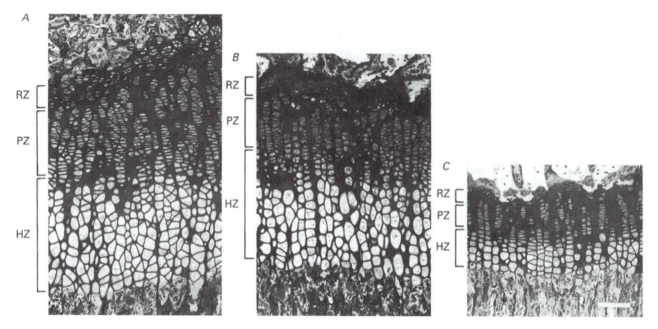

Figure 10-12 The growth plate of the rat proximal tibia is shown at three different ages: 21 days (A), 35 days (B), and 80 days (C), growing at ~275, ~330, and ~85 μm/day, respectively. The extent of the reserve zone (RZ), proliferative cell zone (PZ), and hypertrophic cell zone (HZ) in each growth plate is indicated. Note that, in a very general way, one can correlate rate of growth with cellularity of the growth plate and cellularity of each zone. Bar = 100 μm. (From Hunziker and Schenk 1989; used with permission from Ernst B. Hunziker and from the Physiological Society, Cambridge, UK.)

growth, but also how differential growth is achieved in multiple growth plates.

These data provide some interesting conclusions that might not be, at first, intuitively obvious.

1. Growth plates may have identical cell cycle times, identical numbers of cells, and identical rates of loss of cells at the chondro-osseous junction and still be growing at different rates, depending upon the differential extent of volume and height change during hypertrophy. Additionally, there can be a change in cell cycle time without a change in rate of growth.
2. The shape change is as important as the volume increase. Without a shape change that translates volume increase into height increase in the direction of growth, part of the interstitial growth accompanying hypertrophy will be translated into growth in width of the metaphysis.
3. Cellular proliferation in these growth plates is primarily a mechanism to maintain numbers of chondrocytes participating in the steady state, or to change numbers of chondrocytes as rates of growth are altered.

Hypertrophic chondrocytes in rapidly growing mammalian growth plates undergo a volume increase of well over an order of magnitude (1500 μm³ as a proliferative cell to as much as 30,000 μm³ as a hypertrophic cell). Cellular volume

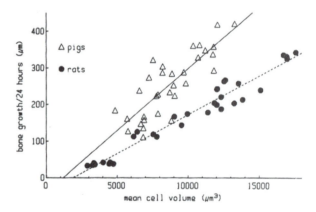

Figure 10-13 A plot of multiple growth plates growing at different rates, from less than 50 μm/day to 400 μm/day, from two different species—rats and miniature Yucatan swine. Note the very strong positive correlation between final hypertrophic cell volume and rate of growth in both species. Note also the different slope to the line in the two species, indicating that, for any given rate of growth, final volume will be greater in rats than in pigs. This indicates that, in rats compared to pigs, a greater percentage of the interstitial elongation achieved by growth plate chondrocytes is during hypertrophy. (Graph modified from Breur et al. 1991 and used with permission from the Orthopaedic Research Society.)

increase during hypertrophy is a highly regulated process; hypertrophy is differential, occurring to the greatest extent in the most rapidly growing growth plates. Therefore, hypertrophy is inversely time dependent—chondrocytes in growth plates growing most rapidly achieve significantly higher final volumes in significantly less time than chondrocytes in slowly growing growth plates. Hypertrophy essentially in-

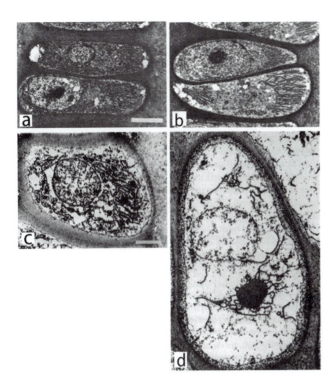

Figure 10-14 Micrographs of proliferative cells (a, b) and hypertrophic cells (c, d) from the proximal tibial growth plate of the rat at two different ages—80 and 35 days. Although proliferative cell volume and height are quite similar at the two ages, hypertrophic cell volume and height are very different. By examining the incremental height difference between proliferative and hypertrophic cells at the two ages, it is clear that the gain in height accompanying the volume and shape change of hypertrophy is significantly greater in the younger animal. The sum of this incremental height increase for all chondrocytes turned over in one day is the major contributor to the growth differential between these two growth plates. Bar = 5 μm. (From Hunziker and Schenk 1989; used with permission from Ernst B. Hunziker and from the Physiological Society, Cambridge, UK.)

volves a rapid increase of cellular fluid intake, since organelle content does not keep pace with the volume increase. However, the regulatory mechanisms involved are not understood (Hunziker et al. 1999; Farnum et al. 2002). If one can generalize from the rat, chondrocytic hypertrophy with appropriate shape change appears to be the primary strategy that rapidly growing mammals use to achieve differential growth during endochondral ossification. As will be discussed below, this is not necessarily true from what we understand from the more limited data from birds, reptiles, fish, and amphibians.

Regulation of Transition Points during Endochondral Ossification

Bone elongation in mammals with well-organized growth plates is achieved through clonal expansion following stem cell division, represented spatially as a column of cells. The spatial alignment of cells within a column mirrors the temporal "life history" of an individual chondrocyte within the differentiation cascade. Hochberg (2002a) presents this concept graphically, showing the same chondrocyte throughout its life history—reserve, proliferative, and hypertrophic (fig. 10.16). The growth plate, in the steady state, should be considered a constantly renewing organ, similar to skin epithelium or gut epithelium. In any given short-term framework, such as 24 hours, cell renewal through proliferation will be offset by chondrocytic loss through death and replacement at the chondro-osseous junction on the metaphyseal side. There is a constant progression of the metaphyseal ossification front, as each cell of the growth plate stays spatially constant while going through its differentiation cascade of proliferation and hypertrophy, and then death by apoptosis (Farnum and Wilsman 1987; Farnum et al. 1990; Hatori et al. 1995; Zenmyo et al. 1996).

The preceding focus was on the kinetics of activity *during* proliferation and hypertrophy that result in quantitative incremental growth. However, transition points within the growth plate can be identified, and it has been demonstrated that the *kinetics of transition* at these points is significant for understanding both rates of growth and how final lengths are achieved. Conceptually one can think of these as points during the differentiation cascade that are controlled as gates: the rate of transition can be modulated. It is presumed that the regulation between stages in the differentiation cascade is not as rigorous in epiphyses that remain cartilaginous and do not show a high level of spatial organization of the chondrocytes (as in figs. 10.3a, b, c), but this has never been studied.

Morphological changes can be used to identify the key transition points (fig. 10.17). The first is the point of initiation of stem cell expansion that can be identified on tissue sections by a change in cellular size, shape, and degree of columnation. The second represents the point at which cellular proliferation ceases and chondrocytes initiate the volume increase and shape change associated with hypertrophy. Conceptually this is considered as occurring in two steps—first, the permanent cessation of cellular proliferation, then initiation of volume increase. These may or may not be linked. This transition can be thought of either as a broad one, in which the signals to end proliferation and those to initiate hypertrophy occur throughout a so-called prehypertrophic zone, or as a transition that has two sharply defined control points—a signal to end proliferation and a signal to initiate hypertrophy, with a transition area of shape change in between. Both views are depicted in figure 10.17.

The combination of immunohistochemistry and *in situ* hybridization, together with transgenic technology, has allowed micro-dissection of these potential points of transition and regulation within the growth plate. The power of these

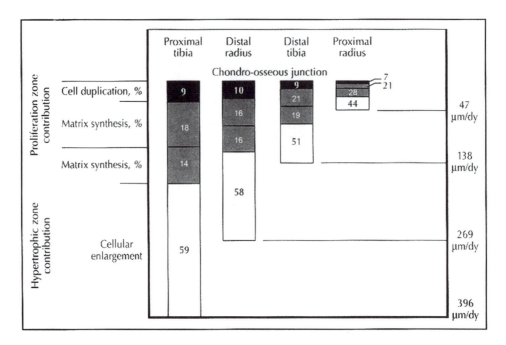

Figure 10-15 This chart summarizes the contributions to growth made at each stage of the differentiation cascade in four growth plates growing at different rates in the 28-day-old rat. Note that (1) at all rates of growth, the most significant contribution is during the hypertrophic stage; (2) the faster the rate of growth, the more is contributed by cellular volume increase and shape change during hypertrophy; and (3) the slower the rate of growth, the more elongation is dependent upon matrix synthesis. This reaches almost 50% in the growth plate growing less than 50 μm/day. (From Wilsman et al. 1996b; used with permission from the Orthopaedic Research Society.)

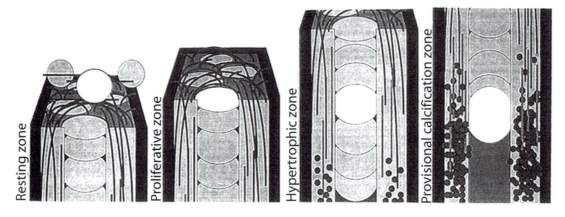

Figure 10-16 As bone elongation occurs, the chondrocyte goes through a differentiation cascade characterized by proliferation followed by terminal differentiation, which is characterized by volume increase during hypertrophy. It is thought that the stem cell population from which clonal expansion of proliferation is initiated is in the resting zone, but there has been limited characterization of these cells. The important point indicated in the figure is that during the differentiation cascade the position of the chondrocyte in absolute space remains the same. By the time the differentiation of a given cell has reached the final stages where the matrix is being calcified, all cells formerly closer to the metaphysis have died. These older cells have been replaced by bone, thus bringing the ossification front to the chondrocyte prior to its own death. (From Hochberg 2002b; used with permission from S. Karger AG, Basel.)

techniques is that analyses can be made at multiple levels of biological organization, and, in many studies, data are provided that link molecular approaches with observations of the whole animal (recently reviewed by Farnum and Wilsman 2001). A similar trajectory of progress in understanding the significance of these transition points has been made

through the molecular analysis of naturally occurring diseases of the skeletal system (International working group 1998; Superti-Furga et al. 2001; Unger 2002). In these diseases, understanding of modification of phenotype is linked to changes at the molecular level (Erlebacher et al. 1995). This has allowed new insights linking the process of endo-

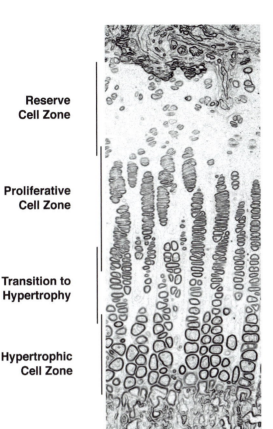

Reserve Cell Zone

Proliferative Cell Zone

Transition to Hypertrophy

Hypertrophic Cell Zone

Figure 10-17 It is becoming increasingly clear that the transitions between populations of chondrocytes in different stages of differentiation are regulated through both positive and negative autocrine and paracrine feedback loops. Important transition zones exist between the stem cell population and the initiation of clonal expansion during proliferation; the termination of proliferation and the initiation of the volume increase and shape change of hypertrophy; and the chondro-osseous junction where chondrocytes die and the space they formerly occupied is invaded by endothelial cells and osteoprogenitor cells. The transition to hypertrophy is sometimes considered to be of long enough duration to warrant the identification of a "prehypertrophic cell zone." (From Farnum and Wilsman 2001; used with permission from Lippincott Williams and Wilkins.)

chondral bone formation to the periosteum, and also illuminating the relationship of endochondral ossification to the developing metaphysis.

If one considers that the growth plate has not only three populations of chondrocytes (reserve, proliferative and hypertrophic) but also checkpoints at the transitions between them, it follows that change in the function of a checkpoint can lead to a dramatic change in phenotype. In some cases this results in chondrodysplasias, some of which are incompatible with life; in others, it can result in a subtle change that ultimately evolves into a distinctly different morphology. In a broad comparative sense, it is only the role of homeobox genes in these processes that has been explored in

any depth (Coates 1994; Sordino et al. 1995; Mortlock et al. 1996; Coates and Cohn 1999; C. Chiu et al. 2000; G. P. Wagner and Chiu 2001; Ahn et al. 2002).

Transition 1: Initiation of Clonal Expansion

The first transition, which is the initiation of stem cell clonal expansion, can be defined morphologically; however, very little is known about regulation of the kinetics at this transition point. The broad question to be asked is this: how much of the activity at this transition, which will be played out in postnatal growth, has been patterned very early in the prenatal environment? If stem cell division and clonal expansion are part of early embryonic patterning, the natural candidates would be some combination of homeobox genes, which have been identified in the early embryonic patterning of all vertebrate limbs studied to date, and of teleost fishes (J. Lewis and Martin 1989; Sordino et al. 1995; G. P. Wagner and Chiu 2001; Ahn et al. 2002). The question is still open as to whether homeobox genes control the number of stem cells and the number of divisions per stem cell, thus controlling total growth potential (Shubin 1995; Campisi 1996; Cohn 2001). *Sox* gene expression, which is known to affect both very early chondrogenesis and growth plate cellular differentiation, is another potential candidate for regulatory influence at this transition (Zhao et al. 1997; W. Huang et al. 2001). Figure 10.18 indicates multiple points in the control spectrum where *sox* genes have been shown to play significant roles, both prenatally and postnatally. How much of the patterning that is influenced by *Sox* in the morphogenesis of mesenchymal condensations and cartilage differentiation is still active in the postnatal animal, specifically at the level of control of chondrocytic clonal expansion (deCrombrugghe et al. 2002; Yan et al. 2002)?

To date no one has been able to use a specific marker to identify on a tissue section the *true stem cells* that create the daughter cells that become the proliferative cells of a given column. No method has been devised to isolate stem cells for study in culture. Therefore, there have been limitations on how the cells of this transition can be studied experimentally. The following very basic questions remain for the role of these cells in postnatal growth. (1) Does a given stem cell have a variable or a fixed number of divisions? (2) To what extent is the *rate* of initiation of stem cell division significant? That is, are there controls that regulate the kinetics of transition at this point? (3) To what extent, and in what combination, do autocrine/paracrine/hormonal and environmental signals influence the initiation of stem cell division? (4) To what extent is the so-called reserve zone the source of stem cells, or is the function of the reserve zone primarily structural (Farnum and Wilsman 1988)?

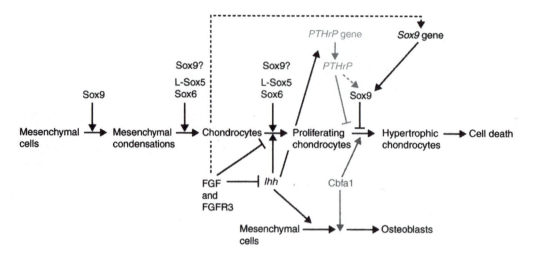

Figure 10-18 Members of the *sox* family of transcription factors, and especially *sox9*, appear to be important regulators of cartilage formation and growth, both prenatally and postnatally. *Sox9* is critical from early time points such as mesenchymal cell condensations and the differentiation of chondrocytes from mesenchymal cells. In addition, members of the *sox* family appear to be involved in regulatory pathways at the transition to proliferation and at the transition to hypertrophy, both in the embryo and during postnatal bone elongation. (From deCrombrugghe et al. 2001; used with permission from Elsevier.)

Transition 2: From Proliferation to Hypertrophy

A "master-gatekeeper" with the function of "putting the brakes on bone growth" within the growth plate was first reported by Vortkamp et al. (1996), who described molecules regulating a gate at this transition between the time that chondrocytes have permanently left the proliferative pool and when they have definitively initiated their terminal differentiation, manifested as the size increase of hypertrophy. The gate of this transition is regulated by a negative feedback loop involving Indian hedgehog (*Ihh*) and PTHrP (parathyroid related protein), their upstream and downstream regulators, and their receptors. Transgenic mice have been created with modifications of the major players of this loop, including knockouts, double knockouts, chimeras, and delayed activations tied to the collagen II-promoter, which ensures activation only in cartilage cells. Double knockouts and partial knockouts have been made in almost every conceivable combination (recently reviewed in de Crombrugghe et al. 2002). Figure 10.19 presents a recent synthesis of the available data indicating the primary regulators in this loop. The control center is in the boxes labeled "maturing chondrocytes" and "commitment to hypertrophy." *Ihh* produced by postproliferative cells in this prehypertrophic zone not only influences proliferations of younger chondrocytes, but also results in communication to the periosteum via *Patched* to stimulated PTHrP, which then provides negative feedback. This loop is thought to be a primary one in regulating the number of cells that will go on to hypertrophy during endochondral ossification (Long and Linsenmayer 1998).

Active research continues into the nature of the multiple

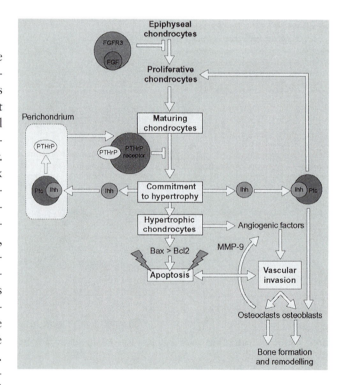

Figure 10-19 One of the best-understood regulatory feedback loops during chondrocytic differentiation in the growth plate involves Ihh, PTHrP, and the PTHrP receptor. Ihh, produced by chondrocytes committed to hypertrophy, interacts through *Patched* in the periosteum to initiate PTHrP production by perichondral cells. This in turn interacts with the PTHrP receptor on growth plate chondrocytes to stop further maturation, effectively "putting the brakes on bone growth" (Vortkamp et al. 1996). As indicated in the diagram, Ihh also has feedback loops to proliferative cells as well as to osteoblasts/osteoclasts involved in bone formation and remodeling. These feedback loops, originally established as important in embryonic growth, are also thought to be active during postnatal bone elongation. (From White and Wallis 2001; used with permission from Elsevier.)

upstream and downstream players involved in this negative feedback loop (Crowe et al. 1999; Minina et al. 2001). At a conceptual level, several questions remain open. First, how do trafficking signals go back and forth between growth plate chondrocytes and the periosteum? The most detailed studies have analyzed mouse embryos starting at day 13.5. Given the small size and the absence of a secondary center in these animals, it seems reasonable to think that diffusion pathways of messengers of this loop could exist between the periosteum, growth plate chondrocytes, and periarticular cells. It is less clear how this might occur with size increases and the development of secondary centers. Second, what is the time frame through which the loop is operating? Is this a control point that acts slowly to regulate cellular numbers, coordinating the balance throughout the growth period between increase in cell numbers through proliferation and loss of cell numbers through death as rates of elongation change? Or is this a gate that acts over very short time frames, such as in response to the near-instantaneous changes in biomechanical stress that cells of the growth plate perceive? An important point to emphasize is how the existence of this loop involves the interdependence of periosteum and endochondral bone formation—an interdependence that has probably existed since the first tetrapods and has evolved into a complex regulatory interplay in mammals and birds.

Transition 3: The Chondro-Osseous Junction

The third transition is at the cartilage/bone junction on the metaphyseal side. Here the terminal hypertrophic chondrocyte dies, most probably by apoptosis (fig. 10.20), and endothelial and osteoprogenitor cells invade the territorial matrix (Farnum and Wilsman 1987, 1989; Farnum et al. 1990; Hatori et al. 1995; Zenmyo et al. 1996; Gibson 1998; Horton et al. 1998). The calcified cartilage of the longitudinal septae is the scaffold for new bone formation by osteoblasts, with different extents of remodeling and loss of the original longitudinal architecture occurring in different species. There is what can be defined as a "kinetics of turnover" at this transition, studied most thoroughly in a serial section analysis of proximal tibial growth plates in swine growing at ~140 μm/day in which ~5.4 hypertrophic chondrocytes per column were lost each day. In this growth plate each chondrocyte spends approximately 4.5 hours as a terminal cell, less than one hour of which it spends in the condensed form that is consistent with apoptosis (Farnum and Wilsman 1989).

Although it is generally agreed that the terminal chondrocyte dies by apoptosis, under a variety of culture conditions it has been demonstrated that growth plate hypertrophic chondrocytes may dedifferentiate and take on an osteoblastic phenotype (Cancedda et al. 1995). The extent to which

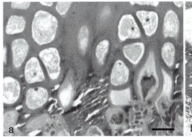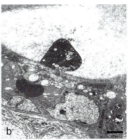

Figure 10-20 There is a general consensus that terminal chondrocytes at the chondro-osseous junction die by rapid cellular condensation with morphology consistent with that of apoptosis (Farnum and Wilsman 1987, 1989). This figure shows the chondro-osseous junction from the distal radial growth plate of a 4-week-old miniature pig, demonstrating the morphology of both fully hydrated terminal chondrocytes and a chondrocyte undergoing condensation. At a higher magnification in panel (b) it can be seen that the chondrocyte maintains an attachment to the last transverse septum while the remainder of its pericellular attachments are lost. The cell fills only a small fraction of the space it previously occupied. Vascular endothelial cells will enter into this space during metaphyseal vascular penetration. (a) Bar = 25 μm; (b) bar = 5 μm.

this occurs *in vivo*, if at all, is still debated. This debate relates to the understanding of the complexities of control of chondrocytic differentiation, including such issues as how terminal differentiation may be linked to proliferation (C. S. Adams and Shapiro 2002). Some suggest that the ultimate fate of hypertrophic chondrocytes may be different in different microanatomical sites within the growth plate based on their proximity to the perichondrium (Cancedda et al. 1995; Bianco et al. 1998). An alternative hypothesis is that apoptosis is only one of several physiological forms of cellular death of hypertrophic chondrocytes (Roach and Clarke 2000). Others hypothesize that the apoptotic program is tied into other feedback loops within the growth plate such as Ihh/PTHrP (Horton et al. 1998), and even that the apoptotic program is involved in the mediation of several other critical processes during endochondral ossification such as intracellular calcium accumulation and release, matrix mineralization, matrix resorption, and attraction of blood vessels at the chondro-osseous junction (Gibson 1998).

Leaving the debate of the fate of the terminal chondrocyte aside, over the last five years, and continuing at an intense rate presently, the complexities of the regulatory mechanisms involved in signaling at the chondro-osseous junction have been studied, by utilizing the power of analysis of multiple perturbations in transgenic mice. A primary role is played by vascular endothelial growth factor (VEGF) produced by hypertrophic chondrocytes, in conjunction with matrix metalloproteinases (MMPs) responsible for matrix degradation (fig. 10.21; Karsenty 1999; Carlevaro et al. 2000). In addition, the transcription factor *cbfa1*, which is critical to osteoblast differentiation, has been shown to be significant in hypertrophic terminal differentiation, thus emphasizing that this transcription factor has multiple poten-

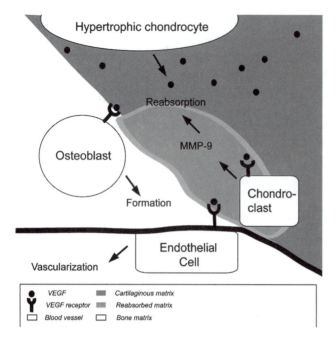

Figure 10-21 The chondro-osseous junction is another important transition region with multiple autocrine and paracrine regulators apparently involved in such events as signaling death of the terminal chondrocyte, activation of metalloproteinases (MMPs) for matrix resorption, and attraction of endothelial cells together with osteoprogenitor cells. Vascular endothelial growth factor (VEGF) produced by hypertrophic chondrocytes interacts with its receptor on endothelial cells, and also is involved with the activation of MMPs on chondroclasts. However, it is presumed that multiple other molecules are involved, some of which trigger apoptosis in the terminal cell, others of which stimulate osteoblasts. (From Karsenty 1999; used with permission from Gerard Karsenty and from Cold Spring Harbor Laboratory Press.)

tial roles in skeletogenesis (Karsenty 1999). The best way to conceptualize the chondro-osseous junction is as an interface with crosstalk between the terminal chondrocytes and vascular endothelial cells, resulting ultimately in apoptotic death of the terminal chondrocyte, angiogenesis, and osteo-progenitor cell recruitment (Farnum and Wilsman 1989; Farnum et al. 1990; Hatori et al. 1995; Zenmyo et al. 1996).

Additional Levels of Regulation

The previous discussion focussed only on one or two important regulators at each transition, thus oversimplifying the complexity of control at each point. Not only are multiple additional growth factors involved in autocrine/paracrine regulation of growth plate function, but systemic hormones also directly affect chondrocytic differentiation kinetics (N. Loveridge and Farquharson 1993; Farnum and Wilsman 1998b, 2002). An example of the former is the family of fibroblast growth factors (FGFs). FGFs comprise a family of 22 known genes that encode structural proteins, and four distinct FGF receptors are activated by most members of the FGF family (Ornitz and Marie 2002). What is particularly

interesting about the FGFs is that they have essential roles at all stages of skeletal development—from positioning and outgrowth of the limb bud, to detailed patterning of limb elements, to regulation at multiple steps of the chondrocytic differentiation cascade both in the embryo and postnatally (Wilkie et al. 2002). It has been hypothesized that the emergence and functional significance of FGF and FGF-R gene-family expansions and divergence occurred very early in vertebrate evolution, and that the gene duplications that occurred in these early stages have been critical determinants of skeletal system formation in vertebrates ever since (Coulier et al. 1997).

Figure 10.22 diagrams the complexity of known interactions of several in the family of FGFs, and their signaling pathways at multiple levels of the growth plate during postnatal development (Z. Liu et al. 2002). A summary would be that FGFs and their receptors are involved with the regulation of intrinsic growth rates in all zones of the postnatal growth plate, perhaps even coordinating proportional bone growth throughout the body (Erlebacher et al. 1995). Thus, FGFs can be considered to be master autocrine/paracrine regulators during postnatal growth, with effects seemingly at

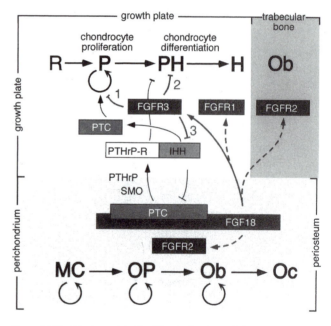

Figure 10-22 Members of the fibroblast family of growth factors have potential regulatory roles at almost every stage of endochondral ossification, effectively maintaining a continuous crosstalk with the surrounding perichondrium and the newly formed trabecular bone with its periosteum. This diagram depicts some of the known interactions of FGFs with the Ihh/PTHrP negative feedback loop. These pathways have been investigated both through the use of transgenic mice and through the analysis of naturally occurring chondrodysplasias in humans. R, reserve cells; P, proliferating cells; PH, prehypertrophic cells; H, hypertrophic cells; SMO, smoothened; PTC, patched; MC, mesenchymal cells; OP, osteoprogenitor cells; Ob, osteoblasts; Oc, osteocytes. (From Z. Liu et al. 2002; used with permission from David Ornitz and from Cold Spring Harbor Laboratory Press.)

every point in the chondrocytic differentiation cascade. Mutations of specific members of the FGF family and their receptors have been documented in multiple chondrodysplasias and other diseases of skeletal growth. As an example, the etiology of most cases of achondroplasia, which is one of the most common dwarfisms, is a glycine-to-arginine substitution in the transmembrane domain of FGFR3 (Ornitz and Marie 2002; Wilkie et al. 2002). Probably no other family of molecules is so key to skeletal regulation from the earliest embryonic stages through to postnatal growth.

It is easy to become entrenched in the understanding of autocrine and paracrine regulatory pathways and to forget that postnatal bone growth occurs in a permissive systemic hormonal environment. Analyses of transgenic mice have greatly augmented knowledge of the significance of key hormones, whose circulation throughout the body is important for generalized levels of control over the rate of bone elongation. A good example of the effect of systemic regulatory control is overall size. There is experimental evidence that, contrary to the original somatomedin hypothesis, growth hormone (GH) has direct effects on growth plate cartilage, and does not need to act through liver-derived insulin-like growth factors (IGFs; fig. 10.23). Chondrocytes in the growth plate express receptors for both growth hormone and for IGFs (Ohlsson et al. 1998). Analysis of transgenic mice with multiple combinations of alteration in growth hormone and/or IGF1 and/or IGF2 and/or their receptors has clearly indicated that growth hormone should be considered the master systemic regulator of postnatal bone elongation in mammals, through multiple direct and indirect actions within the growth plate. All effects of either growth hormone or the IGFs are directed at augmentation of bone elongation (for recent reviews see Lupu et al. 2001; Le Roith et al. 2001).

The following generalizations can be made of the other major systemic hormones and their effects on growth plate cartilage: (1) the major action of thyroid hormone is during chondrocytic hypertrophy, where it acts both to inhibit proliferation and to stimulate terminal differentiation (Ishikawa et al. 1998; Siebler et al. 2001; Harvey et al. 2002); (2) the major actions of glucocorticoids are catabolic, acting to decrease IGF-1 generation, matrix collagen synthesis, mineralization, and vascular penetration (Baron et al. 1994; Hochberg 2002b; Koedam et al. 2002; Smink 2003); and (3) both estrogens and androgens increase bone elongation rate during adolescence, but estrogens are required for the normal timing of growth plate closure and may act by entirely local regulatory mechanisms within the growth plate (Faustini-Fustini et al. 1999; Mauras 2001; van der Eerden et al. 2002; Bilezikian 2002).

Potential Roles of the Extracellular Matrix during Endochondral Ossification

The specific composition of the extracellular matrix affects bone elongation. Growth plate cartilage matrix can be conceptualized as having three different zones (Noonan et al. 1998). There is a pericellular matrix that surrounds each individual chondrocyte and is the interface with the rest of the extracellular matrix (fig. 10.24a; Eggli et al. 1985). Chondrocytes of a given column share a common territorial matrix that is the route of invasion of endothelial cells during capillary penetration at the chondro-osseous junction (fig. 10.24b). Third, there is an interterritorial matrix that separates chondrocytes of different clonal expansions and that is calcified in the very distal hypertrophic cell zone (fig. 10.24b; Farnum and Wilsman 1986). These are shown diagrammatically in figure 10.24c (Hunziker and Herrmann 1990). In the proliferative cell zone, there is a higher *ratio of matrix to cells* than in the hypertrophic cell zone. However, it is important to understand that, as chondrocytes hypertrophy, they produce *more matrix per cell* by a factor of three or more in rapidly growing growth plates (fig. 10.25; Hunziker and Schenk 1989). Thus, matrix production in both cellular zones of the growth plate contributes substantively to longi-

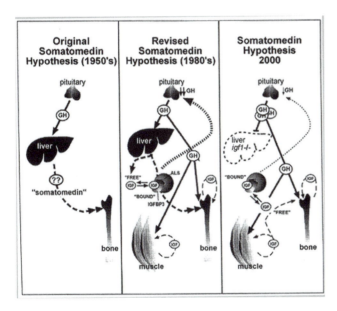

Figure 10-23 The original somatomedin hypothesis (1950s) stated that growth hormone (GH) exerted its peripheral effects through a circulating factor produced by the liver (insulin-like growth factors or IGFs). This hypothesis was revised (1980s) after it was recognized that IGFs synthesized within the growth plate may be involved in autocrine/paracrine regulation. Most recently (2001) this has again been modified, based on multiple experiments with transgenic mice. Results of gene deletion experiments have questioned the role of liver-derived IGF-1 and protein-bound forms circulating in blood in controlling postnatal bone growth, although free IGF may still be involved. (From Le Roith et al. 2001; used with permission from the Endocrine Society.)

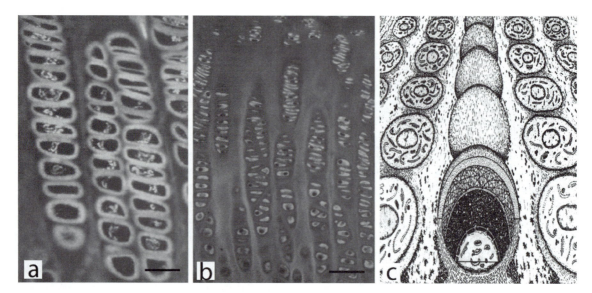

Figure 10-24 Three distinct matrix compartments are present in growth plate cartilage. As shown in panel (a), a pericellular matrix surrounds each individual chondrocyte. This is apparent by staining with the lectin wheat germ agglutinin (WGA), which also stains the Golgi area of individual cells but is negative in the nucleus. Panel (b) shows the distal radial growth plate of a 4-week-old Yucatan pig stained with the lectin concanavalin A (Con-A). Con-A staining is negative in the pericellular matrix, as well as in the territorial matrix that surrounds chondrocytes of a given column. However, the interterritorial matrix between columns of chondrocytes lightly stains with this lectin. The rough endoplasmic reticulum of individual cells is brightly positive, and the nucleus is negative. These matrix compartments are depicted diagrammatically in panel (c), which also demonstrates how the territorial and interterritorial matrices have unique structural properties. (a) Bar = 20 μm; (b) bar = 40 μm. (c modified from Hunziker and Herrmann 1990; used with permission from Ernst Hunziker and from Kluwer Academic/Plenum Publishers.)

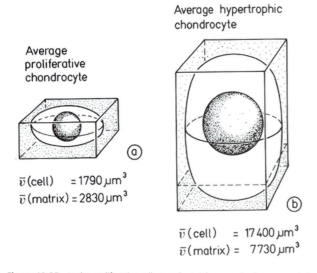

Figure 10-25 In the proliferative cell zone the total area and volume occupied by cells (area fraction of cells [$A_{a\,cells}$] and volume fraction of cells [$V_{v\,cells}$]) are less than in the hypertrophic cell zone. However, the quantity of matrix per cell is significantly greater in the hypertrophic cell zone. Increased matrix synthesis per cell accompanies increase in cellular volume. (From Hunziker et al. 1987; used with permission from the *Journal of Bone and Joint Surgery*.)

tudinal growth. Increased pericellular/territorial matrix volume makes a greater contribution to longitudinal growth than does increased interterritorial matrix volume (Noonan et al. 1998).

The shape change that accompanies chondrocytic hyper-

trophy, effectively changing the direction of the long axis of the cell so that it is aligned with the direction of growth, is critical for converting the volume increase of hypertrophy into longitudinal growth. However, the mechanism is not understood. Buckwalter and Sjolund (1989) have hypothesized that mechanical restraint by the interterritorial matrix between columns of chondrocytes essentially confines expansion in the lateral direction, thus translating the volume increase into increase in the direction of growth (fig. 10.26). This would be equivalent to the way plants grow, with volume increase of apical cells being translated into height increase in the direction of growth due to mechanical restraint by asymmetrical properties of the cell wall (Buckwalter and Sjolund 1989; Noonan et al. 1998). It is interesting to speculate about a commonality between the strategies of plant growth in height and bone growth in length, since both require rapid growth in a structurally solid support system.

Buckwalter's theory would account for the metaphyseal flaring seen in numerous chondrodysplasias. Any abnormality of matrix composition might be expected to alter biomechanical properties of the extracellular matrix and increase lateral growth at the expense of longitudinal growth during the volume increase (see Hochberg 2002a, and Erlebacher et al. 1995 for a recent review of growth plate morphology in human chondrodysplasias). Stereological analysis has shown this to be true in Scottish Deerhounds with naturally occurring pseudoachondroplasia. Final volumes of hypertrophic

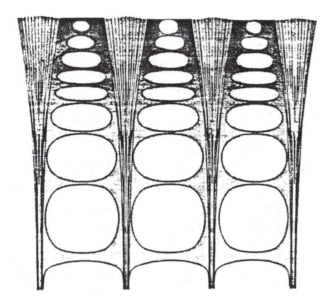

Figure 10-26 A shape change accompanies the volume increase of hypertrophy, translating the size increase into a height increase in the direction of growth. One hypothesis is that the interterritorial matrix restricts expansion of the chondrocyte in width and thus the shape change is primarily a passive one that is dependent upon the structural properties of the matrix. (From Buckwalter and Sjolund 1989; used with permission from the *Iowa Orthopaedic Journal*.)

cells were identical in affected dogs and their normal littermates. The extracellular matrix was highly disorganized in growth plates from affected dogs, and both proliferative cells and hypertrophic cells had a more rounded shape than in the growth plates of normal littermates. The decreased differential height change between proliferative and hypertrophic cells accounted for essentially 100% of the loss of bone elongation in growth plates from affected animals compared to normal littermates (Breur et al. 1992).

Like other hyaline cartilages, the primary collagen of growth plate cartilage is collagen type II, and the primary proteoglycan is aggrecan. Additional components such as collagens types IX and XI and small molecular weight proteoglycans are found throughout the matrix, while others, such as collagen X and cartilage oligomeric matrix protein (COMP), are localized to specific regions (reviewed by Neame et al. 1999). Matrix compartments in the growth plate may sequester growth factors such as transforming growth factor β (TGF-β), which are involved in autocrine regulation during chondrocyte differentiation. Similarly, matrix compartments may contain the enzymes that are required for matrix degradation concomitant with hypertrophy (e.g., matrix metalloproteinases) and matrix mineralization (e.g., alkaline phosphatase). Differences in both organizational and constitutional properties of the territorial and interterritorial matrices may facilitate diffusion of nutrients, growth factors, and ions in certain directions dur-

ing chondrocytic differentiation. It has been emphasized previously that mineralization of the longitudinal septae in the distal hypertrophic cell zone provides a scaffold for subsequent bone formation and that it provides components necessary for the optimum microenvironment needed for hematopoiesis and immune cell function in the bone marrow (Jacenko et al. 2002). Although the spatial heterogeneity of the growth plate matrix has been increasingly studied and understood (see Neame et al. 1999, for a recent review), the significance of the heterogeneity in terms of optimization of growth plate function is clearly in need of further study.

Comparative Observations of Endochondral Ossification in Nonmammalian Epiphyses

The previous examination of our current understanding of the mechanism of endochondral bone elongation in mammals, focusing specifically on what has been learned through analysis of transgenic mice and spontaneous mutants, allows a baseline for comparative observations in animal groups that have been much less well studied. The transgenic mouse system is a good one for coming to understand such questions as the following: (1) What kinds of changes in regulatory pathways (autocrine, paracrine, or systemic) are compatible with life? (2) What kinds of spatial and/or temporal modulations of the chondrocytic differentiation cascade lead to significant modifications that would be consistent with potentially having been involved in dichotomous branching during evolution? or (3) As major structural components of the growth plate evolve and change, what are the effects on bone growth rate? This is not to imply that the full complexity of strategies for bone elongation are present, let alone understood, in murine postnatal bone growth, or that they are more highly evolved in this species than in others. It is only that in this species these issues currently are best understood and so this baseline provides a platform for comparative analyses and/or defining the questions that need to be answered in the analysis of additional species (Enlow 1969; Francillon-Viellot et al. 1990; Tickle and Münsterberg 2001).

Postnatal Bone Elongation through Endochondral Ossification in Fish

As noted by Shubin (1995), a necessary starting point is to clearly define what is meant by a fin. In fish, the endoskeleton forms only a fraction of the surface area of the fin (G. P. Wagner and Chiu 2001). Internal skeletal elements in the proximal region of the fin form through endochondral ossification, and these vary over a very wide range in number and shape. The exoskeleton, consisting of rays of dermal origin, has no counterpart in tetrapod limbs. Fin rays, which are connected by ligaments to the endoskeleton, grow by distal addition of bony segments. Thus, fin radials are not like digits (Shubin 1995; Iovine and Johnson 2000). In the fin-to-limb transition there were major transformations of the endoskeleton, and one of the clearest differences in limb morphology compared to fin morphology is the absence of rays (Coates 1994, 1995; B. K. Hall 1991). I limit consideration to the endoskeletal element.

Multiple studies focus on very early processes of development of fins, looking for analogies to limb development in tetrapods, with some studies extending to experimental analysis of growth of cartilage and bone in teleost fish *in vitro* (Miyake and Hall 1994). One emphasis has been on patterning genes (see Coates 1995; Sordino et al. 1995; Grandel and Shulte-Merker 1998; Coates and Cohn 1999; Chiu et al. 2000; Hinchcliffe 2002; Tanaka et al. 2002 for recent examples) and the T-box gene, *tbx5* (Ahn et al. 2002), using zebrafish as the model. It has been shown that in zebrafish the transcription factor *sox9* is required both for morphogenesis of cartilage condensations and for cartilage differentiation, as it is in mammals (Yan et al. 2002). The primary perspective of these studies is to analyze phylogenetic relationships among fish, the origin of fins, and significant molecular events in the fin-to-limb transition. In contrast, this chapter focuses on postnatal growth of the endoskeletal element, since the overall goal is to make comparative observations with cellular regulation of postnatal endochondral ossification in tetrapods.

Bone growth through endochondral ossification in fish is indeterminate, but with apparent remarkable control over the extent of growth, growth rate, and shape over a wide variety of morphological variants (S. L. Johnson and Bennett 1999). As a generalization, there is no development of a calcified or bony secondary center, or, when one can be found, it has been interpreted as a parallel development to that seen in tetrapods (Haines 1934, 1938). There are zonal divisions of chondrocytes into proliferative and hypertrophic cells based upon morphology and axial ratios, but no regular arrangement into columns (fig. 10.3a; Haines 1934). The chondro-osseous junction also has no regular arrangement, with cartilage remaining in the metaphysis for long periods of time prior to final replacement by bone. Although the role of GH and IGF-1 in growth promotion of multiple organs in fish has been studied (recently reviewed by Mommsen 1998), the role of systemic hormones on endochondral ossification has not been addressed. Fish can display rapid and variable rates of growth, but no study to date has tried to capture the kinetics of this process at the level of cellular dynamics where the chondrocytic differentiation cascade is converted to longitudinal growth.

Postnatal Bone Elongation through Endochondral Ossification in Amphibians

Consideration of postnatal bone elongation in amphibians has many of the same issues as the study of endochondral bone elongation in fish. There is a tremendous variety of epiphyseal structure and in the ratio of permanent cartilage/endochondral bone/periosteal bone among the range of species studied, depending primarily upon the extent of aquatic life (Esteban et al. 1996, 1999). Epiphyseal structure and growth may be very similar to that of fish for primarily aquatic species, while for species spending a large part of their life in a terrestrial habitat there is virtually no endochondral bone and periosteal bone forms a smooth cylinder (Haines 1938).

Although detailed morphological studies of the epiphysis in amphibians in relationship to bone elongation have been made in a few species, most current studies on regulatory mechanisms focus on very early development (Haines 1938; Púgener and Maglia 1997; Sheil 1999; Trueb et al. 2000). The recent study by Bergwitz et al. (1998) demonstrating the significance of PTH/PTHrP receptor isoforms in early *Xenopus* development and that by Hanken et al. (2001) studying *Shh* and *distal-less* protein (*Dlx*) in *Eleutherodactylus coqui* (a direct-developing frog) are the kind of studies needed at all stages of growth in multiple species in order to make comparative studies of basic regulatory mechanisms of bone elongation through endochondral ossification between amphibians and mammals. In an ironic way, the increased incidence and prevalence of malformations in frogs (see Meteyer et al. 2000) are providing opportunities for studying skeletal regulation at multiple levels including roles of GnRH and androgens (Sower et al. 2000), vitamin A (Das and Mohanty-Hejmadi 2000), and FGFs (Loeffler et al. 2001), as examples from the recent literature. As mechanisms involved in limb regeneration are studied more intensively at the molecular level, this may provide further impetus for comparative studies of normal postnatal bone elongation in amphibia (Libbin et al. 1988; Brockes 1997; Gardiner et al. 1999; Wolfe et al. 2000; M. W. King et al. 2003).

A series of recent studies by Felisbino and Carvalho (1999, 2000, 2001, 2002) in *Rana catesbeiana* provided detailed morphological analyses of the "match-stick" type of epiphysis found in *Rana* and some other amphibians, first described as early as the 1910s (A. M. Marshall 1916), but with additional descriptive studies in the 1980s (Dickson 1982) and 1990s (Dell-Orbo et al. 1992). The descriptive analyses by Felisbino and Carvalho are very detailed, and from them hypotheses can be made as to how bone growth is achieved in *Rana* spp. It would be interesting to test these hypotheses using stereological analysis and kinetic markers such as BrdU and fluorochrome labeling.

Figure 10.27 demonstrates the unique features of the epi-

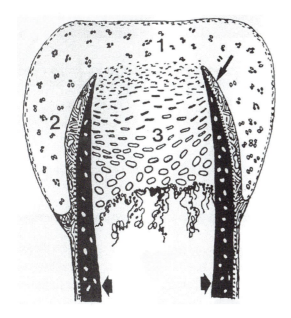

Figure 10-27 The epiphysis of *Rana* spp. has a matchsticklike appearance with three cartilaginous areas: (1) the articular cartilage, (2) the lateral articular cartilage, and (3) the cartilage specifically undergoing endochondral ossification (growth cartilage). Bone elongation is through bone formation at the leading edge of the fibrous periosteum (thin arrowhead). Thick arrows point to periosteal bone. (From Felisbino and Carvalho 1999; used with permission from Elsevier.)

physis in *Rana catesbeiana*. Epiphyseal cartilage is inserted into the end of a tubular bone shaft, therefore delineating three regions: (1) articular cartilage and lateral articular cartilage; (2) growth cartilage; and (3) a fibrous layer of periosteum. The latter is high in blood vessels and is joined to the lateral articular cartilage. Proliferative chondrocytes of the growth cartilage lack any columnar arrangement, and their division is perpendicular to the long axis of the bone. Hypertrophy is not associated with either mineralization or immediate bone formation, and therefore there is no mechanism to translate the cellular volume increase of hypertrophy into longitudinal growth. The conclusion is that periosteal ossification drives bone elongation, that growth cartilage contributes primarily to radial expansion, and that endochondral ossification—the actual replacement of cartilage by bone—is a very late event, probably important as the animal continues to gain weight. The deposition of bone trabeculae is not dependent upon cartilage matrix mineralization (Felisbino and Carvalho 2001).

An osteochondral ligament, perhaps analogous to the periodontal ligament in the jaw, is formed by proliferation of cells at the tip of the periosteal bone (Felisbino and Carvalho 2000). It is a complex, specialized fibrous attachment that essentially guarantees strong and flexible anchorage of the lateral articular cartilage to the periosteal bone. It is hypothesized that the large lateral articular cartilage is an adaptation for loading during high-amplitude movements that occur

during the characteristic limb movements of frog swimming, jumping, and resting. Thus, the interplay between periosteal and endochondral bone formation is highly specialized in these species. However, fundamentally the process of endochondral ossification is the same as has been described before—a combination of chondrocytic proliferation, hypertrophy, and matrix synthesis—even though the temporal juxtaposition of these in relationship to subsequent bone formation is far removed from what was previously described for rapidly growing mammals.

Postnatal Bone Elongation through Endochondral Ossification in Reptiles and Birds

Current research into mechanisms of bone elongations in these animal groups is driven by such diverse goals as understanding the relationship between dinosaurs and modern birds, and maximizing growth rates in the poultry industry. Major themes include the range of bone elongation strategies in modern reptiles, including the presence of both determinate and indeterminate growth; the extreme evolutionary modifications in some groups such as turtles (H. K. Suzuki 1963; Rieppel and Reisz 1999; Sheil 2003); other extreme evolutionary modifications in modern reptiles for limblessness, miniaturization, and/or return to an aquatic environment (Gans 1975; Hanken 1982, 1985, 1993; Raynaud 1990; Cohn and Tickle 1999; Shapiro 2002); and the effect of selective pressures for rapid post-hatching growth in altricial avian species. Although it is beyond the scope of this chapter, the concept that certain dinosaurs grew very rapidly, achieving adult gigantic sizes in as little as seven years through exceedingly rapid periosteal bone formation, is an excellent reminder of the range of strategies that species have evolved for postnatal bone elongation (Horner et al. 1999, 2000, 2001; Padian et al. 2001).

Detailed morphological time-sequenced studies of individual species from several reptilian groups have been made. These studies followed the patterns and timing of chondrification of skeletal elements as well as the pattern and timing of ossification of the same elements (Rieppel 1992a, 1992b, 1993a, 1993b, 1993c; Sheil 2003). A major conclusion drawn from these analyses is that chondrification and ossification often are decoupled, and selection can act on these two phases of limb formation independently. This is one possible explanation for the wide variability of final design. Maisano (2002a), studying the postnatal ontogeny of two species of iguana, also concluded that patterns of prenatal morphogenesis are independent of patterns of postnatal skeletal development, and further suggests that the latter may be very important in defining character data for systematic investigations.

Even within one group of reptiles, there may be significant variations in epiphyseal structure and the extent of indeterminate growth—some have large secondary centers that ossify, in others the secondary center only calcifies, and in still others the epiphysis remains cartilaginous (Haines 1939, 1941). Similarly, the synchronization of perichondral versus endochondral ossification varies in different reptilian groups, with an extreme being crocodiles and alligators. In these animals large masses of hypertrophied cartilage remain isolated in the marrow cavity, developing as cartilage cones that are only slowly replaced by bone. Although the only consistent secondary center in extant birds is in the proximal tibia, Haines (1969) hypothesized that bird ancestors did have secondary centers and that these were secondarily lost, perhaps because birds developed a system of cartilage canals, or perhaps because air sac development and secondary center development were not compatible. Nevertheless, even though secondary centers are lacking, avian bone elongation is considered to be determinate; growth plate cells at some point cease to proliferate and a fibrocartilaginous articular cartilage remains, with variable amounts of endochondral bone replacement within the diaphysis (J. M. Starck 1996).

Two features for understanding differences in rate of bone elongation during post-hatching avian life are (1) final adult size and (2) whether the young are altricial versus precocious. Although more extensive studies are needed across a wide range of avian species, the generalization is that large birds, such as the ostrich, grow by depositing well-developed fibrolamellar bone, a pattern that is considered primitive for birds (Castanet and Smirina 1990; Horner et al. 2001). The faster the postnatal growth rate, the higher the proportion of the skeleton that will be cartilaginous, with replacement of cartilage by bone often lagging well behind the rate of bone elongation per se (J. M. Starck 1996). Some nongiant modern birds, be they precocial or altricial, have very rapid bone elongation rates: in a comparative study of 87 bird species, lengthening of the tibiotarsus ranged from 0.35 to 6.0 mm a day (Kirkwood et al. 1989a). It has been hypothesized that that nonavian dinosaurs never achieved rates as rapid as that of modern altricial birds, and that selection for high rate occurred stepwise after separation from dinosaurs and under very great selection pressure for rapid growth post-hatching and prefledging (Erickson et al. 2001).

In the chick, multiple regulators of postnatal bone elongation have been studied using (1) cultured growth plate cells or sternal cartilage cells (see Grimsrud et al. 1999; Carlevaro et al. 2000; Pateder et al. 2000; LeBoy et al. 2001 as recent examples); (2) beads soaked with a specific growth factor implanted at an early embryonic stage (see Ferguson et al. 1998; Pathi et al. 1999 as recent examples); (3) retroviral misexpression vectors microinjected at different stages of limb de-

velopment (see Crowe et al. 1999; Ferrari and Kosher 2002; Iwamoto et al. 2003 for recent examples); (4) explants of chick embryonic limbs in culture (see Minina et al. 2001 as a recent example); or (5) spatial and/or age-different grafts (Helms et al. 1996 for a recent example). It would probably be fair to say that most of these studies use chick cells because of the ease of access to chick embryos, with the assumption that major prenatal and postnatal regulatory pathways will be similar between birds and mammals. In fact, this has proven to be a reasonable assumption, with no major differences demonstrated for the multiple growth factors and transcription factors studied to date (Farnum and Wilsman 2001). Several important feedback regulatory loops, including that of PTHrP/Ihh, were first described in the chick, which has been an important model for others, such as the role of VEGF and *cbfa1* during vascularization of the chondro-osseous junction (Vortkamp et al. 1996; Pathi et al. 1999; Carlevaro et al. 2000; Iwamoto et al. 2003).

Despite the apparent similarity of signaling pathways during the chondrocytic differentiation cascade, stereological studies of growth plates from chickens and ducklings have demonstrated a significant difference compared to mammals in the emphasis between proliferation and hypertrophy as the chondrocytic differentiation pathway is converted into growth. The morphology of the proximal tibial growth plate of the chick is dominated by the long, rather unorganized columns of proliferative cells, with metaphyseal vessels penetrating well into the hypertrophic cell zone (Lutfi 1970; Howlett 1979, 1980; Hunt et al. 1979; Thorp 1988; Barreto and Wilsman 1994; J. M. Starck 1998). The cytological organization is basically similar across a wide range of altricial, semiprecocial, and precocial neonates (J. M. Starck 1996). Studies of differential growth in multiple growth plates in chicks and ducklings demonstrate the same positive correlations with rate of growth as identified for mammals—that is, that the numbers of proliferative and hypertrophic cells, final hypertrophic cell volume, and cell cycle times all show strong positive correlations with rate of growth. The rate of elongation relates primarily to the cartilaginous volume of the bone, which is larger in altricial hatchlings than in precocial hatchlings. This has been interpreted as meaning that rapid bone elongation is enhanced in altricial species, at a cost of reduced mechanical strength (J. M. Starck 1996, 1998).

However, and importantly, unlike mammalian growth plates, cell cycle time and the number of proliferative cells are significantly more important for understanding differential growth in these avian species than is volume increase during hypertrophy (Kirkwood et al. 1989a, 1989b; Barreto and Wilsman 1994; J. M. Starck 1996). These avian growth plates achieve growth primarily by producing and turning over large numbers of cells per day (Kember and Kirkwood 1987). One study found a range of 6–53 cells/column turned over in growth plates growing at different rates, the latter number being more than five times that ever demonstrated for a mammalian growth plate (Kirkwood et al. 1989b). Chicks produce more cells a day for a given growth rate than does the rat, but the final hypertrophic cell volume is significantly smaller. The distal tibiotarsus of a 14-day-old duckling grows at 318 μm/day and has a final hypertrophic cell volume of 2,710 μm³, compared to a final hypertrophic cell volume in the rat proximal tibia at 21 days of 17,040 μm³ growing at a comparable rate (335 μm/day; Barreto and Wilsman 1994). As emphasized earlier, the cellular activities (proliferation and hypertrophy) are the same; the resultant mix of these is significantly different between chicks/ducks and rats/mice for a given rate of growth. The hypertrophy process is not as efficient in the chick/duck as it is in the rat/mouse, and is not the primary quantitative determinant of bone elongation rate during endochondral ossification. It would be interesting to be able to test the validity of this generalization by looking at a wider variety of species in each group.

Postnatal Bone Elongation by Endochondral Ossification in Didelphids

A separate consideration should be given to marsupials, to emphasize again that there is no "one and only" approach to postnatal endochondral bone formation, even though the chondrocytic differentiation cascade and the fundamental processes are highly conserved in all vertebrates. In marsupials, secondary centers ossify directly from the perichondrium. Haines (1939) argued that cartilage canals, which play a significant role in the formation of secondary centers in most tetrapods, are primitively absent in marsupials. This contrasts with rats and mice, from which cartilage canals have been lost secondarily. Hamrick (1999) analyzed forelimb and hindlimb modifications in differential growth as the newborn opossum goes from the initial need for precocially developing forelimbs to hold on to the mother, to a tetrapod functionally walking on both pairs of limbs at the time of weaning (Maunz and German 1997). To achieve this there is a disassociation of forelimb and hindlimb growth rate; the forelimb develops significantly faster at first, then the hindlimb catches up. Gille and Salomon (1995) noted a similar situation in ducks, where, at hatching, the femur has a higher degree of maturity and faster growth rate than the humerus. This emphasizes the versatility of modular limb design with hierarchical units that can undergo temporal transformations, with different degrees of connectivity or disassociation under different selective pressures—in this example, timing of maturation for functional use (Raff 1996).

The Epigenetic Influence of Biomechanics in Endochondral Bone Elongation

When one tries to devise experiments to directly test hypotheses concerning the effect of the biomechanical environment surrounding a given growth plate on either the rate or the extent of bone elongation, it becomes obvious that bone elongation is so intimately tied to the functioning of multiple systems within the animal that the specific role of biomechanics is almost impossible to disentangle from other factors. A few conclusions are obvious: some loading, minimally from the force of tendons and muscles that cross joints, is required, but the effect of lack of motion such as with denervation or weightlessness is more significant on growth of bone in width and remodeling than on bone growth in length (Sibonga et al. 2000). Although loading is important for cortical bone growth and the acquisition of shape, very little experimental evidence exists to show that *rates* of elongation through endochondral ossification are altered by changes in mechanical loading within physiological ranges (Biewener and Bertram 1993).

Linear growth can be compromised using devices that provide controlled compression across the growth plate, as has been shown in experiments *in vivo* (Stokes et al. 2002). The effects of physiological extremes are more difficult to interpret. As one example, bone maturation diminishes significantly in elite female rhythmic gymnasts, but this is in conjunction with multiple systemic hormonal changes, including a loss of the pubertal growth spurt and nutritional deficits (Georgopoulos et al. 2001, 2002). Interestingly, these young women have a late acceleration of linear growth that continues for chronologically longer periods than normal, so that final height may not be compromised but, rather, reached considerably later than normal because of a delay in growth plate closure. Final height usually is at a level higher than what would be predicted from their genetic disposition alone (Georgopoulos et al. 2001).

Mammalian growth plate cartilage (and probably the epiphyseal cartilages of fish, amphibians, birds, and reptiles) requires motion within a certain range in order to elongate, and this requirement exists from early stages of embryonic development (Sledge 1966). The effect of the radical change in biomechanical environment in the fin-to-limb transition, or the reverse during secondary return to the aquatic environment in multiple species, clearly involved change in form of bone, but the explicit role of this environmental change in affecting mechanisms of regulation of bone elongation during endochondral ossification have not been studied. Cartilage has what some have described as a primary drive for growth that is independent of mechanical stimuli, suggesting a central intrinsically determined growth potential for cartilage that is programmed early in development (Biewener and Bertram 1993). This intrinsic program integrates genetic and epigenetic stimuli over a wide range of normal physical exercise.

Surgeons can modulate rate of elongation using manipulations beyond the normal physiological range for the correction of limb length inequalities, angular limb deformities, and scoliosis, and experimental models of these procedures have been developed (see Alberty and Peltonen 1993; Fjeld and Steen 1990; B. P. Pereira et al. 1997; Farnum et al. 2000; Stokes et al. 2002; as examples). In general, slowing growth has been more successful than accelerating growth, although multiple devices have been created to use controlled physeal distraction (reviewed by Farnum and Wilsman 1998a). Stripping of the periosteum/perichondrium also accelerates growth. This effect has long been interpreted as secondary to the release of tension over the growth plate, since the perichondrium normally functions to create growth-generated stresses that modulate elongation rates (Biewener and Bertram 1993; D. R. Carter et al. 1987, 1998c). Alternatively, the effect of periosteal stripping may be solely through augmenting blood supply and the delivery of systemic growth factors. More mechanistically, periosteal stripping may inhibit the Ihh/PTHrP negative feedback loop that requires communication between growth plate cartilage cells and the periosteum; inhibition of this negative feedback would accelerate growth. This example underscores the complexity of disentangling the multiple influences on bone elongation rate through experiments in *in vivo* systems.

Experimental evidence on the role of mechanobiological influences in bone elongation comes from (1) *in vivo* studies with manipulation during development in chick embryos; (2) *in vitro* studies at the cell, tissue, or organ culture level in which the mechanical environment is manipulated; (3) investigation of existing *in vivo* systems that test the possible limits of mechanical influence; and (4) computational modeling with prediction of outcomes for experimental hypothesis testing. What is clear from these approaches is that cell signaling often involves mechnotransduction, and that these signals can lead to changes in transcriptional regulation. A recently documented example showed that the type X collagen gene of hypertrophic cells is mechanoresponsive under cyclic loading conditions (Q. Wu et al. 2001), that this occurs through the up-regulation of bone morphogenic proteins (BMPs), and that this affects the Ihh/PTHrP feedback loop. One of the multiple future research challenges for understanding the complexity of biomechanical influences over bone growth rate during endochondral ossification will be the investigation of genes involved in chondrocytic differentiation that have mechanosensitive promoter elements.

Carter and colleagues have studied multiple effects of the role of biomechanics on bone growth, from the cellular level to the level of the whole bone, from early stages of embryonic patterning to the timing of secondary centers postnatally, and from fossil evidence to multiple modern species (D. R. Carter et al. 1987; D. R. Carter and Wong 1988; Wong and Carter 1990; D. W. Carter et al. 1998b, 1998c; Henderson and Carter 2002). The basic argument is that the range of skeletal form that is possible in evolution is constrained by developmental rules of construction that reflect biophysical processes associated with tissue mechanical loading. Therefore, mechanical loading plays a major role throughout musculoskeletal development, for both periosteal bone and endochondral bone and for any range of mixed proportions between the two. By extension, growth-generated strains regulate morphogenesis throughout development, including modulating growth rates, rates of differentiation, and directionality of growth. It also is hypothesized that the efficiency of molecular mechanisms of control may differ in different groups of animals as a function of the differences in the sensitivity of their cells (chondrocytes) to mechanical stimuli (Henderson and Carter 2002). These hypotheses should stimulate experimental research into the potential multiple roles of biomechanical regulation of postnatal bone elongation through endochondral ossification.

Nutrition as a Regulator of Rate and Extent of Bone Elongation

It seems obvious that nutritional state has a profound effect on postnatal bone elongation. However, analogous to the situation with biomechanics, it is very difficult to study nutritional effects independently of multiple other epigenetic influences in the postnatal animal. There is a significant role of the IGFs in both prenatal and postnatal bone growth. IGFs in serum are complexed with high-affinity binding proteins (IGFBPs), and receptors for IGFs are present on virtually all cell types, including growth plate chondrocytes (Ohlsson et al. 1998; Smink 2003). In the growth plate, as in most organs, IGFs act by autocrine and paracrine mechanisms, as well as by endocrine mechanisms primarily through growth hormone, although the balance of these three levels of regulation is still under investigation (Thissen et al. 1994). Nutritional status and supply of dietary energy are critical regulators of IGFs and IGFBPs, and IGFs provide an important mechanism linking nutrition and growth, including endochondral bone elongation.

Evidence exists that there is fetal programming of the GH/IGF factor axis, and that there can be reprogramming of this axis under certain kinds of early nutritional stress (Harel and Tannenbaum 1995; Cianfarini et al. 2002; Holt 2002;

Javaid and Cooper 2002). Programming in this sense can be defined as "lasting changes in structure and function caused by environmental stimuli acting at critical periods in early development" (Cooper et al. 2002). Presumably this means that a compromised nutritional environment can act either to permanently alter gene expression or to alter cellular numbers (Javaid and Cooper 2002). In late fetal life, fetal growth rate is determined less by fetal phenotype than by nutrient delivery (glucose, amino acids, oxygen) through the placenta (Holt 2002). Depending upon nutrient level, different levels of programming for the extent of postnatal growth will occur. One hypothesis is that this is a way the fetus can test for environmental growth conditions after birth through the maternal and placental endocrine and metabolic milieu, and respond to potentially prolonged nutritional deficiency by programming for smaller size if indicators are that nutritional will be suboptimal postnatally (Cianfarini et al. 2002; Cooper et al. 2002; Holt 2002). It also is one way that nutrition is linked directly to growth through mechanisms acting centrally.

Postnatally, nutrition can also influence bone growth through the hormone *leptin,* which is produced in adipose tissue and acts as a satiety signal during feeding (reviewed by Schwartz et al. 2000; Margetic et al. 2002; Prentice et al. 2002). Leptin can act as a circulating feedback signal, tied centrally to the GH/IGF axis, to regulate general energy homeostasis by reflecting adipose tissue stores (fig. 10.28; Mauras 2001; Ranke 2002). Although there is much to be learned about the role of leptin in general body metabolism, in rats, maternally delivered leptin reduces skeletal growth in the pups postnatally (Nilsson et al. 2003). Recent evidence suggests that leptin can act through its receptors directly on peripheral tissues, and leptin receptors have been demonstrated on growth plate chondrocytes in the rat (Hoggard et al. 1997; Kume et al. 2002; Phillip et al. 2002). Thus, leptin-mediated nutritional effects on the growth plate, which already are known to act indirectly through GH and thyroxine, may also act directly on the chondrocytic differentiation cascade in the growth plate (Siebler et al. 2001). Leptin has been identified by immunocytochemistry in the stomachs of several nonmammalian vertebrates (trout, frog, lizard, snake; Muruzábal et al. 2002). Future research may add leptin to the already long list of hormonal regulators of post-natal bone growth acting both through central regulating mechanisms and directly on target tissues. Whether ghrelin, a hormone produced by the stomach that has GH-stimulating activity, can affect the growth plate directly has not yet been studied (Broglio et al. 2002).

The complexity of the interplay between nutrition and bone elongation, with potential controls acting both centrally and peripherally, is best demonstrated by *catch-up*

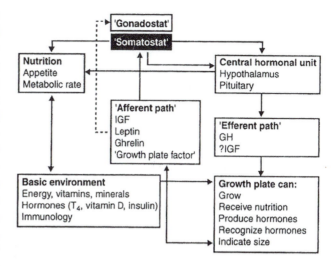

Figure 10-28 Endochondral ossification is influenced not only by the systemic hormonal environment but also by the general metabolic state of the animal, including the level of nutrition. The hormone leptin, produced by adipose cells, may be an important intermediary in the nutritional regulation of longitudinal growth, acting either centrally through the pituitary axis or perhaps with direct actions at the level of the growth plate. (From Ranke 2002; used with permission from Freund Publishing House Ltd.)

growth, first described by Prader et al. in 1963. Catch-up growth has been recently defined as "height velocity above the statistical limits of normality for age and/or maturity during a defined period of time, following a transient period of growth inhibition" (Boersma and Wit 1997, 646). The degree of catch-up following some form of nutritional deprivation depends upon the timing, duration, and the nature of the original deprivation. In rodents, models in which manipulation of nutrition is the primary variable include variations in pre- or postnatal timing of the nutritional deprivation—manipulation of food of the mother during gestation, manipulation of litter size preweaning, or short periods of fasting to the postweaning animal, as examples.

Growth plate cellular function is exquisitely sensitive to nutritional change. As an example, after a three-day fast to four-week-old male rats, the proximal tibial growth plate fell to a rate of bone elongation only 30% of that of the control littermate, with a reduction in all cellular kinetic parameters that promote growth (fig. 10.29a, b; Farnum et al. 2003). Within seven days of return to feeding, 24-hour elongation rate in the proximal tibial growth plate in the previously fasted animals had reached the level of the control littermates, and, more important, it remained elevated *above control level* for the next three weeks of the experiment. Evidence that even individual growth plates may have a "sense of final size" comes from experiments utilizing local delivery of glucocorticoids to an individual growth plate, significantly suppressing growth rate only in the treated growth plate. After cessation of the glucocorticoid treatment, there

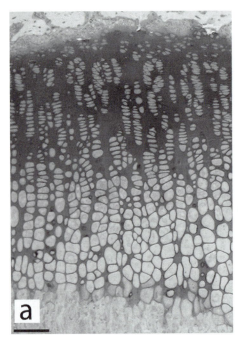
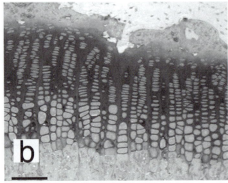

Figure 10-29 The growth plate is very responsive to changes in nutritional state, as shown by this comparison of the proximal tibial growth plate from a 4-week-old control rat (a) compared with that of a littermate that had undergone nutritional deprivation for three days (b). Every parameter of growth plate kinetic cellular activity was affected, including the cellularity of each zone and the final volume of hypertrophic chondrocytes. Seven days after returning to free-choice alimentation, the growth plate had fully recovered. Over the next three weeks it grew at a rate faster than control, indicating the capacity for catch-up growth in this model. Bar = 100 μm. (From Farnum et al. 2003; used with permission from Lippincott Williams and Wilkins.)

was increase in rate of growth only in that particular growth plate, resulting in some degree of catch-up in final length (not as compromised as the initial treatment would have caused; Baron et al. 1994).

The phenomenon of catch-up growth implies that the body has a set point for length of individual bones and that, if the normal growth trajectory is interrupted, faster-than-normal elongation can be achieved under some circumstances so that final length will not be compromised (Harel and Tannenbaum 1995; Ranke 2002). The fact that catch-up can be complete implies that the body has a set point for length of a bone that can be reached only after accelerated elongation at a *rate faster than that expected* at a given bone age. The fact that catch-up sometimes will not be complete, even though growing conditions become optimal, implies that under some circumstances there can be reprogramming of the GH/IGF axis to a new set point (Cianfarini et al. 2002). There are several recent reviews of catch-up growth and the multiple kinds of situations besides nutritional compromise in which it may occur (see Gafni and Baron 2000; Ranke 2002; Wit and Boersma 2002).

Schew and Ricklefs (1998) summarize the issues associated with nutritional effects on avian growth, including bone elongation, by distinguishing between growth (meaning an increase in size) and maturation (meaning changes in an or-

gan that bring it closer to the adult morphology and level of function). They emphasize that total body growth (or growth of a skeletal element) is a function both of the rate of growth and of the period during which growth occurs. The latter is a function of the rate of maturation. A range of responses of the rate of growth/maturation has been seen during different kinds of nutritional stresses in experimental studies of birds. One of the most interesting is that, for many altricial species, a primary response to nutritional stress is to slow both maturation and growth and to fledge at a significantly later time, thus maintaining adult size. They also raise the question of whether examples of acceleration of growth (as during catch-up) are due to qualitatively different processes than seen in normal growth. A start to addressing these kinds of questions could be made through analysis of growth plate cellular kinetic responses during the nutritional stress of these experiments, but no such studies have been done to date.

Two hypotheses have been proposed as explanations for the phenomenon of catch-up growth. The neuroendocrine hypothesis suggests a mechanism where the organism has the ability to recognize the degree of mismatch between its target size and its actual size and to adjust growth rate according to the degree of the mismatch. This implies that there is both a "time-tally" and feedback between the concentration of some hypothetical substance in the periphery and the GH/pituitary axis (Wit and Boersma 2002).

The growth plate hypothesis of catch-up growth is that each growth plate in the body has its own senescence program, and this can run uninhibited under optimal growth conditions (Gafni and Baron 2000; Wit and Boersma 2002). Senescence is not a function of time, per se, but of the cumulative number of divisions stem cells have undergone, which was established during early embryonic programming (Campisi 1996; Gafni et al. 2001; Tyrcha 2001). Information about the number of stem cell divisions completed is retained within the growth plate and influences subsequent growth rates. In this hypothesis, catch-up is considered a manifestation of a delay in senescence; the potential number of stem cell divisions is conserved, but delayed in timing. This mechanism, based on conservation of number of stem cell divisions, fails to explain why there can be a sustained period of growth velocity greater than the normal growth velocity for the corresponding bone age during the catch-up period; it would predict that a given growth velocity would correspond to a given bone age.

The catch-up phenomenon is an experiment of nature that demonstrates an aspect of the intrinsic forces of growth within growing cartilage. Minimally, it suggests both the existence of a sense of final size inherent in bone growth, and the ability of certain external insults, particularly during late fetal and early neonatal growth, to reprogram the target size of the adult organism, including length of individual bones. One could ask whether this "sense of size" extends to the length of individual limbs. Figure 10.30 is a radiograph from a 1.5-year-old dog that had a closed femoral fracture as a puppy. The fracture healed with significant shortening and widening of the femur. However, the magnitude of the tibial compensatory overgrowth was such that, despite severe shortening of the femur, the length of the two hindlimbs was comparable (Schaefer et al. 1995). Equality of limb length was maintained, but at the expense of having femorotibial joints at two quite different levels!

Experimental Approaches to Unanswered Questions

Whether one is considering the full complexity of a mammalian growth plate designed for rapid, efficient elongation of endochondral ossification, or a seemingly more simple structure of the epiphysis of a fish, a number of very basic questions remain unanswered. It is difficult to design experimental approaches for many of these, since they often cannot be addressed satisfactorily using reductionist approaches. Several are considered here, with a summary of recent experimental evidence.

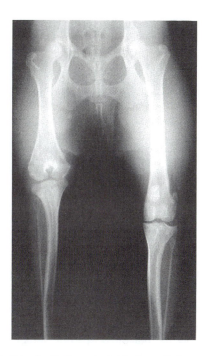

Figure 10-30 This animal had a nontreated fracture, with the result that the femur healed significantly shorter and wider than normal. However, there was a level of tibial overgrowth so that the total length of the limb equaled the normal side, despite having femorotibial joints at two levels. (From Schaefer et al. 1995; used with permission from Schattauer Publications.)

To What Extent Is Cartilage Growth in Postnatal Animals Driven by "Intrinsic" Factors?

It has been argued that the best explanation of catch-up growth is that there exists an intrinsic program within growth plate cartilage based on a finite number of stem cell divisions and memory of how many of these have been completed (Gafni et al. 2001; Tyrcha 2001). Although this may explain intrinsic regulation of the proliferative component of growth plate cellular activity, it ignores the contribution of the hypertrophic stage of chondrocytic differentiation, which, in most mammalian growth plates, contributes the majority of interstitial growth that results in elongation. It would be unlikely that the magnitude and rate of chondrocytic hypertrophy through time also are preprogrammed at a very early stage of development. Experiments in which growth plates have been transplanted between animals of different ages do, however, suggest that ultimate skeletal length is primarily determined by factors inherent in each growth plate. Growth plates from juvenile animals transplanted into older animals grow at the juvenile rate, independent of the age of the recipient (Kline et al. 1990; Stevens et al. 1999). In these experiments bone length was measured, but there was no analysis of cellular kinetics.

Cessation of Growth

In a broad sense, it can be considered that, with the development of a secondary center, determinate growth is assured at the time of growth plate closure. Surprisingly little is known about the mechanism of growth plate closure, other than that it occurs very rapidly and, in epiphyses with secondary centers, closure commences from both the epiphyseal and the diaphyseal side (Haines 1941, 1974, 1975). Timing of closure basically follows the same sequence pattern over a wide range of species (Kohn et al. 1997), implying that it is following the course of a patterning event in the early embryo that has been conserved. The question is whether this is consistent with the hypothesis that the growth plate, an ephemeral organ, has an internal program for senescence (Campisi 1996; Gerhard et al. 2002). Just before closure, most growth plates still have a morphological appearance that would be consistent with the ability to continue growth. That is why it has been pointed out that it is not that growth stops and then bony union occurs, but rather the bony union occurs and stops growth (Parfitt 2002).

Interestingly, it is from identification of a naturally occurring disease in humans that definitive proof has emerged about the role of sex steroids in growth plate closure. In mammals that have a pubertal growth spurt, estrogens in females and androgens in males act on the growth plate to accelerate growth. However, estrogens are the hormones required for growth plate closure, and in males androgens must be converted to estrogens by the action of the enzyme aromatase. Cases have been reported of men either lacking aromatase, or with mutations in aromatase or its receptor, who have open growth plates with continued slow linear growth into their 30s. Their growth plates close not in response to androgen therapy but only in response to estrogen therapy (reviewed by Faustini-Fustini et al. 1999; Bilezikian 2002).

In many rodents such as mice and rats, growth essentially ceases, even though several growth plates do not close with a bony union (A. B. Dawson 1929, 1935). Figure 10.31a shows an example of this in a 2.5-year-old mouse. Bromodeoxyuridine labeling for DNA synthetic activity demonstrates the occasional positive cell (fig. 10.31b). Tetracycline labeling as an indicator of bone growth shows that no growth has occurred in the last 48 hours (fig. 10.31c). Despite the apparent ability of a small number of cells to retain replicative ability, the conclusion is that rodent growth plates cease growing even though a bony union is never formed between the epiphysis and the metaphysis (Roach et al. 2003). The original interpretation was that this is an example of neoteny, with a cessation of growth in the absence of closure or bony union (A. B. Dawson 1929). E. A. Martin et al. (2003) demonstrated

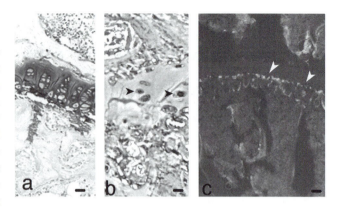

Figure 10-31 Most rodent growth plates stop growing but never form a bony union. In these micrographs from a 2.5-year-old mouse, the cartilaginous interface between epiphyseal and metaphyseal bone is apparent (a). Despite the very low cellularity, occasional proliferating cells can be demonstrated with BrdU immunocytochemistry (arrowheads, b). Nevertheless, localization of OTC 48 hours after injection OTC clearly indicates a lack of any bone elongation in these growth plates, despite the lack of a bony union. White arrowheads, positioned in the growth plate, indicate that all fluorescence remains at the metaphyseal chondro-osseous junction. (a, b) Bar = 20 μm; (c) bar = 40 μm.

using microcomputed x-ray tomography that multiple small bony bridges cross the rat proximal tibial growth plate starting as early as 3.9 months. These continue to form, effectively stopping growth in the male by 8 months and the female by 10 months; the extent of the bony bridging makes the possibility of further growth unlikely, even though extensive cartilaginous areas remain.

Polarity of the Growth Plate and Directionality of Growth

Carter and colleagues provide computational models that predict a given directionality to growth based on the stress/strain environment of the growing bone (Wong and Carter 1990; D. R. Carter et al. 1998c; Henderson and Carter 2002). The form of the woven trabecular bone of the epiphysis is very different from that of the longitudinally directed metaphyseal bone that forms the primary and secondary spongiosa. When experiments have been done to surgically reverse a growth plate by rotating it 180 degrees, the conclusion is that the original polarity remains. Bone with the form of epiphyseal bone grows in the metaphyseal direction and vice versa (Abad et al. 1999; fig. 10.32). This suggests that polarity is intrinsic to the growth plate and that some component of spatial polarity of the growth plate is patterned during embryonic development.

Fracture Repair: A Rerun Governed by the Same Regulatory Mechanisms?

To the extent that it has been studied, cellular kinetic processes during the endochondral stage of fracture repair are

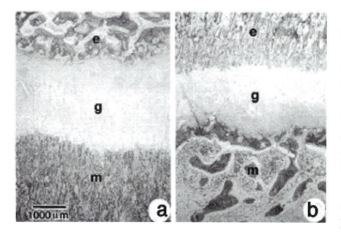

Figure 10-32 These micrographs represent the results of experiments in which the growth plate (g) was isolated and then rotated 180 degrees. Note that the orientation of the epiphyseal bone (e) and its characteristic appearance as woven bone remains even after transplantation. Similarly, the longitudinally orientated trabeculae of the original metaphyseal bone (m) maintain their original metaphyseal alignment, even after growing in the epiphyseal position. This has been suggested as evidence that the growth plate has an intrinsic polarity to its spatial organization. (From Abad et al. 1999; used with permission of the Endocrine Society.)

identical to those seen during bone elongation: mesenchymal cell condensation, differentiation into chondrocytes, and then a differentiation cascade of chondrocytic proliferation, hypertrophy, and matrix production, played out temporally and leading to bony union. In addition, mathematical models have been used to demonstrate that the patterns of tissue differentiation, including the sequence of the chondrocytic differentiation cascade, can be predicted using the same mechanobiological concepts as have been used to understand biomechanical influences on prenatal and postnatal growth, including the timing of the development of secondary centers of ossification (D. R. Carter et al. 1998a; Bailòn-Plaza and van der Meulen 2001). The analysis of molecular regulatory mechanisms during the endochondral stage of fracture repair has, to date, revealed essentially the same autocrine and paracrine regulators as seen in bone elongation—*Ihh*/PTHrP, *Cbfa1*, the BMPs, *VEGF*, and *MMP13*, as examples (Ferguson et al. 1998, 1999; Le et al. 2001; Le Roith et al. 2001; Gerstenfeld et al. 2003). Much less is known about the permissive hormonal environment for fracture repair other than that, in the adult, the general systemic hormonal environment in terms of levels of GH, sex steroids, and thyroxine is very different from that of the postnatal period.

Since bones heal throughout adult life not by forming scar tissue but by generating new bone, this should, most accurately, be called a regenerative process. It is appears that the fundamentals of the regulatory processes involved in fracture healing are the same as those involved in postnatal bone elongation (Gerstenfeld et al. 2003). This regeneration can occur throughout the epiphyseal, metaphyseal, and diaphyseal bone, leading to bony union and remodeling to the correct form. Although there have been reports of growth plate regeneration and bone elongation after experimentally induced mid-diaphyseal amputations (Libbin and Weinstein 1987), in mammals there is essentially no consistent ability for amputated limbs to elongate significantly through a regenerative process. However, there are reports of successful microvascular transplants of growth plates in children (Nunley et al. 1987; Boyer et al. 1994; K. G. Shea et al. 1997).

Chondrocytic Hypertrophy: When and Why Did It First Develop, and How Is It Regulated?

A major unanswered question in the biology of cartilage interstitial growth is the evolution of the functional role of chondrocytic hypertrophy as a major contributor to growth rate and the total amount of growth achieved. Some degree of chondrocytic enlargement is seen in the epiphyseal cartilage of all animals that utilize endochondral ossification, from fish through birds and mammals. Evidence of a hypertrophic zone also is seen in the fossil growth plates of dinosaurs (Horner et al. 2000). But the significance of hypertrophy in the overall growth efficiency of bone elongation in a given group of animals varies widely. Regulated volume increase that brings the cell to a new set point as great as an order of magnitude or more from the previous one is very unusual in the biology of cells, which have powerful regulatory mechanisms to either increase or decrease cell size for the purpose of homeostatic maintenance of original cell volume when exposed to harsh osmotic environments (O'Neill 1999; R. J. Wilkins et al. 2000; Farnum et al. 2002).

The so-called hypertrophy seen in chondrocytes is primarily the result of water imbibition and should therefore be more correctly called cellular swelling (Buckwalter and Sjolund 1986). However, although the increase in intracellular organelles during hypertrophy does not parallel in magnitude the increase in cellular fluid volume, it has been shown that basic organelles such as mitochondria, rough endoplasmic reticulum, and microtubules increase in absolute amount and modulate their rate of activity as the cells increase in size (Farquharson et al. 1999; Hunziker et al. 1999). And, over the last two decades, it has definitely been shown that hypertrophic chondrocytes are active, fully viable cells whose gene expression differs significantly from proliferative cells (Farnum and Wilsman 1987, 2002; Farnum et al. 1990; Nurminskaya and Linsenmayer 1996). It is generally agreed that they die in response to signaling factors, probably from the metaphyseal vasculature, which initiate an apoptotic program (Farnum and Wilsman 1989, 2002; Hatori et al. 1995; Zenmyo et al. 1996).

At least three very basic questions need to be answered about this stage of chondrocytic differentiation. First, did hypertrophy originally evolve as a mechanism of interstitial growth, or, did it evolve in conjunction with mechanisms of mineralization of the cartilage matrix? In mammals these two functions of the hypertrophic chondrocyte can be conceptually considered as separate, since they can occur independently of each other. Has this always been true, and if so, which is the older function? Second, what are the physiological mechanisms that are involved as a chondrocyte rapidly increases its volume? In some mammalian growth plates a regulated volume increase by a factor of 10 to 15 can occur in less than 18 hours. That the increase is regulated and differential is known; what is unknown is the physiological mechanism (Farnum et al. 2002). And third, what is the mechanism of the regulated shape change that occurs during hypertrophy in mammalian growth plates, and when and how did this evolve? As discussed earlier, there are hypotheses that this is caused by passive restraint by the longitudinal septae (Buckwalter and Sjolund 1989; Noonan et al. 1998), but there is no experimental evidence specifically supporting this idea.

Evolutionary Selection Points during Endochondral Ossification: The Mammalian Example

The growth plate is a complex, albeit ephemeral, organ where chondrocytes rapidly differentiate in a cascade of proliferation and hypertrophy. Chondrocytes secrete a complex matrix of multiple compartments, the ratio of cells to matrix changes during differentiation, and the compositional and structural properties of the matrix also change during differentiation. Differentiation is characterized by regulated transitions, much as the cell cycle has checkpoints. Growth plate vasculature enters through two directions—epiphyseal and metaphyseal—and both nutrient and regulatory signals may differ as they enter the growth plate from these two directions (Trueta and Amato 1960). Although the growth plate lacks a direct nervous supply, significant regulatory loops exist and are activated through signals between growth plate cartilage cells and the surrounding perichondrium. Multiple aspects of growth plate function are thought to be patterned during early embryonic development. Nevertheless, the process of endochondral ossification can be modified by changes in the biomechanical environment, the nutritional environment, secondary to changes in any of multiple systemic hormones and secondary to additional epigenetic factors such as pharmaceuticals and disease states.

Although the highly anisotropic growth plate of mammals with its regularly aligned columns of chondrocytes may allow the most efficient conversion of the chondrocytic differentiation cascade into longitudinal growth, in terms of the basics of bone formation through a cartilaginous precursor, the fundamentals are the same as in the cartilaginous epiphysis of fish. Proliferation, hypertrophy, and matrix production are the basic processes. The units are modular, both proximally and distally, and can be changed independently of each other. It is this overall simplicity of the fundamental processes and design units involved in longitudinal growth that has allowed the diversity of limb morphology to evolve (Francillon-Viellot et al. 1990; Shapiro 2002).

Basic patterning genes and regulatory interactions appear to have been conserved over time (Coates 1995; Chiu and Wagner 2000; Cohn 2001; Shubin 2002), and the basic structural materials and their properties also seem to be very ancient (Erickson et al. 2002; Exposito et al. 2002). Limb morphological diversity at the extremes of adaptation—miniaturization, limblessness, return to the aquatic environment, digit reduction—often correlate with simple differences in the pattern of expression of highly conserved regulatory genes with reduction, fusion, or retention of embryological connections being a major cause of morphological change (Felts and Spurrell 1966; Shubin and Alberch 1986; B. T. Shea 1993; Cohn 2001; Shapiro 2002; Shubin 2002). The observation that the sequence of initial chondrification of the multiple cartilaginous anlage segments in the limb does not correspond to the sequence of later ossification has been suggested as important for providing two temporally different stages during which changes in limb morphology may have evolved. Modification during cartilage formation could lead to major reductions as seen in digit reduction in lizards or in ungulates, while modifications during ossification could provide the more specific character changes associated with differences among skeletal features between different species of the same genera, or closely related genera (Rieppel 1992a, 1992b; 1993a, 1993b, 1993c; Shapiro 2002). And selection for shifts in such features as duration of growth and/or time to maturation can also result in complex and novel shape configurations (B. T. Shea 1993).

Since evolutionary changes in morphology can emerge from shifts in growth controls at any of multiple levels, detailed basic understanding learned in any one appropriate model system becomes the basis for informed comparative analyses of multiple other species. Even detailed molecular analysis of the more than 200 recognized forms of human chondrodysplasias (International working group 1998; Unger 2002) is an "experimental" approach to the understanding of the kinds of molecular changes that can have significant influence on both the form and growth pattern of limbs postnatally, including which kinds of molecular changes are compatible with life (Dreyer et al. 1998; International working group 1998; Hochberg 2002b). A specific example is the analysis of the multiple aberrations of the GH/IGF-1 system as manifested in diseases of short or giant stature that has furthered our understanding of the role of this system in postnatal growth at a level parallel with knowledge gained from the creation and analysis of transgenic mice with alterations of this system. Major groups in the molecular-pathogenetic classification of genetic disorders of the skeleton include defects in extracellular structural proteins, defects in metabolic pathways, defects in folding and degradation of macromolecules, defects in hormones and signal transduction mechanisms, and defects in nuclear proteins and transcription factors (Superti-Furga et al. 2001; Unger 2002). The known molecular changes in theses chondrodysplasias demonstrate the multiple levels of biological organization at which selection may act during the endo-

chondral bone elongation process. What is viewed as disease when it occurs in humans, may, in fact be viewed as an opportunity under different selective pressures in other species.

Since selection can act directly on the regulator of, or on any of the upstream regulators of, any growth-altering influence, and since selection can occur at any level of biological organization—molecular, cellular, tissue, organ, behavioral, and so on—or at any point from early development through to the end of postnatal growth, the possibilities of diversity of limb morphology from selective pressures for change in function approach being infinite (A. S. Wilkins 2002). The challenge is to continually synthesize knowledge gained from multiple perspectives into an ever more refined understanding of the regulation of bone elongation during postnatal endochondral ossification.

Acknowledgments

I would like to express thanks to Michelle Lenox, Cherrill Wallen, and Carie Ryan for help in preparation of the manuscript and figures and in gaining the copyright permissions. Michael Simmons, MFA, contributed figures 10.1, 10.2, and 10.8; his creativity is much admired. I would like to thank Dr. Drew Noden for the picture of the 16 mm human embryo in figure 10.1. Original research presented in this chapter was funded through grant NIH AR 35155 to Norman J. Wilsman and Cornelia E. Farnum.

Chapter 11 Paired Fin Repair and Regeneration

Marie-Andrée Akimenko and Amanda Smith

THOMAS HUNT MORGAN is probably better known for his fundamental work on *Drosophila* genetics, but his research interests were first focused on the regenerative capacity of various organisms (Morgan 1898, 1900, 1902, 1904). Morgan chose the term "morphallaxis" to describe the process of regeneration of *Planaria* in which lost body parts are replaced by the remodeling of the remaining part and "epimorphosis" to describe the type of regeneration that, in contrast to morphallaxis, requires active cell proliferation prior to the replacement of the lost body part. Limb regeneration of several urodele amphibians and fin regeneration of teleost fish are examples of epimorphosis. Epimorphic regeneration is characterized by the formation, soon after wound healing, of a blastema: a mass of proliferative, multipotent progenitor cells that will differentiate to give rise to the different cell types and will reconstitute the lost part.

Fin regeneration has been studied in many different teleost fish including tilapia (N. E. Kemp and Park 1970; Santamaría and Becerra 1991; Santamaría et al. 1992), minnows and blennies (Géraudie 1977; Goss and Stagg 1957; Morgan 1900, 1902, 1906; Nabrit 1931; G. P. Wagner and Misof 1992), opaline gouramis (Tassava and Goss 1966), goldfish (Marí-Beffa et al. 1999; Morgan 1902; Nabrit 1931; Santamaría et al. 1992, 1996), trout (Alonso et al. 2000), swordtail fish (Zauner et al. 2003), and zebrafish (Akimenko et al. 1995; Géraudie et al. 1994; S. L. Johnson and Weston 1995; Marí-Beffa et al. 1999; Murciano et al. 2002); see Poss et al. (2003) for a review.

In the last two decades, fin regeneration has received a surge of interest with the introduction of the zebrafish as a model organism for the study of regeneration, genetic analysis of vertebrate embryogenesis, organ development, and disease. The growing interest in zebrafish embryology has stimulated the development of multiple molecular and genetic tools that are also beneficial for investigating regeneration of adult tissue (for reviews, see Akimenko et al. 2003; Poss et al. 2003).

Structure of the Paired Fins of Adult Fish

The development of the fins and the tissue composition of their skeletal elements are described in detail in other chapters. The following is a brief description of the fin structure and, more specifically, of the part of the fin that can regenerate.

The fin skeleton consists of two parts of distinct ontogenetic origin: (1) the *endoskeleton*, located at the base of the fin and made of endochondral bones that supports the exoskeleton; and (2) the *exoskeleton*, which is made of rays of dermal origin and located in the external part of the fin (plate 11.1). Regenerative capacity is limited to the exoskeleton. The bone of each ray, called lepidotrichia, is itself composed of two concave mineralized hemirays facing each other in the shape of parentheses (plate 11.1A, B). These hemirays are made of a successive series of segments linked to each other by nonmineralized ligaments; this segmented structure provides flexibility to the fin (Becerra et al. 1983). Tendons connect the most proximal segment of the lepidotrichia to the striated muscles, the abductors and adductors, which are restricted to the base of the fin. Control of fin movements depends on the contraction of these muscles because the external part of the fin is devoid of muscle. With the exception of the most lateral rays of the caudal fin and the anteriormost rays of the other fins, most of the rays form a few bifurcations or branches along the proximal-distal axis of the ray. The hemirays occupy a subepithelial position and enclose the connective tissue that contains the arterial capillaries and nerve bundles. Venous capillaries are located in the interray connective tissue, which is thinner than the intraray connective tissue. Each ray contains, at its distal tip, an array of unmineralized fibrils called actinotrichia (plate 11.1D). These fibrils are made of elastoidin, a collagen type of protein that is not yet characterized. The actinotrichia line the internal surface of the few distalmost segments of the hemirays and extend distally in the space devoid of lepidotrichial matrix.

Fin Regeneration Capabilities

Morgan (1900, 1902) performed a number of regeneration experiments on the caudal fins of *Fundulus heteroclitus* and *Carassius auratus* as well as many other species. From these studies, he concluded that fin regeneration in *Fundulus heteroclitus* necessitates the presence of a stump of ray tissue; there is no regeneration when the fin ray is totally extirpated. However, other studies demonstrated that ray regeneration following complete ray removal is possible in other species, such as *Tilipia mossambica*, *Carassius*, and *Syngnathus*, and may depend on the developmental stage of the fish (Duncker 1905; Nabrit 1929, 1931; Birnie 1934; N. E. Kemp and Park 1970). For example, N. E. Kemp and Park (1970) reported that following excision of the tail of *Tilapia* to the level of articulation of the lepidotrichia with elements of the endoskeleton, new fins regenerated in some specimens. Nevertheless, the original number and pattern of the rays was not completely restored four to six weeks after total excision.

Despite these conflicting results, it seems that the presence of a stump ray is sufficient to allow regeneration of a ray. This was confirmed by the demonstration that transplantation of fin ray stumps to an ectopic position within the fin can induce the regeneration of fin rays (Birnie 1947; Goss and Stagg 1957; Goss 1969b; Murciano et al. 2002). These results led to the idea that the fin ray is the regenerative unit. More recently, transplantation experiments performed on the caudal fin of zebrafish indicate that a single hemiray can regenerate in the absence of its contralateral counterpart, suggest-

ing that the hemiray may be the regenerative unit (Murciano et al., unpublished data).

Is Fin Regeneration Capability a Property Common to All Teleost Fish?

This question is difficult to answer since, considering the diversity of this group, regeneration capacity has only been examined in a very limited number of teleost species. To try to answer this question, G. P. Wagner and Misof (1992) examined the regenerative ability of the pectoral fins of 14 species from six euteleostean families. Euteleostei form one of the four primary lineages of Teleostei, which are commonly divided into Osteoglossiformes, Elopomorpha, Clupeiformes, and Euteleostei. They chose the pectoral fin as a model system because reports in the literature suggested variability in pectoral fin regenerative capacity among teleosts (see G. P. Wagner and Misof 1992). Another reason was that the number of morphological specializations of the pectoral fins in several groups of fish allowed them to test whether the morphological specialization had an effect on the fin regenerative capability. The results of this study show that, although regeneration was often impaired, the initial stages of regeneration, including wound healing, blastema formation, and blastema growth, were observed in all species tested. Variation existed only in regenerate quality, such as bent fin rays, ray fusions, retarded growth of some rays, and partial regeneration.

Heteromorphic regeneration, which refers to abnormal regeneration, was especially observed in bottom-dwelling species, but not all bottom-dwelling fish showed impaired regeneration, suggesting that loss of regeneration capability is not an adaptive trait for fish living on the bottom but could be a consequence of other adaptive changes.

From this analysis and previous reports from the literature, G. P. Wagner and Misof (1992) proposed that the ability to regenerate pectoral fins is a trait inherited from a common ancestor of euteleostei. They could not make such a conclusion for the whole teleostean group since their analysis only included members of the euteleostei infradivision but no members from more primitive teleostei (Osteoglossiformes, Elopomorpha, and Clupeiformes). Another study, however, reported that the lungfish, *Protopterus*, shows good regeneration of its pectoral fins (Conant 1970).

General Characteristics of Fin Regeneration

Fin regeneration is a rapid, temperature-dependent process (plate 11.2A). In zebrafish, the caudal fin regenerates faster at 33°C than at 25°C. The rapidity with which the fin regenerates is species dependent and also depends on the level at which the amputation is made: the more tissue removed, the faster it is replaced (Tassava and Goss 1966). This is clearly illustrated when "staircase" amputation is performed on a single fin (plate 11.2B–E; Akimenko et al. 1995). Fin regeneration is unidirectional; if a gap is made in a ray by removing a few segments, regeneration will only proceed from the proximal end of the gap (plate 11.1C). The amputated fin grows by the progressive addition of segments and regenerates to its original size, implying a mechanism that controls growth of the regenerate. Although the general pattern of the fin is restored, the regenerated fin differs slightly from the native fin since the position of the first fork of a regenerate appears a few segments distal to its original position (Géraudie et al. 1994, 1995).

Due to the structure of the fin, the regenerate is made of a succession of two different types of blastema: a dense blastema forming at the level of each fin ray surrounded by a loose blastema arising from the regeneration of the interray soft tissue. Results from a few studies aimed at examining regulation of regeneration in the context of the whole fin indicate that the regenerating ray receives influences from the surrounding tissue.

Morgan (1900, 1902) was the first to observe the rate of ray outgrowth on the relative position of blastemas of the surrounding rays. For example, if the lateralmost rays on both sides of a caudal fin are cut at the same level but the neighboring rays are cut more proximally on one side and more distally on the other hand (plate 11.2G), the lateral rays do not regenerate at the same rate. Growth of the lateral ray will be delayed when the adjacent blastemas are more proximal (plate 11.2G, left side of the regenerate), whereas the opposite situation favors growth of the lateral ray (plate 11.2G, right side of the regenerate).

Formation of ray bifurcations also depends on the regulative influence of the surrounding tissue. For example, if a ray is forced to regenerate in the absence of contact with other rays, it will not form bifurcation (Marí-Beffa et al. 1999). On the hand, it is possible to induce the formation of a bifurcation in the lateralmost ray of the zebrafish caudal fin, which normally does not branch, when it is surrounded by interray tissue (Murciano et al. 2002).

Fin regeneration can be divided into successive steps: wound healing, blastema formation, regenerative outgrowth concomitant to differentiation, and patterning of the blastema (plate 11.2F shows the typical structure of a longitudinal section through a fin ray of a growing regenerate). In the following sections, the morphological, histological, and molecular characteristics of each of these steps will be discussed. Most of the studies have been performed on the cau-

dal fin because it is easily accessible for surgery and manipulations and provides a large amount of material to examine compared to the other fins.

Wound Healing

The immediate response to fin injury is wound healing. There is very little bleeding and little inflammation after amputation. Neovascularization of the zebrafish fin regenerate has recently been described. Bleeding terminates within the first hour, and severed blood vessels have healed by 24 hours after amputation (see below and C. C. Huang et al. 2003). The wound closes within 1 to 3 hours in the zebrafish and 7 to 12 hours in other teleosts (Misof and Wagner 1992; Santamaría et al. 1996; Santos-Ruiz et al. 2002).

Wound closure is achieved by rapid apical migration of epithelial cells located lateral to the cut surface without any cellular proliferation of the epithelial cells, as determined by the absence of proliferating cell nuclear antigen (PCNA) immunostaining, which detects cells in S phase of the cell cycle, and bromodeoxyuridin (BrdU) incorporation, a thymidine analog that incorporates into DNA during replication (plate 11.3A; Nechiporuk and Keating 2002; Poleo et al. 2001; Santamaría et al. 1996; Santos-Ruiz et al. 2002).

Following wound closure and during the next 12 to 18 hours, epithelial cells accumulate at the tip of the fin to form a dense multilayered mass of cells called the wound epidermis or apical epidermal cap (Becerra et al. 1996; Bullock et al. 1978; Géraudie and Singer 1992). Again, formation of the epidermal cap occurs without cell proliferation. In contrast, epithelial cells proximal to the level of amputation are strongly labeled with BrdU (plate 11.3B; Poleo et al. 2001; Santos-Ruiz et al. 2002).

Using the fluorescent lipophilic dye DiI as a cell tracer to follow the migration of epithelial cells, the epidermal cap was shown to form by the unidirectional migration of epithelial cells. This migration is essentially observed during the first two days following amputation in zebrafish and involves not only cells located close to the wound but also cells located several segments proximal from the cut surface (at least up to 500 μm in the zebrafish; Poleo et al. 2001). Soon after its formation, the wound epidermis is characterized by two well-organized cell layers: an outer layer, which consists of compact flat cells (Marí-Beffa et al. 1996; Misof and Wagner 1992), and a basal layer of cells recognizable by their cuboidal morphology. The rest of the epithelial cells comprising the epidermal cap form a loose multilayered mass of cells located between the basal and outer cell layers. Mucous cells differentiate within this central epidermal mass of cells (Géraudie and Singer 1992; Marí-Beffa et al. 1996). Soon after their formation, cells of the basal epithelial layer located at the apex of the stump exhibit cytoplasmic protrusions or digitations (Géraudie and Singer 1992; Santos-Ruiz et al. 2002). Such protrusions penetrating among the underlying mesenchymal cells were proposed to reflect intensive exchanges between epithelium and mesenchyme necessary for the formation and the growth of the blastema (Géraudie and Singer 1992).

Poss et al. (2000a) showed that expression of at least two members of the wnt signaling pathway, *wnt5* and *lef1*, a member of the HMG DNA-binding transcription factors that mediate the response to wnt, are activated in epithelial cells lining the mesenchymal cells, before these epithelial cells acquire a cuboidal shape. Development of a number of organs in mammalian species, including teeth, whisker and hair follicles, and gut and mammary glands, are governed by epithelial-mesenchymal interactions (Millar 2002; Pispa and Thesleff 2003) with components of the wnt signal transduction pathway implicated. The early expression of *lef1* and *wnt5* in the wound epidermis of the fin may suggest a role in the differentiation of the epithelial progenitor cells into basal epithelial cells or in signaling to the mesenchymal cells to induce the formation of the blastema. Once the basal layer is formed, *lef1* expression is maintained in these cells during blastema formation. Later, its expression becomes restricted to a subset of these epithelial cells, suggesting that lef1 may play additional roles at later stages of regeneration (see below; Poss et al. 2000a).

Blastema Formation

Epimorphic regeneration is characterized by the formation of a blastema as a mass of proliferative and less differentiated cells located beneath the wound epidermis. In the regenerating limb of urodele amphibians, blastema cell progenitors have been shown to arise through a local dedifferentiation of mature stump cells (Brockes 1997; Brockes and Kumar 2002; Lo et al. 1993). These cells proliferate and differentiate to replace the missing tissue. Cellular dedifferentiation has not yet been addressed experimentally in the regenerating fins of fish.

Histological analyses indicate that, as the wound epidermis is forming, fibroblast-like cells of the connective tissue (which represent the main cell population within the fin ray) become disorganized and go from stellate-shaped to rounded (Becerra et al. 1996). In addition, these cells start to present signs of BrdU incorporation, indicating cell proliferation (plate 11.3C; Nechiporuk and Keating 2002; Poleo et al. 2001; Santos-Ruiz et al. 2002). Rapidly, the number of proliferating cells increases to form a proliferative zone that cov-

ers up to two segments proximal to the amputation level. In zebrafish, the first signs of proliferation are detectable between 18 and 24 hours postamputation, and high proliferation rate is observed during the second day of regeneration (Nechiporuk and Keating 2002; Poleo et al. 2001; Santos-Ruiz et al. 2002). DiI labeling and BrdU pulse chase experiments have shown that the fibroblast-like cells rapidly become mobile and migrate toward the amputation plane, where they participate in blastema formation. Indeed, when intraray cells are labeled with DiI prior amputation, two segments (160 μm) proximal from the amputation level, staining is later confined to the proximal and differentiated part of the 7-day fin regenerate indicating their early participation to the blastema (plate 11.3D, E; Poleo et al. 2001).

Proliferating cells are essentially fibroblast-like cells of the connective tissue, but some endothelial cells and scleroblasts or bone-forming cells—recognizable by their elongated shape and their position along the inner and outer surfaces of the lepidotrichia—incorporate BrdU (Poleo et al. 2001; Santos-Ruiz et al. 2002). In addition, histological analysis indicates that scleroblasts lining the hemirays form a cap over the surface of the amputated hemirays, suggesting a role in bone repair (Misof and Wagner 1992). Similarly, endothelial cells of the blood vessels of the stump seem to be involved in revascularization of the regenerate (C. C. Huang et al. 2003). However, there is to date no indication of the participation of scleroblasts and endothelial cells in the regeneration of other structures. Altogether these data indicate that amputation triggers the entry of local mature cells into the cell cycle and support the hypothesis that, in the fin as in the urodele limb, the blastema emerges through a process of cell dedifferentiation.

Stem Cells?

Besides a putative dedifferentiation process, blastema cells could originate from quiescent stem cells. So far, only pigment cells of the zebrafish caudal fin have been proposed to differentiate from a population of stem cells distributed along the intraray tissue. It has been shown that, during regeneration, the fin stripe of melanocytes is reestablished from the *de novo* differentiation of unpigmented precursors. Furthermore, analysis of temperature-sensitive zebrafish mutants revealed that melanocyte differentiation in adults depends on the function of *kit* receptor tyrosine kinase (Rawls and Johnson 2000, 2001; Rawls et al. 2001). The white cells characteristic of the long rays may also originate from unpigmented progenitors; grafting an unpigmented fragment of a long ray in an ectopic interray gives rise to a fin ray containing white cells in an environment normally devoid of such cells (Murciano et al. 2002). However, there is no demonstration that pigment stem cells can give rise to other cell types.

Besides the local origin of cells forming the blastema, DiI cell-labeling experiments also show that mesenchymal cells can migrate from distant locations (up to seven segments or 750 μm) and become incorporated into the blastema of a two-day regenerate (Poleo et al. 2001). In the absence of the characterization of the identity and of the fate of these cells, one can only speculate on their role in the regenerating fin. The movement of these distant cells may maintain integrity of the tissue by compensating for the migration of cells closer to the wound. Another possibility is that these cells are reserve stem cells that may be specifically recruited following amputation to participate in blastema formation.

The conclusion we draw is that blastema formation could be the result of two cellular mechanisms, not mutually exclusive, cell dedifferentiation and activation of quiescent reserve cells.

Factors Involved in the Formation and Growth of the Blastema

In urodele amphibians, the apical epidermal cap plays an instructive role for initial blastema formation and further outgrowth; removal of the wound epidermis inhibits limb regeneration (Goss 1956; Mescher 1976; for a review see Nye et al. 2003).

In regenerating fins, a similar function of the wound epidermis is suspected but has not been demonstrated. Although the identity of the molecular factors that trigger formation of the blastema is not yet known, it is interesting to note that a number of the players known to be involved in the development of organs through epithelial-mesenchymal interactions are up-regulated in blastemata. For example, as mentioned above, the early expression of *lef1* and *wnt5* in the basal epithelial cell layer of the wound epidermis suggests a potential role in blastema formation (Poss et al. 2000a).

Fgf

Development of many different ectodermal organs, including limbs/fins, lungs, and teeth and hair follicles, relies on epithelial-mesenchymal interactions involving members of the fibroblast growth factor (FGF) signaling pathway (Crossley et al. 1996; Fischer et al. 2003; Mariani and Martin 2003; Millar 2002; Ohuchi et al. 1997; Pispa and Thesleff 2003; Vogel et al. 1996; Warburton et al. 2000). Results from numerous studies support a role of the Fgf signaling pathway in initiation of amphibian limb regeneration (Boilly et al. 1991;

Christen and Slack 1997; D'Jamoos et al. 1998; Giampaoli et al. 2003; Mullen et al. 1996; Poulin et al. 1993). Similarly, recent studies highlight the requirement of Fgf signaling for the formation of the blastema and its growth in regenerating zebrafish fins.

Poss et al. (2000b) showed that an Fgf receptor, Fgfr1, is expressed in the fibroblast-like cells at the onset of blastema formation, and this expression is maintained during blastema outgrowth. To investigate the role of Fgf signaling, these authors used a pharmacological inhibitor of Fgfr1, SU5402, and showed that incubation with SU5402 immediately following amputation prevents blastema formation without affecting wound healing (plate 11.4A–D). This inhibition is accompanied by an absence of cell proliferation, as shown by BrdU incorporation analysis. However, absence of FGF signaling does not prevent the presumptive blastema cell population from adopting a disorganized aspect, suggesting that these cells can respond to an early signal distinct from Fgf.

In a search for a candidate FGF that would signal to the Fgfr1-expressing mesenchymal cells, they found that a factor called wound fgf (wfgf), later renamed fgf24 (Fischer et al. 2003), is expressed at levels only detectable by reverse transcriptase-polymerase chain reaction (RT-PCR) during blastema formation (Poss et al. 2000b). *Wfgf* expression becomes detectable by *in situ* hybridization in the epidermal cap during blastema outgrowth.

In a different study, Tawk et al. (2002) addressed the role of Fgf signaling in the fin regenerate using an electrotransfer approach that consists of electroporation of the plasmid construct expressing the gene of interest following its injection into fin tissue (Tawk et al. 2002). They ectopically expressed a dominant-negative form of the *Xenopus* Fgfr1, which lacked the intracellular tyrosine kinase domain and had previously been shown to strongly compromise Fgf signaling in *Xenopus* embryos (Amaya et al. 1991). Electrotransfer of the construct prior amputation of the caudal fin completely inhibits initiation of regeneration. When the electrotransfer is performed during fin regeneration, ectopic expression of the defective Fgfr1 blocked further outgrowth in the majority of the fin regenerates analyzed (Tawk et al. 2002).

Msx

Another important family of genes involved in epithelial-mesenchymal interactions is the *msx* homeobox gene family. Msx gene products act as transcriptional repressors that appear to orchestrate the inductive events essential to organogenesis (for reviews, see Alappat et al. 2003; Bendall and Abate-Shen 2000). Most notably, *Msx1* expression in the progress zone of the limb buds of mouse embryos has been related to the maintenance of an undifferentiated state of the Msx-expressing cells (Song et al. 1992; Woloshin et al. 1995). In the regenerating urodele limb, *Msx2* is rapidly induced in the wound epidermis as well as in the subjacent stump tissue; *Msx1* is restricted to the blastema cells (M. R. Carlson et al. 1998; Koshiba et al. 1998; Simon et al. 1995; Yokoyama et al. 2001). In fetal mice, the capacity to regenerate the digit tips correlates with the presence of the nail organ and, more precisely, corresponds to the domain of expression of *Msx1* (Reginelli et al. 1995). Recently, Han et al. (2003) showed that fetal mouse digits of *Msx1* mutant display regenerative defects, further linking *Msx* gene function to the regenerative process. Interestingly, in support of a role of Msx1 in enhancing cellular plasticity expression of *Msx1* in mature C2C12 myotubes can induce a dedifferentiation response producing multipotential progenitor cells (Odelberg et al. 2000).

The zebrafish *msx* gene family consists of five members, *msxa, msxb, msxc, msxd,* and *msxe* (Akimenko et al. 1995) With the exception of *msxe,* all other *msx* genes are upregulated during fin regeneration with distinct spatiotemporal patterns of expression. During blastema formation, *msxa* and *msxd* transcripts are restricted to the epithelial cap, whereas *msxb* and *msxc* transcripts are localized in the proliferating cells of the blastema as early as 18 hours postamputation (Akimenko et al. 1991; Poss et al. 2000b). Consistent with a potential function of the *msx* genes in blastema formation, treatment with the inhibitor of Fgf signaling, SU5402, which prevents blastema formation and cell proliferation, also inhibits the induction of *msxb* and *msxc* expression in the cells of the connective tissue (plate 11.4E, F; Poss et al. 2000b).

Msx proteins are repeatedly used in multiple growth factor signaling pathways, such as the bone morphogenetic factor (Bmp), Fgf, and sonic hedgehog (Shh) signaling pathways during organogenesis. For example, during tooth development, Bmp2, Bmp4, Fgf2, and Fgf4 expression in the dental epithelia induce Msx1 expression in the underlying mesenchyme (Alappat et al. 2003; Pispa and Thesleff 2003). In turn, mesenchymal expression of growth and transcription factors, such as *Bmp4, Fgf3, Dlx2,* and *Patched* (*Ptc,* the membrane receptor of Shh protein), depends on *Msx1* expression.

In the fin regenerate, once the blastema is formed, most probably through inductive signals from the wound epidermis, the blastema is thought to send inductive signals to the epidermis.

The hypothesis of inductive interactions between the blastema and the overlying epidermis was tested by analysis of *msxd* and *msxa* expression following single ray ablation.

Under such condition, the gap left by the ablation of the ray is rapidly healed and filled by the migration of epithelial cells. Therefore, the ray often regenerates within a well-differentiated epithelial tissue.

Murciano et al. (2002) showed that *msxd* and *msxa* expression is activated in the epidermis overlying the blastema. Expression of *msx* genes in the epidermis is accompanied by the expression of the bone morphogenetic factor bmp4 in the blastema cells, a situation similar to that observed during normal fin regeneration (Murciano et al. 2002). The functional relationship between the msx proteins and bmp4 in the fin regenerate remains to be determined. *Msx* genes are expressed all along the regeneration process. However, the domain of expression of *msxb* and *msxc* changes during outgrowth of the fin regenerate, as described below (Akimenko et al. 1995).

It is clear from the results of a very limited number of studies that the nervous system plays a critical role in fin regeneration as in urodele limb regeneration. Teleost fins fail to regenerate if deprived of their nerve supply (Géraudie and Singer 1977, 1985; Goss 1969b). Indeed, resection of the three nerves of the brachial plexus of the pectoral fin of *Fundulus* prevents blastema formation and fin outgrowth but does not affect wound healing (Géraudie and Singer 1977; Géraudie and Singer 1985). However, the factors mediating this dependency are still unknown.

Outgrowth of the Regenerate

Following blastema formation, outgrowth of the regenerate is marked by changes in proliferation, morphological, and molecular profiles. During outgrowth, the regenerate comprises, underneath the epithelial layers, a mature blastema in its distal portion and a differentiating region in its proximal portion.

During blastema formation, all blastema cells proliferate at a similar rate (Nechiporuk and Keating 2002; Poleo et al. 2001; Santamaría et al. 1996). During regenerative outgrowth the blastema becomes divided into two distinct populations with different cell cycle kinetics (Nechiporuk and Keating 2002; Santamaría et al. 1996). The proximal blastema (PB) cells, defining the PB region, proliferate at a faster rate than during blastema formation (the length of the G2 phase changes from one hour to six hours as determined in zebrafish at 33°C; Nechiporuk and Keating 2002). In contrast, the distalmost blastema (DB) cells (in the DB region) have a very low cycling to nonproliferating rate. Consequently, only cells of the PB are expressing the proliferating marker, proliferating cell nuclear antigen (PCNA). Results from BrdU pulse-chase experiments indicate that the slow cycling cells of the DB originate from the highly proliferative PB zone and segregate into the DB at the end of blastema formation. According to Nechiporuk and Keating (2002), this finding does not support the hypothesis of a slow cycling stem cell population origin of the DB cells.

Nechiporuk and Keating (2002) showed that the transition to a different mode of proliferation is associated with a modification of *msxb* expression. While *Msxb* transcripts are found in a diffuse pattern during blastema formation, they become restricted to the DB cells during the outgrowth phase (plate 11.5).

As *Msx2* expression has been correlated with a reduced proliferation state and undifferentiated state in other systems of higher vertebrate species (Hamrick 2001), Nechiporuk and Keating (2002) proposed that *msxb* expression may be necessary to slow cell cycling during blastema formation and to inhibit cell division in the DB during regenerative outgrowth. The functional basis for the subdivision of the blastema into two cell populations is not clear. However, they suggest that the msxb-expressing cells would establish a boundary of proliferation and provide a direction for regenerative outgrowth.

Whatever the basis of such transition, the molecular mechanisms underlying the establishment and maintenance of these cell populations remain to be determined. As during blastema formation, *wfgf* is expressed in the epidermal cap and *Fgfr1* expression is now confined to the DB zone of the blastema. Furthermore, brief treatment with SU5402 downregulates *msxb* expression and prevents further growth of the regenerate, suggesting that *msxb* expression is maintained via Fgf signaling (Poss et al. 2000b). Several other factors such as members of the even-skipped family (*eve1, evx1,* and *evx2*), the retinoic acid receptor *rarγ, lef1,* and *bmp4* are confined to the DB during outgrowth of the regenerate (Borday et al. 2001; Brulfert et al. 1998; Murciano et al. 2002; Poss et al. 2000a; J. A. White et al. 1994). Further analyses are required to decipher the function of these factors during this phase.

In a screen for temperature-sensitive mutants using the strategy shown in plate 11.5, Poss et al. (2002) isolated the *nightcap* (*ncp*) mutation that specifically affects proliferation of proximal blastema cells at the onset of the regenerative outgrowth phase. Positional cloning of the mutation revealed that it affects the gene *mps1*, encoding a kinase required for the mitotic checkpoint. During blastema formation *mps1* is coexpressed with *msxb* in the proliferating cells. During regenerative outgrowth, *msxb* and *mps1* transcripts segregate in two separate domains, the poorly proliferative DB and the highly proliferative PB, respectively (plate 11.6;

Poss et al. 2002), suggesting that mps1 function is required to establish or to maintain intense proliferation in the proximal blastema during outgrowth. Interestingly, mps1 function seems to be dispensable during blastema formation, suggesting a possible mechanism of compensation by another factor to control cell proliferation at the beginning of the fin regeneration process.

As mentioned above, the growth rate of the regenerate varies according to the level of amputation along the proximal-distal fin axis; it is faster in its proximal part than in its distal part (Akimenko et al. 1995; Tassava and Goss 1966). A staircaselike amputation of the zebrafish caudal fin revealed a correlation between *msxb* expression with the growth rate of fin regenerates (plate 11.2B–E; Akimenko et al. 1995). Thus, *msxb* expression progressively decreases from the proximal to the distal cut. As the cell number is roughly the same in the three regenerates, it appears that the hybridization signal, reflecting the expression level of *msxb*, is stronger in individual cells of the proximal regenerate. Such variations of expression along the length of the fin ray are not observed for *msxc*, the other *msx* gene expressed in the blastema cells. Therefore, it appears that msxb function may regulate growth of the regenerate. Alternatively, *msxb* level of expression may depend on positional cues along the proximal-distal axis of the ray.

Based on these results, Zauner et al. (2003) investigated the role of *msx* genes in the development and regeneration of the sword of *Xiphophorus* caudal fins. During sexual maturation, males from some *Xiphophorus* species develop a sword that is an extension of the ventral rays of the caudal fin (plate 11.7A). They characterized the ortholog of *msxc* in this species and another gene they named *msxE/1* because of sequence similarity with the zebrafish *msxe* and the *Xenopus msx1* genes. Gene expression analysis revealed a specific correlation between sword development and regeneration and the upregulation of *msxc* expression in the fin rays of the sword compared to the other fin rays (plate 11.7B–D). This result argues for a role for msxc in the growth control of this sexually selected morphological trait.

Surprisingly, expression of these two *msx* genes in regenerating fins of swordtails differs from that observed in zebrafish (Zauner et al. 2003). Indeed, *msxE/1* is induced in the distal blastema of swordtails (plate 11.7F, H), whereas it is not in zebrafish (Akimenko et al. 2003; Poss et al. 2003), and *msxc* shows an additional pattern of expression in swordtails compared to zebrafish (plate 11.7E, G). Indeed, in addition to being expressed in the distal blastema, *msxc* transcripts are also found in a region of the distal-lateral mesenchyme that may correspond to cells destined to become scleroblasts. The function of msxc in these cells is unknown, but, based on the known role of the msx in other systems, Zauner et al. (2003) suggest that msxc may be required to maintain the expressing cells undifferentiated. However, since this expression does not correlate to any identified zebrafish *msx* gene, it implies that either the true zebrafish ortholog of the *Xiphophorus msxc* gene is not yet identified, or the *msx* gene family is evolving particularly rapidly in teleosts compared to other branches of vertebrates, with paralogs acquiring distinct functions.

Differentiation and Pattern Formation

As the blastema is still in its outgrowth phase, cells located in the proximal part of the blastema start the differentiation program. The zone of differentiation and patterning is located proximal to the PB zone. Most of the studies regarding cell differentiation and patterning during fin regeneration have focused on the analysis of the regeneration of the skeletal elements with an emphasis on the lepidotrichia.

Actinotrichia

During embryonic development, actinotrichia appear before lepidotrichia (Géraudie 1977; Geraudie and Landis 1982; A. Wood and Thorogood 1984; see also Polly, chap. 15 in this volume). However, the timing of regeneration of these two skeletal elements is still controversial. Many studies present evidence that lepidotrichia are forming in the blastema prior to actinotrichia (Goss and Stagg 1957; Marí-Beffa et al. 1989; Misof and Wagner 1992; Santamaría and Becerra 1991), while scanning electron microscopy of fin regenerates in *Fundulus heteroclitus* and *Lebistes reticulates* seems to indicate the reverse (Géraudie and Singer 1992).

The proteins comprising the actinotrichia are still to be identified, but many studies agree that actinotrichia are made of collagen and/or of a collagen-like protein called elastoidin (Chandross and Bear 1979; Garrault 1936; Géraudie and Landis 1982; Gross and Dumsha 1958). The actinotrichia appear in the blastema extracellular matrix beneath the basal lamina in the distal part of the regenerate, and the fibrils will keep an apical position all through the regeneration process (plate 11.1D). Using a pulse-chase analysis of ³H-proline incorporation during regeneration of *Xiphophorus* fin rays, Marí-Beffa et al. (1989) investigated the mechanism by which the actinotrichia conserve their apical position. Length and distribution of the actinotrichia during regeneration seem to be under the balanced control of their synthesis and degradation, supporting the hypothesis of a turnover mechanism of elastoidin synthesis rather than an apical transport of the fibril by the growing regenerate. The identity of the cells synthesizing the actinotrichia is not yet known.

Lepidotrichia

The matrix comprising the lepidotrichia is secreted by scleroblasts or lepidotrichia-forming cells (LFCs). These blastema cells lining the epithelial tissue and located at the limit between the PB and the patterning zone differentiate into scleroblasts.

Lepidotrichia will mature as the regenerate grows (the less mature form of the lepidotrichia being located in the distal part of the regenerate and the more mature being in the part closest to the stump; plate 11.2F). These cells start to release the lepidotrichial matrix in the epithelial-mesenchymal interface. As the regenerate grows, some scleroblasts migrate around the newly deposited bone matrix and become localized between the basement membrane along the basal epithelial cell layer and the newly formed lepidotrichia. As these scleroblasts continue to synthesize and secrete the bone matrix on the outer surface of the lepidotrichia, they contribute to its thickening. So, in the proximal part of the regenerating bone, the hemiray is surrounded by scleroblasts (Santamaría and Becerra 1991). The bone matrix will then mineralize in a proximal-distal direction.

Lepidotrichial matrix is made of glycosaminoglycans and type II collagen, and its composition changes as the bone matures (Montes et al. 1982; Santamaría and Becerra 1991; Santamaría et al. 1992). Drugs affecting collagen metabolism in mammals can impair the deposition and organization of the lepidotrichia and of the extracellular matrix of the connective tissue (Bechara et al. 2000). The type II collagen gene *col2a1* is one of the earliest genes coding for a protein involved in bone synthesis in the regenerating fin of zebrafish (S. L. Johnson and Weston 1995). It is first expressed one day after amputation in blastema cells close to the stump of the bones and later in the scleroblasts of the regenerate on both sides of the lepidotrichia.

As during development, the bones regenerate by the successive addition of new segments. The transcription factor *evx1* is probably involved in joint formation, since *evx1* gene expression in scleroblasts correlates with segment boundaries during fin development and regeneration (plate 11.8I; Borday et al. 2001). The expression pattern of two members of the hoxa gene complex, hoxa11b and hoxa13b, in differentiating scleroblasts suggests that they may also be involved in dermal bone cell differentiation or ray patterning (Géraudie and Borday Birraux 2003).

Shh and BMP

A number of lines of evidence suggest that the sonic hedgehog (shh) and bone morphogenetic factor (bmp) signaling pathways are implicated in bone morphogenesis and patterning.

The signaling molecule, *sonic hedgehog (shh)*, involved in the patterning and morphogenesis of many organs, is expressed in a subset of cells of the basal epithelial layer located at the limit between the PB and the patterning zone, just at the level of new matrix bone deposition in each fin ray (plate 11.8A, D; Laforest et al. 1998). Transcripts of its receptor, the transmembrane protein, *patched (ptc1)*, are found in a large domain that overlaps the domain of *shh*, and are also found in the adjacent scleroblasts, some of which are proliferating (plate 11.8B, E). The bone morphogenetic protein, *bmp2b*, which in many developing systems has been shown to be a secondary signal activated by shh signaling, presents a pattern of expression similar to that of *ptc1* (plate 11.8C, F).

A role for shh signaling pathway in bone patterning was recently supported by the demonstration that ectopic expression of *shh* following the direct microinjection of *shh*-expressing constructs in the tissue separating the branches of the fin ray can induce ectopic bone formation leading to ray fusion (plate 11.8G, H; Quint et al. 2002). This was accompanied by induction of *ptc1* and of *bmp2b* in this interray tissue. Similarly, injection of *bmp2b*-expressing construct can also induce bone fusion but, in this case, without upregulation of *shh*. In addition, ectopic expression of the *bmp* inhibitor, *chordin*, can antagonize bone fusion induced by shh misexpression.

These results suggest that, in the fin regenerate, bmp2b is a downstream target of shh signaling. How shh signaling is acting to determine proper bone deposition is still unclear. It could stimulate either the differentiation of new scleroblasts or the proliferation and recruitment of already existing scleroblasts. Inhibition of shh signaling using cyclopamine, a steroidal alkaloid that interrupts shh signaling by acting on smoothened (smo), a component of the receptor complex present on the surface of the target cell, leads to the inhibition of cell proliferation in the blastema, which ultimately leads to an arrest of fin growth (Quint et al. 2002). Nevertheless, cyclopamine treatments do not stop bone matrix deposition by already differentiated scleroblasts, suggesting that *shh* has no effect on bone matrix synthesis and release.

Interestingly, the phenotype of cyclopamine-treated fins (Quint et al. 2002) is similar to that observed following treatment with the inhibitor of Fgfr1 signaling, SU5402 (Poss et al. 2000b). In both cases, there is a significant decrease in cell proliferation accompanied by an arrest of fin growth, a continued bone deposition, and a melanocyte accumulation in the distal part of the regenerate. Treatment with SU5402 entails the down-regulation of *shh* expression. Reciprocally, *Fgfr1* expression is abolished following cyclopamine treatment (M.-A. Akimenko, unpublished data). At that stage of the regeneration process, *fgfr1* is expressed in the DB cells as well as in the basal layer of the epidermis in a domain that overlaps that of *shh* (Poss et al. 2000b). Further experiments are required to understand the exact function of *shh* signaling and its relationships with *FGF* signaling.

Revascularization of the Regenerate

The transparency and easy accessibility of zebrafish embryos, combined with the possibility of producing transgenic fish, make the zebrafish an ideal organism in which to follow organ formation in live embryos/adults using transgenic lines expressing the Enhanced Green Fluorescent Protein (EGFP) driven by tissue-specific gene regulatory regions.

N. D. Lawson and Weinstein (2002) established a zebrafish transgenic line expressing EGFP under the control of the regulatory region of the *fli1* gene, an endothelial cell marker, and used the *TG(fli1:EGFP)^{y1}* line to visualize the development of the embryonic vasculature *in vivo*. Using this transgenic line, C. C. Huang et al. (2003) examined the neovascularization in regenerating caudal fins of zebrafish (plate 11.9; C. C. Huang et al. 2003).

Each fin ray is irrigated by one artery running in a central position in the intraray mesenchyme between the hemirays and is closely flanked by two veins located in the interray mesenchyme. The artery and the veins of a ray and the veins of adjacent rays are connected by intervessel commissures that are more densely distributed in the distal portion of the fin. In wild-type fish, fin amputation triggers very minimal bleeding.

The vessels' ends are healed within the first day postamputation. The artery and veins of a ray are then reconnected through anastomosis of sprout extensions from the healed vessel ends. Two days postamputation, blood circulation resumes in the connecting bridges (plate 11.9B). Next, blood vessel regeneration can be divided into two phases. The early phase, observed between 2 and 8 days postamputation when zebrafish are kept at 25°C, is characterized by the formation of a plexus made of numerous thin vessels branching from the connecting bridges and making connections with neighboring vessels (plate 11.9C).

As the regenerate grows, the excess intervessel commissures are removed by pruning starting from the proximal part of the regenerate, and this blood vessel remodeling leads to the definitive arteries and veins. During the late phase of regenerative angiogenesis, growth of the blood vessels continues by sprouting from their distal ends without formation of a plexus intermediate. The basis for the use of these two growth mechanisms is unknown yet. Nevertheless, evidence that the two phases of vessel growth involve distinct genetic programs comes from the analysis of the temperature-sensitive *reg6* mutation, which displays severe blood vessel blisters at the restrictive temperature of 33 °C (C. C. Huang et al. 2003; S. L. Johnson and Weston 1995).

C. C. Huang and colleagues (2003) examined the *reg6* mutation into the *TG(fli1:EGFP)^y1* background. Using temperature shift experiments at various time points following amputation, they could pinpoint the critical period to the early phase of blood vessel regeneration. They found that *reg6* mutants present defects in anastomosis and in the branching process of the blood vessel commissures leading to plexus formation. The few vessel extensions and branches that are formed become swollen. In contrast, when temperature is shifted to 33 °C during the late phase of regenerative angiogenesis, vascularization is not affected. Characterization of the *reg6* mutation may help to determine whether the defective vascularization is the primary cause of the regenerative defect.

The origin of the cells forming the regenerating blood vessels has not been directly addressed in this study. Although the data indicate that new blood vessels seems to derive from preexisting vessels by angiogenesis, they do not exclude the possibility that blood vessel cells may originate from the differentiation of undifferentiated cells of the blastema. Further analysis using, for example, cell lineage tracing is required to examine the possible contribution of the blastema to the regenerating vessels.

Concluding Remarks

Significant progress has been made during the last 15 years in our knowledge of the structure of the fin regenerate and in our understanding of some of the mechanisms controlling fin regeneration. One idea that emerges from the numerous histological and molecular studies is that the blastema is a more complex structure than a simple population of undifferentiated cells. It also appears, based on the gene expression patterns reported thus far, that the growth, differentiation, and patterning of the regenerate involve molecular mechanisms similar to those involved in development of the fin rays during the larval stage. It is important to note that the initiation of fin regeneration has to be considered separately and may involve factors and genetic pathways that are not required during larval stages.

One of the important questions that remain to be answered is the nature of the molecular mechanisms that underlie initiation of regeneration. If cell dedifferentiation is the main cellular mechanism contributing to the formation of the blastema, what are the factors that control this response to amputation? What are the contributions from the nervous system?

Most of the recent progress in our understanding of fin regeneration is attributable to the use of zebrafish as a model system, which enabled combined genetic and molecular approaches; also, genetic screens specifically adapted to the analysis of fin regeneration have been developed (S. L. Johnson and Bennett 1999; S. L. Johnson and Weston 1995; Poss et al. 2002). Since fin regeneration often involves genes that are expressed during embryonic or larval development of the fin buds and of their skeleton, mutations that influence fin regeneration could affect fish development in a way that may impede the analysis of their effects on the regeneration process.

To circumvent this potential problem, S. L. Johnson and Weston (1995), and later Poss et al. (2002), generated temperature-sensitive mutants of fin regeneration. Characterization of some of these mutants has already brought significant insights into some aspects of the genetic program of fin regeneration (Nechiporuk et al. 2003; Poss et al. 2002; Rawls and Johnson 2001). However, one limitation of screens based on temperature-sensitive mutants is that only a small percentage of ethylnitrosourea-induced mutations are temperature-sensitive. Therefore, additional screens have to be developed that would reveal mutations specific to the regeneration process—that is, for factors that are dispensable or not involved during embryogenesis or larval development but that would be essential for the initiation of the regeneration of adult organs.

Direct perturbation of the fin regenerate might constitute a useful approach to complement those mentioned above. Microinjection of plasmid constructs in the fin regenerate already proved to be an efficient technique to misexpress secreted factors locally and thus perturb the normal regeneration process (Quint et al. 2002). However, this approach may have some limitations when applied to nonsecreted molecules, such as transcriptional regulators, because of the relatively low number of cells that would receive the construct. Coupling microinjection to electroporation as shown by Tawk et al. (2002) may increase the efficiency of gene delivery to a larger number of blastema cells.

Transgenic lines of zebrafish have been produced in which the green fluorescent protein (GFP) is expressed under the control of gene regulatory sequences that target expression of this reporter to specific cell populations (Higashijima et al. 1997; Ju et al. 1999; Z. Gong et al. 2001; Gibbs and Schmale 2000). The use of such transgenic lines to follow the lineage of GFP-expressing cells should help answer some key questions of fin regeneration, as exemplified by the recent analysis of the revascularization of the fin regenerate using the *TG(fli1:EGFP)*^{y1} transgenic line (C. C. Huang et al. 2003).

The identification of genes differentially expressed during the different steps of the regeneration process will also be instrumental in deciphering the underlying molecular mechanisms. This is presently carried out via cDNA subtraction and differential display reverse transcriptase (DDRT)-PCR approaches (M.-A. Akimenko, unpublished data) and shall soon be complemented with screening of zebrafish microarrays as these become more widely used.

Molecular mechanisms of fin regeneration have been investigated in a limited number of species. At this time, besides zebrafish, the medaka is the other teleost species amenable to embryological and genetic studies for which a number of molecular tools, including a complete genome sequence, are either readily available or expected to be so in the near future. Most observations reported thus far suggest that the general framework of fin regeneration is probably similar in the different teleosts. However, some differences have also been observed, for example, in *msx* gene expression during fin regeneration in the sword-bearing *Xiphophorus* fish and in zebrafish (Akimenko et al. 1995; Nechiporuk and Keating 2002; Zauner et al. 2003). Therefore, mechanisms of fin regeneration may show important variations among teleosts and valuable insights into the regeneration process may be gained by examination of additional species. The identification of which molecules and mechanisms are part of a general framework and which constitute species-specific adaptations will enrich our understanding of the complex and fascinating phenomenon of epimorphic regeneration not only of fins but also of other organs.

Acknowledgments

The authors thank C. Murciano for sharing unpublished results. This work was supported by funding from Canadian Institutes of Health Research (CIHR) and Natural Sciences and Engineering Research Council (NSERC) of Canada.

Chapter 12 Tetrapod Limb Regeneration

David M. Gardiner and Susan V. Bryant

SALAMANDERS ARE unique among tetrapod vertebrates in their ability to regenerate their limbs as adults. As a consequence, salamander limb regeneration has been extensively studied by several generations of biologists, and a wealth of cellular and molecular data have provided valuable insight into the mechanisms controlling tissue and organ regeneration. Until recently, progress lagged behind studies of limb development; however, advances in molecular biology, bioinformatics, and functional analysis have stimulated a renewed interest in regeneration that has led to exciting new opportunities to understand regenerative mechanisms and, ultimately, to stimulate regenerative responses in humans.

Animals that can regenerate have long stimulated the imagination of researchers and reinforced a belief that we will eventually be able to coax humans to regenerate. Successive generations of regeneration biologists, each with the acquisition of new technologies, have declared that human regeneration is imminent, yet such optimism has thus far proved to be unfounded. As will become evident from this review, we argue once again that the time has arrived when we can finally understand the biology of regeneration. We are at the beginning of a regeneration renaissance, in terms of both our understanding of tetrapod limb regeneration and, more important, an understanding of regeneration as a fundamental process of all organisms. Thanks to the progress made in sequencing the genomes of numerous organisms, we are no longer limited to studying the few model organisms that lend themselves to genetic analysis. Through

the power of comparative biology, we can begin to understand which aspects of regeneration are conserved and which are novel. Only time will tell whether we are just another generation who thinks it sees the top of the mountain, or whether we will at last reach the top and be in a position to help realize the potential of humans for regeneration.

In this review, we summarize the most recent findings regarding tetrapod limb regeneration in the context of more classical studies. We emphasize what we consider to be important principles of regeneration to emerge from this synthesis of old and new. At the end, we briefly discuss tetrapod limb regeneration in the context of our knowledge of regenerative processes in other organisms, with particular attention to regeneration of fish fins, the topic of the previous chapter in this book.

Tetrapod Limb Regeneration

Of all model systems for studying regeneration, none has received more attention than the tetrapod limb. Given the strikingly conserved anatomy of the limb among all extant tetrapods, the regenerative ability of urodele limbs causes one to wonder if similar regenerative abilities would be possible for all tetrapods, including humans. As a result of decades of research driven largely by this question, we have acquired a considerable knowledge of cell- and tissue-level phenomena associated with limb regeneration. In recent years advances in molecular biology and somatic cell transgenesis have created opportunities for studies of the mechanisms controlling this remarkable ability.

Many excellent reviews of tetrapod limb regeneration have been published over the years. We refer the reader to *Vertebrate Limb Regeneration* (Wallace 1981) as a comprehensive review of research through the late 1970s. Discoveries based on studies of cell biology and cell contribution over the next 20 years are reported and discussed in reviews from a number of laboratories (Brockes 1997; Gardiner and Bryant 1996; Maden 1997; Mescher 1996; Muneoka et al. 1989; Stocum 1996). The molecular and functional analyses that inform current thinking about the mechanisms controlling regeneration are reviewed in the context of the older literature (Brockes and Kumar 2002; Bryant et al. 2002; Gardiner and Bryant 1998; Gardiner et al. 1999, 2002; Géraudie and Ferretti 1998; T. L. Muller et al. 1999; Nye et al. 2003).

We present here a brief overview of the chronological events of limb regeneration as a reference for the topics discussed in this review. In addition, we review studies that in our view are most relevant to the key principles of tetrapod limb regeneration.

Overview of the Events of Tetrapod Limb Regeneration

Regeneration of a salamander limb results in a quantitatively, qualitatively, and functionally perfect replacement of the portion removed by amputation (fig. 12.1). Thus, a limb that is amputated through the distal humerus replaces the distal humerus, the zeugopod, and autopod, whereas amputation at the wrist results in regeneration of only the elements of the autopod. These and other observations lead to the

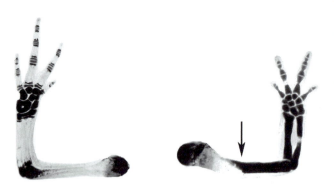

Figure 12-1 A mature urodele (*Notophthalmus viridescens*) limb (left) has an anatomy that is typical of all tetrapods, including humans. When amputated proximally (arrow, right), a urodele limb regenerates to re-form a limb with a structure that is identical to what was removed by amputation. These limbs have been stained for cartilage with Victoria blue (Bryant and Iten 1974). At late stages of limb development and regeneration, the cartilage ossifies and does not stain (clear regions of the limb on the left and the mature, proximal stump tissues on the right). The regenerated elements of the limb on the right have not yet begun to undergo ossification. (Modified from Gardiner and Bryant 1998.)

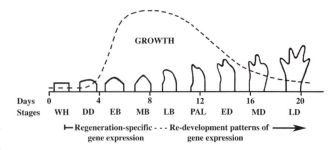

Figure 12-2 Sequence of events during urodele limb regeneration. The stages of regeneration are based on morphological criteria (Iten and Bryant 1973; Tank et al. 1976) and are referred to as WH (wound healing), DD (dedifferentiation), EB (early bud blastema), MB (medium bud), LB (late bud), PAL (palette), ED (early digits), MD (medium digits), and LD (late digits). In a small axolotl (5–7 cm total body length), this sequence is complete within about 4 weeks. Growth of the regenerate (cell proliferation as measured by either labeling index or mitotic index) is low during the first few days following amputation, and is maximal at the blastema and predigit stages (Wallace 1981). At the later stages, the patterns of gene expression are comparable to those observed in developing limb buds. In contrast, the temporal and spatial patterns of gene expression during the early stages of regeneration often are different than during limb development. (Modified from Gardiner and Bryant 1998.)

conclusion that cells have information about their location relative to one another within the limb. This positional information is inferred to be a stable property of cells in the mature limb that can be accessed during regeneration to guide replacement of the missing information (Nye et al. 2003; Stocum 1984, 1996).

The sequence of regeneration can be broken into a series of stages, progressing from amputation through blastema formation, growth, and differentiation to the fully regenerated limb (fig. 12.2). The time needed to complete this development sequence varies between species and as a function of animal size, with larger animals progressing more slowly through the sequence. As a guide, axolotls (*Ambystoma mexicanum*) ranging in size from 5 to 10 cm will complete regeneration in about one month.

Within hours of amputation, the epidermis at the periphery of the wound migrates to cover the wound surface. Over the next several days this wound epidermis (WE) thickens to form the apical epidermal cap (AEC) of the regenerate. As discussed below, formation of the blastema and proliferation of blastema cells are dependent on factors produced by the WE and the AEC (Stocum and Dearlove 1972; Thornton 1968), as well as by the limb nerves (Singer 1952, 1978). Within the first few days, the tissues near the wound become disorganized, and cells from the periphery begin to migrate under the WE and accumulate at the center of the stump (Gardiner et al. 1986). It is at this early phase of regeneration that cells in the stump begin to give rise to the blastema cells that will eventually form the new limb. The process by which blastema cells are generated is classically referred to as dedifferentiation, and it is an important topic that we address in this review. Proliferation rates are low during the early, preblastema stages of regeneration, and then increase as the migrating stump-derived cells begin to accumulate under the WE (early bud blastema; see Wallace 1981). As the blastema increases in size, it flattens along the dorsal-ventral axis (late bud). Differentiation progresses from proximal to distal, and from anterior to posterior (Palette). Continued growth of the regenerate to normal size is dependent on an adequate supply of nerves; however, differentiation is not, and limbs that are denervated at late stages form normally patterned miniatures (Singer and Craven 1948).

Principles of Tetrapod Limb Regeneration

We focus the discussion around the three issues that we consider central to an understanding of tetrapod limb regeneration.

All Tetrapod Limbs Regenerate at Some Stage in the Life History of the Organism

Although no adult vertebrates, with the exception of urodeles, can regenerate limbs, there is evidence of regenerative ability in the developing limb buds of all vertebrate species that have been tested. This issue has been most extensively studied in the frog, *Xenopus laevis*, an important model organism for studies of vertebrate limb regeneration.

Early-stage (predifferentiation) limb buds of *Xenopus* can regenerate perfectly. As *Xenopus* larvae progress through metamorphic climax, regenerative ability declines as differentiation progresses (Dent 1962; Muneoka et al. 1986a). The sequence of differentiation in the developing limb is from proximal to distal and from posterior to anterior. Proximal amputation prior to the onset of differentiation leads to a fully formed regenerate. At later stages when proximal levels no longer regenerate, distal levels are still undifferentiated and remain able to regenerate. Eventually, even distal amputations yield incomplete regenerates, and ultimately regenerative ability is lost.

Presence or absence of regenerative ability is an intrinsic property of the limb cells, and not a consequence of hormonal or other changes in the animal. Young limb buds grafted to regeneration-incompetent older hosts retain their intrinsic regenerative ability, whereas older limb buds grafted to regeneration-competent younger hosts remain unable to regenerate (Sessions and Bryant 1988). One noteworthy observation is that repeated amputation prolongs the window of regeneration in animals as they continue their progression to a nonregenerative state (Kollros 1984). In these experiments, the contralateral control limb loses its ability to regenerate at the normal time. These results indicate that

repeated injury can maintain limb cells in a regeneration-competent state.

Loss of regenerative ability in *Xenopus* is associated with a progressive loss of the ability to reactivate expression of genes involved in growth and pattern formation—for example, *Shh* (T. Endo et al. 1997). Treatment of amputated late-stage regeneration-incompetent *Xenopus* limb buds with FGF10, normally expressed in the mesenchyme of young limb buds, induces expression of a number of genes, including epidermal *fgf8*, and partially rescues regenerative ability (Yokoyama et al. 2001). A comparable result has been observed in regeneration-incompetent amputated chick limb buds that are supplied with either FGF or a grafted (FGF-expressing) apical ectodermal ridge (AER; T. L. Muller et al. 1999). Exogenous FGFs appear to substitute for the endogenous FGFs normally produced by the apical epidermis of young limb buds (Christen and Slack 1997; Yokoyama et al. 2000). The partial rescue of regeneration by FGFs is confined to stages in which intrinsic regenerative ability is declining. Once all regenerative abilities have been lost, exogenous FGF is no longer effective in stimulating a regenerative response. Although relatively little is known about the expression and function of FGFs in the limbs of either developing or regenerating urodeles, *fgf8* and *fgf10* are expressed in both the apical epidermis and the mesenchyme of regenerating urodele limbs (Christensen et al. 2002; M. Han et al. 2001), and it is reasonable to assume that their functions in FGF signaling pathways are conserved in regeneration.

Information about the regenerative ability of developing mammalian limbs is limited to mice and rats. An earlier report that opossum hindlimb buds can regenerate at birth (Mizell 1968) was subsequently determined to be incorrect, and apparently had resulted from a misinterpretation of the fate map of the limb bud at that stage (Fleming and Tassava 1981). Studies of rodent limb buds *in vitro* in organ culture provided some evidence of a regenerative response, but the short time period available for maintenance, coupled with limited progression of limb development under culture conditions (Chan et al. 1991; Deuchar 1976; Lee and Chan 1991), prevented further exploration of regeneration *in vitro*.

Equally challenging are *in vivo* experiments, since early limb bud stages, those most likely to be able to regenerate, occur *in utero*, and very young embryos do not survive the trauma of amputation (Muneoka et al. 1986b; Wanek et al. 1989). At the youngest stage at which the embryo does survive the surgery, the limb bud is at a relatively advanced stage of development; however, it can still regenerate from amputations at distal levels and replace nearly entire digits (Wanek et al. 1989). This degree of regenerative ability is comparable to that observed in equivalently developed late-stage *Xenopus* tadpoles, and thus it is likely that mammalian limb buds possess more extensive regenerative abilities at earlier stages of development. At even later stages *in utero*, during digit formation, the proximal limit of regenerative ability corresponds to the proximal boundary of expression of *msx1*. As discussed below, expression of *msx1* has been implicated in maintaining cells in an undifferentiated, proliferative state in a number of developing systems, and it is considered to be one of the genes with a critical functional role in the initiation of a regenerative response.

Mouse limbs, as well as the limbs of other mammals, including humans, retain the ability to regenerate the most distal structures, the terminal phalanx of digits. Digit tip regeneration was first discovered in humans as a consequence of the fact that fingertip amputation is a common trauma (T. L. Muller et al. 1999). If a digit amputation is treated conservatively by preventing infection of the wound and allowing it to heal without suturing, the digit tip will regenerate. Although originally reported to occur only in children, digit tip regeneration also occurs in adults. If the wound surface is sutured closed, as is often the practice, the regenerative response is inhibited. Similar surgical closure of the wound in amputated urodele limbs is one of the very few ways to effectively inhibit natural limb regeneration (Wallace 1981).

Digit tip regeneration has been studied experimentally in rodents. Amputation of the digit tips at levels distal to the nail matrix in juveniles and adults results in regeneration. As in urodele regeneration, it appears that connective tissue fibroblasts (discussed below) are the cells that participate in re-forming the lost structure. However, during mouse digit tip regeneration, the bone re-forms directly within the connective tissue matrix (T. L. Muller et al. 1999), rather than from an intermediate cartilaginous model as in endochondrial skeletal formation in regenerating urodele limbs. This pattern of bone formation appears more comparable to that of dermal bone, such as fish fin rays (discussed in the previous chapter), which also regenerate in the adult organism.

Unlike *Xenopus* limb buds, chick limbs are not able to regenerate after a simple amputation at any stage of development. However, in all developing or naturally regenerating limbs, an outgrowth-permissive epidermis is required (AER in chick and mouse, AEC in amphibians). Even in animals that can regenerate, any treatment that inhibits the formation or function of the apical epidermis also inhibits limb development (Tschumi 1957) or regeneration (Stocum and Dearlove 1972). When this epidermis is removed from amphibian limb buds or blastemas, it is regenerated, and in these animals an early response to amputation is regeneration of the apical epidermis. In contrast, the AER of chick limb buds is not re-formed, and removal results in truncated limbs. Therefore, it follows that at least one barrier to limb bud regeneration in the chick is failure to re-form the AER. If

the AER is replaced surgically postamputation (Hayamizu et al. 1994) or its function is replaced by exogenous FGF (Kostakopoulou et al. 1996, 1997; G. P. Taylor et al. 1994), outgrowth and regeneration proceed. As in *Xenopus*, regenerative ability is restricted spatially along the proximal-distal axis, such that at stage 24–25 the distal region of the limb bud (progress zone) can regenerate, but the proximal region cannot (G. P. Taylor et al. 1994).

As we have seen, studies of vertebrate limb buds indicate that all vertebrates tested show some capacity for regeneration during development. These findings suggest that during development, limb cells exist in at least two states with regard to their ability to regenerate. Cells that are located more distally, and have neither differentiated nor made an irreversible commitment to differentiate, have the ability to regenerate, provided that either an outgrowth permissive epidermis or equivalent factors are present. Cells in more proximal regions of the limb bud, where differentiation of mature tissues is occurring, have lost the ability to respond to signals for regeneration, even in the presence of permissive factors. Urodeles are the only exception to this generalization, because they have the unique ability to regenerate *after* differentiation as well as before.

Regeneration Is a Two-Step Process

Given that vertebrates can regenerate their limbs during embryonic development, it follows that if mature limb cells were able to revert to a stage equivalent to the early limb bud, they might then be able to bring about regeneration. A number of discoveries suggest that in fact this is the mechanism whereby urodele limbs regenerate. Successful limb regeneration is a two-step process involving a preparation phase (phase I) that establishes a blastema that is equivalent to a limb bud, which then progresses through a redevelopment phase (phase II). Evidence for these two phases in urodele limb regeneration comes from classical and molecular studies.

The most direct demonstration that developing limb buds and regenerating limb blastemas use the same molecular mechanisms to control growth and pattern formation comes from classical tissue-grafting experiments. These studies are based on the induction of supernumerary growth and pattern in response to grafting that creates new cellular interactions (Bryant et al. 1981; French et al. 1976). When cells from normally nonadjacent positions (e.g., anterior and posterior) are grafted adjacent to each other, growth is stimulated and a new pattern is formed. The new pattern is that which would normally occur between the interacting cells. This experimental paradigm has been used to discern the rules governing pattern formation and to isolate and test the function of numerous molecules that both encode positional information and control the interactions that stimulate extra growth and pattern formation. Because the forelimb and hindlimb of the axolotl develop asynchronously, it is possible to test whether developing limb bud cells and regenerating blastema cells interact in the same way that each would interact with their own type. Reciprocal grafting of developing and regenerating tissues does in fact yield the same patterns of supernumerary outgrowths and patterns of cell contribution to the outgrowths that are observed after limb bud–limb bud interactions or blastema-blastema interactions (Muneoka and Bryant 1982, 1984). Thus, developing limb bud cells and regenerating blastema cells communicate with one another and use the same mechanisms for the control of growth and pattern formation. It follows from this conclusion that the potential to regenerate a limb from a blastema is present in all vertebrates. The challenge is to understand how to form a blastema.

In recent years, analysis of gene expression has confirmed that patterns seen in the later (blastema) stages of regeneration (phase II) are comparable to those observed during similar stages of limb development. However, this is not the case during the early (phase I) preblastema stages. In hindsight, these findings are perhaps to be expected since no stage in development corresponds to the preblastema stage of regeneration, in which cells of the mature tissue of the stump give rise to the limb bud–like cells of the blastema.

The first molecular evidence for the occurrence of unique patterns of gene expression during phase I of regeneration came from an analysis of the expression of the HoxA genes.

There is considerable molecular and genetic evidence that the most 5′ members of the HoxA complex (*Hoxa9–13*) function in the control of pattern formation during limb development (A. P. Davis et al. 1995; Fromental-Ramain et al. 1996b; Small and Potter 1993). By generating a Hox code of overlapping expression domains, the HoxA genes function in the specification of segmental identity along the proximal-distal limbs axis (fig. 12.3A). During limb development in all tetrapod limbs examined, including urodele limbs, the more 3′ genes are expressed earlier, and in an expression domain that extends further proximally than the more 5′ genes that are expressed progressively later, and in progressively more distally restricted regions of the limb bud. The proximal boundaries of each more 5′ gene correspond to progressively more distal limb segments (Yokouchi et al. 1991; Haack and Gruss 1993; Gardiner et al. 1995).

In contrast to the conserved spatial and temporal patterns of *HoxA* expression observed in developing limbs, during limb regeneration, 5′ *HoxA* genes are expressed synchronously in stump cells regardless of the level of amputation along the proximal-distal limb axis (Gardiner et al. 1995). Expression of both *Hoxa9* and *Hoxa13* is detected about 24 hours postamputation. A few days later, all expressing cells

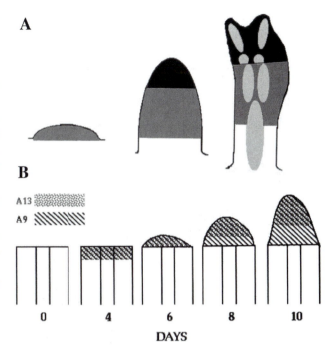

Figure 12-3 Expression of *Hoxa9* and *Hoxa13* during axolotl limb development and limb regeneration. The spatial and temporal patterns of expression in developing axolotl limb buds (A) are comparable to other tetrapods. *Hoxa13* is expressed later and in a more distally restricted domain (black) as compared to *Hoxa9* (gray). During regeneration (B), both genes are re-expressed at the same time in cells adjacent to the amputation plane (4 days) and in the early bud blastema (6 days). As the blastema grows, expression patterns that are comparable to the limb bud arise (8–10 days) in which *Hoxa13* is expressed in a more distal domain relative to *Hoxa9*. (Modified from Gardiner and Bryant 1996.)

are found in the newly formed blastema. Future lineage studies will determine whether the expressing cells migrate to form the blastema, as we suspect. After another few days, as the blastema grows, a region of cells expressing *Hoxa9* but not *Hoxa13* is generated at the base of the blastema, leaving cells expressing both markers at the tip. When the regenerated skeletal elements begin to differentiate, both *Hoxa9* and *Hoxa13* are expressed in the autopod, whereas *Hoxa9*, but not *Hoxa13*, is expressed in the zeugopod. This final spatial expression pattern is the same as in developing limbs of all vertebrates, including urodeles. In summary, the later (phase II) expression pattern is comparable to that seen in limb development, whereas phase I of regeneration is unique to regeneration, and the expression pattern of HoxA genes is not comparable to any stage of limb development.

Historically, the issue of the relationship between regenerative and developmental mechanisms has been the subject of speculation and debate (Bryant and Muneoka 1986; Ferretti and Brockes 1991; Muneoka and Sassoon 1992; Stocum 1991). As discussed, it now appears that regeneration is both different from and similar to limb development. What is less clear is the point at which the blastema acquires the properties and behaviors of a developing limb bud. Early grafting studies were typically performed with medium to late bud blastemas, and results indicated that at that stage the blastema is more similar to than different from a limb bud. An extensive experimental literature indicates that the regeneration blastema changes its requirement for nerves at about this same stage in regeneration (Singer 1952). Regeneration up to this point is absolutely dependent on a nerve supply, such that denervation of the limb inhibits regeneration. At the medium to late bud blastema stages, the regenerate is no longer completely dependent on nerves, and will complete the process of differentiating the missing structures without a nerve supply, although the final growth of the regenerate is reduced, resulting in a normally patterned but small regenerate. Since developing limb buds are not dependent on nerves, it is likely that the transition from the regeneration-unique phase to the redevelopment phase occurs when the blastema becomes independent of the nerve for further development. It is important to recognize that the transition between these two phases is unlikely to be abrupt and that there is a period when cells at the proximal base of the blastema are still more stumplike than blastema-like, and at the same time, cells in the distal region of the blastema have acquired properties of limb bud cells.

The Key to Understanding Regeneration Is in Understanding the Mechanisms Controlling Phase I Events

It follows from the above that since the early events of regeneration are unique, these could be the barriers to regeneration that have been hypothesized to exist (T. L. Muller et al. 1999). It is possible that assisting an amputated limb to progress through the early stage and form a blastema would be sufficient to enable it to access the redevelopment phase leading to complete limb regeneration. If this is the case, then the first challenge to inducing regeneration is to understand the mechanisms controlling phase I. At least two critical issues need to be understood: the origin of the blastema cells that carry on during phase II to form the regenerated limb, and the cellular and molecular mechanisms that generate this population of blastema cells.

Origin and Genesis of Blastema Cells

Given the amount of effort that has gone into identifying the source of regeneration progenitor cells and how blastema cells interact during phase II to control growth and pattern formation, it is surprising how little is known about the mechanisms whereby blastema cells are generated from progenitor cells. As discussed below, limb connective tissues, particularly the dermis of the skin, are a major source of blastemal cells, although cells from other tissues also contribute to the blastema cell population.

Origin of Blastema Cells

As a starting point, we consider one series of experiments particularly noteworthy, since it established that the source of blastema cells is within a few millimeters of the amputation plane (fig. 12.4).

When the body of a salamander is shielded, while the limb is x-irradiated to inhibit limb cells from proliferating, the limb fails to regenerate when amputated through the x-irradiated tissues (Wallace 1981). Thus, regeneration cannot be rescued by circulating cells from sites that are remote from the limb, or by activated stem cells from a location that is distant from the amputation site. In contrast, when the distal portion of the limb is shielded and the proximal regions are x-irradiated, limbs that are amputated through the distal unirradiated tissues do regenerate without contribution from the more proximal irradiated tissues.

Given that blastemal cells arise from cells at or near the amputation plane, it is noteworthy that after a limb has regenerated, it can be re-amputated and the process of regeneration is repeated exactly as before. Thus, limb progenitor cells persist locally within the regenerated limb, and by definition are self-renewing, making blastema precursor cells in the mature urodele limb at least functionally equivalent to limb stem cells (Blau et al. 2001; Flake 2001; Marshak et al. 2001). What is learned from studies of mammalian stem cells, particularly those present in adults, almost certainly

Figure 12-4 The cells that form the blastema and regenerated limb arise from tissues localized near the amputation plane. If a limb is x-irradiated locally (shaded circular region at the elbow), with the rest of the limb and body of the animal shielded, the limb will regenerate when amputated through a shielded region located either proximal or distal to the irradiated tissues. Limbs that are amputated through the irradiated region fail to regenerate, and thus regeneration is not rescued by cells located at a distance from the amputation plane (see Wallace 1981).

will accelerate progress toward understanding urodele limb regeneration. Conversely, understanding the biology of urodele blastema cells likely will contribute to a further understanding of stem cell biology.

If not the most critical issue, certainly a place to start is the identification of the source(s) of blastema cells. At the outset, it is important to recognize that, as in limb development, there are expected to be two distinct populations of blastema cells. One population consists of cells equivalent to the undifferentiated mesenchymal cells in the distal region of the limb bud (progress zone [PZ]), which function in the control of growth and pattern formation during limb development (Summerbell et al. 1973). The other population consists of cell types that migrate into the regenerating limb in response to the positional clues provided by the first population. A number of studies of developing limbs have demonstrated that precursors for blood vessels and muscle (summarized below), as well as nerve fibers, are in the second category. The cells of the PZ, once their patterning role is complete, contribute primarily to connective and skeletal tissues but not to blood vessels or muscles (S. Li and Muneoka 1999). Similar studies have yet to be conducted in regenerating limbs, but we anticipate that the results will be equivalent to those in developing limbs. The fact that a normally patterned limb can be regenerated without muscle (Dunis and Namenwirth 1977; Holder 1989; Lheureux 1983) is comparable to results from developing limbs (Kieny and Chevallier 1979) and lends support to this idea.

The question of where the cells of the blastema come from has attracted considerable attention; however, for a variety of reasons, mostly involving technical limitations, it remains largely unanswered. It is unlikely that significant progress in mechanistic studies can be made until the critical cell populations have been unambiguously identified and characterized. At this point, the evidence suggests that all tissues of the mature limb contribute cells to the regenerate to some extent; no data indicate extensive metaplasia or transdifferentiation during normal limb regeneration (M. R. Carlson 1998; Gardiner and Bryant 1998; Mescher 1996). Most studies of the origin of blastema cells have involved the grafting of marked tissues or cells into limbs or blastemas, and then following the fate of these cells into the regenerated tissues. This approach is limited by the fact that most tissue grafts are composed of multiple cell types. In particular, most tissues contain connective tissue fibroblasts, making it very difficult to identify with certainty the cell type that was the source of the labeled cells observed in the regenerated limb. The earliest studies on this topic did not utilize cell autonomous markers to determine lineage relationships, but rather deduced relationships based on histological observations of the cells in the stump. In the most detailed of these studies, it was inferred that tissues contribute cells to the blastema in direct proportion to their abundance in the stump (Chalkley 1954; Hay and Fischman 1961). Subsequent lineage studies demonstrated that, despite appearances, this is not the case. Some cell populations, dermal fibroblasts in particular, overcontribute. Others, such as chondocytes and perichondrial cells, significantly undercontribute (Muneoka et al. 1986b). Taking all available data into account, cellular contribution to the blastema is as follows.

Nerves and Blood Vessels. The regenerate contains both nerves and blood vessels that are continuous with more proximal structures in the stump. Several studies have reported that the early blastema is poorly vascularized, with blood vessels eventually growing distally into later-stage blastemas from preexisting vessels in the stump. However, a recent study demonstrated that revascularization occurs much earlier during the regeneration process, by in-growth of blood vessels whose cells are derived from severed vessels at more proximal levels in the stump (Rageh et al. 2002).

Preexisting axons in the stump, severed at amputation, initially regress for a short distance proximally before regenerating rapidly to innervate the early blastema and overlying epidermis. Since the cell bodies of axons in the blastema reside within the central nervous system, these cells obviously are not considered part of the blastemal cell population. The source of the regenerated ensheathing Schwann cells is unclear, but they are presumed to arise from preexisting Schwann cells in the stump that accompany the regrowing axons into the blastema. When the nerves initially regress, they leave behind a connective tissue sheath composed of fibroblasts and Schwann cells. These cells appear to proliferate, to contribute to the blastema (Chalkley 1954, 1959; Hay and Fischman 1961), and to associate with the regenerating axons to re-form the nerve sheath.

As noted earlier, the first phase of regeneration is dependent on the presence of an adequate nerve supply. If the nerve supply is severed proximally, at the base of the limb, regeneration at a distant limb site is inhibited. It is has been proposed that proximal denervation results in the release of inhibitory factors, perhaps from the Schwann cells that remain after the axon dies (Irvin and Tassava 1998), which prevent regeneration. In proximal denervation experiments, the cut nerve eventually regenerates to the amputation surface, at which time the proposed inhibitory influences would be neutralized, allowing delayed regeneration to resume. According to this view, regeneration inhibitory factor(s) would also be released after simple amputation as a result of the limited local regression that occurs normally at amputation, and such inhibition would be quickly overcome by the regrowth of the axons.

As in most cases, whether a process requires activation or release from inhibition (or both) is difficult to determine. In contrast to the disinhibition view just presented, an argument could be proposed that nerves provide a stimulatory factor(s) needed for regeneration. In the regeneration literature on the role of nerves, activation is in fact the dominant hypothesis, and has led to the identification of several candidate neurotrophic factors. Since it is known that growth is controlled by interactions between blastema cells, such a factor(s) could function as a growth-permitting factor. Below a threshold level of nerves, growth does not occur; above that threshold, normal growth occurs. The amount of growth above the threshold is not influenced by variation in abundance of the nerve supply (Muneoka et al. 1989).

The target cells of the neurotrophic factor have not been extensively characterized, though at least one population of responsive cells can be identified immunohistochemically (Fekete and Brockes 1987). The nerve-dependency of muscle is well known and has been studied in several systems. Denervated and injured muscle fibers undergo extensive degeneration, which is presumed to account for the muscle fiber degeneration observed when amputation both severs and locally denervates muscle fibers (Jo-Ann Cameron, Department of Cell and Structural Biology, University of Illinois, pers. comm.; Echeverri et al. 2001). In the induction of accessory limbs (discussed below), nerves are required for the accumulation of cells with the potential to form an accessory limb, and therefore connective tissue cells are also a likely target for a positive-acting factor derived from nerves.

Skeletal Tissues. The skeletal elements of the limb stump (depending on position within the element) consist of chondrocytes and perichondrial cells or osteocytes and periosteal cells, as well as adherent connective tissue cells (fibroblasts). In response to amputation there is histolysis of bone and cartilage. Early descriptive studies concluded that the cells contributing to the blastema were derived from the peripheral tissues, the periosteum or perichondrium, rather than from the centrally located chondrocytes or osteocytes. However, when adherent connective tissues are removed from skeletal elements (perichondrium and chondrocytes) prior to grafting into limb stumps to assess their relative contributions to the blastema, skeletal tissue contribution was far less than would be expected based on the proportion of cells they represent in the stump (Muneoka et al. 1986c).

The fact that the limb skeleton can be regenerated entirely from cells derived from connective tissue fibroblasts is further evidence that skeletal tissues have a relatively minor role in contributing cells to the blastema and subsequent regenerate. Limbs in which contribution from all tissues other than the dermis has been prevented by x-irradiation still regenerate a normal skeletal pattern (Dunis and Namenwirth 1977; Holder 1989; Lheureux 1983). Limbs that have had a bone removed prior to amputation still form a normal skeleton distal

to the amputation, even though the missing skeletal element is not re-formed in the stump proximal to the amputation plane (Goss 1969a). However, it is also the case that grafting an additional skeletal element into the stump induces the regeneration of an additional element in the regenerate that is continuous with the extra element proximally (Goss 1969a). The skeletal tissues of the mature limb are thus not entirely without influence on the regenerated skeletal pattern.

Muscle. The fate of skeletal muscles has been a subject of great interest over the decades, particularly in the past few years. Understanding the behavior of muscle-associated cells during regeneration has been complicated by the fact that muscle is a complex tissue composed of multiple cell types, including myofibers, nerve-associated Schwann cells, vascular tissues, connective tissue fibroblasts, and muscle stem cells (satellite cells/postsatellite cells). This issue has attracted attention recently because of several studies of the behavior of cells from a newt muscle cell line *in vitro,* as well as after grafting into the blastema in a fashion analogous to earlier contribution studies (Brockes 1998; Kumar et al. 2000; Lo et al. 1993; E. M. Tanaka et al. 1997). This cell line, newt A1, was derived from explants of mature limb tissues and can form myotubes *in vitro.* These multinucleate myotubes can contribute to regenerated limbs if they are grafted into a blastema, where they reenter the cell cycle and fragment to give rise to mononucleated cells.

The behavior of newt A1 cells *in vitro* is similar to that of the well-characterized mouse C2C12 myogenic cell line. These cells were originally derived from satellite cells, which are myogenic stem cells associated with mature muscle fibers (Lattanzi et al. 2000). Although the origin of the newt A1 cells is not known, they may be derived from postsatellite cells, which are the myogenic stem cells in adult urodeles hypothesized to be equivalent to satellite cells in mammal muscle (see J. A. Cameron et al. 1986; B. M. Carlson 2003). Myogenic newt A1 cells can be induced to form myotubes *in vitro* that will then respond to serum stimulation by entering S phase. *In vitro* these cells arrest in the cell cycle prior to mitosis; however, when implanted into blastemas, they give rise to mononucleated cells by an unknown mechanism that appears not to require cytokinesis (Velloso et al. 2000). When implanted into blastemas, mononucleate cells contribute mostly to regenerated muscle; however, some labeled cells also become incorporated into regenerated cartilage, suggesting that a limited degree of transdifferentiation can occur (Lo et al. 1993).

An essential caveat to the observation of infrequent transdifferentiation or stem cell conversion is that the biological relevance of such observations is uncertain (Blau et al. 2001). Additionally, recent studies of multipotent cells from adult mammalian tissues have demonstrated that cell-cell fusion can occur to form heterokaryons that appear to be evidence that transdifferentiation has occurred (Blau 2002). It is possible that cell fusion in response to injury may be a normal occurrence in tissues (Blau 2002), particularly in the case of muscle regeneration, where fusion of mononucleate cells (myoblasts) to form multinucleate cells (myotubes/myofibers) is an essential feature of the developmental program. Thus, newt myofibers appear to respond to amputation by giving rise to myogenic cells that constitute one source of progenitor cells for muscle regeneration.

In addition to mononucleate cells derived via fragmentation of mature myofibers, regenerated muscle also forms from the progeny of postsatellite cells. Satellite cells are myogenic stem cells present beneath the external lamina of skeletal muscle fibers in many vertebrates, as well as in larval urodeles (see J. A. Cameron et al. 1986). Satellite cells in mammals are the source of myogenic stem cells in adult muscle (M. Li et al. 2000; Pastoret and Partridge 1998) and the source of cells for the mouse myogenic C2C12 cell line (Lattanzi et al. 2000). During metamorphosis of larval urodeles, it appears that satellite cells become enveloped in their own external lamina adjacent to the external lamina of the myofibers to give rise to the postsatellite cells, which are considered to be functionally equivalent to the satellite cells of other adult vertebrates (see J. A. Cameron et al. 1986). Postsatellite cells respond to injury by incorporating 3H-thymidine, proliferating in culture, and fusing to form new myotubes, in a fashion comparable to newt A1-derived myotubes. Myotubes *in vitro* derived from postsatellite cells express both blastema cell and myoblast-specific antigens, consistent with what is observed *in vivo*. As in other vertebrates, a major source of regenerated muscle in urodeles is a population of stem cells that are intimately associated with differentiated myofibers.

Regardless of the relative contribution of cells from these two sources for generating muscle precursor cells, it is evident that regenerated muscle arises from cells present in pre-existing muscle tissues (see B. M. Carlson 2003). Thus, during regeneration, muscle arises from a discrete lineage that is separate from other limb tissues, as it does during limb development. During development, limb myoblasts originate in the somites and migrate into the limb bud after it is relatively well formed. Experiments that inhibit the migration of muscle progenitor cells result in the development of muscle-less limbs (Kieny and Chevallier 1979). These limbs are equivalent to the regeneration of muscleless limbs resulting from x-irradiation of the limb to block cellular contribution from stump tissues. Grafts of skin from unirradiated limbs will rescue regeneration; however, the limbs that form lack muscle, even though they have a normal skeletal pattern and

contain tendons, connective tissues, nerves, and blood vessels (Dunis and Namenwirth 1977; Holder 1989; Lheureux 1983).

Connective Tissue Fibroblasts. Cells that form the connective tissue of the dermis and surround muscles, nerves, and blood vessels are collectively referred to as fibroblasts. During regeneration, growth and pattern formation are coordinately regulated by interactions between cells derived from connective tissues of the amputated stump, which are presumed to be the fibroblasts. Of all cell types, limb fibroblasts have the greatest influence on regeneration both in terms of contribution of cells to the blastema and in the control of growth and pattern formation in the regenerating limb (see Bryant et al. 2002; Mescher 1996).

As discussed above, when examined histologically, all tissues of the stump appear to contribute cells in proportion to their availability in the stump (Chalkley 1954, 1959; Hay and Fischman 1961); however, cell lineage data indicate this is clearly not the case (Muneoka et al. 1986c). Whereas some tissues contribute relatively few cells to the blastemas (e.g., cartilage), dermal fibroblasts significantly overcontribute. The progeny of dermal fibroblasts account for between 19% and 78% of the cells of the early blastema (42% on average), even though dermal fibroblasts represent less than 20% of the cells of the stump, suggesting that this population is subject to selective expansion in the blastema. Since dermal fibroblasts account for about half of all fibroblasts in the limb (Tank and Holder 1979), it is possible that essentially all of the early blastema cells are derived from fibroblasts, half from the dermis and half from elsewhere in the stump, even though fibroblasts in total account for less than half of the cells of the mature limb. As in the PZ of developing limb buds (S. Li and Muneoka 1999), it is likely that the early blastema and the distal region of the later-stage blastema are populated primarily by cells that are actively involved in growth and pattern formation. These cells would be the progeny of connective tissue fibroblasts, and their progeny would contribute to the connective tissues and skeleton of the regenerated limb. Cells that contribute to regenerated muscle would be derived from injured muscle fibers in the stump and would be expected to migrate distally within the blastema.

In addition to overcontributing to the blastema, fibroblast-containing tissues are the only mature limb tissues that influence growth and pattern formation during regeneration. It also appears that among the fibroblast-containing tissues, the dermis has a particularly dominant effect, which presumably is a consequence of fibroblasts being the predominant cell type in this tissue. Supernumerary outgrowths can be induced by grafts of skin (Bryant et al. 1987), and the pattern of the final regenerate is determined by the orientation of the grafted skin, rather than that of the stump (see Muneoka et al. 1986c). In addition, as discussed above, studies of muscleless limbs have demonstrated that a normal limb pattern composed of all limb tissues (other than muscle) can be regenerated from grafts of skin as the sole source of progenitor cells for the blastema. We conclude that the fibroblast-derived cells of the blastema form a blueprint that guides the migration and growth of nerves, blood vessels, and myogenic cells. A comparable blueprint in the interdigital connective tissue has been hypothesized to control digit identity in the developing chick wing (Dahn and Fallon 2000). Pattern formation in other organs of the body may also be controlled by the connective tissue matrix during both development and regeneration.

The recognition of fibroblasts, and dermal fibroblasts in particular, as the cell type from which the majority of the early blastema cells are derived is of considerable consequence to the design of future experiments to study the mechanisms controlling regeneration. At present it is unclear whether there is a population of quiescent stem cells in the dermis that are activated during regeneration, or whether dermal fibroblasts lose their differentiated phenotype and give rise to the relatively undifferentiated blastema cells (discussed in the next section). If there is a quiescent stem cell population for limb regeneration, it can be isolated from the dermis. Likewise, if dedifferentiation is the mechanism for induction of limb progenitor cells, then connective tissue fibroblasts must be responsive to the growth and differentiation factors that are produced in response to amputation of the limb.

Mechanisms of Genesis of Blastema Cells

Cells of the connective tissues, the fibroblasts, form the major source of blastema cells. Fibroblasts do not have a highly specialized appearance and have not been well characterized in any system. However, recent gene expression studies in mammals have begun to provide insights into the complexity and diversity of fibroblast differentiation and function (H. Y. Chang et al. 2002). In urodele amphibians, it is not clear whether all fibroblasts are capable of contributing to the blastema, or whether there is a subpopulation that function as blastema cell progenitors. Despite this paucity of data regarding fibroblast phenotypes and function, it has been assumed historically that blastema cells arise as a result of dedifferentiation, both of fibroblasts and of the other cell types that eventually contribute to the blastema. The term "dedifferentiation" has been used loosely, and appears to mean different things to different authors (M. R. Carlson 1998). It is frequently used to refer to the reversal of differentiation and reversion to a less differentiated state, as if it is known with certainty that this process occurs, though in reality it is inferred but not proven. The fact that a quiescent

stem cell population in the connective tissue has not been identified is neither evidence against the existence of such cells nor evidence for the existence of dedifferentiation, because there has not yet been a concerted effort to look for such a population. "Nonseeing is not the same as nonbeing" (Blau et al. 2002).

The recent recognition of the plasticity of differentiated cells and the existence of multipotent cells in adult tissues have not yet been integrated into the regeneration field as alternatives to the idea that dedifferentiation is the dominant developmental mechanism in the genesis of blastema cells. When dedifferentiation was first proposed as a mechanism for generating blastema cells (see Wallace 1981), cells in adult tissues were considered to be terminally differentiated, and the developmental fate of their nuclei to be committed and irreversible. Although it has been known for some time that this is not the case, the extent to which cells in differentiated tissues are able to exhibit dynamic cellular behaviors and patterns of gene expression in response to changes in the extracellular environment they encounter has been appreciated only recently. C2C12-derived myotubes provide a well-characterized example of such developmental plasticity. In addition to dedifferentiating to give rise to mononucleate myogenic progenitor cells, C2C12 myotubes can also give rise to cartilage and adipocytes depending on exposure to different differentiation factors (see Odelberg et al. 2000). Another example is the fibroblast phenotype of skin fibroblasts, which can be altered in response to stress or injury to give rise to myofibroblasts that assume a myocyte phenotype and express some smooth muscle marker genes (see Lorena et al. 2002).

Aside from the issue of developmental plasticity of cells in adult tissues, it is clear that blastema cells are not a homogeneous population of undifferentiated cells. Although they exhibit a homogeneous cytology, their heterogeneity can be detected immunohistochemically using monoclonal antibodies as cell markers (see Géraudie and Ferretti 1998). In addition, classic studies involving induction of supernumerary growth and pattern have repeatedly demonstrated that cells from different regions of a blastema have distinctly different positional properties relative to each other and to the limb stump. This positional information is an encoding of relative spatial order along the proximal-distal, dorsal-ventral, and anterior-posterior axes of the limb, and is inherited from precursor cells in the stump. The positional information that characterizes different populations of blastema cells determines their behavior and, as a result, controls the growth and pattern formation that are essential for regeneration.

The existence of positional information is demonstrated by induction of growth and pattern formation in response to interactions between cells from different positions within the

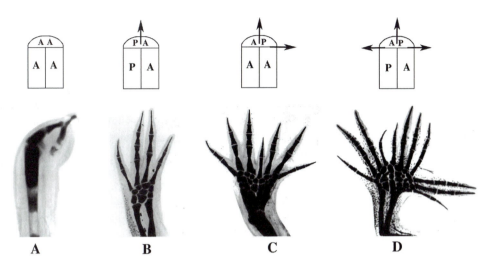

Figure 12-5 Growth and pattern formation are coordinately regulated. Cartoons at the top indicate the relationship between cells derived from anterior (A) and posterior (P) positions within the limb. In each situation, a blastema is located distal to the stump tissues, and the arrows indicate the positions where outgrowth occurs. The corresponding limb patterns (Victoria blue staining) that are formed are illustrated below each cartoon. The normal limb (B), when amputated, forms a single outgrowth giving rise to a normal limb pattern. When a blastema is grafted to the contralateral limb so as to appose A and P axes (D), two supernumerary outgrowths are induced at the blastema-stump boundary, resulting in three times as much growth and pattern. Limbs can be created surgically that contain less than the normal amount of positional information (e.g., a double anterior limb in A). These limbs exhibit less growth and less pattern formation. An intermediate amount of growth and pattern formation is induced when a normal blastema is grafted on a double anterior limb stump (C), which induces two centers of outgrowth. (Modified from Gardiner and Bryant 1998.)

limb. This growth and patterning response is referred to as intercalation, and it can occur along the proximal-distal and transverse limb axes, resulting in either restoration of positional continuity along the proximal-distal axis (regeneration) or induction of supernumerary pattern along the transverse axes (Bryant et al. 1981; French et al. 1976). In both instances, growth and pattern formation come about as a consequence of short-range interactions between cells with different positional information. Interactions between cells that express the same positional information do not stimulate growth or pattern formation (fig. 12.5). Although the molecules that confer these positional properties have not been identified, there are some candidates of interest. Cell-sorting experiments demonstrate that cells from different limb positions express different cell surface molecules. This has been directly demonstrated for cells along the proximal-distal axis (da Silva et al. 2002; Nardi and Stocum 1983), and similar properties have been attributed to the transverse axes of the limb stump (Stocum 1996). If proximal and distal blastema cells are juxtaposed in culture, the proximal partner engulfs the distal. This behavior is encoded, at least in part, by cell surface, GPI-linked molecules (da Silva et al. 2002), as has been demonstrated for equivalent adhesive interactions in developing chick limb bubs (Wada et al. 1998). One candidate molecule, Prod 1, the newt ortholog of CD59, has been isolated from and characterized in blastemas, and appears to have an appropriately graded distribution relative to the proximal-distal limb axis (da Silva et al. 2002).

Fibroblasts have not received as much research attention as more exotic cell types, maybe because they have a generic, nondistinguishing morphological appearance (spindle-shaped) and do not express markers for other cell lineages (H. Y. Chang et al. 2002). As an appreciation develops that this simple appearance masks a rich diversity of specializations, including the encoding of unique information about position within the body, this situation will change. Advances in gene expression profiling have recently enabled a much richer portrait of fibroblasts as distinctly different cells from different tissues, as well from different spatial locations within a given tissue (H. Y. Chang et al. 2002). Thus, fibroblasts are differentiated cells that exhibit an impressive degree of developmental and physiological plasticity. By this view, fibroblasts are not stem cells, nor do they dedifferentiate to give rise to blastema cells; rather, fibroblasts are blastema cells. They have the inherent ability to give rise to many of the cell types of the limb in response to appropriate developmental signals. For example, in the classic studies of ectopic bone induction by implantation of BMP into connective tissues, the target cells are presumed to be fibroblasts that are responsive to the BMP signal as a differentiation factor. It is has not been presumed that fibroblasts in the connective tissue first dedifferentiate and then transdifferentiate into chondrocytes and osteocytes. Similarly, dissociated cartilage can yield undifferentiated fibroblastic cells that express type I collagen, a marker for connective tissue fibroblasts. These cells can be induced to give rise to chondrocytes expressing type II collagen, which is a cartilage marker, and ultimately the hypertrophic cartilage marker, type X collagen, and the osteoblast marker, Cbfa1 (Reiter et al. 2002). We presume that urodele fibroblasts would exhibit a similar diversity of responses to signaling molecules that control cell fate and differentiation.

We emphasize regarding the genesis of blastema cells that currently there is a critical lack of understanding of the biology of fibroblasts in general and urodele limb fibroblasts in particular. Regardless of the specific molecular events involved, fibroblasts are the critical target population of the signals controlling the genesis of the limb blastema. The key to identifying these signals and understanding how those signals might be used to induce human limb regeneration lies in understanding the response of limb fibroblasts to those signals. A critical area of needed research thus involves the isolation, purification, and characterization of this population (these populations) of cells.

Signals Controlling the Genesis of Limb Blastema Cells

Several experimental models provide an opportunity to identify and test the function of the signals controlling the early events of regeneration leading to formation of the blastema. Some are based on relatively recent discoveries, others on classical studies that can be combined with modern techniques of gene discovery and functional analysis. There is a wealth of data from classical studies that provide a number of fascinating experiments that can be used to collect important molecular data. Such studies have already led to new insights about mechanisms of regeneration, particularly with regard to the transition from limb cells to blastema cells.

Although the earliest events after amputation are common to both regeneration and normal wound healing, within hours after amputation there begins a cascade of events that is specific to limb regeneration. Collectively these events result in the generation of the population of proliferating blastema cells that is usually described as being generated by "dedifferentiation." Although dedifferentiation may or may not represent a specific biological mechanism for the genesis of blastema cells (discussed above), it is a useful descriptive term for this phase of regeneration that represents the transition from the mature limb tissues to the blastema. An early molecular marker associated with dedifferentiation, as distinct from wound healing, is the re-expression of *Hoxa9/Hoxa13*, which begins at about 24 hours postamputation. This molecular event precedes by many days the morphologically observable changes in stump tissues that will eventually lead to the accumulation and proliferation of blastema cells. These subsequent events include tissue histolysis, accumulation of immunoreactive cells, degradation of

the extracellular matrix, fragmentation of myotubes to give rise to myogenic mononucleate cells, activation of myogenic postsatellite cells, migration of connective tissue fibroblasts leading to accumulation of cells distally, thickening of the WE to form the aAEC, and the onset of blastema cell proliferation. Each of these events has been studied to some degree, and the eventual understanding of regeneration will require an integrated model to account for all these critical events.

Signals from the Apical Epidermis

Although epidermal cells do not contribute directly to the blastema, a covering of a specialized apical epidermis is critical for the success of regeneration (Wallace 1981). Epidermal cells function to enable outgrowth and may also function in the control of pattern formation, as is the case during limb development. Treatments that affect formation of the WE or the AEC alter the course of regeneration. For example, if formation of the WE is inhibited by a graft of mature skin over the amputation surface, regeneration is inhibited. Removal of the WE or AEC after it has formed also inhibits regeneration. Conversely, experiments that relocate the position of the AEC induce limb outgrowth at the new position (Thornton 1968).

It is presumed that critical interactions occur between the WE and the underlying stump cells. Expression of several genes by stump cells provides visual evidence for the necessity of such interactions (fig. 12.6). When mature skin is grafted over the end of the amputated stump, re-expression of a number of regeneration genes does not occur. Typically, there are regions where a WE forms adjacent to the grafted skin, and the genes are re-expressed in a localized region of the stump that is in contact with the WE. Since gene expression is not induced in adjacent cells that are in contact with mature skin, the interactions between the WE and stump cells are mediated by short-range signals and require a close approximation between these two cell types. A new basal lamina is not re-formed until relatively late during regeneration, which would allow for critical epithelial-mesenchymal interactions during all the early stages of regeneration.

Although the signals mediating interactions between stump and epidermis are not well understood, it is likely that FGFs are important during regeneration, as they are during limb development. Several *fgfs* have been shown to be expressed in regeneration (Géraudie and Ferretti 1998), including *fgf4, fgf8,* and *fgf10* (Christensen et al. 2002; Han et al. 2001). In addition to functioning in the control of cell proliferation, FGFs also function as chemoattractants and stimulate migration of limb bud cells toward the apical epidermis (S. Li and Muneoka 1999). Since one of the early cellular events of regeneration is the migration of dermal fibroblasts

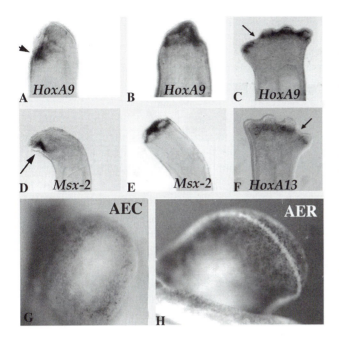

Figure 12-6 Interactions between the apical epidermis and underlying mesenchymal cells. *Hoxa9*, *Hoxa13* and *Msx2* are expressed in stump tissues underlying the apical (wound) epidermis (B, F, and E respectively). If mature skin is grafted over the end of the amputated stump, expression of these genes is not induced distally (A and D, data for *Hoxa13* not shown), but is induced where there are gaps between the grafted skin and the stump (arrows). Similarly, when limbs are amputated near the base of the digits, there often are regions where the mature skin is not removed by amputation, and expression of these genes is not induced (arrows in D and F). The apical epidermis is distinct from the adjacent epidermis in that the cells are not proliferating, and thus do not incorporate BrdU. Nonproliferation is observed in both regenerating limbs (G, white area in this view of the distal end of a regenerating limb prior to blastema formation) and developing limbs (H, white stripe representing the apical ectodermal ridge of a chick limb bud).

and other stump cells distally toward the WE (Chalkley 1959; Gardiner et al. 1986), FGF could be an early signal controlling this cell migration event. Cell migration precedes onset of cell proliferation, and it has been hypothesized that proliferation occurs as a result of cells with different positional properties contacting one another under the WE (Gardiner et al. 1986). Cells of the apical epidermis itself do not proliferate (fig. 12.6G), which may be related to their unique function in signaling to the underlying stump cells. Since changes in the length of the cell cycle have been shown to be able to regulate patterns of gene expression (Ohsugi et al. 1997), the lack of proliferation may be critical to the new function of the epidermal cells.

Signals Associated with Establishing the Distal Tip
Observations of *HoxA* expression indicate that one of the first events of regeneration is the respecification of the distal tip of the limb, as discussed earlier (fig. 12.3). Since these genes are not expressed in lateral, nonregenerating wounds,

we hypothesize that they are expressed in response to signals from the WE. Based on molecular and genetic evidence in other vertebrates, specification of the distalmost region of the pattern (autopod) is a consequence of the coexpression of 3′ and 5′ members of the HoxA and HoxD complexes (A. P. Davis et al. 1995; Fromental-Ramain et al. 1996b; Gardiner et al. 1995; Haack and Gruss 1993; Small and Potter 1993; Yokouchi et al. 1991). Early coexpression of *Hoxa9* and *Hoxa13* in stump cells indicates that one of the earliest events in regeneration is the reestablishment of the distalmost part of the limb, regardless of the level of amputation. The more proximal regions of the pattern arise subsequently as a consequence of growth and pattern formation induced by interactions between the distal tip and the cells of the more proximal stump via a mechanism referred to as intercalation (see Gardiner and Bryant 1998). Early establishment of the distal tip ensures that the regenerated tissues will always be an exact replacement of the portion of the pattern that has been removed by amputation.

Intercalation has been studied extensively in classic grafting and cell contribution studies. It is stimulated by interactions between cells with different positional identities, and results in the replacement of the tissues that are missing between the two interacting populations of cells. A basic feature of intercalation is that the interacting cells both signal and respond to each other, although not necessarily equally. In the context of the intercalation that occurs during normal limb regeneration, the most relevant experiments involve interactions between cells from different levels along the proximal-distal limb axis (fig. 12.7). A graft of a distal blastema (amputation at the level of the autopod) onto a proximal stump (amputation at the level of the stylopod) induces the formation of the missing zeugopod in a polarized fashion. All the intercalated tissues are derived from cells of the proximal stump, which is consistent with the principle of regeneration that limb cells can only generate new pattern that is more distal than their level of origin (see Wallace 1981). The distal blastema thus signals the proximal stump cells to proliferate and regenerate (intercalate) the missing intermediate structures. The presence of such a signal is confirmed by grafting a very late-stage distal blastema or a mature hand onto a proximal amputation stump (fig. 12.7B), in which case intercalation by proximal stump cells is not stimulated. Since the proximal stump cells have the ability to respond to distal signals when they are present, it is evident that such signals are produced early on and throughout much of regeneration but cease as regeneration is completed and are not present in mature tissues. At this point, it appears that the induction of distal HoxA expression by the WE is the earliest indication of the establishment of a distal fate, and hence the earliest time at which distal signals are present.

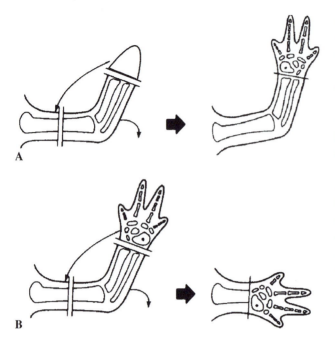

Figure 12-7 Signals from distal limb tissues induce growth and pattern formation in proximal cells. (A) When a distal blastema is grafted to a more proximal level (arrow indicates that the blastema is removed surgically, the limb is amputated proximally, the amputated piece is discarded, and the blastema is grafted onto the stump), the proximal tissues regenerate the missing structures. (B) When a late digit blastema or mature hand is grafted proximally (as described for panel A), the stump tissues do not regenerate the missing structures.

Signals from the Nerves

The importance of nerves in regeneration has long been recognized, but the molecular mechanisms mediating neuronal influences are still largely unknown. Nerves are severed during amputation; after a brief period of regression, they begin to regenerate rapidly into the stump tissues at the amputation plane, and they subsequently innervate the blastema and overlying epidermis. If the limb is denervated during the early stages of regeneration, the limb fails to regenerate. This observation has led to the hypothesis that nerves produce a neurotrophic factor required for the initiation and progression of the early stages of regeneration. Denervation of regenerating limbs at later stages inhibits further growth of the regenerate, but not the differentiation of a normally patterned, but diminutive, limb.

The production of factors required for regeneration may be a normal function of nerves. Alternatively, this function of nerves may be acquired in response to amputation and subsequent interactions with cells that are mobilized during regeneration, such as fibroblasts and the wound epidermis. Regenerating axons grow distally, innervate the apical epidermis during the early stages of regeneration, and presumably respond to signals in both the blastema and the wound epidermis that guide their pattern of regrowth and innerva-

tion. Since the behavior of nerves during regeneration has not been well characterized at the molecular level, the nature of such interactions cannot be evaluated at the present time.

Understanding the signals produced by nerves is critical to devising strategies to induce regeneration in humans, since nerves are the source of factors that are required for the recruitment and proliferation of cells that generate the blastema. Progress in this area of research has been limited by the lack of efficient and appropriate functional assays. Typically experiments on signals produced by nerves involve denervation of the limb, which inhibits regeneration, followed by application of test treatments for their ability to rescue regeneration. Rescue is generally limited in nature and rarely results in normal regeneration (however, see Mullen et al. 1996). Several factors have been identified that have effects that are characteristic of the neurotrophic factors (reviewed extensively by Mescher 1996).

Signals from the Extracellular Matrix (ECM)

During the early stages of regeneration, the level of ECM degrading enzymes increases (Dresden and Gross 1970; Grillo et al. 1968; Miyazaki et al. 1996; E. V. Yang et al. 1999). Presumably this is required to release cells from the matrix to enable them to migrate distally to form the blastema. A number of matrix molecules have been identified from urodele limb tissues and from blastemas (reviewed extensively by Géraudie and Ferretti 1998; Mescher 1996). In addition to providing a substrate for cell migration and a matrix for holding tissues together, the ECM also is a reservoir of matrix-associated signaling molecules that could function in controlling regeneration. For example, in developing limbs, the ECM plays an active role in modulating signals controlling growth and pattern formation. The positional identity of anterior limb bud cells is stabilized by ECM-associated members of the FGF family of growth factors; when those signals are inhibited, anterior cells are converted to posterior cells (Schaller and Muneoka 2001). Experimental implantation of ECM components that interfere with normal FGF signaling in the anterior induces conversion to a posterior positional identity, leading to growth and formation of supernumerary structures along the anterior-posterior limb axis.

Although the role of the ECM in modulating signaling during regeneration has not yet been investigated, results from the developing chick limb suggest a possible role for the ECM in encoding positional memory in the mature urodele limb. As discussed earlier, blastema cells express different positional identities in accordance with their position of origin in the limb. This information controls growth and pattern formation in the blastema, and is subsequently encoded in the limbs cells that differentiate at the end of regeneration. This positional memory is a stable property of limb cells, and is recalled if and when the limb is amputated again. The mechanism for storing positional signals could involve the ECM, since the ECM is a product of the connective tissue fibroblasts that build the limb pattern. If positional memory is encoded by growth and differentiation factors sequestered in the ECM, these signals would be released when the ECM is degraded during the initial stages of regeneration to provide stump cells with local information about their position.

Regeneration-Specific Signals

The expression of several genes in limb regeneration has been characterized (see Géraudie and Ferretti 1998). Typically, genes expressed during limb regeneration are expressed during limb development. In the case of genes that are expressed during the later, redevelopment phase of regeneration (phase II), the spatial and temporal patterns of expression are similar to those observed during development. For genes expressed during the unique preparation phase (phase I), their spatial and temporal patterns of expression may differ, but the function appears to be conserved (Gardiner et al. 1998).

To date, no regeneration-specific genes (i.e., not expressed at any other time in the life of the organism) have been identified. There is, however, at least one example of a gene (*Hoxc10*) that appears to respond to regeneration-specific signals, and there likely will be more, since the early stages of regeneration are quite different from development. Genes within the HoxC complex are involved in specification of positional identity along the rostral-caudal axis of vertebrate embryos. *Hoxc10* is expressed in developing hindlimbs and tails, but not in the forelimbs either of urodele larvae or of other vertebrate embryos (M. R. Carlson et al. 2001; Peterson et al. 1994). *Hoxc10* is however expressed at high levels in response to forelimb amputation in axolotls. Presumably there are elements in the promoter region of the axolotl *Hoxc10* gene that are responsive to signals present in forelimb regeneration but not in forelimb development. An analysis of the promoter of this gene would provide an opportunity to isolate and identify those signals.

Signals Controlling Myogenesis *in vitro* and *in vivo*

A model system that in recent years has proved informative involves the response of myogenic cells to regeneration signals *in vitro*. These studies have utilized both the mouse C2C12 cell line and the newt A1 cell line. In culture, the differentiation of C2C12 cells into myotubes can be reversed by ectopic expression of the transcription factor, *msx1* (Odelberg et al. 2000). *Msx1* expression induces multinucleated C2C12 myotubes to fragment, giving rise to proliferating mononucleated cells. These mononucleated cells can be induced to enter multiple differentiation pathways, as could

the original C2C12 cells, including myogenesis, adipogenesis, chondrogenesis, or osteogenesis depending on exposure to a variety of well-characterized growth and differentiation factors. This ability of myotubes to revert to a mononuclear phenotype is similar to that reported for newt A1 cells (discussed in detail above).

Signals are present in a protein extract of newt limb blastemas that are able to induce the same transition to proliferating mononucleate cells from C2C12 myotubes as does ectopic expression of *msx1* (McGann et al. 2001). The blastema extract can also induce entry into S phase in newt A1 myotubes *in vitro*, where they arrest. Newt A1 myotubes also enter S and arrest in the presence of serum. The response of C2C12 myotubes to blastema extract is significant in demonstrating the presence of a blastema-derived factor(s) that is involved in the control of the differentiated state. Since *msx1* is known to be involved in controlling this response, it is likely that the blastema-derived signals are operating upstream of the *msx* transcription factors.

Other studies have suggested an important role for *msx* in the regulation of regeneration in urodeles (M. R. Carlson et al. 1998; Crews et al. 1995; Koshiba et al. 1998; Simon et al. 1995) as well as mammals (Reginelli et al. 1995; Y. Wang and Sassoon 1995), and *msx* genes are expressed early in regeneration (M. R. Carlson et al. 1998; Koshiba et al. 1998). Several signaling molecules have already been identified as regulators of *msx* expression, including members of the FGF and Wnt signaling pathways (Bang et al. 1999; Bushdid et al. 2001; Yokoyama et al. 2001). In addition, both *msx1* and *msx2* are known indicators of BMP signaling (M.-F. Lu et al. 1999; Pizette et al. 2001; Scaal et al. 2002; Vainio et al. 1993). The C2C12 myotube assay has the potential to identify other signaling molecules from blastemas involved in the control of differentiation and growth. Members of the FGF, Wnt, and BMP signaling pathways are obvious candidates for further studies.

Strategies for Identifying Phase I Regeneration Signals

With the availability of new technologies in functional genomics that can be used in a wide range of organisms, it may finally be possible to discover the so-far elusive mechanisms that enable tetrapod limb regeneration. As discussed earlier, the important entry points are the early unique parts of the process, and the biology of the fibroblasts that generate the blastema. The goal of future studies is to understand how fibroblasts generate the limb blueprint that other cell types, such as muscle, nerve, and blood vessel, can respond to. Since each of the nonfibroblast components of the limb can regenerate individually, they cannot be the limiting step in successful regeneration. We hypothesize that once the blueprint is established, muscle, nerve, and blood vessels will respond appropriately to make a complete regenerate. The strategy for understanding how the blueprint is established and functions will require identification of all the molecular components of the regeneration process, and the ability to test the function of key components.

Microarray Analysis and Gene Expression Profiling

In recent years, molecular analyses have begun to provide insights into the mechanisms controlling regeneration. It is already clear that regeneration involves complex molecular interactions between multiple tissues, and thus we can expect that there will be many intersecting pathways involved, rather than just a small number of "regeneration genes." The challenge of understanding the biology of complex systems is hardly unique to regeneration biology. In recent years, techniques have become available to identify all the molecular components of a system or process, and to study the interactions between those components. Key to the success of such system approaches is the ability to combine the molecular catalogue with assays for relevant cell- and tissue-level properties.

Unlike most developmental processes, which are part of a continuum that starts at fertilization, regeneration is initiated at a specific moment by an external event. From an experimental point of view, this makes it possible to define a temporal time frame for regeneration, beginning at the moment of amputation, and to clone and identify all genes expressed in regeneration limbs. By identifying changes in expression at each stage of regeneration, it should be possible to identify the signals involved in each transition. To date, the axolotl expressed sequence tag (EST) project has cloned and identified nearly 7,000 nonredundant regeneration genes (unpublished data; Putta et al. 2004), a number that will continue to increase over the next few years. Consequently, the availability of cloned regeneration genes will not be a factor that will limit progress in understanding limb regeneration. The challenge will entail drawing on the wealth of information from classical studies to guide the identification of the function of this large set of genes.

Functional Analysis

Although it has become relatively routine to identify large numbers of genes associated with particular biological processes, it remains a considerable challenge, especially in nongenetic organisms, to test the function of a gene. The regeneration field has struggled to optimize techniques for introducing transgenes into urodeles. In spite of much effort, to date there have been few successes, and it has been difficult to alter expression in regenerating limbs so as to induce the dramatic phenotypic results comparable to those

achieved in organisms such as the fly, worm, chick, or mouse. The list of less successful methods includes injection of DNA into the blastema, lipofection, *in vivo* electroporation (but see below), and adenovirus and bacculovirus viral vectors. More successful techniques include biolistics, direct intracellular microinjection of plasmid DNA, and infection with replication-incompetent retrovirus. These techniques have been used to test the function of the different isoforms of the retinoic acid receptor, and most recently to identify retinoic acid target genes (Cash et al. 1998; Pecorino et al. 1996). In addition, vaccinia viral vectors have been used to test the function of *Shh* and members of the Wnt family of signaling molecules (Roy et al. 2000 and unpublished results). Vaccinia vectors introduce the transgene into virtually all cells in the vicinity of the injection site, and expression persists for at least one week in blastema cells and nearly two weeks in mature limb cells (fig. 12.8A, B). More recently, lentiviral vectors have been used to drive transgene expression in urodele limb and blastema cells (Ghosh et al. 2001) and offer the advantage of long-term expression without death of the host cell (fig. 12.8C). Finally, techniques for electroporation have improved dramatically in that past few years, and this technique is likely to become one of the most common method for regeneration studies *in vitro* and *in vivo* (fig. 12.8D).

Accessory Limbs as an Assay System for Regeneration Signals

Until recently, most attempts to alter regeneration have relied on methods to inhibit the process (e.g., by denervation of the limb) and then test for the ability of a particular agent to rescue inhibited regeneration. Interpretation of negative results in such inhibition experiments is challenging at best. Ideally, a preferred approach would enable the induction of regeneration in a regeneration-competent animal. The ability of urodeles to form accessory limbs from wounds on the sides of limbs, rather than the ends, provides this opportunity. The seminal experiments involving the induction of accessory limbs were performed by Bodemer in the mid-2oth century (Bodemer 1958, 1959) and have been expanded upon in recent years (Egar 1988; Maden and Holder 1984; Reynolds et al. 1983). As an assay for the signals that control wound healing, dedifferentiation, growth, and pattern formation, this model system offers the important advantage of testing for a positive response in an organism in which all the necessary components for limb regeneration are known to be present. This model system is also important in that it allows for the experimental dissection of the multiple steps occurring during the early stages of regeneration.

Wound Healing. If a piece of skin (epidermis and underlying dermis) is removed carefully so as not to induce damage to the underlying muscle and connective tissue, a WE forms and the skin is regenerated. The genes expressed in lateral wound healing are also expressed in regeneration and are part of the common early part of the pathway (fig. 12.9). Although lateral wounds do not give rise to outgrowths, the missing skin, including skin appendages such as glands, regenerates perfectly. The origin of the cells for skin regeneration is not known; however, we presume that they are dermal fibroblasts that migrate from the wound periphery, as occurs in early regeneration (Gardiner et al. 1986). Since these wounds do not form the scar tissue that would result from equivalent wounds in other adult vertebrates, lateral wound healing is comparable to the scarless wound healing observed in embryonic vertebrates, including mammals (P. Martin 1997). Understanding the mechanisms of urodele skin regeneration will be useful in guiding efforts to induce skin regeneration or to engineer a skin replacement in humans.

Genesis of Blastema Cells. If a nerve is surgically deviated to the site of lateral limb wound, an outgrowth is induced. Under the influence of appropriate additional signals (fig. 12.10A, B, C), the cells in this outgrowth are capable of forming an entire new limb. Thus, a lateral wound plus a deviated nerve provides an experimental model to study the origin of the blastema. We do not yet know which cells of the mature limb contribute to nerve-induced lateral blastemas, but assuming the outgrowths are equivalent to the early, nerve-dependent blastema, it is likely that fibroblasts from the dermis surrounding the wound contribute a large proportion of the cells. Without additional signals, nerve-induced lateral blastemas persist for a few weeks but eventually regress. It is possible that regression occurs because the epidermis fails to progress from a WE to the specialized, thickened AEC that is required for limb outgrowth during limb regeneration.

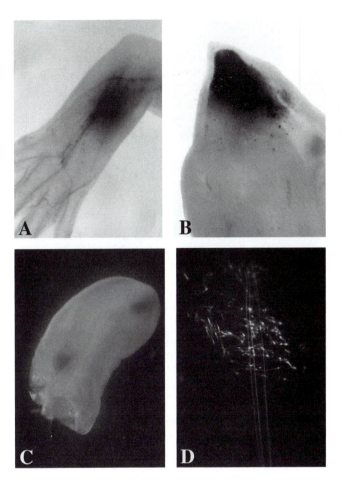

Figure 12-8 Functional analysis in urodele limbs *in vivo.* Vaccinia beta-gal expression (black regions) in a mature limb (A) and in a medium bud blastema (B). Lentiviral beta-gal expression (two black spots in C) in a distal blastema (upper right) and in proximal stump tissues (lower left) of a limb that has been removed from the animal (end of the humerus is visible at the lower left). Electroporation of a GFP expression plasmid into a mature limb results in expression of the transgene (white areas in D) in a large number of cells, including myofibers (vertical lines).

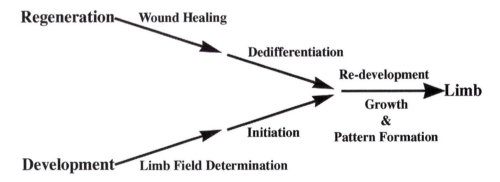

Figure 12-9 Convergence of pathways for regeneration and development. The point of convergence likely occurs at the late bud blastema stage of regeneration, and from that point onward, regeneration appears to be a recapitulation of the events associated with the late stages of limb development. Prior to the point of convergence, the events of regeneration are associated with the transition from a mature limb to a blastema that is equivalent to a limb bud. (Modified from Bryant et al. 2002.)

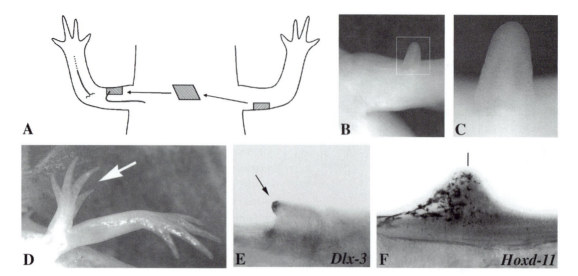

Figure 12-10 The induction of lateral limbs as a model system for the induction of limb formation. There are three different types of wounds that can be generated, and each induces a different growth and pattern formation response. If a piece of skin is removed, the wound heals and the skin is regenerated (not illustrated). A nerve can be surgically deviated to the site of the wound and a piece of skin from the opposite circumferential position can be grafted to the wound site (posterior to anterior in A). If no skin is grafted, a symmetrical outgrowth (bump) is induced (B and at higher magnification in C) that eventually stops growing and regresses. If skin is grafted, induced bumps continue to grow and eventually form an accessory limb (D) and express genes that are expressed during both limb development and limb regeneration from an amputated stump. *Dlx-3* is expressed at the distal tip (arrow in E) and *Hoxd11* is expressed in the posterior half of the outgrowth (to the right of the vertical line in F; the anterior tissues were derived from a pigmented animal and pigment cells can be seen to the left of the vertical line). (Modified from Gardiner et al. 2002.)

Redevelopment of the Limb. If in addition to a deviated nerve, a piece of skin from the opposite side of the limb is grafted to the lateral wound site (fig. 12.10A, D), an accessory limb is induced to form at the site of the wound (Maden and Holder 1984; Reynolds et al. 1983). These outgrowths express several of the genes that are characteristic of the redevelopment phase of regeneration, including *Dlx-3* and *Hoxd11* (fig. 12.10E, F). It is seen repeatedly that interactions between cells from disparate positions within the limb are required for normal outgrowth and pattern formation during regeneration (Bryant et al. 1981; French et al. 1976). Nerve-derived signals are required to induce an outgrowth from lateral limb wounds, but formation of an entire limb requires signals associated with the diversity of positional information that is provided by skin grafting. In the absence of a skin graft, the cells that form the lateral blastema are derived from a limited region of the limb and have limited positional information. Surgically created limbs that are symmetrical, and thus are limited in their diversity of positional information, similarly form symmetrical, growth-inhibited outgrowths (Bryant et al. 1981; fig. 12.5A), which we hypothesize to be equivalent to the nerve-induced outgrowths from lateral limb wounds. Since lateral limb outgrowths continue to grow and form normally patterned limbs, the interactions between the cells presumably lead to the generation of signals that stimulate the epidermis to form an AEC, and this in turn provides permissive signals for the formation of a limb.

Comparative Aspects of Regeneration and the Evolution of Regenerative Mechanisms

The ability to discover the underlying molecular mechanisms of regeneration in many organisms and tissues is allowing for new comparative studies of the phenomenon of regeneration. These studies will eventually lead to the identification of what is common, and thus likely to be essential to regenerative abilities, and what is different, providing information about novel ways to achieve successful regeneration.

A comparison of regeneration among vertebrates, including urodele limbs, fish fins, mouse digit tips, and larval anuran limbs, reveals many similarities (for recent reviews, see Akimenko et al. 2003; Brockes and Kumar 2002; Bryant et al. 2002; Gardiner et al. 2002; T. L. Muller et al. 1999; Nye et al. 2003; Poss et al. 2003). Among these similarities are a requirement for nerves and for interactions between a WE and underlying stump tissues. Where tested, denervation inhibits regeneration, and it has been suggested that the nerve is involved in signaling by members of the FGF family of growth factors. Studies of the role of the WE have implicated a number of signaling molecules as potentially important, among them members of the FGF family. Another common molecular feature of regeneration is expression of members of the *msx* family of transcription factors (M. R. Carlson et al. 1998; Crews et al. 1995; Koshiba et al. 1998; Reginelli et al. 1995; Simon et al. 1995; Y. Wang and Sassoon 1995). These molecules appear to function by maintaining cells in an undifferentiated and proliferative state, and thus are thought to be important in the transition of cells from the mature limb into blastema. Members of the FGF family of growth factors can regulate *msx* expression, suggesting a link between nerves, epidermis, and FGF. Other signaling pathways involved in the regulation of *msx* expression include the Wnts and BMPs, which likely will prove to be involved in the control of dedifferentiation, growth, and differentiation during regeneration in most, if not all, systems.

One noteworthy observation is the occurrence of populations of cells in regenerating tissues that are not in the cell cycle or have slow cell cycle kinetics. Alterations in cell cycle kinetics can alter gene expression and pattern formation, and it has been hypothesized that a long cell cycle is characteristic of cells that function as signaling centers during embryonic development (Ohsugi et al. 1997). During regeneration of both urodele limbs and fish fins, cells of the apical epidermis have a very long cell cycle, or may even be arrested in the cell cycle (Hay and Fischman 1961; Poleo et al. 2001). Similarly, the distalmost blastema cells in regenerating urodele limbs have a long cell cycle compared to the cells at more proximal levels (Chalkley 1959; Connelly and Bookstein 1983), and a comparable population at the distal tip of regenerating fish fins has also been observed (Nechiporuk and Keating 2002). It is unclear whether cell cycle kinetics plays a causal role in specification of the distal tip of the pattern; however, it is likely that distal tip cells exhibit a unique pattern of gene expression as a consequence of their unique growth characteristics (Ohsugi et al. 1997).

One important aspect of urodele limb regeneration, also a feature of other regenerating systems, is establishment of the distal tip first, followed by intercalation of the intermediate missing parts of the pattern. This has not been reported as such for fish fin rays; however, this phenomenon has been reported during planarian regeneration, where entire animals can be regenerated from small fragments. When this occurs, the first structures to re-form are those at the ends of the animal with the intermediate structures being formed by intercalation (Agata et al. 2003). The mechanism for generation of the distal tip of the pattern is not known; however, the processes of intercalary growth and pattern formation have been well characterized and are conserved between arthropods and vertebrates (Bryant et al. 1981; French et al. 1976). It therefore is likely that the cells of all successfully regenerating organisms have the ability to establish a distal tip and to engage in positional interactions between the tip and the stump and form new structures via the process of intercalation. Cases where intercalation is defective make the point that the first step is formation of the distal tip. Mouse digits that have been amputated *in utero* often form a complete dis-

tal tip in spite of incomplete intercalation of more proximal structures (Wanek et al. 1989). When regeneration of late stage *Xenopus* limb buds is stimulated by FGF, distal structures are formed but more proximal structures are not intercalated (Yokoyama et al. 2001). Therefore, an understanding about how the distal tip is generated in response to amputation is the critical first step in understanding how to induce regeneration, followed by a molecular understanding of intercalation.

Along with the similarities in regenerative mechanisms between organisms, there are also some notable differences. Digit tip regeneration in adult rodents and primates is limited to the distal phalangeal bone in regions that include the nail matrix, and appears to occur via the direct activity of fibroblasts rather than through formation of a blastema, as in urodeles. Another unique feature of digit tip regeneration is that it involves the direct formation of bone from connective tissue, rather than through the formation of a cartilaginous model that is subsequently replaced by bone. Similarly, regeneration of fish fins is restricted to the distal dermal skeleton, the fin rays (Géraudie et al. 1998). An amputation at the level of the more proximal endochondrial skeleton, the radials, does not lead to regeneration. As in regenerating mammalian digit tips, the bony rays form directly in the regenerating connective tissues, without a cartilage anlagen. The pattern of gene expression during fin ray regeneration is similar to that observed during fin ray development, but it differs from that observed during the early stage of development of the endochondrial skeleton (Géraudie et al. 1998).

Conclusions about the relationship between urodele limb and fish fin regeneration are difficult to draw since there is no direct homology between the elements of these two appendages (G. P. Wagner and Chiu 2001), and they are of different developmental origin (endochondrial bone in urodeles and dermal bone in fish fins). Fin regeneration appears to progress through a stage that is comparable to the redevelopment phase (phase II) in urodeles; however, it is unclear whether it goes through an early preparation stage comparable to phase I in urodeles. The similarity between normal growth and regenerative growth of fin rays may be a refection of the fact that fins are continuously developing throughout the life of the adult fish. Thus, an early preparative phase such as occurs in urodele limbs may not be required for regeneration. Similarly, a preparative phase is not required for regeneration of predifferentiated developing *Xenopus* limb buds, and if the limb bud cells are caused to remain in an extended embryonic state (by continual re-amputation), the period of regenerative ability is extended (Kollros 1984). The lack of a regenerative response from proximal amputations through the endochondrial skeleton of fish fins, and from proximal amputations through differentiated regions of

Xenopus limb buds, may be a consequence of the absence of a preparative phase needed to return cells that have completed development back to a regeneration-competent state. These observations suggest that a third target for study, along with formation of the distal tip and the stimulation of intercalation, is the preparation phase needed in order to initiate regeneration from mature structures that are no longer in development.

Finally, some studies indicate that the endochondral skeleton of lungfish fins can regenerate (Conant 1970), which may indicate that appendage regeneration, with all of the necessary components discussed above, was an ability present in the ancestors of the tetrapods.

Part III
Transformation

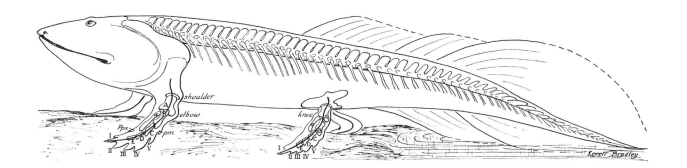

Chapter 13 Evolution of the Appendicular Skeleton of Amphibians

Robert L. Carroll and Robert B. Holmes

The Ancestry of Tetrapods

Between obligatorily aquatic fish and the dominant terrestrial vertebrates of the modern fauna (the reptiles, birds, and mammals) there lies a large assemblage of tetrapods that have a biphasic life history. The capacity for frogs and salamanders to live both in the water and on land was first recognized by the Greeks, who termed them *amphibios* ("living a double life"). With the rise of evolutionary thought, it was recognized that living amphibians represent the descendants of much earlier animals that were phylogenetically intermediate between primitive fish and amniotes. Fossils representing nearly a hundred families that had a biphasic life history are now known from the Paleozoic and Mesozoic eras. Although they are admittedly a paraphyletic assemblage, they too will be referred to as amphibians in this chapter. Unfortunately, there remains a vast gap in both time and morphology between Paleozoic and modern amphibians. As a result, very few studies of extant amphibians have included information regarding the first 100 million years of their ancestry, when amphibians were the dominant land vertebrates. Knowledge of these primitive tetrapods, and especially the nature of the appendicular skeleton, is also vital for understanding the origin of amniotes. In the absence of any prior publications that summarize the evolution of the girdles and limbs of the major groups of primitive tetrapods that lived from the Late Devonian into the Cretaceous, this subject will be discussed in considerable detail in this chapter.

The appendicular skeleton of terrestrial vertebrates generally occupies a much greater proportion of the body mass and provides a much larger contribution to locomotion than that of fish. Fortunately, knowledge of the transition between osteolepiform sarcopterygian fish (see Coates and Ruta, chap. 2 in this volume) and early amphibians documents the primitive anatomical pattern from which that of tetrapods evolved (Ahlberg and Johanson 1998; Clack 2000; Coates et al. 2002). Direct homologies can be recognized for most of the skeletal elements, but their functional roles differ significantly from those of their obligatorily aquatic ancestors (Andrews and Westoll 1970a; Clack 2002b).

Superficially, the well-known choanate sarcopterygians, *Eusthenopteron* (fig. 13.1A) and *Panderichthys*, resemble other Paleozoic bony fish in the external appearance of their paired fins. The bones of the dermal pectoral girdle appear as essentially a continuation of the skull roof and operculum, and are structurally and functionally integrated with the feeding and respiratory activities of the skull. On the other hand, the strong bony axis of the paired fins (also evident in the median fins) and the presumed presence of intrinsic musculature clearly distinguish the endochondral skeleton from that of any actinopterygian fish and suggest the capacity for their use in pushing against the substrate, but provide no evidence for a role in supporting the body out of the water. As in actinopterygians, all the fins are sheathed with dermal lepidotrichia, extending well beyond the limits of the endochondral skeleton.

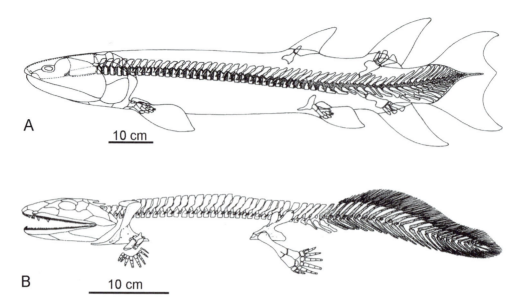

Figure 13-1 (A) Skeleton of the tristichopterid fish *Eusthenopteron* from the Upper Devonian (after Andrews and Westoll 1970a). (B) Skeleton of the Upper Devonian tetrapod *Acanthostega* (after Coates and Clack 1995).

In contrast, the appendicular skeletons of the earliest adequately known tetrapods, the Upper Devonian genera *Acanthostega* and *Ichthyostega*, differ dramatically from those of even the most closely related choanate fish in the separation of the dermal skeleton of the pectoral girdle from the skull by the loss of the most posterior bones of the skull roof and most of the opercular elements, and the vast increase in the proportional size of the endochondral skeleton of both the pectoral and pelvic appendages (fig. 13.1B). Numerous large digits have evolved, and the lepidotrichia have vanished.

However, both *Acanthostega* and *Ichthyostega* retain a large, fishlike tail, with variable expression of supraneural vertebral supports and dermal lepidotrichia, indicating the retention of effective aquatic locomotion. The presence of lateral-line canals in the skull and the configuration and retention of the hyoid apparatus into the adults support the arguments of Clack and Coates (1995) that these animals were primarily aquatic in their lifestyle, but the general structure of the appendicular skeleton indicates a major change in locomotion relative to their sarcopterygian ancestors. Taken as a whole, the appendicular skeleton of *Acanthostega*, *Ichthyostega*, and *Tulerpeton* (Coates 1996; Lebedev and Coates 1995) is much more similar to that of Carboniferous amphibians than it is to that of *Eusthenopteron* and *Panderichthys*. Although the structural hiatus evident between these well-known genera may be bridged by additional data from the several less well-known taxa of putative tetrapods from the Upper Devonian (*Ventastega*, *Hynerpeton*, *Metaxygnathus*, *Elginerpeton*, and *Obruchevichthys* [Clack 2000]), there was nevertheless a fundamental change in the structure of the ap-

pendicular skeleton during the Upper Devonian that led to the pattern of all subsequent land vertebrates.

Classification

Analysis of the patterns of evolution of the appendicular skeleton of amphibians should be based on a reliable understanding of the probable relationships of this assemblage. Unfortunately, establishing a convincing scheme for classifying the amphibians is very difficult because of two long gaps in our knowledge of their fossil record. Coates and Clack (1995) and Clack (2002a) pointed out a critical period of approximately 30 million years at the beginning of the Carboniferous during which the initial radiation of terrestrial vertebrates must have occurred, but for which there is very little fossil evidence.

By the first appearance of the major lineages of late Paleozoic tetrapods in the Viséan (mid-Carboniferous), all were already very distinct from one another, and few synapomorphies can be recognized that permit the determination of reliable sister-group relationships (Rolfe et al. 1994; Clack 2001). However, one can recognize three clearly distinct morphotypes among Paleozoic tetrapods:

1. A primitive assemblage, termed the "labyrinthodonts," of relatively large size (up to a meter or more in length) that retained many of the primitive character states of their plausible osteolepiform sister taxa. Well-known members of this assemblage, the anthracosauroids and temnospondyls, dominated the Upper Carboniferous and continued as a highly

diverse group into the Triassic (Heatwole and Carroll 2000). The Devonian tetrapods and four less well-known families first appearing in the Lower Carboniferous—the Crassigyrinidae, Baphetidae, Whatcheeridae, and Colosteidae—appear as an evolutionary grade bridging the structural gap between *Eusthenopteron* and *Panderichthys* and later labyrinthodonts (Clack 2002a). The labyrinthodonts have long been assumed to include the ancestors of all other tetrapods, and so are a conceptually paraphyletic assemblage. However, it remains convenient to retain this term, at least for informal usage, until their specific affinities with other groups of terrestrial vertebrates can be more reliably established.

2. The term "lepospondyl" (R. L. Carroll 2000b) has long been applied to an assemblage of generally much smaller Paleozoic tetrapods that differ from labyrinthodonts in the presence of a number of derived characters of the skull and vertebrae, most of which may be associated with their small size and precocial ossification (R. L. Carroll 1999). Their vertebrae form as complete cylinders, in contrast with the multipartite centra of labyrinthonts. Neither the monophyly nor the specific relationship(s) of lepospondyls to labyrinthodonts has yet been convincingly demonstrated (see alternative hypotheses of relationships discussed in Anderson 2001).

3. Unquestioned amniotes appear by the Upper Carboniferous, with plausible stem taxa from the Viséan (R. L. Carroll 1964a; Smithson et al. 1994; Paton et al. 1999). They have been postulated as having a sister-group relationship either with the anthracosaurian labyrinthodonts or with one or another lepospondyl order (R. L. Carroll 1991b; Laurin and Reisz 1997). The only way that these major questions of interrelationships can be convincingly resolved is through the discovery of intermediate forms from the Lower Carboniferous (Clack 2002a).

However, in contrast with the difficulty of establishing relationships between these large assemblages, most species of Paleozoic tetrapods can be assigned to a number of readily recognizable smaller clades (at the level of families and orders) on the basis of numerous unique derived characters of the skull and vertebral column. Although subject to some degree of variability, the distribution of these characters is sufficiently conservative that most of these taxa have been accepted for more than 100 years, supported by numerous detailed anatomical studies summarized in Heatwole and Carroll (2000). These clearly definable taxa, each of which probably had a monophyletic origin, include two large clades of labyrinthodonts, the anthracosauroids (including seymouriamorphs) and temnospondyls (including stereospondyls), and four orders of lepospondyls: aïstopods, microsaurs, nectrideans, and adelogyrinids. These taxa will serve as the primary units for the discussion of the evolution of the appendicular skeleton in this chapter.

The second major hiatus in our knowledge of the history of amphibians is an 80 to 100 million year gap between the first occurrence of the modern amphibian orders in the Early and Middle Jurassic and putative plesiomorphic sister taxa in the Carboniferous and Lower Permian (R. L. Carroll 2001a). This gap in time and morphology has led to a longstanding controversy as to the origin and nature of interrelationships of the frogs, salamanders, and caecilians. Recent papers have argued alternatively for their common ancestry from lepospondyls (Laurin and Riesz 1997; Laurin 1998a, 1998b) or temnospondyl labyrinthodonts (A. R. Milner 1993, 2000), or separate origins for frogs and salamanders from labyrinthodonts, and caecilians from lepospondyl microsaurs (R. L. Carroll 2000c, 2001b; R. L. Carroll et al. 2004; Schoch and Carroll 2003).

Although it is not yet possible to establish a fully resolved phylogeny for amphibians, this investigation of the evolution of their appendicular skeleton provides much new information that can be compared with previous databases that relied primarily on the skull and vertebral column. It thus provides a valid test for previous hypotheses of relationships, a procedure recently discussed by Rieppel (2003).

A further problem regarding the systematics of amphibians is that of restricting terminology to monophyletic groups, as defined by Hennig (1966) to include ancestral species *and all of their descendants*. The Amphibia, as that term has been used historically (e.g., Duellman and Trueb 1986), is clearly a paraphyletic assemblage since it explicitly excludes their probable descendants, the amniotes. Common usage of the Amphibia is as a grade, not a clade, that includes animals that descended from among obligatorily aquatic sarcopterygian fish and encompasses ancestors of the lineage leading to amniotes. Although amphibians, as so recognized, are unquestionably a paraphyletic group, it is nevertheless convenient for them to be considered as a unit in terms of investigating the structure and function of the tetrapod appendicular skeleton from its origin among obligatorily aquatic fish through its evolutionary radiation in land vertebrates that have not evolved the skeletal and physiological characteristics of amniotes. This usage concurs with the initial concept of amphibians as vertebrates having a biphasic way of life, including both aquatic and terrestrial stages, which has been a critical factor in the mode of development and pattern of evolution of their appendicular skeleton.

Whatever the specific ancestry of amniotes and the three living amphibian orders, their recognition as monophyletic groups leaves behind a number of paraphyletic assemblages among the previously named Paleozoic clades. These include some or all of the following taxa: labyrinthodonts, temnospondyls, anthracosaurs, lepospondyls, and microsaurs. Until such time as these relationships are much better established, it remain convenient, for the sake of communication, to retain the well-established terminology.

Labyrinthodonts, Lepospondyls, and the Ancestors of Amniotes

Fish-Amphibian Transition

The genera *Ichthyostega, Acanthostega,* and the less well-known *Tulerpeton* (Coates 1996; Clack 2000, 2002b) are unquestionably tetrapods, as that term is typically used for animals with feet, rather than fishy fins. Whatever specific patterns of relationship may emerge among the many lineages of later Paleozoic amphibians and amniotes, the general configuration of the appendicular skeleton of these Upper Devonian tetrapods can be traced with unquestioned continuity into the stem taxa of all later land vertebrates.

Pectoral and Pelvic Girdles

Changes in the structure and function of the girdles and limbs between *Eusthenopteron* and early tetrapods were discussed in detail by Andrews and Westoll (1970a), and more recent data were provided by Coates (1996) and Coates et al. (2002). As in their fish ancestors, the pectoral and pelvic girdles of early tetrapods differed significantly in the presence of both dermal and endochondral elements in the shoulder girdle, but only endochondral elements in the pelvis (figs. 13.2–13.7). The most primitive tetrapods still retained a vestige of the bones that in osteolepiforms joined the dermal shoulder girdle and the skull. These include the anocleithrum dorsally, present in *Acanthostega* and *Tulerpeton* (fig. 13.2; but not in *Ichthyostega*) and in the Carboniferous anthracosaur *Pholiderpeton* (Clack 1987) and the Lower Permian seymouriamorph *Discosauriscus* (Klembara and Bartik 2000). This distribution suggests that this bone was lost separately in at least three lineages: one or more times among Devonian and Lower Carboniferous tetrapods, and later among anthracosaurs and seymouriamorphs.

The subopperculum is reported in *Ichthyostega*, as is a bone termed the preopercular. The latter bone is also present in *Acanthostega* and the Lower Carboniferous *Crassigyrinus,* but despite its name, it is functionally associated with the skull, not the operculum (Clack 2002b). Despite the retention of these small elements, the skull would have been essentially uncoupled from the trunk so that the head could have moved independently in terrestrial feeding rather than requiring reposition of the entire body, as would have been necessary in choanate fish. However, the configuration of the anterior edge of the clavicle and cleithrum in *Acanthostega* has been interpreted by Clack and Coates (1995) and Clack

(2002b) as being homologous with the postbranchial lamina that in fish helps to direct water flow out of the gill chamber and to form a seal against which the operculum closes following respiration. Its retention implies continued close association of the shoulder girdle with the gill chamber in this genus and perhaps some other early tetrapods.

The remaining dermal elements of the shoulder girdle, cleithrum, clavicle, and interclavicle, present in tristichopterids (including *Eusthenopteron*) and panderichthyids, are retained in almost all Paleozoic tetrapods, but their configuration and relative size vary from group to group. In the fish, the cleithrum is a wide, strap-shaped bone extending nearly to the ventral midline, anterior to which there is a small triangular exposure of the clavicle that extends medially beneath the small oval interclavicle. The size of the cleithrum is progressively reduced in early tetrapods, revealing the extent of the previously underlying endochondral scapulocoracoid. In most Carboniferous tetrapods, the cleithrum is reduced to a small dorsal blade and a narrow stem running along the anterior edge of the scapula and inserting behind the stem of the clavicle. The primitive dermal sculpturing of the cleithrum is lost in most Carboniferous tetrapods. The clavicle evolves a relatively narrow dorsal stem and a broader, distinctly sculptured, ventral blade, whose extent is highly variable within the various clades.

The interclavicle is greatly expanded from its dimensions in advanced choanate fish, perhaps to protect the ventral surface of the body as it was dragged across the substrate by the sprawling limbs (Clack 2002b). Its shape is extremely variable both between and within major groups, from diamond shaped (as in *Tulerpeton* and *Acanthostega*), to having a very long and narrow posterior stem (in *Ichthyostega*; fig. 13.4A). The portions not underlaid by the clavicular blades are variably sculptured. The anterior margin of the interclavicle may extend far anteriorly, beneath the back of the lower jaws. In numerous specimens belonging to various clades, the anterior margin is formed by a dense series of small anteriorly projecting processes that may have functioned for the attachment of hyoid musculature. This is especially conspicuous in small, immature specimens.

The endochondral shoulder girdle of choanate fish and primitive tetrapods consists primitively of a single paired element, the scapulocoracoid, composed of laterally and ventrally facing surfaces that are broadly comparable to the scapula and coracoid bones, which are variably distinct from one another in later tetrapods. Both groups have a conspicuous glenoid depression for articulation with the head of the humerus. This faces largely posteriorly in the fish, but posterolaterally in tetrapods. The articulating surface is a relatively simple concavity in the fish, but in tetrapods it has a complex, helical configuration that controls the orientation

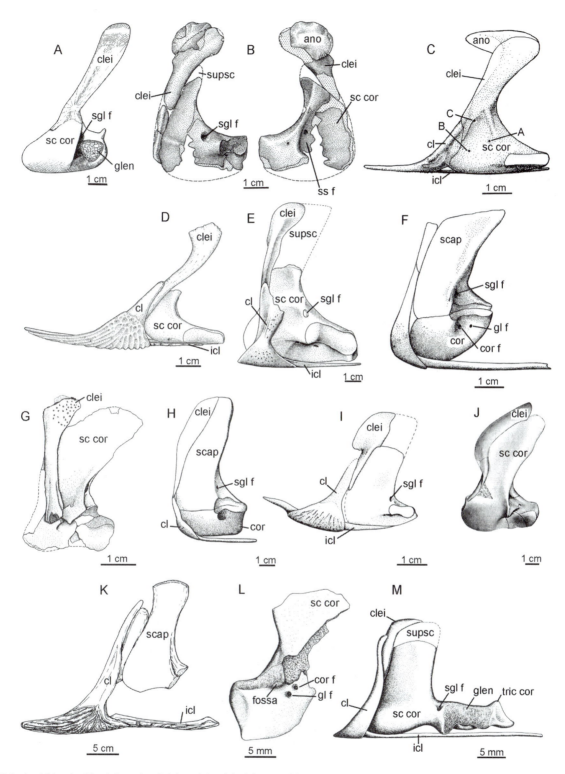

Figure 13-2 Amphibian shoulder girdles, primarily in lateral view. (A) *Ichthyostega* (after Jarvik 1980). (B) *Tulerpeton*, lateral and medial views (after Lebedev and Coates 1995). (C) *Acanthostega* (after Coates 1996). (D) *Greererpeton* (after S. J. Godfrey 1989). (E) The anthracosaur *Proterogyrinus* (after Holmes 1984). (F) The seymouriamorph *Seymouria* (after Romer 1956). (G) The Carboniferous limnoscelid diadectomorph *Limnostygis* (after R. L. Carroll 1967b). (H) *Diadectes* (after Romer 1956). (I) The Upper Pennsylvanian temnospondyl *Dendrerpeton* (after Holmes et al. 1998). (J) The Lower Permian temnospondyl *Dissorophus* (after Case 1910). (K) The Triassic stereospondyl *Siderops* (after Warren and Hutchinson 1983). (L) The Upper Pennsylvanian microsaur *Trachystegos* (after R. L. Carroll and Gaskill 1978). (M) The Lower Permian microsaur *Pantylus* (after R. L. Carroll and Gaskill 1978). Abbreviations: A, B, C, D, E, arbitrary designation of foramina in the scapulocoracoid of *Acanthostega*; see appendix for other abbreviations.

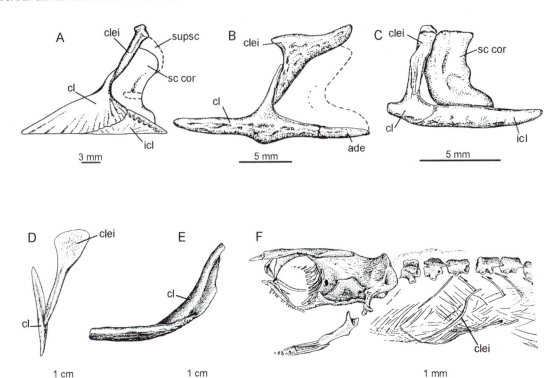

Figure 13-3 Lateral views of the shoulder girdle of lepospondyls. (A) The urocordylid nectridean *Urocordylus*. (B) The diplocaulid nectridean *Diceratosaurus*. (C) The scinosaurid, *Scincosaurus*. (D) The clavicle and cleithrum of the adelogyrinid *Adelogyrinus*. (E) Posterior view of the clavicle of the adelogyrinid *Adelospondylus*. (F) Lateral view of the anterior portion of the body of *Phlegethonia*, showing the position of what may be the cleithrum, the only element of the dermal shoulder girdle known in the family Phlegethontiidae. See appendix for abbreviations. (After R. L. Carroll et al. 1998.)

of the humeral shaft as it is moved during terrestrial locomotion (Holmes 1980). Andrews and Westoll (1970a) argued that this pattern was already initiated in *Eusthenopteron*.

Numerous small opening extend from the medial to lateral surfaces of the scapulocoracoid of *Acanthostega*. Coates (1996) refers to them by the letters A through E because of uncertainty of homology with the smaller number in later tetrapods. A is in the position of the supraglenoid foramen in later tetrapods, and D and E are in the general position of the glenoid and the supracoracoid foramina that pierce the coracoid plate.

In *Eusthenopteron* (Andrews and Westoll 1970a), the lateral surface of the scapulocoracoid is attached to the medial surface of the cleithrum at three points, separated by wide channels. In *Panderichthys*, considered closer to tetrapods in the loss of the dorsal and anal fins, the scapulocoracoid is enlarged and most of its lateral surface is in contract with the cleithrum. Although the scapulocoracoid and cleithrum are firmly attached in these fish, there remains a clear distinction between the endochondral and dermal bones. This distinction is lost in the Upper Devonian amphibians *Ichthyostega*, *Acacanthostega*, and *Hynerpeton* (Daeschler et al. 1994), in which their line of contact cannot be seen superficially. In

contrast, *Tulerpeton* (Lebedev and Coates 1995) and all later Paleozoic tetrapods have a clear distinction between the cleithrum and scapulocoracoid. The later bone becomes fully exposed laterally as the cleithrum becomes further reduced.

In most Paleozoic amphibians, the scapular and coracoid portions of the scapulocoracoid appear as a single area of ossification, more or less centered on the glenoid. Its expansion from the fish condition appears to have occurred via progressive increase in the extent of ossification along the dorsal and posteroventral margins. Growth of the scapulocoracoid in *Greererpeton* (S. J. Godfrey 1989) clearly extends from a single area of ossification some distance above the glenoid and progresses both dorsally and posteroventrally to the limits of the bone, without any evidence of separate areas of ossification as occur in some later tetrapods. The coracoid portion may be poorly ossified or appear missing in small and secondarily aquatic species. Lower Permian relatives of anthracosaurs, the discosauriscids and *Seymouria* (fig. 13.2), have evolved a sutural separation between the scapula and coracoid, as is also the case among some lineages of primitive amniotes, including primitive synapsids (pelycosaurs), captorhinomorphs, and procolophonids (Romer 1956). As the cleithrum was reduced in Upper Devonian amphibians, a

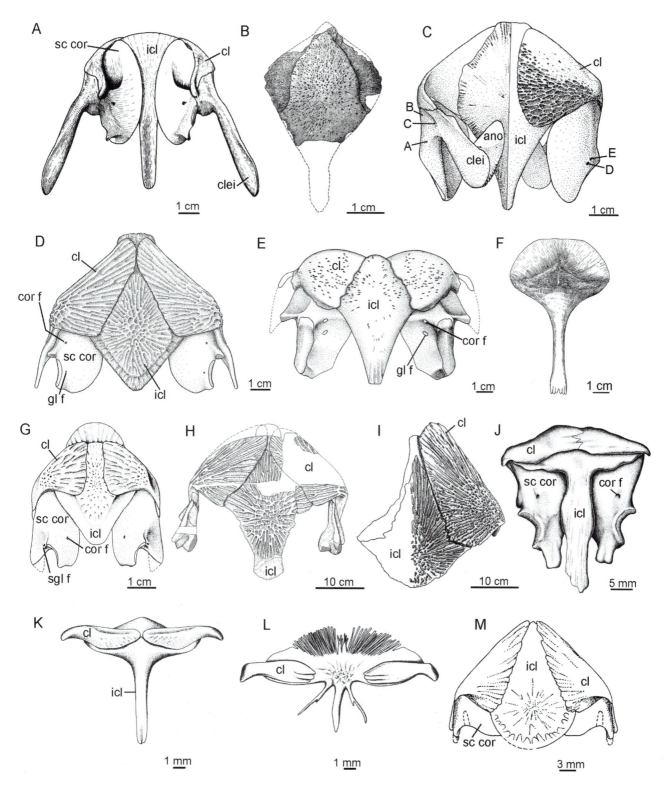

Figure 13-4 Dorsal and ventral views of the shoulder girdle. (A) Dorsal view of the shoulder girdle of *Ichthyostega* (after Jarvik 1980). (B) Ventral view of the interclavicle in *Tulerpeton* (after Lebedev and Coates 1995). (C) *Acanthostega*: left side shows the shoulder girdle in dorsal view, and the right side shows it in ventral view (after Coates 1996). (D) Ventral view of *Greererpeton* (after S. J. Godfrey 1989). (E) Ventral view of the anthracosaur *Proterogyrinus* (after Holmes 1984). (F) Ventral view of the interclavicle of *Seymouria* (after T. E. White 1939). (G) Ventral view of the Upper Pennsylvanian temnospondyl *Dendrerpeton* (after Holmes et al. 1998). (H) Ventral view of the Triassic stereospondyl *Siderops* (after Warren and Hutchinson 1983). (I) Ventral view of the dermal shoulder girdle of the stereospondyl *Metoposaurus* (after Sawin 1945). (J–L) Ventral views of microsaur interclavicles (after R. L. Carroll and Gaskill 1978); (J) *Pantylus*; (K) *Tuditanus*; (L) *Microbrachis*. (M) The nectridean *Urocordylus* (after R. L. Carroll et al. 1998). Abbreviations: A, B, C, D, E, arbitrary designation of foramina in the scapulocoracoid of *Acanthostega*; see appendix for other abbreviations.

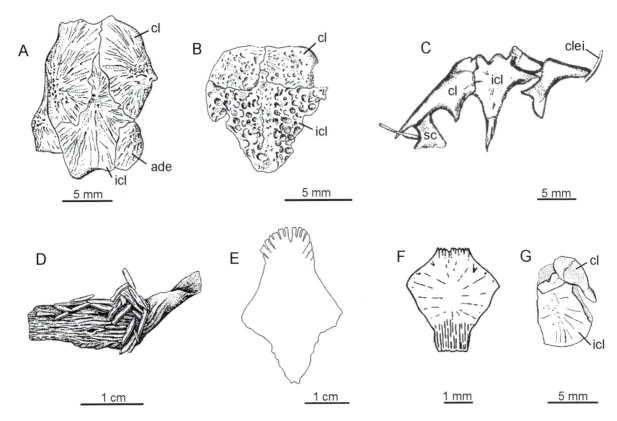

Figure 13-5 Ventral or dorsal views of the shoulder girdle of lepospondyls. (A) The diplocaulid *Diceratosaurus*. (B) The scincosaurid *Scincosaurus*. (C) The lysorophid *Pleuroptyx*. (D) Right clavicle of the adelospondyl *Adelospondylus*. (E) Outline of the interclavicle of *Adelogyrinus*. (F) Putative interclavicle of the aïstopod *Ophiderpeton*. (G) Clavicle and interclavicle of *Acherontiscus*. See appendix for abbreviations. (After R. L. Carroll et al. 1998.)

gap appears above the dorsal margin of the scapulocoracoid that is assumed to have been occupied by a cartilaginous suprascapula, as in modern amphibians.

Reduction and nearly complete loss of the shoulder girdle occur in several lepospondyl groups.

The pelvic girdle is even more dramatically altered than the shoulder girdle in the transition between osteolepiform fish and early tetrapods (fig. 13.6). In *Eusthenopteron*, it consists of a single paired element, attached neither at the midline nor to the vertebral column (fig. 13.6A, M). It is roughly triangular in shape, with an anteromedially facing pubic process, a dorsoposterior iliac process, and a posteromedially positioned acetabulum in the form of a cup-shaped depression facing primarily posteriorly. A very similar structure occurs in the distantly related rhizodontid fish *Gooloogongia loomesi* (Johanson and Ahlberg 2001), but the pelvis of *Panderichthys* is small and lacks an iliac ramus (Boisvert 2005).

The pelvic girdle in *Acanthostega* and *Ichthyostega* differs dramatically in having a shape and relative size obviously comparable to those of most other Paleozoic tetrapods (fig. 13.6B, C, N). The three regions common to later genera, the pubis, ischium, and ilium, are clearly defined, although

in *Acanthostega* no sutures are recognizable between them. In *Ichthyostega*, the suture lines are almost obliterated, as in very mature specimens of later temnospondyls, but those demarcating the pubis are evident in one specimen (Coates 1996) and would seem to mark the line of fusion on the medial surface between the ilium and ischium in the specimen illustrated as Jarvik's (1980) figure 160B.

Without intermediates (aside from a tantalizing fragment described by Ahlberg 1998, his fig. 12), it is difficult to imagine how the simple triradiate structure of *Eusthenopteron* could have evolved into the complex, tripartite bone of early tetrapods. Panchen and Smithson (1990) pointed out that *Eusthenopteron* has no area of ossification behind the acetabulum. They suggested that the more posterior element, the ischium, is a neomorph, perhaps derived from the dermal pelvic scute (which shares a similar topographic position; see fig. 13.13A later in this chapter). Although dermal and endochondral bones generally have distinct developmental and evolutionary histories, there is a close connection between the dermal cleithrum and the endochondral scapulocoracoid in the pectoral girdle of Upper Devonian tetrapods. However, it seems highly unlikely that a superficial dermal bone

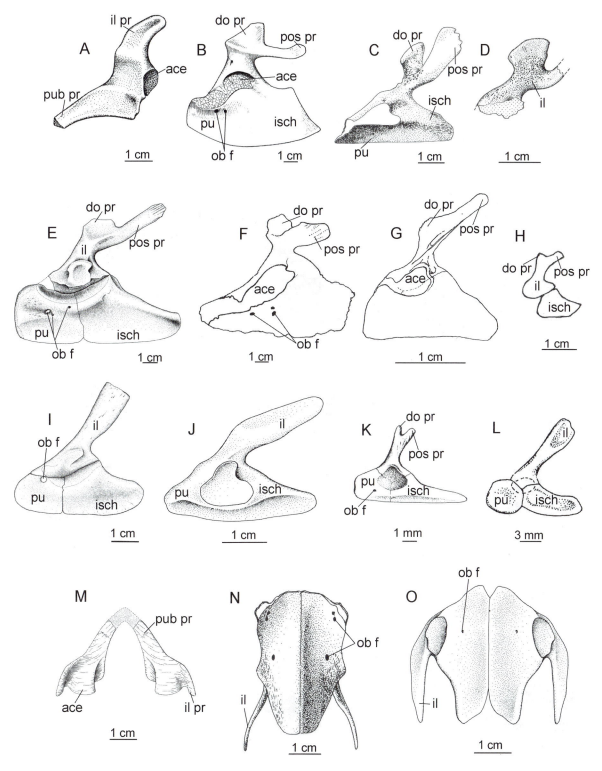

Figure 13-6 Pelvic girdles. (A–L) Lateral views: (A) The tristichopterid fish *Eusthenopteron* (after Andrews and Westoll 1970a). (B) *Ichthyostega* (after Coates 1996). (C) *Acanthostega* (after Coates 1996). (D) *Tulerpeton* (after Lebedev and Coates 1995). (E) *Proterogyrinus* (after Holmes 1984). (F) *Whatcheeria* (after Lombard and Bolt 1995). (G) *Caerorhachis* (after Ruta et al. 2002). (H) *Eucritta* (after Clack 2001). (I) *Greererpeton* (after S. J. Godfrey 1989). (J) *Dendrerpeton* (after Holmes et al. 1998). (K) The microsaur *Tuditanus* (after R. L. Carroll and Gaskill 1978). (L) The nectridean *Urocordylus* (after R. L. Carroll et al. 1998). (M) Dorsal view of the pelvic girdle of *Eusthenopteron* (after Andrews and Westoll 1970a). (N) Ventral view of *Acanthostega* (after Coates 1996). (O) Ventral view of *Dendrerpeton* (after Holmes et al. 1998). See appendix for abbreviations.

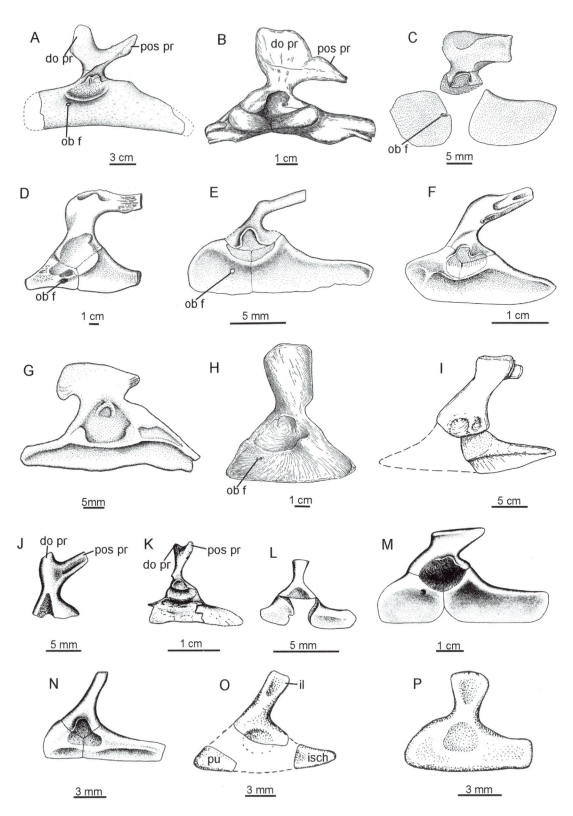

Figure 13-7 Pelvic girdles in lateral view. (A–E) Anthracosauroids: (A) The primitive anthracosauroid *Eoherpeton* (after Smithson 1985). (B) *Seymouria* (after T. E. White 1939). (C) *Discosauriscus* (after Klembara and Bartik 2000). (D) *Limnoscelis* (after Romer 1956). (E) *Westlothiana* (after Smithson et al. 1994). (F) The early amniote *Hylonomus* (after Smithson et al. 1994). (G–I) Temnospondyls: (G) *Parioxys* (after R. L. Carroll 1964c). (H) *Eryops* (after Gregory 1951). (I). *Siderops* (after Warren and Hutchinson 1983). (J–N) Microsaurs: (J) Ilium of the Carboniferous *Asaphestera* (after R. L. Carroll and Gaskill 1978). (K) *Sparodus* (after R. L. Carroll 1988). (L) *Hapsidopareion* (after R. L. Carroll and Gaskill 1978). (M) *Micraroter* (after R. L. Carroll et al. 1998). (N) *Rhynchonkos* (after R. L. Carroll et al. 1998). (O, P) Nectrideans: the diplocaulid *Keraterpeton* and the scincosaurid *Scincosaurus* (after R. L. Carroll et al. 1998). See appendix for abbreviations.

could have become integrated with the pelvis without interfering with posterior muscles inserting on the femur. Panchen and Smithson (1990) further suggested that the area of the pubis may have evolved from a cartilaginous anterior extension of the original fish pelvis. Its elaboration may have formed the anterior margin of the obturator foramen (multiple in *Ichthyostega*, *Acanthostega*, *Whatcheeria*, and *Proterogyrinus*).

An analogous sequence of evolutionary changes may have affected the endochondral bones of the pelvic and pectoral girdles. The fossil record shows that the scapulocoracoid expanded dorsally and ventroposteriorly to form the large, complex bone seen in early tetrapods, and remains as a single area of ossification in most Paleozoic amphibians. In contrast, the scapulocoracoid in early synapsids and some anaspids developed as a bipartite or tripartite structure, and sutural separations are retained during early growth.

The same pattern of evolution and developmental change may have occurred in the pelvic girdle. The small element present in *Eusthenopteron* could have increased in size in its descendants by growth along the anterior and posterior margins, presumably in response to the forces of expanding muscle masses acting to depress and adduct the femur. In the primitive genera *Ichthyostega*, *Acanthostega*, *Whatcheeria*, *Caerorhachis*, and *Eoherpeton* the girdle showed little evidence of subdivision. Only in more derived taxa are immature specimens known that show initial separation of the three elements. This is evident in most anthracosauroids, *Dendrerpeton*, *Eucritta*, and *Crassigyrinus* (the latter a very large but secondarily aquatic species).

The acetabulum was modified from facing primarily posteriorly in *Eusthenopteron* to primarily laterally facing in tetrapods. In this process, the supra-acetabular buttress was elaborated to resist the upward thrust of the proximal end of the femur. A feature of uncertain significance in Upper Devonian and the Lower Carboniferous genera *Whatcheeria* and *Caerorhachis* was the extension of the unfinished surface of the acetabulum to the anterior margin of the pubis. This was lost in all other Paleozoic tetrapods.

As seen in figure 13.1, the iliac process in *Eusthenopteron* stops well below the vertebral column, and there is no evidence of modification of either the ribs or vertebrae for an incipient sacral attachment. The most dorsal portion of the ilium (showing the iliac canal present in *Ichthyostega*) may be represented by a fragment attributed to the Frasnian genus *Elginerpeton*, closely resembling the anterior dorsal process of *Ichthyostega* (Ahlberg 1998). The latter genus and other Upper Devonian tetrapods, *Acanthostega* and *Tulerpeton*, also have a long posterior iliac process that is angled more laterally, beyond the plane of the anterior process. These processes were presumably deeply embedded in the axial musculature.

A well-defined sacral rib is known in *Acanthostega*, which has a configuration similar to that of thoracic ribs, with a moderately elongate shaft and the distal end compressed mediolaterally and expanded dorsoventrally that attached to the inner surface of the iliac blade, just beneath the base of the anterior dorsal processes. Coates (1996) identified a similar-shaped sacral rib in *Ichthyostega*.

In their relative size, general proportions, and attachment of the iliac blade to the vertebral column via a sacral rib, the pelvises of Upper Devonian tetrapods are comparable to those of Carboniferous and Permian genera that are otherwise thought to be primarily terrestrial. Of particular significance is the great expanse of the puboischiadic plate in a manner comparable to those of primitive synapsids and archosauromorphs. In these taxa, in common with living lizards and crocodiles, this area presumably served for attachment of large ventral muscles—the puboischiofemoralis externus, ischiotrochantericus, and the adductor femoris—that serve to lower and adduct the distal end of the femur and so lift the body from the substrate. Such a role for the ventral surface for the pelvis was neither necessary nor possible in *Eusthenopteron*.

It is almost impossible to conceive of the pelvic structure of *Ichthyostega* and *Acanthostega* having evolved except to support and move their bodies on land, without the buoyancy of water (R. L. Carroll et al. 2005). These genera were almost certainly capable of effective aquatic locomotion, as evidence by the structure of their tail, but there can be no reason for having an extensive, co-ossified puboischiadic plate if it were not used for the attachment of muscles capable of lifting the body off the ground. There is an extensive fossil record of later tetrapods having reverted to life in the water, as exemplified by the ichthyosaurs, mosasaurs, and cetaceans. However, their return to the water is evidenced by traits very different from those seen in the Devonian amphibians: reduced ossification of the girdles and limbs, loss of sacral attachment, shortening of proximal limb elements, and elimination of the hinging function of the carpus and tarsus. All of these changes were essentially in the opposite direction from those seen in the origin of tetrapods, with the transformation of fins to limbs.

Limbs

In contrast with the many differences between the configuration of the pectoral and pelvic girdles of both sarcopterygian fish and early tetrapods, the skeletal elements of the pectoral and pelvic fins of choanate fish and the fore- and hindlimbs of early tetrapods appear much more similar to one another. Both appendages have a single proximal element, the humerus anteriorly and the femur posteriorly, that articulates with the girdle via a joint that allows some degree of movement in all planes. Each of these bones are succeeded by two elements, the radius and ulna in the forelimb and the tibia and fibula in the hindlimb, with which they have a hinge like articulation. Further bifurcations proceed posteriorly, beyond the ulna and fibula.

Because of their positional similarity, the units of the fore- and hindlimbs are frequently referred to by a common set of terms. The humerus and femur are together designated the stylopod; the radius and ulna and the tibia and fibula are termed the zeugopod; and the more distal bones are grouped as the autopod. The bones of the stylopod and zeugopod are readily recognizable across the fish-amphibian transition, but the autopod is drastically remodeled.

The paired fins in tristichopterids and panderichthids served as paddles for steering and propulsion. Although they were flexible in all directions, there were no clearly defined joints or lines of flexion below the zeugopod. Neither the carpus nor the tarsus forms a recognizable region of the fins in *Eusthenopteron*. Rather, the bones of this region overlap in a manner that restricts bending, while retaining the limited mobility typical of a fish fin.

Despite the similar sequence of branching of the bones distal to the humerus and femur in the best-known choanate fish, *Eusthenopteron*, the configurations of the humerus and ulna are quite different from those of their serial homologues of the hindlimb. This suggests functional differences in their use in osteolepiforms that may presage the fundamental differences in the nature of the elbow and wrist joints and those of the knee and ankle in tetrapods.

As discussed by Coates (1996), one may trace a progressive series of changes through the humeri of *Eusthenopteron, Panderichthys, Acanthostega,* and *Tulerpeton,* leading to the general pattern of Carboniferous tetrapods (figs. 13.8–13.12). In *Eusthenopteron*, the bone is in the form of a short, stout column, with a conspicuous posterodistal process. That of *Panderichthys* (Vorobyeva 1992) is much more flattened, and the posterior process initiates the appearance of the tetrapod entepicondyle. The proximal articulating surface remains a broad convex arch extending across the entire proximal end of the bone throughout this transition. The articulating surfaces for the radius and ulna shift from facing essentially distally, to achieve a more ventral exposure. The various foramina that are present in osteolepiform fish, *Ichthyostega,* and *Acanthostega* (Coates 1996) are reduced to a single entepicondylar foramen in most Lower Carboniferous tetrapods.

The humeri of Upper Devonian and Lower Carboniferous tetrapods have an essentially L shape in dorsal view, with the entepicondyle extending posteriorly. Below the deltopector crest, common to most Paleozoic tetrapods, is an anterior crest that forms the margin of the bone. The anterior and posterior articulating surfaces are in nearly the same plane,

with no evidence of a shaft between the proximal and distal areas of expansion. A major change that occurs in many later lineages is the loss of the anterior crest, resulting in the appearance of a distinct shaft. This is accompanied by the "twisting" of the proximal and distal areas of expansion so that they are at about a 90 degree angle to one another.

Other trends can be noted both between and within specific clades. Smaller species have relatively longer shafts and smaller distal extremities and in some clades lose the entepicondylar foramen. Larger animals exaggerate the extent of the extremities and the areas of articulation, and are more likely to retain the entepicondylar foramen. The numerous differences that evolved within lineages overshadow the features that distinguish the most basal members of different clades.

The femur of *Eusthenopteron* (fig. 13.10A) differs significantly from the humerus in lacking a posterodistal process and in dorsal view broadly resembles that of primitive tetrapods in having a stout, columnar shaft and clearly distinct distal condyles for articulation with the tibia and fibula. However, the ventral surface of the femur, as illustrated by both Jarvik (1980) and Andrews and Westoll (1970a), is much simpler than those of any early tetrapod in lacking the conspicuous ridge system that presumably served for the insertion of the large muscles that adducted and retracted the femur in terrestrial vertebrates. All that was recognized by Coates (1996) was a faint, obliquely oriented ridge, vaguely comparable with the adductor blade and adductor crest in tetrapods. There is no trace of an intertrochanteric fossa or trochanters.

All early tetrapods differ significantly from *Eusthenopteron* in the greater overall size of the femur and its elongation relatively to the width of the anterior and posterior extremities. A strong adductor blade and associated ridge have evolved in *Ichthyostega,* but not the more proximal trochanters seen in later tetrapods. The shaft is narrower than that of *Eusthenopteron* but remains much wider than in any other early tetrapods. In strong contrast, the femora of *Acanthostega* and *Tulerpeton* much more strongly resemble those of amphibian groups known from the Carboniferous in the relative proportions of the shaft and articulating surfaces and the nature of the areas for muscle attachment on the ventral surface. All have highly distinctive adductor crests and a variable expression of trochanters at its proximal extremity. However, the femur of *Acanthostega* differs from that of all other early tetrapods in the strong anteroventral rotation of the shaft so that the anterior (tibial) condyle is largely ventral to the fibular condyle. Coates (1996) argued that this configuration was correlated with a paddlelike stroke of the limb for aquatic locomotion, but it does not preclude the use of strong adductor and retractor muscles, indicated by the

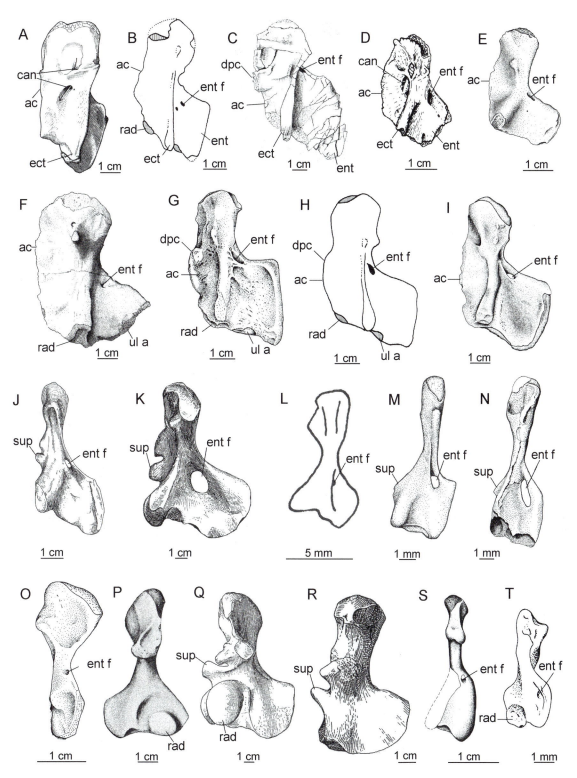

Figure 13-8 Left humeri, primarily in dorsal view. (A) *Ichthyostega* (after Jarvik 1980). (B) *Tulerpeton* (after Lebedev and Coates 1995). (C) *Whatcheeria* (after Lombard and Bolt 1995). (D) *Crassigyrinus* (after Panchen 1985). (E) *Greererpeton* (after S. J. Godfrey 1989). (F) *Baphetes* cf. *kirkbyi* (after A. C. Milner and Lindsay 1998). (G) A primitive anthracosaur from the Tournaisian of Nova Scotia (after Clack and Carroll 2000). (H) *Eoherpeton* (after Smithson 1985). (I) *Proterogyrinus* (after Holmes 1984). (J) *Seymouria* (after T. E. White 1939). (K) *Limnoscelis* (after Williston 1911). (L) *Casineria* (after Paton et al. 1999). (M) *Westlothiana* (after Smithson et al. 1994). (N) *Hylonomus*, the oldest known amniote (after Smithson et al. 1994). (O) The temnospondyl *Dendrerpeton* (from Holmes et al. 1998). (P) *Dissorophus* ventral view (after Case 1910). (Q) *Eryops* ventral view (after Case 1911). (R) *Metoposaurus* (after Sawin 1945). (S) The microsaur *Asaphestera* (after R. L. Carroll et al. 1998). (T) The nectridean *Scincosaurus* ventral view (after R. L. Carroll et al. 1998). See appendix for abbreviations.

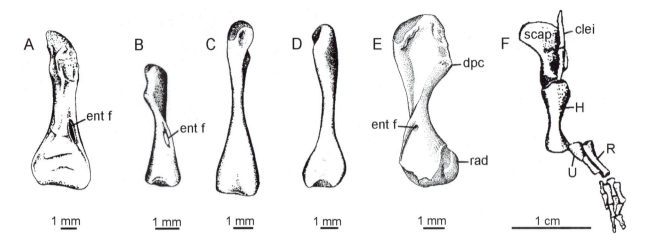

Figure 13-9 Humeri of lepospondyls. (A–E) Microsaurs: (A) *Tuditanus*. (B) *Saxonerpeton*. (C) *Leiocephalikon*. (D) *Rhynchonkos*. (E) *Batropetes*. (F) Shoulder girdle and forelimb of the Carboniferous lysorophid *Brachydectes newberryi*. (A–D, F after R. L. Carroll et al. 1998; E after R. L. Carroll 1991a). See appendix for abbreviations.

ventral ridge system, to pull the body forward, using the well-developed foot as a point of contact with a terrestrial substrate.

Coates noted that the posterior end of the adductor blade became progressively more proximal in position from *Ichthyostega* through *Acanthostega* and *Tulerpeton*, suggesting a more rapid but less powerful movement resulting from the contraction of the medially positioned adductors and retractors, but the general configuration of the femur had reached a roughly similar level of advancement to that of the earliest known members of the more derived clades, as shown by the many femora illustrated by Coates (1996, his fig. 36) and Ruta et al. (2002, their fig. 12). Within the Carboniferous and Permian, differences in the configuration of the femora are primarily in the proportions of the bone, which may be attributed largely to variably body size and the degree of terrestrial adaptation, rather than to taxonomic affinities. Ossification of the articulation surfaces is typically reduced in aquatic genera, in which the limb bones are commonly shortened relative to the length of the trunk.

Primitive tetrapods retain the limited hingelike articulation between the stylopod and zeugopod seen in their aquatic ancestors, although the surface of articulation at the end of the humerus and femur has extended more ventrally, elaborating the elbow and knee joints. In *Eusthenopteron*, the ulna is a short, columnar bone, resembling the femur in terminating in two condyles at right angles to the shaft that articulate with two elements of the autopodium, the intermedium and ulnare (fig. 13.11A). The fibula also articulated with two bones of the autopodium but also had an extensive posterodistally directed flange, resembling that of the humerus and ulnare of the forelimb (fig. 13.13A; see also fig. 13.14).

Acanthostega appears to retain the most primitive config-

uration of these elements among tetrapods. The bones of the zeugopod remain relatively flat, with the radius and tibia conspicuously longer than the ulna and fibula, precluding the hingelike articulation between the stylopod and the autopod common to later tetrapods. A distinct olecranon was not yet evident. In *Ichthyostega* (fig. 13.11C), the shafts of the radius and ulna are of approximately the same length and the olecranon is highly developed. In *Tulerpeton*, the greater length of the ulna and radius relative to the width of the articulating surfaces approaches that of Carboniferous tetrapods, and the ulna has a distinct olecranon.

In *Ichthyostega* and *Acanthostega*, the tibia and fibula broadly resemble their counterparts in Carboniferous tetrapods, although they are shorter, relative to the femur, and the tibia remains somewhat longer than the fibula. In *Tulerpeton*, the shafts of these bones are longer, relative to the width of the articulating surfaces, and they become separated from one another distally, as in Carboniferous tetrapods.

No autopodial bones of the forelimb are known in *Ichthyostega*. In *Acanthostega*, the only recognizable carpal bone is the intermedium that parallels the distal end of the radius to augment the upper line of flexure between the stylopod and the autopod. Despite the high degree of ossification of these proximal elements and the more distal bones of the autopodium, no other carpal bones have been recognized, suggesting that the remainder of the carpus was unossified, as is the case in numerous later tetrapods, especially among the anthracosauroids. In marked contrast, the proximal units of the autopod are ossified in *Tulerpeton* and appear to show a comparable line of flexion between it and the stylopod, as seen in later tetrapods. However, more detailed comparison shows that the clear distinction between the mesopodium (carpals and tarsals) and metapodium

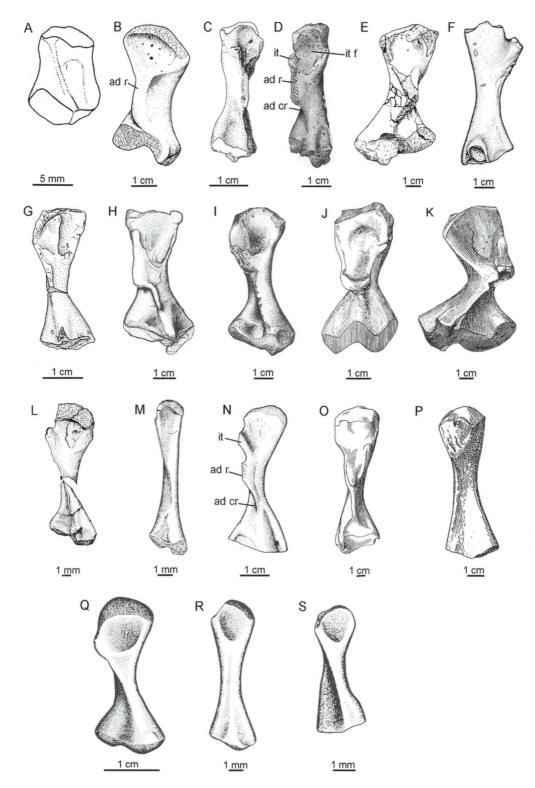

Figure 13-10 Right femora, all in ventral view. (A) *Eusthenopteron* (after Coates 1996). (B) *Ichthyostega* (after Jarvik 1980). (C) *Acanthostega* (after Coates 1996). (D) *Tulerpeton* (after Lebedev and Coates 1995). (E) *Whatcheeria* (after Lombard and Bolt 1995). (F) *Crassigyrinus* (after Panchen and Smithson 1990). (G) *Greererpeton* (after S. J. Godfrey 1989). (H) Putative anthracosaur femur from the Tournaisian of Horton Bluff (after Clack and Carroll 2000). (I) *Proterogyrinus* (after Holmes 1984). (J) *Seymouria* (after T. E. White 1939). (K) *Limnoscelis* (after Williston 1911). (L) *Westlothiana* (after Smithson et al. 1994). (M) The early amniote *Paleothyris* (after R. L. Carroll 1969a). (N) The temnospondyl *Dendrerpeton* (after R. L. Carroll 1967a). (O) *Eryops* (after Case 1911). (P) *Metoposaurus* (after Sawin 1945). (Q–S) Microsaurs (after R. L. Carroll and Gaskill 1978): (Q) *Trachystegos*. (R) *Rhynchonkos*. (S) The aquatic genus *Microbrachis*. See appendix for abbreviations.

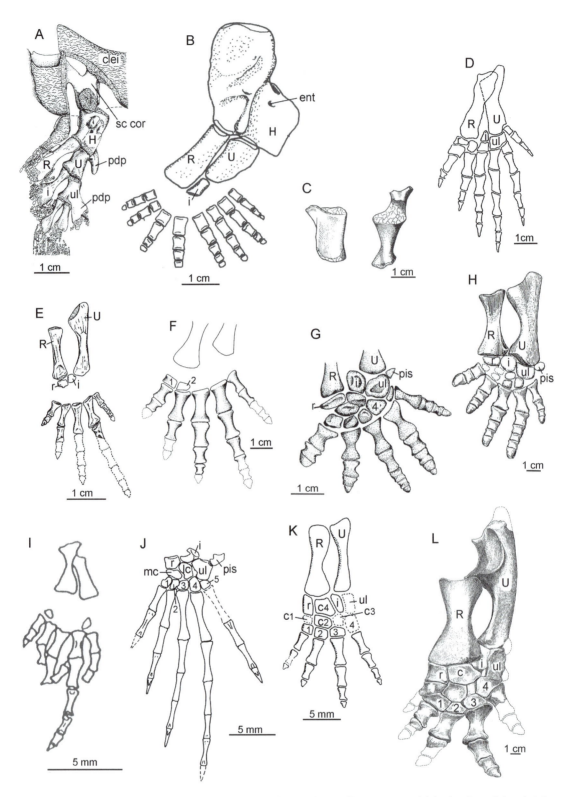

Figure 13-11 Forelimb. (A) *Eusthenopteron* (after Andrews and Westoll 1970a). (B) *Acanthostega* (from Coates 1996). (C) Left radius and ulna of *Ichthyostega* (after Jarvik 1980). (D) *Tulerpeton* (after Lebedev and Coates 1995). (E) *Gephyrostegus* (after R. L. Carroll 1970). (F) *Proterogyrinus* (after Holmes 1984). (G) *Seymouria sanjuanensis* (after Berman et al. 2000). (H) *Limnoscelis* (after Williston 1925). (I) *Casineria* (after Paton et al. 1999). (J) The early amniote *Paleothyris* (after R. L. Carroll 1969a). (K) The earliest known temnospondyl, *Balanerpeton* (after A. R. Milner and Sequeira 1994). (L) *Eryops* (after Gregory et al. 1923). Abbreviations: 1–5, distal carpals; ent, entepicondylar foramen; see appendix for other abbreviations.

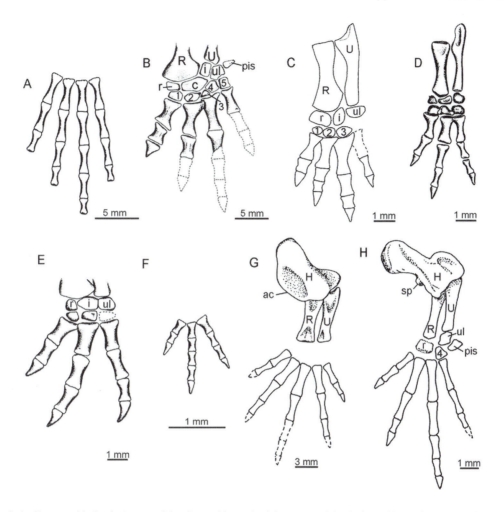

Figure 13-12 Forelimb of lepospondyls. (A–F) Microsaurs: (A) *Tuditanus*. (B) *Pantylus*. (C) *Batropetes*. (D) *Hyloplesion*. (E) *Microbrachis*. (F) *Odonterpeton*. (G, H) Nectrideans: *Urocordylus* and *Sauropleura scalaris*. Abbreviations: 1–5, distal carpals; sp, supinator process; see appendix for other abbreviations. (From R. L. Carroll et al. 1998.)

(metacarpals and metatarsals) has not yet reached the level seen in well-known Carboniferous tetrapods.

Only two proximal tarsal bones are known in *Acanthostega,* the intermedium and the tibiale, with the rest of the tarsus apparently remaining cartilaginous. The entire area of the tarsus is ossified in *Ichthyostega* and *Tulerpeton,* but the specific homology of the more distal elements remains uncertain.

Functionally, the wrist and ankle joints are neomorphs of tetrapods. Although some bones were formed from preexisting elements, the specific history of this transformation has been difficult to establish. Comparison of currently available information from both sides of this transition does suggest a fairly consistent scenario (figs. 13.11–13.13). *Eusthenopteron,* as described by Andrews and Westoll (1970a), shows a nearly comparable pattern of the sequence of bifurcation of the major elements of the fore- and hindlimbs. Posteriorly, the ulna and fibula each articulate with two more distal ele-

ments, homologized with the ulnare and intermedium of the carpus, and the fibulare and intermedium of the tarsus. The ulnare and fibulare, in turn, articulate with two additional, elongate elements, whose specific homologies are less obvious. Accepting the homology of the ulnare and fibulare and making comparison with Upper Devonian tetrapods, these may be compared with metacarpals and metatarsals, based on their elongate appearance, or with distal carpals and tarsals, which occupy this position relative to the ulnare and intermedium in Carboniferous tetrapods. In the absence of knowledge of the ulnare in *Acanthostega* and of the entire manus in *Ichthyostega,* those genera cannot be included in this comparison.

Aside from the intermedium, the specific homologies of the elements of the autopodium in *Acanthostega* are difficult to establish on the basis of current evidence. As the best-known specimen (individual X of block MGUH field no. 1227) is preserved, each digit appears in continuity with a

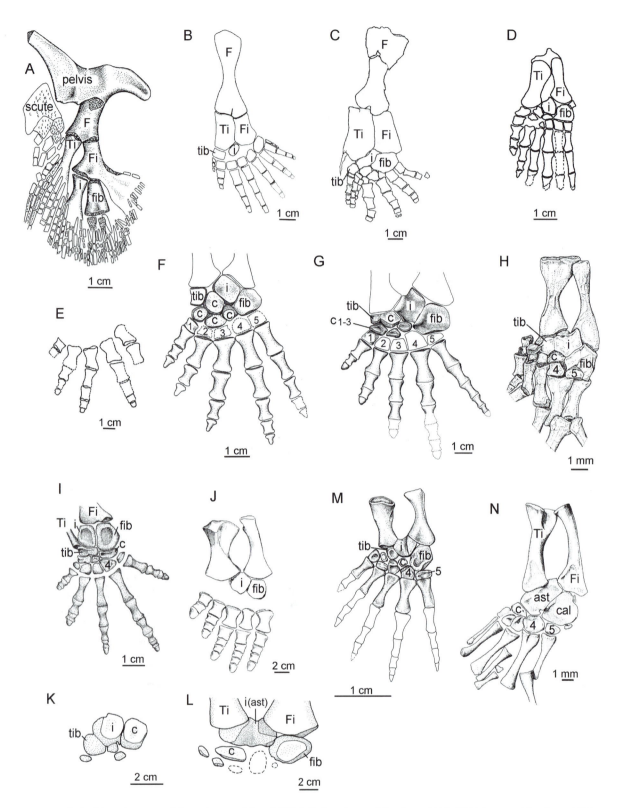

Figure 13-13 Rear limbs. (A) *Eusthenopteron* (after Andrews and Westoll 1970a). (B) *Acanthostega* (after Coates 1996). (C) *Ichthyostega* (after Coates 1996). (D) *Tulerpeton* (after Coates et al. 2002). (E) *Pederpes* (after Clack 2002a). (F) *Greererpeton* (after S. J. Godfrey 1989). (G) *Proterogyrinus* (after Holmes 1984). (H) *Gephyrostegus* (R. L. Carroll 1970). (I) *Seymouria sanjuanensis* (after Berman et al. 2000). (J) *Limnoscelis* (after Williston 1925). (K, L) Undescribed new diadectid and *Diadectes* sp. showing that co-ossification of the intermedium and tibiale occurred within the Diadectidae (after Berman and Henrici 2003). (M) Lower rear limb of *Westlothiana* (after Smithson et al. 1994). (N) Lower rear limb of *Hylonomus* (after Smithson et al. 1994). Abbreviations: 1–5 distal tarsals; see appendix for other abbreviations.

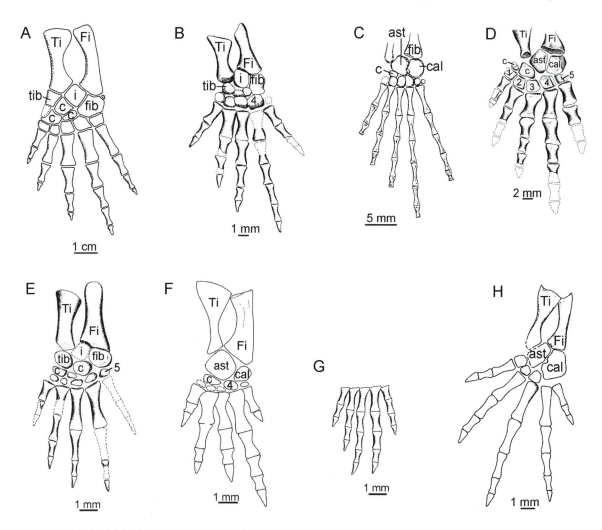

Figure 13-14   Lower hindlimbs. (A) The oldest known temnospondyl *Balanerpeton* (after A. R. Milner and Sequira 1994). (B–G) Microsaurs: (B) *Hyloplesion*. (C) *Tuditanus*. (D) *Pantylus*. (E) *Rhynchonkos*. (F) *Batropetes*. (G) *Microbrachis*. (H) The nectridean *Sauropleura scalaris*. (B–H after R. L. Carroll et al. 1998.) Abbreviations: 1–5 distal tarsals; see appendix for other abbreviations.

metacarpal and perhaps a distal carpal. If the distal carpals are not ossified (as was assumed by Coates 1996), the phalangeal count would be 3-3-3-3-4-4-4-3. But, if the carpals are ossified and remain in association with the corresponding metacarpals, the count would be one less for each digit. The former would seem the more probable in view of the common lack of ossification of distal carpals in anthracosauroids.

In all Carboniferous tetrapods for which the carpus and tarsus are adequately known, the metacarpals and metatarsals each articulate with a single, more proximal element, identified as a distal carpal or tarsal. In this respect, the feet of *Acanthostega* and *Ichthyostega* appear anomalous, for what appear to be two (or even three) metatarsals articulate with the fibulare. The forelimb of *Tulerpeton* is informative in that the length of the bones that articulate with the ulnare are of different lengths, the more lateral being short, like that

of a distal carpal, and the more medial being longer, resembling metacarpals. The tarsus, in contrast, resembles that of more derived tetrapods, in which the two bones articulating with the fibulare both have the proportion of distal tarsals.

If this is the correct determination of homologies, it implies that the distinction of the lateral distal carpals and tarsals from metapodials, common to Carboniferous tetrapods, had evolved in *Tulerpeton* but not in *Acanthostega* and *Ichthyostega*. Hence, direct comparisons of phalangeal counts between the latter genera and Carboniferous tetrapods are questionable (see also Coates et al. 2002). The elongation of the lateral portion of the carpus and tarsus may explain the apparent difference between tetrapod footprints described from the Upper Devonian (Clack 2000) and those from the Tournaisian of Horton Bluff (R. L. Carroll et al. 2005).

If the patterns of the hands and feet of *Tulerpeton* are

accepted as being the most similar of any Devonian tetrapods to those of Carboniferous amphibians, they can form the basis for understanding the subsequent evolution of the limbs among various groups of Paleozoic tetrapods. There is no direct evidence for more than five digits, fore or aft, in any post-Devonian tetrapods, although some footprints from the Tournaisian are equivocal. Current evidence can be interpreted as supporting the common ancestry of the foot structure in all adequately known Carboniferous tetrapods.

With some hesitation regarding the identity of distal carpals, the phalangeal count of the manus of *Tulerpeton* is 2-3-4-5-4-2. From its anomalous articulation with the ulna, the most lateral digit is the one presumed lost. The number of phalanges in the first five digits of the manus resembles most closely that of anthracosaurs, among Carboniferous tetrapods. The first two digits of the foot have the count of 2 and 3 common to nearly all early tetrapods. The count of the lateral digits cannot be established. The most conspicuous subsequent change is the further reduction in the number of digits of the manus to four in temnospondyls and most specimens of colosteids. This may have happened later in microsaurs and nectrideans.

Origin of Tetrapod Limbs

Although this chapter is primarily concerned with anatomical descriptions, it is impossible to understand the origin of the tetrapod appendicular skeleton without consideration of the adaptive nature of the shift from aquatic to terrestrial locomotion.

The clearly amphibious nature of the skeleton of *Acanthostega* and *Ichthyostega* emphasizes the basically biphasic life history of the amphibians since their origin in the Upper Devonian. Nearly every feature of the limbs and girdles of *Acanthostega* and *Ichthyostega* is derived relative to those of *Eusthenopteron* and *Panderichthys* in its potential for support of the body above the substrate and locomotion on land. In contrast, the tail is clearly suited for aquatic locomotion in its extensive caudal fin. Seen together, these seem to indicate very different adaptive strategies, but they also reflect the temporal progression from an obligatorily aquatic way of life as hatchlings to a facultatively terrestrial way of life as adults that must have been faced during the ontogeny of most early amphibians.

Eusthenopteron, which has served as the model for the ancestry of amphibian anatomy, is known to have had a long period of growth without evidence of metamorphosis (K. S. Thomson and Hahn 1968; Schultze 1984; Cote et al. 2002). Without evidence to the contrary, it may be assumed that the early development of *Acanthostega* and *Ichthyostega* was initially comparable, but, as in modern salamander larvae, limb development was delayed and locomotion in the water

was initially based on the undulation of the posterior trunk and tail. As in most primitive salamanders, early tetrapods almost certainly continued to use the tail for aquatic locomotion even as adults. In fact, even into the amniote level of evolution, the tail continued to be important in both aquatic and terrestrial locomotion because of the great importance of the caudofemoralis muscle, which is used for both flexure of the tail and retraction of the femur. It would not be surprising if the gill supports, together with their arterial circulation and relicts of the operculum, were functional, at least through the period of obligatorily aquatic development and perhaps into the adult stage.

One is, however, still faced with explaining the selective pressure for terrestrial locomotion. The most striking problem is that of diet. The dentition of *Acanthostega* and *Ichthyostega*, like that of *Eusthenopteron*, can be assumed to be adapted to feeding on relatively large prey, of which little was apparently available on land in the Late Devonian and Early Carboniferous, aside from the early amphibians themselves. In fact, the richest food resources almost certainly continued to be in the water. These included a varied assemblage of fish, as well as larger shallow-water invertebrates such as arthropods. What then was the impetus for the ancestors of tetrapods to move onto land? (See Clack 2002b.)

While aquatic prey could only have been caught in the water, there would have been a physiological advantage for early amphibians to have basked on land when not feeding, as is the case in the modern crocodiles and the marine iguana. Even in tropical regions, the water temperature in marginal marine environments would almost always have been below the body temperature that could have been achieved by land animals from the radiant heat of the sun (R. L. Carroll et al. 2005).

Carboniferous through Jurassic Labyrinthodonts

Lower Carboniferous Amphibians

The rarity of amphibian fossils from the Lower Carboniferous results in a gap in continuity between the Upper Devonian genera and major groups that do not appear until the Viséan. According to Clack (2001, 2002a), a number of families that are first recognized in the Lower Carboniferous appear to represent an intermediate grade that radiated prior to the appearance of the temnospondyl and anthracosaurian lineages that dominated the later Paleozoic and early Mesozoic. These include the Baphetidae, Whatcheeridae, Crassigyrinidae, Colosteidae, and the genus *Caerorhachis*. They share a number of primitive characters and others that are exhibited in a mosaic fashion by later labyrinthodonts, but few that support specific interrelationships with one another.

Crassigyrinus, the only known member of the Crassigyrinidae (Panchen 1985; Panchen and Smithson 1990; Clack 1998a), is unique in the extremely primitive configuration of the palate and the great reduction in the size of the limbs, especially the forelimb. The dermal shoulder girdle is coarsely sculptured, the interclavicle roughly diamond shaped, and the clavicular blade extensive. The scapulocoracoid is not known and may have been unossified; the humerus (fig. 13.8D) is very incompletely so. It appears to be pierced by foramina on the upper anterior surface, as in *Ichthyostega* and *Acanthostega*, in addition to the entepicondylar foramen. An anterior crest is retained. The pelvis retains the dorsal and posterior iliac blades present in Devonian tetrapods, but the posterior one is quite short. The pubis is not ossified, but there is a well-developed sacral rib. The femur lacks the prominent adductor crest and intertrochanteric fossa seen in *Acacanthostega* and *Tulerpeton*.

Whatcheeria (Bolt and Lombard 2000) and its Tournaisian antecedent *Pederpes* (Clack 2002a; Clack and Finney 2005) are plausible sister taxa of the anthracosauroids. However, the resemblances are primarily of a primitive nature, such as the extensive anterior crest of the humerus and the presence of both anterior and posterior processes of the ilium, although the latter is much shorter in the whatcheerids. Clack suggests that the manus of *Pederpes* may have had more then five digits, but not all are preserved. The metatarsals (fig. 13.13E) are clearly distinguishable and are bilaterally and proximodistally asymmetrical, in contrast with the pattern seen in Devonian genera. This suggests that the digits pointed anteriorly, rather than laterally, as is the case for *Acanthostega*. The phalangeal count of the pes, restored as 2-3-4-4-2+, is similar to that of the colosteid *Greererpeton*.

The Baphetidae (previously termed the Loxommatidae) have long been known from a number of genera represented primarily by their skulls, which are unique in the presence of a keyhole-shaped orbit (Beaumont and Smithson 1998). Only recently have specimens been described that include significant elements of the postcranial skeleton (A. C. Milner and Lindsay 1998; Clack 2001). What elements are known reflect a primitive level of evolution among Carboniferous labyrinthodonts, as in whatcheerids, but without obvious affinities with other clades.

In the most complete skeleton known, that of an immature specimen of *Eucritta* (Clack 2001), only the scapular portion of the scapulocoracoid is preserved, but it shows no taxonomical informative features. The clavicular blades are narrow and fairly widely separated from one another ventrally by an area of medial sculpturing on the interclavicle. The latter bone has a broadly tapering stem. The cleithrum is unusually robust, but essentially straight, with a slightly expanded dorsal blade. The ilium, which is only loosely articu-

lated with the ischium, has dorsal and posterior processes of nearly equal length. The pubis was apparently unossified.

The humerus, better known in *Baphetes* cf. *kirkbyi* (A. C. Milner and Lindsay 1998; fig. 13.8F), has an anterior crest resembling that of early anthracosauroids. Nothing is known of the manus. The femur of *Eucritta* shows no surface details. The zeugopodial elements are approximately 60% the length of the stylopod. Nothing is known of the anterior autopod. Surprisingly in such an immature specimen, four poorly ossified tarsals are present in *Eucritta*. The phalangeal count of the pes 2-2-3-4-? resembles that of temnospondyls. Clack (2001) suggests that *Eucritta*, the oldest known baphetid, probably retained a more terrestrial way of life than its later relatives, despite its low degree of ossification.

The Colosteidae were extremely common in some Lower Carboniferous localities and lingered on into the Upper Carboniferous. Although their anatomy is extremely well known, their specific affinities with other labyrinthodont clades remain unresolved (Panchen 1975; Hook 1983; S. J. Godfrey 1989). The best-known genus, *Greererpeton*, has an elongate trunk but relevatively small and poorly ossified limbs indicative of a secondarily aquatic way of life.

The pectoral girdle (figs. 13.2D, 13.4D) shares some features with temnospondyls, specifically the course "pit and ridge" ornamentation of the dermal elements and the diamond-shaped outline of the interclavicle, but these may represent primitive characters. In contrast, the poorly ossified scapulocoracoid, bearing both supracoracoid and glenoid foramina, is actually more comparable to the configuration exhibited by anthracosaurs and *Tulerpeton*.

The pelvic girdle (fig. 13.6I) resembles that of generalized temnospondyls such as *Dendrerpeton* and *Balanerpeton* in that the ilium bears a simple posterodorsally directed process but otherwise resembles that of other primitive tetrapods. The pubis is, at best, poorly ossified.

The humerus (fig. 13.8E) retains the primitive L shape of Upper Devonian tetrapods. The major difference from most other archaic tetrapods and anthracosaurs is its relatively low degree of ossification (S. J. Godfrey 1989). Humeralzeugopodial length ratios are similar to those of anthracosaurs. In contrast with most primitive tetrapods, however, at least some carpal elements appear to be well ossified, although no known specimens are sufficiently well articulated to permit reconstruction. Although originally described with a four-digit manus (S. J. Godfrey 1989), one specimen has recently been discovered with five digits (Coates 1996). No specimens of *Greererpeton* are sufficiently well articulated to permit an accurate phalangeal count, but *Colosteus* (Hook 1983) has a count of 2-2-3-3, similar to that of temnospondyls. The femur (fig. 13.10G) bears a prominent adductor ridge that extends far out onto the shaft. At its proximal end, just distal to the femoral head, it extends ventrally as a very prominent internal trochanter. The tarsus of *Greererpeton* (fig. 13.13F) includes a proximal row comprised of a fibulare, intermedium, large fourth (proximal) centrale, and tibiale. As in the temnospondyl *Balanerpeton*, the fourth centrale does not appear to contact the tibia. Three additional centralia are also present. The phalangeal formula of the pes is 2-2-3-4-3+, within the range seen in temnospondyls, but lower than that of anthracosaurs.

A final Lower Carboniferous labyrinthodont of uncertain taxonomic position is *Caerorhachis bairdi*, known from a single fairly complete specimen (Holmes and Carroll 1977; Ruta et al. 2002). The skull generally resembles that of later temnospondyls, but unlike nearly all members of that assemblage, the vertebral pleurocentra are medial, high-sided crescents, rather than small, paired elements.

All that is known of the shoulder girdle is the fimbriated anterior margin of the interclavicle and a fragment of a clavicular blade. The pelvis is primitive in the presence of only faint sutures separating the three elements and the extension of the unfinished bone surface of the acetabulum to the anterior margin of the pubis, as in Upper Devonian tetrapods. In contrast, the ilium is dominated by a posterodorsally oriented blade, resembling that of primitive temnospondyls and early amniotes. There is only a small remnant of an anterior process.

Of the forelimb, only a fragment of the humerus and the proximal ends of the radius and the ulna (showing a well developed olecranon) are known. What is preserved suggests a small limb. No carpals are preserved, and the hand is not sufficiently complete to establish either the digital or phalangeal count. The femur is preserved in primarily dorsal view, but shows the adductor crest extending about half the length of the bone. There is a prominent internal trochanter. The tibia and fibula are approximately 40% the length of the femur. The tarsus appears to have been completely ossified, although it cannot be reconstructed with assurance. It has been restored as having three proximal elements (the tibiale, intermedium, and fibulare), four centralia, and five distals. The phalangeal formula is probably close to 2-3-3-3-3 (similar to that of temnospondyls) but unlikely to be as high as that of anthracosaurs.

Anthracosauroids

All other Carboniferous and later labyrinthodonts can be assigned with assurance to two very diverse clades, the anthracosauroids (Smithson 2000) and the temnospondyls (Holmes 2000). The skull of anthracosauroids, which include the predominately aquatic embolomeres, the more terrestrial gephyrostegids, and the seymouriamorphs, is characterized by the posterolateral extension of the parietal to reach the tabular but the retention of a closed palate and primitively by a line of mobility between the skull roof and cheek. The primitively paired pleurocenta fuse ventrally to form a crescentic structure, which in more advanced genera become cylindrical and firmly attached to the neural arches. Of all Paleozoic tetrapods, they are most often implicated in the origin of amniotes (e.g., R. L. Carroll 1987–1988). As in most Paleozoic tetrapods, the dermal pectoral girdle of anthracosauroids (figs. 13.2, 13.4) is extensive. In embolomerous anthracosaurs (e.g., *Proterogyrinus, Archeria, Pholiderpeton*), the interclavicle is kite shaped, with a distinct posteriorly directed triangular parasternal process. In the one known exception, *Eldeceeon* (Smithson 1994), the interclavicle consists of a fan-shaped anterior plate and a long, narrow, parallel-sided parasternal process. The latter morphology is present in the archaic tetrapods *Whatcheeria, Ichthyostega*, some microsaurs, and the seymouriamorphs *Seymouria* (fig. 13.4F) and *Discosauriscus*. It is difficult to establish which of the two patterns is more primitive since both occur among Upper Devonian amphibians. The ventral surface of the interclavicle is broadly overlapped anteriorly by the triangular ventral plates of the clavicles. At the lateral apex of the plate, each clavicle turns dorsally as an attenuated stem. Extending dorsally from the latter is the splintlike cleithrum that expands slightly at its dorsal end. Together with the clavicular stem, it forms the anterior margin of the girdle. Surprisingly, the anocleithrum is retained in two anthracosauroids, the embolomere *Pholiderpeton* (Clack 1987) and the seymouriamorph *Discosauriscus* (Klembara and Bartik 2000), suggesting that this element may have been lost more than once in this assemblage. Whether this indicates a bony connection between the pectoral girdle and skull is doubtful, since anthracosaurs have a distinct, presumably functional neck separating the skull and girdle.

The scapulocoracoid is pierced ventrally by both a supracoracoid and glenoid foramen in all taxa in which this element is well preserved, and above the anterior extremity of the glenoid by the supraglenoid foramen. The glenoid is elongate and "strap shaped" but does not appear to be strongly twisted along its length, suggesting that retraction of the humerus was accompanied by relatively little rotation (Romer 1957; Holmes 1980).

Seymouriamorphs, including the terrestrial *Seymouria* (T. E. White 1939) and the aquatic, larval discosauriscids (Klembara and Bartik 2000), comprise a distinct anthracosauroid radiation. As in amniotes, the cleithrum is small and lacks a dorsal expansion. The interclavicle resembles that of *Eldeceeon, Whatcheeria*, and *Ichthyostega* in having a long, parallel-sided parasternal process. However, the scapula and coracoid each forms a discrete ossification, although the coracoid is not divided into anterior and posterior components (fig. 13.2F).

The ilium of early anthracosauroids (figs. 13.6 and 13.7)

retains the pattern of bifurcation seen in Upper Devonian tetrapods, with a long, posterodorsally directed process and a short, broad dorsal process of variable length that articulated with the sacral rib. In some taxa, the pubis may be absent (e.g., *Silvanerpeton, Eldeceeon,* and small individuals of *Gephyrostegus* and *Proterogyrinus*), because it either remained unossified or was only weakly attached to the rest of the girdle and was lost before preservation. Pubes are known in some specimens of *Archeria* and *Proterogyrinus.* In the former, the bone is pierced by a single obturator foramen as in most other primitive tetrapods. The pubis of *Proterogyrinus* is more primitive in being pierced by a second foramen near its suture with the ischium as in a number of archaic tetrapods including *Whatcheeria* (Lombard and Bolt 1995), *Ichthyostega* (Jarvik 1996), and *Acanthostega* (Coates 1996).

The ilium of *Seymouria* resembles that of earlier anthracosaurs in possessing both dorsal and posterior processes, although the dorsal process is relatively larger, and has partly coalesced with the posterior process. An expansion of the dorsal process at the expense of the posterior process also occurs in more derived temnospondyls and in the large lepospondyl *Micraroter.* The pubis is pierced by a single obturator foramen. In *Discosauriscus,* the posterior process is relatively broader, parallel-sided, and finished in cartilage posteriorly, suggesting that it was significantly longer in life. The ilium of *Discosauriscus* is remarkably similar in shape to that of *Limnoscelis* (Romer 1956). In contrast with anthracosaurs and most other primitive Paleozoic tetrapods, in which ossification of the pubis occurs relatively late, the pubis of *Seymouria* is well ossified, and pubic ossifications are known even in premetamorphic individuals of *Discosauriscus.*

The humeri of anthracosauroids (fig. 13.8G–I) retain the primitive L-shaped outline seen in archaic Devonian tetrapods (Clack 2002a) and colosteids (S. J. Godfrey 1989). A large entepicondylar foramen pierces the anteroproximal corner of the large rectangular entepicondyle in all known genera. A prominent crest, absent in most other late Paleozoic tetrapods, projects from the anterior surface of the humerus, precluding the expression of a well-defined shaft. In contrast with most later tetrapods, in which the humerus is markedly "twisted" so that proximal and distal heads are at approximately 90 degrees to one another, the broad, rectangular distal head is set off from the plane of the proximal head at an angle of only about 50 degrees.

Another important feature shared with most Devonian and Lower Carboniferous amphibians, but not seen in later tetrapods, is that the bicondylar axis (connecting the centers of the ulnar and radial facets), as viewed dorsally, does not intersect the long axis of the humerus at right angles but forms an angle of approximately 60 degrees (Holmes 1980). In addition, the axis of the ulna, when in articulation, is not

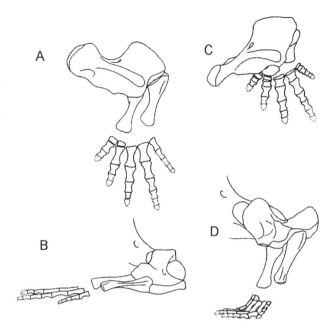

Figure 13-15 Reconstruction of the position of the pectoral limb of the anthracosaur *Proterogyrinus.* (A, B) Beginning of power stroke in dorsal and lateral views. (C, D) End of power stroke in dorsal and lateral views. (After Holmes 1980.)

oriented at approximately right angles to the plane of the distal humeral head, but projects, when fully extended at the beginning of the stance phase, almost directly anteriorly, and when fully flexed at the end of the stance phase, forms an angle of no more than 15 degrees with the plane of the entepicondyle (fig. 13.15). The radial articulation (capitulum) is poorly developed and, rather than facing directly ventrally as in most primitive tetrapods, faces anterodistally. This unusual morphology indicates that unlike later primitive tetrapods (e.g., Holmes 1977), not only did the zygopodium of anthracosaurs projected more or less anteriorly rather than ventrally from the humerus throughout the stance phase, but the capacity of the ulna and radius to flex on the humerus was limited. A similar arrangement is seen in the archaic Devonian tetrapods *Ichthyostega* (Jarvik 1996), *Acanthostega* (Coates 1996) and *Tulerpeton* (Lebedev and Coates 1995), and may represent a primitive condition.

The humerus of *Seymouria* (fig. 13.8J) shares a number of primitive features with Carboniferous anthracosaurs, including a low degree of torsion (the proximal and distal heads form significantly less than a 90 degree angle with one another), a large entepicondyle and prominent, dorsally directed ectepicondyle producing an L-shaped outline, and a distal bicondylar axis that is set off at an angle of much less than 90 degrees to the long axis of the humerus. However, the anterior crest has been lost, and a short but distinct shaft is present. At the point where the ectepicondyle emerges from the humeral shaft, the anterior surface of the humerus bears a stout supinator process. The humerus of the larval seymouriamorph *Discosauriscus* is not as well ossified but resembles that of *Seymouria* in being L shaped. However, it lacks an anterior flange.

The zeugopod of Carboniferous anthracosaurs, *Seymouria*, and *Discosauriscus* is approximately 60% of the stylopod. The carpus of anthracosauroids (fig. 13.11) is poorly known. The first and second distal carpals and a small, poorly ossified element that might represent an intermedium are preserved in one large specimen of *Proterogyrinus* (Holmes 1984). Two bones, tentatively identified as a radiale and intermedium, are present in one specimen of *Gephyrostegus* (R. L. Carroll 1970). An incomplete carpus has been described in *Archeria* (Romer 1957), but owing to its disarticulated condition, a reconstruction was not attempted. Well-preserved articulated front limbs are known in *Eusauropleura* (R. L. Carroll 1970), *Silvanerpeton* (Clack 1994), and *Eldeceeon* (Smithson 1994), but none shows a trace of carpal elements. This suggests that ossification of the carpus was either delayed ontogenetically or, in some taxa, may not have occurred at all.

In contrast with embolomeres, a well-ossified, articulated carpus is known in *Seymouria* (Berman et al. 2000). The proximal row consists of an ulnare, intermedium, and radiale, and a small bone identified as a pisiform (fig. 13.11G). But only two centralia (a proximal and a distal) are present, resembling the condition in primitive amniotes. The carpus of *Discosauriscus* is not ossified.

A well-preserved, articulated, although commonly incomplete, manus is known in several taxa. The phalangeal count appears to vary little within the group. In *Eusauropleura*, a count of 2-3-4-5-3, the same as in basal amniotes, has been established based on complete material. In *Gephyrostegus*, the most complete manus has been badly scattered, but has been plausibly reconstructed with the same count (R. L. Carroll 1970). Complementary evidence from several incomplete specimens permitted Romer (1957) to reconstruct the first four digits of *Archeria* as 2-3-4-5, and while he had firm evidence of at least three phalanges in the fifth, he opted to reconstruct it with four, by analogy with the pes. This formula has also been reconstructed for both *Eldeceeon* and *Silvanerpeton*, although in neither case is the manus complete. The phalangeal count for the manus of *Seymouria baylorensis* (T. E. White 1939) and *Discosauriscus* is 2-3-4-5-3, as in primitive amniotes, but there is one less phalanx in the fourth digit of *Seymouria sanjuanensis* (Berman et al. 2000).

The femora of *Proterogyrinus*, *Archeria*, and *Seymouria* are generally primitive in structure, bearing an adductor crest that extends far distally. The elements of the stylopod are stout but relatively short in most taxa, with the length of the tibia being between 55 and 60% of the length of the fe-

mur in Carboniferous anthracosauroids and about 75% in *Seymouria*, but approaching the length of the femur in *Limnoscelis* (Williston 1911).

The most complete tarsus is preserved in *Proterogyrinus* (Holmes 1984) and *Gephyrostegus* (R. L. Carroll 1970; fig. 13.13). In *Proterogyrinus*, 12 tarsal elements are present: a proximal row comprising a large medial fibulare, an intermedium that articulates with both the tibia and fibula, and laterally a large fourth (proximal) centrale and small tibiale, both articulating with the tibia. In one large specimen, the tibiale has partially fused with the fourth centrale located immediately lateral to it (Holmes 1984). Immediately distal to the tibiale, fourth centrale, and intermedium is a series of three subequal-sized centralia, and, finally, five distal tarsals, the fourth of which is, as in most primitive tetrapods, the largest. Ten elements are present in the tarsus of *Gephyrostegus* (R. L. Carroll 1970). The proximal row consists of a fibulare and a larger element articulating with both the fibula and tibia that appears to represent the fusion of the intermedium and the tibiale to form a compound element analogous to the astragalus of amniotes. However, in contrast with amniotes, the proximal centrale remains as a separate element. The tarsus of *Silvanerpeton* is not ossified, and in *Archeria* it is disarticulated and incomplete. It is well ossified and appears to be complete in *Eldeceeon*, but has not been completely described.

The tarsus of *Seymouria baylorensis*, as described by T. E. White (1939), has the pattern common to Carboniferous anthracosaurs, with three proximal elements—radiale, intermedium, and tibiale—distal to which is a wide medial centrale. In marked contrast, Berman et al. (2000), reconstructed the tarsus of the European species *S. sanjuanensis* as having only two large elements in the proximal row, identified as the fibulare and intermedium (fig. 13.13I). The tibiale is described, not in its normal position medial to the intermedium, but distal to it. As a consequence, the tibia appears to have little or no contact with the intermedium. The next row of tarsal elements, comprising three centralia, is also unusual in that they occupy the entire width of the tarsus rather than being restricted to the central and medial portions. The significance of such an arrangement is unknown. The tarsus of *Discosauriscus* is not ossified.

Silvanerpeton is the most primitive anthracosauroid in which a complete, articulated pes is preserved. The phalangeal count is unequivocally 2-3-4-5-4, as in *Seymouria* and *Limnoscelis* and in primitive amniotes (Clack 1994). The same counts are confirmed for digits I and V in *Gephyrostegus*, and digits I, II, and tentatively V (it is almost certain that only the terminal phalanx is missing from the last digit) in *Proterogyrinus*. This count was also plausibly reconstructed for *Archeria*, although the material is badly disarticulated.

Sister Taxa of Amniotes

It has long been thought that anthracosauroids gave rise to amniotes. In the early years of the 20th century, attention was focussed on *Seymouria* as an intermediate between embolomeres and reptiles, based on such features of the appendicular skeleton as the presence of a long-stemmed interclavicle, similar numbers of digits and phalanges on the hands and feet, and the partial incorporation of a second pair of sacral ribs. The length of the limbs suggested a terrestrial way of life. In contrast, the skull was primitive in the absence of a transverse flange of the pterygoid and the retention of an intertemporal bone. More importantly, the discovery of the closely related discosauriscids demonstrated that seymouriamorphs had gilled aquatic larvae and so were certainly not physiologically amniotes (S̆pinar 1952; Boy and Sues 2000).

The description of numerous species from well down in the Carboniferous with typical amniote cranial and postcranial features (R. L. Carroll 1964a, 1969a; Reisz 1972) shows that amniotes, including both synapsids and the ancestors of diapsids, had evolved long before the appearance of seymouriamorphs in the fossil record, suggesting a far earlier origin. Few fossils from the Lower Carboniferous show specific links between early anthracosaurs and amniotes. One possibility is *Casineria* from the Lower Carboniferous of Great Britain (Paton et al. 1999). It is represented by a single specimen lacking a skull. As in the oldest known amniotes, it is both very small and highly ossified. As in *Seymouria* and *Discosauriscus* and some early amniotes, the endochondral shoulder girdle is formed by a scapula and a single coracoid ossification, and the splintlike cleithrum lacks a dorsal expansion. As in primitive amniotes, the proximal and distal heads of the humerus are set at a full 90 degrees from one another. The entepicondyle, the length of which in anthracosauroids and archaic Devonian taxa represents a full 50% of the total length of the humerus, is equivalent to only 33% of humeral length in *Casineria*, resulting in the presence of a distinct shaft, lacking an anterior crest. These features combine to produce a shape similar to that of the stem amniote *Westlothiana* (Smithson et al. 1994) and primitive amniotes (fig. 13.8L–N). A partial phalangeal count of the Lower Carboniferous *Casineria* has been established as follows: digit I, 2; digit IV, 5 (fig. 13.11I). The remaining three digits are incomplete, but are compatible with an amniote count of 2-3-4-5-3. The pelvis resembles that of the early amniote *Hylonomus lyelli* (fig. 13.7F) in having a narrow, posteriorly directed iliac blade, without a trace of an anterior process. The zygopod is only slightly more than 50% of the length of the stylopod. The tarsus and pes cannot be reconstructed.

The slightly younger, but still Lower Carboniferous, genus *Westlothiana* was originally described as the oldest known amniote (Smithson 1989) based on the apparent possession of an astragalus. A more complete description of the specimens (Smithson et al. 1994) revealed a more conventional primitive anthracosauroid tarsus, with a large fibulare, intermedium, and tibiale in the proximal row, as well as a proximal centrale (fig. 13.13M). *Westlothiana* also lacks the transverse flange of the pterygoid that characterizes early amniotes (R. L. Carroll 1991b). On the other hand, the vertebrae are very similar to those of primitive amniotes, specifically *Captorhinus* (Smithson et al. 1994).

Little of the dermal shoulder girdle is known. The scapulocoracoid appears to be ossified as a unit, without distinction between the scapula and coracoid, but is otherwise poorly known. The pelvis has a long but slender posterior process, with only a slight extension anteriorly, in the position of the dorsal process. The ischium extends far behind the acetabulum (fig. 13.7E).

The structure of the humerus (fig. 13.8M) closely resembles that of the early amniote *Hylonomus* in the presence of a long shaft and a short supinator process, but it is remarkably short relative to the length of the trunk or of individual trunk vertebrae—only about half the relative length in *Hylonomus*. It is about 60% the length of the femur. The ulna and radius are similarly short, but the olecranon is very well developed. At least two carpals are ossified, but their identity cannot be established. The manus cannot be reconstructed. The femur (fig. 13.10L) has a very broad proximal expansion and a clearly defined internal trochanter, but the adductor crest is not extended far distally. The tibia and fibula are approximately 70% the length of the femur. As in the embolomere *Proterogyrinus*, the tibiale and intermedium are not integrated with a proximal centrale to form an astragalus. The phalangeal count is 2-3-4-5-4, as in early amniotes.

Westlothiana may be viewed as a close sister taxon of amniotes, but is precluded from a potentially ancestral position by the apomorphies of an increased number of presacral vertebrae—approximately 36 versus about 25—and the great reduction in the relative length of the forelimb.

Three families from the Upper Carboniferous and Lower Permian, the Limnoselidae, Tseajaiidae, and Diadectidae, united in the Diadectiomorpha, may be phylogenetically even closer to the amniotes, but they lack significant shared derived characters with that group and appear far too late in time to be their actual antecedents. *Limnoscelis*, from the Lower Permian, is the best-known primitive member of this assemblage and lacks many features that amniotes had evolved 40 million years earlier, notably the consolidation of the the tibiale, intermedium, and proximately centrale in an astragalus (fig. 13.13J). All known members of this group are much larger than Carboniferous amniotes, with consequent changes in the structure of both the vertebrae and the limbs, making direct comparison difficult.

Among the diadectomorphs, the endochondral girdle expands far dorsally. In *Diadectes* the scapula and coracoid are separated by a suture, although this is missing in the much earlier limnoscelid *Limnostygis* (fig. 13.2G, H). The sculpturing of the dermal elements is lost. The humerus becomes especially massive, with the greatly expanded articulating areas much restricting the extent of the shaft (fig. 13.8K). A supinator process evolves, and the entepicondylar foramen becomes greatly enlarged. The carpus of *Limnoscelis* (fig. 13.11H) consists of the usual proximal bones (radiale, intermedium, and ulnare), but there is an additional lateral element identified as a pisiform, which may be homologous with that bone in amniotes (Williston 1912). However, the convergent origin of a pisiform-like element in nectrideans and microsaurs suggests that this bone may have evolved independently among the diadectomorphs as well. Three distal elements are also preserved, but their specific identity is uncertain. The phalangeal count is 2-3-4-5-3, but the phalanges are much shortened, except for the broadened terminal elements.

The iliac blade is expanded above the acetabulum (fig. 13.7D). However, it does not divide into a bifurcate structure, as in most anthracosauroids, but extends primarily posteriorly. The proximal and distal extremities of the femur are greatly expanded, to appear like two converging equilateral triangles. Zygopodials are correspondingly stout. In *Limnoscelis,* the tarsus remains unossified except for two large proximal elements most plausibly interpreted as a fibulare and intermedium, and two smaller bones in the position of centralia 3 and 4 (fig. 13.13J). The phalangeal formula is 2-3-4-5-5.

Berman and Henrici (2003) have ably demonstrated that fusion of the proximal tarsals into a unified astragalus occurred within the family Diadectidae, subsequent to divergence from more primitive diadectomorphs. This process can be observed to occur ontogenetically within a primitive but as yet unnamed sister taxon of *Diadectes,* as well as in some immature specimens of *Diadectes.* Ossification is much reduced in the more distal parts of the tarsus (fig. 13.13K, L). The convergent origin of a bone with the structural and developmental characteristics of an amniote astragalus, at least 25 million years after this element had evolved in the common ancestors of mammals and diapsids, implies the even earlier divergence of amniotes and diadectomorphs. The phalanges of the manus and pes are further shortened in *Diadectes* but retain the same count as in *Limnoscelis.*

Temnospondyls

The other major group of labyrinthodonts is the temnospondyls, which first appear in the fossil record in the Viséan and continue into the Cretaceous (Holmes 2000). Their skulls are derived in the firm attachment of the cheek and skull table and the large size of the interpterygoid vacuities, but primitive in retaining contact between the postparietal and the supratemporal. Most temnospondyls are characterized by vertebrae in which the intercentrum is the dominant central element. They are generally considered to be ancestral to some, if not all, living amphibians. The temnospondyls underwent two periods of major radiation, the first principally within the Carboniferous and early Permian, which included a large number of primarily terrestrial lineages. The second, beginning in the Late Permian and continuing into the Cretaceous, consisted largely of secondarily aquatic families, referred to collectively as the stereospondyls (Warren 2000).

The pectoral girdle of primitive temnospondyls (figs. 13.2I–J, 13.4G) resembles that of anthracosaurs in general form. The lozenge or diamond-shaped interclavicle forms a broad, midventral plate, but a posterior stem, if present at all, is less well defined than in anthracosaurs. Dorsally, the cleithrum expands to form a spoon-shaped plate that forms part, and in a few cases all (e.g., *Cacops* and *Dissorophus,* Williston 1910; and *Eryops,* Miner 1925), of the dorsal margin of the girdle. The girdle is positioned immediately behind the skull, with the interclavicle projecting ventrally between the rami of the lower jaws, possibly restricting movement of the head relative to the trunk. However, none of the extracleithral or extrascapular elements present in sarcopterygian fish and in part retained in at least some Devonian tetrapods, *Pholiderpeton,* and *Discosauriscus* have been reported in temnospondyls.

As in the majority of Paleozoic amphibians, sutures marking the boundaries between the presumed scapular and coracoid ossifications of the scapulocoracoid are absent. The vertical scapular blade is well developed in all temnospondyls, but is only fully ossified to its dorsal extremity in large or terrestrial taxa. Ventrally, the scapulocoracoid turns medially to form a horizontal coracoid plate. At the junction of the scapular and coracoid surfaces, a horizontally elongate glenoid forms a complex surface that faces posteroventrally at its anterior end and twists to face laterally at its posterior termination. This shape suggests that as the humerus was retracted during the stance phase, it was forced to rotate considerably about its proximodistal axis. The anterior end of the glenoid is supported dorsally by a stout buttress pierced by a supraglenoid foramen (Borsuk-Białynicka and Evans 2002). The coracoid plate is pierced anterior to the glenoid by a supracoracoid foramen, and in some taxa (e.g., *Sclerocephalus,* Meckert 1993), medial to the glenoid, by a glenoid foramen.

The dermal girdle of stereospondyls is similar in structure to that of Paleozoic temnospondyls (fig. 13.4H, I). The interclavicle is diamond shaped, although the relative size of the anterior and posterior halves is variable. The ventral process of the clavicle is extremely variable in form (Hellrung 2003). In *Siderops* (Warren and Hutchinson 1983), it is narrow at the base and tapers to a point as in Paleozoic taxa, but is much shorter in most other taxa, and can be very broad. The scapulocoracoid is generally small and poorly ossified, both dorsally and ventromedially. In *Metoposaurus*, both scapular and coracoid portions are ossified (although not separated by a suture), but in most stereospondyls, little if any of the coracoid is ossified (e.g., *Paracyclotosaurus*, *Siderops*, fig. 13.2K).

The ilium of primitive temnospondyls such as *Balanerpeton* (A. R. Milner and Sequeira 1994) and *Dendrerpeton* (Holmes et al. 1998) typically bears a long posterodorsal process, but without expression of the anterior dorsal process of Devonian tetrapods and anthracosauroids (figs. 13.6 and 13.7). Although this pattern is retained in most later temnospondyls, in a few of the more terrestrial taxa such as *Eryops*, *Cacops*, and *Trematops* it is in the form of a short, dorsally directed blade that expands little beyond what is required to articulate with the sacral rib. However, that of *Parioxys* (fig. 13.7G) expands both anteriorly and posteriorly, perhaps for the attachment of additional musculature to move the limb anteriorly and posteriorly. In amniotes, most of the lateral surface of the ilium provides attachment points for limb muscles, but in urodeles, the pelvic limb muscles originate only from the base of the ilium (E. T. B. Francis 1934). Most of the lateral surface of the blade is occupied by the origin of the iliocostalis, and its dorsal edge, by the longissimus (Romer 1922). A similar arrangement has been hypothesized for *Eryops* (E. C. Olson 1936) and probably could be extended to other temnospondyls. The platelike ischium is typically triangular in outline, with its apex directed posteriorly. In taxa for which there is at least a partial growth series known (e.g., A. R. Milner and Sequeira 1994), the pubis is relatively late to ossify. When it is preserved, it is invariably pierced by a single, small obturator foramen.

The ilium, ischium, and pubis of stereospondyls (fig. 13.7I) are always preserved as separate elements (Warren and Snell 1991). The ilium, which is usually the best preserved, resembles that of many Paleozoic temnospondyls in bearing an expanded dorsal process separated from the acetabular area by a distinct waist. The ischium is variable in shape, suggesting that it was normally incompletely ossified. Pubes are unknown in most stereospondyls, presumably because they remained cartilaginous.

In temnospondyls, the L-shaped humerus seen in Devonian tetrapods, colosteids, anthracosaurs, and some nectrideans is modified by the absence of the anterior crest and a realignment of the ectepicondyle, which diverge anterodis-

tally from the central shaft rather than lying parallel to the proximodistal axis of the humerus (fig. 13.8O–R). The deltopectoral crest, located approximately midway between the proximal and distal ends of the bone in archaic Devonian forms (Clack 2002b) has migrated toward its proximal end. This results in a distinct, albeit short, shaft separating broad, flattened proximal and distal humeral heads that are set at an angle of about 90 degrees from one another, producing the "tetrahedral" configuration common in more derived Paleozoic tetrapods. The capitulum for reception of the radius is larger and faces more directly ventrally than in anthracosaurs and Devonian tetrapods, indicating the ability to flex the elbow more completely. The capitulum is also more distally located. As a result, the bicondylar axis (connecting the centers of the radial and ulnar articular surfaces) is set at close to 90 degrees to the long axis of the humerus. In contrast with anthracosaurs and Devonian stem tetrapods, in which the axis of the zeugopod lies more or less parallel with the distal humeral expansion, the ulna and radius of temnospondyls are directed more ventrally, lying approximately at right angles to the distal humeral expansion.

The entepicondylar foramen, present in early temnospondyls (*Dendrerpeton*, *Balanerpeton*), is lost in later genera. Humeri are known from a number of presumed aquatic taxa (e.g., *Cheliderpeton*, Werneberg and Steyer 2002; *Trimerorhachis*, Williston 1915; *Dvinosaurus*, Bystrow 1938). They agree in general form with that of *Eryops*, but incomplete ossification of their distal and proximal ends makes detailed comparison difficult. There is a general trend for the largest members of individual clades to expand the relative proportions of the end of the limb and to add a suppinator process (e.g., *Edops*, *Eryops*, and *Trematops*, and the stereospondyl *Metoposaurus*), versus more lightly built members of the same families.

The humeri of Late Permian and early Mesozoic stereospondyls resemble those of Paleozoic temnospondyls in general form, although the distal expansion is set at an angle considerably less than 90 degrees to the proximal head (Warren and Snell 1991). Only in metoposaurs does the humerus approach the robust construction seen in the largest Paleozoic temnospondyls. In other stereospondyls, the proximal and distal ends are incompletely ossified and rugosities and processes for muscle attachment are not as well developed, presumably as a consequence of obligatorily aquatic habits.

As in most Paleozoic tetrapods, the zeugopod of temnospondyls is usually between 50 and 60% of the humeral length, although it approaches 65% in the presumed terrestrial dissorophoid *Amphibamus*. The ulna is slightly longer that the radius, largely due to the presence of an olecranon, which is normally much better ossified and prominent in large or more terrestrial taxa (fig. 13.11). In stereospondyls, the ulna and radius, other than tending to be less well ossified, are similar to those of Paleozoic temnospondyls.

The carpus of most temnospondyls is poorly ossified, so our knowledge of this aspect of their anatomy is limited to a few taxa. In *Balanerpeton* (fig. 13.11), the earliest known temnospondyl, the carpus is well ossified but incomplete. Three proximal elements, interpreted as a radiale, a large proximal centrale, and intermedium, are present, as are three distal carpals and an additional element interpreted as a distal centrale. As reconstructed, there would have been four centralia in total. Some temnospondyls have been described with fewer than four centralia (e.g., *Dendrerpeton;* Holmes et al. 1998), but this may simply reflect incomplete ossification or incomplete material. The carpus of most stereospondyls is unknown (and presumably unossified). Only in *Mastodosaurus* (Fraas 1889) and *Uranocentodon* (van Hoepen 1915) are a few ossified elements recorded.

The carpus of the large, amphibious *Eryops* is known from at least one well-ossified, articulated specimen (Miner 1925). Eleven elements appear to be present: an ulnare, intermedium, and radiale, four centralia, and four distal carpals. The ulnare, the most lateral of the proximal elements, articulates with the ulna; the radiale, the most medial, articulates with the radius; and the intermedium articulates with both epipodials. The fourth (proximal) centrale, which is the largest, articulates with the radius between the intermedium and the radiale. The remaining three centralia are arranged in a transverse series, beginning immediately distal to the fourth centrale and terminating at the medial side of the carpus. Each of the four distal carpals articulated with a metacarpal. Gregory et al. (1923), based on embryology of extant amphibians, comparison with urodeles, and what they interpreted as articular facets on the preserved carpal elements, reconstructed in addition a prepollex, vestigial fifth distal carpal, and a postminimus. However, the presence of these structures has never been confirmed in any *Eryops* specimen or, for that matter, any temnospondyl.

Phalangeal counts have been reconstructed in few temnospondyls and confirmed in fewer (figs. 13.11 and 13.14). A count of 2-2-3-3 has been reconstructed in *Balanerpeton* (although digit I is incomplete), *Amphibamus* (R. L. Carroll 1964b), and *Micromelerpeton* (Boy and Sues 2000), but the counts differ in other genera. In *Trimerorhachis* (Case 1935) it has been reconstructed as 2-2-3-2, and *Tambachia* (Sumida et al. 1998), 2-3-3-3. *Eryops* has been reconstructed with a count of 2-2-3-2, apparently based on incomplete fossil material and comparison with extant urodeles (Gregory et al. 1923). The same count has been given for *Apateon*, based on many well-articulated specimens (Boy and Sues 2000). In no known case does the count of an individual digit exceed the range presented above. With the possible excep-

tion of *Paracyclotosaurus* (D. M. S. Watson 1958), which has been restored with a five-toe manus, all stereospondyls resemble Paleozoic temnospondyls in having only four digits in the manus. The phalangeal formula varies but falls within the same range.

The femur of early temnospondyls (fig. 13.10), exemplified by *Balanerpeton* and *Dendrerpeton,* is little modified from that of archaic Devonian tetrapods in the possession of a prominent adductor blade that extends far distally onto the femoral shaft, and appears to bear a distally placed fourth trochanter. In more derived temnospondyls including *Eryops*, the crest still extends far out onto the femoral shaft but is most prominent at its proximal end, suggesting a proximal migration of the fourth trochanter, possibly correlated with increasingly terrestrial habits (Clack 2002b). The femora of stereospondyls, although generally not as well ossified, are otherwise similar to those of Paleozoic temnospondyls. The adductor crest and fourth trochanter, however, have migrated onto the proximal femoral head.

As in the front limb, the tibia and fibula of Paleozoic temnospondyls and stereospondyls are significantly shorter than the femur, being between 50 and 60% of its length, although this ratio approaches 70% in the dissorophoid *Eoscopus* (Daly 1994). The tarsus is only rarely adequately known among either temnospondyls or stereospondyls, because the elements either remained cartilaginous throughout life or were lost from the specimens. The best-preserved tarsus is exhibited by *Balanerpeton*. Ossification in this terrestrial temnospondyl (A. R. Milner and Sequeira 1994) is advanced, leaving little doubt as to its basic structure (fig. 13.14A). A fibulare, intermedium, and tibiale comprise the proximal row of elements. A large proximal (fourth) centrale wedges between the intermedium and tibiale, but does not articulate with the tibia. Three additional centralia are located distal to the fourth, extending from the middle of the tarsus to its medial edge. Evidence from three different specimens suggests that they are variable in size and shape (A. R. Milner and Sequeira 1994). Five distal tarsals are present, the fourth being the largest. One specimen exhibits a small ossification on the lateral edge of the tarsus lying between the fibula and fibulare that has been identified as a postminimus.

The phalangeal count of the pes in *Balanerpeton* is 2-2-3-4-3. This is a common count in temnospondyls (e.g., *Palatinerpeton*, Boy 1996; *Dissorophus, Eoscopus,* and *Amphibamus grandiceps*, Daly 1994; probably *Tambachia*, Sumida et al. 1998), but it is not universal. For example, a count of 2-2-3-4-4 has been reported for *Amphibamus lyelli* (Daly 1994), and 2-3-3-3-2 for *Trimerorhachis* (Case 1935). Scattered elements of the pes are preserved in *Paracyclotosaurus* (D. M. S. Watson 1958), but in no case can a phalangeal count be established in this or any other stereospondyls.

Lepospondyls

In contrast with the generally large-bodied and anatomically stereotyped labyrinthodont that dominated the amphibian fauna from the Upper Devonian through the Triassic were a number of highly distinct lineages of generally much smaller tetrapods, grouped as lepospondyls (R. L. Carroll et al. 1998). This includes five clades classified as orders—the Microsauria, Nectridea, Lysorophia, Adelospondyli, and Aïstopoda—and a single genus, *Acherontiscus*. The microsaurs and nectrideans generally have fairly well-developed appendicular skeletons, but the others show variable degrees of limb reduction and loss, accompanied by significant elongation of the trunk. The aïstopods are the most highly derived in the loss of nearly the entire appendicular skeleton, but were the first to appear in the fossil record (mid-Viséan). There is little convincing evidence for a particular pattern of relationships among these clades, which exhibit very distinct patterns of cranial morphology. None show obvious affinities with any of the labyrinthodont clades. Microsaurs and nectrideans show a mosaic of features that are otherwise present among either anthracosaurs or temnospondyls.

Microsaurs

Microsaurs appear to be the most conservative of lepospondyls in their overall skeletal anatomy, but they also show the most taxonomic diversity. Eleven families are recognized (R. L. Carroll and Gaskill 1978). The earliest known microsaur (unnamed) is from the Elvirian of Goreville, Illinois, equivalent to the Namurian E2 of Europe, slightly younger than the East Kirkton locality, near the base of the Upper Carboniferous (Lombard and Bolt 1999). Eight individuals are preserved in a single nodule, but their close juxtaposition has made thorough preparation, especially of the appendicular skeleton, very difficult. The anatomy of the skull and vertebral column show well-established microsaurian features, indicating a significant prior history since this lineage diverged from other tetrapod lineages.

The next oldest microsaur is *Utaherpeton*, from the Mississippian-Pennsylvanian boundary of North America, equivalent to the lowermost Namurian B (basal Bashkirian) of Europe (R. L. Carroll et al. 1991; R. L. Carroll and Chorn 1995). Two specimens of differing degrees of maturity show most of the skeleton and provide a basis for determining the primitive configuration of the appendicular skeleton. By the Westphalian A of Joggins, Nova Scotia, a host of lineages had arisen, outlining the diversity of the group that continued until its extinction after the Lower Permian.

As in early labyrinthodonts, the dermal elements of the shoulder girdle were extensive, especially in more primitive species (figs. 13.2–13.4). In the Joggins genus *Asaphestera*, the clavicular blades were large, triangular structures with the stem located posteriorly—the interclavicle a wide, diamond-shaped plate. The clavicular blades are much narrower in other genera. The anterior margin of the interclavicle is smooth in most adults but is fimbriated in immature animals and in at least one aquatic genus. As in labyrinthodonts, the configuration of the posterior portion of the interclavicle varied extensively from genus to genus. Most genera had a narrow process, like that of *Ichthyostega*, but of different lengths. The most distinctive interclavicle is that of *Microbrachis* (fig. 13.4L), with a persistently fimbriated anterior margin and small, paired processes extending obliquely posteriorly from the base of the posterior stem. The cleithra are narrow and angle to a variable degree over the top of the unossified suprascapula. The dermal bones of the shoulder girdle never show the deep sculpturing evident in most labyrinthodonts. This may be attributed to their small size, but this is contradicted by the deep sculpturing seen on all dermal elements of the shoulder girdle among equally small nectrideans.

The endochondral shoulder girdle is first adequately known in the Joggins genera (fig. 13.2). In large, mature animals it is extensive and well ossified. There is never a trace of subdivision between the scapular and corocoid portions, but in small and/or immature individuals, the coracoid portion fails to ossify. There are commonly two foramina ventral to the glenoid, which may be comparable to those seen in labyrinthodonts. At least one large genus, *Pantylus*, has a supraglenoid foramen, but only a single foramen in the coracoid. These openings are difficult to identify in small forms (R. L. Carroll 1968).

The glenoid had a helical surface for articulation with the humerus, as in temnospondyls. Several genera, including *Asaphestera* and *Trachystegos* from Joggins, and *Batropetes,* from the Lower Permian of Germany, have a distinct pit, of no obvious function, anteroventral to the glenoid. A similar depression in the scapula occurs in the Lower Jurassic caecilian *Eocaecilia* (Jenkins et al. 2006). A gap between the ossified dorsal surface of the scapulocoracoid and the cleithrum is assumed to have been occupied by a cartilaginous suprascapula, as in labyrinthodonts.

In contrast to the most primitive labyrinthodonts, the pelvis of microsaurs is lightly built and shows clear separation of the three elements (figs. 13.6, 13.7). The ilia of Carboniferous genera are notable for the presence of both an anterior dorsal and a posterior process, of approximately equal length. Later species belonging to several lineages independently reduced and eventually lost the distinction between the two processes and formed a single posteriodorsally extending blade, resembling that of early temnospondyls and amniotes. The very large genus *Microroter* has a single anteroposteriorly expanded plate, which was supported by three sacral ribs. The pubis is only unossified in the smallest specimens.

The earliest known microsaur humerus is that of an unnamed genus from the late Chesterian, of Goreville Illinois. It was described as having a distinct shaft, expanded proximal and distal ends that were at a nearly 90 degree angle to one another, and an entepicondylar foramen. Neither this nor any of the other early microsaurs has the anterior crest of the humerus that is characteristic of early anthracosauroids. This general pattern is retained thought the history of the order (fig. 13.9). The entepicondylar foramen is lost in a few species with relatively small limbs (e.g., *Rhynconkos*) and/or aquatic habits (*Odonterpeton*). The main variation of the humerus that can be seen among microsaurs is in the relative extent of the articulation surfaces, which is conspicuously greater in taxa with larger-sized individuals.

In the earliest microsaur known from an articulated skeleton, *Utaherpeton*, the very lightly built ulna (with a considerable olecranon) and the radius are nearly the length of the humerus. In general, the proportions of the individual bones correspond with the mass of the body, and the olecranon is lost in small and poorly ossified forms.

The carpus is unossified in the very small specimens of *Utaherpeton*, as well as in some larger and later taxa. It is best known in the massive *Pantylus*, in which there are the usual proximal row of ulnare, intermedium, radial, and a large centrale, but also a clearly defined pisiform (fig. 13.12B). No more than one centrale is known in other microsaurs. Surprisingly, there are five carpals in articulation with the metacarpals, although there are only four digits. This suggests that a fifth digit may have been present in the immediate ancestors of microsaurs. *Microbrachis* has three proximal carpals, and probably three distals, to support three digits. There is the same number of elements in *Hyloplesion*, but the central distal carpal has two lateral processes, which appear as if they were the fused remnants of once-separate bones.

No more than four digits are known in the manus of any microsaur. The complete phalangeal count is known in only a few: *Utaherpeton*, 2-3-3-3, and *Tuditanus*, 2-3-4-3 (exceptional in the unguals being expanded at their tips). The Lower Permian genus *Baptropetes* has 2-3-3-2. Among three-toed microsaurs, the count in *Microbrachis* and *Hyloplesion* is 2-3-3, and in the tiny *Odonterpeton*, 2-4-3.

The larger and more completely ossified femora (fig. 13.10Q–S) resemble those of small to medium-sized temnospondyls. In animals of smaller size the degree of distinction of the adductor crest and the internal trochanter is diminished. The ends are incompletely ossified in the smallest and most aquatic. The tibia and fibula of *Utaherpeton* are

about 45% the length of the femur, but in the more gracile *Tuditanus*, approximately 55%, and in the elongate Lower Permian *Goniorhynchus*, 47%.

The most primitive tarsal pattern known in microsaurs is that of *Hyloplesion*, from the Upper Carboniferous, with two to three centralia, in addition to the proximal tibiale, intermedium, fibulare, and five distal tarsals (fig. 13.14). This is close to the number in the Lower Carboniferous temnospondyl *Balanerpeton*, with three or possibly four centralia, and the anthracosaur *Gephyrostegus*, with three. The microsaurs *Pantylus*, *Tuditanus*, and *Batropetes* all have a large element medial to the calcaneum that has been identified as an astragalus. This element is certainly not homologous with the astragalus in amniotes (in the sense of having had a common phylogenetic ancestry of this element), but it may have had a comparable developmental origin through fusion of the intermedium with the tibiale and a centrale, as seems to be the case for *Tuditanus*, judging by the retention of suture lines between the elements (R. L. Carroll and Baird 1968).

Utaherpeton has a phalangeal count of the pes of 2-3-4-5-3, while *Tuditanus* has 2-3-4-5-4; *Microbrachus*, 2-3-4-4-3; *Hyloplesion*, 2-3-4-5-3?; and *Batropetes*, 2-3-4-4-1. The primitive count for the manus is close to that of temnospondyls, while that of the pes resembles anthracosauroids and amniotes.

Nectrideans

The nectrideans are readily defined by the highly conspicuous configuration of the caudal vertebrae in which the neural and hemal arches are nearly symmetrical, extensive, and compressed to form an effective sculling appendage. In the most primitive members of this order, only a single skull bone, the intertemporal, has been lost from the complement common to primitive temnospondyls and anthracosaurs.

Three families of nectrideans are recognized: the primitive, aquatic urocordylids; the diplocaulids, with highly derived skulls; and the small, apparently terrestrial scincosaurids (Bossy and Milner 1998). All have extensive and conspicuously sculptured elements of the dermal shoulder girdle (figs. 13.3–13.5). The interclavicle may be pointed or rounded posteriorly. This bone is marked by very large pits in *Scincosaurus*. The cleithra are narrow and unsculptured in urochordylids and *Scincosaurus*, but are greatly expanded dorsally as well as sculptured in diplocaulids. Among the diplocaulids, *Diceratosaurus* has an additional paired element posterior to the clavicular blade and lateral to the posterior end of the interclavicle. The scapulocoracoid is only well ossified in *Scincosaurus*, but in no genus is it well enough preserved to determine which of the foramina common to microsaurs and labyrinthodonts are present.

In contrast with anthracosaurs and early microsaurs, the

iliac blade is never bifurcate (figs. 13.6 and 13.7). In most genera it is a slender, dorsoposteriorly directed process. In *Scincosaurus* it extends dorsally and is expanded at its extremity. In the latter genus it appears fused to the pubis and ischium, but in the other genera the bones are suturally distinct. The pubis and ischium are slow to ossify or may remain cartilaginous in diplocaulids. Where known, the sacral rib (never more than a single pair) is typically a long, slender structure, with little terminal expansion, except for *Sauropleura scalaris* and *Scinosaurus*, in which it is shorter and more expanded distally.

All nectrideans have small but well-defined limbs; none have more than 26 presacral vertebrae. The shape of the humerus varies extensively within the nectrideans. That of *Urocordylus* has a long anterior crest and relatively little torsion, as in anthracosaurs, but lacks an entepicondylar foramen (fig. 13.12G). *Sauropleura scalaris* has lost the anterior crest but retains a supinator process in its place. The humerus of diplocaulids is small and poorly defined. In *Scincosaurus*, there is no anterior crest, the ends of the bone are twisted at approximately 90 degrees, as in temnospondyls and amniotes, and an entepicondylar foramen is present. A. C. Milner (1980) used the configuration of the humerus in scincosaurids to argue for their more terrestrial nature, which may have been primitive for nectrideans.

The ulna and radius are slender and generally poorly ossified except for *Scincosaurus*, in which they are robust, with a well-defined olecranon. In most species the carpus is unossified. *Sauropleura scolaris* is exceptional in having four elements, including one identified as a pisiform. *Scincosaurus* has a number of carpals as well, but their arrangement cannot be restored. *Urocordylus wandsfordi* appears to have a phalangeal count of the manus of 2-3-2+-3-2, but other genera certainly have only four digits. The count of *Ptyonius* and *Sauropleura sclaris* is 2-3-4-3.

Except for *Scinosaurus*, in which the femur is not well known, this bone is poorly ossified and shows little of the ventral ridge system. The tibia and fibula are short but well defined in *Urocordylus* and *Scincosaurus*, but more slender and with less well-defined articulating surfaces in other genera. *Sauropleura* is distinctive in having two large proximal tarsals, one identified as an astragalus, and the other, with the margin of a perforating foramen, termed the calcaneum, as well as three anterior distal tarsals (fig. 13.14H). *Scincosaurus* also has a number of tarsals, but their individual identity cannot be determined. The highest phalangeal count is that of *Lepterpeton*, which appears to be 2?-3-4?-4-3. The hindlimb of this genus is 50% longer than the forelimb. The phalangeal count of *Urocordylus* and *Sauropleura scalairs* is 2-3-4-4-2. *Ptyonius* and *Ctenerpeton* have only four digits, with a count of 2-3-4-3. *Keraterpeton* is the only diplocaulid

in which the phalangeal count can be established with assurance, as 2-3-3-3-3. *Scincosaurus* is thought to have had a count of 2-3-4-3-2.

Lysorophids, Adelogyrinids, and Aïstopods

Lysorophids, adelogyrinids, and aïstopods show progressive loss of girdle and limb elements, which is in reverse order to the timing of their first appearance in the fossil record (R. L. Carroll et al. 1998). The lysorophids are not known until the Westphalian B. Well-known specimens from the Westphalian D through the Lower Permian show elongation of the vertebral column from 69 to 97 presacrals and reduction in the relative size and number of elements of the girdles and limbs, but these are already so far reduced in the Westphalian as to preclude specific comparison with those of any other Carboniferous clades (Wellstead 1991).

The best-known pectoral girdle is that of *Pleuroptyx* from the Upper Carboniferous (fig. 13.5C). The dermal elements are nearly smooth. The interclavicle has a sharply tapering stem and clearly defined embayments for overlap of the broad clavicular blades. The cleithra are very slender rods. Only the scapular portion of the endochondral girdle is ossified. The interclavicle of the Lower Permian *Brachydectes elongatus* is a small, transversely expanded plate, the blade of the clavicle has been reduced, and the cleithrum may have been lost. The ilium, ischium, and pubis form a simple, triangular plate with a modest iliac blade, but do not co-ossify.

The limbs are tiny and poorly ossified. The humeri and femora are respectively the length of one and two trunk centra. In both limbs, the zygopodial elements are about 50% the length of the stylopod. There is no evidence of an entepicondylar foramen, and no carpal ossification is known. The Carboniferous *Brachydectes newberryi* probably has a phalangeal formula of 3-?3-3-2 in the manus. The femur has a short adductor crest and an internal trochanter. Three proximal tarsal are ossified, and the phalangeal formula is 2-3-3-3-2.

The Adelospondyli is known from four nominal genera from the Lower Carboniferous of Scotland (Andrews and Carroll 1991). Adelospondyls have at least 70 precaudal vertebrae, but there is no fossil evidence for the presence of the endochondral shoulder girdle, pelvis, or any limb elements in any specimen from the short duration of this assemblage. The dermal shoulder girdle of adelogyrinids broadly resembles that of anthracosauroids and temnospondyls, but without evidence for specific affinities with either (figs. 13.3D, E and 13.5D, E). What little is known of their appendicular skeleton provides no evidence for their relationships.

The snakelike aïstopods, from their first appearance in the Viséan, had already lost most of their appendicular skeleton (Anderson 2001, 2002, 2003; Anderson et al. 2003). None shows any trace of the limbs, the scapulocoracoid, or the pelvic girdle. The only bone of the appendicular skeleton that is common in the two major clades, Ophiderpetontidae and Phlegethontiidae, is a sigmoidal-shaped element the level of the third to fourth cervical vertebrae that is apparently a cleithrum (fig. 13.3F). A rhomboidal interclavicle with a fimbriated margin has been reported in *Ophiderpeton*, from the Westphalian B (fig. 13.5F).

The family Acherontiscidae is represented by a single specimen from the Lower Carboniferous of Scotland (R. L. Carroll 1967b). It is long bodied, but has two central elements per segment that vaguely resemble those of embolomeres, but with a skull and dentition superficially resembling those of gymnarthrid microsaurs. The only elements of the appendicular skeleton that are known are a rhomboidal interclavicle with a delicate pattern of radiating sculpture, and a narrow-bladed clavicle (fig. 13.5G).

As far as we know, none of the lepospondyl groups survived the Permian. The only taxon that may have relationships with the modern amphibian orders is the microsaur *Rhynchonkos,* which has been hypothesized as a sister taxon of the caecilians.

Modern Amphibian Orders

In contrast with the strong evidence for the common ancestry of Carboniferous tetrapods from among choanate sarcopterygians, there is continuing debate as to whether the frogs, salamanders, and caecilians shared an immediate common ancestry from among a single clade of Paleozoic tetrapods or evolved from two or three separate lineages. Three morphologically and phylogenetically distinct clades continue to be discussed as possible sister taxa for one or more of the modern orders: the temnospondyl labyrinthodonts, specifically the superfamily Dissorophoidea (A. R. Milner 1993; Trueb and Cloutier 1991); the lepospondyl order Lysorophia (Laurin and Reisz 1997); and the microsaur family Goniorhynchidae (R. L. Carroll 2000a).

Because of the great anatomical and adaptive differences between the Paleozoic and extant amphibian orders, whose fossil record is separated by 80 to 100 million years, it is difficult to begin a discussion of the evolutionary changes in the appendicular skeleton on the basis of what is known of the Paleozoic genera. No known Carboniferous or Permian amphibians show specific characteristics in the adult stage that are unequivocally comparable with those seen in the living orders. Rather, it is necessary to work back from the highly derived features of the anatomy of the living groups and attempt to recognize incipient stages in the early Mesozoic and late Paleozoic.

The appendicular skeletons of frogs, salamanders, and caecilians are remarkably divergent from one another (fig. 13.16). Salamanders are clearly the most primitive, in the retention of many features common to most Paleozoic amphibians. In most (including the probable stem taxa) the fore- and hindlimbs are well developed and of similar size and function.

In frogs, the pelvis and rear limb, as well as the axial skeleton, are always highly specialized in relationship to their saltatory locomotion. In most genera, this is associated with symmetrical use of the rear limbs for both terrestrial and aquatic locomotion. Living caecilians are unique among amphibians in having no trace of girdles or limbs in either adult stages or during development. These highly divergent features of the appendicular skeleton suggest a long period of divergence of the living orders, but must be considered in evaluating their probable ancestry.

Caecilians

Modern caecilians or gymnophonians, including approximately 175 species grouped in six families, are poorly known limbless forms now restricted to the damp tropics (M. H. Wake 1993a, 2003). Elements of the appendicular skeleton are, however, known from Jurassic and Cretaceous fossils. The most recent vestige is the presence of a femur that is apparently associated with unquestionably caecilian jaw bones and vertebrae from the Lower Cretaceous of Morocco (S. Evans and Sigogneau-Russell 2001).

Much more is known of the appendicular skeleton of *Eocaecilia* from the Lower Jurassic of Arizona (Jenkins and Walsh 1993; Jenkins et al. 2006). Described material includes scapulocoracoids, humeri, radii, ulnae, femora, tibiae, and fibulae. The proximal articulating surface of the humerus has a spiral shape like that of Paleozoic amphibians. The pelvic girdle has not yet been found, but two contiguous vertebrae can be recognized as sacrals. The proximal portion of the femur resembles that of the Lower Cretaceous caecilian. It has a prominent trochanter, strongly resembling that of salamanders. Digits are associated with both the fore- and hindlimbs, with a maximum of at least three associated with the latter. The exact vertebral count is not known, but there must have been approximately 70, within the range of the smallest number known in living species.

Eocaecilia is unquestionably identifiable as a caecilian on the basis of detailed study of the skull and lower jaws. The small size of the skull, the anterior position of the jaw articulation, the pattern of the bones of the skull roof, and the presence of medial rows of teeth on both the palate and the inside surface of the lower jaw most closely resemble the anatomy of the goniorhynchid microsaur *Rhynchonkos* among Paleozoic tetrapods (R. L. Carroll and Currie 1975;

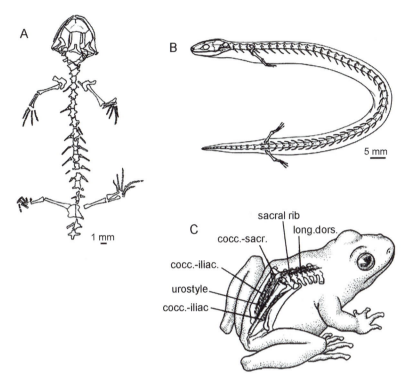

Figure 13-16 Modern amphibian orders. (A) *Valdotriton*, the oldest known representative of a metamorphosed salamandriform from the Lower Cretaceous (from S. E. Evans and Milner 1996). (B) Reconstruction of *Eocaecilia*, the oldest known caecilian from the Lower Jurassic of Arizona (after Jenkins and Walsh 1993). (C) The anuran caudopelvic mechanism for saltation. In stationary posture the ilio-sacral and sacro-urostylic joints are flexed, resulting in angulation between the ilia + urostyle and the sacrum + presacral vertebrae. During the early phase of a jump, this angle increases as the trunk extends. Extension of the trunk occurs through the action of the longissimus dorsi and coccygeo-sacralis muscles. This movement entails rotation at the ilio-sacral joints and the extension at the sacro-urostylic joint (after Jenkins and Shubin 1998). See appendix for abbreviations.

R. L. Carroll and Gaskill 1978; R. L. Carroll 2000a). The most conspicuous resemblance of *Rhynchonkos* to primitive caecilians is the presence of 36 or more presacral vertebrae and much-reduced limbs. Such a great degree of trunk elongation and limb reduction is also exhibited by other lepospondyls, but not among any labyrinthodonts. However, the specific configurations of the girdles and limbs (figs. 13.7N, 13.9D, 13.10R, 13.14E) are similar to those of other elongate microsaurs, and do not specifically point to caecilian affinities. The only exception is the common presence of a circular fossa in the scapulocoracoid anteroventral to the glenoid in *Eocaecilia* in several genera of microsaurs (fig. 13.2L), but this is not known in *Rhynchonkos*.

No convincing synapomorphies of the appendicular skeleton have been recognized that support a specific sister-group relationship between caecilians and either Salentia or Caudata on the basis of current knowledge of those taxa.

Salamanders

In contrast with caecilians, there is strong evidence that both frogs and salamanders evolved from among labyrinthodonts

(R. L. Carroll 2004; R. L. Carroll et al. 2004). The most general similarity is that most primitive members of both modern groups, in common with labyrinthodonts, reproduce via aquatic larvae with conspicuous external gills. Lepospondyls, in contrast, appear to have developed directly, for there is no evidence of a distinct gilled larvae stage in any of the orders, all of which are known from very small specimens that already show highly ossified cylindrical centra (R. L. Carroll et al. 1998).

Evidence from both adult and larval anatomy enables us to trace the divergence of the caudate and salientian lineages back to the Upper Carboniferous, within the temnospondyl superfamily Dissorophoidea (Holmes 2000). The primitive members of this group are known from relatively small animals, with a snout/vent length of the adults of approximately 10 cm. The best-known families from the Upper Carboniferous are the Amphibamidae, Micromelerpetontidae, and Branchiosauridae. The Amphibamidae is known from both adults and larvae, while the latter two families are known almost exclusively from larvae, suggesting at least facultative neoteny. It is this early dichotomy in life history traits that eventually led to the distinctive anatomies of the modern phyla. This is specifically associated with the divergence of the appendicular skeleton.

The most direct evidence for the origin of salamanders is the unique degree of similarity in the sequence of ossification of the skull bones in the temnospondyl labyrinthodont genus *Apateon* and that of primitive living salamanders (Schoch and Carroll 2003). However, many points of resemblance can also be seen in the appendicular skeleton. Our understanding of the common pattern of the appendicular skeleton of salamanders has been based on the more derived families, notably the Salamandridae, as illustrated by E. T. B. Francis (1934) and Duellman and Trueb (1986). That of the stem taxon, Hynobiidae, however, provides a more direct comparison with putative ancestors (figs. 13.17 and 13.18). An important feature of the species *Hynobius nigrescens* is its greater degree of ossification, which provides a more direct mode of comparison with Paleozoic tetrapods.

All salamanders have lost all three dermal bones of the shoulder girdle. Their reduction may already have been initiated in the Upper Carboniferous temnospondyl *Apateon*, in which the clavicular blades are reduced in size and the interclavicle is very slow to ossify. The endochondral shoulder girdle of *Hynobius nigrescens* resembles that of primitive labyrinthodonts in being ossified as a single element, with a conspicuous supracoracoid foramen. As in most salamanders, there is an extensive area of cartilage surrounding the bony portion, with three prominent areas termed the suprascapula, procoracoid, and coracoid, the latter of which overlap ventrally. Posteriorly, the coracoid cartilages articulate with the anterior margin of a triangular cartilaginous sternum, which is a neomorph.

The humerus has a long shaft, with proximal dorsal and ventral crests. There is no entepicondylar foramen, but a large radial condyle. The pattern of the carpals in *H. nigrescens* resembles that of Paleozoic tetrapods in the large number of independent elements, including the ulnare, intermedium, centrale, and radiale in the proximal row. An element below the radiale in modern salamanders is termed the prepollex but is presumably comparable to a bone recognized as a further centrale in Paleozoic amphibians. An additional centrale occupies a position in the middle of the carpus. The most striking difference between the carpus of all limbed salamanders and that of all Paleozoic amphibians is the presence of only a single element proximal to the base of digits I and II. This is referred to as the basale commune. E. T. B. Francis (1934) provided several lines of evidence that it develops from and presumably evolved from two separate bones, as seen in all Paleozoic tetrapods. The carpus of *Salamandra* is simplified by fusion of the ulnare, the intermedium, and possibly the proximal centrale into a single element.

The pelvis of salamanders broadly resembles that of Paleozoic amphibians in consisting of three elements. The pubis,

however, tends to be abbreviated and may be primarily cartilaginous. The femur has a very prominent proximal trochanter but a short adductor ridge. The tarsus of *H. nigrescens* resembles the carpus in the retention of three elements in the area of the centralia of Paleozoic temnospondyls, but a single distal tarsal articulates with the first two digits.

The most dramatic change in the appendicular skeleton among modern salamanders is the complete loss of the hindlimbs in sirenids and the great reduction of both fore- and hindlimbs in amphiumids. On the other hand, there is a striking conservatism in the phalangeal count among most members of the families Cryptobranchidae, Ambystomatidae, and Salamandridae, as well as primitive plethodontids. The phalangeal count of these salamanders is surprisingly close to that of most frogs, and also to dissorophoid temnospondyl labyrinthodonts from the Permo-Carboniferous (table 13.1).

Alberch and Gale (1985) made an extensive study of the degree of variability of the phalangeal count of the hindlimb in frogs and salamanders. They found that frogs were more conservative, with almost all species having a count of 2-2-3-4-3. Only two species had completely lost one toe (digit I). Three species had lost one phalanx, in all cases from the first toe. Two showed loss from digits I and V. Two genera had lost phalanges in digits I, V, and III. Urodeles showed somewhat more variability. Four-toed species occur in seven genera within the Hynobiidae, Salamandridae, Proteidae, and Plethodontidae. The highest count documented was 2-3-3-4-2 (*Ambystoma mabeei*), but the most common ambystomatid count is 2-2-3-4-2. A count of 2-2-3-3-2 is common to Cryptobranchidae, Hynobiidae, Dicamptodontidae, and some salamandrids. Among genera retaining five toes, the greatest phalangeal loss was recorded in *Botioglossa*: 1-2-1-1-1.

The striking conservatism of phalangeal counts in most adult salamanders is in strong contrast with the evidence for marked differences in the pattern of limb development that appears to distinguish salamanders from all other tetrapods (Shubin and Alberch 1986; D. B. Wake and Shubin 1994; Hinchliffe and Vorobyeva 1999). This subject is treated in more detail by Wagner and Larsson (chap. 4 in this volume).

Surprisingly, the differences in early stages of development—tissue condensation and early chondrification—do not appear to influence the appearance of the pattern of final adult ossification. On the other hand, in spite of having nearly the same phalangeal count, the sequence of ossification of the phalanges does differ. The sequence of phalangeal ossification was first studied by Erdmann (1933). Work now in progress shows considerable variation, but commonly ossification of one or more of the distal phalanges precedes those in a more proximal position in the same digit. This commonly occurs in the first and second digits of living sala-

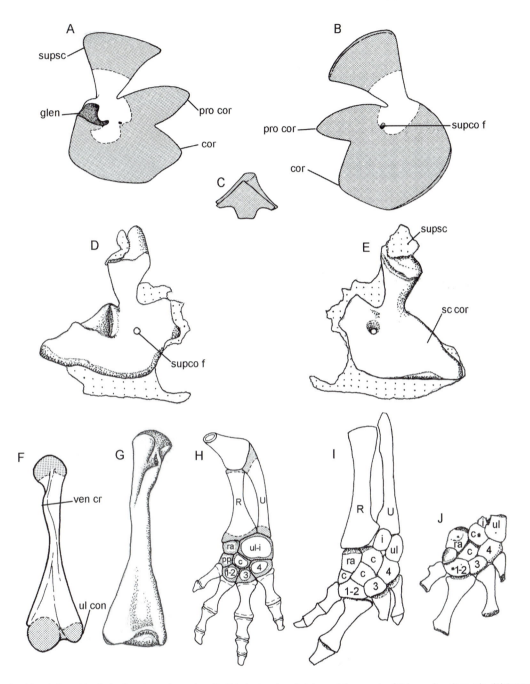

Figure 13-17 Shoulder girdle and forelimb of modern salamanders. (A, B) Lateral and medial views of the scapula of *Salamandra salamandra*. (C) Sternum of *S. salamandra*. (D, E) Lateral and medial views of the scapulocoracoid of *Hynobius nigrescens fuscus*. (F) Humerus of *S. salamandra* in ventral view. (G) Humerus of *Hynobius nigrescens*. (H) Left lower forelimb of *S. salamandra* in ventral view. (I) Left lower forelimb of *Hynobius nigrescens* in dorsal view. (J) Right lower carpus of *Hynobius nigrescens* in ventral view. Abbreviations: 1–2, basal commune, an element formed by the fusion of distal carpals 1 and 2; 3, 4, distal carpals; ra, radiale; see appendix for other abbreviations. (*Salamandra salamandra* after Duellman and Trueb 1986; *Hynobius nigrescens fuscus* based on specimen 22513 from the Herpetology Collection of the Museum of Comparative Zoology, Harvard.)

manders, but it is even more marked among the distal phalanges of all digits in the Permo-Carboniferous branchiosaurid *Apateon* (Schoch 1992).

Considering all the skeletal features of adult urodeles and the sequence of their ossification, their only plausible ancestry among Paleozoic tetrapods is among the branchiosaurid

dissorophoids, within the temnospondyl labyrinthodonts. Within the appendicular skeleton, the only major change is the loss of the dermal cleithrum, clavicle, and interclavicle, and the appearance of the neomorphic cartilaginous interclavicle. Although minor variation occurs, the same phalangeal count occurs in *Apateon* and both Mesozoic and ex-

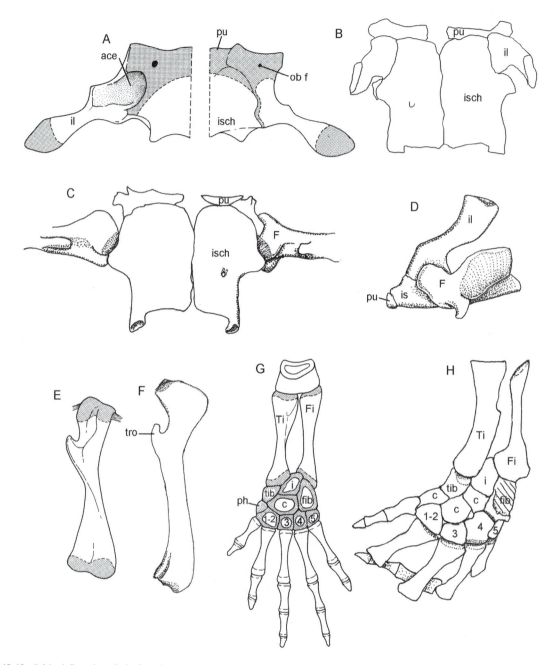

Figure 13-18 Pelvic girdle and rear limb of modern salamanders. (A) Ventral (left) and dorsal (right) views of the pelvis of *Salamandra salamandra*. (B–D) Dorsal, ventral, and lateral views of the pelvis of *Hynobius nigrescens fuscus*. (E, F) Ventral views of the femora of *Salamandra salamandra* and *Hynobius nigrescens*. (G) Left lower hindlimb of *Salamandra salamandra*. (H) Left lower hindlimb of *Hynobius nigrescens*. Abbreviations: 1–2, basal commune, an element formed by the fusion of distal tarsals 1 and 2; 3–5, distal tarsals; is, isch, ischium; see appendix for other abbreviations. (*Salamandra salamandra* after Duellman and Trueb 1986; *Hynobius nigrescens fuscus* based on specimen 22513 from the Herpetology Collection of the Museum of Comparative Zoology, Harvard.)

tant crown-group salamanders. The only difference in the carpus and tarsus of primitive salamanders is the fusion of the first and second mesopodials to form the basale commune. The great conservatism of the appendicular skeleton, except among the sirenids, amphiumids, and derived plethodontids, may be attributed to a similar biphasic life style, with recurrent neoteny.

Frogs

The appendicular skeleton of frogs is the most highly derived but also the most stereotyped of any order of terrestrial vertebrates, in association with their primarily saltatory mode of terrestrial locomotion (Vial 1973). The capacity for jumping has evolved in many vertebrate groups, but the anatomi-

Table 13-1 Phalangeal count of temnospondyl labyrinthodonts, salamanders, and frogs, showing high level of consistency

	Manus	Pes	Source
Most primitive known temnospondyls			
Balanerpeton woodi	2-2-3-3	2-2-3-4-3	Milner and Sequeira 1994
Superfamily Dissorophoidae			
"Branchiosauridae"	2-2-3-3	2-2-3-4-3	Nyřany (unpublished data)
Amphibamus grandiceps	2-2-3-3	2-2-3-4-3	(authors' observations)
Eoscopus lackardi	—	2-2-3-4-3	Daly 1994
Micropholis	2-2-3-3	2-2-3-4-3	Broili and Schröder 1939
Micromelerpeton credneri	2-2-3-3	2-2-3-4-3	Boy and Sues 2000
Apateon	2-2-3-3	2-2-3-4-3	Royal Ontario Museum, no. 44276
Apateon pedestris	2-2-3-2	2-2-3-4-3	Boy and Sues 2000
Apateon caducus	2-2-3-2	2-2-3-4-3	Boy and Sues 2000
Jurassic and Cretaceous salamanders			
Chunerpeton tianyiensis	2-2-2-?-2	2-2-3-4-3	Gao and Shubin 2003
Karaurus sharovi	2-2-3-2	2-2-3-4-3	Ivachnenko 1978
Jeholotriton paradoxus	2-2-3-2	2-2-3-3-2	Wang 2000
Valdotriton gracilis	2-2-3-2	2-2-3-4-2	Evans and Milner 1996
Mesozoic and Tertiary frogs			
Vieraella herbsti	2-2-3-3	—	Roček 2000
Notobatrachus degiustoi	2-3-3-3	2-2-3-4-3	Roček 2000
Notobatrachus degiustoi	2-2-3-3	2-2-3-4-3	Sanchiz 1998
Eodiscoglossus santonjae	2-2-3-3	2-2-3-4-3	Roček 2000
Palaeobatrachus grandipes	2-2-3-3	2-2-3-4-3	Sanchiz 1998

cal specializations that evolved during the ancestry of frogs are unique. As recently discussed by Jenkins and Shubin (1998), these involved elongation of the hindlimbs, with the proximal tarsals elaborated as a separate segment, strengthening of the zeugopod of both the fore- and hindlimbs by radio-ulnar and tibio-fibular fusion, reduction in the number of trunk vertebrae to between 6 and 10, and consolidation of the anterior caudal skeleton into a rodlike urostyle.

In contrast with all other terrestrial vertebrates, there is not a fixed connection between the sacrum and the ilium. Rather, the ilia lie beneath the sacral diapophyses, and can rotate around that joint, which is well anterior to the acetabulum. In addition, a hinge joint has developed between the urostyle and the sacrum, which are tightly linked by the coccygeo-sacralis muscle.

At rest, in a crouched position, the ilio-sacral and sacro-urostylic joints are flexed, with an angle of approximately 135 degrees between the trunk and the urostyle (fig. 13.16C). Extension of the trunk results from actions of the longissimus dorsi and the coccygo-sacralis muscles, assisted by the thrust of the forelimbs. This movement involves rotation at the ilio-sacral joint and extension at the sacro-urostylic joint.

In contrast with salamanders, frogs retain both the cleithrum and the clavicle, although they do lose the interclavicle. The ventral portion of the shoulder girdle is much modified (fig. 13.19). The coracoid is clearly distinct from the

scapula, and the unique omosternum and sternum have evolved anterior and posterior projections from the midline. A great variety of specific patterns of the ventral portion of the shoulder girdle have evolved in relationship to resisting the force of striking the ground (Emerson 1988b). In addition to great elongation of the iliac blade, the pelvis of all frogs is derived in the posterior rather than ventral position of the pubis and ischium, with the pubis frequently cartilaginous (fig. 13.19B).

All of the fundamental characteristics of the appendicular skeleton of living anurans were achieved by the first appearance of anurans in the fossil record in the Early Jurassic (Jenkins and Shubin 1998; Sanchiz 1998; Roček 2000; Gao and Wang 2001). The most important fossil for establishing the ancestry of frogs is the Lower Triassic genus *Triadobatrachus massinoti* (fig. 13.20B). The ilium of *Triadobatrachus* is extended far anteriorly, relative to the acetabulum, and the number of presacral vertebrae is reduced to 14. Only six caudal vertebrae are preserved, but their diameter quickly reduces posteriorly, suggesting that the tail was originally not much longer. The tibiale and fibulare are also somewhat elongated in *Triadobatrachus*. The fusion of the frontal and parietal bones is a synapomorphy shared with modern frogs, and the deeply embayed squamosals suggest support of a froglike tympanum.

Important advances in the evolution of the anuran shoulder girdle are also evident in another Lower Triassic salientian, *Czatkobatrachus* (Borsuk-Białynicka and Evans 2002). Notable is the apparent transformation of the supraglenoid foramen, common to Paleozoic labyrinthodonts, to the

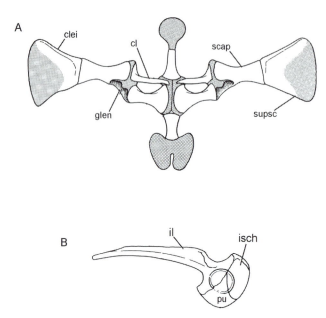

A

B

Figure 13-19 Pectoral and pelvic girdles of *Rana esculenta*. (A) Ventral view of pectoral girdle. (B) Lateral view of left pelvic girdle. See appendix for abbreviations. (After Duellman and Trueb 1986.)

scapular cleft that divides the base of the scapula in modern anurans.

The ancestry of frogs is less securely traced back to the Carboniferous and Permian temnospondyls with small body size and elongate limbs. The family Amphibamidae, within the Dissorophoidea, shows the greatest resemblance to anurans (Bolt 1969; 1991). The skull of *Doleserpeton*, from the Lower Permian, has a superficially froglike skull—very wide and low, with extremely wide interpterygoid vacuities. More specifically, the otic notch extends the entire height of the cheek, suggesting the presence of a large tympanum. The quadrate has a large dorsal process, which in modern frogs supports the tympanic annulus, and ossifies at very small body size. The stapes is small and light, without evidence for the distal attachment to the cheek that is present in salamanders and caecilians. The marginal teeth are pedicellate, as in frogs (but also caecilians and salamanders). The atlas vertebrae are uniquely anuran among Paleozoic temnospondyls in having a unipartite structure with widely spaced double condyles.

Reconstructions of the appendicular skeleton of *Doleserpeton* have not been published, but individual elements bear strong resemblance to those of the Upper Pennsylvanian genera *Amphibamus* (R. L. Carroll 1964b) and *Eoscopus* (Daly 1994) and the Lower Permian *Micropholis* (fig. 13.20A). All of these animals are relatively small but highly ossified, showing fully ossified carpals and tarsals. The tarsals of

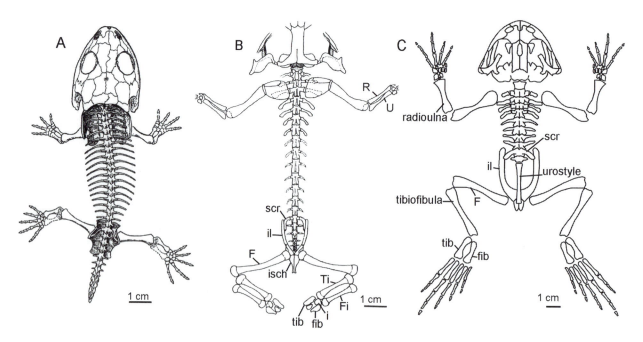

Figure 13-20 Skeletons of a primitive anuran and putative Mesozoic sister-taxa. (A) Lower Triassic amphibamid *Micropholis* (after Broili and Schröder 1937). (B) The Lower Triassic salentian *Triadobatrachus* (after Roček and Rage 2000). (C) The Upper Jurassic anuran *Notobatrachus* (after Sanchiz 1998). See appendix for abbreviations.

Eoscopus and *Doleserpeton* resemble those of *Triadobatrachus* in the degree of elongation of the tibiale and fibulare. The phalangeal counts of the amphibamids *Amphibamus, Eoscopus,* and *Micropholis* are identical with those of early frogs, and very similar to those of salamanders, a resemblance not shared by any groups of Paleozoic amphibians other than temnospondyls.

Although both frogs and salamanders can be traced to a single superfamily of temnospondyls, the Dissorophoidea, and putative sister taxa of both orders are known to have evolved pedicellate teeth by the Late Pennsylvanian or Early Permian, the two lineages may be distinguished by the Late Pennsylvanian by clearly divergent lifestyles and associated skeletal modification. Both are known to have had gilled, aquatic larvae, of similar general appearance. However, all members of the family Branchiosauridae, which include the putative sister taxa of salamanders, had a much extended period of growth within an aquatic environment, retaining external gills, gill rakers of a configuration suited for suck and gap feeding, and greatly delaying ossification of the carpals and tarsals. Formation of a distinct otic notch is also much delayed. In contrast, members of the Amphibamidae, including *Doleserpeton,* ossify their carpals and tarsals at very small size, and quickly develop the characteristic of an impedance matching middle ear. This suggests an early and rapid switch from an aquatic to a terrestrial way of life, as is characteristic of the majority of the crown group Anura.

Appendix: List of Abbreviations

ac	anterior crest
ace	acetabulum
ad cr	adductor crest
ade	accessory dermal element
ad r	adductor ridge
ano	anocleithrum
ast	astragalus
c	centrale
cal	calcaneum (equivalent of fibulare)
can	canals present in some Upper Devonian and Lower Carboniferous tetrapods
cl	clavicle
clei	cleithrum
cocc.-iliac	coccygeo-iliacus
cocc.-sacr.	coccygeo-sacralis
cor	coracoid
cor f	coracoid foramen
do pr	dorsal process of ilium
dpc	deltopectoral crest
ect	ectepicondyle
ent	entepicondyle
ent f	entepicondylar foramen
F	femur
Fi	fibula
fib	fibulare
glen	glenoid fossa
gl f	glenoid foramen
H	humerus
i	intermedium

icl	interclavicle
il	ilium
il pr	iliac process
isch	ischium
it	internal trochanter
it f	intertrochanteric fossa
lc	lateral centrale
long.dors.	longissimus dorsi
mc	medial centrale
ob f	obturator foramen
pdp	posterodistal process
ph	prehallux
pis	pisiform
pos pr	posterior process of ilium
pp	prepollex
pro cor	procoracoid cartilage
pu	pubis
pub pr	pubic process
R	radius
ra	radiale
rad	radial condyle
scap	scapula
sc cor	scapulocoracoid
scr	sacral rib
sgl f	supraglenoid foramen
ss f	subscapular fossa
sup	supinator process
supco f	supracoracoid foramen
supsc	suprascapula

Ti	tibia		ul	ulnare
tib	tibiale		ul a	articulation for ulna
tric cor	process for attachment of coracoid head of triceps		ul con	ulnar condyle
tro	trocanter		ven cr	ventral crest
U	ulna			

Acknowledgments

We wish to thank Mary Ann Lacey for assembling, labeling, and scaling all the figures. This research was supported by a succession of grants from the Natural Sciences and Engineering Research Council of Canada.

Chapter 14 Limb Diversity and Digit Reduction in Reptilian Evolution

Michael D. Shapiro, Neil H. Shubin, and Jason P. Downs

THE STUDY OF morphological rules, or trends, offered classical biologists the opportunity to address the mechanisms underlying the evolution of anatomical designs. Regularities in evolution suggested that common functional or developmental rules governed the transformation of structures. Parallelism is one such example. If different groups share a similar set of developmental mechanisms, then the parallel evolution of similar designs will be a frequent event in the history of life. Similarly, convergent adaptations can arise from the biomechanical or kinematic demands placed by selection on different taxa. Indeed, the selective and developmental explanations for trends are not mutually exclusive.

The reptilian limb offers an excellent opportunity to explore the persistence of, and mechanisms behind, morphological trends. The origin of new adaptive designs, from flight to cursoriality, involves changes in the shape, proportion, and number of bones in the limb. More often than not, bones are lost during the adaptive evolution of new designs. In fact, among living amniotes, only *Sphenodon* and some lizards retain the ancestral phalangeal formula (Romer 1956; Hopson 1995). Reduction is a common mode of limb evolution in other groups as well, being the predominant pattern of morphological change not only in reptiles but also in salamanders, frogs, and synapsids (including mammals). Indeed, this trend in the evolution of tetrapods is so common that it has been called a rule of morphological evolution (Sewertzoff 1931).

Here, we present an analysis of digital patterns in a phylogenetically and functionally diverse set of reptiles. These data

can serve to elucidate the possible functional and developmental mechanisms behind major transformations in reptilian limb structure, and our analysis of this database reveals regularities in the patterns of phalangeal and digital loss. This analysis further reveals qualitative differences between the reduction of digits and the loss of an entire digit. Patterns of digit reduction and loss can be clade specific and/or correlated to major differences in functional design, as revealed by the parallel acquisition of similar patterns in distantly related taxa.

What Is a Reptile?

The use of "Reptilia" to designate a monophyletic group is a recent development in a long taxonomic history marked by an ever-changing circumscription. Though Linnaeus (1758) earns credit for the name "Reptiles" and Laurentus (1768) for raising "Reptilium" to class status, both were building on the Aristotelian tradition to classify all nonmammal, nonbird tetrapods together. With the systematization of phylogenetic taxonomy (de Queiroz and Gauthier 1990, 1992, 1994), "Reptilia" was applied to the clade of amniotes stemming from the most recent common ancestor of turtles and saurians (lepidosaurs and archosaurs; Gauthier et al. 1988a, 1988c). By including Aves and excluding all of Synapsida, this usage makes "Reptilia" a monophyletic group, the most exclusive to include crown groups Testudines, Lepidosauria, and Archosauria.

Following Gauthier et al. (1988b), our usage of "reptile"

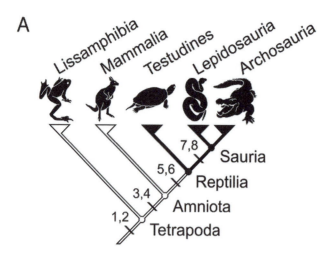

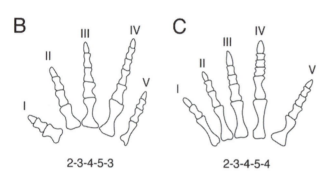

Figure 14-1 (A) Cladogram showing phylogenetic relationships of the Tetrapoda as proposed by Gauthier et al. (1988b). All names refer to a crown group. The eight numbers refer to the following characters: (1) pentadactyly; (2) dissociation of the pectoral girdle from the skull; (3) internal fertilization; (4) amniotic egg; (5) β-keratin in integumentary structures; (6) uricotelic metabolism; (7) large retroarticular process; (8) impedance-matching auditory system. Representative ancestral digit configurations of reptiles, based on the (B) manus and (C) pes of the basal amniote *Labidosaurus* (after Sumida 1997).

here refers to the species of Reptilia, the clade stemming from the most recent common ancestor of Testudinata and Sauria (fig. 14.1A). Although Aves is nested within Reptilia, this chapter will focus more specifically on Testudinata, Lepidosauria, and nonavian Archosauria. "Testudinata" refers to crown turtles and their turtle-shelled out-groups (Joyce et al. 2004); "Lepidosauria" refers to the most exclusive clade that includes Squamata (lizards, snakes, and amphisbaenians) and *Sphenodon* (Gauthier et al. 1988b); and "Archosauria" refers to the most exclusive clade that includes Crocodylia (crown crocodilians) and Aves (crown birds), and therefore also includes pterosaurs and nonavian dinosaurs (Gauthier et al. 1988b).

Primitive phalangeal formulae for testudinates, lepidosaurs, and archosaurs were assumed to match the basal amniote condition: 2-3-4-5-3 (digit I to digit V) for the manus and 2-3-4-5-4 for the pes (Romer 1956; fig. 14.1B, C).

Construction of Database and Mosaic Plots

A database of tetrapod phalangeal formulae was compiled from the neontological and paleontological literature and from museum specimens (see appendix table 14A.1). Additionally, an unpublished analysis of testudinate limb diversity was kindly made available by C. Crumly (later published in Crumly and Sánchez-Villagra 2004). We were unable to use many studies of digital formulae because the absence or reduction of a digit was not distinguished. For example, some authors use 0 in a digital formula to denote either reduction of a digit to a metacarpal or metatarsal only, or complete loss of phalanges and their supporting metapodial; such data were not useful for this analysis. As a familiar example, a phalangeal formula of 0-0-3-0-0 does not adequately describe the elements present in the manus of the modern horse because it implies that digit I, which is completely absent, and digit II, represented by a splintlike metacarpal, are equivalent. In our analysis, absent elements are denoted with X, those with metapodials only are denoted with 0, and those with phalanges are denoted by the number of phalanges present. Hence, we would have scored a horse as X-0-3-0-X.

Only cases of reduction from ancestral formulae (see below) were analyzed, and limbs with hyperphalangy of any digit, or lacking digits altogether, were excluded. Fossil forms were also excluded in cases of questionable completeness or homology. These criteria placed an emphasis on only utilizing cases in which issues of phylogenetic relationships and homology were well established. These criteria are discussed in the next section.

The frequency of different patterns of reduction is presented graphically as a set of charts known as mosaic plots (fig. 14.2). The appendage of each higher-level taxon is figured as five columns, with each digit a subdivided column. The length of the subdivisions within each column corresponds to the frequency of each kind of reduction seen in that digit. This type of plot provides a rapid visual assessment for the way in which digital reduction patterns differ between taxa and among digits in a particular taxon. Configurations listed in the appendix were reduced to a data matrix that tabulated kind of reduction with higher-level taxon. Reductions in each digit were coded in the following way (fig 14.2): one phalanx lost, all phalanges lost, entire digit lost, and unmodified.

Differences in the width of the columns in the mosaic plots are the product of an unavoidable artifact of homology assessment; not all types of variation could be factored into our coding scheme. Two kinds of coding difficulties exist, one where a single type of reduction can be coded in multiple ways, the other where a transformation cannot be coded at all. In the first instance, in a hypothetical lineage with only a

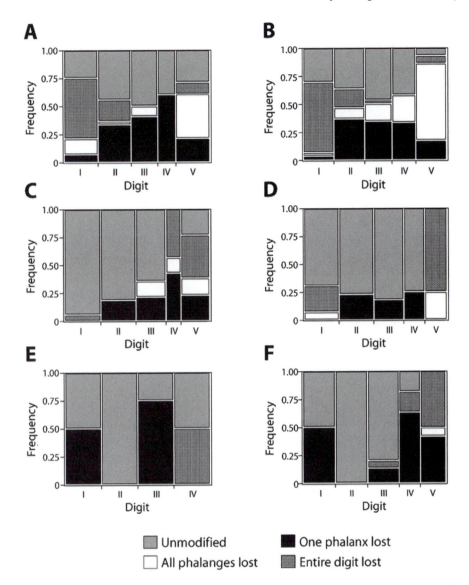

Figure 14-2 Mosaic plots of digit reduction patterns in (A) lepidosaur manus, (B) lepidosaur pes, (C) archosaur manus, (D) archosaur pes, (E) caudate manus, and (F) caudate pes. Each digit is presented as a separate column. The frequency of different reduction states (unmodified, one phalanx lost; all phalanges, but not metacarpal/metatarsal, lost; entire digit lost) is depicted for each digit.

single phalanx in the ancestral state, the loss of this phalanx would also involve loss of all phalanges—two different states in our coding scheme. The second type of difficulty is exemplified by testudinates. Digit IV of many testudinate configurations loses two phalanges relative to the ancestral amniote formula, a type of transformation that cannot be coded. Since our goal is to compare digit reduction patterns within and between taxa, we adopted a coding scheme that could be applied to all digits. Hence, the loss of two phalanges could not be coded for digit IV because such a loss is synonymous with the loss of all phalanges in digit I. Similarly, we could not score the loss of three phalanges in digit IV because digits III and V (manus only) have only three phalanges in the ancestral state. If we were to apply our mosaic

plot analysis to testudinates, therefore, the column for digit IV would be very narrow because our usable sample of digit IV reductions in this group would be small. Variation among the digital patterns of reptiles is so extreme that no single coding scheme could account for their patterns of reduction. The approach that we chose maximized the comparisons between homologous elements, as discussed below.

Digit Homology and Identity

The limbs of reptiles whose digital homologies are ambiguous were excluded, and only ossified elements in adult limbs were considered; however, adult configurations may not always reflect embryonic condensation patterns (see the section "Discussion"). We did not assign identities to individual phalanges, but present the phalangeal counts for each digit in the limb. For example, using this scheme, a digit I with a single phalanx would be scored as having lost a single, unidentifiable phalanx relative to the ancestral amniote condition, not as having lost specifically the first or second phalanx of that digit. This criterion allowed comparisons between taxa.

Digit homologies are sometimes difficult to establish when fewer than five digits are present. We assigned digit homologies and identities in reduced configurations based principally on patterns of loss within a clade, adult morphology, and developmental origins of digits. In the absence of developmental data, for example, intra- and interspecific variation revealed which digit was missing in four-digit turtle limb configurations. Within several species (*Chersina angulata*, *Geochelone elephantopus*, *Homopus areolatus*, and *Testudo* spp.), pes configurations vary between 2-2-2-2-1 and four digits with two phalanges each (Zug 1971). In these cases, we determined that the four-digit formula was 2-2-2-2-X for two reasons: (1) the remaining digits had the phalangeal formula of digits I–IV in the five-digit morph, and (2) this formula was more parsimonious than the alternative configuration of X-2-2-2-2. Assuming that the ancestor of these species had five digits, the X-2-2-2-2 configuration would necessitate both loss of the first digit and hyperphalangy of the fifth.

Other configurations were difficult to evaluate without making assumptions based on known developmental data. In pentadactyl amniotes, for example, the condensation of digit IV is the first to appear along the "primary axis" (Burke and Alberch 1985; Shubin and Alberch 1986; Müller and Alberch 1990; Shapiro 2002). Hence, in lepidosaurs with only one digit, we scored digit IV as present (e.g., *Lerista* and *Lialis*, X-X-X-0-X).

Results

Testudinata: Manus

Digit I is variable among turtle manus configurations examined. This digit is absent from one-third of configurations, while others lose one (42%; 5/12) or two (25%; 3/12) phalanges.

Digits II–V tend to be reduced to similar numbers of phalanges in all configurations. A complete (three phalanges) digit II is present in only two configurations (2-3-3-3-3 and 2-3-3-3-2), and two phalanges are lost in the most reduced configuration encountered (X-1-2-1-1; *Testudo*). More com-

monly, one phalanx is lost from digit II (75%; 9/12), yielding a total of two.

Likewise, two or fewer phalanges are typically retained by digits III–V. Digits II, III, and IV are reduced to two phalanges in all but two configurations (2-3-3-3-3 and 2-3-3-3-2), and digit V is similarly reduced in all but one configuration (2-3-3-3-3). Digit III never bears less than two phalanges, but digits IV and V are reduced to one phalanx in 33% (4/12) and 58% of (7/12) configurations, respectively.

Testudinata: Pes

Digit I is not subject to reductions in the turtle pes and always bears two phalanges.

Digit II loses a single phalanx from the ancestral three in one-third of configurations but is otherwise unaltered. Reductions of digit II occur in a variety of taxa, including *Proganochelys* (2-2-2-2-2), the earliest well-known testudinate (Gaffney 1990).

No turtle pes bears the ancestral amniote condition of four phalanges on digit III. Two-thirds of configurations lose a single phalanx, while the other one-third loses two.

Digit IV has four phalanges in the testudinate configuration with the least reductions (2-3-3-4-3, restricted to trionychids only) but is otherwise reduced to three or two phalanges.

The phalangeal count for digit V is the most variable in the testudinate pes, ranging from four phalanges to absent. Only *Dermatemys* (2-3-3-3-4) bears the ancestral four phalanges, while two configurations (2-3-3-4-3, trionychids only; and 2-3-3-3-3, many taxa) bear three. Most configurations bear two or fewer phalanges on digit V. Whether or not digit V is ever completely lost depends on the assignment homologies. According to Joyce (2000; pers. comm. 2004), the fifth metatarsal is rectangular in shape and, consequently, misidentified as a tarsal. This identification further implies that the first phalanx, when present, is commonly misidentified as the metatarsal. The phalangeal formula of 2-2-2-2-X, indicating the loss of digit V, would be revised to 2-2-2-2-0 under this interpretation. We retain the traditional assessment, pending the results of additional phylogenetic, morphological, and embryological studies on testudinate limb diversity.

Lepidosauria: Manus

Digit I loses at least one element in 75% (21/28) of reduced lepidosaurian configurations (fig. 14.2A). Most commonly, however, digit I is either complete (21%; 6/28) or absent (54%; 15/28). In 11% (3/28) of configurations, digit I is reduced to a metacarpal only, while a single phalanx remains in only two configurations (1-3-4-5-3, *Heterodactylus;* and 1-2-3-3-2, *Bachia*). Digit I reductions are not necessarily correlated with reductions of other digits: the full range of digit I reductions is observed without any alterations of digits II–V (1-3-4-5-3, 0-3-4-5-3, and X-3-4-5-3).

Digit II reductions tend to be bimodal: either one or all phalanges are lost. The only lepidosaur encountered that loses two phalanges on digit II is the skink *Anomalopus* (X-1-2-2-0). Of the six configurations with no digit II phalanges, five also lack a metacarpal. In all but one configuration (2-2-3-3-2, *Moloch*), digit II reductions always occur with reductions of digit I only, or digit I and another digit.

Digit III retains the ancestral phalangeal count of four in 36% (10/28) of reduced configurations, and loses a single phalanx in 29% (8/28). The eight configurations in which digit III is reduced to two or fewer phalanges also exhibit severe reductions to digit I (i.e., loss of all phalanges or the entire digit) and usually to digit V. Digit III is lost entirely only in lepidosaurs lacking a manus.

Digit IV loses at least one phalanx in most reduced configurations. Nevertheless, digit IV bears phalanges in all configurations not lacking a manus. A single phalanx is lost in 29% (8/28) of reduced configurations. When two are more phalanges are lost, however, all other digits are usually reduced as well. An exception to this trend is the gekkonid *Nephrurus* (2-3-3-3-3), in which only digits III and IV are reduced. Furthermore, in configurations with three or fewer digit IV phalanges, 64% (9/14) are missing digit I and all phalanges of digit V.

Digit V undergoes frequent reductions: at least one phalanx is missing in 71% (20/28) of reduced configurations. No configuration loses two digit V phalanges, but half (14/28) lose all three. Loss of the fifth metacarpal is comparatively infrequent, however, occurring only in three of the most reduced configurations (X-X-2-3-X, X-X-0-3-X, and X-X-0-2-X). Loss of a single phalanx from digit V is not necessarily correlated with losses from other specific digits, but loss of all three phalanges is correlated with the complete loss of digit I in all but two cases (2-3-4-4-0, *Chalcides;* and 0-3-4-4-0, *Hemiergis*).

Lepidosauria: Pes

Digit I is the most frequently reduced in the lepidosaurian pes (fig. 14.2B), with 70% (21/30) of configurations exhibiting loss of at least one phalanx. Digit I is completely absent from 63% (19/30) of reduced configurations, and two other cases lose either one (1-3-4-5-0, *Anotosaura*) or both (0-3-4-4-0, *Hemiergis*) phalanges.

As in the manus, digit II reductions in the pes are bimodal, with either one or all phalanges lost. Of the latter

cases, most lose digit II entirely. Digit II reductions only occur when either digit I or V is also reduced, but reductions of the outer digits do not necessarily affect digit II. For example, the ancestral phalangeal counts of digits II–V are unaffected in the configuration X-3-4-5-0 (*Hemiergis*).

Twenty-seven percent (8/30) of reduced configurations lose a single phalanx from digit III, while 30% (9/30) lose two or more. All of the latter cases are highly reduced forms in which at least two digits other than digit III have lost all phalanges. Digit III phalanges are lost only when at least one digit—but usually more—is reduced. Metacarpal III is highly stable and is lost only in monodactyl (X-X-X-0-X) and digitless configurations.

Digit IV is the last to be lost in its entirety, but it is subject to frequent reductions. One or more phalanges are lost in 67% (20/30) of configurations, and two or more are lost in 40% (12/30). When two or more phalanges are lost from digit IV, all other digits are usually reduced as well. Three configurations restricted to two genera are the exceptions, and all retain a complete digit I: 2-2-2-0-0 and 2-2-3-3-0, *Bachia;* and 2-2-3-3-2, *Moloch*.

Digit V reductions are the most frequent in the lepidosaurian pes, characterizing 90% (27/30) of configurations. All phalanges are typically lost when this digit is reduced, but one or two are occasionally lost as well. Digit V reductions can occur independently, but the loss of all digit V phalanges co-occurs with the complete loss of digit I in 73% (16/22) of cases. Although digit V is subject to frequent reductions, it is rarely lost entirely: only the two most reduced lepidosaurian configurations (X-X-0-0-X and X-X-X-0-X; and limbless forms) lose the fifth metatarsal.

Archosauria: Manus

Digit I is remarkably stable among archosaurs (fig. 14.2C). Hadrosaurs lose this digit entirely (X-3-3-3-3), but it otherwise retains the ancestral phalangeal count of two in all configurations.

Digit II also shows little variability, with phalangeal losses occurring in 24% (4/17) of reduced configurations. These reductions occur in the graviportal stegosaurs (2-2-2-2-1) and sauropods (2-2-1-1-1 and 2-1-1-1-1), and the bipedal theropod *Compsognathus* (2-2-0-X-X).

Digit III is never lost but exhibits a full range of phalangeal counts. At least one phalanx is lost in 47% (8/17) of configurations, and most of these lose at least two phalanges. In all but one of the latter cases (2-3-0-X-X, *Tarbosaurus*), digit II is also reduced.

Digit IV is missing at least one phalanx in all reduced configurations. Two phalanges are lost in 35% (6/17) of cases, while three or more are lost in 47% (8/17). In all five theropod configurations, digit IV either has one phalanx, has no phalanges (but retains a metacarpal), or is absent. These reductions are also accompanied by severe reductions of digit V.

Digit V is highly variable. At least one phalanx is lost in 82% (14/17) of configurations, and half of these lose all digit V phalanges. Five configurations lose digit V entirely, and all but one of these (2-3-4-4-X, numerous pterosaurs) represent theropods.

No independent reductions of individual digits occur in the archosaur manus. That is, at least two digits are affected by reductions when reductions occur. However, when only two digits are affected by reductions, digits IV and V are always involved.

Archosauria: Pes

In the archosaur pes (fig. 14.2D), digit I has either all or none of its phalanges. One-third of reduced configurations have no digit I phalanges, and most of these also lose the first metacarpal. Digit I reductions are always correlated with the loss of digit V.

Digit II is never lost in the pes and has either two or three phalanges. Two of three configurations with two phalanges show reductions in all five digits; only *Camarasaurus* (2-2-2-1-1) reduces digit II and retains the ancestral two phalanges in digit I.

Digit III bears four phalanges in most configurations, but 15% (2/13) of cases lose a single phalanx, and the same proportion loses two. Phalangeal loss in digit III always entails reductions in at least three other digits.

Digit IV varies in phalangeal number from the ancestral five to a highly derived one; however, at least four phalanges are retained in most configurations. Only large dinosaurs such as sauropods (2-3-4-2-1, 2-3-3-2-1, and 2-2-2-1-1) and stegosaurs (X-2-3-3-X and X-2-2-2-X) are exceptions to this trend.

Digit V loses phalanges in all reduced pes configurations, and only rhamphorhynchoid pterosaurs (2-3-4-5-2) have more than a single phalanx. In 62% (8/13) of cases, all phalanges are lost, and most of these involve the loss of the entire digit. Digit V reductions can occur independently of reductions in other digits, as evidenced by several configurations with reductions in this digit only: 2-3-4-5-2, 2-3-4-5-1, 2-3-4-5-0, and 2-3-4-5-X.

Discussion

This study reveals different modes of digit reduction and loss that occur at distinct levels of organization. First, at the level of a single digit, some or all phalanges (but not the supporting metacarpal or metatarsal) may be lost. Alternatively, the

entire digit may be lost. As we discuss below, partial and complete losses of the digit skeleton represent qualitatively different developmental and evolutionary phenomena. Certain digit reductions or losses tend to occur in concert with others. Some of these correlations differ among groups, whereas others occur independently in unrelated lineages. Another type of change involves alteration of the absolute size and proportions of the autopod. This type of modification was addressed by Holder (1983) and cannot be inferred directly from our data. Second, at a higher level of organization, the integration of digit reductions or losses influences the configuration of the autopod as a whole, including its overall shape. For example, trends of limb reduction that emphasize the loss of external digits often yield more narrow or elongate autopodia than those that consist of uniform reductions across all digits.

The distinct trends we observe within and between clades invite further exploration of potential common mechanisms underlying the evolution of limb reduction in reptiles. Specifically, we will address the role of shared external deterministic agents, namely selection for a particular limb function, and the internal developmental parameters that may characterize each group.

Digit Reduction versus Loss

Although the distinction is seldom emphasized, the loss of all phalanges in a digit (a type of digit *reduction*) is not the same as the loss of all phalanges plus their supporting metacarpal or metatarsal (digit *loss*). Lepidosaurs, for example, frequently reduce manual and pedal digits I and V, but patterns of complete loss differ dramatically. In the lepidosaurian manus, most (15/19) of the configurations without phalanges on digit I also lose the first metacarpal. This frequency of loss differs considerably from that of digit V, however, in which less than one-quarter (3/14) of configurations lacking phalanges also include loss of the metacarpal. The disparity is even greater in the pes, wherein all cases but one (19/20) of complete phalangeal loss in digit I also exhibit loss of the whole digit, but only a small proportion of configurations (2/22) show the same pattern in digit V. In both the manus and pes, digit V is lost only in highly reduced configurations that also lose at least two other digits, whereas digit I can be lost independently. These results demonstrate that digit V is subject to frequent *reductions* but is highly resilient to complete *loss*, whereas *loss* of digit I is more common than *reduction* by one or more phalanges only.

From a developmental perspective, the loss of phalanges of a digit is qualitatively different from the loss of the entire digit. When phalanges are lost, a developing digit primordium segments into fewer elements than in the ancestral form (Storm and Kingsley 1996). Absence of a digit, on the other hand, implies that the digit never forms in the first place (but for discussions of limb element "loss" through nonossification, fusion, or resorption see Ewart 1894; Mettam 1895; Hinchliffe and Johnson 1980; Rieppel 1992b; Galis et al. 2001; Larsson and Wagner 2002; Kundrát et al. 2002). This distinction is important because the absence of all phalanges (but presence of a metapodial) and the complete absence of a digit are frequently synonymized in the literature; however, these two phenomena are not developmentally or phylogenetically equivalent.

With this distinction in mind, we can evaluate differences in manual and pedal evolution among reptiles. For example, whereas digits tend to be *lost* from the periphery of the autopod in lepidosaurs, digits may be frequently *reduced* in the center. Indeed, in the manus, digit IV is reduced by at least one phalanx in 82% of configurations, more than any other digit. Likewise, in the pes, digit IV is reduced more often than either digit II or digit III. The frequency of loss of *all* phalanges in a digit, however, follows the order I > V > II > III > IV, a sequence often called "Morse's Law" (Morse 1872). Morse based this generalization primarily on birds, but it has since been co-opted to include lizards and other tetrapod groups. Morse's Law is usually considered to be the order of loss of complete digits, but our analysis suggests otherwise. Based on the configurations we examined, loss of entire digits, seen most commonly and completely in lepidosaurian limbs, follows the sequence I > II > V > (III, IV) in the manus, and I > II > V > III > IV in the pes. In the context of complete digit loss, therefore, Morse's Law is not upheld. The observed order of loss is, however, more reflective of the sequence of digit chondrogenesis in amniotes (Mathur and Goel 1976; Burke and Alberch 1985; Shubin and Alberch 1986; Müller and Alberch 1990; Shapiro 2002). That is, digits are lost in the reverse order of digit primordia appearance. Patterns of chondrogenesis may, therefore, be a predictor of the sequence of digit loss, although not necessarily of individual digit reduction.

In summary, patterns of digit reduction and loss in lepidosaurs demonstrate that these two phenomena are not equivalent. Digit V is frequently reduced, indicating a high degree of evolutionary plasticity in its phalangeal number. In contrast, digit V is rarely completely lost, thus highlighting its remarkably conserved presence in different types of configurations.

Integrated and Correlated Digit Reductions: Shaping the Hand and Foot

Preferential reduction and loss of the outer digits is a principal mode of pedal evolution among lepidosaurs and archosaurs. Among archosaurs, for example, two pedal configurations representing theropods and bipedal ornithischians

(0-3-4-5-X and X-3-4-5-X) involve reduction or loss of outer digits with retention of the ancestral formula in the central ones. However, a different pattern of loss emerges in the archosaurian manus: reductions and losses are biased toward the postaxial digits, with only modest alterations of the preaxial digits. Digits I and II are nearly always present in these groups, whereas digits IV and V are the most frequently absent.

Another mode of reduction contrasts with the loss of digits and instead produces uniform phalangeal formulae across all digits. Such is the case among testudinates, in which phalangeal counts tend to be either two or three. Furthermore, in no turtle manual configuration does the phalangeal count of one digit differ from another by more than one (except when digit I is entirely missing). A similar trend emerges among certain archosaurs in both the manus and pes. These include the sauropods (manus: 2-2-2-2-1, 2-2-1-1-1, and 2-1-1-1-1; pes: 2-2-2-1-1) and several ornithischians (manus: 2-3-3-3-3, 2-3-3-3-2, 2-2-2-2-1, and X-3-3-3-3; pes: X-2-3-3-X and X-2-2-2-X).

Thus, similar reduction trends tend to occur repeatedly among reptiles, and some of these trends are independent of phylogeny. These trends yield combinations of individual digit reductions and losses, but we have yet to examine why such correlated digit reductions are indeed correlated. The convergent appearance of reduction trends in distantly related taxa suggest that limb function may be a common factor. On the other hand, the limited number of observed trends suggests developmental constraints on the ways a limb can be reduced. Below, we consider the role of function as a target of selection in certain reduction trends, following which we discuss the potential underlying developmental mechanisms.

Functional Correlates of Digit Reduction

Reduction Patterns among Archosaurs Correlate with Functional Differences Between and Within Organisms

A single monophyletic group, Archosauria, contains taxa with three major trends in digit reduction. These different patterns are especially pronounced in the manus, where modes of reduction are seen in different functional settings.

First, in the manus of theropod dinosaurs, digits I–III are typically retained with full complements of phalanges, whereas digits IV and V are highly reduced or absent (fig. 14.3A–D). All theropods are bipedal, and their hands are often well developed with an opposable first digit, suggesting a grasping function (Romer 1956). Evolutionary retention of the preaxial digits and loss of the postaxial digits appear to preserve this specialization (Sereno 1997). Interestingly, similar digit reductions, localized to the postaxial aspect of the autopod, are also observed among salamanders (fig. 14.2E,

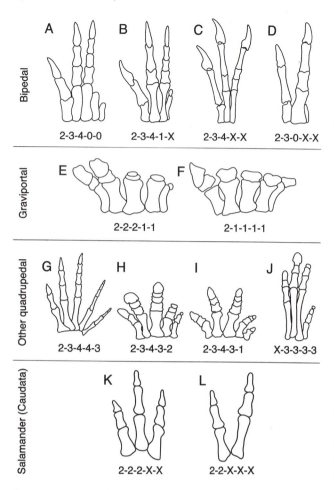

Figure 14-3 Patterns of manual digit loss and reduction among archosaurs. In all diagrams, distal is up and anterior is to the left. Phalangeal formulae are indicated below each manus. (A–D) Manual digits of the bipedal theropods: (A) *Eoraptor*, (B) *Syntarsus*, (C) *Harpymimus*, and (D) *Tarbosaurus*. The principal reduction trend within theropods involves phalangeal and digit losses from the postaxial side of the autopod. Manual digits of (E) the graviportal ornithischian dinosaur *Stegosaurus* and (F) the sauropod *Diplodocus*. In these quadrupeds, phalanges are lost essentially uniformly across all digits but the first. (G–J) Manual digits of other quadrupedal or facultatively quadrupedal archosaurs. The digits of (G) *Crocodylus*, (H) *Centrosaurus*, (I) *Leptoceratops*, and (J) *Anatosaurus* feature modest reductions, with the central digits emphasized. (K, L) Pedal digits of the salamanders *Amphiuma* (C) and *Proteus* (D). (A, E after Sereno 1997; B after Raath 1985; C after Barsbold and Osmólska 1990; D, F after Norman 1985; G after Romer 1956; H after Lull 1933; I after Brown and Schlaikjer 1940; J after Lull and Wright 1942; K, L after Shubin et al. 1995.)

F). All salamanders are obligate quadrupeds that employ a sprawling gait, so their manus and pes have a fundamentally different function from that of the manus of bipedal dinosaurs. The most reduced-limbed forms, such as *Amphiuma* (2-2-2-X-X) and *Proteus* (2-2-X-X-X; hindlimbs; fig. 14.3K, L), are elongate, undulatory swimmers with drastically shortened limbs. Despite these functional differences, salamanders and theropods are the only tetrapod groups in which the loss of postaxial digits (i.e., digits IV and V) is not accompanied by the loss of anterior digits as well.

In a second trend, exemplified by transformations in the

forelimbs of sauropod and stegosaur dinosaurs, convergent evolution of similar phalangeal formulae is seen in a different functional category. Unlike theropods, sauropod and stegosaur mani tend to be reduced to only one or two phalanges on each digit, and no digits are lost (fig. 14.3E, F). These two clades are not closely related (Sereno 1997), but both include massive, quadrupedal animals that likely assumed a graviportal posture and gait (Romer 1956). Hence, unlike theropods, their mani were regularly in contact with the ground and were primarily weight-bearing, not grasping. The uniform trend of digit reduction shared by these groups, therefore, appears to be correlated with obligate quadrupedality and a graviportal posture.

A third archosaur reduction pattern comprises intermediate functional regimes in the manus of facultative quadrupeds, nongraviportal obligate quadrupeds, and sprawling or semi-aquatic quadrupeds (fig. 14.3G–J). These taxa are characterized by modest reductions compared to the graviportal dinosaurs, and central digits tend to be the longest or bear the most phalanges, unlike the theropods. Some of these mani are clearly used for weight bearing, as evidenced by hooves on the central digits of hadrosaurs (fig. 14.3J).

Notably, only one of these three trends is also observed in the archosaurian pes. With the exception of *Apatosaurus* (2-3-4-2-1), graviportal dinosaurs tend to have nearly uniformly reduced digits. Digit losses, however, characterize the other major trend in archosaur pes reduction, occurring primarily among bipeds. Nearly two-thirds (62%) of the archosaurian pes configurations considered show loss of all phalanges from digit V, and nearly half (46%) show loss of the entire digit. Digit I can also be greatly reduced in length relative to digits II–IV, show a reduced phalangeal count, or be entirely lost (fig. 14.4A–C). Digits II–IV retain their ancestral phalangeal numbers and are likely the only functional digits in the foot. Hence, through localized reductions of the outer digits, the central digits are emphasized in these forms. Their elongate foot morphology and reduced outer digits invite functional interpretations of cursoriality (see Carrano 1999). Cursorial mammals follow a similar reduction trend. In mammals, however, the most reduced forms retain a complete digit III (the modern horse) or digits III and IV (artiodactyls, such as antelope and deer) as the functional digits, rather than digits II–IV (fig. 14.4D–F).

Localized Outer Digit Reductions in Lepidosaurs

Other reptilan groups converge upon each of the three archosaurian trends described above. First, like the hindlimbs of reduced-limbed archosaurs such as theropod dinosaurs, most reduced-limbed lepidosaurs preserve the central digits (i.e., digits II–IV; fig. 14.4G–J), but functional considerations differ. Unlike dinosaurs, which have a parasagittal posture, lizards typically maintain a sprawling posture and gait. Nev-

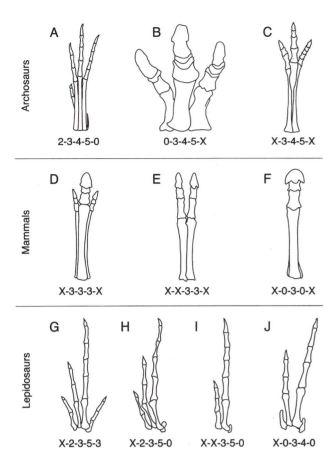

Figure 14-4 General patterns of outer digit loss in the pes of archosaurs, mammals, and lepidosaurs. In all diagrams, distal is up and anterior is to the left. Phalangeal formulae are indicated below each pes. (A–C) Pedal digits of the dinosaurs (A) *Compsognathus*, (B) *Iguanodon*, and (C) *Struthiomimus*. Each pes exhibits loss or severe reduction of one or more outer digits. Metatarsal 1 (described, but not figured, by Norman 1990) is present in *Iguanodon* (B) but is not visible in this drawing. (D–F) Pedal digits of (D) the cursorial perissodactyl *Miohippus*, (E) the artiodactyl *Poebrotherium*, and (F) the perissodactyl *Equus*. Pedal digits of (G–I) the scincid lepidosaurs *Lerista* and (J) *Hemiergis*. Both genera contain species with multiple limb morphologies (see appendix table 14A.1). In panels (G–I), digit IV is not reduced, whereas all others are either reduced or lost. In panel (J), only two digits with phalanges remain, and both are reduced in phalangeal count. (A after Ostrom 1978; B after Norman 1985; C after Osborn 1916; D after Osborn 1918; E after Scott 1940; F after Romer 1966; G–I after Greer 1990.)

ertheless, digits are typically lost or severely reduced such that digits III and IV are conserved. However, in contrast with archosaur pedal configurations, in which reductions are usually restricted to digits I and V, lepidosaurs are subject to frequent reductions of the central digits. This trend is especially notable in the Australian scincid *Lerista*, which provides the best example of graded limb reduction in any tetrapod (Greer 1987, 1989, 1990; fig. 14.4G–I). Many cases of digit reduction and loss in lizards are accompanied by body elongation and a de-emphasis on limb-powered locomotion (Gans 1975). *Lerista* species with more than three manual digits tend to move above the ground surface using their limbs, whereas forms with fewer phalanges or digits—and

typically shorter limbs—are subsurface foragers that move principally by lateral body undulation (Greer 1989).

In contrast, the phalangeal formulae of the functional digits of cursorial archosaurs are typically unaltered and are the distal components of elongate, but not shortened, limbs. Hence, the patterns of reduction shared by archosaurs (hindlimbs only) and lepidosaurs (forelimbs and hindlimbs) are superficially similar with respect to the retention of central digits, yet different with respect to the details of phalangeal loss. The retention of complete central digits in archosaurs may be related to functional constraints of parasagittal locomotion, whereas in lepidosaurs, which typically employ a sprawling gait and often utilize undulatory trunk locomotion in forms with highly reduced limbs, functional constraints for the maintenance of complete digits are likely weaker or absent. Unlike the pedes of some lepidosaurs, the reduced pedes of archosaurs are necessarily used in locomotion, and thus strongly developed central digits are retained.

Uniform Reductions in Turtles

The testudinate manus and pes both undergo uniform phalangeal reductions similar to those seen in a second functional class of archosaurs, the graviportal dinosaurs (fig. 14.5). In the manus, most (10/12) reduced turtle configurations bear only one or two phalanges on each digit. In the pes, with only two exceptions, phalangeal counts are subequal and range between one and three. While the general reduction patterns in turtles are similar to those of graviportal archosaurs, drawing functional analogies between them remains difficult; however, Zug (1971) notes that the extreme reductions in testudinoids (pond turtles and terrestrial tortoises) are associated with a shift to terrestrial locomotion.

In summary, the reduction patterns of distantly related clades—such as cursorial dinosaurs and lepidosaurs, or testudinates and graviportal dinosaurs—may show broad morphological similarities. These commonalities, however, are not always easily reconciled by functional interpretations. For example, the mode of digit loss shared between theropods and lepidosaurs is not explained by a functional similarity. Intrinsic or developmental constraints ("generative" constraints of Richardson and Chipman 2003) may also play a role in reduction patterns; therefore, we will now consider the role of development in the evolutionary transformation of digits.

How to Deconstruct a Limb: Developmental Origins of Limb Reduction

Among reptiles, and amniotes in general, digits I and V are most susceptible to complete loss or severe reductions (e.g., the loss of all phalanges). However, as described above,

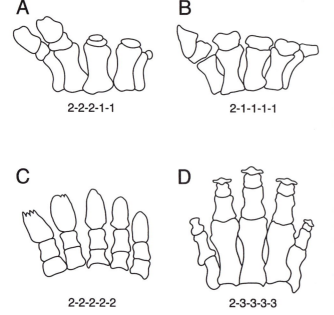

Figure 14-5 Uniform manual digit reductions among reptiles. In all diagrams, distal is up and anterior is to the left. Phalangeal formulae are indicated below each drawing. (A, B) Manual digit morphologies of the archosaurs (A) *Stegosaurus* and (B) *Diplodocus*. (C) Manual digits of the testudinate *Testudo*, which shows a broadly similar pattern of uniform reductions across all digits. (D) Manual digits of the graviportal mammal *Elephas*. In contrast with (A) and (B), the elephant does not lose phalanges from any digit relative to the ancestral mammalian formula (2-3-3-3-3; Flower 1870; Jenkins 1971), and the central manual digits are the most prominent. (Credits for A, B listed in figure 14.3. C after Williston 1925; D after Cornwall 1956.)

digit I is lost much more frequently than is digit V, and this morphological pattern has developmental implications. First, this pattern implies that digit I either condenses and develops two phalanges, condenses and soon regresses, or does not condense at all. This tendency contrasts with patterns in other digits, especially digits II–IV, which often exhibit modest reductions of one or two phalanges. Second, digit V nearly always develops, but subsequently does not segment to form phalanges in many taxa. As a result, only a metapodial persists into adulthood. The mechanism underlying nonsegmentation of digit primordia is unknown but may involve localized tissue shortages (i.e., not enough mesenchyme to support additional chondrogenesis; see Alberch and Gale 1983; 1985; Raynaud 1990; Shapiro et al. 2003) or changes in the regulation of joint formation (see Storm and Kingsley 1996, 1999; Hartmann and Tabin 2001).

The evolutionary conservation of digit V is intriguing, and its developmental persistence may be even greater than adult limb skeletal morphologies indicate. For example, the presence of small splints fused to the postaxial aspects of metatarsal 4 in single specimens of the dinosaurs *Iguanodon*

(fig. 14.4B) and *Stegosaurus* (Galton 1990b) hint at the transient presence of digit V in these species. This phenomenon may extend beyond reptiles as well. Although most birds (pes only) and mammalian ruminants lack any trace of digit V postnatally, a digit V metapodial appears in the embryos of some species but later fuses to digit IV (Mettam 1895; Romanoff 1960; Hinchliffe 1977). The frequency and phylogenetic distribution of this phenomenon is difficult to assess, however, due the paucity of basic embryological studies of reduced-limbed species. The frequency is even more difficult to assess in fossil forms, which usually provide very few ontogenetic data.

What are the changes in development that bring about the patterns of reduction observed in reptiles? Among living taxa, reptiles offer a multitude of candidate model systems to study the developmental origins of limb reduction. For example, lizards exhibit varying degrees of evolutionary limb reductions, ranging from the loss of a single phalanx to complete limblessness, even at low taxonomic levels (Greer 1991). While the adult morphologies of many reduced-limbed lizards have been studied in detail, the developmental and molecular mechanisms producing these morphologies have not been explored, with a few exceptions.

In some limbless reptiles, for instance, embryonic cessation of limb outgrowth appears to result from a loss of distal signals, which in turn leads to an arrest of mesenchymal proliferation and patterning. In particular, limb bud degeneration in some serpentiform lizards (Raynaud 1962; 1963; Rahmani 1974; Vasse et al. 1974; Raynaud et al. 1975) and a snake (Cohn and Tickle 1999) is known to result from early breakdown of the apical ectodermal ridge (AER), an ectodermal thickening in the distal limb bud that is critical for its outgrowth (Saunders 1948). AER breakdown, in turn, leads to a loss of normal patterning and proliferation of the underlying mesenchyme, resulting in a dramatically truncated (or absent) limb (Raynaud 1985; 1990; Cohn and Tickle 1999).

Early decline or arrest of limb bud proliferation can also result in modest limb reductions, including the loss of digits. Raynaud and colleagues (Raynaud 1986, 1989, 1990, 1991; Raynaud and Clergue-Gazeau 1986; Raynaud and Kan 1988) show that experimental treatment of lizard embryos with the mitotic inhibitor cytosine-arabinofuranoside decreases cell proliferation in the limb buds and can lead to a loss of digits. Moreover, these experimental digit reductions and losses occur in the same order as evolutionary digit loss in many lepidosaurs, with peripheral digits lost before central ones (Morse 1872; Sewertzoff 1931; Greer 1987; Raynaud 1987; Greer 1989; 1990; 1991). Similarly, Alberch and Gale (1983, 1985; Alberch 1985b) demonstrate that experimentally induced (also by mitotic inhibition) patterns of salamander digit loss mimic naturally occurring ones. The striking similarities between experimental and evolutionary patterns of digit loss suggest a common mechanism in both phenomena (Raynaud 1990), and this mechanism may be common to multiple tetrapod lineages. Like the complete loss of limbs in serpentiform reptiles, the loss or reduction of individual digits may be linked with a shortage of tissue in the developing limb.

In both lizard and salamander experiments, the loss of a single digit is usually not accompanied by widespread reductions of the remaining digits. Similarly, some natural populations and species of lepidosaurs and archosaurs in this study show single digit losses without major (if any) reductions of other digits. Based on known chondrogenesis sequences, these reductions do not represent simple truncations of pentadactyl skeletal developmental programs. If they did, we would expect them to match, or at least closely resemble, intermediate embryonic configurations of pentadactyl developmental sequences of other amniotes (table 14.1); instead, they represent novel configurations.

This observation is further supported by developmental studies of limb reduction among natural populations of reptiles. The Australian skink genus *Hemiergis*, for example, provides an ideal model organism in which to study developmental aspects of evolutionary digit reduction and loss. Shapiro (2002; Shapiro et al. 2003) detailed the limb skeleton development sequences of *Hemiergis* populations and species with two, three, four, and five digits, and concluded that limbs with fewer than five digits did not result from simple developmental truncations of a pentadactyl (or any other) chondrogenesis sequence (fig. 14.6). Rather than yielding fewer numbers of complete digits, truncations of a five-digit developmental program in *Hemiergis* would yield a series of incomplete digits (fig. 14.6, intermediate configurations).

The archosaur manus, especially that of theropods, follows an evolutionary pattern of reduction that is superficially similar to the common lepidosaurian mode seen in *Hemiergis*. Nevertheless, the sequence of digit loss in the theropod manus differs from that of other reptiles and, indeed, all other amniotes. Theropods lose digits IV and V but never completely lose digits I–III (the enigmatic monodactyl theropod *Mononykus* was excluded from our analysis due to ambiguous digit homologies). Hence, theropods appear to defy the digit loss trend seen in all other reptile groups. This difference may reflect an anterior shift in the primary axis to digit III, which could potentially increase the stability of the anterior digits. Alternatively, digit IV—the "primary axis" digit that is first to appear in known amniote developmental sequences—may have been the first to form in these organisms, only to later obtain the identity of a digit III during osteogenesis (G. P. Wagner and Gauthier 1999), or to regress

or fuse with digit III. The latter scenario is observed in the "missing" metatarsals 2 and 5 of bovid mammals, which fuse embryonically with metatarsals 3 and 4, respectively (Mettam 1895). Similarly, in the avian foot a cartilaginous metatarsal 5 fuses with metatarsal 4 (Hinchliffe 1977).

Notably, the identity of the earliest digits to develop differs among tetrapod lineages. Specifically, salamanders differ from all other groups for which data are available (Shubin and Alberch 1986; table 14.1). In amniote (and anuran) sequences, metapodials generally appear in the order IV > (III, V) > II > I. Phalanges are added essentially uniformly across all digits, beginning with the digits that develop first, until the adult configuration for each digit is attained. Digits that appear first may add phalanges before other digits begin chondrogenesis, but phalangeal addition is completed nearly simultaneously across all digits. In salamanders, however, digits appear in the order (I, II) > III > IV > V (Shubin and Alberch 1986; Blanco and Alberch 1992). Moreover, in contrast to amniotes, chondrification of the phalanges is not uniform across the autopod, and in some cases each digit completes phalangeal condensation before the next one begins (Blanco and Alberch 1992). Thus, salamander digits emerge from the limb and chondrify one at a time (or nearly so), whereas the digits of other groups develop essentially simultaneously. The earliest-appearing digits—I and II in salamanders, III and IV in all other groups— are also the most evolutionarily stable. The theropod manus is a possible exception to this rule. In general, however, patterns of digit stability and loss are dependent on developmental properties of each group (Alberch and Gale 1983).

In contrast to the localized reduction and loss of outer digits described above, relatively uniform reductions occur across all digits in several lepidosaurs (e.g., *Nephrurus* manus, 2-3-3-3-3; *Moloch* manus, 2-2-3-3-2), graviportal archosaurs (e.g., *Stegosaurus* manus, 2-2-2-2-1), and most reduced testudinate configurations. In many of these configurations, all five digits are retained, and some of these formulae closely resemble intermediate stages of pentadactyl chondrogenesis sequences. Hence, since the last phalanges to form are the first to be lost, these configurations may indeed represent truncations of ancestral chondrogenetic sequences (also see fig. 10.10 of Shapiro and Carl 2001; Shapiro 2002). This trend is distinguished from those described earlier because all five digits typically persist. Condensation of all five digit primordia, an early event in digit morphogenesis, often occurs in taxa showing uniform reductions, but all phalanges do not necessarily undergo segmentation within the digital ray, a later event in morphogenesis. Therefore, the morphogenetic mechanisms that lead to loss of a digit must act earlier than those that result in a reduction in the number of phalanges. Some uniform truncations of this type are

Table 14-1 Pentadactyl developmental sequences for several tetrapod species

Taxon	Manus	Pes	Reference
Testudinata (Chelydridae)			
Chelydra	2-3-**3-3**-3[a]		Burke and Alberch 1985
	0-0-1-1-1		
	X-X-0-0-0		
Testudinata (Emydidae)			
Chrysemys		2-3-**3-3**-4[a]	Burke and Alberch 1985
		0-0-1?-1-1?	
		X-X-0-0-0	
Lepidosauria (Agamidae)			
Calotes	2-3-4-5-3	2-3-4-5-4	Mathur and Goel 1976[b]
	2-3-4-4-3	1-2-3-4-3	
	1-2-4-3-2	1-2-3-3-3	
	0-1-3-2-1	1-2-2-2-2	
	0-1-2-2-1	0-1-1-1-1	
	0-0-1-1-0	0-0-1-1-1	
Lepidosauria (Scincidae)			
Hemiergis	2-3-4-**4**-3	2-3-4-**4-3**	Shapiro 2002
	1-3-3-3-2	1-3-3-3-2	
	1-2-3-3-2	1-2-3-3-2	
	1-2-2-2-2	1-2-2-2-2	
	0-1-1-1-1	0-1-1-2-1	
	X-0-0-1-0	X-0-0-1-0	
	X-X-0-0-X	X-X-0-0-X	
Archosauria (Crocodylidae)			
Alligator	2-3-4-5-*4*	2-3-4-5-**0**	Müller and Alberch 1990
	1-2-2-2-1	2-3-3-2-0	
	1-1-1-1-1	1-1-1-1-0	
	0-1-1-1-0	X-0-1-0-0	
	X-1-1-1-0	X-X-X-0-X	
	X-X-X-0-X		
Archosauria (Phasianidae)			
Gallus		2-3-4-5-**X**[c]	Hinchliffe 1977
		0-1-2-2-0	
		0-1-2-1-0	
		0-1-1-1-0	
		X-1-1-1-0	
		X-X-0-0-0	
Archosauria (Anatidae)			
Anas		2-3-4-5-**X**[c]	Romanoff 1960
		1-2-3-4-0	
		1-2-3-3-0	
		0-1-1-1-0	
		X-X-0-0-X	
Mammalia (Hominidae)			
Homo	2-3-3-3-3[d]	2-3-3-3-3[d]	O'Rahilly et al. 1957
	2-2-2-2-2	2-2-2-2-2	
	0-1-1-1-1	1-1-1-1-1	
	X-X-0-0-X	0-0-1-1-0	
		0-0-0-0-0	
		X-0-0-0-0	
Caudata (Abystomatidae)			
Ambystoma		2-2-3-4-2	Alberch and Gale 1985
		2-2-3-4-1	
		2-2-3-3-0	
		2-2-3-1-X	
		2-1-1-0-X	

Table 14-1 (continued)

Taxon	Manus	Pes	Reference
		1-0-0-X-X	
		0-0-X-X-X	
Anura (Pipidae)			
Xenopus		2-2-3-4-3	Alberch and Gale 1985
		2-2-3-3-2	
		1-1-3-2-1	
		0-1-2-2-1	
		0-0-1-1-0	
		X-0-1-1-0	
		X-0-0-0-0	
		X-X-0-0-X	

NOTE: Adult phalangeal formulae are in the top row for each sequence, with progressively earlier (principally cartilaginous) embryonic configurations listed below. Boldface numbers in adult formulae indicate reductions relative to ancestral formulae, and italics indicate hyperphalangy.
[a]Adult formulae from Zug 1971.
[b]From Greer 1991.
[c]Metatarsal 5 never supports phalanges at any embryonic stage and fuses with metatarsal 4 while still cartilaginous; metatarsal 5 thus appears to be missing in the adult.
[d]These formulae represent the ancestral configurations for Mammalia but reduced configurations for Synapsida (manus: 2-3-4-5-3; pes: 2-3-4-5-4).

coincident with evolutionarily miniaturized limbs (Lande 1978), suggesting that this mode of reduction, too, has a tissue availability component. Miniaturization is not an issue with the gigantic sauropods that follow this trend, however.

Molecular Control of Limb Reduction

Our understanding of the genetic control of limb development has increased manifold in recent years (for reviews, see Tickle 1995; R. L. Johnson and Tabin 1997; Shubin et al. 1997), but the molecular mechanisms underlying limb reduction remain poorly understood. In amniotes, the zone of polarizing activity (ZPA) and the AER coordinate mesenchymal proliferation and patterning through a positive feedback loop involving *Sonic hedgehog* (*Shh*), which encodes a secreted intracellular signal expressed in the ZPA, and fibroblast growth factors (FGFs), expressed in the AER (Laufer et al. 1994; Niswander et al. 1994; Zúñiga et al. 1999). Notably, reduced-limbed morphologies in *Hemiergis* are associated with decreased Shh protein expression (plate 14.1; Shapiro et al. 2003). Furthermore, these naturally occuring digit configurations are strikingly similar to some experimental morphologies generated by Raynaud's mitosis inhibition experiments. Such similarities are not surprising, however, since mitotic inhibition of limb mesenchyme and decreased Shh expression should have similar effects: both decrease proliferation and limit the quantity of tissue available for digit condensations. Consequently, differences in proliferation mediated by Shh (or another signal in its feedback loop with the AER, including FGFs) or changes in digit identity and quantity regulated by Shh (Harfe et al. 2004) and Shh-Gli3 interactions (Aoto et al. 2002; Litingtung et al. 2002; te Welscher et al. 2002a, 2002b) may regulate digit number. The prevalence of outer digit reduction and loss among reptiles, and tetrapods in general, further suggests that this may be a common developmental mechanism of limb reduction across several major clades.

"Natural experiments" such as *Hemiergis,* in which a variety of morphologies occur among closely related taxa, are underrepresented in the evolutionary studies of vertebrate development. A comparative approach complements experimental work on traditional model species: experimental manipulations using chicks, mice, and frogs can help unravel the mechanisms and pathways of development, but they cannot always predict how evolution modifies these mechanisms to produce novel morphologies. While important for understanding the molecular basis for limb reduction, studies of nontraditional model organisms tend to be correlative. In the case of *Hemiergis,* for example, Shh is implicated in the loss of digits. However, based on available evidence, we cannot determine whether a regulatory or coding change in *Shh* is a primary cause of limb reduction, or whether *Shh* is downregulated in response to another molecular cue. A crucial next step in studies of evolution and development will entail the integration of genetics in studies of morphological diversity among natural populations of vertebrates. To date, no such studies focus on reptile limb diversity, but limb (fin) studies in other vertebrates suggest the feasibility of such experiments in the near future (Peichel et al. 2001; Shapiro et al. 2004).

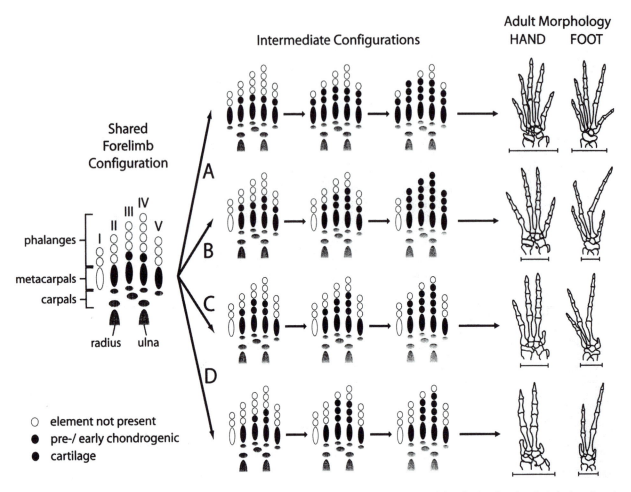

Figure 14-6 Limb skeletal development in *Hemiergis*. In all diagrams, distal is at the top and anterior is to the left. Following shared, early skeletal configurations (left), the developmental trajectories of (A) *H. initialis* (5/5), (B) *H. peronii* (4/4), (C) *H. peronii* (3/3), and (D) *H. quadrilineata* (2/2) autopodia diverge, culminating in different adult morphologies (right). The shared and intermediate stages depict forelimb configurations only, but hindlimb data are virtually identical. Intermediate configurations are based on data from whole mounts and serial sections, and do not necessarily represent identical embryonic stages among all four morphs. Scale bars = 1 mm for adults. (After Shapiro et al. 2003.)

Reducing the Limb: A Two-Tiered Hierarchy

As outlined above, tetrapod limb reduction operates at distinctly different levels of organization. Individual digits can be either reduced in phalangeal count or lost altogether, and correlated patterns of reduction and loss determine the overall autopod shape and phalangeal configuration. Limbs are functional organs used by organisms to move about in their environments and to manipulate their surroundings. Thus, we should not be surprised to find functional correlates of limb reduction trends among, for example, cursorial or graviportal tetrapods in our analysis. That these trends span phylogenetic boundaries lends support to the notion that function is a primary determinant of overall autopod shape.

If function determines *what* a reduced limb looks like, then development determines *how* reductions are effected. Developmental processes dictate the order of digit formation

and therefore which digits are most susceptible or resistant to reduction and loss. The last digits to form are typically the first to be lost, but these losses are not consistent with a simple model of heterochronic truncation at the level of limb chondrogenesis (for an extended discussion, see Shapiro and Carl 2001; Shapiro 2002). Nevertheless, studies of comparative patterns of limb skeletal chondrogenesis provide insight and predictability regarding the sequence of reductions. Amniotes form the most preaxial digits last, and these are the first digits to be lost (except, apparently, in theropod dinosaurs). In cases of relatively uniform reductions across all digits, the resulting phalangeal formulae often resemble intermediate developmental configurations, and thus phalanges that fail to form are those that would otherwise form last. Hence, this type of reduction is predictable based on known developmental sequences. Importantly, uniform reductions across multiple digits are limited to amniotes (and anurans) because they assemble all digits essentially simultaneously. Truncations at intermediate developmental stages, therefore, may produce a series of incomplete digits. Salamanders, on the other hand, typically complete development of one digit before beginning the next, and thus any simple truncation would produce fewer, but more compete, digits.

Reptiles are a group of organisms whose phylogenetic, anatomical, and functional diversity can be used to assess the mechanisms behind morphological trends. Some trends in reptilian limb morphology are unique to specific groups, while others are more general and appear to evolve convergently. Reptilia exhibits a morphological diversity that makes the clade an especially attractive subject for the study of the developmental and genetic bases of morphological change. Despite this fact, reptiles have not received a level of

attention to match their high potential for discovery. A major challenge for the future of reptilian limb studies will be to build upon the molecular, developmental, and genetic tools and techniques that are widely used in studies of "traditional" model organisms (e.g., the laboratory mouse, the chicken, and the frog *Xenopus laevis*). In doing so, we can hope to determine whether convergence in reptilian limb morphology is the product of common developmental and genetic mechanisms. Are convergent morphologies in different taxa caused by changes in the same genes? Are similar numbers of genetic changes required to effect similar morphological changes among different lineages of reptiles? Answers to these questions will only come from the comparative molecular analysis of reptile limb evolution and development.

Appendix

Table 14A-1 Phalangeal formulae for reptiles and salamanders used in this study

Configuration	Taxon	Source
Testudinata: manus		
2-3-4-5-3	*Captorhinus*	Gaffney 1990
2-3-3-3-3	*Macroclemys*	Gaffney 1990
	Podonemis	Gaffney 1990
2-3-3-3-2	*Thalassochelys*	Romer 1956
2-2-2-2-2	*Chersina*	Crumly and Sánchez-Villagra 2004
	Geochelone	Crumly and Sánchez-Villagra 2004
	Gopherus	Crumly and Sánchez-Villagra 2004
	Homopus	Crumly and Sánchez-Villagra 2004
	Indotestudo	Crumly and Sánchez-Villagra 2004
	Kinixys	Crumly and Sánchez-Villagra 2004
	Malacochersus	Crumly and Sánchez-Villagra 2004
	Meiolania	Gaffney 1990
	Proganochelys	Gaffney 1990
	Pyxis	Crumly and Sánchez-Villagra 2004
	Stylemys	Crumly and Sánchez-Villagra 2004
	Testudo	Williston 1925
2-2-2-2-1	*Geochelone*	Crumly and Sánchez-Villagra 2004
	Gopherus	Crumly and Sánchez-Villagra 2004
	Indotestudo	Crumly and Sánchez-Villagra 2004
	Kinixys	Crumly and Sánchez-Villagra 2004
	Malacochersus	Crumly and Sánchez-Villagra 2004
	Manouria	Crumly and Sánchez-Villagra 2004

continued

Table 14A-1 (continued)

Configuration	Taxon	Source
	Psammobates	Crumly and Sánchez-Villagra 2004
	Pyxis	Crumly and Sánchez-Villagra 20043
	Testudo	Crumly and Sánchez-Villagra 2004
2-2-2-1-1	Kinixys	Crumly and Sánchez-Villagra 2004
1-2-2-2-2	Testudo	Crumly and Sánchez-Villagra 2004
1-2-2-2-1	Gopherus	Crumly and Sánchez-Villagra 2004
	Testudo	Crumly and Sánchez-Villagra 2004
1-2-2-1-1	Gopherus	Crumly and Sánchez-Villagra 2004
X-2-2-2-2	Homopus	Crumly and Sánchez-Villagra 2004
X-2-2-2-1	Homopus	Crumly and Sánchez-Villagra 2004
	Testudo	Crumly and Sánchez-Villagra 2004
X-2-2-1-1	Testudo	Crumly and Sánchez-Villagra 2004
X-1-2-1-1	Testudo	Crumly and Sánchez-Villagra 2004
Testudinata: pes		
2-3-4-5-4	Captorhinus	Gaffney 1990
2-3-3-4-3	Lissemys	Zug 1971
	Pelochelys	Zug 1971
	Trionyx	Zug 1971
2-3-3-3-4	Dermatemys	Zug 1971
	Malaclemys	Zug 1971
	Chrysemys	Zug 1971
2-3-3-3-3	Carettochelys	Zug 1971
	Casichelydia	Gaffney 1990
	Claudius	Zug 1971
	Clemmys	Zug 1971
	Damonina	Zug 1971
	Kinosternon	Zug 1971
	Macroclemys	Gaffney 1990
	Podocnemis	Gaffney 1990
	Staurotypus	Zug 1971
	Terrapene	Zug 1971
2-3-3-3-2	Casichelydia	Gaffney 1990
	Claudius	Zug 1971
	Platysternon	Zug 1971
	Rhinoclemys	Zug 1971
2-3-3-2-2	Terrapene	Zug 1971
2-3-3-2-1	Terrapene	Zug 1971
2-2-2-2-2	Proganochelys	Gaffney 1990
2-2-2-2-1	Chersina	Crumly and Sánchez-Villagra 2004
	Geochelone	Zug 1971
	Gopherus	Zug 1971
	Homopus	Crumly and Sánchez-Villagra 2004
	Kinixys	Zug 1971
	Malachochersus	Crumly and Sánchez-Villagra 2004
	Manouria	Crumly and Sánchez-Villagra 2004
	Psammobates	Crumly and Sánchez-Villagra 2004
	Testudo	Crumly and Sánchez-Villagra 2004
2-2-2-2-X	Chersina	Crumly and Sánchez-Villagra 2004
	Geochelone	Crumly and Sánchez-Villagra 2004
	Gopherus	Zug 1971
	Homopus	Crumly and Sánchez-Villagra 2004
	Kinixys	Crumly and Sánchez-Villagra 2004
	Meiolania	Gaffney 1990
	Pyxis	Crumly and Sánchez-Villagra 2004
	Testudo	Crumly and Sánchez-Villagra 2004
Lepidosauria: manus		
2-3-4-5-3	Ancestral saurian	Romer 1956
2-3-4-5-2	Chamaesaura	Greer 1991

Table 14A-1 (continued)

Configuration	Taxon	Source
2-3-4-4-3	*Anotosaura*	Kizirian and McDiarmid 1998
	Chalcides	Caputo et al. 1995
	Ctenophorus	Greer 1989
	Hemiergis	Choquenot and Greer 1989
	Moloch	Greer 1989
	Nephrurus	Stephenson 1960
	Rhynchoedura	Greer 1989
2-3-4-4-2	*Chalcides*	Caputo et al. 1995
2-3-4-4-0	*Chalcides*	Caputo et al. 1995
2-3-3-4-3	*Hemidactylus*	Haake 1976
2-3-3-3-3	*Nephrurus*	Stephenson 1960
2-2-3-3-2	*Moloch*	Greer 1989
1-3-4-5-3	*Heterodactylus*	Kizirian and McDiarmid 1998
1-2-3-3-2	*Bachia*	Kizirian and McDiarmid 1998
0-3-4-5-3	*Colobodactylus*	Kizirian and McDiarmid 1998
	Colobosaura	Kizirian and McDiarmid 1998
0-3-4-4-3	*Bachia*	Kizirian and McDiarmid 1998
0-3-4-4-0	*Hemiergis*	Choquenot and Greer 1989
0-2-2-2-2	*Bachia*	Kizirian and McDiarmid 1998
X-3-4-5-3	*Hemiergis*	Choquenot and Greer 1989
X-3-4-5-0	*Hemiergis*	Choquenot and Greer 1989
X-2-4-5-3	*Lerista*	Greer 1989
X-2-3-4-2	*Lerista*	Greer 1990
X-2-3-4-0	*Lerista*	Greer 1989
X-2-3-3-0	*Coeranoscincus*	Greer and Cogger 1985
	Chalcides	Caputo et al. 1995
X-2-3-2-0	*Anomalopus*	Greer and Cogger 1985
X-2-2-2-0	*Chalcides*	Caputo et al. 1995
X-1-2-2-0	*Anomalopus*	Greer and Cogger 1985
X-0-3-4-0	*Hemiergis*	Choquenot and Greer 1989
X-X-2-3-0	*Chalcides*	Caputo et al. 1995
X-X-2-3-X	*Lerista*	Greer 1989
X-X-1-3-0	*Chalcides*	Caputo et al. 1995
X-X-0-3-X	*Lerista*	Greer 1989
X-X-0-2-X	*Lerista*	Greer 1989
Lepidosauria: pes		
2-3-4-5-4	Ancestral saurian	Romer 1956
2-3-4-5-3	*Anotosaura*	Kizirian and McDiarmid 1998
	Chalcides	Caputo et al. 1995
	Ctenophorus	Greer, 1989
	Heterodactylus	Kizirian and McDiarmid 1998
	Rankinia	Greer, 1989
2-3-4-5-0	*Heterodactylus*	Kizirian and McDiarmid 1998
	Chalcides	Caputo et al. 1995
2-3-4-4-4	*Nephrurus*	Stephenson 1960
2-3-4-4-3	*Agama*	Greer 1991
	Chalcides	Caputo et al. 1995
	Hemiergis	Choquenot and Greer 1989
	Moloch	Greer 1989
2-3-4-4-0	*Bachia*	Kizirian and McDiarmid 1998
2-3-3-4-3	*Hemidactylus*	Haake 1976
	Stenodactylus	Haake 1976
2-2-3-3-2	*Moloch*	Greer 1989
2-2-3-3-0	*Bachia*	Kizirian and McDiarmid 1998
2-2-2-0-0	*Bachia*	Kizirian and McDiarmid 1998
1-3-4-5-0	*Anotosaura*	Kizirian and McDiarmid 1998
0-3-4-4-0	*Hemiergis*	Choquenot and Greer 1989

continued

Configuration	Taxon	Source
X-3-4-5-3	*Hemiergis*	Choquenot and Greer 1989
X-3-4-5-0	*Hemiergis*	Choquenot and Greer 1989
X-3-4-4-0	*Hemiergis*	Choquenot and Greer 1989
X-2-4-5-4	*Lerista*	Greer 1989
X-2-4-5-0	*Lerista*	Greer 1987
X-2-3-5-3	*Lerista*	Greer 1990
X-2-3-5-0	*Lerista*	Greer 1989
X-2-3-4-0	*Chalcides*	Caputo et al. 1995
X-2-3-3-0	*Coeranoscincus*	Greer and Cogger 1985
	Chalcides	Caputo et al. 1995
X-2-2-2-0	*Chalcides*	Caputo et al. 1995
X-2-2-0-0	*Anomalopus*	Greer and Cogger 1985
X-0-3-4-0	*Hemiergis*	Choquenot and Greer 1989
	Chalcides	Caputo et al. 1995
X-0-2-2-0	*Alcys conciana*	Greer 1989
X-0-0-0-0	*Delma*	Greer 1989
	Pygopus	Stephenson 1962
X-X-3-5-0	*Lerista*	Greer 1989
X-X-0-3-0	*Lerista*	Greer 1989
X-X-0-0-0	*Paradelma*	Greer 1989
X-X-0-0-X	*Anomalopus*	Greer and Cogger 1985
X-X-X-0-X	*Lerista*	Greer 1990
	Lialis	Stephenson 1962
Archosauria: manus		
2-3-4-5-3	Ancestral saurian	Romer 1956
2-3-4-4-3	*Crocodylus*	Romer 1956
2-3-4-4-2	*Anchisaurus*	Carroll 1987
	Geosaurus	Romer 1956
2-3-4-4-X	Numerous pterosaurs	Wellnhofer 1991
2-3-4-3-2	*Centrosaurus*	Lull 1933
	Heterodontosaurus	Norman 1985
	Plateosaurus	Galton 1990a
	Protoceratops	Brown and Schlaikjer 1940
2-3-4-3-1	*Leptoceratops*	Brown and Schlaikjer 1940
2-3-4-3-0	*Hypsilophodon*	Norman 1985
2-3-4-1-X	*Ceratosaurus*	Norman 1985
	Syntarsus	Raath 1985
2-3-4-0-0	*Eoraptor*	Sereno 1997
2-3-4-X-X	*Allosaurus*	Molnar et al. 1990
	Deinonychus	Ostrom 1969
	Dromeceiomimus	Barsbold and Osmólska 1990
	Scipionyx	Dal Sasso and Signore 1998
	Struthiomimus	Barsbold and Osmólska 1990
2-3-3-3-3	*Iguanodon*	Norman and Weishampel 1990
2-3-3-3-2	*Camptosaurus*	Norman and Weishampel 1990
	Pinacosaurus	Coombs and Maryanska 1990
2-3-0-X-X	*Tarbosaurus*	Norman 1985
2-2-2-2-1	*Shunosaurus*	McIntosh 1990
	Stegosaurus	Galton 1990b, Sereno 1997
2-2-1-1-1	*Apatosaurus*	McIntosh 1990
2-2-0-X-X	*Compsognathus*	Norman 1990
2-1-1-1-1	*Brachiosaurus*	McIntosh 1990
	Diplodocus	Norman 1985
X-3-3-3-3	Hadrosauridae	Weishampel and Horner 1990
Archosauria: pes		
2-3-4-5-4	Ancestral saurian	Romer 1966
2-3-4-5-2	Most Rhamphorhynchoidea	Wellnhofer 1991

Configuration	Taxon	Source
2-3-4-5-1	*Pterodactylus*	Wellnhofer 1991
	Anchisaurus	Romer 1956
2-3-4-5-0	*Alligator*	Kuhn-Schnyder and Rieber 1986
	Camptosaurus	Norman and Weishampel 1990
	Centrosaurus	Lull 1933
	Ceratosaurus	Norman 1985
	Compsognathus	Ostrom 1978; Norman 1990
	Deinonychus	Ostrom 1969
	Dromeceiomimus	Barsbold and Osmólska 1990
	Plateosaurus	Galton 1990a
	Protoceratops	Brown and Schlaikjer 1940
	Protosuchus	Romer 1956
	Most Pterodactyloidea	Wellnhofer 1991
	Syntarsus	Rowe and Gauthier 1990
2-3-4-5-X	*Allosaurus*	Molnar et al. 1990
	Heterodontosaurus	Norman 1985
	Hypsilophodon	Norman 1985
	Leptoceratops	Brown 1914
	Tarbosaurus	Norman 1985
2-3-4-4-0	*Geosaurus*	Romer 1956
2-3-4-4-X	*Talarurus*	Coombs and Maryanska 1990
2-3-4-2-1	*Apatosaurus*	Gilmore 1936
	Diplodocus	McIntosh 1990
2-3-3-2-1	*Diplodocus*	McIntosh 1990
	Janenschia	McIntosh 1990
2-2-2-1-1	*Camarasaurus*	Norman 1985
0-3-4-5-X	*Iguandodon*	Norman and Weishampel 1990
X-3-4-5-X	*Edmontosaurus*	Norman 1985
	Euoplocephalus	Coombs and Maryanska 1990
	Struthiomimus	Osborn 1916
X-2-3-3-X	*Huayangosaurus*	Galton 1990b
X-2-2-2-X	*Stegosaurus*	Galton 1990b
Caudata: manus		
2-2-3-2-X	*Cryptobranchus*	Cope 1889
	Karaurus	Ivachnenko 1978
	Lipoxitriton	Shubin and Wake 2003
	Laccotriton	Gao and Shubin 2001
2-2-2-2-X	*Siren*	Shubin and Wake 2003
2-2-2-X-X	*Pseudobranchus*	Shubin and Wake 2003
1-2-3-2-X	*Taricha*	Shubin and Wake 2003
	Triturus	Shubin and Wake 2003
	Sinerpeton	Gao and Shubin 2001
1-2-2-X-X	*Amphiuma*	Cope 1889
Caudata: pes		
2-2-3-4-3	*Karaurus*	Ivachnenko 1978
2-2-3-4-2	*Ambystoma*	Alberch and Gale 1985
	Laccotriton	Gao and Shubin 2001
2-2-3-3-2	*Ambystoma*	Shubin et al. 1995
	Cryptobranchus	Alberch and Gale 1985
	Dicamptodon	Alberch and Gale 1985
	Hynobius	Alberch and Gale 1985
	Liua	Shubin et al. 1995
	Tylototriton	MVZ 219764
2-2-3-3-1	*Echinotriton*	Alberch and Gale 1985
2-2-3-3-0	*Echinotriton*	Alberch and Gale 1985

continued

Table 14A-1 (continued)

Configuration	Taxon	Source
2-2-3-3-X	*Echinotriton*	Alberch and Gale 1985
2-2-3-2-X	*Batrachuperus*	CAS 152088
	Hynobius	Alberch and Gale 1985
	Necturus	Shubin et al. 1995
2-2-2-X-X	*Amphiuma*	Shubin et al. 1995
2-2-X-X-X	*Proteus*	Shubin et al. 1995
1-2-3-4-2	*Sinerpeton*	Gao and Shubin 2001
1-2-3-3-2	*Taricha*	Shubin et al. 1995
	Most Plethodontidae, some Salamandridae	Alberch and Gale 1985
1-2-3-3-X	*Batrachoseps*	Alberch and Gale 1985
	Eurycea	Alberch and Gale 1985
	Hemidactylium	Alberch and Gale 1985
1-2-3-3-1	*Thorius*	Alberch and Gale 1985
1-2-3-2-2	*Bolitoglossa*	Alberch and Gale 1985
1-2-3-2-X	*Salamandrina*	MVZ 184845
	Batrachoseps	Alberch and Gale 1985
	Thorius	Shubin et al. 1995
1-2-2-1-1	*Bolitoglossa*	Alberch and Gale 1985
1-2-1-1-1	*Bolitoglossa*	Alberch and Gale 1985

Note: In many cases, formulae listed are present in additional taxa as well. Institutional abbreviations: CAS, California Academy of Sciences, San Francisco, CA; MVZ, Museum of Vertebrate Zoology, Berkeley, CA.

Acknowledgments

We thank A. W. Crompton, J. Hanken, J. A. Hopson, F. A. Jenkins Jr., W. Joyce, N. Rosenthal, S. Scott, and C. Tabin for helpful comments and discussion on earlier drafts of the manuscript. Portions of this work were supported by an R. A. Chapman Memorial Fellowship, a Helen Hay Whitney postdoctoral fellowship, and grants from the National Science Foundation, the Society for Integrative and Comparative Biology, and Sigma Xi to Michael D. Shapiro.

Chapter 15 Limbs in Mammalian Evolution

P. David Polly

THE MORE THAN 4,000 living species of mammal have infiltrated almost every habitat in the world. From alpine mountaintops to plains grasslands, from aerial heights to the depths of the ocean, from slender forest branches to narrow subterranean burrows, mammals occupy an extraordinary variety of substrates. Some are capable of running, swimming, or swinging at great speeds, while others creep slowly along limbs or push laboriously through the soil. The limbs of mammals reflect the diversity of their habitats (fig. 15.1). The long, slender legs and two-toed feet of the antelope allow it to survey the plains for predators and bound off at their approach. The short, muscled arms and broad, thick-clawed hands of the mole scrape through soil as the animal searches for worms and grubs. Spidery, elongated hands and fingers with webs of skin between them propel bats on their erratic chases after night-flying insects. Dense, paddlelike fins steer whales through the watery course on which they are propelled by their massive tail flukes. Limbs are crucial for mammalian locomotion, social behavior, and feeding.

The functional diversity of mammal limbs is facilitated by sometimes subtle structural differences. A small discrepancy in the proportion of one limb segment and its distal neighbor can translate into significant disparity in running speed. The position of a muscle insertion along a long bone shaft can make the difference between a mammal that can tear its way through thick turf and one that cannot. The fusion of two bones in the forearm or wrist may mean that one animal cannot turn its palm to grasp a limb as it tries to climb, while another species can easily wrap its forelimbs around a tree trunk and scamper to safety. A large part of a mammal's lifestyle can, therefore, be read in the structure of its limbs.

This chapter first reviews the anatomical structure of the mammalian limb. Some brief mention of comparative differences is made in relation to structure, but those are reserved for the most striking ones in which the number of elements differs or the structure of homologous elements is particularly diverse. Special emphasis is given to the structures that are most obviously related to function. The chapter then reviews the functional diversity of the mammalian limb from the perspective of gross ecomorphological categories, groups that are primarily locomotory in nature. The chapter then reviews aspects of variation, genetics, and development of the mammalian limb. Finally, the early evolution of the mammalian limb is reviewed.

This chapter concentrates on terrestrial mammals. Bats and gliding mammals are covered by Gatesy and Middleton (chap. 16 in this volume), and whales and aquatic mammals by Thewissen and Taylor (chap. 18 in this volume).

General Considerations

Mammalian Taxonomy

On all points except the most major, the taxonomy in this chapter follows McKenna and Bell's (1997) *Classification of Mammals above the Species Level*. The name "Mammalia"

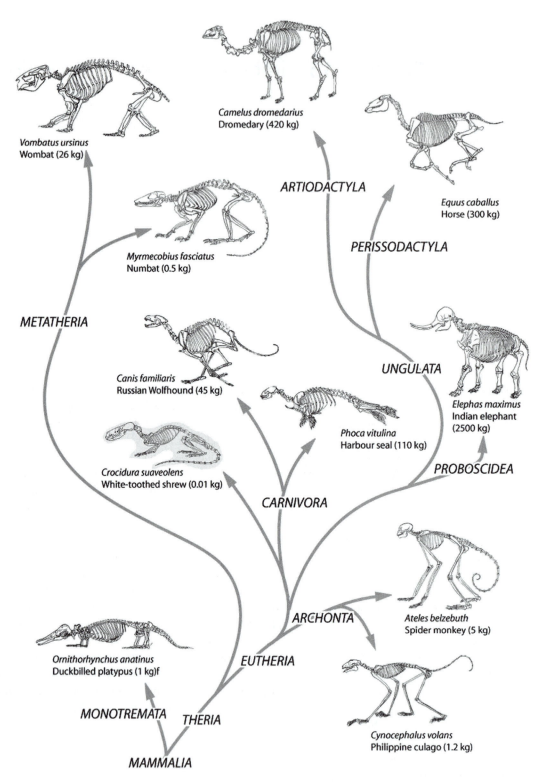

Figure 15-1   A sample of the diversity of limb skeletons of living mammals. Monotremes have a sprawling posture, large fibulae, and multiple elements in their pectoral girdles, all of which are ancestral traits for mammals. Marsupials have epipubic bones, which are a primitive trait for mammals, and relatively large fibulae. Living placentals have lost the epipubic bones and severely reduced the size of the fibulae. Many specializations are not shown, including volant bats and aquatic whales. Note the relationship between the orientation of the pelvis and body mass. The largest mammals have vertically oriented pelvises that are perpendicular to the vertebral column, and the smallest ones have horizontally oriented pelvises that parallel the column. (*Elephas maximus* and *Camelus dromedarius* after Young 1981; all others after Gregory 1951.)

is used here to designate the earliest common ancestor of living mammals to have an exclusively dentary-squamosal jaw joint and all descendants of that ancestor (cf. Simpson 1945; Z. Luo 1994). This traditional usage differs from that of some authors who have restricted the name "Mammalia" to the crown group—the last common ancestor of living monotremes, marsupials, and placentals and the descendants of that ancestor—and used the name Mammaliaformes for the more inclusive group of animals with a dentary-squamosal jaw joint (Rowe 1988, 1993; McKenna and Bell 1997). "Metatheria" and "Eutheria" are used in this paper for marsupial and placental mammals, respectively, rather than McKenna and Bell's use of "Marsupialia" and "Placentalia."

Several higher taxonomic categories of mammal are frequently referred to in this paper. Therians are the group that includes metatherians and eutherians, their last common ancestor, and the descendants of that ancestor (fig. 15.1). These mammals share the tribosphenic molar pattern and many specializations in the skeleton. Monotremes are nontherian mammals that live today in Australia and New Zealand. Multituberculates were nontherian mammals that are now extinct, but were diverse and geographically widespread during the Mesozoic and early Cenozoic. Archonta is the group of eutherian mammals that includes primates, bats, dermopterans (flying lemurs), and scandentians (tree shrews). Ungulata is the group of eutherian mammals that includes perissodactyls (horses, rhinos, and tapirs) and artiodactyls (pigs, camels, deer, cattle, antelope, etc.).

Anatomical Terminology

Anatomical structure names follow standard paleontological and zoological usage, which is essentially that of Romer and Parsons (1977). This usage differs in details from standard usage in anthropological, medical, and veterinary disciplines. The greatest discrepancies are in the names of carpal and tarsal bones.

Many anatomies have been published that include the limbs of one or more species of mammal. Two of the most detailed are *Gray's Anatomy* of the human (Williams and Warwick 1980) and *Miller's Anatomy of the Dog* (H. E. Evans 1993). The general anatomical structure of the limbs presented here owes a great debt to these works.

Structure and Function of Mammalian Limbs

General

The mammalian appendicular skeleton differs from that of other vertebrates in several respects (fig. 15.2A). Mammals have fewer bony elements than most other vertebrates, though the evolutionary reduction of the number of bones is compensated for by an increase in the number of muscles. With the exception of monotremes, the mammalian pectoral girdle has been reduced to a single major bone, the scapula (which incorporates a remnant of the coracoid). The number of carpals has been reduced to nine or fewer elements and the tarsals to seven or fewer. Mammals typically have an upright posture, with the humerus and femur oriented almost vertically. Compared to other tetrapods, limb movement during terrestrial locomotion is roughly in a parasagittal plane (see Jenkins and Camazine 1977 for a nuanced discussion; and see Inuzuka 1984 for an argument that desmostylians, an extinct group of manatee-like eutherian mammals, had a sprawling posture). The rotation of the limbs into an upright stance was correlated with changes in the configuration of distal limb elements, notably the repositioning of the tarsals so that the astragalus sits partially atop the calcaneum (O. J. Lewis 1989; Szalay 1994). Bony processes for the insertion of extensor muscles are also especially well developed in mammals, particularly the olecranon process of the ulna, the greater trochanter of the femur, and the tubercle of the calcaneum. The pelvis of mammals is also reorganized, with a long ilium that positions the pelvis caudoventrally to its sacral attachment. Mammals and many of their synapsid relatives also have a ventral contact between the pubis and ischium, which encloses the obturator foramen. Most of these skeletal changes are associated with changes in the posture of mammals and correlated muscular reorganizations that give mammals very fast and efficient modes of locomotion.

A large part of the diversity of mammalian limb morphology can be summarized by their stance. Habitual standing limb postures have been categorized as plantigrade, digitigrade, and unguligrade (fig. 15.2B; Howell 1944). Plantigrade mammals stand with the carpals and tarsals in contact with the substrate. Digitigrade mammals stand on the ends of their metapodials, with the carpus and tarsus raised above the substrate. Unguligrade mammals carry the digitigrade stance further and stand on the tips of their digits. These categories form a continuum, especially between plantigrade and digitigrade, and some mammals have multiple stances depending on the situation. Each stance position has its consequences for locomotion. Plantigrade mammals are able to use the feet for forward propulsion to a greater extent than digitigrade or unguligrade mammals (Clevedon Brown and Yalden 1973). Digitigrade mammals gain an extra limb segment and a proportionally longer distal limb, which increases the length of stride and, thus, the speed of movement, but they are more reliant on the proximal limb for forward momentum. Unguligrade mammals gain yet more segments and distal length, enabling them to move quickly and efficiently.

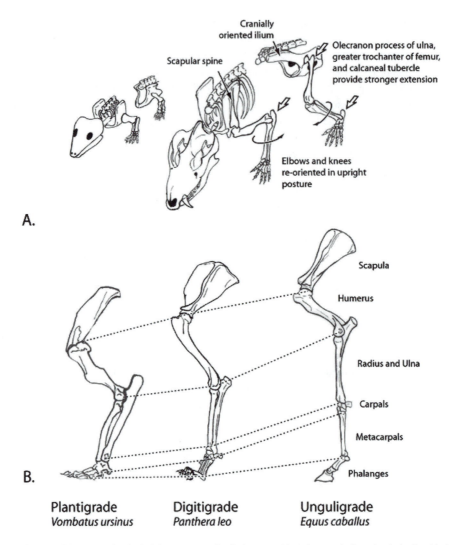

Figure 15-2   (A) Specialized features of the mammalian limb skeleton. Mammalian limbs are positioned parasagitally under the body, with the elbow rotated posteriorly and the knee rotated anteriorly. The number of elements in the pectoral girdle is reduced, and the scapula is often highly mobile. The ilium of the pelvis is reoriented anteriorly. A number of bony processes—notably the olecranon, greater trochanter, and calcaneal tuber—add leverage to limb extension (after Radinsky 1987). (B) Typical mammalian limb postures. Plantigrade mammals rest their entire foot on the substrate. Digitigrade mammals stand on their toes, with the wrist and ankle elevated above the substrate. Unguligrade animals carry the digitigrade posture even further, standing on the tips of the digits. Digitigrade and unguligrade postures are often associated with cursorial locomotion. (*Vombatus* after Gregory 1951; others after E. E. Thompson 1896.)

Pectoral Girdle

The pectoral girdle of mammals is light and mobile. In therians, the girdle is composed of a large scapula, a small coracoid (fused to the scapular head), and, often, a clavicle. The girdle is embedded in a muscular sling that provides scapular mobility and bodily support. When present, the clavicle connects the scapula and the sternum, but otherwise the pectoral girdle has no bony connection to the rest of the skeleton. Body weight in mammals is transmitted to the forelimb via the *serratus* muscle complex, which suspends the thorax from the dorsal margin of the scapula. Weight is transmitted

through the scapula to the limb via the glenoid fossa of the scapula, which forms the proximal part of the shoulder joint. The pectoral girdle also serves as a point of origin for muscles of the arm. The pectoral girdle thus provides support, offers propulsive power, and helps absorb the impact of the forelimb during locomotion.

In monotremes and many extinct mammalian groups the pectoral girdle is more bulky and retains bones that were lost in therians, including the anterior clavicle (or proclavicle) and interclavicle. These animals retain a sprawling posture, and the humerus transmits a medial as well as vertical force vector onto the pectoral girdle. Consequently, the left and

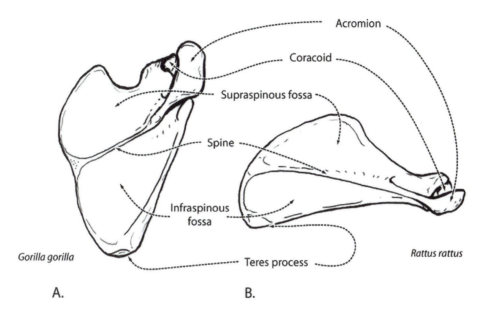

Acromion

Coracoid

Supraspinous fossa

Spine

Infraspinous fossa

Gorilla gorilla

Teres process

Rattus rattus

A.   B.

Figure 15-3 The mammalian scapula. Both elements are in life position. (A) Right scapula of a gorilla in dorsal view. The gorilla is a very large ape, with typical body masses ranging between 90 and 200 kg. Gorillas are partially arboreal, and have a semiupright stance on the ground, where they move by knuckle-walking. The vertical orientation of the glenoid cavity and scapular spine are related to the dual function of the forelimb during brachiation and quadrupedal walking. The large acromion is associated with massive deltoideus and trapezius muscles, which provide strength of movement to the scapula and arm. The long, acute angle of the teres process provides an advantageous moment arm for the teres major, strengthening adduction of the arm (after Raven 1950). (B) Right scapula of a rat in lateral view. The rat is a medium-sized rodent, with typical body masses around 0.15 kg. Rats are terrestrial quadrupeds, with good scansorial abilities. The horizontal orientation of the glenoid cavity and scapular spine is common in mammals with small body masses (after E. C. Greene 1935).

right girdles are braced at the midline via the clavicle, anterior coracoid, and interclavicles. The buttressing prevents collapse of the shoulders (Howell 1937a).

Scapula

Structure

The scapula of therian mammals is a large, flat, triangular bone that tapers distally into a neck supporting the articular region (fig. 15.3). The lateral surface is bisected by a raised spine that divides the supraspinous and infraspinous fossae. The spine terminates distally of the acromion process, which overhangs the proximal humerus. The distal scapula has a round depression called the glenoid cavity, which receives the humeral head to form the shoulder joint. The coracoid process, composed of the vestigial coracoid bone, projects around the cranial side of the glenoid. The cranial border of the scapula is usually gently curved, while the dorsal and caudal borders come together in an angle known as the teres process. The costal surface of the scapula is shallowly concave and almost completely occupied by the subscapular fossa. In some mammals, notably perissodactyls and artiodactyls, the dorsal portion of the scapula is unossified and is composed in the adult of cartilage.

Comparative Function

The shape and size of the scapula and its muscular attachments reflect the locomotory habit of the animal (fig. 15.3). The primary functional components of the scapula are (1) the relative breadth of the proximal scapula blade from the teres process to the cranial border, which determines the moment arms of flexors and extensors of the shoulder; (2) the orientation of the scapular axis, which relates to whether the scapula contributes substantially to limb flexion and extension; and (3) the size and shape of the acromion and coracoid processes, which are related to the size and moment arms of shoulder muscles. Cursorial (running) mammals, for example, usually have a long, narrow, vertically oriented scapular blade that helps increase stride length. Ambulatory (unspecialized terrestrial) mammals may also have a narrow scapula, but it is usually more horizontally oriented. Fossorial (digging) and natatorial (swimming) mammals usually have a triangular blade with a large, robust teres process that provides the teres major muscle with more efficient leverage for powerful adduction of the forelimb. In some fossorial and natatorial animals a second low spine divides the infraspinous and teres fossae.

Clavicle

Structure

The clavicle is a long bone that connects the scapula and sternum, but which is rudimentary or absent in many mammalian groups. When present the bone usually has a bowed shaft that attaches at its medial end to the manubrium of the sternum via a saddle-shaped synovial joint, and at its distal end to the acromion of the scapula via a flat joint. The clavicle is a dermal or intermembranous bone and is the only such bone to be retained in the therian pectoral girdle. Despite this developmental categorization, the clavicle develops in part from cartilaginous precursors, and it has multiple primary and secondary ossification centers (Williams and Warwick 1980). In humans, two primary intermembranous ossification centers form and quickly fuse as ossification proceeds. True cartilage is formed at the medial and lateral ends, which are engulfed by expanding ossification. Further endochondral growth proceeds from the ends, and there are also two endochondral epiphyses that form. A secondary ossification center later appears at the sternal end. The two primary ossification centers correspond to the attachment points of the sternocleidomastoid and trapezius muscles (Howell 1937b). The clavicle is the first postcranial element to ossify in therians (Sánchez-Villagra 2002), but its ossification may not be complete until after sexual maturity.

Several muscles may attach to the clavicle, leaving their imprint on the bone. When the clavicle is fully developed, as in humans, the clavicular head of the deltoid muscle is often attached to the ventral border of the lateral end, and the trapezius to the dorsal. The sternocleidomastoid (or its equivalent) inserts on the cranial border of the medial end, and the subclavius muscle originates from the ventral border. The pectoralis major originates from the ventral border of the medial end. In groups with a reduced clavicle, such as dogs, the sternocleidomastoid and the clavicular head of the deltoid form a more or less continuous muscle with the splinterlike clavicle embedded in a tendinous sheath at the level of the shoulder.

Absence of the clavicle can occur individually in normally claviculate species. In humans, for example, the clinical condition of cleidocranial dysplasia may present symptoms of reduced or absent clavicles. Cleidocranial dysplasia can be caused by a mutation in one or both copies of the *cbfa1* gene, which has its most profound effects on the ossification of membraneous bone (Mundlos 1999). In domestic dogs (*Canis familiaris*), which have small clavicles unattached to other bones, the clavicle is variable both in size and degree of ossification (H. E. Evans 1993). Among living mammals, monotremes have an exceptional clavicle, which is solidly attached to the rest of the pectoral girdle and has a much dif-

ferent function than in other groups. The emballonurid Ghost Bats, *Diclidurus*, also have an unusual clavicle, which is flattened and triangular with pronounced process midshaft (G. S. Miller 1907; Howell 1937b; N. B. Simmons, American Museum of Natural History, New York, pers. comm.).

Comparative Function

The clavicle has two major functions: it participates in shoulder movement and it maintains a fixed distance between the shoulder joint and the sternum. The clavicle's exact role depends on the presence of the bone itself and the configuration of the muscles attached to it. In humans, for example, the clavicle plays an important role in lifting the shoulder, where it acts as a lever with the fulcrum at the manubrium (Williams and Warwick 1980). The trapezius, which inserts at the distal end, pulls the clavicle cranially and thus raises the shoulder. This movement is facilitated by caudal translation of the proximal end of the scapula across its cartilaginous disc, powered by the pectoralis major. In humans, the bony clavicle adds stability to movements when the forelimb is used for manipulation and contributes to an arcuate movement of the shoulder (Jenkins 1974). In the domestic dog, where the clavicle is reduced to a splint of bone, the bone serves as an interface between the cleidocervicalis and cleidobrachialis muscles (H. E. Evans 1993). The bone's ligamentous attachments to the sternum and the scapular head prevent the scapula from moving too far caudally and allow it to be pulled cranially by the clavicular muscles.

Evolutionary loss of the clavicle allows the shoulder to move parallel to the thorax, an adaptation for cursoriality (Chubb 1932; Jenkins 1974), but its absence in species that are normally claviculate results in shoulder collapse. Those mammals that retain the clavicle—such as most marsupials, primates, and bats—often use the manus to climb, fly, or manipulate objects (Howell 1937b).

Humerus

Structure

The humerus is the only bone of the brachium. It supports the weight of the anterior body in quadrupeds, serves as the insertion point for muscles moving the brachium, and is the point of origination for muscles moving the forelimb and manus. In most groups the humerus has a long, cylindrical shaft with a rounded proximal head and a flared distal condylar region (fig. 15.4).

At its proximal end, the head of the humerus articulates with the glenoid fossa of the scapula. The proximal humerus usually has four tubercles (or tuberosities, depending on their size and shape) for muscle attachments: two immedi-

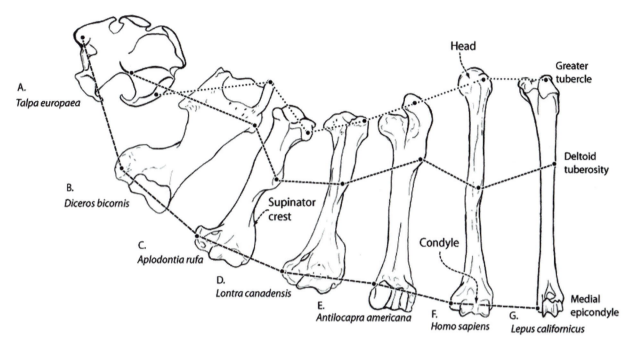

Figure 15-4 The mammalian humerus. Comparative cranial views of the left humerus of seven mammalian species (not to scale). Dotted lines show the highlight of different positions, sizes, and orientations of the functionally important greater tubercle, deltoid tubercle, and medial epicondyle. (A) Left humerus of the European mole, a small (0.1 kg) fossorial lipotyphlan insectivore. The trochanters and epicondyles are greatly enlarged to provide longer lever arms for muscles of the forelimb, particularly those involved in extension, adduction, and supination (after Gregory 1951). (B) Left humerus of the Black rhinoceros, an extremely large (800–2,000 kg) perissodactyl ungulate (after R. Walker 1985). (C) Left humerus of the Mountain beaver, a medium-sized (1.3 kg) subterranean rodent. The broad epicondyles and pronounced deltoid tuberosity are typical of fossorial mammals (after B. M. Gilbert 1973). (D) Left humerus of the North American river otter, a medium-sized (5 kg) natatorial carnivoran. Note the similarities with the humerus of the fossorial Mountain beaver (after B. M. Gilbert 1973). (E) Left humerus of the Pronghorn antelope, a large (60 kg) artiodactyl cursor. Note the similarities in shape to the humerus of the Black rhinoceros, which differs mainly in the thickness required to support the much larger body mass. (F) Left humerus of a human, a large (40–100 kg) bipedal ape. Note the broad epicondyles, which provide moment arms for supination of the forearm, and the wide humeral head, which allows a wide range of movements of the arm (after Gregory 1951). (G) Left humerus of a Black-tailed jackrabbit, a medium-sized (2 kg) cursorial lagomorph. Note the narrow epicondyles and strongly hinged condyle, which stabilize the forelimb (after B. M. Gilbert 1973).

ately adjacent to the head and two further down the shaft. The greater tubercle is located at the craniolateral margin of the head and is the insertion point of the supraspinatus and infraspinatus muscles. The lesser tubercle lies on the medial margin of the head and is the insertion point of the subscapularis muscle. The muscles of the greater and lesser tubercle serve primarily to stabilize the shoulder joint and secondarily to move the brachium. Further distally, on the lateral side of the shaft, the deltoid tuberosity marks the insertion of the deltoideus muscle, a major extensor of the brachium. The teres major tuberosity is usually visible on the medial side of the shaft and is the insertion point of the teres major, a major flexor of the brachium.

At the distal end of the humerus, a condyle articulates with the radius and ulna to form the elbow joint. The condyle is divided into a medial trochlea, which is a grooved structure that articulates with the ulna, and a lateral capitulum, which is convexly rounded and articulates with the fovea of the radial head. Immediately above the condyle are the medial and lateral epicondyles, which are the points of origin for muscles of the antebrachium. The medial epi-

condyle is the point of origin for the pronator teres muscle, which pronates the forearm, and the common flexors of the carpus and digits. The lateral epicondyle is the origin of the common extensors of the carpus and digits. Just proximal to the lateral epicondyle is the lateral supracondylar ridge, or supinator crest, from which the supinator muscle arises. In many mammals, an entepicondylar foramen runs through the medial epicondyle. The entepicondylar foramen serves as a retinaculum for the median nerve and sometimes the brachial artery (Landry 1958). Presence of the canal is the ancestral condition for mammals, but it has been lost independently in many groups that have reduced the epicondyle in association with restricting the ability to abduct the humerus and supinate the forearm.

Comparative Function

Many of the morphological differences among mammalian humeri have direct functional correlations. The size, shape, and orientation of the tubercles, the orientation of the head, the width of the epicondyles, and the shape of the condyle are often indicative of the range of movements in the forearm and the locomotory style of the animal (fig. 15.4). For example, in the pronghorn antelope (*Antilocapra americana*), a cursor, the deltoid and teres major tubercles are proportionally located about a quarter of the way down the shaft. The moment arm for the deltoid and teres major muscles is consequently short, allowing rapid but relatively weak flexion and extension of the forelimb. The same tubercles in the natatorial otter (*Lutra canadensis*) are located nearly halfway down the shaft, providing for proportionally more powerful flexion and extension. The proportions of the epicondylar region in the same species are also quite different. The pronghorn, whose stabilized forearm is incapable of supination, has narrow epicondyles and consequently short moment arms for the pronator and supinator muscles. The otter has wide epicondyles and a pronounced supinator crest, providing long moment arms for powerful pronation and supination abilities that are used by the animal in swimming and food manipulation. The same phenomenon is seen in the condyles. In the pronghorn, both trochlea and capitulum are hingelike, but in the otter the capitulum is broadly rounded for rotation of the radial head in supinating movements. In carnivores, cursors without the ability to supinate the forelimb apparently have a restricted maximum body mass, whereas generalists that retain supination can reach very large body sizes (K. Andersson and Werdelin 2003).

Radius and Ulna

Structure

The radius and ulna make up the forearm, or antebrachium (fig. 15.5). The radius provides the primary support for anterior body mass, and the ulna stabilizes the elbow joint and provides a point of insertion for elbow extensors. In some mammals, particularly cursorial ones, these bones are fused and the ulna reduced.

The radius has a head at its proximal end, a long shaft, and an enlarged distal extremity with articulation surfaces for the ulna and carpals. The radial head is usually oval in outline with a concave proximal end, the articular fovea. The fovea contacts the capitulum of the humerus. The surface around the margin of the head makes rotational contact with the radial notch of the ulna. On the medial side of the proximal shaft is the radial tuberosity, the insertion point of the radial head of the biceps brachii muscle. The radius usually broadens at its distal end where it articulates with the carpals. On the cranial side, grooves are often visible for the tendons of the abductor pollicis longus, the extensor carpi radialis, and the extensor digitalis. On the lateral surface of the extremity there is usually a concave ulnar notch that provides the distal point of rotation of the radius against the ulna. The distal end has a concave articular surface for the

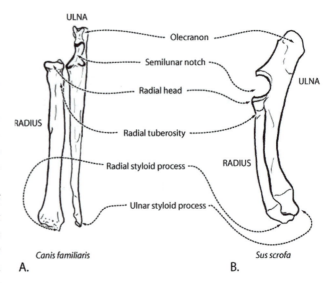

Figure 15-5 The mammalian radius and ulna. (A) Disarticulated left radius (caudal view) and ulna (cranial view) of a dog (after H. E. Evans 1993). (B) Articulated left radius and ulna of a pig in lateral view. The radius and ulna are normally fused in adult pigs (after Sisson and Grossman 1938).

scaphoid and lunate bones of the carpus. A radial styloid process projects from the medial side of the distal end.

The ulna has a large proximal end, a tapering body, and usually a distal end with an articular surface for the carpals. The proximal end has a crescent-shaped semilunar notch that receives the trochlea of the humerus. The large olecranon process projects from the proximal end of the ulna and receives the insertions of the extensor muscles, most notably the triceps brachii. The distal border of the trochlear notch terminates in the coronoid process of the ulna. Distal to the coronoid process of the trochlear notch is a concave surface, the radial notch, which receives the articular circumference of the radial head. The distal extremity thickens into the pointed styloid process and has a facet called the ulnar articular circumference that contacts the distal head of the radius. The styloid processes of ulna and radius provide lateral support for the carpals, and are homologous to the radiale and ulnare carpal bones of ancestral tetrapods (Cihak 1972).

Comparative Function

The function of the forelimb bones has been extensively studied (Howell 1944; Jenkins 1973). Important factors for comparative forelimb function include (1) the degree of fusion of the radius and ulna; (2) the shape of the radial head and corresponding ulnar articular surface; (3) the proportional length of the olecranon process; and (4) the proportional position of the radial and ulnar tuberosities. The degree of fusion and the shape of the radial head help determine the range of pronation-supination movement possible at the manus. Cursorial mammals often have restricted pronation-supination, whereas scansorial mammals can usually completely supinate the manus. Pronation-supination movements are accomplished by flipping the distal radius over the distal ulna, rotating the radius about its long axis. As the radius rotates, the radial head rolls within the radial notch of the ulna. Round radial heads roll easily, whereas flattened heads do not. Morphometric ordination of the outline of the radial head alone does a good job of discriminating scansorial, fossorial, ambulatory, and cursorial species (MacLeod and Rose 1993). The length of the olecranon determines the moment of effort for forelimb extension. The olecranon is relatively longer in fossorial and natatorial animals, and shorter in cursorial ones. The position of the tuberosities affects the moment arm of effort for forearm flexion.

These differences are illustrated by a comparison of the forelimb of a dog with that of a pig (fig. 15.5). Dogs are relatively agile, medium-weight digitigrade cursors, while pigs are less agile, heavier unguligrade cursors. The olecranon process, which is the in-lever of the system used to extend the forelimb, is proportionally longer in the pig than in the dog. This gives the pig a more efficient moment arm for extension, which is related partly to the larger mass of the pig but also to differences in distal limb morphology (pigs have a longer carpus and manus, which compensate for the relatively short radius and ulna in providing overall length to the limb). Consequently, the pig has a slower, more powerful extension than the weaker but rapid extension of the dog. Pigs

have a close fit between radius and ulna, which provides stability and completely restricts pronation-supination of the manus. Dogs have a looser fit between the two elements, although supination is restricted compared to some other mammals. Though it is not visible in the figure, the radial head of the dog is ovate, whereas that of the pig has an irregular shape that locks it into the ulnar notch, preventing supination.

Manus

Structure

The manus consists of the wrist, or carpus, the metacarpus, and the digits (fig. 15.6). The carpus is composed of a dual series of carpal and sesamoid bones. The proximal series articulates with the radius and ulna, whereas the distal one articulates with the metacarpals. The number and shape of the carpal bones vary among mammalian groups, with both evolutionary fusions and losses of bones having occurred. In the ancestral therian there were three proximal carpals, which from medial to lateral are the scaphoid, lunate, and triquetral. The first two articulate with the radius and are fused in many groups (notably in Carnivora), and the last articulates with the ulna. The pisiform, a sesamoid bone with an important function in flexion of the carpus, lies lateral and caudal to the triquetrum. The distal carpals are, from medial to lateral, the trapezium, trapezoid, capitate, and

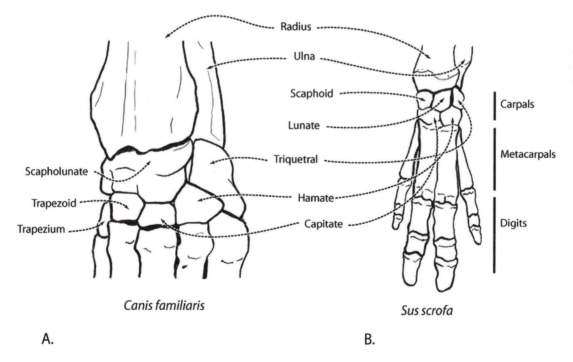

Canis familiaris

A.

Sus scrofa

B.

Figure 15-6   The mammalian carpus and manus. (A) Left carpus of a dog in cranial view (after H. E. Evans 1993). (B) Left manus of a pig in cranial view. Note the reduction in digits and associated carpal bones (after Sisson and Grossman 1938).

hamate. The first three articulate individually with the first three metacarpals, while the fourth carpal articulates with both the forth and fifth metacarpals. The triquetral often contacts the fifth metacarpal as well.

One to five metacarpal bones (depending on the group) extend distally from the distal carpals. They each have a blocky base with an articular face for the carpals, a long body, and a rounded head with a joint surface for articulation with the digits. Usually, the number of metacarpals corresponds directly to the number of digits. Some mammalian groups, such as equids, have reduced the number of metacarpals to only one, while others retain the ancestral number of five. Interestingly, the first metacarpal has a growth plate at the proximal and distal ends (as does the first metatarsal), whereas the other four have growth plates only at their distal ends (Shively 1978).

Distal to the metacarpals are the digits (one per metacarpal bone), most of which are each composed of three phalanges (proximal, middle, and distal). The first digit, the pollex, has only two phalanges. The distal phalanx is usually specialized for supporting a nail, claw, or hoof.

Comparative Function

The manus of mammals differs in extreme and remarkable ways (fig. 15.1). The number of digits ranges from one (in Equidae, the horses) to five (the ancestral condition, which is retained in many living groups). Cursorial mammals often have a reduced number of digits, though the number and amount of reduction varies considerably. Canids, for example, have reduced the first digit so that they retain four fully functional ones; tapirs have reduced the first and fifth, leaving three large ones; and cervids have reduced digits one, two, and four, functioning with only two primary digits. Arboreal animals often have the full complement of five, usually with an opposable pollex and long metacarpals and phalanges. Bats have extremely elongated digits that support their wing membranes. Seals and sea lions have a broad manus with paddlelike proportions.

Most marsupials retain all five digits of their manus, regardless of the locomotory specializations of the hindlimb. The ubiquity of generalized, grasping hands in marsupials may be due to constraints imposed by metatherian reproduction. Metatherian young are born at a very early embryonic stage and continue development attached to the nipple after climbing there with precocially developed forelimbs (Lillegraven 1975; L. S. Hall and Hughes 1987; Szalay 1994). Neonatal marsupials are effectively "scansorial" in that they have to climb hand over hand from vulva to nipple, requiring them to have well-developed forelimbs and the ability to grasp and pull. Eutherians, with their intrauterine development, are not so constrained and have a much wider diversity of forelimb specialization.

Pelvis

Structure

The pelvis (fig. 15.7) anchors the hindlimb to the axial skeleton, transmits weight from the body onto the hindlimb, and supports and protects the internal organs in the posterior part of the body cavity. The left and right halves of the pelvis are known as innominate bones, or *os coxae*. Each innominate is made up of the fusion of the ilium, ischium, pubis, and acetabular bone. The pelvis is attached to the sacrum at the iliac crest, and the two halves are attached to one another at the pubic symphysis. On the lateral faces of the pelvis is the round depression known as the acetabulum that receives the ball of the femur. The obturator foramen lies on the ventral aspect of the pelvis between the pubis and ischium.

Many mammals, especially living metatherians, have epipubic bones. These slender bones extend cranially into the abdominal wall from the pubis. Whereas textbooks often associate epipubic bones with support of a pouch, or marsupium, they are present in both pouched and pouchless metatherians, and they are found in monotremes and early eutherian groups.

Comparative Function

The shape of the pelvis is closely associated with the body mass and locomotory repertoire of the animal. Differences are most notable in the orientation of the iliac crests and in the shape and depth of the acetabulum. The orientation of the iliac crest is related to the habitual posture of the animal. Bipedal mammals, such as humans, have an upright ilium that parallels the vertical orientation of the sacroiliac joint, whereas quadrupedal mammals usually have a horizontally oriented ilium that parallels their horizontal vertebral column (Schultz 1936). Orientation is also related to body mass, most notably among quadrupedal species. Heavy mammals tend to have a more upright pelvis than do light ones (compare the shrew, *Crocidura suaveolens,* the dog, *Canis familiaris,* the horse, *Equus caballus,* and the elephant, *Elephas maximus,* in fig. 15.1). The vertical orientation is capable of supporting greater weight without dislocating the sacroiliac joint or putting undue torsion on the vertebral column. The shape of the acetabulum is also related to locomotory style (Jenkins and Camazine 1977). Ambulatory mammals have a wide range of hip movements, including significant abduction of the femur. The acetabulum in these animals is more shallow and open than that of cursorial mammals. The lengths and angles of the ilium and ischium relative to the acetabulum also have important functional consequences because the surfaces of these bones are the origination points for most of the extensors and flexors of the hips. The primary extensors of the hip arise from the lateral face of the ilium (the gluteus muscles) and from the ischiatic tuberosity

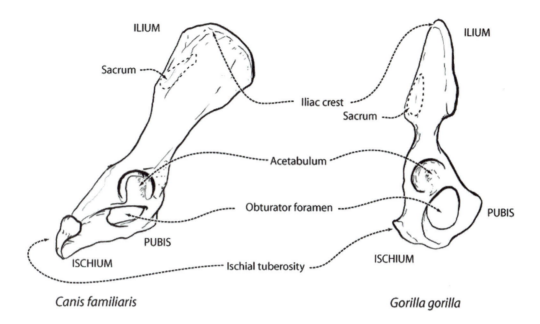

Figure 15-7 The mammalian pelvis. (A) Right innominate of a dog. Note the anterodorsal angle of the ilium, which intersects the vertebral column at an oblique angle. This morphology is typical of terrestrial quadrupeds (after H. E. Evans 1993). (B) Right innominate of the gorilla. Note the rounded iliac crest and the upright orientation (after Raven 1950).

(the semimembranosus and other muscles). The lengths and angles of these areas of the pelvis thus determine the moment arms for hip extension, which provides forward momentum during locomotion.

Most likely, the epipubic bones have multiple functions, including support of the ventral body wall, influencing movements of the pelvis during locomotion, and support of young (Szalay 1994). Interestingly, T. D. White (1989) found in pouchless marsupials a direct relationship between litter mass on the one hand and size of the epipubic bones and their degree of sexual dimorphism on the other; however, among pouched species the relationship was not so strong, suggesting that the litter-supporting function of the bones is greater in species without a pouch and that other functions predominate in pouched species.

Femur

Structure

The femur is a long bone with a distinctive ball-like head at the proximal end and a large, triangular distal end with a groove on the cranial side for the patella and patellar tendon (fig. 15.8). The head is usually located on a restricted neck and is surrounded by three large processes, or trochanters, whose positions help determine the lever advantage for the flexors and extensors of the hip. The greater trochanter arise lateral to the femoral neck and projects proximally. Three ex-

tensors of the hip—the gluteus medius, gluteus profundus, and piriformis muscles—insert on it. The lesser trochanter lies distal to the head on the caudomedial margin of the femur. The iliopsoas muscle, a flexor of the hip, inserts on the lesser trochanter. The third trochanter lies more distally down the body of the femur and is more variable in size from group to group. Another extensor of the hip, the gluteus superficialis muscle, inserts on the third trochanter. The body of the femur is usually long with a rounded cross section. At the distal end, two condyles curve caudally, with the patellar groove lying between them on the cranial face and the intercondyloid fossa between them on the caudal face. The length and depth of the patellar grove are often correlated with locomotory type. On the caudal side, just above the condyles, are the lateral and medial supracondylar tuberosities, which are the origin of the gastrocnemius muscle.

Comparative Function

Some functionally important characters of the femur are (1) the length and orientation of the greater trochanter; (2) the size of the third trochanter; (3) the shape of the head and position of the fovea; and (4) the depth of the patellar groove. The greater trochanter functions as a primary lever for extension of the hip. Consequently, it is often long and robust in cursorial mammals. The muscles that insert on the greater trochanter run caudally from the anterior pelvis, pulling the trochanter forward, which helps extend the limb by pivoting

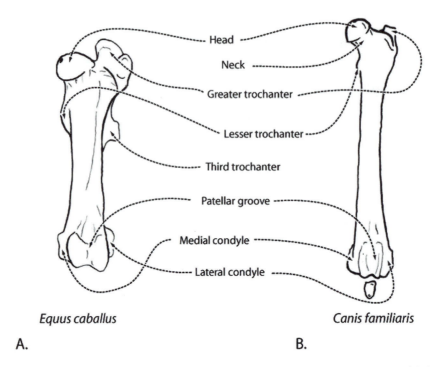

Head
Neck
Greater trochanter
Lesser trochanter
Third trochanter
Patellar groove
Medial condyle
Lateral condyle

Equus caballus

A.

Canis familiaris

B.

Figure 15-8 The mammalian femur. (A) Left femur of a horse. Note the especially large third trochanter, on which insert extensors of the hip that provide important leverage for large cursors (after Sisson and Grossman 1938). (B) Left femur of a dog (after H. E. Evans 1993).

it around the head. The longer the trochanter, the more effi-cient the moment arm for extension. The third trochanter is well developed in cursorial mammals because it is also the insertion point of hip extensors. These muscles run in the op-posite direction from the posterior part of the pelvis, from where they pull the distal femur caudally by tugging on the third trochanter. The head of the femur is broader and more proximally directed in species that significantly abduct the femur during locomotion, such as ambulatory mammals (Jenkins and Camazine 1977). The patellar groove is longer and deeper in cursorial and saltatory mammals than it is in ambulatory or scansorial ones. The patella, or kneecap, itself is a sesamoid bone embedded in the tendon of the quadriceps femoris muscle. The patella rides in the patellar groove along with the tendon. The extra depth of the groove provides sta-bilization for the knee via the patellar tendon.

Crus

Structure

The lower hindlimb between knee and ankle is the crus, and is made up of the tibia and fibula, plus several sesamoid bones (fig. 15.9). In most mammals, the tibia is the larger bone. It lies medial to the fibula and supports most of the body weight. The proximal end flares at the knee joint, and tapers distally into a long body. The proximal articular sur-face is divided into a lateral and medial condyle, with an in-

tercondyloid eminence between them. Just distal to the artic-ular surface on the cranial border is the tibial tuberosity, which is the insertion point of the quadriceps femoris and the sartorius muscles. A small articulation surface for the fibular head is located on the lateral side of the proximal end. The position of the contact between tibia and fibula varies considerably among mammal groups. In placentals the artic-ulation is just distal to the margin of the lateral condyle, whereas in monotremes and many marsupials the contact is at the margin itself, often with the head of the fibula extend-ing proximally to the level of the distal femur (Szalay 1994). At the distal end the tibia flares again, with an articular sur-face for the ankle joint. The articular surface is often grooved, sometimes quite deeply, to receive the ridges of the astragalus. A distal projection on the medial side forms the medial malleolus, which wraps around the medial face of the astragalar trochlea. In some taxa the articular surface and malleolus tightly envelop the astragalus, restricting move-ment to parasagittal plane, whereas in others, especially marsupials, the articulation is more open, allowing more freedom of movement for the foot (Szalay 1994). The lateral side of the distal tibia has an articular surface for contact with the fibula. The fibula is usually thin (especially in euthe-rians) and serves mostly for muscle attachment and to stabi-lize the lateral ankle joint. In some taxa the fibula is fused to the tibia, and in some it is extremely reduced. The proximal head is usually enlarged and has a small articulation surface

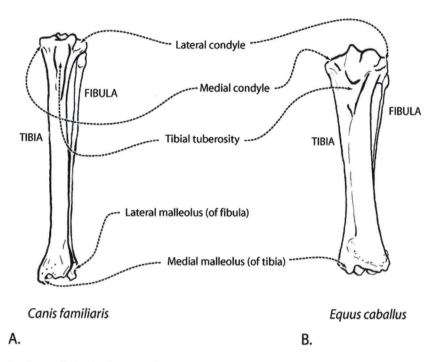

Figure 15-9 The mammalian tibia and fibula. (A) Left tibia and fibula of a dog (after H. E. Evans 1993). (B) Left tibia and fibula of a pig. Note the reduced fibula (after Sisson and Grossman 1938).

for contact with the tibia. The distal end is enlarged into the lateral malleolus, which usually participates in forming the ankle joint.

Comparative Function

Functional variety in the tibia and fibula is not as visually obvious as it is in other limb structures. The most conspicuous functional differences are seen in the distal articular surface of the tibia. This joint surface can be deeply grooved in cursorial mammals, spiraled in species with significant limb abduction (some ambulatory mammals) or those capable of hindfoot reversal (scansorial mammals), or comparatively flat (some ambulatory and scansorial species). The shape of the surface contributes to the degree of ankle stabilization and parasagittal movement. The degree of fusion of the fibula also varies among mammals, but mostly in relation to body mass. The tibia and fibula are often fused at the distal end in the smallest mammals, and occasionally in saltatory or cursorial mammals. Freedom of the fibula enhances abduction/adduction of the ankle, which can be important for scansorial mammals, especially marsupials, which normally have a proportionally large fibula.

Pes

Structure

The ankle, or tarsus, consists of several blocky bones (fig. 15.10). The ancestral therian had seven tarsal bones, a num-ber that has been reduced in some cases by loss or fusion of elements. The largest tarsals are the calcaneum and astragalus (also known as the talus in medical terminology). These bones, especially the astragalus, articulate with the crus at the tibiotarsal, or upper ankle joint. The astragalus is divided into a body, neck, and distal head. The proximal (or dorsal) surface of the body forms the tibial articular surface, or trochlea. The shape of the trochlear surface mirrors the shape of the distal articular surface of the tibia. In most groups, the trochlea has articular surfaces on its medial and lateral faces, which articulate with the inner surfaces of the malleoli of the tibia and fibula, respectively. Facets for articulation with the calcaneum are found on the plantar surface of the body. The neck and head lie distal to the body. The head articulates with the navicular bone, and the neck (and sometimes head) contacts the calcaneum (and sometimes the cuboid bone). The calcaneum is elongate, with the long axis paralleling the long axis of the foot. A large calcaneal process, or tuber, projects caudally and is the point of insertion for the large tendons of the calf muscles. The calcaneum has a large facet for articulation with the astragalus on its dorsal side and for the cuboid on its distal end. In many mammals, especially eutherians, a large process called the sustentaculum astragali projects from the medial side of the calcaneum to support the neck of the astragalus. The joint between astragalus and calcaneum is the astragalocalcaneal, or lower ankle joint. The navicular bone lies distal to the astragalar head and the cuboid distal to the calcaneum. The joint be-

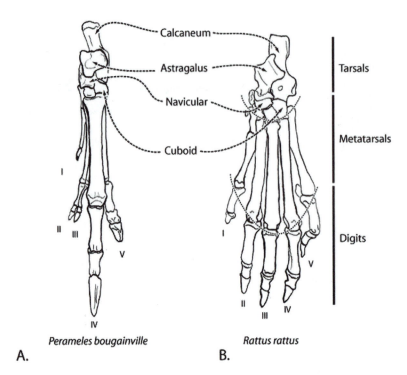

Figure 15-10   The mammalian pes. (A) Left pes of a Western barred bandicoot. Note the reduction in digits and associated carpals, and the syndactylous condition of digits II and III (after Szalay 1994). (B) Left pes of a rat (after E. C. Greene 1935).

tween the proximal and distal tarsals is the transverse tarsal joint, which is composed of the astragalonavicular joint and the calcaneocuboid joint. Distal to the navicular and medial to the cuboid are the lateral, intermediate, and medial cuneiforms. These and the cuboid collectively articulate with the metatarsals.

Comparative Function

Like the manus, the functional morphology of the pes is complex and varied, so only a few aspects will be considered here. Comparative movements of the pes can be most easily understood by considering several major joints and axes of rotation (Szalay 1984, 1994). The upper ankle joint is the primary joint for dorsiflexion and plantarflexion, and it is the point where force is transmitted between crus and pes. The shape of the upper ankle joint surface has both taxonomic and functional variety. The upper ankle joint of metatherians has a relatively smooth, simple structure where some abduction and adduction occurs along with dorsi- and plantarflexion. In didelphids, the astragalus is mediolaterally expanded so that it has three facets, two for the tibia and one for the fibula, which is a weight-bearing bone (Jenkins and Mc-Clearn 1984; Szalay 1994). The upper ankle joint of eutherians usually has a sculpted surface, often with two ridges along the edges of the astragalar trochlea with correspond-

ing grooves in the distal tibial articulation. These ridges and grooves have the greatest relief in cursorial and saltatorial mammals, where they restrict plantarflexion to a parasagittal plane and provide stabilization for the ankle. The upper ankle joint of multituberculates is markedly different from the same structure in therians in that the distal tibia bears two condyles rather than a single concavity and the astragalus has dual grooves rather than ridges (Jenkins and Krause 1983). The shape of these surfaces is spiraled to facilitate hindfoot reversal. The lower ankle joint is also functionally diverse, and movement there may occur in many circumstances. In some taxa the lower ankle joint is especially important in hindfoot inversion (useful to scansorial mammals) because specialized facets allow the calcaneum to slide around the astragalus and so twist the distal foot into an inverted position (Jenkins and McClearn 1984). The lower ankle joint is tightly locked in cursorial mammals to prevent inversion or other distal foot rotation. The transverse tarsal joint varies similarly, with considerable rotation in mammals capable of hindfoot inversion, but tight interlocking in cursorial mammals.

The observation is worth making that pronation and supination of the manus are accomplished by movement in the forelimb, but that inversion and eversion of the pes are accomplished at tarsal joints.

Ecomorphologic Diversity of Mammalian Limbs

Mammalian limbs are often categorized into locomotory or ecomorphologic types (fig. 15.11). Locomotory classifications have been based on gaits (Heglund et al. 1974; Alexander and Jayes 1983), on combinations of musculoskeletal features (Dublin 1903; Gregory 1912; Maynard Smith and Savage 1956; M. E. Taylor 1974; Van Valkenburgh 1985), on limb ratios (Gambaryan 1974; Gonyea 1976; Hildebrand 1980; Christiansen 2002), and on combinations of these criteria. Systems have also been proposed for specific taxonomic groups. Many authors have lamented that locomotory classification schemes are not specific enough to be useful (e.g., Clevedon Brown and Yalden 1973), but in most cases increased specificity inhibits broad applicability. In this chapter only a few general ecomorphologic categories will be discussed, primarily because only very loose groupings can be applied across the whole of Mammalia.

Overviews of limb mechanics and functional diversity are found in Flower (1870), Muybridge (1893), Howell (1944), Tricker and Tricker (1967), Gambaryan (1974), Oxnard (1984), Hildebrand (1988), and Alexander (1983a, 2002).

Generalized or Ambulatory Mammals

Some mammals do not have obvious specializations for any particular locomotor style, substrate, or activity, and are often categorized as "generalized" or "ambulatory." The raccoon, *Procyon lotor,* and the opossum, *Didelphis virginiana,* are examples of these generalized or ambulatory mammals (Jenkins and Camazine 1977). Walking, running, climbing, and manipulating objects are all possible for ambulatory mammals because they do not have skeletal specializations that limit any particular activity. Ambulatory mammals have relatively mobile joints, the ability to pronate-supinate the manus, five digits (often with opposable hallux and pollex), and a plantigrade to semi-digitigrade posture. The proportions of the three major limb segments are roughly equal (see *Marmota, Spermophilus,* and *Rattus* data in plate 15.1). Ambulatory mammals range in body mass from small (e.g., *Mus musculus*) through medium (e.g., *Didelphis virginiana*) to large (e.g., *Ursus arctos*).

Several skeletal traits are associated with an ambulatory mode of locomotion. The scapula is often triangular, providing a powerful moment arm for the flexors of the forearm. The deltoid crest of the humerus is often prominent and positioned well down the length of the shaft, almost to the midpoint in some cases. Ambulatory mammals often have a clavicle (Jenkins 1974). The supinator crest is large, the capitulum is rounded, the radial head is ovate to circular in outline (MacLeod and Rose 1993), and the radius and ulna are not fused, allowing supination of the manus. Ambulatory species have a wide range of hip movements, including significant abduction, and their acetabulae are accordingly open, the balls of their femora are broadly rounded, and the angle of the neck is obtuse relative to the axis of the femoral shaft (Jenkins and Camazine 1977). The tarsal joints permit a variety of movements, including some abduction and adduction at the upper ankle joint in marsupials (Jenkins and McClearn 1984).

Cursorial Specializations

Many ungulates and carnivores have specializations that allow greater running speeds, specializations that are called "cursorial" (fig. 15.11B). Horses, *Equus caballus,* goitered gazelles, *Gazella subgutturosa,* and cheetahs, *Acinonyx jubatus,* are examples of mammals with well-developed cursorial traits. Cursorial animals typically have relatively long limbs with distal segments proportionally longer than proximal ones (e.g., see *Gazella, Dama, Equus, Acinonyx,* and *Lepus* species in plate 15.1). The posture of cursors is usually digitigrade or unguligrade. Limb joints, especially in the hindlimb, are often structured to permit only parasagittal motion, which stabilizes them against mediolateral collapse during running. The scapula of cursors tends to be long, with a short teres process. The reduction or loss of the clavicle further enhances scapular mobility and parasagittal flexion and extension of the forelimb (Jenkins 1974). The deltoid and teres tuberosities of the humerus are proximally located. Restriction of the ability to supinate the manus and invert the pes contributes to joint stabilization. The pelvis and thigh of cursors are organized to maximize parasagittal movement, which is most obvious in the cylindrical nature of the acetabulum and the horizontal orientation of the femoral neck (Jenkins and Camazine 1977). The carpals and tarsals are often closely packed and structured to inhibit movements among them. The tibiotarsal joint has a hinge/groove structure that stabilizes it and restricts flexion and extension of the foot to a parasagittal plane (Van Valkenburgh 1985; Van Valkenburgh 1987). The metapodials and proximal phalanges are often greatly elongated, which increases the length of stride. The peripheral digits of cursors are often reduced or lost, the most extreme example being in the horses, where the number is reduced to one.

The shapes of bones and the positions of muscle origins and insertions give cursors a high "gear ratio" (Hildebrand 1988). The combination of short in-levers and long out-levers produces rapid translation of the manus and pes, but comparatively little power. Consequently, cursors also have specializations that lighten the distal limb. The number of digits is often reduced, with the single digit found in the horse

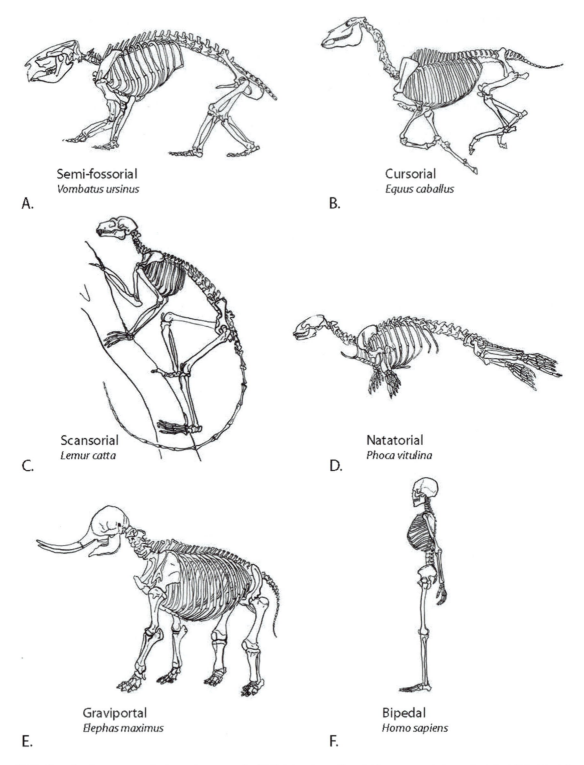

Figure 15-11 Examples of some mammalian locomotory ecomorphs. (A) Skeleton of a semifossorial Common wombat, a medium-sized (26 kg) Australian marsupial with badgerlike burrowing habits. Note the high scapular spine, the long teres process, the pronounced deltoid tuberosity, and the long olecranon process (after Gregory 1951). (B) Skeleton of a horse, a large (300 kg) cursorial ungulate. Note the combination of cursorial (elongated distal digits and narrow, vertical scapula) and graviportal (vertical pelvis) features (after Gregory 1951). (C) Skeleton of a ring-tailed lemur, a medium-sized (3 kg) scansorial primate. Note the triangular scapula, long tarsus, and opposable pollex and hallux (after Gregory 1951). (D) Skeleton of a harbor seal, a large (100 kg) natatorial carnivoran. Note the very large teres process, the short humerus, and the long digits (after Gregory 1951). (E) Skeleton of an Indian elephant, an extremely large (2,500 kg) graviportal proboscidean. Note the long proximal and short distal limb segments, the vertical orientation of the pelvis, and the digitigrade stance (after Young 1981). (F) Skeleton of a human, a large (40–100 kg) bipedal ape. Note the unusual shape of the pelvis, the long proximal limb segments, and the pronounced plantigrade stance (after Young 1981).

being an extreme example. Muscles in the distal segment are also reduced. Centrifugal force, elongated tendons, and spring ligaments often combine to assist with segment extension (Camp and Smith 1942; C. R. Taylor et al. 1980; Alexander and Jayes 1983). The limbs of cursors may not be optimized for maximum speed, as suggested by Howell (1944), but for minimization of locomotory energy costs (Gambaryan 1974; Garland and Janis 1993; Christiansen 2002).

Saltatory Specializations

Some desert and grassland mammals have specializations for jumping, or saltation. Many saltatory specializations are simply exaggerations of cursorial ones, such as long distal limb segments, high gear ratios, digit reduction, and hinged joints (Hall-Craggs 1965; Clevedon Brown and Yalden 1973). However, bipedal saltators, such as the red kangaroo, *Macropus rufus,* or the jerboa, *Jaculus jaculus,* differ markedly from their quadrupedal counterparts in that their fore- and hindlimbs have dissimilar specializations because saltation is performed bipedally. The forelimbs of saltatory animals often resemble those of ambulatory or scansorial species, particularly the mobile elements and grasping hands. Consequently, saltators can be recognized by their limb ratios because of their proportionally shorter manus than pes (e.g., *Allactaga, Eremodipus, Jaculus, Pygeretmus, Gerbillus,* and *Macropus* species in plate 15.1). Because saltators move bipedally, they must balance their weight over the hindlimb. In many species, a long, heavy tail counterbalances the weight of the forelimbs and head.

Saltators use their hindlimbs for propulsion, weight support, and maneuverability, which requires a stable hindfoot. Many saltators have a tridactyl foot, which provides stability and allows adjustment to direction of movement.

Despite claims that bipedal saltation might be more energy efficient for covering the distances necessary to forage in the desert, some studies have found that bipedal and quadrupedal locomotion are similar in energy costs at typical foraging speeds (S. D. Thompson 1985).

Scansorial Specializations

Scansorial, or climbing mammals, characteristically have mobile limbs and the ability to grasp with their hands and often their feet (fig. 15.11C). The terms "scansorial" and "arboreal" have an overlapping usage. Here "scansorial" is used for the general climbing locomotory category, whereas "arboreal" is used only in reference to species that live in trees. Scansorial locomotion is found in many mammal groups, and examples include such divergent species as the red squirrel, *Sciurus vulgaris,* the howler monkey, *Alouatta alouatta,* the three-toed sloth, *Bradypus tridactylus,* the kinkajou, *Potos flavus,* and the tree kangaroo, *Dendrolagus bennettianus.* Most primates, all scandentians (tree shrews), and many rodents, didelphids, and carnivores are scansorial to a greater or lesser degree. However, there is considerable diversity in the amount of time spent climbing, the method and speed of climbing, and the morphological structures that can clearly be associated with a scansorial lifestyle (Dublin 1903). Some climbers suspend themselves from tree limbs, such as sloths or gibbons, and have elongated limbs; others, such as tree squirrels, scamper on limbs and climb by clinging to projections and irregularities. And yet others, such as tarsiers, spring from place to place in the trees. Some scansors spend their entire lives in trees, while some climb only occasionally. Universal scansorial features are thus difficult to identify.

Caveats aside, some features are stereotypical of scansors (Dublin 1903). Most are plantigrade and have grasping feet and hands. In some cases the manus and pes have become elongated, either at the tarsus for jumping scansors, or at the metatarsus and digits in suspensory and brachiating mammals. Suspensory species, such as sloths, have highly modified distal phalanges with long, curved claws for grasping. Many scansorial species have long, dorsoventrally thickened claws (Cartmill 1974). In almost all cases, the forelimb is capable of extensive pronation and supination, and the radius is mobile with a rounded head. The clavicle is often present, which stabilizes the shoulder, and the scapula is often triangular. The limbs as a whole are often elongated, the forelimb more than the hindlimb in brachiating species (Erikson 1963). Many scansorial mammals have prehensile tails.

Fossorial Specializations

Some mammals have fossorial or digging adaptations (fig. 15.11A). Fossorial mammals include those that dig occasionally in pursuit of food or in the construction of burrows, such as badgers, *Meles meles,* and wombats, *Vombatus ursinus,* and those that live most of their lives below the surface, such as mole rats (e.g., *Cryptomys damarensis*) and moles (e.g., *Talpa europea*). Many insectivorous mammals, especially those that eat ants, also have fossorial specializations, including armadillos, pangolins, and anteaters.

Fossorial specializations are most pronounced in the forelimbs, which are the primary tools of excavation (Shimer 1903; Reed 1951). In all fossorial mammals, the limbs have a "low gear ratio" (Hildebrand 1988), which emphasizes strength of extension rather than speed. In semifossorial burrowers, such as the wombat, the low gear ratio is manifested in the long teres process of the scapula, which increases the moment arm of the teres major muscle, the strong deltoid and teres tubercles, and the long olecranon process of the ulna. In subterranean mammals such as moles the differences are more extreme. The manubrium of the sternum is long, which moves the pectoral girdle forward so that it is at the very front of the animal; the clavicles are short, which pull the shoulders close to the body; the scapula is extremely long and narrow; the humeri are bizarrely shaped with large tubercles for muscles of flexion, extension, abduction, and adduction (fig. 15.4A); the radius and ulna are tightly interlocked, the olecranon process is long and robust, and the distal articular surfaces of both bones are flared to provide a massive contact with the carpus; and the manus is broad and inflexible, with interlocking carpals, extremely short metacarpals and phalanges, and long, thick claws (Reed 1951; Yalden 1966). Adaptations to digging are covered in more detail elsewhere (see Kley and Kearney, chap. 17 in this volume).

Natatorial Specializations

Some mammals are adapted to life in water, and those that use the limbs as their primary means of propulsion are called "natatorial." The most extreme examples are cetaceans, or whales, and pinnipeds, or seals, sealions, and walruses; however, many other animals swim regularly, such as beavers, otters, nutrias, water voles, and water shrews (Osburn 1903). Depending on the degree of specialization, mammals may be classified as natatorial (swimming) or aquatic.

Natatorial features of the forelimb are similar to those of semifossorial mammals, but in swimmers the hindlimbs are also specialized (fig. 15.11D). Swimmers require powerful forelimb extension for pushing through the dense medium of the water. Accordingly, natatorial mammals have large teres processes, short humeri with large deltoid and teres tubercles, and long ulnar olecranon process. Unlike diggers, swimmers have an elongated manus with especially long digits. The proximal hindlimb is sometimes reduced in length, but the crus is often long, and the pes paddlelike. In many cases, the toes are webbed (like in the beaver) or have become flippers (as in seals and sea lions). The sternum of natatorial mammals is often reduced, and the clavicles are usually absent (Osburn 1903). Aquatic and natatorial adaptations are covered in detail elsewhere (see Thewissen and Taylor, chap. 18 in this volume).

Graviportal Specializations

Mammals with extremely large body masses require special features simply to support their weight during locomotion (fig. 15.11E). Animals that show such skeletal specializations are said to be graviportal. The most obvious graviportal features are the diameter of the limb bones, which are disproportionately large to support a body mass that has increased as the cube of body length (Schmidt-Nielsen 1984). The orientation of the pelvis is also a notable graviportal feature. The orientation of the ilium is highly correlated with body mass, and it becomes more vertical as body mass increases (fig. 15.1). The vertical orientation reduces the torque placed on the pelvis and sacrum under large loads. It is often said that decreased length of the distal limb segments, especially the metapodials and digits, is a graviportal specialization (Hildebrand 1988), but close inspection of plate 15.1 shows that this is not the case. In the plate, the standardized lengths of the proximal, middle, and distal limb segments of both the forelimb and hindlimb are shown. They have been ordered by the relative length of the pes, with the longest at the top of the graph and the shortest at the bottom. Elephants, which are the very largest of living land mammals, are indeed found at the bottom of the graph because of the reduced length of their manus and pes. However, the next heaviest mammal, the giraffe, *Giraffa camelopardalis*, is at the top because of its proportionally long manus and pes. Other large mammals, such as the bison, are found scattered through, suggesting that the short manus and pes of the elephants is a feature specific to the Proboscidea and not an adaptation to large body mass.

Bipedality

An unusual locomotory specialization of mammals is non-saltatory bipedalism (fig. 15.11F). Humans are the only habitual nonjumping bipeds, although some other species, mostly primates and bears, are facultative bipeds. Bipeds are typically plantigrade, have short tails, and vertically oriented pelvises. The human hindlimb has unique specializations for bipedalism, which include a short, flared ilium, an extremely long femur, a down-turned tuber of the calcaneum, and a large, elongated pollex (O. J. Lewis 1989).

Variability, Genetics, and Development in Mammalian Limbs

The interplay between structure, function, and development makes mammalian limbs a particularly interesting system, one that remains fertile despite its long history of study (Mariani and Martin 2003). The mammal limb skeleton has a particularly high genetic component. The size, shape, and structure of limb bones are sufficiently distinctive as to reliably indicate the species from which it came (B. M. Gilbert 1973), and the phylogenetic continuity of limb structure is remarkable (Szalay 1994), even though superficial homoplastic resemblances develop as rapid evolution scampers across the same substrates again and again. In another sense, though, limbs are improbably plastic. Bones remodel rapidly in response to the stress and strain they encounter. Habitual exertion can leave a marked impact on the structure of limbs, from bowlegged cowboys, to asymmetrically armed Anglo-Saxon bowmen, to the eerily humanlike pelvis of Slijper's bipedal goat (Slijper 1946). Excision and tendon-severing experiments suggest that the structure of limb bones is due more to the forces impinging on them than to anything fundamentally heritable. The ecophenotypic plasticity of the limb skeleton suggests a lack of heritable variance. Some have suggested that the latter has wrongly encouraged systematists to reject limb characters from phylogenetic analyses (Szalay 1977). Almost in contradiction, patterns of expression of intrinsic skeletal genes are known to influence morphology and have been used to explain evolutionary transformations (Lovejoy et al. 2000). The evolutionary transformations of the limb are clearly a fuzzy combination of the genetic and epigenetic, the selective and the ecophenotypic (nongenetic, life historical), the developmental and the functional. Mammal limbs provide a potentially fruitful venue for exploring the interaction among these factors because they have a greater structural complexity in terms of muscles and bony features than those of other vertebrates, allowing a greater diversity of approach to functional interpretation; mammals have evolved a greater functional diversity than other vertebrate groups; and the mammalian fossil record is particularly rich in its postcrania.

Selected literature on variability, genetics, and development in mammalian limbs is reviewed here. In considering variability, it is important to distinguish between ontogenetic variation, population variation, and interspecific or evolutionary variation. Ontogenetic and population variation are discussed in this section and related to intraspecific variation, which was discussed above. It is also important to distinguish among variation in different kinds of traits: variation can be in the dimensions of bones (either uniformly as "size" or nonuniformly as "shape"), variation can be in the structure of limb elements (such as the surface shape of a joint articulation, for example), and variation can be polymorphic (such as polydactly). These different types of trait probably have radically different genetic and epigenetic underpinnings. Consequently, quantitative or experimental data on one type of trait may have little explanatory power over another. Trait types may be functionally integrated, however, as the previous discussion of limb segment ratios and skeletal structure demonstrates.

Quantitative Variability

Variation in the dimensions of limb bones has been studied since Galileo noted in 1637 that the diameter of long bones is disproportionately large compared to their lengths in mammals with large body mass (Galilei 1914, as described by Schmidt-Nielsen 1984). The functional correlates of quantitative variation in mammal limbs have been the subject of many recent interesting studies (Schmidt-Neilsen 1984; Van Valkenburgh 1985, 1987; Ruff and Runestad 1992; Heinrich and Biknevicius 1998; Polk et al. 2000; Ruff 2000), and the relationship is becoming reasonably well understood. But what is the genetic and developmental context of the evolution of quantitative aspects of limb morphology?

One fundamental issue is the extent to which limb size traits are heritable. It is well known that environmental factors such as nutrition and body mass can affect limb dimensions, so it is interesting to know how much of the variance in limb elements is genetic. There have been few studies of heritability in limb elements (see reviews by Cock 1966; Thorpe 1981) except for a series of papers on the quantitative genetics of skeletal traits in mice by Leamy and coworkers (Leamy 1974, 1975, 1977, 1981; Leamy and Bradley 1982; Leamy and Sustarsic 1978). These studies considered osteological traits of the skull and limbs, along with measures of body size. Several limb traits were included: innominate length, ilium length, obturator foramen (all pelvis traits), femur length, tibia length, scapula length, humerus length, and radio-ulna length. These traits collectively represent all of the segments of both limbs, excluding the feet themselves.

Heritability of limb element lengths in mice was high, on average slightly higher than skull or body traits (Leamy 1974). This result dispels the notion that limb traits exhibit greater nongenetic, ecophenotypic variation than do dental or skull traits. Heritability of forelimb elements was generally higher than for those of the hindlimb, but the two were comparable. A multivariate component analysis grouped osteological traits into related factors, including a skull factor, a limb factor, a pelvic factor, and a width factor (Leamy 1975). The pelvis thus appears to be partly dissociated from other limb traits, including the scapula.

Even though the heritability of limb traits is high, they do not appear to evolve as quickly on microevolutionary time scales as do molar or skull traits. Leamy and Sustarsic (1978) looked at variation between different inbred lines of mice, finding that limb traits did not differ as much as did tooth traits. Despite their possibly slower rate, limb traits do evolve on very small evolutionary timescales. Schnell and Selander (1981) found that pelvic length showed greater differentiation among populations of *Mus musculus musculus* and *M. musculus domesticus* than did ulna length or hindfoot length. Means of ulna length, condylar width of the humerus, pelvic length, and hindfoot length were significantly different in the two subspecies.

A high heritability does not in itself explain the mechanism by which the heritable variance is transmitted. Limb traits have long been known to be pleiotropic, or influenced by many genes (Lande 1978), and now a bewildering array of genes have been identified that are active during various stages of limb development (Karsenty 2003; Mariani and Martin 2003). These genes interact in complex, dynamic ways to produce skeletal structures, and their contribution to gross heritability of limb measurements is not straightforward. For example, Hox genes are involved in patterning the segments of limbs (A. P. Davis et al. 1995), but so does pleiotrophin, an extracellular growth-differentiation factor, which is expressed long after limb patterning has taken place (Tare et al. 2002). Both genes act directly on skeletal development, but there are other sources of heritable variance that act indirectly through nonskeletal tissues. The muscular and vascular systems are the most obvious sources of such epigenetic interactions, but less obvious sources, such as the nervous system, also have an effect on skeletal development (Wermel 1934). Heritable variance may also be transmitted through interactions among elements of the limb skeleton themselves, as early excision experiments indicate. For example, the removal of part of either the radius or ulna results in an increase in thickness and decrease in length of the other bone (Wermel 1934; Murray 1936). Habitual behaviors can also contribute to heritable variation, or at least to species-specific limb morphology. Modified behaviors have a demonstrable effect on the limb skeleton, as observation of habitually bipedal rats and goats has shown (H. S. Colton 1929; Slijper 1946). Whereas bipedal goats and rats are so extreme as to be unlikely in the "natural" world, learned behaviors such as gait or resting posture may have a substantial effect on limb bones and may contribute to the shared morphology of a population or species. Some journals discourage morphometric studies of the limb skeleton of zoo animals because of the systematic effects that captive life can exert on limb morphology.

Polymorphic Variability

Limb elements may exhibit polymorphic variability, or multiple discontinuous manifestations of a trait within the same population (Yablokov 1974). Examples of polymorphic variation include polydactyly, supernumerary bones or muscles, or achondroplastic dwarfism.

The most common polymorphisms involve missing bones (or parts of bones) associated with otherwise normal tissues. Most of these types of polymorphism are caused by events that happen after the basic limb pattern is specified. Recently, Packard et al. (1993) argued that such malformations are caused primarily by teratogenic effects on the arterial system rather than the skeletal system per se. Development of the arterial system is closely associated with the development of the skeletal system (Karsenty 2003), and its malformation impedes the normal development of the bone. The nonheritable condition of mice in which fibulae and feet are both missing but the rest of the skeleton is normal (Grüneberg 1952) is probably an example.

Not all losses of limb elements are teratogenic, however. The mouse polydactylous condition known as "luxate" produces a reduced or absent tibia, which is replaced by a ligament. The fibula is also malformed, usually bent, and medial digits are polydactylous. The luxate condition has a semi-dominant inheritance pattern (Grüneberg 1952) and is caused by a mutation that reduces the size of the domain of Fgf-8 expression in the preaxial (radial) side of the limb at the initial stage of limb development (Yada et al. 2002; see also T. C. Carter 1954 for a similar argument that an anterior shift in the apical ectodermal ridge was responsible for the luxate condition). The reduction in the Fgf domain allows Shh to be expressed more anteriorly, which expands the expression domains of other genes normally confined to the posterior bud. The result is polydactyly of the anterior (radial) digits. The gene involved in the mouse luxate mutation is now known to be a member of a large family of luxoid genes. Interestingly, Grüneberg (1952) reported that there are several types of inherited preaxial polydactylism in mammals, but postaxial polydactylism is much rarer.

The example of the luxate condition illustrates the possibility that some common genetic limb polymorphisms may have few evolutionary consequences. Luxate limbs, though heritable, are so universally detrimental that they are almost certain to be removed from a population by selection, even though mutations leading to the condition are common enough for luxoid conditions to be taxonomically widespread (Grüneberg 1952; Pucek 1965; Herreid 1958).

Other limb polymorphisms exist that may be of evolutionary significance, however. For example, Yablokov (1974) reported that the pisiform and some distal carpals were variably absent in *Delphinapterus leucas*, the Beluga whale, and that extra elements were commonly present between the proximal and distal carpals. Whales as a group are characterized by hyperphalangy, or the possession of extra phalanges in the digits, and the variation described by Yablokov is probably not detrimental for whales as it would be for horses. The condition of having extra elements was apparently heritable in the belugas because similar carpal patterns were found in mothers and calves (Belkovich and Yablokov 1965). Another limb polymorphism with interesting patterns of variation was also found in belugas, whose fourth digit is often forked (a comparatively rare postaxial polydactyl condition). In a few populations, however, the fifth digit rather than the fourth is bifid (Yablokov 1974).

Alberch (1985a) argued that the distribution of supernumerary and vestigial digits among breeds of domestic dog is due to a completely unrelated phenomenon: the number of cells in the developing limb buds. Alberch argued that the number of digits was a function of the size, or number of cells in, the developing limb bud. He argued that a critical mass of cells was required before the condensation of an anlage could occur. Small breeds fell below the required threshold and often lacked a digit, whereas large breeds had more than the requisite number of cells and sometimes had an extra digit. This mechanism of polymorphic variation is an example of an interactive effect between size and number of digits. Lande (1978) provided an interesting worked example from a population genetics perspective of how such interactive polymorphisms might evolve.

An interesting limb polymorphism with widespread pathological occurrence, but which is the norm in some mammal groups, is syndactly. A large clade of Australian marsupials, the Syndactyla, share an unusual hindfoot morphology in which digits II and III are comparatively gracile and are bound together as a single toe (fig. 15.10A). This syndactylous condition is found homologously in peramelids, notoryctids, phalangerids, vombatids, and macropodids (Szalay 1994). In these marsupials, syndactyly is the normal condition, it is heritable, and it is not associated with other major skeletal abnormalities. Syndactyly is a widespread pathology in other groups of mammals, however, where it is usually accompanied by other defects. Grüneberg (1952) reported a bilateral defect of the radius, in which carpals, metacarpals, and digits on the radial side of the hand were also missing, and digits III and IV were syndactylous. He also reviewed a defect in mice in which both hindlimbs were completely missing, the pelvis had associated defects, and digits II and III of the forelimb were syndactylous. These conditions were probably not heritable, and both were extremely detrimental to the individuals that had them. Other examples of syndactyly are heritable, but also have associated detrimental effects, such as oculo-dento-digital dysplasia in humans (Ioan et al. 2002). The disjunction between syndactyly and other skeletal (and nonskeletal) effects suggests that the molecular developmental pathway leading to marsupial syndactylism is different from in commonly studied pathologies. In the marsupials, the syndactylous condition was fixed in the ancestral population only through the conjunction of heritability, disassociation with other detrimental effects, and functional suitability.

These examples indicate that polymorphic variation can be either genetic or ecophenotypic, and that the ultimate cause of polymorphisms may be associated directly with skeletal developmental system (as in the luxate condition) or indirectly with another system, such as the arterial one.

Early Diversification of the Mammalian Limb

Most evolutionary scenarios put the ancestors of placental and marsupial mammals in the trees. The view that ancestral therian mammals were arboreal has been around since the early 20th century when Matthew (1904) argued that ancestral placental mammals may have had opposable digits on the manus and pes. Matthew was extending to placentals an earlier argument that marsupials had an arboreal ancestry (Huxley 1880; Dollo 1899), saying that the "Cretaceous ancestors of Tertiary mammals were small arboreal animals of very uniform skeletal characters, but probably somewhat differentiated in dentition according as fruit, seeds and nuts, or insects formed the staple of their diet" (Matthew 1904, 816). Matthew's argument was based on an analysis of skeletal characters, primarily from the placental mammals in the early Cenozoic fossil record.

The arboreal origin of mammals has not gone unquestioned, however. Haines (1958) presented detailed comparative anatomical data on the feet of mammals and concluded that the earliest then known placentals were probably terrestrial and that, in fact, Cretaceous forests would have had little to offer mammals. Szalay (1984, 1994) reiterated the possibility that the earliest placental mammals were terrestrial and argued that scansorial adaptations in placentals (especially archontans) might not be homologous with scansorial features in the ancestral marsupial. A lively literature has grown up around the debate, which obviously remains open (Jenkins 1974; Cartmill 1974; O. J. Lewis 1989; Bloch and Boyer 2002, 2003; Kirk et al. 2003).

Recent finds show that early mammals were more diverse in their postcranial skeleton and locomotory habits than Matthew imagined, however (fig. 15.12). New limb morphologies have been found in eutherians in other groups of Mesozoic mammals. It is now clear that early mammals, far from being a homogeneous group of arboreal specialists, inhabited a broad range of ecological niches, partitioned by substrate, diet, body mass, and geography.

Two triconodonts, members of a paraphyletic group of mammals from the Late Jurassic and Early Cretaceous, are now known from their postcranial skeletons: the late Jurassic *Jeholodens jenkinsi* from China (Ji et al. 1999) and the Early Cretaceous *Gobiconodon ostromi* from North America (Jenkins and Schaff 1988). *Jeholodens* was most likely a terrestrial, plantigrade animal. The arrangement of tarsals is

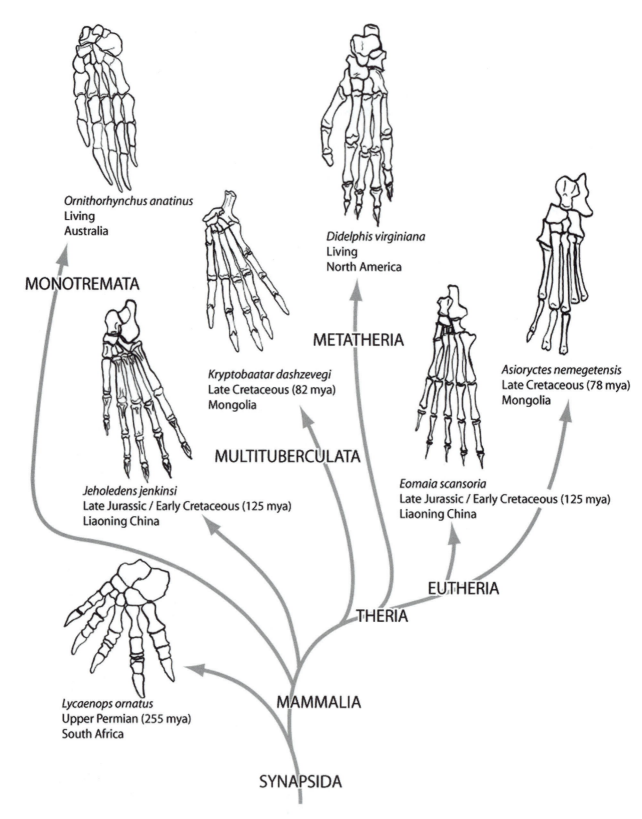

Ornithorhynchus anatinus
Living
Australia

MONOTREMATA

Didelphis virginiana
Living
North America

METATHERIA

Kryptobaatar dashzevegi
Late Cretaceous (82 mya)
Mongolia

MULTITUBERCULATA

Asioryctes nemegetensis
Late Cretaceous (78 mya)
Mongolia

Jeholodens jenkinsi
Late Jurassic / Early Cretaceous (125 mya)
Liaoning China

Eomaia scansoria
Late Jurassic / Early Cretaceous (125 mya)
Liaoning China

EUTHERIA

THERIA

Lycaenops ornatus
Upper Permian (255 mya)
South Africa

MAMMALIA

SYNAPSIDA

Figure 15-12 Evolution of the mammalian tarsus. Left tarsus of selected living and extinct taxa showing the diversity and evolutionary transitions of major mammalian groups. Therian mammals share an astragalus that sits partially or fully over the calcaneum. (*Lycaenops ornatus* after Gregory 1951; *Jeholodens jekinsi* after Ji et al. 1999; *Eomaia scansoria* after Ji et al. 2002; *Kryptobaatar dashzevegi* after Kielan-Jaworowska et al. 2000; *Ornithorhynchus anatinus* after Szalay 1994; and *Asioryctes nemegetensis* and *Didelphis virginiana* after Kielan-Jaworowska 1977.)

like nonmammalian cynodonts, montotremes, and multituberculates in having the axis of rotation of the upper ankle joint at an oblique angle to the main axis of the foot, a condition that suggests a posture that was sprawling. The calcaneum had a tubercle, but the astragalus was not subtended by the calcaneum as it is in therians.

One of the most surprising features of *Jeholodens* and *Gobiconodon* is that they had a scapular spine. Unlike other scapular features, the spine is a specialized structure not found in monotremes, multituberculates, or *Morganucodon* (Jenkins and Parrington 1976; Krause and Jenkins 1983). Traditionally, the spine was thought to have evolved from the cranial border of the scapula in nonmammals. It was thought that the spine had evolved by the addition of a supraspinous portion to the blade leaving a spine running along the junction of the neomorph and the "old" scapula (Romer and Parsons 1977). Paleontological and developmental information now suggests that the spine itself is the novelty, formed (developmentally and evolutionarily speaking) as part of the reorganization of the trapezius and deltoid musculature associated with the evolution of upright posture and a more mobile pectoral girdle. The spine of *Jeholodens* and *Gobiconodon* suggests that these triconodonts had a muscular anatomy that permitted extensive movement of the pectoral girdle during locomotion.

Developmental studies support the distinction between spine and the rest of the scapula. The spine develops in one of three ways: by apposition to the acromion, as an independent condensation, or as an intermuscular ossification (Sánchez-Villagra and Maier 2002; Grossman et al. 2002). In all cases, the spine has a developmental origin separate from the acromion process, which has its own condensational history and a distinctive pattern of *Hox5a* and *Pax1* coexpression (Timmons et al. 1994; Aubin et al. 2002).

Other important discoveries have come from the postcrania of multituberculates. Multituberculates were a diverse group of rodentlike animals that flourished in the Cretaceous and early Tertiary, but their postcrania have been almost completely unknown. A skeleton of *Ptilodus kummae* from the Paleocene of North America demonstrated specialized arboreal features, including a unique upper ankle joint that facilitated squirrel-like hindfoot reversal (Krause and Jenkins 1983; Jenkins and Krause 1983). Subsequent discoveries indicate that multituberculates retained an interclavicle (Sereno and McKenna 1995), that the humerus had significant torsion, that the humerus had no trochlea, that the calcaneum contacted the fifth metatarsal (compare the therian and nontherian feet in fig. 15.12), and that multituberculates had a posture that was much more sprawling than that of therians (Kielan-Jaworowska and Gambaryan 1994; Gambaryan and Kielan-Jaworowska 1997). The tarsus of *Kryptobaatar dashzevegi* and *Ptilodus kummae* is shorter and broader than therian mammals, with a wide astragalus and a calcaneum whose tuber is angled posterolateral to the main axis of the metatarsus (fig. 15.12). Most likely, the parasagittal stance (fig. 15.2) is a specifically therian condition not found more broadly among mammals. Considerable locomotory diversity existed among multituberculates, as indicated by the humeral differences among *Bulganbaatar, Lambdopsalis,* and *Kryptobaatar* (Kielan-Jaworowska et al. 2000).

Several new skeletons have also expanded our picture of eutherian limb diversity. One of the most striking is the 125-million-year-old *Eomaia scansoria* from Early Cretaceous of China. A contemporary of *Jeholodens, Eomaia* had many of the features associated with the therian and eutherian skeleton. It had a mobile scapula with well-developed spine; a well-developed ulnar olecranon and calcaneal tuber; an upright, parasagittal posture; a fibula that was significantly more gracile than the tibia; an astragalus subtended by the calcaneus; an astragalar trochlea; a calcaneum that was separated from the metatarsals by the navicular; and a long axis of the metatarsus that was roughly perpendicular to the axis of rotation in the upper ankle joint. *Eomaia* has been interpreted as being arboreal, though that interpretation is not universally accepted (Weil 2002). Other important eutherian limb skeletons that have been described recently include *Ukhaatherium nessovi,* an asiorychtithere from the Late Cretaceous of Mongolia (Horovitz 2000), and *Deccanolestes hislopi* from the Late Cretaceous of India (Prasad and Godinot 1994).

Whereas the diversity in mammal limb structure today evolved over the last 140 million years since the last common ancestor of marsupials and placentals, limb and substrate diversity was already varied in the Mesozoic. The earliest known placental, *Eomaia,* was a scansorial tree dweller; *Zhangheotherium,* a therian contemporary of *Eomaia,* was terrestrial (Hu et al. 1997), as was *Jeholodens.* Because of structural correlates to limb function, many locomotory aspects of a mammal can be reliably interpreted from its skeleton, even when it is extinct with no exact analogues among the living (Szalay 1994).

The evolutionary migration of mammals through changing environments and habitats can be traced using the bones of the same legs that carried them on daily forays during their lives. The constant requirements of support and movement mean that habitat, behavior, and growth are intertwined with bone, muscle, and tendon in phylogenetic transformations (Szalay 1981). One aspect can hardly change without the others doing likewise.

Acknowledgments

The author would like to thank Jason Head, Christine Janis, Marcelo Sánchez-Villagra, Heidi Schutz, Nancy Simmons, Fred Szalay, and Mark Uhen, who answered questions and provided useful discussion. Jason Head provided many useful suggestions on the text. I would also like to specially thank Brian Hall for inviting this contribution and patiently editing its content.

Chapter 16 Skeletal Adaptations for Flight

Stephen M. Gatesy and Kevin M. Middleton

V ERTEBRATES AROSE IN an aquatic environment dominated by the dynamics of fluid flow. With the advent of terrestriality, tetrapods moved into a substrate-based domain in which gravity and inertia became most significant for locomotion. Among the amniotes, three lineages once again exploited fluid forces to conquer the aerial realm through powered flight. The ancestors of birds, bats, and pterosaurs lengthened and broadened their forelimbs into aerodynamically proficient appendages that we collectively call "wings." Such convergence in overall shape and movement presumably stems from the presence of stringent constraints on the aerodynamic performance of flapping forelimbs. Wings are thus a classic example of vertebrate limbs evolving common solutions to a shared mechanical demand. Yet these volant forms would never be mistaken for one another; each clade arose from a different preflight condition and exhibits unique specializations in airfoil design.

Many authors have addressed vertebrate flight from the viewpoint of aerodynamics, scaling, ecology, and evolution (e.g., Pennycuick 1975; Rayner 1979, 1981, 1988; Norberg 1981, 1985, 1990). In this chapter we focus on wings from an osteological perspective. We begin by introducing the phylogeny of powered fliers (fig. 16.1). Because flapping flight originated in both Reptilia (pterosaurs, birds) and Synapsida (bats), the most recent common ancestor of these groups lies at the root of the clade Amniota (e.g., Gauthier et al. 1988b). Following Sumida (1997), we use closely related fossil taxa to reconstruct the forelimb skeleton and pectoral girdle in the hypothetical ancestral amniote. This allows us to character-

ize wing osteology in each group relative to a common reference point in their evolutionary history.

We then delineate the disparity of wing skeletal design by constructing two theoretical morphospaces, within which we plot select forelimb dimensions of birds, bats, pterosaurs, their close nonvolant relatives, and taxa near the base of Amniota. The distribution of these data reveals patterns of "the range of anatomical design" (Gould 1991, 412) and is used to explore very basic questions about wing evolution. For example, how disparate are the skeletal and segmental proportions of vertebrate wings? Are wings highly constrained by aerodynamic requirements compared to nonflying forelimbs? Have pterosaurs, birds, and bats converged on a single design, or does each clade show singular proportions commensurate with their unique solutions to supporting the flight surface? Are some flying clades more disparate than others, and if so, why? How might variation in wing proportions relate to body size, flight style, and the forelimb's role in behaviors other than flight? Finally, how do wing skeletons in each group differ from the limb skeletons of their close, nonvolant relatives, and what might these differences reveal about the origin of flight? Our aim is to begin to discern which character states preceded, coincided with, and followed the three transitions from limb into wing.

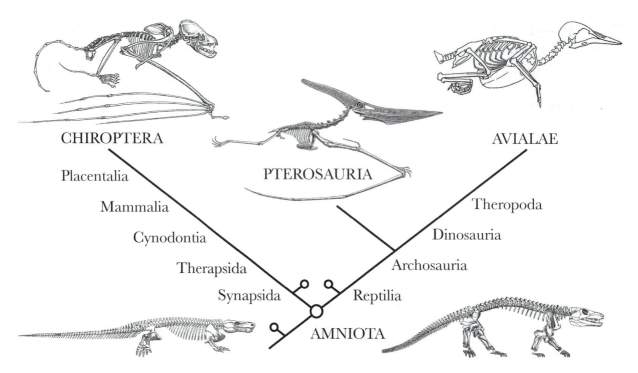

Figure 16-1 Simplified cladogram of amniote phylogeny showing three origins of powered flight. The highly derived wings of bats, pterosaurs, and birds evolved from the terrestrial forelimb of an ancestral amniote (large circle) that we reconstruct based on closely related taxa (small circles) such as *Limnoscelis* (lower left) and *Seymouria* (lower right). (Cladogram based on Laurin and Reisz 1995. Drawings modified from Eaton 1910; Williston 1911; T. E. White 1939; Jepsen 1970; R. L. Carroll 1987–1988; Jenkins et al. 1988.)

Overview and Wing Osteology

Ancestral Amniote

During flight the musculoskeletal elements of the forelimb support and deform the wing's aerodynamic surface—a membrane in pterosaurs and bats and feathers in birds (fig. 16.2). Such highly derived morphologies arose independently from the less specialized forelimb found in their most recent common ancestor at the base of the clade Amniota. Based on taxa near the origin of amniotes (e.g., *Limnoscelis*, *Seymouria* [Williston 1911; T. E. White 1939]; see Reisz 1997 and Sumida 1997 for further discussion), we briefly describe the forelimb skeleton of a hypothetical ancestral amniote so that it can serve as a reference for appreciating the modifications of its volant descendants.

Primitively, the amniote shoulder girdle (fig. 16.3A) consisted of paired scapulae, coracoids, clavicles, and splintlike cleithra, as well as an unpaired, median interclavicle (Sumida 1997). The scapula and coracoid both contributed to a "screw-shaped" glenoid fossa that formed the shoulder joint with the humeral head. The humerus had a tetrahedral organization (Romer 1956) with proximal and distal articular surfaces oriented approximately 90 degrees to one another and pronounced processes for muscle attachment. The radius

and ulna articulated with a distinct capitulum and trochlea on the distal humerus; pronation and supination were likely limited (Sumida 1997). Insufficient specimens with complete carpals make details of the wrist region unclear, but flexibility was likely spread across multiple joints rather than concentrated at a single articular axis (Holmes 1977; Sumida 1989). The manus was pentadactyl (fig. 16.3B), with digit IV longest (Sumida 1997).

This basal form can be reconstructed as an obligate quadruped (Sumida 1997) that lived in the Early Carboniferous approximately 340 million years ago (Paton et al. 1999). Its pectoral appendages were short relative to its trunk and bore comparatively homogeneous digits. Yet from such unspecialized forelimbs evolved three unique wing morphologies capable of powered flight.

Pterosaurs

Pterosaurs were the first amniotes to evolve flapping flight. Based on cranial morphology, pterosaurs have been reconstructed as piscivores, insectivores, filter feeders, and scavengers (Wellnhofer 1978, 1991), presumably filling many of the aerial niches now occupied by birds. The pterosaur fossil record extends back at least 210 million years to the Late Triassic (Wellnhofer 1978), and even the oldest and most primi-

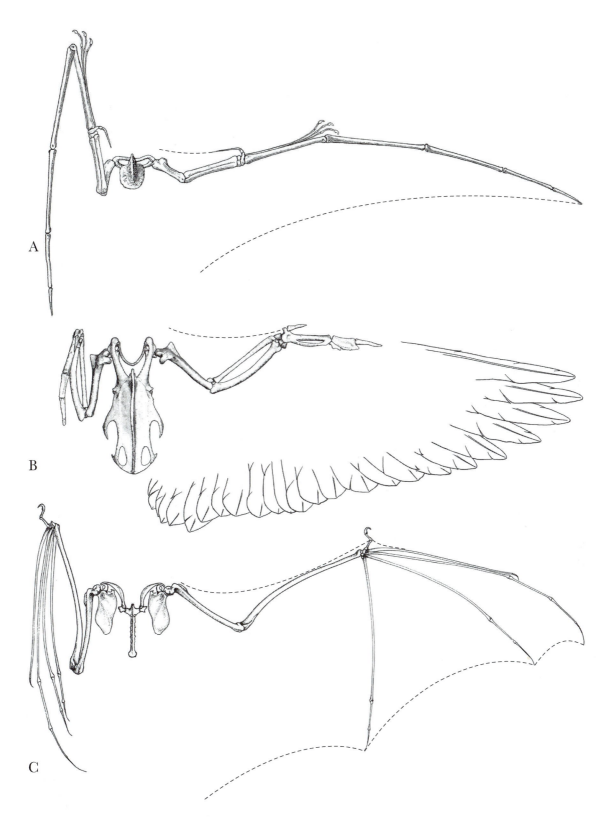

Figure 16-2 Wing osteology in vertebrate powered fliers, ventral view. (A) The pterosaur, *Pteranodon* (modified from Eaton 1910). The main wing membrane is primarily supported by the hypertrophied fourth digit. Mobility was greatest at the shoulder, elbow, and fourth metacarpophalangeal joint. (B) An extant pigeon, *Columba* (modified from N. S. Proctor and Lynch 1993). Primary feathers extending from the manus significantly increase wingspan. Note the proximity of the wrist to the shoulder when folded. (C) An extant bat (modified from Hill and Smith 1984). Four elongated digits support the main wing membrane, which is also attached to the hindlimb.

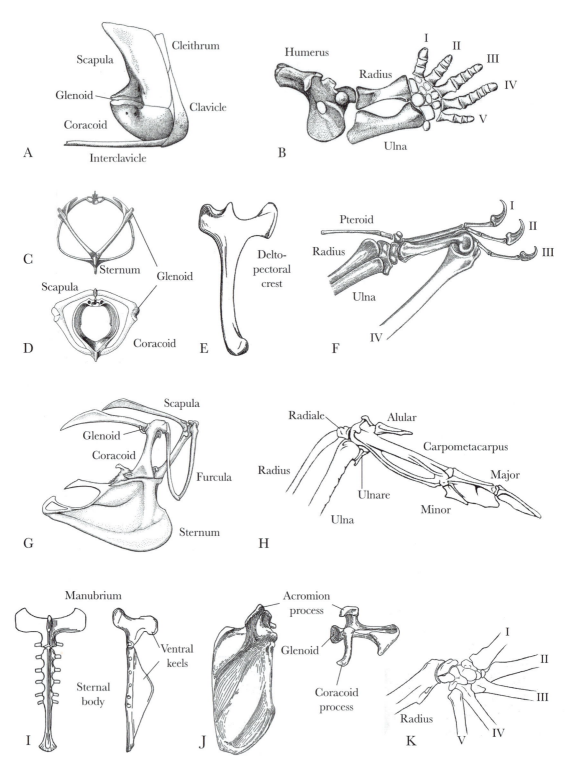

Figure 16-3 Osteology in taxa close to the origin of amniotes (A, B), pterosaurs (C–F), birds (G, H), and bats (I–K). (A) Right lateral view of the pectoral girdle of *Seymouria* (modified from Romer 1956 after T. E. White 1939). (B) Dorsal view of right forelimb of *Limnoscelis* (modified from R. L. Carroll 1987–1988 after Williston 1925), showing relatively short, stout elements and the full complement of digits. (C) Anterior view of thorax and pectoral girdle of a *Rhamphorhynchus* (modified from Wellnhofer 1991). (D) In pterodactyloids such as *Pteranodon*, the scapulae articulate with fused vertebrae, which form a notarium (modified from Bramwell and Whitfield 1974). (E) Right humerus and (F) hand of *Rhamphorhynchus* in dorsal view (modified from Wellnhofer 1978, 1991). (G) Anterolateral view of the pectoral girdle of a pigeon, *Columba* (modified from N. S. Proctor and Lynch 1993). The sternum bears a prominent ventral keel, and the clavicles have fused into a furcula, or wishbone. (H) Wrist of *Columba* in dorsal view showing extensive fusion and reduction in element number (modified from Vasquez 1994). (I) A chiropteran sternum in ventral and right lateral view. (J) Dorsal and anterior views of a right scapula. (K) Right wrist in dorsal view showing the distal radius, complexly faceted carpals, and five proximal metacarpals (bat figures modified from Vaughn 1959).

tive pterosaurs are easily recognizable as volant organisms (Wild 1978, 1984a, 1984b; Jenkins et al. 2001). Despite the fragility of pterosaur bones, many are so distinctive (e.g., humeri, wing-finger phalanges) that even very fragmentary fossils can be confidently assigned to Pterosauria. General information on pterosaur anatomy and evolution has been reviewed by Eaton (1910), Romer (1956), Wellnhofer (1978, 1991), and R. L. Carroll (1987–1988).

Although no good transitional forms are known, pterosaurs are thought to be closely related to archosaurs near the base of Dinosauria (Sereno 1991; Sereno and Arcucci 1993, 1994; Benton 1999). Several phylogenetic analyses have found the small, bipedal archosaur, *Scleromochlus taylori,* from the Late Triassic of Scotland (Woodward 1907) to be the sister taxon of pterosaurs (Sereno 1991; S. C. Bennett 1996; Benton 1999). The monophyly of Pterosauria has never been questioned, and the clade has been traditionally divided into two groups, rhamphorhynchoids and pterodactyloids (Wellnhofer 1978). Rhamphorhynchoids, known from the Late Triassic to the Cretaceous, had relatively small heads, short necks, and long tails (Wellnhofer 1975a, 1975b, 1975c, 1991). The more derived pterodactyloids, known from the Jurassic through the end of the Cretaceous, were characterized by having relatively larger heads, longer necks, and shorter tails (Wellnhofer 1970; S. C. Bennett 2001a, 2001b). Pterodactyloids include the largest flying animals to have ever lived, reaching an estimated wingspan of 11–12 meters in *Quetzalcoatlus* (Lawson 1975). A recent cladistic study by Unwin (1995) supported pterodactyloid monophyly but revealed "rhamphorhynchoids" to be paraphyletic.

Pterosaur flight has received attention from a variety of researchers (Brower 1982; Padian 1985, 1991; Rayner 1988; Hazlehurst and Rayner 1992; Padian and Rayner 1993a, 1993b; Alexander 1994; Marden 1994; Unwin and Bakhurina 1994; S. C. Bennett 1997). Most modern workers concur that pterosaurs were capable of flapping flight, although the largest pterodactyloids are thought to have primarily harnessed winds and/or thermals for soaring (Bramwell and Whitfield 1974). A more contentious issue has been terrestrial locomotion. Padian (1983a, 1983b) and Padian and Olsen (1984a, 1984b) reconstructed pterosaurs as bipedal animals, whereas others have argued for quadrupedal progression (S. C. Bennett 1990; J. M. Clark et al. 1998; Unwin 1987, 1988a, 1988b, 1997, 1999; Unwin and Henderson 2002; Wellnhofer 1988). Recent analyses of footprints (e.g., Mazin et al. 1995; J. L. Wright et al. 1997) have convinced many workers that pterosaurs habitually used all four limbs when on the ground.

We characterize the pectoral girdle and wing of pterosaurs based largely on Wellnhofer (1978) and S. C. Bennett (2001a). The shoulder girdle is simplified relative to the ancestral amniote condition (figs. 16.3C, D). Dermal elements are absent, leaving only an ossified sternum and paired scapulae and coracoids (Romer 1956). The fused sternal plates bear a large, anteroventrally projecting keel to accommodate enlarged flight musculature. Each scapula and coracoid is co-ossified into a V-shaped scapulocoracoid, which forms a saddle-shaped glenoid. In large pterodactyloids, such as *Azhdarcho, Dsungaripterus, Nyctosaurus, Pteranodon, Quetzalcoatlus,* and *Santanadactylus* (S. C. Bennett 2001a), the distal end of the scapula articulates with a series of three to eight fused thoracic vertebrae, which form a notarium (fig. 16.3D).

Many of the wing bones are pneumatic and thin-walled, and in large pterosaurs, even the distal phalanges are pneumatic (Bramwell and Whitfield 1974). The humerus's saddle-shaped proximal surface articulates with the glenoid to form a sellar joint. A large, linguiform deltopectoral crest projects forward for the insertion of the pectoral musculature (fig. 16.3E). The forearm is dominated by the ulna, and the elbow is a simple hinge joint (Bramwell and Whitfield 1974). The pterosaur wrist originally contained five carpal bones in two rows. The two proximal carpals fuse in all but the most primitive pterosaurs. In some derived forms the two distal carpals associated with metacarpal IV fuse as well. The remaining distal carpal articulates with a hollow spur of bone called the pteroid (fig. 16.3F), a structure unique to pterosaurs. The pteroid, which likely is bone rather than calcified tendon (Unwin et al. 1996), supports the edge of an anterior wing membrane spanning from neck to wrist (fig. 16.2A).

Digit IV dominates the hand of all pterosaurs. The fourth metacarpal is extremely robust but varies in length, being relatively short in rhamphorhynchoids (fig. 16.3F) and longer in pterodactyloids (figs. 16.1, 16.2A). The distal articular surface of metacarpal IV is a trochlea with offset condyles, allowing the wing finger to automatically supinate during upstroke (Padian 1983a; Jenkins et al. 2001) and to tuck alongside the body when on the ground (Bramwell and Whitfield 1974). Distally, the wing membrane is supported only by the four elongated phalanges of digit IV, and no ungual is present. Metacarpals I–III are slender and bear digits with a phalangeal formula of 2-3-4 (fig. 16.3F); the fifth metacarpal and digit are lost. Wellnhofer (1991) noted that the unguals of digits I–III show well-developed flexor tubercles and hypothesized that these digits had strong grasping ability.

Birds

The extant avifauna is globally distributed and includes over 9,000 species exhibiting a wide spectrum of body size, locomotor style, and diet. Birds are first found in the fossil record approximately 145 million years ago in the form of *Archaeopteryx lithographica* from the Late Jurassic Solnhofen limestones of southern Germany (von Meyer 1861a, 1861b, 1862). During the last 20 years an influx of new fossil material, primarily from China, Spain, and South America, has substantially increased our understanding of Mesozoic bird diversity (for reviews see Chiappe and Witmer 2002 as well as Chiappe and Dyke 2003). In addition to ornithological texts (e.g., Proctor and Lynch 1993) there are several scientific volumes about avian anatomy and evolution (K. E. Campbell 1992; Baumel et al. 1993; Mindell 1997; Olson 1999; Gauthier and Gall 2001; Chiappe and Witmer 2002; Zhou and Zhang 2002c). General descriptions of avian osteology are relatively common (e.g., Fürbringer 1888; Fisher 1946; Owre 1967; Raikow 1985), with the *Handbook of Avian Anatomy: Nomina Anatomica Avium* (Baumel et al. 1993) providing standardized nomenclature.

The evolutionary relationships of birds to other amniotes have been a persistent question for over 100 years (reviewed in Gauthier 1986; Sereno 1991; Witmer 1991, 2002). Workers using cladistic techniques have unanimously favored a theropod ancestry of birds (Cracraft 1986; Gauthier 1986; Norell et al. 2001; J. M. Clark et al. 2002), a conclusion reached by previous authors based on comparative anatomy (Huxley 1868, 1870a, 1870b; Ostrom 1973, 1974, 1975, 1976a, 1976b; Bakker and Galton 1974). In his phylogenetic analysis of Diapsida, Gauthier (1984, 1986) found strong support for birds being nested within maniraptoran coelurosaurs, with dromaeosaurs such as *Velociraptor* and *Deinonychus* as close sister taxa. More recent discoveries, in particular spectacular specimens preserving soft tissues from China (e.g., P.-J. Chen et al. 1998; Ji et al. 1998; Xu et al. 1999a, 1999b, 2000, 2001, 2003; Norell et al. 2002), have begun to blur the distinction between feathered theropods and primitive birds (Witmer 2002). Avian systematics has a rich and diverse history summarized by Sibley and Ahlquist (1990). Intraordinal relationships are being addressed using both morphological (Ericson 1997; Livezey and Zusi 2001; Cracraft and Clarke 2001; Dyke et al. 2003) and molecular approaches (contributions in Mindell 1997; Groth and Barrowclough 1999; van Tuinen et al. 2000; Ericson et al. 2001), but many nodes remain poorly resolved.

The mechanics of bird flight is addressed in an expansive and rapidly growing literature. Over the past 25 years, studies have elucidated flight from the perspectives of aerodynamics (e.g., Pennycuick 1975, 1986; Rayner 2001b; Hedrick et al. 2002, 2003; Spedding et al. 2003), kinematics (e.g., Jenkins et al. 1988; Tobalske and Dial 1996), and muscle activity/mechanics (e.g., Dial et al. 1987, 1988, 1991; Dial 1992a, 1992b; Biewener et al. 1992). Birds are the only group of powered fliers in which this ability has been secondarily lost. Flightlessness has evolved in at least 34 separate families

of birds, including auks, cormorants, dodos, ducks, grebes, and parrots (Livezey 1995).

The avian sternum (figs. 16.1, 16.2B, 16.3G) bears a ventrally expanded, midline keel to accommodate the hypertrophied flight musculature (supracoracoideus and pectoralis muscles). This keel is absent in the flightless ratites. Paired scapulae and coracoids articulate with the fused clavicles, which constitute the furcula or wishbone. Ventrally, the flared ends of the elongated coracoids fit into anterior sulci on the sternum. The coracoids meet the scapulae at an acute angle, together forming the saddle-shaped glenoid cavity (Jenkins 1993). An acrocoracoid process projects anteriorly, forming a pulley to deflect the tendon of the supracoracoideus muscle and allowing it to supinate and elevate the wing (Poore et al. 1997). The avian scapula has a very characteristic strap shape, thinning caudally as it lies along the ribcage roughly parallel to the vertebral column (fig. 16.3G).

The proximal humerus bears an ovoid articular head, a deltopectoral crest that is less prominent than that of pterosaurs, as well as dorsal and ventral tubercles (fig. 16.2B). The shaft is pneumatized in most species. Distally, the humerus bears articular condyles for the ulna and radius. The ulna is bowed caudally and bears quill knobs for the attachment of secondary flight feathers, whereas the radius is straighter and smaller in diameter. The ulna and radius articulate with a free pair of carpal bones, the ulnare and radiale (fig. 16.3H). These carpals articulate with the carpometacarpus, a complex co-ossification of three distal carpals and three metacarpals. The identity of the three digits has received a great deal of attention (Hinchliffe 1977; Hinchliffe and Hecht 1984; Burke and Feduccia 1997; G. P. Wagner and Gauthier 1999; Feduccia and Nowicki 2002; Wagner and Larsson, chap. 4 in this volume). Herein we follow Baumel and Witmer (1993) by referring to the digits as the alular, major, and minor rather than by number. Alular and minor digits typically have one phalanx, whereas the major digit has two (fig. 16.3H). Ungual phalanges are found on the alular and/or major digits in many birds (Fisher 1940).

Bats

With over 900 extant species, bats (Chiroptera) comprise approximately one-quarter of present mammalian diversity. Bats do not reach the size of the largest birds or pterosaurs; all bats retain the ability to fly. Diets include insects, vertebrates, fruit, pollen/nectar, and blood. Extant bats are found in most tropical and temperate regions except certain remote islands (Nowak 1991). Although tooth fragments from the Late Paleocene have been referred to Chiroptera (Gingerich 1987), the first definitive fossils are about 53 million years old from Early Eocene deposits (Simmons and Geisler 1998). Fossil bats offer few clues to the order's phylogenetic position among mammals or the origin of chiropteran flight. Even the earliest known forms, such as *Icaronycteris*, have forelimbs modified into wings (Jepsen 1970) and ears specialized for echolocation (Novacek 1985).

Conventionally, bats have been divided into two suborders that form a monophyletic group closely allied with dermopterans ("flying lemurs"; Altringham 1996). Megachiropterans, or megabats, consist of a single family of Old World fruit bats, including the "flying foxes." Microbats, suborder Microchiroptera, encompass all other families. Over the past decade these viewpoints have been challenged on several fronts. First, Teeling et al. (2000, 188) found no molecular evidence for close relationship between Chiroptera and Dermoptera and remarked that any "presumed shared derived characters for flying lemurs and bats are convergent features that evolved in association with gliding and flight, respectively." Second, based on brain morphology, J. D. Pettigrew et al. (1989) suggested that megabats and microbats may have evolved separately from nonchiropteran ancestors. Later studies supported bats as a natural group (e.g., Bailey et al. 1992; Honeycutt and Adkins 1993). Finally, microchiropteran monophyly has also been questioned (Stanhope et al. 1998; Springer et al. 2001; Teeling et al. 2000, 2002). In these molecular phylogenies Megachiroptera remained monophyletic, but several microchiropteran families (Hipposideridae, Megadermatidae, Nycteridae, and Rhinolophidae) were found to be more closely related to Megachiroptera than to other microchiropterans. Some recent morphologic studies, however, have supported microbat monophyly (e.g., Novacek 1992; Shoshani and McKenna 1998; Simmons and Geisler 1998; K. E. Jones et al. 2002). This issue remains to be resolved.

Most bats flap continuously during flight, although some megachiropterans also soar (Norberg et al. 2000). Many species can hover and are extremely aerobatic. The wing's aerodynamic surface is divided into regions between the digits, body wall, hindlimb, and tail. The axial skeleton of bats has been stiffened in association with flight. In some families, the final cervical vertebra fuses with the first and sometimes second thoracic vertebra. Functional studies of bat flight include descriptions of kinematics, muscle activity, bone stresses, and aerodynamic performance (e.g., Norberg 1972, 1976; Hermanson and Altenbach 1983; Rayner and Aldridge 1985; Rayner 1987; Swartz et al. 1992; Swartz 1998), as well as wing allometry (Norberg 1981; Swartz 1997) and ecomorphology (Norberg and Rayner 1987).

Our osteological description is largely based on the work of Vaughan (1959), Walton and Walton (1970), Strickler

(1978), Hill and Smith (1984), and Koopman (1984). The chiropteran sternum is T-shaped and composed of two segments (figs. 16.2C, 16.3I). Ventral keels project from the manubrium and sternal body in some forms (Cuvier 1805), but in others the enlarged pectoralis musculature meets at a midline tendon sheet (Altringham 1996). Large lateral processes of the manubrium articulate with the relatively massive clavicles. Each clavicle articulates with its scapula's acromion process, coracoid process, or both and likely functions as a spoke guiding scapular rotation during flight. The rectangular or oval scapular blades lie on the dorsal surface of the ribcage roughly parallel to the vertebral column. A scapular spine separates the small supraspinous fossa from the larger infraspinous fossa. As in other mammals, the coracoid is represented by the coracoid process but in bats is extremely large (up to half the length of the scapula), ventrally directed, and laterally curved (fig. 16.3J). The shallow glenoid socket faces laterally and in some bats is augmented by a dorsal facet that articulates with the greater tuberosity of the humerus.

Bats have non-pneumatized, marrow-filled longbones (Nowak 1991). Although not air-filled, the humerus and radius are relatively thin-walled (Swartz et al. 1992). The humerus has a straight or slightly sigmoid shaft and a head ranging "from nearly round to elliptical to roughly tear-shaped" (Strickler 1978, 43). A prominent pectoral crest similar to that in birds projects from the cranial border, merging with the greater tuberosity. In some species the greater tuberosity projects proximally beyond the humeral head far enough to articulate with the scapula, forming a "scapulo-humeral lock" (Vaughn 1959; Altenbach and Hermanson 1987). Distally, the olecranon fossa, which in other mammals accepts the olecranon process of the ulna, is rudimentary or absent. The radius is the dominant forearm element (figs. 16.1, 16.2C). The olecranon portion of the ulna is fused to the radius. Distally the ulna thins or becomes cartilaginous, and it may not reach the wrist. The carpal bones of bats are arranged in two rows, and the fusion pattern of some elements can be used to determine an animal's age. Grooves on the distal radius interlock with the proximal carpals to restrict wrist movement to flexion and extension (Grassé 1955a).

The hand is composed of five digits, with the metacarpals and phalanges of digits II–V supporting the wing membrane. When the wing is fully extended, the five metacarpals fan out almost 180 degrees with metacarpal V trailing chordwise behind the wrist (fig. 16.3K). Digit I is short and typically bears a strong claw for clinging (Nowak 1991). Other digits lack unguals in most bats, although some megabats retain an ungual on digit II. Compared with nonflying mammals, bat hands have relatively long metacarpals, phalanges with tapering shafts, and a specialized proximo-distal gradient in cortical thickness and mineralization (Swartz 1997; Papidimitriou et al. 1996). Terminal phalanges can lack a marrow cavity and consist only of cartilage (Papidimitriou et al. 1996), a morphology thought to promote, rather than resist bending (Swartz 1997, 1998).

Wing Disparity

How Should Wing Skeletons Be Configured?

The skeletal elements of any wing must perform multiple tasks during flapping flight. Many of these functions are primitive, such as providing structural support against muscular, gravitational, and inertial forces, or articulating to allow changes in limb length and position. Other functions, however, are novel. For example, wing skeletons bear aerodynamic loads from the flight surface (feathers or wing membrane) far greater than their terrestrial ancestors ever encountered. The forelimb skeleton also dictates, to varying degrees, the overall size and shape of the airfoil. This is especially true in bats, where bones delineate the wing's length, chord (breadth), and tip shape. In pterosaurs, the forelimb skeleton determined wing length but not wing chord, which depended on the soft-tissue membrane and its attachment to the hindlimb (see Padian and Rayner 1993b; Unwin and Bakhurina 1994). A bird's wing is the least prescribed by bones because feathers contribute to both its length and its chord. As in the other two clades, however, the length and orientation of skeletal elements is critical for establishing the wing's internal points of mobility and the mechanics of these lever systems.

Wings, however, do not function exclusively for flight. Bats and pterosaurs (based on referred footprints) walk quadrupedally, demanding that their airworthy wings also operate as supportive limbs when on substrates. Birds, as obligate bipeds, are less hindered in this respect, but their wings are often used in nonflight behaviors such as display, brooding, defense, and predation. Bat wings can also have roles in feeding, either to trap flying insects or to manipulate fruit, and in thermoregulation. Finally, winged vertebrates must also manage their hypertrophied forelimbs when not in use. Compact folding may act as an additional constraint on skeletal proportions and joint mobility (Middleton and Gatesy 2000).

Given these basic demands, how should wing skeletons be configured? Most studies of vertebrate wing design use basic parameters of fixed wing aircraft, such as aspect ratio and wing loading (e.g., Pennycuick 1975; Norberg and

Rayner 1987). Such static representations of wing shape and size often correlate with flight performance and ecology but are insufficient for asking many basic morphological questions. How much of the wing should be supported by the arm, forearm, and hand skeleton? How should each of these elements be oriented within the wing? Where should the main joints be positioned for folding? How should the relative segment lengths vary in wings of different flight surface, size, and performance? In the absence of human-made flapping machines to use as paradigms for optimal design (e.g., Lauder 1996), it has been difficult to know how to interpret articulated biological wings from a mechanistic perspective. In the next section we carry out a simple analysis of skeletal proportions in pterosaurs, birds, and bats to begin to explore patterns of wing disparity (e.g., Gould 1991). These patterns may reveal some of the rules governing the construction of vertebrate wings.

How Are Wing Skeletons Proportioned?

The disparity of forelimb elements can be studied using a proportion morphospace (Gatesy and Middleton 1997; Middleton and Gatesy 2000). We visualize this morphospace as a ternary diagram (triangular graph) on which we plot the relative contribution of three variables to a whole. This is a theoretical morphospace, encompassing all possible combinations of three elements rather than an empirical morphospace created only from sampled data (McGhee 1999). Forelimbs with similar proportions will be restricted to one area of the ternary diagram, whereas more disparate limbs will be spread out into a larger point cloud. First, we calculated the relative length of bones representing the arm, forearm, and palm to create a skeletal proportion morphospace (fig. 16.4). For each specimen, the length of the humerus, radius, and metacarpal (IV in pterosaurs, carpometacarpus in birds, III in bats) was divided by the summed length of all three bones to yield percentages. These data are plotted on a ternary diagram in which the lower left, lower right, and upper vertices correspond to 100% humerus, 100% radius, and 100% metacarpal, respectively (fig. 16.4A). Our data set includes 113 specimens from 9 genera of pterosaurs, 554 specimens from 266 genera of volant birds, and 79 specimens from 45 genera of bats (see the appendix for references).

Our sample of vertebrate wings is neither widely nor evenly distributed in ternary morphospace (fig. 16.4A). Wings are restricted to the middle region, whereas the corners, representing limbs with one extremely long element, are entirely empty. The contribution of the humerus varies 24%, from 21% (pterosaur and bat) to 45% (bird). Radial variation is similar, spanning 26% between two pterosaurs at 28% and 54%. Metacarpal percentage is most variable

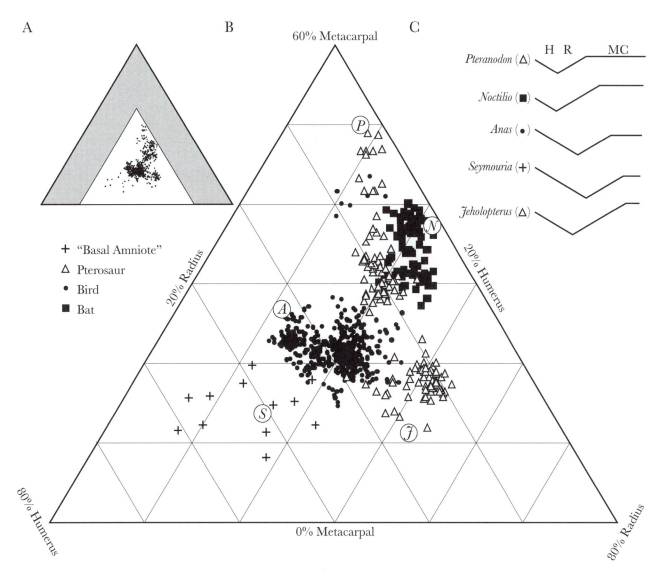

Figure 16-4 Forelimb skeletal proportions in ternary morphospace. (A) The relative lengths of the humerus, radius, and longest metacarpal are plotted for pterosaurs, birds, bats, and taxa near the origin of Amniota ("basal amniotes"). Wings are restricted to the middle of the morphospace. (B) A subset of ternary morphospace enlarged to show the distribution of each group. "Basal amniotes" fall to the lower left of birds, which cluster around the midline. Pterosaurs and bats have relatively shorter humeri than most birds, with the exception of hummingbirds and swifts, which have elongated carpometacarpi. (C) Diagrams of skeletal proportions drawn to identical total length to show the range of anatomical design. Figured specimens are shown in (B) with letters within open circles.

(38%), with pterosaurs again having both the lowest (12%) and highest (50%) values. Only a subset of birds crosses the midline by having the humerus longer than the radius. The radius is never the shortest element of the three, although all other combinations of rank order have evolved. Following Middleton and Gatesy (2000) we can use a disparity index (DI: 100 times the distance between points divided by the maximum possible distance) to quantify interpoint difference. The most disparate wings belong to two pterosaurs (figs. 16.4B, C). The lowermost (*Jeholopterus*) and uppermost (*Pteranodon*) have a DI of 34, representing a divergence in proportions of one-third the maximum theoretically possible. Flying birds have a disparity index of 25, while in bats it is 11.

Bird and bat wings occupy distinct regions of the morphospace, but pterosaurs fill this gap and overlap with both groups. Each clade is further divided into clusters, which we discuss in more detail later. The pterosaur point cloud is long and narrow compared to the more globular distributions of birds and bats. A simple way to estimate the area colonized is to subdivide the morphospace into 400 triangular cells (Gatesy and Middleton 1997). All vertebrate wings occupy 43 cells, of which birds are present in 24, pterosaurs in 21, and bats in just 10.

Basal amniotes and close nonamniote relatives (14 specimens of 12 genera) differ from most volant forms in their relative bone lengths, falling below and to the left of the main bird cloud. Although there is variation among these taxa, they can be characterized as having a humerus longer than the radius and relatively short fourth metacarpals of less than 20%.

How Are Functional Wing Segments Proportioned?

The proportions of the wing's three major longbones allow one aspect of anatomical disparity to be assessed, but func-

tional implications of these patterns are difficult to interpret. To help compare wing skeletons from different clades as articulating structures, we divided the entire wing into proximal, middle, and distal segments that move as units during flight. We plotted these data to create a segmental proportion morphospace (fig. 16.5). Bats primarily bend the wing skeleton at the elbow and wrist, so the three segments are the humerus, radius, and metacarpal + phalanges of digit III (79 specimens of 45 genera). For pterosaurs, most flexion/extension of the distal wing took place at the fourth metacarpophalangeal joint rather than at the relatively immobile wrist (Padian 1983a). Therefore, the three segments

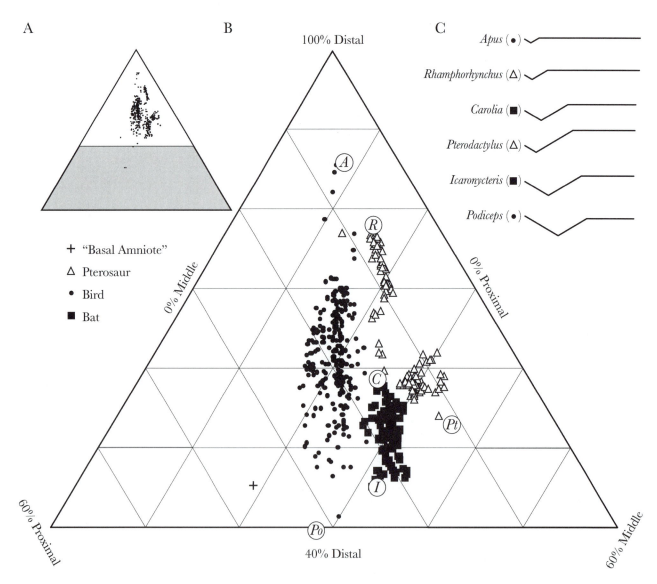

Figure 16-5 Forelimb segmental proportions in ternary morphospace. (A) The relative lengths of the proximal, middle, and distal wing segments are plotted for pterosaurs, birds, bats, and taxa near the origin of Amniota ("basal amniotes"). Wings are restricted to the upper portion of the morphospace because the distal segment is always longest. (B) A subset of ternary morphospace enlarged to show the distribution of each group. Birds, bats, and pterosaurs fill distinct regions with little overlap. (C) Diagrams of segmental proportions drawn to identical total length to show the range of anatomical design. Figured specimens are marked in (B) by circled letters.

in pterosaurs are the humerus, radius + metacarpal IV, and phalanges of digit IV (113 specimens from 9 genera). For birds, we measured the humerus and radius for the first two segments. Because the major digit of birds does not reach the wing tip, we used study skins to measure the length from the wrist to the tip of the longest primary feather (known as the "wing chord" in field ornithology) for the distal functional segment. Few museum preparations preserve measurable wing bones with primary feathers still attached. Consequently, we combined humeral and radial data with wing chord data from different specimens to make composite percentages for 268 specimens from 216 genera. The ternary diagram axes are organized such that the lower left, lower right, and upper vertices correspond to 100% proximal segment, 100% middle segment, and 100% distal segment, respectively.

Within this more functionally-based theoretical morphospace, wings are restricted to a narrow wedge near the top of the ternary (fig. 16.5A). The distal segment, representing the "handwing," is always the longest. Humeral contributions span 26% between two birds at 6% and 32%. The middle segment differs similarly (28%) from a bird (8%) to a pterosaur (36%). Variation is greatest (46%) in the distal segment's contribution, ranging from 40% to 86% of total wing length in two birds. This grebe and swift have the most disparate wings in our sample, with a DI of 38 (figs. 16.5B, C).

Each clade occupies a distinct region of morphospace (fig. 16.5B). Bats form a small, ovoid cluster nestled between the elongated bird and pterosaur clouds making up the left and right sides of the wedge, respectively. The most disparate pterosaurs have a DI of 24, whereas bats show a more conservative DI of only 12. Wings are present in 44 of the 400 possible cells (11%). Birds are present in 25 cells, whereas pterosaurs occupy 14 and bats only 6.

Our sample of taxa near the origin of amniotes is extremely limited by the small number of specimens with complete hands. One taxon, *Crassineria*, has an unusually long manus (Paton et al. 1999), but others fall lower with distal segments about one-third of limb length (fig. 16.5A).

Pterosaurs

The skeletal proportions of the two main groups of pterosaurs, pterodactyloids and rhamphorhynchoids, form loose upper and lower clouds, respectively (fig. 16.6A). These clusters correspond with the classic characterization of pterosaurs as having either short or long metacarpals. Basal pterosaurs cluster with the rhamphorhynchoids but tend to have relatively longer humeri. Members of the genus *Pteranodon*, a large, toothless, crested pterodactyloid from the Late Cretaceous, has the most extreme proportions with an elongated fourth metacarpal approaching the length of the

humerus and radius combined. Unfortunately, presumptive close relatives of pterosaurs such as *Scleromochlus* are too fragmentary to measure reliably.

The distribution of taxa changes significantly on the segmental ternary diagram (fig. 16.6D). Rhamphorhynchoids, which were lowest on the skeletal plot because of their relatively short metacarpals, are now highest with hypertrophied fourth digit phalanges making up 69–78% of the wing. In contrast, pterodactyloid wing fingers only contribute 53–61%; basal forms have distal segments of intermediate proportion. All pterosaurs have an exceptionally short humeral segment, which contributes only 6–15% to total length.

Birds

The majority of birds form a large, ovoid cloud near the bottom of the skeletal ternary diagram (fig. 16.6B). In contrast, swifts and hummingbirds are above the main cluster separated by a DI of 10. These are the only birds in which the carpometacarpus is longer than the humerus. As pointed out previously (Middleton and Gatesy 2000), birds with relatively short humeri (swifts, hummingbirds, swallows, martins) are regarded as highly maneuverable. Birds with the relatively longest humeri (alcids, loons, cuckoos, grebes, and albatrosses) are considered poorly maneuvering fliers. Six specimens of *Archaeopteryx*, the most primitive bird known, are located along this lower left edge of the distribution (43%:38%:19%) coproportional with grebes. Other Mesozoic birds (e.g., *Concornis, Confuciusornis, Jeholornis, Yanornis*) are scattered within the middle of the lower point cloud. Nonavian theropods considered closely related to birds have relatively longer humeri, but recently described forms with feathered forelimbs (*Caudipteryx, Microraptor, Sinornithosaurus*) are coproportional with many extant birds.

On the segmental ternary diagram (fig. 16.6E), birds form a tall cloud along the midline. The addition of phalanges and primary feathers to the carpometacarpus makes the distal segment the longest and again the most variable. As before, hummingbirds and swifts top the distribution, with wing chords as long as 86%. Below this, with handwings making up 74–79% are a storm petrel, a swallow, and two martins. A diversity of birds show a distal segment of ca. 70%, including pigeons, terns, caprimulgiforms (nighthawks and whip-poor-wills), parrots, and various passerines (songbirds). Most of these taxa are small and considered adept fliers; many are able to hover and/or feed on the wing. By contrast, birds with the relatively shortest wing chords (less than 50%; grebes, albatrosses, and pelicans) are larger and less maneuverable. All other birds in our sample have distal segment proportions of 50–70%, including *Archaeopteryx* (23%:20%:57%). *Confuciusornis* (16%:13%:71%), an Early Cretaceous bird from China, appears among relatively adept extant aerialists.

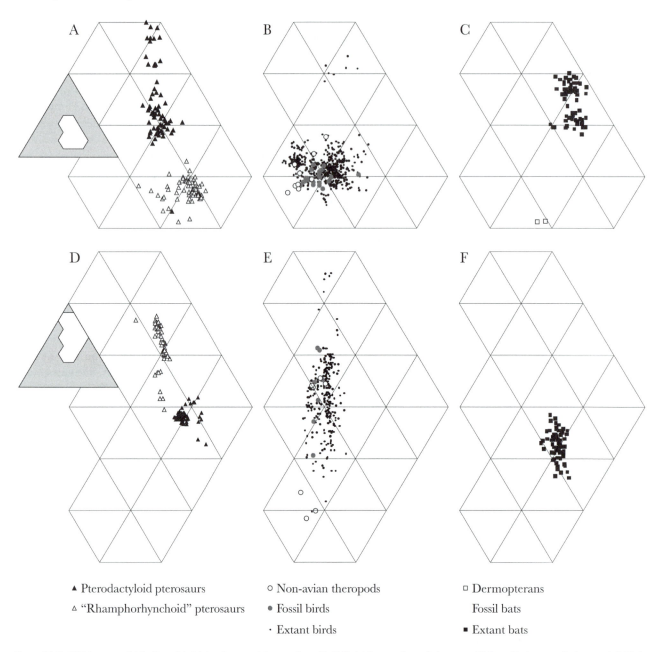

▲ Pterodactyloid pterosaurs   ○ Non-avian theropods   □ Dermopterans

△ "Rhamphorhynchoid" pterosaurs   ● Fossil birds   Fossil bats

· Extant birds   ■ Extant bats

Figure 16-6 Within-group distribution of skeletal and segmental proportions. (A–C) Skeletal proportions of pterosaurs, birds, and bats, respectively, are subdivided into phylogenetic groups (pterosaurs) and into extant and fossil taxa. Note the overlap between nonavian theropods and fossil and extant birds in (B). Dermopteran gliders, shown in (C), are widely separated from both fossil and extant bats. While no longer believed to be closely related to bats, dermopterans are still used as a proxy for the intermediate gliding morphology presumably present during chiropteran evolution. Similarly, (D–F) show segmental proportion in pterosaurs, birds, and bats. Note the distal elongation in "rhamphorhynchoid" pterosaurs, which reverses their position relative to pterodactyloids (A). In (E), nonavian theropods are coproportional with extant birds.

Surprisingly, nonavian theropods with preserved feathers (two specimens each of *Caudipteryx* and *Microraptor*) have distal segment proportions of ca. 65%. Three genera of closely related nonavian theropods (*Bambiraptor, Deinonychus, Sinornithosaurus*) are proportioned much like grebes despite lacking preserved hand feathers.

Bats

On the skeletal ternary graph (fig. 16.6C) bats form two closely spaced clusters with relatively low humeral percentages (20–30%) and average radial and metacarpal values compared with other fliers. The upper cloud contains microchiropterans having a third metacarpal relatively longer than the other bats in our sample. The lower cloud is primarily composed of megachiropterans, but also includes several microchiropterans. Interestingly, these genera are members of those microchiropteran families (Hipposideridae, Megadermatidae, Nycteridae, and Rhinolophidae) that Teeling and colleagues (2000, 2002) found to be more closely related to megabats in their phylogenetic analyses. Most fossil bats have proportions intermediate between these two clusters, although the extinct megabat, *Archaeopteropus,* is an exception with the lowest metacarpal percentage in the sample. Dermopteran gliders (genus *Cynocephalus*) exhibit some metacarpal elongation relative to arboreal rodents but fall short of chiropteran values.

The division among clusters is lost when the phalanges of digit III are added to the third metacarpal to yield a distal segment length (fig. 16.6F). Megabats and microbats form a single cloud with very little variation in segmental proportions. Bats with the longest distal segments (56–58%) are all phyllostomids (leaf-nosed bats). Fossil forms fall to the bottom of the distribution, with handwings of only 45–48%, although *Archaeopteropus* is slightly higher. Despite the relative homogeneity of bats in our segmental plot (fig. 16.6C), correlations between other aspects of wing morphology (wing loading, aspect ratio, tip shape) and flight style, feeding strategy, and ecology have been identified (see Norberg and Rayner 1987 and references therein).

Functional and Evolutionary Insights

Sampling Artifacts

How accurately do the skeletal and segmental ternary diagrams reflect disparity in wing proportions? One concern is sampling bias, which distorts the actual distribution of each clade to different degrees. Birds are best represented in terms of the number of specimens and genera, but we still only have data from a small fraction of the ca. 9,000 extant species. On the other hand, we specifically sought out unusual birds to try to delineate their distribution in morphospace as well as possible. We suspect that the large gap separating swifts and hummingbirds from other birds is not spanned by extant forms. Fossils could bridge these clusters, and extinct taxa could also expand the bird cloud in other directions. In contrast, few of the ca. 900 species of bats are in our plots, but it is unlikely that we have overlooked an extant form with wings grossly different than those sampled. Pterosaurs are a different case yet again. Our sample is surprisingly disparate given that the specimens are 210 to 65 million years old and extremely fragile. Therefore, it is quite possible that we

are missing a significant fraction of true pterosaur wing disparity.

Specializations

What factors might be responsible for differences in morphospace distribution among volant clades? One likely candidate is body size. Bats could be less disparate simply because of their smaller size range. Even the largest flying fox (1,400 g) is small compared to the largest extant flying bird (12,000 g) and minute compared to the largest pterosaur (Norberg 1981). What is striking, however, is the relative uniformity of bats compared to birds and pterosaurs of similar size.

Another obvious difference among vertebrate wings is the nature of the flight surface. One might predict that membranous wings would evolve to be more similar to each other than they are to feathered wings, but a tight clustering of bats and pterosaurs to the exclusion of birds is not obvious on either ternary graph. On the contrary, bats occupy relatively exclusive regions of skeletal morphospace, whereas some pterosaurs and birds converge on similar proportions. Segmental proportions are remarkably clade-specific, with little overlap among groups. Basal birds, bats, and pterosaurs had unique combinations of wing segment lengths. Flying descendents in each group pioneered new regions of segmental morphospace but without converging on common proportions (fig. 16.5). Why is there so little homoplasy in relative segment lengths among volant clades? Differences in handwing support may be partially responsible. The mechanical and aerodynamic properties of the flexible wing tips of bats might be different enough from the stiffer fourth digit of pterosaurs and primary feathers of birds to preclude similar proportions. Further quantitative analyses are warranted.

The forelimb's role in terrestrial locomotion is likely to be important as well. Although there has been controversy (e.g., Padian 1983a), pterosaurs appear to have been competent quadrupeds, contacting the substrate with digits I–III and the fourth metacarpophalangeal joint. Few bats are agile on the ground (but see Schutt et al. 1997), but movement on surfaces and through branches involves all four limbs. In contrast, birds walk, run, and hop exclusively on their hindlimbs. Could quadrupeds be restricted to distinct regions of morphospace because their forelimbs must function as both wings and legs? In pterosaurs the middle segment (radius + metacarpal IV) is always much longer than the proximal segment, ranging from 163 to 321% of humeral length. Unfortunately, we are unable to explain why walking would favor these proportions over those found in birds.

Alternatively, avian wings could be prohibited from colonizing pterosaur and bat segmental morphospace because of bipedalism. When not in flight, birds compactly fold their wings by retracting the humerus against the body and flexing the elbow and wrist joints (fig. 16.2B). This Z-configuration allows the primary feathers to cross over the back, irrespective of their length. Folding is most effective when the proximal (humerus) and middle (radius) segments are of comparable length, so that upon elbow flexion the wrist lies adjacent to the shoulder (Middleton and Gatesy 2000). If the middle segment was exceedingly long, as in pterosaurs (figs. 16.1, 16.2A), the wrist would project forward beyond the contour of the thorax (up to two humeral lengths) and interfere with neck and head movement. Bird wings avoid this conflict by being centered along the midline of the segmental ternary (radius = humerus). Taxa in our sample have radii ranging from 72 to 148% of humeral length; more than half fall within ±10% of parity.

The relevance of folding for wing design in birds has been questioned. Workers consider the relative lengths of the proximal and middle wing segments to be dominated by selection for flapping flight (Dyke and Rayner 2001; Rayner and Dyke 2003; Nudds et al. 2004). These authors have found that the ratio of humeral length to radial or ulnar length, known as the brachial index (BI; Howell 1944), correlates with several measures of flight morphology (Rayner and Dyke 2003). In particular, bird wings with low BI tend to have low moments of inertia, which should reduce power requirements. But Rayner and Dyke (2003) were unable to conceive of selective pressures favoring aerial wings with high BI. To put avian BI values in perspective, intersegmental ratios can be superimposed on a ternary diagram. Two patterns are apparent. First, BI exhibits the least amount of variation of any such ratio among birds. BI only varies over a twofold range, whereas humerus/handwing and radius/handwing both vary by a factor of more than seven. Such dramatic differences in relative primary length likely have greater aerodynamic significance than do minor inequities in arm and forearm size. Second, birds have higher BI than all but a few bats and pterosaurs. Reducing BI likely benefits flight performance and cost, but for some reason birds have failed to realize the lower ratios achieved by other fliers. Folding is likely a contributing factor constraining birds to remain close to a BI of 1.

Additional support for this folding hypothesis may come from a correlation between feeding ecology and radio-humeral proportions. Birds that feed while their wings are folded should have radii short enough to prevent the wrist from obstructing the head. Indeed, taxa falling on the left side of the segmental ternary (radius < humerus) are almost entirely ground-dwelling galliforms (chickenlike birds) and various waterbirds (ducks, geese, swans, grebes, loons) that feed with their wings furled. Birds on the right side of the

ternary (radius > humerus) either catch their prey on the wing (swifts, swallows, flycatchers), are plunge-divers (kingfishers, pelicans), or have long necks (storks, vultures), thereby minimizing any reduction in cranial mobility. In bats the radius varies from 131 to 201% of humeral length, but the projecting wrist (figs. 16.1, 16.2C) remains clear of the head in aerial insectivores/nectivores or is used for climbing and manipulating in frugivores. Given their more extreme proportions, pterosaurs must have been even more prone to interference between the neck, head, and wing while not in flight.

Convergent Similarities

Based on our comparison of flying forms with their most recent common ancestor at the base of Amniota, several generalizations can be made about skeletal evolution. First, the origin of flight entailed not only hypertrophy of the pectoral appendage and girdle, but also a change in forelimb proportions. All clades have converged on elongated handwing segments, albeit by different solutions. Pterosaurs are the most dramatic in this respect; the four phalanges of digit IV account for up to 78% of total segment length. Most birds show the least derived skeletal proportions, presumably because feathers make up a significant portion of the handwing. Second, the number of independent components was reduced by loss and fusion. Bird wings are most derived, such that living taxa typically have only seven functional elements making up the wrist and hand skeleton compared to a primitive count of about two dozen. Other fliers are less reduced but still show digit loss and carpal fusion (pterosaurs) or radio-ulnar fusion (bats). Finally, each clade independently acquired humeral shapes to counter torsion. Bird and bat humeri experience high torsional and bending stresses during flight (Biewener et al. 1992; Swartz et al. 1992); pterosaurs are thought to have been similar. A structural solution to this type of loading, which can cause failure due to shear, is to increase bone diameter while keeping wall thickness low to minimize mass (Currey and Alexander 1985).

Comparisons to such a deep ancestor can be deceiving, however. Some of the differences between volant forms and the ancestral amniote condition are present in more closely related taxa that do not fly. For example, many features of extant birds traditionally interpreted as flight adaptations arose well before the origin of flight. Maniraptoran theropods possessed a tridactyl manus, feathers, fused clavicles, hollow bones, and carpal fusion before the origin of flight (although birds later fuse and reduce further). Unfortunately, the ancestry of bats and pterosaurs is more ambiguous, making the sequence of spectacular hypertrophic changes in their hand morphology less well clarified.

Appendix: Data Sources

Basal Tetrapods: Benton 1999; R. L. Carroll 1969a; Colbert and Kitching 1975; Heaton and Reisz 1980; Holmes 1984; Langston and Reisz 1981; Paton et al. 1999; Reisz et al. 1984; Sumida 1989; Watson 1957; T. E. White 1939.

Bats: Personal measurements (Museum of Comparative Zoology, Cambridge, MA); Habersetzer and Storch 1987; Jepsen 1970.

Birds: Personal measurements (Museum of Comparative Zoology, Cambridge, MA); Dong 1993; Lacasa Ruiz 1989; Middleton and Gatesy 2000; Perle et al. 1994; Sanz et al. 1995; Wellnhofer 1974a, 1988, 1993; Zhou 1992; Zhou and Zhang 2001, 2002a, 2002b.

Nonavian Theropods: Burnham et al. 2000; Hwang et al. 2002; Middleton and Gatesy 2000; Ostrom 1976c; Xu et al. 1999b, 2003; Zhou and Wang 2000; Zhou et al. 2000.

Pterosaurs: S. C. Bennett 2001a; Dalla Vecchia 1998; Jenkins et al. 2001; Unwin 1988b; X. Wang and Lü 2001; X. Wang et al. 2002; Wellnhofer 1970, 1974b, 1975a, 1991; Wild 1978, 1984a, 1984b.

All data are available from the authors.

Acknowledgments

We would like to thank Brian Hall for inviting us to write this chapter and for his patience while we completed it. For access to specimens in their care, we thank Douglas Causey and Judith Chupasko of the Museum of Comparative Zoology. This chapter benefited greatly from discussions with Sharon Swartz.

Chapter 17 Adaptations for Digging and Burrowing

Nathan J. Kley and Maureen Kearney

DIGGING IS A relatively common activity among vertebrates. It is an important component in the feeding repertoires of some species, being used either as a method of foraging or as a means by which to conceal stores of food. Also, many oviparous species dig underground nests in which to deposit their eggs. In general, however, the most extensive digging activities are those associated with the construction of patent underground tunnel systems, or burrows.

Although burrowing is most common among highly fossorial species, it is also a crucial behavior in the life histories of many terrestrial, aquatic, and arboreal vertebrates. In some species, burrows may be built on a daily or nightly basis to provide temporary shelter from predators or from extreme environmental conditions. Other species burrow only seasonally to hibernate or aestivate, or to deliver and rear their young. Some truly subterranean forms build burrows as more or less permanent underground homes. Such species often feed and even breed within the confines of their burrows.

Vertebrates dig and burrow in a variety of different ways. Digging (i.e., the breakup and removal of soil) is performed most frequently with the hands or feet, which are often equipped with robust claws, nails, or tubercles (Hildebrand 1985a). However, the head is sometimes used for digging as well, and in some taxa the snout is highly modified for this purpose (e.g., hog-nosed snakes [*Heterodon*]; D. D. Davis 1946). In the broadest sense of the word, "digging" may also be done with the mouth (e.g., Nile crocodiles [*Crocodylus niloticus*]; Pooley 1969, 1982) or even the neck (e.g., bullsnakes, pinesnakes and gophersnakes [*Pituophis*]; C. C.

Carpenter 1982; Zappalorti et al. 1983; Himes 2000), but such cases are relatively rare and involve few, if any, structural specializations. (Adaptations for scraping are seen in the dentition of chisel-tooth digging rodents [Hildebrand 1985a], but these species generally remove soil from their burrows with their feet rather than their mouths [e.g., Jarvis and Sale 1971].)

Digging is perhaps the most common method of burrow construction among terrestrial vertebrates, but several other techniques may be used. Foremost among these are a variety of head-first burrowing mechanisms used by several different groups of slender, elongate reptiles and amphibians (e.g., amphisbaenians, uropeltid snakes, and caecilians). In forms such as these, initial penetration into the substrate and subsequent tunnel extension occur through forward pushing movements of the head. Tunnels are then widened by compaction, either through up-and-down or side-to-side movements of the head, or through lateral undulation or regional expansion of the trunk (Gans 1974).

In this chapter we examine the digging and burrowing mechanisms of tetrapods, focusing in particular on specializations of the appendicular skeleton that have evolved in association with these mechanisms. We exclude other craniates from our overview because hagfishes, lampreys, and burrowing fishes rarely use their fins to construct their burrows; instead, they rely mainly on oral excavation techniques or undulatory movements of the body and tail (Atkinson and Taylor 1991). Similarly, we also omit birds from our survey. Although a small number of birds do nest in burrows (most notably a variety of colonial seabirds; Furness 1991), none are known to exhibit structural adaptations for digging (Hildebrand 1982). Moreover, several burrow-nesting species commonly nest in burrows dug by other animals (e.g., burrowing owls; Bent 1938; D. J. Martin 1973; MacCracken et al. 1985). Thus, in the present account we focus on adaptations for digging and burrowing in amphibians, nonavian reptiles, and mammals.

Amphibians

Salamanders

Among the hundreds of species of fossorial and semifossorial amphibians, salamanders are the least specialized diggers. Although many terrestrial species frequently take shelter underground, few seem capable of constructing their own burrows in even moderately compacted soils (Chadwick 1940; Heatwole 1960; Brandon 1965; Semlitsch 1983; Parmelee 1990). Instead, some simply seek refuge in the burrows of other animals, such as rodents and crayfishes (e.g., Trenham 2001; Neill 1951; Petranka 1998). Others use shovel-like movements of the head and undulatory movements of the trunk to enlarge and extend preexisting crevices or holes in the ground (Heatwole 1960; Brandon 1965; Semlitsch 1983; Parmelee 1990). The latter technique may be especially important among the elongate, limb-reduced, aquatic forms, *Amphiuma, Siren* and *Pseudobranchus*, all of which are frequently found in deep underground burrows (Holbrook 1842; Brimley 1920; Cockrum 1941; Cagle 1948; Knepton 1954; Duellman and Schwartz 1958; J. R. Freeman 1958; Gehlbach et al. 1973).

Among the few species of terrestrial salamanders known to build their own burrows, the tiger salamander (*Ambystoma tigrinum*) appears to be the most capable digger (Semlitsch 1983; Gruberg and Stirling 1972; Bishop 1941, 1943). This large species uses rapid backward thrusts of its muscular forelimbs to excavate its burrows, alternating from one arm to the other every 3 to 10 strokes. During digging, soil is often thrown into the air, gradually accumulating in piles along the sides of the animal's trunk. At irregular intervals throughout the excavation process, the salamander stops digging and uses alternating movements of its hindlimbs to plow these piles of displaced soil backward (Gruberg and Stirling 1972).

Although *Ambystoma tigrinum* appears to be exceptional among salamanders in its digging abilities, this species exhibits no obvious morphological adaptations for burrowing. Its heavy build, stout limbs, and short, broad-based digits (Cope 1889) are doubtless advantageous for digging, but

other species of *Ambystoma* with similar morphologies are relatively poor burrowers (although the closely related *A. californiense* may be an exception; Jennings 1996). Moreover, the smallest juvenile *A. tigrinum* show the same strong burrowing tendencies as do the larger adults (Semlitsch 1983; R. G. Webb 1969). Thus, the tiger salamander's digging capabilities appear to be attributable to behavioral, rather than anatomical, specializations. Based on the hypothesis that ambystomatids originated in the Late Mesozoic or Early Tertiary of western North America (Tihen 1958), Semlitsch (1983) suggested that digging in *A. tigrinum* may represent the retention of an ancestral behavioral trait that facilitated survival in dry, desertlike conditions. Although more recent phylogenetic analyses (Shaffer et al. 1991; Jones et al. 1993) cast some doubt on this scenario, it is evident that digging is important in allowing *A. tigrinum* to survive in extremely arid regions within its current geographic range (R. G. Webb 1969).

Frogs

In contrast to salamanders, many species of frogs are accomplished diggers, capable of constructing their own burrows in a variety of different soil types and for a number of different purposes. In general, the drive to burrow is strongest among terrestrial anurans that inhabit hot, arid environments: myobatrachids in Australia (Main et al. 1959; Main 1968; Sanders and Davies 1984); *Breviceps, Hemisus,* and numerous ranids in southern Africa (Poynton 1964; Poynton and Pritchard 1976; J. P. Loveridge 1976); and *Rhinophrynus* and several pelobatids in Mexico and the southern United States (Fouquette and Rossman 1963; Bragg 1965; McClanahan 1967; Ruibal et al. 1969). However, burrowing is also relatively common in a variety of other frogs (e.g., bufonids, leptodactylids, microhylids, pelobatids, ranids) that inhabit less extreme environments (Emerson 1976; Dutta and Pradhan 1985). Most fossorial frogs feed and breed above ground, but an almost completely subterranean existence has evolved several times within Anura (e.g., *Myobatrachus gouldii, Hemisus marmoratus, Breviceps adspersus, Rhinophrynus dorsalis*). Forms such as these exhibit the most conspicuous adaptations for burrowing.

Feet-First Burrowers
The vast majority of fossorial and semifossorial anurans burrow hindfeet-first (Emerson 1976), submerging either in a circular, corkscrew fashion, or by sliding directly backward and downward into the ground (Sanders and Davies 1984). Accounts of burrowing in pelobatids (e.g., R. M. Savage 1942; Jansen et al. 2001), hylids and myobatrachids (Sanders and Davies 1984), and microhylids (Emerson 1976; Mukerji

1931) suggest that digging behavior in feet-first burrowers is highly stereotyped. Burrows are excavated using alternating scraping movements of the hindfeet. Starting from a resting position, the anterior portion of the body is first raised up off of the ground by partial extension of the forelimbs. One leg is then fully flexed at the hip, knee, and ankle joints, and the tibiofibula is raised slightly above the femur (Sanders and Davies 1984). Following this preparatory phase, the inner metatarsal tubercle, projecting from the ventromedial surface of the foot (see below), is carefully positioned against the substrate. This positioning phase involves a lateral rotation of the femur about its own long axis and an eversion of the foot. Then, during the digging stroke, the femur is rotated medially and the knee partially extended (Emerson 1976). Through this action, soil immediately behind and beneath the plantar surface of the foot is scraped away, and is either swept lateroposteriorly (Emerson 1976; Sanders and Davies 1984) or scooped directly onto the frog's back (Jansen et al. 2001; R. M. Savage 1942). Throughout the digging process, the hands are braced firmly against the ground to resist forward movement generated through knee extension during the digging stroke (Emerson 1976). In addition, in some instances the forelimbs are used to actively propel the frog through the soil that has already been loosened by the hindfeet (Mukerji 1931; Ball 1936; Sanders and Davies 1984).

Three functionally significant structural modifications of the hindlimb evolved independently in several lineages of feet-first burrowing frogs, and these modifications are es-

pecially pronounced in taxa that spend most of their lives underground. The first is an enlargement of the inner metatarsal tubercle, an integumentary projection located ventromedially on the plantar surface of the anuran foot. In many burrowing forms—perhaps most notably in the spadefoot toads, *Scaphiopus*, *Spea*, and *Pelobates*—the inner metatarsal tubercle is greatly expanded and heavily keratinized, often bearing a sharp cutting surface along its free margin (e.g., Boulenger 1896; Cope 1889; Noble 1931). During burrowing, this structure thus functions as a broad and highly effective digging "spade" (e.g., Mukerji 1931; Emerson 1976; Sanders and Davies 1984; Jansen et al. 2001).

A second modification, strongly related to the first, is an increase in the size and robusticity of the prehallux, the skeletal structure that directly supports the inner metatarsal tubercle. With the exception of some pipids, the prehallux is a universal feature of the anuran pes (Howes and Ridewood 1888; Fabrezi 1993, 2001). In most nonburrowing forms, it is composed of between one and four cylindrical or cuboidal elements arranged in series, thus appearing superficially like an additional preaxial digit (fig. 17.1A). In most burrowing forms, however, the prehallux nearly always consists of two elements, the distal of which is often flattened and greatly expanded (fig. 17.1B). In some highly fossorial taxa, such as pelobatids, *Rhinophrynus*, and *Breviceps*, the enlarged distal element of the prehallux is shaped somewhat like the head of an axe, the curved distal margin of which supports the cutting edge of the spadelike inner metatarsal tubercle

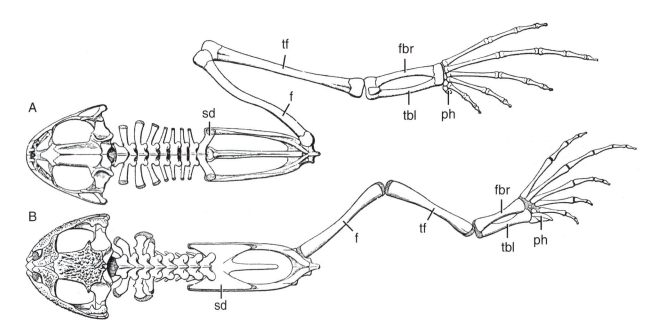

Figure 17-1 Dorsal views of the skeletons of (A) *Rana temporaria*, a jumping frog, and (B) *Pelobates fuscus*, a feet-first burrowing frog. The forelimbs, pectoral girdles, and left hindlimbs have been removed. Note the relatively short tibiofibula (tf) and enlarged prehallux (ph) in the burrowing species. f, femur; fbr, fibulare; ph, prehallux; sd, sacral diapophysis; tbl, tibiale; tf, tibiofibula. (A modified from Boulenger 1897; B modified from Boulenger 1896.)

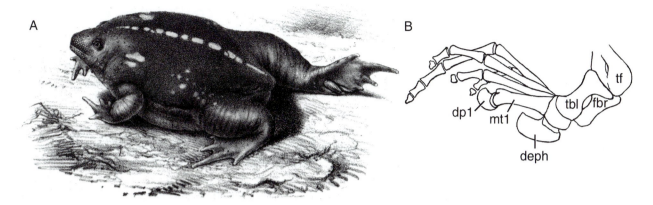

Figure 17-2 Adaptations for burrowing in *Rhinophrynus dorsalis*, a highly fossorial feet-first burrowing frog. (A) Dorsolateral view of a living specimen, with the right hindlimb partially extended and the right foot everted. Note the large inner metatarsal tubercle on the medial edge of the plantar surface of the foot, and the smaller adjacent tubercle on the tip of the first toe. (B) Medial view of the right pes, showing the enlarged distal element of the prehallux (deph) and the highly modified single phalanx of digit I (dp1). deph, distal element of the prehallux; dp1, distal phalanx of digit I; fbr, fibulare; mt1, metatarsal of digit I; tbl, tibiale; tf, tibiofibula. (A reproduced from Günther 1858a; B modified from Noble 1931.)

(Fabrezi 2001; fig. 17.2B). Based on the presence of specialized muscles inserting onto the prehallux in some burrowing species (e.g., *M. adductor praehallucis*, *M. abductor praehallucis*; Burton 2001), it is likely that at least some spade-footed anurans have the ability to exert relatively fine-scale control over the positioning of the prehallux (and thus the inner metatarsal tubercle) during digging.

A third major morphological specialization that characterizes nearly all backward-burrowing frogs relates to the relative proportions of the long bones of the hindlimb. One of the most functionally significant trends in the early evolution of Anura was a marked increase in the relative length of the hindlimbs. Longer legs allow frogs to transfer more kinetic energy to the body during the take-off phase of a jump, thus facilitating more efficient saltatory locomotion (Gans and Parsons 1966; J. Gray 1968). Indeed, among extant frogs and toads, jumping performance is generally highest among species that have relatively long legs (relative to the length of the trunk), and in most such taxa the tibiofibula is longer than the femur (Zug 1972, 1978; Dobrowolska 1973; Emerson 1976, 1988a; fig. 17.1A). In frogs that burrow feet-first, however, most of the force applied to the substrate during digging is generated through extension of the knee. The tibiofibula is thus the out-lever in this system, and the length of this element is therefore a major factor in determining the amount of force that can be generated during the digging stroke. Consequently, most feet-first burrowing anurans have relatively short hindlimbs, in which the tibiofibula is often shorter than the femur, in many instances accounting for less than one-quarter the total length of the hindlimb (Emerson 1976; Dobrowolska 1973; Zug 1972; Başoğlu and Zaloğlu 1964; fig. 17.1B).

In addition to these three specializations, the Mexican

burrowing toad (*Rhinophrynus dorsalis*) exhibits two further adaptations for burrowing in the structure of its posterior limbs. The first is a loss of one phalanx from the first digit (not the fifth, as mistakenly reported by Trueb and Gans [1983] and followed by Dutta and Pradhan [1985]), resulting in the unique phalangeal formula 1-2-3-4-3 in the pes (C. F. Walker 1938; Duellman and Trueb 1986). In addition, the single remaining phalanx in the hallux is highly modified, bearing a strong resemblance to (but remaining smaller than) the greatly expanded distal element of the adjacent prehallux (Noble 1931; Fouquette 1969; Trueb 1996; fig. 17.2B). Thus, *Rhinophrynus* is equipped with two spadelike digging instruments on each of its feet, rather than the single structure present in other burrowing taxa. Secondly, the integument underlying these digging spades is unusual, and very much unlike the hard, smooth, keratinized metatarsal tubercles commonly seen in other digging forms. The thickened areas of skin surrounding these structures in *Rhinophrynus* are distinctly striated, bearing a series of strong transverse ridges across their surfaces (Günther 1858b; fig. 17.2A). Consequently, each of the two digging spades has an almost rasplike surface texture. These features, together with the extreme reduction in length of the tibiofibula (Emerson 1976), render *Rhinophrynus* one of the best-equipped diggers among all anurans (Kellogg 1932; Fouquette and Rossman 1963; M. S. Foster and McDiarmid 1983).

Unfortunately, despite the extreme degree of morphological specialization exhibited by *Rhinophrynus*, detailed studies of its burrowing behavior are lacking. It is known, however, that this species occasionally digs with synchronous movements of its hindlimbs (M. S. Foster and McDiarmid 1983), a pattern that is apparently unique among anurans. In addition, when searching for food (termites and ants) underground, it is believed to burrow forward, using its spatulate forelimbs and tuberculate hands (Trueb and Gans 1983). Consistent with this belief is its solid skull structure (Trueb and Cannatella 1982), anteriorly placed pectoral girdle, and specialized integument overlying the snout (Trueb and Gans 1983), features commonly seen in head-first burrowing frogs (see below).

Somewhat surprisingly, fossorial anurans that dig with their hindfeet appear to exhibit no clear-cut burrowing adaptations in the morphology of their pelvises and associated structures. Although it has long been recognized that many burrowing frogs possess greatly expanded sacral diapophyses (e.g., Günther 1858b; H. W. Parker 1931), detailed morphological studies have demonstrated that these are associated with two very different types of iliosacral articulations in different taxa (Emerson 1979, 1982). The first—articulation type I of Emerson—facilitates anteroposterior translation of the pelvic girdle. The functional utility of this type of sliding iliosacral joint during swimming has been demonstrated in the aquatic pipid *Xenopus* (Videler and Jorna 1985; Palmer 1960), but it seems unlikely that such an articulation would be particularly advantageous for digging in the burrowing taxa that possess it (e.g., pelobatids). The second type of iliosacral articulation associated with expanded sacral diapophyses—type IIA of Emerson—functions quite differently, permitting the pelvis to rotate laterally in the horizontal plane. Such an articulation could conceivably function to increase the length of the digging stroke in burrowers that possess it (e.g., *Rhinophrynus*, bufonids, microhylids, leptodactylids). However, a more likely functional explanation is that this latter type of articulation has evolved (repeatedly) as an adaptation for increasing stride length in frogs and toads that rely more heavily on walking than jumping for locomotion (Emerson 1979).

Head-First Burrowers

Although the great majority of burrowing anurans dig feet-first into the ground, head-first burrowing also has evolved in several different lineages of frogs. The best-known examples are *Myobatrachus gouldii* and *Arenophryne rotunda* in Australia (Myobatrachidae; Main et al. 1959; M. J. Tyler et al. 1980); *Pseudacris illinoensis*, *P. ornata*, and *P. streckeri* in the United States (Hylidae; L. E. Brown et al. 1972; L. E. Brown and Means 1984; L. E. Brown 1996); and *Hemisus marmoratus* in Africa (Hemisotidae; Emerson 1976). Several microhylids from New Guinea and Brazil are also believed to be head-first burrowers (Menzies and Tyler 1977; A. L. De Carvalho, qtd. in Emerson 1976).

Most of what is known about head-first burrowing mechanisms in frogs derives from descriptions of burial behavior in *Hemisus marmoratus* (Emerson 1976) and *Arenophryne*

rotunda (M. J. Tyler et al. 1980). In both of these species, burrowing is initiated by flexing the head ventrally about the cranio-vertebral joint and thrusting the snout into the soil. Once the snout is positioned firmly in the ground, the frog then begins to use its hands to excavate the earth surrounding its head. During digging, the forelimbs may be used in an alternating fashion (*Hemisus;* Emerson 1976), synchronously (*Pseudacris;* L. E. Brown et al. 1972; L. E. Brown and Means 1984), or unilaterally, switching sides every three to four strokes (*Arenophryne;* M. J. Tyler et al. 1980).

As explained by Emerson (1976), the wide-arcing movements of the forelimbs serve two functions. During the early phase of the digging stroke, the lateral component of the force vector generated by the digging hand is greater in magnitude than that of the posteriorly directed component. As a result, soil is scooped from in front of the frog's head and displaced to the side. As the stroke continues, however, lateral forces diminish and posterior forces increase. Consequently, toward the end of the digging stroke, most of the force generated by the arm is contributing to forward propulsion of the body into the excavated hole (Emerson 1976). At least in *Arenophryne*, once the body is largely buried, the hindlegs may also become involved in propelling the frog forward (M. J. Tyler et al. 1980). Also, throughout the excavation process, *Hemisus* uses shovel-like movements of its head to compact and further penetrate the soil (Emerson 1971; J. C. O'Reilly, Department of Biology, University of Miami, pers. comm.).

Although some head-first burrowing frogs exhibit modifications of the skull and the integument overlying the snout (Menzies and Tyler 1977; M. J. Tyler et al. 1980; Davies 1984), the most conspicuous adaptations associated with this burrowing style are often found in the structure of the forelimbs. Externally, the arms are short, broad, and muscular, often having an almost spatulate appearance, and the fingers of the hand are typically very short and stout (L. E. Brown et al. 1972; Davies 1984; Dutta and Pradhan 1985; figs. 17.3D and 17.4B). Internally, modifications of the forelimb skeleton are extensive as well, involving changes in both the size and the shape of elements in the stylopodium, zeugopodium, and autopodium.

The humerus in head-first burrowers is more robust than that of closely related nonburrowers, and the crista ventralis humeri is greatly expanded (Emerson 1976; Chantell 1968). This enlargement of the ventral humeral (= deltoid) crest provides an increased area of attachment for the large muscles powering the digging movements of the arm (e.g., M. pectoralis, M. latissimus dorsi, M. deltoideus). It also lengthens the moment arm about the longitudinal axis of the humerus, thereby facilitating long-axis rotation of that bone during the latter phases of arm retraction (Emerson 1976).

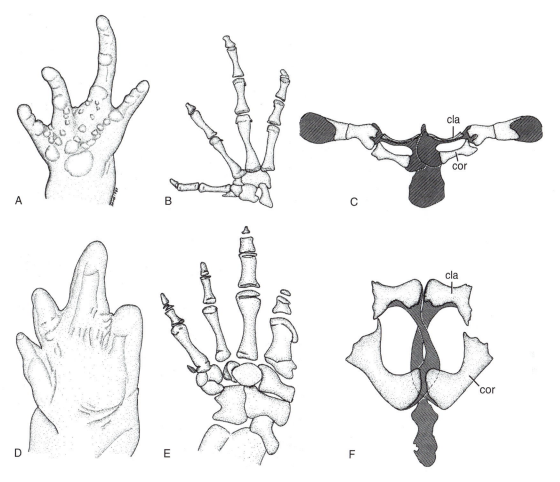

Figure 17-3 Morphology of the hand and pectoral girdle in feet-first burrowers (A–C) versus a head-first burrower (D–F) in the anuran family Myobatrachidae. (A) Palmar view of the left hand of *Uperoleia talpa*. (B) Dorsal view of the right manus of *Pseudophryne douglasi*. (C) Ventral view of the pectoral girdle of *P. douglasi*. (D) Palmar view of the left hand of *Myobatrachus gouldii*. (E) Ventral view of the left manus of *M. gouldii*. (F) Ventral view of the pectoral girdle of *M. gouldii*. Note the increased robusticity of the hand in the forward-burrowing species (*Myobatrachus*, D–F), as well as the increased size and angle of the coracoids (cor). cla, clavicle; cor, coracoid. (Modified from Davies 1984.)

Like the humerus, the radioulna has also become some-what modified in head-first burrowers. It is typically quite short and stout, and at least in *Pseudacris illinoensis* and *P. streckeri*, the olecranon process is markedly enlarged (Chantell 1968). Presumably, this latter specialization provides an increased surface area for the insertion of the main elbow extensor, the *M. anconeus*, and may also increase the force advantage of this muscle by lengthening its in-lever.

Several adaptations for digging have been identified in the structure of the manus of forward-burrowing frogs, but different structural changes have evolved in different taxa. In the myobatrachids *Arenophryne* and *Myobatrachus*, one phalanx in digit IV has been lost, resulting in the reduced phalangeal formula 2-2-3-2 in the manus. In addition, the re-maining phalanges in the hand of *Myobatrachus* are impressively stout relative to those seen in closely related species (Davies 1984; fig. 17.3). These modifications result in shorter,

stouter digits, which contribute to the formation of a more robust digging "shovel."

Digital reduction has also occurred in the burrowing hylids *Pseudacris illinoensis*, *P. streckeri*, and *P. ornata*, but in a very different way. Like all other hylids, these species possess intercalary cartilages between the subterminal and terminal phalanges of their digits. These cartilages, which have evolved independently in several families of arboreal anurans, are more or less cuboidal in shape in most taxa, and are positioned such that the terminal phalanx of each digit lies ventral to the subterminal one (fig. 17.4C). It is believed that this arrangement increases the rotational freedom of the distal phalanges, thereby enhancing the functional efficiency of the adhesive toe pads that characterize most tree frogs (Noble and Jaeckle 1928; Emerson and Diehl 1980; Green 1981). In the forward-burrowing species *Pseudacris illinoensis*, *P. streckeri*, and *P. ornata*, however, the toe pads are lost

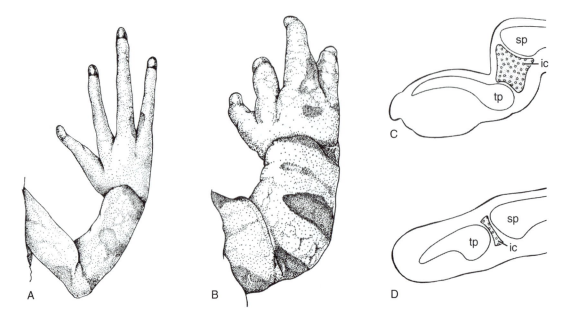

Figure 17-4 Morphology of the forelimbs and digits in terrestrial (A, C) versus head-first burrowing (B, D) species of chorus frogs (*Pseudacris*). (A) Dorsal view of the right forelimb of *P. triseriata*. (B) Dorsal view of the right forelimb of *P. illinoensis*. (C) Sagittal section through a toe of *P. nigrita*. (D) Sagittal section through a toe of *P. illinoensis*. Note the robust forelimb (B) and greatly reduced intercalary cartilage (D) in *P. illinoensis*, a head-first burrower. ic, intercalary cartilage; sp, subterminal phalanx; tp, terminal phalanx. (A and B reproduced from L. E. Brown et al. 1972; C and D modified from Paukstis and Brown 1991.)

(A. H. Wright and Wright 1949) and the intercalary cartilages are reduced to thin, waferlike elements (Paukstis and Brown 1987, 1991). As a result, the terminal phalanx of each finger lies in the same plane as the rest of the digit (fig. 17.4D). This reduces mobility at the distal interphalangeal joints, and thus presumably renders the fingers as more sturdy digging instruments.

Finally, several modifications of the pectoral girdle have been noted in head-first burrowers, and these are believed to be related directly to the digging mechanisms used by these frogs. In both *Hemisus* and *Myobatrachus* there has been a reorientation of the coracoids, such that these elements have become longer and more obliquely positioned relative to the long axis of the body (Emerson 1976; van Dijk 2001; Davies 1984; figs. 17.5 and 17.3F). In addition, in head-first burrowing myobatrachids and hylids, the coracoids are significantly more robust than in other closely related species (Davies 1984; Chantell 1968; fig. 17.3). This enlargement of the coracoids provides an increased surface area for the attachment of the hypertrophied forelimb retractor muscles (e.g., *M. pectoralis pars sternalis*, *M. coracobrachialis*; Emerson 1976). In *Hemisus*, elongation of the coracoids has also resulted in a forward shift of the suprascapulae (fig. 17.5), the anterior margins of which overlap the posterolateral portions of the skull. Emerson (1976) placed great emphasis on the functional significance of this feature, which she believed was crucial in permitting the skull to be rotated ventrally about

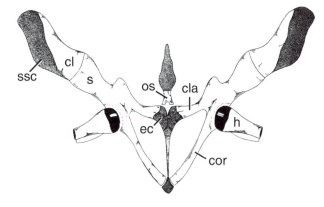

Figure 17-5 Dorsal view of the pectoral girdle in a forward burrowing frog, *Hemisus marmoratus*. Note the long, obliquely oriented coracoids (cor) and the anteriorly placed suprascapulae (ssc). cl, cleithrum; cla, clavicle; cor, coracoid; ec, epicoracoid cartilage; h, humerus; os, omosternum; s, scapula; ssc, suprascapula. (Modified from Engler 1929, after Braus 1919.)

the atlas. Interestingly, a similar anterior shift of the suprascapulae has been noted in *Rhinophrynus*, a predominantly feet-first burrower, but one that is believed to burrow head-first underground (Trueb and Gans 1983).

Caecilians

Caecilians are the most highly adapted burrowers among living amphibians. Nearly all of the more than 160 species contained within Gymnophiona are fossorial, and many lead an almost exclusively subterranean existence (e.g., Duellman and Trueb 1986; M. H. Wake 1993b; Himstedt 1996). Even the few species of Typhlonectidae that have become secondarily aquatic have retained strong fossorial tendencies, often burrowing into the mud along the banks of rivers and lakes (Moodie 1978; Wilkinson and Nussbaum 1999).

Burrowing behavior is unknown for most species of caecilians. However, during the last few decades, gymnophione burrowing mechanisms have received increasing attention from functional morphologists (e.g., Gaymer 1971; Gans 1973a; Renous and Gasc 1986; Renous et al. 1993; M. H. Wake 1993b; Ducey et al. 1993; O'Reilly et al. 1997; Summers and O'Reilly 1997; Teodecki et al. 1998). In all taxa that have been studied, the head is used as the primary burrowing instrument. Accordingly, the skull in most caecilians is heavily ossified and, in many species, quite rigidly constructed to resist the large compressive forces acting on the head during tunnel construction (e.g., Wiedersheim 1879; E. H. Taylor 1969; M. H. Wake and Hanken 1982). However, an even more striking specialization for tunneling in Gymnophiona has been the reduction and subsequent loss of the limbs. Although the Early Jurassic form †*Eocaecilia micropodia* retained small forelimbs and hindlimbs († = extinct taxon; Jenkins and Walsh 1993), the appendicular skeleton has been lost entirely in all, or nearly all, extant caecilians (e.g., E. H. Taylor 1968; Duellman and Trueb 1986). A single possible exception to this is *Ichthyophis glutinosus*, which may retain microscopic vestiges of the pectoral apparatus (Renous et al. 1997).

Reptiles

Turtles

Digging is an important component in the life histories of many terrestrial, aquatic, and marine turtles. In general, burrowing is most prevalent among land tortoises (Testudinidae), some of which are known to excavate burrows in excess of 10 m in length (e.g., *Gopherus;* Ernst and Barbour 1989; Woodbury and Hardy 1948). Digging is also relatively common in other terrestrial forms, such as the North American box turtles (*Terrapene;* Allard 1948; L. F. Stickel 1950; Claussen et al. 1991). However, digging activities are not confined to the terrestrial environment. Underwater digging has been observed both in sea turtles (e.g., *Caretta;* Limpus et al. 1994) and in highly aquatic freshwater taxa (e.g., trionychids; Graham and Graham 1991). Moreover, some sea turtles are known to occasionally overwinter in shallow coastal areas by partially burrowing into the seabed and entering into a prolonged state of torpor (Felger et al. 1976;

Carr et al. 1980; Ogren and McVea 1982). In contrast, some semiaquatic forms follow an alternative strategy, leaving the water entirely during periods of extreme cold or heat in order to dig underground shelters in which to hibernate or aestivate (e.g., kinosternids; D. H. Bennett et al. 1970; D. H. Bennett 1972; Moll 1979). Finally, all of the nearly 300 extant species of Testudines are oviparous, and the vast majority of living species deposit their eggs on land in self-constructed underground nests (Pough et al. 1998; Zug et al. 2001).

The most detailed descriptions of digging in turtles have come from behavioral studies of nesting females (e.g., Kenefick 1954; Auffenberg and Weaver 1969; Burger 1977; Congello 1978; Alho and Pádua 1982; Linck et al. 1989). Nest excavation has been particularly well documented in sea turtles, owing to their large size and predictable nesting habits (e.g., Carr 1952; Hendrickson 1958; Carr and Ogren 1959; Carr et al. 1966; Bustard et al. 1975; Hailman and Elowson 1992). In general, nesting behavior is highly stereotyped; egg chambers are invariably dug with alternating scooping movements of the hindfeet. In some taxa, however, a shallow body pit is also dug just prior to excavation of the egg chamber itself. Excavation of the body pit may involve synchronous or alternating movements of the forelimbs, diagonally alternating movements of all four limbs, alternating movements of the hindlimbs, or a sequential progression through several of these movement patterns.

In contrast to nest digging, burrowing in turtles is nearly always done with the forelimbs (e.g., Allard 1948; Bramble 1982). Morphological adaptations for forelimb digging are rare, however, and are only known to occur in taxa that do extensive tunneling, such as the North American gopher tortoises (*Gopherus;* Bramble 1982; Hildebrand 1985a).

The burrowing abilities of gopher tortoises have been enhanced significantly by several modifications of the autopodia that have resulted in a broadening, shortening, and stiffening of the forefeet (Bramble 1982; Hildebrand 1982, 1985a). The most obvious adaptation for digging is the marked expansion of the distal phalanges (unguals; fig. 17.6). These robust, spatulate ungual phalanges support the broad, flat, keratinous nails that the tortoises use to scrape away earth during the excavation of their tunnel systems.

In addition to this expansion of the unguals, there has also been a reduction in the length of the digits, an adaptation that has evolved numerous times in a wide variety of fossorial quadrupeds (Hildebrand 1985a). In *Gopherus,* digital reduction has evolved in two ways. First, there has been a reduction in the number of phalanges in digit V (and apparently also in digit I in some instances; Crumly 1994), resulting in the reduced phalangeal formula 2-2-2-2-1 or 1-2-2-2-1 in the manus (Bramble 1982; Crumly 1994; but see Auffenberg 1966, 1976). (The most common phalangeal formula among

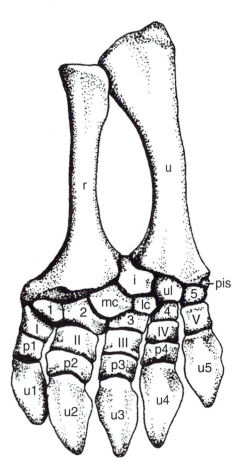

Figure 17-6 Morphology of the forefoot of the gopher tortoise (*Gopherus polyphemus*), a testudinid with strong burrowing tendencies. Note the enlarged, spatulate ungual phalanges (u1–u5), the short proximal phalanges (p1–p4; lost in digit V) and metacarpals (I–V), and the flat articular surfaces of the tightly packed carpal elements, which collectively serve to broaden, shorten, and rigidify the manus. Homologies of carpal elements after Bramble 1982. 1–5, distal carpals; I–V, metacarpals; i, intermedium; lc, lateral centrale; mc, medial centrale; p1–p4, proximal phalanges; pis, pisiform; r, radius; u, ulna; ul, ulnare; u1–u5, ungual phalanges. (Modified from Bellairs 1969.)

ments are packed together quite tightly, and little movement is possible between them. This arrangement aids in rigidifying the carpus, thereby rendering it more resistant to torsional forces incurred by the forefoot during digging (Bramble 1982). In *G. polyphemus* the hand is further rigidified through the effective loss of the mesocarpal joint. In this species (and in *G. flavomarginatus* to a lesser extent), flexion and extension of the forefoot occur almost exclusively at the brachiocarpal (wrist) joint (Bramble 1982). These specializations of the carpus, together with shortening of the digits and expansion of the ungual phalanges, render the forefeet highly effective digging tools, allowing gopher tortoises to burrow much more extensively than other turtles.

Crocodilians

All living crocodilians are predominantly semiaquatic in their habits (Pough et al. 1998; Zug et al. 2001). Most species spend a considerable proportion of their time in water and limit their terrestrial activities largely to thermoregulatory basking along the banks of the rivers, lakes, and swamps in which they live (e.g., McIlhenny 1935; Cott 1961; Neill 1971; Ross 1989). Consequently, digging and burrowing are not as common among alligators, caimans, crocodiles, and gavials as these activities are among many other lineages of tetrapods. Nevertheless, all crocodilians are oviparous, and all of the 23 extant species are known to bury their eggs on land in self-constructed hole or mound nests (Greer 1971; H. W. Campbell 1972). Several species also commonly assist their young at hatching by digging them out of their nests (e.g., Pooley 1977; Deitz and Hines 1980; Whitaker and Whitaker 1984). In addition, numerous species of crocodilians burrow underground during periods of extreme cold or drought (Hornaday 1904; Neill 1971; Richardson et al. 2002), and at least one species (*Crocodylus palustris*) may bury surplus food during times when prey are particularly abundant (Abercromby 1922).

Nesting crocodilians exhibit a wide array of digging behaviors. In both hole nesters and mound nesters, the hindlimbs, forelimbs, and snout all may be used for digging (e.g., Modha 1967; Chu-chien 1982; Lang 1987; Waitkuwait 1989; Whitaker and Whitaker 1989). In grassy areas, the powerful jaws are also sometimes used to tear into the earth (Modha 1967; Pooley 1982). Most commonly, however, egg chambers are dug with alternating scooping movements of the broad, webbed hindfeet (e.g., McIlhenny 1935; Compton 1981; Pooley 1982; Waitkuwait 1985; Whitaker and Whitaker 1989). In contrast, alternating strokes of the forefeet are usually used to excavate nests at hatching time (Modha 1967; Pooley 1977; Whitaker and Whitaker 1984; Thorbjarnarson 1989). Like nest construction, however, nest opening may involve a vari-

nonfossorial testudinids is 2-2-2-2-2 [Crumly 1994], and the primitive formula for the clade containing all living turtles [Casichelydia] is believed to be 2-3-3-3-3 [Gaffney 1990; Zug 1971; W. F. Walker 1973].) Second, especially among the most highly fossorial members of the genus (*G. polyphemus* and *G. flavomarginatus*), the proximal phalanges and metacarpals of all digits are shortened significantly, often being wider than they are long (fig. 17.6). Together, these features result in an overall shortening of the digits, allowing them to better withstand the high forces generated during the digging stroke.

Adaptations for digging in *Gopherus* are also evident in the structure of the carpus. Most notably, nearly all the carpal elements are more or less cuboidal in shape, bearing rather flat articular surfaces (fig. 17.6). As a result, these ele-

ety of behaviors, including biting (Herzog 1975; Pooley and Gans 1976; Whitaker and Whitaker 1984), vigorous thrashing of the body and tail (Cott 1961), and digging with the snout or hindlimbs (Pooley 1977; Crawshaw and Schaller 1980; Whitaker and Whitaker 1984; Messel and Vorlicek 1989).

Digging is also used by many crocodilians to escape from extreme environmental conditions (E. N. Smith 1979). Available accounts of tunneling behavior in several species of *Alligator* and *Crocodylus* suggest that the methods used for den construction are nearly as diverse as those used in nest building. For instance, the Chinese alligator (*A. sinensis*) uses scratching strokes of its forelimbs and ramming thrusts of its head to create elaborate dens that may extend over 50 m in length (Chu-chien 1982). The mugger crocodile (*C. palustris*) also digs with its forefeet, but uses its hindfeet and tail as well to disperse the excavated soil (Whitaker and Whitaker 1984). In contrast, Nile crocodiles (*C. niloticus*) depend largely on their powerful jaws to bite their way through the banks of rivers, lakes, and ponds. Using this technique they are able to construct large tunnel systems that may extend more than 10 m long and 3 m deep (Pooley 1969, 1982; Gadow 1901). All of these techniques are apparently used at different times by the American alligator (*A. mississippiensis*), a species well known for its propensity for digging "gator holes" (McIlhenny 1935; Neill 1971).

Although many researchers have focused considerable attention on the function of the limbs and limb girdles during walking (e.g., Schaeffer 1941; Brinkman 1980; Parrish 1987; Gatesy 1991; Reilly and Elias 1998), galloping (Zug 1974; G. J. W. Webb and Gans 1982), and breathing (Carrier and Farmer 2000; Farmer and Carrier 2000), the functional morphology of digging in crocodilians has not yet been investigated. Based on previous descriptive anatomical studies of the appendicular skeleton (e.g., Mook 1921; Knüsel 1944; Romer 1956), there appear to be no obvious structural specializations for digging. All morphological features that might potentially aid in burrowing are believed to be either plesiomorphic (i.e., robust claws) or adaptive for a semi-aquatic existence (i.e., webbed feet). Apparently, the relatively short limbs and well-developed appendicular muscles of crocodilians (Fürbringer 1876; Gadow 1882; Romer 1923) equip them adequately for the limited amount of digging that they do. Nevertheless, given the remarkable volumes of earth that large crocodilians must excavate to create their dens (nearly 10 m³ in the case of adult mugger crocodiles, based on tunnel dimensions reported by Whitaker and Whitaker 1984), their digging mechanisms clearly warrant further study. Ontogenetic studies of digging mechanics and performance may be particularly enlightening given the allometric growth of the limbs in crocodilians (Dodson 1975; Meers 2002).

Tuataras, Lizards, and Snakes

Lepidosauria includes more digging and burrowing species than any other major clade of vertebrates. Burrowing is, of course, most prevalent among truly subterranean forms, such as amphisbaenians and uropeltid snakes, which spend most of their lives underground (Gans 1974). However, a wide variety of other lepidosaurs, including both quadrupedal and limbless taxa, regularly engage in burrowing for a number of different reasons (e.g., concealment, refuge from extreme environmental conditions, foraging, nesting, etc.).

Although tuataras, lizards, and snakes frequently seek shelter in the burrows of other animals (e.g., Witz et al. 1991; Greer 1989; A. H. Wright and Wright 1957a, 1957b; Robb 1977), most quadrupedal forms exhibit at least some capacity for digging in loose or moderately compact soils. Even arboreal chameleons, with their highly specialized zygodactylous feet, are able to dig relatively long underground tunnels in which to deposit their eggs (Schmidt and Inger 1957; Menzies 1958; Shaw 1960). Most quadrupedal species dig their burrows with their forelimbs. In some taxa with relatively weak forelimbs, however, the head may be used as the primary digging instrument (e.g., *Lanthanotus;* B. Harrisson 1961; T. Harrisson and Haile 1961). In addition, numerous desert-dwelling species bury themselves by literally diving head-first into the sand (e.g., E. N. Arnold 1995).

The vast majority of quadrupedal lepidosaurs excavate their burrows with repetitive scraping strokes of their clawed forefeet (e.g., Schmidt and Inger 1957; Steyn 1963; Steyn and Haacke 1966; Haacke 1975, 1976a, 1976b, 1976c; Auffenberg 1981; J. M. Walker et al. 1986; A. P. Russell and Bauer 1990). Typically, several digging strokes are performed with one arm before switching to the other. Most oviparous species use the same technique for nest digging (e.g., Asana 1941; Deraniyagala 1958; Blair 1960; V. A. Harris 1964; C. C. Carpenter 1966; Rand 1968; Propper et al. 1991; Auffenberg 1981), a rare phenomenon among other reptiles. However, morphological specializations for limb-based digging appear to be relatively uncommon. Although the robusticity of the limbs is occasionally somewhat greater in burrowing forms than in closely related nonburrowing taxa (i.e., barking geckos [*Ptenopus*]; Brain 1962; Haacke 1975), more radical modifications of the appendicular skeleton, such as those seen in many fossorial mammals, have evolved only rarely in Lepidosauria.

Adaptations for limb-based digging appear to be most common in dune-dwelling geckos, especially those that burrow into the compacted sand along the windward sides of dunes (A. M. Bauer and Russell 1991). Among these, by far the most specialized digger is the Namibian web-footed gecko, *Palmatogecko rangei* (A. M. Bauer and Russell 1991;

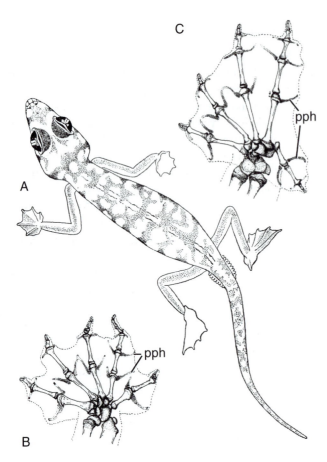

Figure 17-7 Adaptations for digging in the Namibian web-footed gecko, *Palmatogecko rangei*. Dorsal views of (A) the entire body, (B) the right manus, and (C) the right pes. Note the cartilaginous paraphalanges (pph) that support the extensive webbing in the hands and feet of this burrowing lizard. (A modified from J. B. Proctor 1928; B and C modified from Haacke 1976d.)

Haacke 1976a; Mertens 1960). The most striking morphological features of this desert lizard are its strongly webbed hands and feet (L. G. Andersson 1908; J. B. Proctor 1928; fig. 17.7). Early workers believed that the webbed feet of *Palmatogecko* might have evolved as "sand shoes" to facilitate locomotion across desert sands (Barbour 1926; J. B. Proctor 1928). However, recent behavioral and mechanical analyses strongly suggest that they evolved primarily as adaptations for digging (A. M. Bauer and Russell 1991; A. P. Russell and Bauer 1990). Indeed, the forefeet are used in a shovel-like fashion during digging. First, the sand is broken with several loosening strokes. Then the loosened sand is scooped out and transported backward. Finally, the bolus of excavated sand is deposited alongside the trunk. As in many other forelimb diggers, the hindlimbs are used at variable intervals during burrowing to push back the piles of excavated material that accumulate around the trunk (A. P. Russell and Bauer 1990).

In addition to digital webbing, several other specializa-

tions for digging are also evident in the structure of the manus and pes in *Palmatogecko*. In *P. rangei*, digital webbing is partially reinforced by cartilaginous paraphalanges that extend laterally from the interphalangeal joints (Haacke 1976d; A. P. Russell and Bauer 1988; fig. 17.7B, C). The presence of paraphalanges per se cannot be considered an adaptation for burrowing in *P. rangei* because these structures are present in a wide variety of gekkonids, especially in arboreal taxa (A. P. Russell and Bauer 1988). However, in climbing geckos, paraphalanges are confined largely to the distal portions of the digits, and are believed to function in association with the adhesive subdigital pads that characterize most tree-dwelling forms (A. P. Russell 2002; A. P. Russell and Bauer 1988; Mohammed 1991). In contrast, the adhesive toe pads have been lost in *P. rangei*, and the paraphalanges are located more proximally. This arrangement may provide added structural rigidity to the webbing, thereby making the hands sturdier for digging. More importantly, however, the paraphalanges provide skeletal support for specialized interparaphalangeal muscles, which run alongside each digit and permit relatively fine-scale control over movements of the digital webbing during digging (A. P. Russell and Bauer 1988; A. M. Bauer and Russell 1991). Curiously, the paraphalanges are highly reduced or absent in the only other member of the genus (*Palmatogecko vanzyli*, also a web-footed burrowing species), but several of the phalanges themselves are expanded laterally at their distal ends (Haacke 1976d; A. P. Russell and Bauer 1988).

In contrast to *Palmatogecko*, other fossorial and semifossorial desert lizards dig without the benefit of webbed feet. Instead, many are equipped with comblike arrays of enlarged, spinous scales that project medially and laterally from the digits. These toe fringes, which have evolved numerous times within Squamata (Mosauer 1932; Bellairs 1969; Luke 1986), serve to entrap particles of sand during each digging stroke, thereby effectively broadening the toes. This increases the amount of sand that can be excavated with each stroke, and thus enhances overall digging efficiency (Mosauer 1932; Brain 1962; Haacke 1975). However, toe fringes also enhance the sprinting abilities of lizards on loose, shifting sands (Carothers 1986). Moreover, many fringe-toed species do not regularly dig with their limbs, but instead rely on various sand-diving mechanisms to bury themselves (e.g., *Uma*; Stebbins 1944; Pough 1970). Thus, it is likely that toe fringes have evolved most frequently in desert lizards as adaptations for surface locomotion rather than for digging.

The construction of patent tunnels is nearly impossible in dry, loose sand (Mosauer 1932). Therefore, many lizards that live in arid deserts do not commonly dig burrows. Instead, they bury themselves in the sand by using a variety of different "sand diving" or "shimmy burial" techniques (E. N.

Arnold 1995; Halloy et al. 1998; Louw and Holm 1972; Pough 1970; Stebbins 1944). Most such lizards exhibit a suite of morphological adaptations that have evolved in association with these behaviors: a prominent, wedge-shaped snout (Mosauer 1932); specialized supralabial scales that form a sharp cutting surface along the upper lip (Stebbins 1944; E. N. Arnold 1995); and a number of modifications to facilitate subarenaceous respiration (Stebbins 1943; Pough 1969; Louw and Holm 1972). In general, however, sand diving and shimmy burying do not necessitate adaptive modifications of the limbs. In fact, in most cases the forelimbs are tucked tightly against the body once the snout begins to penetrate into the sand (Stebbins 1944; Pough 1969, 1970; E. N. Arnold 1995; Halloy et al. 1998).

Although the head is sometimes used extensively during sand diving, it has become the principal tool for burrowing in several lineages of elongate squamates. This is particularly true of amphisbaenians (Gans 1960, 1968, 1969, 1974) and uropeltid snakes (Gans 1973b, 1976; Rajendran 1985), which spend most of their lives underground, and which exhibit numerous adaptations for head-first burrowing. Many of these adaptations are seen in the structure of the skull (e.g., Zangerl 1944; Gans 1974), vertebral column (Gans 1958), axial musculature (Gans 1974; Gans et al. 1978), and integument (Gans and Baic 1977). However, modifications of the appendicular skeleton are also common; nearly all head-first burrowing squamates lack external limbs, and in some taxa (e.g., uropeltids) the appendicular skeleton has been lost altogether.

Elongate, Head-First Burrowers

Amphisbaenians live mainly in sandy soils in tropical to temperate regions of the Americas, Africa, Europe, and the Middle East, and have undergone major structural modifications related to fossoriality (Gans 1978; Kearney 2003). With the exception of three species within the genus *Bipes*, all amphisbaenians are completely limbless externally. *Bipes* is exceptional in retaining functional forelimbs (but no external hindlimbs; fig. 17.8).

Amphisbaenians are head-first burrowers. Various cranial shapes among amphisbaenians are functionally correlated with specific tunneling behaviors to some degree (Gans 1968, 1974). Some taxa have a so-called shovel-headed morphology, with a dorsoventrally flattened snout and a strong craniofacial angle. These forms generally burrow by thrusting the head forward and slightly downward, then lifting the head dorsally to compress soil into the roof of the tunnel. The tunnel floor is smoothed with the pectoral region, which often bears enlarged scales. In the "spade-headed" (*sensu* Kearney 2003) trogonophids there is also a strong craniofacial angle, but the organization of underlying bones and overlying scales is quite different (Gans 1960; Kearney 2003).

In trogonophids, a unique oscillatory tunneling behavior occurs during which the head is rotated about its long axis in alternating directions so that soil is shaved off the end of the tunnel with the sharp edges of the face. The shaved soil is then compacted into the tunnel walls with the smooth sides of the head and body. The "keel-headed" forms (found only within the family Amphisbaenidae) have heads that are laterally compressed; these burrow by driving the head alternately left and right and smoothing the tunnel walls with the sides of the head. Most amphisbaenians are "round-headed." These species burrow by using the head essentially as a battering ram, thrusting forward and using irregular motions of the head to compress the soil.

The limbed bipedids exhibit a unique combination of burrowing behaviors, including both forelimb-based digging and head-first burrowing. On the surface *Bipes* moves awkwardly, using its short forelimbs to steer the trunk and using either rectilinear locomotion or lateral undulation to propel the body forward (Bogert 1964; Gans 1974). Initial entry into the soil is accomplished by scraping the earth with the robust and heavily clawed forelimbs. During digging, the head is moved to one side and the opposite arm is extended forward past the face. Then the claws are engaged with the surface of the ground and the arm is retracted (Gans 1974). In contrast, when burrowing deep underground, *Bipes* holds its arms close against the body and uses its head to burrow forward (Bogert 1964; Gans 1969, 1974).

Bipes has complete forelimb bones: a carpus comprising nine ossified elements; a broad, flat sesamoid plate on the palmar side of the carpus; short, robust phalanges; and claws (Castañeda and Alvarez 1968; fig. 17.8B). In comparison to those of other lizards, the forelimbs of *Bipes* are relatively wide for their length, relatively short in relation to the greatly elongated body, and relatively large in relation to the diameter of the body (Kearney 2002). The zeugopodium is shortened relative to the stylopodium; the radius and ulna are approximately one-half the length of the humerus (Kearney 2002). The phalanges are uniformly stout, and the distal phalanges support large claws. The three species exhibit different phalangeal formulae: 3-3-3-3-3 for *B. biporus*; 3-3-3-3-2 for *B. canaliculatus*; and 0-3-3-3-2 for *B. tridactylus* (Castañeda and Alvarez 1968; Papenfuss 1982; Crumly 1990). These phalangeal formulae differ from the primitive lepidosaurian condition of 2-3-4-5-3 (Cope 1892a; Romer 1956), especially in the higher number of phalanges in the first digit. Hyperphalangy of the first digit is relatively rare among squamates (occurring elsewhere only in some geckos; Haacke 1976d; Kluge and Nussbaum 1995), and is rare even among vertebrates in general (Romer 1956). Gans (1978) suggested that hyperphalangy in *Bipes* is directly related to selection for tunnel creation and widening.

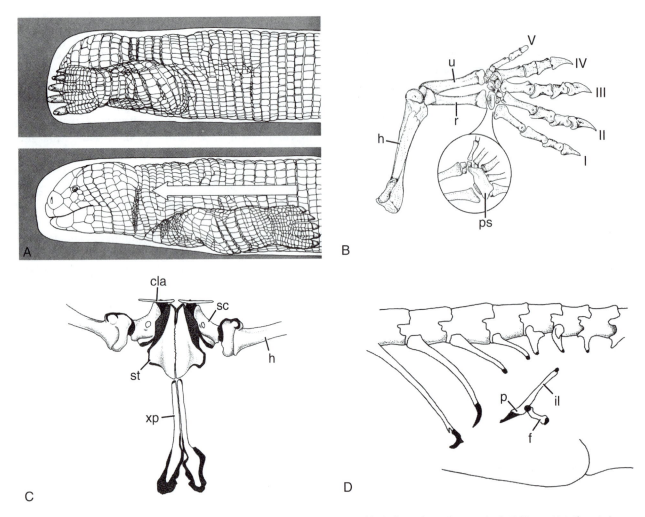

Figure 17-8 Morphology of the appendicular skeleton in the amphisbaenian genus *Bipes*. (A) Digging and tunnel progression in *B. biporus*. Note the anterior placement of the forelimbs, which allows the broad hands to be extended forward past the tip of the snout. (B) Dorsal view of the right forelimb in *B. canaliculatus*. Note the short antebrachium, robust digits, and well-developed interphalangeal joints. In the inset, a ventral view of the carpus, showing the large palmar sesamoid (ps). (C) Ventral view of the pectoral girdle in *B. biporus*. (D) Left lateral view of the pelvic apparatus in *B. biporus*. I–V, digits I–V; cla, clavicle; il, ilium; f, femur; h, humerus; p, pubis; ps, palmar sesamoid; r, radius; sc, scapulocoracoid; st, sternum; u, ulna; xp, xiphoid process. (A modified from Gans 1974; B modified from Castañeda and Alvarez 1968; C and D modified from Kearney 2002.)

Although *Bipes* has no external hindlimbs, an internal element of the stylopodium is retained (the femur; fig. 17.8D), as it is in one other amphisbaenian genus, *Blanus* (Renous et al. 1991; fig. 17.9B). The femur is a small but clearly recognizable element that articulates with the pelvis but does not extend outside the body wall (Zangerl 1945; Kearney 2002). Among other "limbless" elongate squamates, a small hindlimb element is retained in pygopodids, dibamids, and some snakes (e.g., Fürbringer 1870; Greer 1985, 1989). The anguid *Ophisaurus* also retains internal hindlimb elements (e.g., Cope 1892a; Sewertzoff 1931; F. Tiedemann 1976).

Modifications of the limb girdles (including complete loss) are prevalent among amphisbaenians. *Bipes* exhibits the most complete pectoral girdle, retaining all basic elements (fig. 17.8C). The pectoral girdle is shifted anteriorly

and is positioned unusually close to the head (Kearney 2002), at the level of the third cervical vertebra. This is in contrast to the condition in most lizards, in which the pectoral girdle occurs at the level of the sixth cervical vertebra or more posteriorly. This anterior displacement of the girdle in *Bipes* allows the forelimbs to be extended beyond the snout during digging (fig. 17.8A).

The relatively basal amphisbaenian *Blanus* exhibits a less complete pectoral girdle than that found in *Bipes* (fig. 17.9A). In the limbless trogonophids, unique modifications to the pectoral girdle are found. In most members of this family a broad, thin, platelike, calcified cartilaginous sternum is positioned beneath the skin, superficial to the hypaxial musculature and loosely joined to the overlying muscles by connective tissue (Gans 1960; fig. 17.9C). Typically, paired ossified

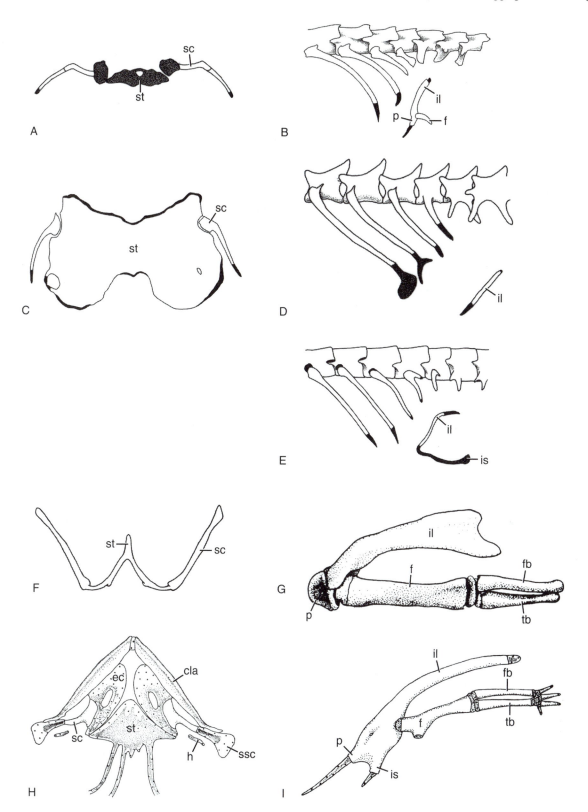

Figure 17-9 Morphology of the appendicular skeleton in a variety of elongate, limb-reduced squamates. Ventral views of the pectoral apparatus (A, C, F, H) and left lateral views of the pelvic apparatus (B, D, E, G, I) in the amphisbaenians *Blanus* (A, B), *Agamodon* (C, D), and *Rhineura* (E); the dibamid *Dibamus* (a male; F, G); and the pygopodid *Delma* (H, I). The amphisbaenians and *Dibamus* are burrowers. *Delma* is largely terrestrial. All traces of the pectoral apparatus have been lost in *Rhineura*. cla, clavicle; ec, epicoracoid cartilage; il, ilium; is, ischium; f, femur; fb, fibula; h, humerus; p, pubis; sc, scapulocoracoid; ssc, suprascapular cartilage; st, sternum; tb, tibia. (A-E modified from Kearney 2002; F and G modified from Greer 1985; H and I modified from Stephenson 1962.)

lateral elements (most likely scapulocoracoids) are joined to this plate anterolaterally. In other amphisbaenians, the pectoral girdle occurs in various stages of degeneration, including complete loss (e.g., *Rhineura* and some amphisbaenids; Zangerl 1945; Gans 1978; Kearney 2002).

In contrast to the pectoral girdle, the pelvic girdle in amphisbaenians is never completely lost (Kearney 2002). The most basal members of the group (e.g., *Bipes, Blanus*) retain a reduced pelvic girdle as well as an internal vestige of the hindlimb (Zangerl 1945; Renous et al. 1991; Kearney 2002; figs. 17.8D, 17.9B). Other amphisbaenians typically retain a single girdle element, the ilium, which may be modified in terms of position relative to the vertebral column, size, and shape (fig. 17.9D). *Rhineura* exhibits an additional cartilaginous pelvic element, which presumably represents the ischium (Kearney 2002; fig. 17.9E). Interestingly, most limbless lizards and primitive snakes also retain at least one pelvic element. Thus, despite the frequent loss of the entire pectoral girdle in some lizards, some amphisbaenians, and all snakes, and despite the frequent loss of the hindlimb, at least one pelvic element is retained in all amphisbaenians, all limbless lizards, and most basal snakes (Kearney 2002). The fact that pelvic girdle elements are so widely retained and variously modified among species suggests possible new functional roles beyond the primitive one of supporting the hindlimb.

Among other squamate reptiles, the striking morphological transformation from a lizardlike body form to a snakelike (elongate, limbless) body form has occurred independently many times (e.g., Greer 1991; Pough et al. 1998) with concomitant reductions and modifications of the limbs and limb girdles. Various patterns of reduction and modification exist, all of which ultimately lead to a superficially similar end result.

Limb reduction and loss in squamates seem to be correlated with body elongation (Camp 1923; Gans 1975; Lande 1978; Wiens and Slingluff 2001). Although many burrowing species are limb-reduced and elongate, these conditions are not linked exclusively to fossoriality in squamates. Camp (1923) showed that the evolution of a serpentiform body among squamate reptiles occurs in species that are adapted to burrowing, grass-dwelling, or surface habitats. Gans (1974) suggested that the common link to limb loss among tetrapods is body elongation and that the adaptive significance of this morphology may be for utilizing undulatory locomotion in particular habitats. Lateral undulation is common among snakelike squamates (Camp 1923; Gans 1974); elongation of the body and loss of the limbs can be viewed as advantageous for this mode of locomotion since limbs may act as a hindrance to locomotion through soil or dense grass.

The diversity of patterns of limb and girdle reduction and loss among squamates is extensive. We briefly describe some of this diversity below; for primary references on the morphology of appendicular skeleton in specific squamate groups, see table 17.1.

The anniellids of southern California and northwestern Baja California are sand swimmers that are slender, limbless, long-bodied, and short-tailed (Coe and Kunkel 1906; C. M. Miller 1944). They have no external limbs, but they retain a single rodlike pelvic element in close proximity to the cloacal opening (Cope 1892a; Kearney 2002). Some workers have described a vestigial shoulder girdle in *Anniella pulchra* (Camp 1923; Stokely 1947a), but others found no trace of the pectoral apparatus (e.g., Cope 1892a; Coe and Kunkel 1906; Kearney 2002). Such conflicting findings are difficult to interpret based on the limited evidence currently available, but they do suggest that the pectoral girdle may be only variably present in members of this family.

The glass lizards (*Ophisaurus*) of North America, Europe, Africa, and Asia are terrestrial and cryptic, often inhabiting tallgrass prairie habitats and "swimming" through dense vegetation (e.g., Fitch 1989). Some species exhibit occasional burrowing tendencies as well (e.g., Gans and Gasc 1990; E. Frey 1982b). These lizards are completely limbless externally and have long trunks and extremely long tails (up to two and a half times the length of the trunk; Fitch 1989; Wiens and Slingluff 2001). Most species retain pectoral and pelvic girdles including all elements, but these are often in reduced form. Additionally, at least some species retain internal rudiments of the hindlimb (Cope 1892a; Fürbringer 1870; Sewertzoff 1931; F. Tiedemann 1976).

Dibamids are small, snakelike lizards whose evolutionary origins are uncertain. *Anelytropsis* is native to northeastern Mexico, while its close relatives of the genus *Dibamus* occur in the Philippines, New Guinea, and Malaysia. Dibamids are mainly fossorial, found beneath detritus or in burrows or crevices. They possess no forelimbs, but males do retain small, flaplike external hindlimbs (Greer 1985). The rudimentary pectoral girdle consists of a reduced sternum and scapulocoracoids only (Camp 1923; Gasc 1968; Greer 1985; fig. 17.9F). The pelvic girdle is somewhat reduced, with the pubis and ischium being much smaller than the bladelike ilium (Gasc and Renous 1979; Greer 1985; fig. 17.9G). In males, the hindlimb consists of a femur, tibia, and fibula. In the externally limbless females, however, only the femur is retained. This sexual dimorphism in limb development is an unusual condition among squamates, and is known elsewhere only in some macrostomatan snakes (e.g., Hoge 1947). The hindlimbs of male dibamids are more similar to the flaplike hindlimbs of pygopodids than they are to the styliform limb rudiments found in other limb-reduced species.

The pygopodids of Australia and New Guinea are elongate

Table 17-1 Selected references on the morphology and evolution of the appendicular skeleton in elongate, limb-reduced squamates

"Lizards"

Pygopodidae
Fürbringer 1870
Cope 1892a
M. Müller 1900
Stokely 1947b
Underwood 1957
Stephenson 1962
Gasc 1968
Kluge 1976
Greer 1989

Scincidae
J. Müller 1831
Heusinger 1833
Fürbringer 1870
W. C. H. Peters 1882
Cope 1892a
M. Müller 1900
Camp 1923
Essex 1927
Sewertzoff 1931
Stokely 1947a
Heyer 1972
Gasc and Renous 1974
M.-A. Tiedemann and Tiedemann 1975
Vasse et al. 1974
Lande 1978

Renous and Gasc 1979
Raynaud 1985
Greer 1987, 1989, 1990, 1991, 1997
Choquenot and Greer 1989
Caputo et al. 1995
Greer et al. 1998
Kearney 2002
Shapiro 2002
Whiting et al. 2003

Cordylidae
Cope 1892a
Duerden 1922
Essex 1927

Gymnophthalmidae
Cope 1892a
Camp 1923
Presch 1975, 1980
Lande 1978
Kearney 2002

Anniellidae
Cope 1892a, 1900
Baur 1894
Coe and Kunkel 1906
Camp 1923

Stokely 1947a
Kearney 2002

Anguidae
Mayer 1825
J. Müller 1831
Heusinger 1833
W. K. Parker 1868
Fürbringer 1870
Cope 1892a, 1900
M. Müller 1900
O. Müller 1913
Sewertzoff 1931
Stokely 1947a
F. Tiedemann 1976
Raynaud 1985
Wiens and Slingluff 2001

Dibamidae
Cope 1892a, 1900
Camp 1923
Gasc 1968
Gasc and Renous 1979
Greer 1985
Kearney 2002

Amphisbaenians

Bipedidae
J. Müller 1831
W. K. Parker 1868
Cope 1892a, 1900
Fürbringer 1900
Zangerl 1945
Castañeda and Alvarez 1968
Renous-Lécuru 1974
Gans 1978
Greer and Gans 1983
Kearney 2002

Rhineuridae
Cope 1892a, 1900
Fürbringer 1900

Zangerl 1945
Gans 1978
Kearney 2002

Trogonophidae
W. K. Parker 1868
Smalian 1885
Fürbringer 1900
M. Müller 1900
Gans 1960, 1978
Kearney 2002

Amphisbaenidae
Mayer 1825
J. Müller 1831

Heusinger 1833
W. K. Parker 1868
Fürbringer 1870, 1900
Smalian 1885
Cope 1892a, 1900
M. Müller 1900
Zangerl 1945
Gans 1978
Renous et al. 1991
Kearney 2002

Snakes

†Pachyrhachis
Caldwell and Lee 1997
M. S. Y. Lee and Caldwell 1998

†Haasiophis
Tchernov et al. 2000
Rieppel et al. 2003

†Eupodophis
Rage and Escuillié 2000

Leptotyphlopidae
Fürbringer 1870
W. C. H. Peters 1882
Duerden and Essex 1923a

Essex 1927
List 1955, 1966
Fabrezi et al. 1985

Typhlopidae
J. Müller 1831
Fürbringer 1870

(continued)

Table 17-1 (continued)

W. C. H. Peters 1882	Fürbringer 1870	**Boidae**
Duerden and Essex 1923b	Mlynarski and Madej 1961	Mayer 1825
Essex 1927		Heusinger 1833
Mahendra 1936	**Aniliidae**	Fürbringer 1870
Evans 1955	Mayer 1825	Beddard 1905
List 1958, 1966	Fürbringer 1870	W. H. Stickel and Stickel 1946
	Mlynarski and Madej 1961	Hoge 1947
Anomalepididae		
List 1966	**Pythonidae**	**Tropidophiidae**
	D'Alton 1836	Bellairs 1950
Anomochilidae	Gasc 1966	
Brongersma and Helle 1951	Raynaud 1972, 1985	
	Renous et al. 1976	
Cylindrophiidae	Cohn and Tickle 1999	
Mayer 1825		

lizards, ranging from terrestrial to fossorial in their habits (Greer 1989). External forelimbs are completely lacking, and the external hindlimbs are small and flaplike. The pectoral girdle is well-developed in most pygopodids (fig. 17.9H) but highly reduced in the fossorial genus *Aprasia* (Stephenson 1962; Kluge 1976; Greer 1989). A tiny, unarticulated, ossified rod in some species has been interpreted as a vestige of the humerus (Stephenson 1962; Greer 1989). The pelvic girdle in pygopodids retains all basic elements (i.e., ilium, ischium, and pubis), but these nearly always become fused in adults (Stephenson 1962; fig. 17.9I). The hindlimbs in pygopodids vary in degree of development among species, being the best developed in *Aclys, Pygopus,* and *Delma,* and the most reduced in *Lialis* and *Pletholax.* In the most well-developed forms, the hindlimb consists of a shortened femur, tibia, fibula, several carpal elements, and four digits (Greer 1989; Stephenson 1962). There is no evident sexual dimorphism in these elements, unlike in dibamids.

Scincidae is the most speciose family of lizards (Pough et al. 1998). Within this clade alone, many independent instances of limb reduction and loss have occurred (Greer 1970). For example, the sand-swimming scincine *Neoseps,* which is confined to sandy soils in central Florida, has a single digit on the forelimb and two on the hindlimb, and exhibits elongation of the trunk. Other skinks found in Africa and Asia exhibit virtually all stages of reduction in digits and limbs (e.g., Essex 1927; Caputo et al. 1995; Whiting et al. 2003; Heyer 1972). Acontines are uniformly limbless, fossorial forms inhabiting sandy soils in southern Africa. Feylinines are also completely limbless, burrowing in the forests of western and central Africa. Lygosomines include strongly fossorial forms, many of which retain small appendages, although both digit and limb reduction occurs repeatedly within the group (Greer 1970). Some limbless scincids ex-

hibit relatively well-developed pectoral and pelvic girdles, but many stages of degeneration are found within the group (Essex 1927). The pelvic girdles often consist of a single pair of ossified rods in the most modified forms.

In all of the above-described nonamphisbaenian squamates, rudiments of the hindlimbs typically persist after the forelimbs disappear. Only some small members of the family Gymnophthalmidae occurring in Central and South America exhibit a pattern similar to that found in the amphisbaenian genus *Bipes*—that is, having the hindlimbs in a more advanced stage of reduction than the forelimbs (e.g., *Bachia;* Presch 1975). Most gymnophthalmids are small and terrestrial, found within the detritus of the forest floor. The pectoral and pelvic girdles are well developed in gymnophthalmids, even among limb-reduced forms (Presch 1975; Kearney 2002).

Whereas the tendency to lengthen the body and reduce or lose the limbs is widespread among reptiles, snakes have committed to this strategy most consistently, regardless of habitat. Snakes are uniformly long-bodied and short-tailed, and the vast majority are completely limbless. The forelimbs and pectoral girdle are completely absent in all known fossil and extant snakes. (Although a pectoral girdle was once believed to be present in anomalepidids [Dunn and Tihen 1944], this has since been shown to be false [Tihen 1945; Warner 1946].) However, a number of fossil snakes are known to have retained well-developed hindlimbs, each consisting of a femur, tibia, fibula, astragalus, and calcanuem (Caldwell and Lee 1997; Tchernov et al. 2000; Rage and Escuillié 2000; fig. 17.10A); in one fossil form (†*Haasiophis*) metatarsals (and possibly phalanges) were retained as well (Tchernov et al. 2000). Many basal extant snakes (some scolecophidians, anilioids, and booids) also retain elements of the pelvic girdle and hindlimb (e.g., Mayer 1825; Bellairs 1950; List 1955; Mlynarski and Madej 1961; fig. 17.10B–D). The hindlimb elements are often highly modified in these forms.

Mammals

Like amphibians and reptiles, mammals exhibit a wide variety of digging mechanisms, several of which involve extensive modifications of the appendicular skeleton. However, in contrast to amphibians and reptiles, digging in mammals has been studied rather extensively, and several excellent reviews are available (e.g., Hildebrand 1985a; Dubost 1968; Shimer 1903; Nevo 1979, 1999; Reichman and Smith 1990). For this reason, we make no attempt here to present an exhaustive systematic overview of digging and burrowing adaptations in Mammalia. Instead, we limit our discussion to a general overview of the three basic types of limb-based digging mechanisms used by mammals: scratch digging, hook-and-pull digging, and humeral rotation digging (Hildebrand 1985a).

Scratch Diggers

Mammalian scratch diggers use vigorous scraping movements of their clawed forefeet to excavate their burrows (Hildebrand 1985a). In this respect, their general method of digging is somewhat similar to that used by a wide variety of quadrupedal reptiles (see above). However, mammalian scratch digging differs in two important ways from the superficially similar digging techniques used by many lizards, crocodilians, and turtles. First, most scratch-digging mammals dig with rapid, alternating strokes of their forefeet (e.g., Holliger 1916; Hickman 1985; Camin et al. 1995; Vassallo 1998). In contrast, most reptiles that dig with their forefeet perform numerous strokes with one arm before switching to the other, and these strokes are typically executed at a relatively slow and deliberate pace. Second, due to the orientation of the limbs in mammals, scratch diggers use predominantly parasagittal movements of the forelimbs for digging (B. Campbell 1938; Reed 1951; Gasc et al. 1986; Hildebrand 1985a). As a result, their digging strokes are generally less intricate than those of reptilian diggers, which typically involve more complex three-dimensional movements of the arm (e.g., long-axis rotation of the humerus, adduction of the humerus, extension of the antebrachium).

Although scratch diggers rely mainly on their forefeet for digging, other parts of the body may be used during the course of tunneling activities. For instance, many scratch-digging rodents use their large, procumbent incisors to assist in loosening hard soils (e.g., Stein 2000; Ebensperger and Bozinovic 2000; Vassallo 1998; Lessa and Stein 1992; Lessa and Thaeler 1989; Agrawal 1967). Some rodents have aban-

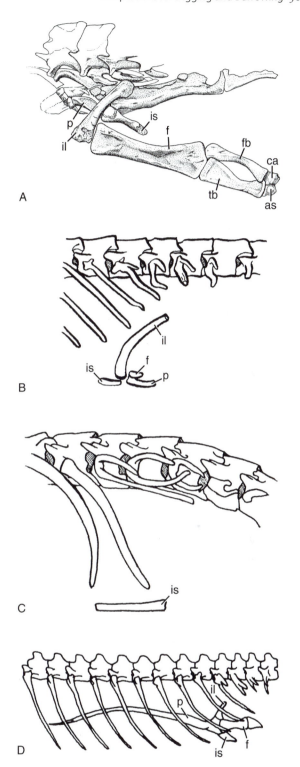

Figure 17-10 Morphology of the pelvic apparatus in a variety of basal snakes. (A) †*Pachyrhachis problematicus*, a marine snake from the Late Cretaceous of Israel. (B) *Leptotyphlops kafubi*, a fossorial leptotyphlopid. (C) *Typhlops vermicularis*, a fossorial typhlopid. (D) *Python reticulatus*, a terrestrial pythonid. All images are shown as left lateral views. as, astragalus; ca, calcaneum; il, ilium; is, ischium; f, femur; fb, fibula; p, pubis; tb, tibia. (A modified from M. S. Y. Lee and Caldwell 1998; B and C modified from List 1966; D modified from M. S. Y. Lee 1997, after Renous et al. 1976.)

doned scratch digging almost completely in favor of this "chisel-tooth digging" technique (e.g., Hildebrand 1982, 1985a; Laville 1989; Jarvis and Sale 1971; Reed 1958). Other scratch diggers augment limb-based digging with powerful lifting movements of the head (e.g., Puttick and Jarvis 1977; Hildebrand 1985a; Gasc et al. 1986; Laville et al. 1989; Gambaryan and Gasc 1993), either to compact the roof of the tunnel (in deep burrows) or to expand the tunnel upward (in shallow runways). Finally, many scratch diggers use their hindfeet in a variety of ways to remove soil already loosened by the forefeet. Most commonly, the hindfeet are used to disperse soil that accumulates underneath the animal during digging (e.g., Vassallo 1998; Gambaryan and Gasc 1993; Gasc et al. 1985; Hickman 1985; Casinos et al. 1983; Howe 1975; Taber 1945; Holliger 1916). However, many fossorial mammals also use backward kicking movements of their hindfeet to transport loosened earth out of their burrows (e.g., Jarvis and Sale 1971; Hickman 1985; Camin et al. 1995). In contrast, others turn around within their tunnels and push the piles of excavated soil forward, using some combination of their forefeet, chest, and head (e.g., Gambaryan and Gasc 1993; Jarvis and Sale 1971; W. H. Lehmann 1963; Nevo 1961).

Scratch digging is the most common method of digging among mammals (Reichman and Smith 1990). It is used not only by many fossorial species, but also by a variety of semi-fossorial and terrestrial forms as well. Some scratch diggers exhibit few, if any, structural adaptations for digging (Hildebrand 1985a). Familiar examples include a variety of carnivores (e.g., dogs, bears), rodents (e.g., marmots and prairie dogs), lagomorphs (e.g., many rabbits) and insectivores (e.g., tenrecs). In contrast, many other mammals are highly specialized for scratch digging: the aardvark (*Orycteropus*); ground squirrels (e.g., *Spermophilus, Spermophilopsis*); pocket gophers (e.g., *Geomys*); tuco-tucos (*Ctenomys*); degus, rock rats, and the cururo (Octodontidae); dune mole-rats (*Bathyergus*); some pangolins (*Manis*); armadillos (Dasypodidae); and the marsupial mole (*Notoryctes*). In taxa such as these, adaptations for digging are often numerous, involving changes in the structure of the pectoral girdle, the forelimbs, and the pelvis (e.g., Dubost 1968; Hildebrand 1985a).

Adaptive modifications of the pectoral girdle in scratch diggers are most evident in the shape and size of the scapula. In most taxa, the acromion process is quite long, often extending well beyond the glenoid fossa (e.g., Carlsson 1904; Hill 1937; Maynard Smith and Savage 1956; W. H. Lehmann 1963; fig. 17.11A). This may increase the area of origin for the *Mm. acromiodeltoideus et spinodeltoideus* (extensors of the humerus), but it may also stabilize the shoulder in some species by restricting lateral rotation of the humerus (Hildebrand 1985a). In addition, in many species the posteroventral portion of the scapula has become significantly enlarged,

forming a distinct postscapular fossa for the origin of the hypertrophied *M. teres major,* one of the main retractors of the forelimb (e.g., Carlsson 1904; Maynard Smith and Savage 1956; W. H. Lehmann 1963; fig. 17.11A).

The humerus is also commonly modified in scratch diggers. The deltoid tubercle is usually enlarged and positioned relatively far from the shoulder joint (e.g., Vizcaíno and Milne 2002; Hildebrand 1985a; W. H. Lehmann 1963; Holliger 1916; Carlsson 1904). This increases the insertional area and the mechanical advantage of the deltoid muscles. Also enlarged is the median epicondyle (Hildebrand 1985a; Goldstein 1972), the site of origin for *M. pronator teres* (and portions of some of the digital and carpal flexors). In general, the humerus is shorter and stouter in diggers than in nondiggers (e.g., Fariña and Vizcaíno 1997; Casinos et al. 1993; Bou et al. 1987; Hill 1937). Moreover, in at least some taxa there has been a thickening of the cortical bone along the diaphysis, rendering the humerus more resistant to bending and twisting forces during digging (e.g., *Ctenomys*; Biknevicius 1993).

In the forearm, specializations for scratch digging are commonly seen both in the structure of the ulna and the radius. Most notably, the olecranon process is usually quite long, in some taxa accounting for nearly half the total length of the ulna (e.g., some dasypodids; Vizcaíno et al. 1999; Hildebrand 1985a; Kühlhorn 1984; Miles 1941; fig. 17.11A). This elongation of the olecranon provides longer in-levers and larger areas of insertion for the main elbow extensors (e.g., *M. triceps brachii, M. anconeus*), thus facilitating more powerful extension of the antebrachium during the digging stroke. It also greatly increases the surface area available for attachment of the carpal and digital flexors that originate from the medial surface of the olecranon. In some scratch diggers, elongation of the olecranon is accompanied by reduction in the length of the radius (e.g., Carlsson 1904; Hildebrand 1982; fig. 17.11A). This further increases the mechanical advantage of the elbow extensors by shortening their out-lever.

In golden moles (Chrysochloridae), modifications of the antebrachium are even more extensive (fig. 17.11B, C). In these highly specialized burrowing insectivores, the large tendon of the *M. flexor digitorum profundus* has been replaced largely by bone (Flower 1876). After originating from the median epicondyle of the humerus, this tendon becomes heavily ossified, extending toward the carpus as a robust "flexor bone." At its distal end near the wrist, the bone once again becomes tendinous before bifurcating to insert ventrally onto the distal phalanges of digits III and IV (Puttick and Jarvis 1977). Because the median epicondyle arches ventroposteriorly in chrysochlorids, the tendinous origin of the flexor lies posterior to the elbow joint. Consequently, when the arm is flexed at the elbow during the early phase of the

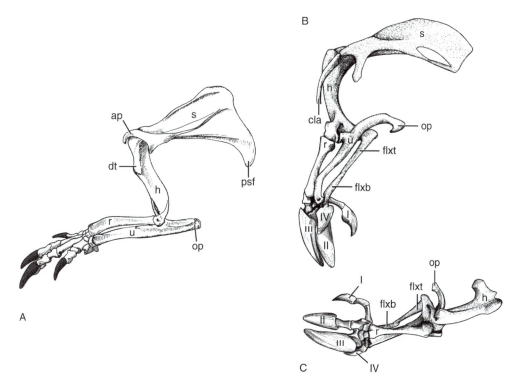

Figure 17-11 Adaptations for scratch digging in mammals. (A) Lateral view of the pectoral apparatus of an armadillo (*Dasypus*), showing the well-developed postscapular fossa (psf), the origin of the *M. teres major*. Note also the long olecranon process (op) and the short radius (r), features that facilitate forceful extension of the antebrachium during the digging stroke. (B) Lateral view of the left forearm and the pectoral girdle in the Namib Desert golden mole (*Eremitalpa granti namibensis*), a highly specialized subterranean scratch digger. (C) Dorsal view of (B) with the pectoral girdle removed. Note the long, curved olecranon process (op) and the ossified distal portion of the digital flexor tendon (flxb). I–IV, claws of digits I–IV; ap, acromion process; cla, clavicle; dt, deltoid tubercle; h, humerus; flxb, flexor bone; flxt, flexor tendon; op, olecranon process; psf, postscapular fossa; r, radius; s, scapula; u, ulna. (A modified from Hildebrand 1982; B and C modified from Gasc et al. 1986.)

digging stroke, the flexor is placed under tension, thereby re-sulting in the passive flexion of the large third and fourth dig-its (Hildebrand 1985a; Gasc et al. 1986).

As in many other fossorial tetrapods, some of the most obvious adaptations for burrowing in scratch diggers are found in the structure of the manus. In general, the forefoot is shorter in diggers than in closely related nondiggers, due largely to a reduction in the lengths of the metacarpals and all but the distalmost phalanges (Hildebrand 1985a; fig. 17.11). This increases the effective force of the digital flexors by shortening their out-levers. In some taxa, digital flexion is further facilitated by the presence of vertically oriented keel-and-groove articulations at the metacarpo-phalangeal and interphalangeal joints (Hildebrand 1985a). These joints per-mit free rotation of the phalanges in the sagittal plane (i.e., flexion/extension) but limit all other movements, thereby re-ducing the risk of dislocation during vigorous digging. The distal phalanges are typically long and robust, providing strong internal support for the enlarged claws. In addition, in some species the tips of the distal phalanges are markedly bi-fid, forming V-shaped notches that embrace internal vertical

ridges within the keratinous claws (Gambaryan 1960; Hilde-brand 1985a). In most scratch diggers the claws are large and fast-growing. Moreover, their shapes and patterns of growth are often highly specialized for the specific ways in which they are used during digging (Gambaryan 2002).

Although scratch diggers rely mainly on their forelimbs for digging, they usually brace themselves within their bur-rows by anchoring their hindfeet to the tunnel walls (Hilde-brand 1985a). In many taxa, the crucial importance of this bracing behavior is reflected in the morphology of the pelvic apparatus. Specifically, in most burrowing rodents, insecti-vores, and marsupials, the pelvis is more or less horizontally oriented, lying nearly parallel to the vertebral column. More-over, the acetabula are often positioned relatively high, in some instances lying nearly adjacent to the vertebral column (Chapman 1919). This allows the hindlegs to push laterally against the walls of the tunnel without inducing torsion about the vertebral axis. Most scratch diggers also have an expanded sacrum, consisting of between four and six verte-brae (Chapman 1919; Hildebrand 1985a). Presumably, this elongation of the sacrum provides added stability for the pelvis, thereby allowing greater forces to be generated by the hindlimbs during bracing.

Hook-and-Pull Diggers

In contrast to scratch digging, which is used by a wide variety of mammals, hook-and-pull digging (Hildebrand 1985a) is a much more specialized method of excavation that is used exclusively by anteaters (Myrmecophagidae). It is not used for burrowing, however. Instead, anteaters use this technique while foraging to rip holes into termite mounds, ant hills, and rotting logs (e.g., B. K. Taylor 1978; Hildebrand 1982; Nowak 1991). These holes then provide a passage for the long, protrusible tongue, which is used to extract the insects from their nests (Naples 1999).

As the name implies, hook-and-pull digging is a two-step process. First, the enormous, curved claws on the second and third digits of the forefoot (fig. 17.12A) are hooked into a preexisting crack, crevice, or hole. Then the digits are strongly flexed and the forefoot is pulled back toward the body. Retraction of the forefoot during the pulling phase involves flexion of the elbow and retraction of the humerus, both of which are carried out with tremendous force (B. K. Taylor 1978; Hildebrand 1985a). Throughout the digging process, the forefeet are used one at a time (Naples 1999).

Hook-and-pull diggers exhibit numerous specializations for digging, nearly all of which are seen in the structure of the pectoral girdle and forelimbs (B. K. Taylor 1978; Hildebrand 1985a). The most noteworthy modification of the pectoral girdle is a ventroposterior expansion of the scapula (fig. 17.12D). This is achieved through an enlargement of the postscapular fossa, the site of origin for one of the main forelimb retractors, the *M. teres major*. In both the giant anteater (*Myrmecophaga*) and the lesser anteaters (*Tamandua*), the postscapular fossa is relatively large (more than half the area of the infraspinous fossa; Gaudin and Braham 1998) and extends ventrally to the approximate level of the shoulder joint (B. K. Taylor 1978; Maynard Smith and Savage 1956). As a result, the moment arm of the teres major is increased significantly, thereby facilitating more powerful retraction of the arm during the digging stroke.

In addition to retraction of the arm (i.e., "pulling"), powerful flexion of the enlarged third finger (i.e., the "hook") is critically important for hook-and-pull diggers. Accordingly, anteaters have a highly specialized mechanism for flexing the central digits of their forefeet, reflected in both the osteology and myology of the forelimbs (B. K. Taylor 1978; Hildebrand 1985a). In both *Myrmecophaga* and *Tamandua*, the distal tendon of the medial head of the *M. triceps brachii* bypasses the olecranon process of the ulna to merge with the tendon of the deep digital flexor (*M. flexor digitorum profundus*).

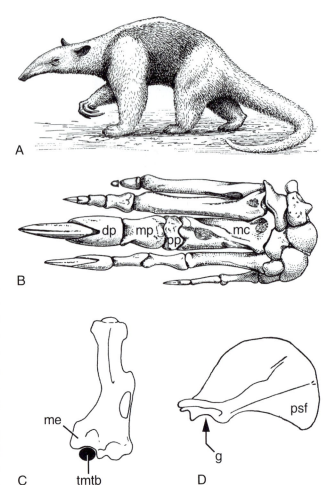

Figure 17-12 Morphological specializations for hook-and-pull digging in myrmecophagine anteaters. (A) Left lateral view of a giant anteater (*Myrmecophaga*) showing the large, curved claws on the second and third digits of the forefeet. (B) Dorsal view of the right manus in *Myrmecophaga*. Note the robust structure of digit III, the main axis of the anteater's "hook." (C) Ventral view of the left humerus in a lesser anteater (*Tamandua*), showing the concave distal margin of the enlarged median epicondyle (me), forming a notch through which passes the robust tendon of the medial head of the triceps (tmtb). (D) Lateral view of the left scapula of *Myrmecophaga*. Note the large, ventrally projecting postscapular fossa (psf), the area of origin for one of the main retractors of the forelimb, the *M. teres major*. dp, distal phalanx of digit III; g, glenoid cavity; mc, metacarpal of digit III; me, median epicondyle of humerus; mp, middle phalanx of digit III; pp, proximal phalanx of digit III; psf, postscapular fossa; tmtb, tendon of the medial head of the *M. triceps brachii*. (A and B modified from Grassé 1955b; C modified from B. K. Taylor 1978; D modified from Maynard Smith and Savage 1956.)

Consequently, this largest head of the triceps muscle functions no longer as a primary extensor of the antebrachium, but rather as a flexor of the hypertrophied third digit. This mechanism of digital flexion is further enhanced by the structure of the humerus. In particular, the median epicondyle is expanded significantly (fig. 17.12C), thereby providing an increased area of attachment for the epicondylar head of the *M. flexor digitorum profundus* (as well as a greater moment arm for the *M. pronator teres*). Moreover, the distal margin of the median epicondyle bears a prominent concavity that serves as a pulley for the robust tendon of the medial head of the triceps (fig. 17.12C). This allows the forces generated by that muscle to be redirected along the long axis of the antebrachium, thus augmenting the forces produced by the digital flexors.

Digital flexion in myrmecophagids is also strongly influenced by the morphology of the large central digit of the manus (B. K. Taylor 1978; Hildebrand 1985a; fig. 17.12B). In both *Myrmecophaga* and *Tamandua* (but not in the arboreal silky anteater, *Cyclopes*; Grassé 1955b), this digit consists of a long metacarpal, relatively short proximal and middle phalanges, and a robust distal phalanx that supports the large, hooklike claw. Both the distal end of the proximal phalanx and the proximal end of the middle phalanx have relatively flat, faceted articular surfaces. As a result, little movement is possible at the proximal interphalangeal joint. In contrast, the metacarpo-phalangeal joint and the distal interphalangeal joint are highly mobile. However, the articulations at these joints are specialized to allow flexion and extension only. The articular surfaces on the distal ends of the metacarpal and the middle phalanx are spool-shaped, bearing strong vertical ridges along their medial and lateral edges. These ridges fit into deep, vertically oriented concavities on the proximal ends of the proximal and distal phalanges. A similar (but single) articulation is also present at the center of the metacarpo-phalangeal joint. These vertical keel-and-groove articulations allow extensive flexion and extension but prevent movements in all other planes. In addition, joint stability is further increased by well-developed collateral ligaments that run along the sides of each joint. This arrangement prevents both lateral and torsional movements within the digit, thus making the anteater's "hook" highly resistant to dislocation during digging (B. K. Taylor 1978; Hildebrand 1985a).

Humeral Rotation Diggers

Long-axis rotation of the humerus is a common component of the digging mechanisms of many quadrupedal amphibians, reptiles, and mammals (e.g., Emerson 1976; A. P. Russell and Bauer 1990; Thewissen and Badoux 1986). In some mammals, however, digging strokes of the forefeet are powered almost exclusively by such movements (Hildebrand 1982, 1985a). Among extant taxa, humeral rotation digging is best known in moles and shrew moles (Talpidae), and the following account is based entirely on these highly derived burrowing insectivores. However, similar digging mechanisms may also be used by monotremes, in which humeral rotation plays a key role in both terrestrial and aquatic loco-

motion (e.g., Howell 1937c; Jenkins 1970; Hildebrand 1985a; Pridmore 1985).

Moles use powerful lateral thrusts of their broad forefeet (fig. 17.13A–C) to dig into and through the ground (e.g., Hisaw 1923; Skoczen 1958), and these digging thrusts are generated almost entirely through long-axis rotational movements of the humeri (Reed 1951; Yalden 1966). In loose soil both arms may be used synchronously in a breaststroke fashion (Skoczen 1958). However, in more compact soil only one forefoot is used at a time, while the other remains braced firmly against the opposite side of the tunnel (Hisaw 1923; Arlton 1936; Skoczen 1958). Usually several digging strokes are performed with each arm before switching sides (Skoczen 1958). In general, digging movements are much slower than those exhibited by scratch diggers (Hildebrand 1985a).

Moles dig two different types of tunnels: shallow surface runways and deep burrows (G. Godfrey and Crowcroft 1960; Gorman and Stone 1990). Shallow runways are dug just beneath the surface, usually as a means by which to locate food. In building these shallow tunnels, moles do not remove the soil that they have loosened. Instead, they forcefully press it against the roof of the tunnel, resulting in the formation of a surface ridge along the overlying ground. This compaction is done with strong lateral thrusts of the forefeet, requiring the mole to rotate the anterior portion of its body before each tamping stroke in order to properly position its forefeet (Hisaw 1923; Skoczen 1958). In contrast, when building deep tunnels, moles remove the soil that they have loosened, thereby forming prominent "mole hills" outside the entrances to their burrows. In such instances, loosened earth is thrown backward by the digging forefeet, and occasionally pushed further back by the hindfeet. Then, after a certain amount of soil has accumulated behind the mole, it somersaults between its spread hindlegs to reverse direction and then pushes the pile of earth out of the tunnel. As in scraping and tamping, this pushing is done with lateral thrusts of the forefeet. Therefore, before starting to move the pile of earth, the mole first turns its head and shoulders to the side in order to properly position the forefoot that will be used for pushing. The other three feet are then used to drive the mole forward toward the tunnel exit (Hisaw 1923; Skoczen 1958).

Morphological adaptations for humeral rotation digging in moles are numerous, involving extensive modifications of both the pectoral apparatus and the pelvis (e.g., Slonaker 1920; Reed 1951). In the pectoral girdle, the most functionally significant of these modifications relate to the structure and orientation of the scapula. In contrast to other mammals, the talpid scapula is oriented almost horizontally, and it extends anteriorly to the level of the occiput (Reed 1951; fig. 17.13B). This carries the glenoid cavity far forward, resulting in an anterior displacement of the forelimbs, thereby

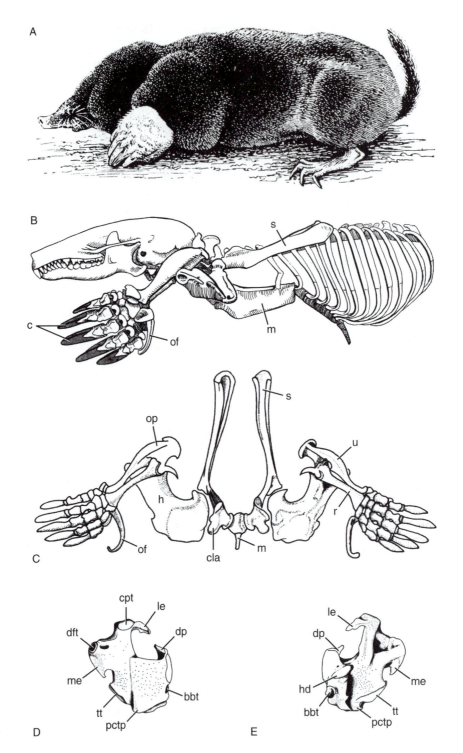

Figure 17-13 Adaptations for humeral rotation digging in moles. (A) Left lateral view of a European mole (*Talpa*), showing the out-turned orientation of the forefeet and the enormous development of the forelimbs relative to the hindlimbs. Note also that the stylopodial and zeugopodial portions of the limbs are contained largely within the contour of the body. (B) Left lateral view of the anterior portion of the skeleton of a western American mole (*Scapanus*), showing the greatly enlarged manus, robust *os falciforme* (of), horizontally oriented scapula (s) and keeled manubrium (m). Note the anterior position of the hands. (C) Frontal view of the pectoral girdle and forelimbs of *Talpa*, showing the steeply angled orientation of the broad humeri (h). Note also the relatively short antebrachia. (D) Anterolateral view of the right humerus of *Scapanus*. (E) Posteromedial view of the right humerus of *Scapanus*. Note the wide separation between the teres trochanter (tt) and the head of the humerus (hd), which significantly increases the moment arms of the *M. teres major* and *M. latissimus dorsi*, the main rotators of the humerus. bbt, tunnel for tendon of the *M. biceps brachii*; c, claws; cla, clavicle; cpt, capitulum of humerus; dft, fossa for origin of digital flexor tendon; dp, deltoid process; h, humerus; hd, humeral head; le, lateral epicondyle; m, manubrium; me, medial epicondyle; of, *os falciforme*; op, olecranon process; pctp, pectoral process; r, radius; s, scapula; tt, teres trochanter; u, ulna. (A modified from Grassé 1955b; B modified from Hildebrand 1982; C modified from Dubost 1968, after Vialleton 1924; D and E modified from Dubost 1968, after Reed 1951.)

allowing the forefeet to be extended forward beyond the tip of the snout (Hildebrand 1985a). In addition, elongation of the scapula increases the area of origin for one of the main rotators of the humerus, the *M. teres major*, which has hypertrophied in moles to become the largest muscle in the body (Reed 1951; B. Campbell 1939; R. A. Freeman 1886). Like the scapula, the manubrium is also greatly elongated in talpids. Moreover, it bears a strong, ventrally projecting keel along most of its length (fig. 17.13B), thus providing an exceptionally broad area of origin for the enlarged pectoral muscles (Slonaker 1920; Reed 1951; Yalden 1966). Finally, the glenoid cavity is strongly modified as well, forming an elliptical articulation with the humeral head (R. A. Freeman 1886). Due to the shape and orientation of this articulation, movement of the humerus about the shoulder joint can occur only in the horizontal plane (Reed 1951).

Like the pectoral girdle, the talpid humerus is highly specialized for its role in digging. Rather than being positioned more or less vertically as it is in most other mammals, the humerus is oriented obliquely, projecting dorsolaterally from the glenoid cavity at a rather steep angle (e.g., B. Campbell 1939; Reed 1951; Yalden 1966; fig. 17.13C). Consequently, the elbow joint lies above the shoulder, and the forefoot is positioned laterally rather than ventrally. Equally as important, however, are structural specializations that facilitate the forceful rotation of the humerus during digging. Foremost among these is an increase in humeral width (e.g., B. Campbell 1939; Reed 1951). Particularly significant from a biomechanical perspective is the wide separation between the head of the humerus and the teres tubercle (fig. 17.13E), the point of insertion for the main rotators of the humerus, the *M. teres major* and the *M. latissimus dorsi*. As a result of this arrangement, these muscles operate with a tremendous mechanical advantage when rotating the humerus during the digging stroke (B. Campbell 1939; Hildebrand 1985a).

In addition to its orientation and overall shape, the talpid humerus exhibits two other morphological specializations for humeral rotation digging. The first of these is a transversely oriented tunnel passing through the proximalmost portion of the bone (fig. 17.13D, E). This tunnel transmits the long tendon of origin of the *M. biceps brachii* from the medial to the lateral side of the humerus (e.g., R. A. Freeman 1886; B. Campbell 1939; Whidden 2000). Due to this deflection, contraction of the biceps not only flexes the elbow but also counterrotates the humerus during the recovery stroke (Reed 1951; Hildebrand 1985a). Also of functional significance is a large fossa at the distal end of the medial epicondyle (fig. 17.13D), which serves as the site of origin for the tendinous deep digital flexor (*M. flexor digitorum profundus*). Due to the peculiar orientation of the brachium, this fossa lies far lateral to the axis of rotation of the humerus. Therefore, when the humerus is rotated during the

digging stroke, the flexor tendon is placed under tension, resulting in a passive flexion of the manus about the wrist joint (D'A. W. Thompson 1884; Reed 1951; Yalden 1966).

Adaptations for humeral rotation digging are also evident more distally, in both the antebrachium and manus. Both the radius and ulna are exceptionally robust, strengthened to withstand the tremendous forces generated during the digging stroke (e.g., Slonaker 1920; Reed 1951). In addition, these bones have become much shorter in moles than in their more surface-active relatives (e.g., shrew moles, shrews; Reed 1951). This, together with a marked shortening of the metacarpals and proximal phalanges (Slonaker 1920), greatly decreases the length of the distal portion of the forelimb, and hence the length of the out-lever in the mole's digging system. However, in contrast to the proximal phalanges, the distal phalanges are greatly elongated to support the large, spatulate claws that project from all five of the mole's digits (Reed 1951; fig. 17.13B, C). One of the most distinctive features of the manus in moles is the large radial sesamoid (the *os falciforme*; fig. 17.13B, C) that runs along the inner edge of the forefoot medial to digit I. This robust sesamoid widens the hand significantly (especially when actively abducted), thereby allowing more soil to be excavated with each digging stroke (e.g., Reed 1951; Hildebrand 1985a).

Although moles dig with their forelimbs, they brace themselves in their tunnels with their hindlimbs (e.g., Howell 1923; Skoczen 1958; Yalden 1966). Accordingly, the talpid pelvic apparatus is highly modified for this function. Most importantly, the pelvis has grown considerably in length and is oriented nearly parallel to the vertebral column (Chapman 1919; Slonaker 1920). As a result, the laterally facing acetabula are raised nearly to the level of the vertebral axis (Reed 1951; Hildebrand 1985a). This arrangement allows the hindlimbs, which are splayed laterally against the tunnel walls during burrowing, to transmit forces to the sacrum in a more direct manner (i.e., without torsion). In addition, the long pelvis is solidly ankylosed to the sacrum at as many as three different places, along both the iliac and ischiadic portions of the innominate bone (Slonaker 1920; Reed 1951; Hildebrand 1985a). Thus, the pelvis is more solidly anchored to the vertebral column in moles than in any other mammals.

In general, the structure of the talpid hindlimb is unremarkable. However, the limb has grown significantly shorter in the most fossorial taxa (e.g., *Talpa, Scapanus, Scalopus*), mainly through a shortening of the tibiofibula and the pes (Reed 1951). This increases the force with which moles can extend their hindlegs, important not only in bracing, but also in pushing soil out of their tunnels. Presumably, this morphology also increases overall locomotor efficiency within the confines of narrow tunnels.

The radically modified appendicular musculoskeletal system of moles renders these small mammals among the most accomplished digging tetrapods. This is illustrated not only by the great lengths of their tunnel systems (commonly extending for hundreds of meters, and sometimes over 1,000 m; Arlton 1936; Hickman 1983, 1984), but also in the strength of their digging strokes. With each lateral thrust of a single forefoot, a mole may generate a force equivalent to more than 30 times its own body weight (Arlton 1936). By comparison, the strongest human weight lifters are unable to lift even three times their own weight when pushing with two hands. Moreover, moles are able to maintain their strength over long periods of time. For instance, one 80 g European mole (*Talpa europaea*) observed by Skoczen (1958) formed four molehills weighing a total of 15.5 kg in 90 minutes. To equal this feat, a 70 kg human would have to excavate nearly 10 tons of earth in a single hour.

Summary

In this chapter we have examined many of the diverse ways in which the limbs of tetrapods have become adapted for digging and burrowing. We have done this in large part by focusing on selected features of the appendicular skeleton in a relatively small number of highly fossorial amphibians, reptiles, and mammals. However, when considered collectively, these specific examples provide a more generalized "thumbnail sketch" of digging and burrowing in tetrapods, one that serves to bring several broader evolutionary patterns into sharper focus:

1. Most fossorial quadrupeds dig with their forefeet. Those that do not burrow with their hindfeet (e.g., most fossorial frogs), head (e.g., some amphibians and a few small mammals), or jaws (e.g., many rodents and some crocodilians).

2. Among quadrupedal amphibians and reptiles that dig with their forefeet, adaptations for digging are rarely evident in the morphology of the appendicular skeleton. Noteworthy exceptions include the hypertrophied forelimbs of *Myobatrachus*, the spatulate forefeet of *Gopherus*, and the webbed feet of *Palmatogecko*.

3. Among mammals that dig with their forefeet, adaptations for digging are often seen in the structure of both the pectoral apparatus and the pelvis.

4. The most common modifications of the appendicular skeleton in digging tetrapods include an increase in the overall robusticity of the bones of the forelimb; expansion of the deltoid tubercle of the humerus; a decrease in the length of the antebrachium; an increase in the length of the olecranon process of the ulna; an increase in the width of the manus; a decrease in the length of metacarpals and all but the distal phalanges; a decrease in the number of phalanges; modifica-

tions of the metacarpo-phalangeal and interphalangeal joints, either to prevent movement or to restrict movement to a single plane; an increase in the length and rate of growth of the claws on the forefeet; and an increase in the number of articulations between the pelvis and the sacrum.

5. Mammalian scratch digging differs from the forelimb-based digging mechanisms of quadrupedal amphibians and reptiles in two significant ways. First, scratch-digging mammals generally dig with alternating strokes of their forefeet, whereas most amphibian and reptile diggers use repetitive, unilateral strokes of a single forefoot before switching sides. This is likely related to the importance of side-specific lateral bending during digging in most quadrupedal amphibians and reptiles, which necessitates a repositioning of the body with each alternation of the forelimbs. Second, the digging strokes of most mammalian scratch diggers are much more rapid than those of amphibians and reptiles. This suggests that the momentum of the forefeet may assist scratch diggers in loosening hard soils.

6. Head-first burrowing is extremely common among fossorial amphibians and reptiles. In most head-first burrowers, the limbs are reduced or lost, the trunk is greatly elongated, and forward thrusts of the head are powered exclusively by movements of the axial skeleton. Such trends toward elongation, limb reduction, and axial propulsion do not occur among burrowing mammals, even among the few taxa in which the head is used for digging. The absence of elongate, head-first burrowing mammals may be due to physiological constraints associated with endothermy (placing an upper limit on surface-to-volume ratios, thus preventing extensive elongation of the body), and/or morphological constraints inherent to the mammalian locomotor apparatus (facilitating sagittal flexion of the spine, but limiting lateral flexion).

Digging Deeper: Directions for Future Research

Rigorous study of morphological adaptation requires an integrative approach that incorporates data on structure, function, performance, natural history, and phylogeny (e.g., Gould and Vrba 1982; H. W. Greene 1986; Coddington 1988; Baum and Larson 1991; Larson and Losos 1996). However, most of the adaptive hypotheses presented in this chapter derive largely from comparative anatomical studies that have been interpreted in the light of available behavioral and/or ecological data. This clearly underscores the need for an increased emphasis on function, performance, and phylogeny in future studies of digging and burrowing.

Functional studies of digging and burrowing are particularly challenging due to the inherent difficulties associated with observing subterranean behaviors. For most fossorial taxa, the precise ways in which the limbs, head, and trunk are used during burrowing remain poorly known. Consequently, assessing the functional utility of morphological characters believed to be adaptive for burrowing is in many cases difficult, if not impossible. In the last two decades, however, functional morphologists have begun to use X-ray cinematography to study burrowing behavior directly in a variety of fossorial mammals (Casinos et al. 1983; Gasc et al. 1986; Laville 1989; Laville et al. 1989; Gambaryan and Gasc 1993). This technique has also been used with great success to elucidate the various mechanisms of tunnel progression used by limbless amphibians and reptiles (Gaymer 1971; Gasc 1982; Summers and O'Reilly 1997). Our understanding of tetrapod burrowing mechanics could be improved dramatically through the increased use of X-ray videography in future studies.

Burrowing performance can be assessed in a variety of ways, but this has been done only rarely. Available performance estimates are limited largely to scattered reports of tunneling speed (e.g., Arlton 1936; Camin et al. 1995; Vassallo 1998). Estimates are also sometimes given concerning the force generated during the digging stroke. This is most commonly measured by an animal's ability to move or lift heavy objects or volumes of earth (e.g., Arlton 1936; Skoczen 1958; Bateman 1959; Gambaryan 1960), but more recent studies have used force platforms to more accurately quantify the forces that fossorial animals exert on the environment during burrowing (O'Reilly et al. 1997, 2001). One of the simplest ways to assess burrowing performance is to test an animal's ability to penetrate soils of known levels of compaction (e.g., Ducey et al. 1993). Moreover, the use of soil penetrometers to precisely quantify soil compaction has the added advantage of making the results of independent behavioral studies more readily comparable, since burrowing behavior is often modulated according to how loose or compact the soil is (e.g., Hisaw 1923; Gans 1974; Vassallo 1998). Regardless of the method(s) used, a greater emphasis on performance in future studies of digging and burrowing would greatly facilitate the assessment of the adaptive significance of the many anatomical specializations that have evolved among fossorial tetrapods.

Finally, the conspicuous absence of cladograms in this chapter reflects one of the greatest areas of inadequacy in our current understanding of the evolution of tetrapod digging and burrowing mechanisms. Phylogeny is rarely considered in anatomical studies of fossorial taxa, and we are aware of no comparative studies to date that have attempted to correlate burrowing performance with morphology or natural history within a phylogenetic framework. Consequently, although most of the adaptive hypotheses presented in this chapter have relatively long histories and are generally accepted, few if any have been tested in a rigorous fashion. The use of modern comparative methods—phylogenetically independent contrasts, squared-change parsimony, concentrated-changes test, and the like—in future studies of digging and burrowing would represent a significant advance in this field.

Chapter 18 Aquatic Adaptations in the Limbs of Amniotes

J. G. M. Thewissen and Michael A. Taylor

THE KEY INNOVATION that characterized amniote origins is the amnion. It allowed early amniotes to become independent from life in water. In that light, it seems paradoxical that a great many amniotes have returned to the water for much or all of their lives. The ubiquity of secondarily aquatic amniotes is evolutionarily important. Even a terrestrial amniote can live amphibiously, and slight aquatic adaptations can greatly facilitate life in water. This leads to evolutionary intermediates that are fit in aquatic and terrestrial environments. As a result, there is great variation in the degree of aquatic adaptations among aquatic amniotes. In some, modifications are limited to one system; in others, all organ systems are affected and life in water is obligate. The benefits of living amphibiously without the need for strong aquatic adaptations probably facilitated the multiple origins of aquatic adaptations. In contrast, active flight requires extreme morphological adaptations, and intermediate morphologies may not offer some of the advantages of active flight. As a result, active flight evolved only rarely in amniotes.

Aquatic adaptations offer a unique opportunity to study the evolutionary process because of the wide range of aquatic adaptations in unrelated amniote lineages. Overall, limb specializations for aquatic behavior pertain to locomotor specializations for swimming, although occasionally they also affect other functions, most commonly reproduction and feeding. For amphibious taxa, the limbs show not only adaptations for life in the water, but also features that make land locomotion possible.

The purpose of this chapter is to catalogue the diversity of aquatic amniotes and identify parallels between swimming modes among the amniotes. Aquatic adaptations can be subtle, and taxa that swim readily may appear to lack any morphological modifications. Such is the case, for instance, with several species of cottontail rabbits (*Sylvilagus palustris* and *S. aquaticus*), which are adept swimmers but lack morphological changes. Other aquatic amniotes lack adaptations for swimming, even though they spend most of their time in the water. Hippos (Hippopotamidae, Artiodactyla; Eltringham 1999) are simply bottom walkers. Others, such as wading birds and moose (*Alces*, Cervidae, Artiodactyla), wade in water with emerged head and lack adaptations for swimming. Some feed but do not locomote in water (ospreys, fishing bats).

These may be excellent models describing the earlier stages of evolution in modern aquatic vertebrates, and they are thus important evolutionarily. We discuss some of these, but a complete discussion would run to many pages, and we do not attempt to elucidate the evolution of the locomotor patterns that led to the most aquatic forms.

In evolving the body form of a swimmer, tetrapods had to adapt their bodies to several different physical properties of water. They had to change their propulsive forces, alter their buoyancy control mechanism, and modify their steering and balance organs. A discussion of the theoretical basis and the practical solutions found by aquatic amniotes is beyond the scope of thischapter. For general discussion of differences between water and air with respect to physics and physiology, see Denny (1993) and Elsner (1999). P. W. Webb and Blake (1985), Fish (1996, 1998b), and T. M. Williams and Worthy (2002) discussed some more general issues relating to the forces and control surfaces important in swimming vertebrates. P. W. Webb and de Buffrénil (1990), M. A. Taylor (1994), and Fish et al. (2002) gave an evolutionary overview of buoyancy control mechanisms. Buchholtz (1998) discussed locomotor similarities in the axial skeleton among aquatic amniotes.

In this chapter, we divide the aquatic amniotes into three groups: mammals, birds, and nonmammalian, nonavian amniotes ("reptiles"). For each, we present phylogenetic trees with time axes. These diagrams summarize the diversity of aquatic representatives and emphasize the multiple origins of aquatic adaptations. The sections of the text that accompany these figures summarize the existing diversity functionally, organizing aquatic amniotes on the basis of observed or inferred locomotor behaviors, not on particular anatomical features. Our interest is in identifying locomotor similarities among the secondarily aquatic amniotes.

Mammals

The phylogeny of aquatic and semiaquatic mammals is summarized in figure 18.1. Aquatic adaptations (fig. 18.2) were discussed in detail by Howell (1930), and important functional reviews on the locomotion of aquatic mammals include those of Tarasoff et al. (1972), Fish (1993, 1996, 2001), Pabst et al. (1999), E. M. Williams (1999), and T. M. Williams and Worthy (2002). Based on studies of modern mammals in flow tanks, Fish (1993, 1996, 1999, 2001) developed a conceptual model that describes evolutionary transitions between swimming modes. This model arranges predominant swimming modes along trajectories of increasing efficiency, and predicts stages of locomotor evolution along these trajectories (fig. 18.3).

Wading

Wading probably represents the simplest way to live in water and retain a mostly terrestrially adapted body. Shallow-water waders, such as tapirs, whose head remains emergent, show no or few aquatic adaptations. Deeper-water waders may have very heavy bones (pachyostosis, swollen bones, and osteosclerosis, thick cortical layers) to counteract buoyancy. Aquatic locomotion in *Hippopotamus* is mostly restricted to wading in deep water (Eltringham 1999) with submerged head. Hippos have osteosclerotic bones as a buoyancy control mechanism (Wall 1983).

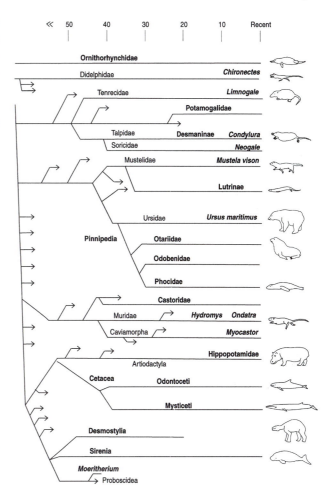

Figure 18-1 Phylogenetic tree showing the diversity of aquatic mammals with approximate relationships and divergence times in millions of years before present. Clades with largely or wholly aquatic members are shown in boldface (most terrestrial mammals are not indicated). It is apparent that aquatic adaptations arose independently many times.

Quadrupedal Paddling

Terrestrial and some semiaquatic mammals usually swim by quadrupedal paddling (Fish 2001). The limb specializations displayed by these mammals are usually very minor or absent altogether. In minks (Mustelidae, *Mustela vison*) the hands and feet are slightly webbed, but otherwise their limbs are morphologically not different from terrestrial relatives (T. M. Williams 1983). The insectivore *Micropotamogale* paddles with hands and feet, which are webbed in *M. ruwenzorii* but not in *M. lamottei* (Kingdon 1974).

Alternating Pelvic Paddling

Alternating pelvic paddling is an aquatic locomotor mode that is more efficient than quadrupedal paddling (Fish 2001). In alternating pelvic paddling, the swimmer propels its body

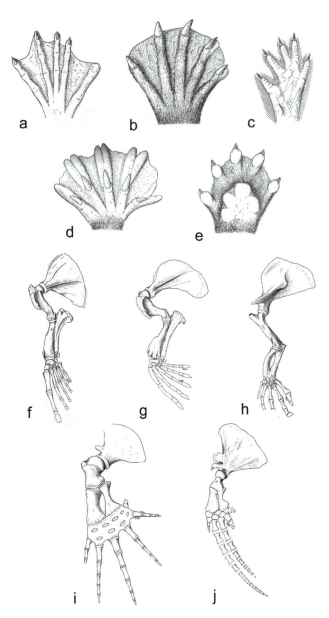

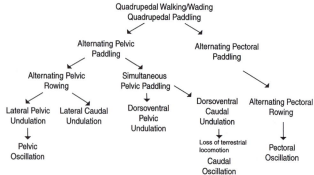

Figure 18-3 Model for the evolution of aquatic locomotor modes in mammals (slightly modified from Fish 1993, 1996, 2001).

Figure 18-2 Aquatic adaptations of the limbs of mammals. Left hindfeet of the marsupial *Chironectes* (a), the beaver *Castor* (b), the muskrat *Ondatra* (c). Left forefeet of the platypus *Ornithorhynchus* (d) and the otter *Lutra* (e). Bones of the left forelimb of the sealion *Zalophus* (f), the seal *Phoca* (g), the dugong *Dugong* (h), the Right Whale *Balaena* (i), and the pilot whale *Globicephala* (j). (From Howell 1930.)

by flexion and extension movements of the hindlimbs in a more or less vertical plane. The left and right limb move out of phase. Mammals that use this locomotor method usually display morphological adaptations for swimming. Drag forces propel these swimmers (Fish 1996), and these forces can be increased by enlarging the size of the foot, which is the propelling surface. To this end, webbing is common (e.g., the beaver *Castor*, the water opossum *Chironectes*; Fish 1993; fig. 18.2a and b), although not universal; for example,

it is poorly developed in the water rat *Hydromys* (Fish and Baudinette 1999).

In several genera of water shrews (e.g., *Sorex palustris* and *Nectogale*), aquatic specializations to increase the propelling surface include a frill of stiff hairs that broadens the feet (Howell 1930). The surface area of the feet can also be enlarged by lengthening the toes, and pelvic paddlers—*Castor, Hydromys,* and *Chironectes*—commonly have proportionally longer feet than their terrestrial relatives (Howell 1930). There are also differences in the myology of aquatic taxa (such as *Chironectes*) and their terrestrial relatives (opossum *Didelphis*; Stein 1981). *Chironectes* uses its forelimbs in procuring food underwater, and this may have been an added selection pressure for shifting the propulsion function completely to the hindlimbs. The feet remain relatively symmetrical in alternating pelvic paddlers, with the hallux commonly reduced and not involved in locomotion.

Alternating Pelvic Rowing

In alternating pelvic rowing, sequential beats of the left and right hindlimb move in a horizontal rather than a vertical plane (Fish 1996). Limb specializations that enlarge the propelling surface (the foot) in these rowers are similar to those in alternating pelvic paddlers. Muskrats (*Ondatra*) use this swimming mode while swimming underwater (Mizelle 1935; Fish 1982, 1984a). Howell (1930) noted that the foot of muskrats (fig. 18.2c) is everted (lateral edge faces more dorsally than the medial edge). This would change the plane of the foot by directing it more perpendicularly to the plane of hindlimb movements, in contrast to the foot of pelvic paddlers.

Lateral Pelvic and Caudal Undulation

In lateral undulation, the body is propelled by sinuous movements of the vertebral column. Intrinsic muscles of the hindlimbs play a secondary role in propulsion. These vertebral movements may power propelling surfaces on the feet (pelvic undulation) or the tail (caudal undulation). The otter shrew *Potamogale* uses its flattened tail in caudal undulation (Kingdon 1974; Nowak 1999). Because the limbs are not used in the most common mode of aquatic locomotion, the feet of *Potamogale* actually display fewer aquatic specializations than those of its paddling relatives (*Micropotamogale*). *Potamogale* lacks webbing but has a skin flap along its feet that allows the animal to hold its legs along its body without disrupting its streamlined contour. The tail is mediolaterally flattened.

Pelvic Oscillation

Lateral pelvic oscillation involves moving the feet in a horizontal plane with the plane of the feet held vertically and most of the power provided by the vertebral column. Phocids (true seals) and odobenids (walruses) swim in this way (Tarasoff et al. 1972; Fish et al. 1988).

Phocids (fig. 18.2g) usually swim by abducting the toes in the vertical plane and forcefully adducting the feet with alternate beats. Anatomically, phocid hindlimbs display many modifications. In the innominate, the ilium is short and its anterior part laterally deflected, whereas ischium and pubis are elongated (Howell 1930; J. E. King 1983). The femur is very short but remains robust in areas where muscles attach. The tibia and fibula are long, and there is usually a synostosis proximally. The first and fifth toes are longer and much more robust than toes II–IV, but the foot remains more or less symmetrical.

Propulsion in seals is mainly provided by the long back muscles attaching to deflected ilia (iliocostalis; J. E. King 1983). This muscle pulls the pelvis and hindlimb medially during each power stroke. The foot of phocids (but not odobenids) is permanently extended, and the sole faces dorsally when the animal is at rest. As a result, phocids cannot fold their feet under the body. This position of the foot is caused by a long proximal process on the astragalus over which the large flexor hallucis longus runs (J. E. King 1983). This mechanism prevents passive dorsiflexion of the foot by reactive forces during the power stroke. However, this mechanism also prevents phocids from using their hindlimbs in land locomotion.

The forelimbs of phocids differ among the subfamilies but always have a short humerus and forearm, and the hand is long. The thumb is robust and long, and the remaining fingers are slender and decrease in length from medial to lateral (digits II–V). Land locomotion in phocids is provided either by sinuous movement of the vertebral column in which the limbs are not important (lobodontines; O'Gorman 1963; J. E. King 1983) or by dragging the body with the forelimbs (phocines; Backhouse 1961). Phocines anchor their hands by flexing their fingers, digging them into the substrate, and then pulling their body forward by elbow and shoulder flexion. The hands of lobodontines and phocines reflect these differences in land locomotion. Lobodontine hands are finlike, with poorly pronounced individual fingers with full interdigital webbing, and small nails (J. E. King 1983). Phocine hands have robust fingers and clawlike nails, large sesamoids embedded in the tendons of the flexors, and less pronounced interdigital webbing (Backhouse 1961). Forelimb morphology in phocids reflects their mode of land locomotion, not that of swimming.

Simultaneous Pelvic Paddling

Simultaneous pelvic paddling differs from alternating pelvic paddling in that left and right hindlimbs are now in the same phase of the stroke cycle (Fish 1993, 1996). Both modes of swimming display similar morphological adaptations. With both hindlimbs in the same phase, it is possible for the swimmer to recruit the back muscles in powering the feet; modifications in the vertebral column may result. The only mammals known to locomote in this fashion are the generalized otters (*Lutra* and *Lontra*). In these otters, the tail assumes some role in providing propulsion (Tarasoff et al. 1972; Fish 1994).

Dorsoventral Pelvic Undulation

Dorsoventral pelvic undulation is used by the seaotter *Enhydra* (T. M. Williams 1989). This is a lift-based mode of locomotion, where thrust is provided at both upstroke and downstroke. Each foot is a highly asymmetrical hydrofoil, with the digit V being the longest (Thewissen and Fish 1997). The thigh (femur) and leg (tibia and fibula) are short, providing the hindlimb muscles with short lever arms that enhance torque. Fish and Stein (1991) discussed pachyostosis in the mustelids.

Dorsoventral Caudal Undulation

In dorsoventral caudal undulation, the tail provides more of the propulsive power than the feet. The giant freshwater otter *Pteronura* swims in this way (Fish 1993 and pers. comm.). It has a flattened tail and webbed fingers and toes (Redford and Eisenberg 1992), but the fore- and hindlimbs are relatively generalized, as propulsive function has shifted to the tail.

Caudal Oscillation

Caudal oscillation is practiced by two orders of mammals that are obligately aquatic, Cetacea and Sirenia (Fish and

Hui 1991; T. M. Williams et al. 1992; Fish 1998a). They have very specialized morphologies; their tails are modified into a flat paddle shape (manatees) or fluke (dugongs and cetaceans). There are no external hindlimbs. The forelimbs of cetaceans (fig. 18.2i and j) are greatly modified, whereas in sirenians the bones retain more land mammal–like morphologies (fig. 18.2h). The only fully functional synovial joint in cetaceans is the shoulder joint; elbow, wrist, and fingers are relatively immobile (Benke 1993; A. G. Watson et al. 1994). The hand has interdigital webbing, and individual fingers are not recognizable externally. The fingers are highly asymmetrical, with the leading edge of the flipper being much longer (hyperphalangeous) than the trailing edge, forming a hydrofoil. The forelimb is mainly used in steering and stabilization, and the tail is the main propulsive organ.

The evolutionary hypothesis presented in figure 18.2 is based on experimental data (Fish 1993, 1996, 2001). For cetaceans, this hypothesis is corroborated by independent evidence from the fossil record. The limb morphology of the most archaic cetaceans, pakicetids, is suggestive of a terrestrial-wading (clearly nonswimming) lifestyle (Thewissen et al. 2001). In the fossil record, pakicetids are followed by ambulocetids, which were probably pelvic paddlers or dorsoventral undulators (Thewissen and Fish 1997). Ambulocetids are followed by remingtonocetids, in which the limbs are reduced and the tail lengthened, features of committed dorsoventral undulators (Bajpai and Thewissen 2000). Caudal undulation is practiced by modern whales, but was already prevalent in late Eocene basilosaurids and dorudontids (Uhen 1998; Buchholtz 1998).

Alternating Pectoral Paddling

Alternating pectoral paddling with the hindlimbs passively trailing is used by polar bears (Flyger and Townsend 1968), which differ little from other bears in swimming morphology. As in pelvic paddlers, the propelling surface (the hand) may be widened to increase its area. Ewer (1973) hypothesized that the forelimb morphology of terrestrial bears is similar to that of polar bears because similar limb movements can be used to climb and swim.

Alternating Pectoral Rowing

Alternating pectoral rowing, involving the movement of the forelimbs in a horizontal plane, is practiced by a single mammal, the platypus, *Ornithorhynchus*, which swims by alternating rowing movement of the forelimbs, which are oriented in a horizontal plane (Fish et al. 1997). It is not clear how these movements are produced, but terrestrial locomotion is well studied (Pridmore 1985). The main movement in

Ornithorhynchus forelimbs (fig. 18.2d) during walking is caused by rotation of the humerus. A similar locomotor mode is used by walking echidnas, *Tachyglossus* (Jenkins 1970), and by digging placental moles (Talpidae; Yalden 1966; Kley and Kearney, chap. 17 in this volume).

Animals that swim using humeral rotation share remarkable locomotor specializations (Hildebrand 1985b). The scapula is long and narrow in order to provide a long lever arm for the humeral rotators (*teres major, latissimus dorsi, supraspinatus, infraspinatus*). The humerus is short and squat, with large medial and lateral epicondyles. Pronation-supination is limited, and the olecranon is long. In moles, as in *Ornithorhynchus*, the hand is wide, providing a large propelling surface during swimming. As a result of a general adaptation for humeral rotation locomotion (in swimming and digging), the forelimb of *Ornithorhynchus* resembles that of moles (Hildebrand 1985b).

Pectoral Oscillation

In pectoral oscillation, the forelimbs provide lift (Fish 1996) during the entire locomotor cycle (upstroke and downstroke). Otariids (sea lions and fur seals) use this mode of locomotion (English 1976a; Feldkamp 1987). Their forelimbs are situated relatively far back on the body (due to the long neck) and are powerful, with short arm and forearm and a large hand (Howell 1930; English 1976b; fig. 18.2f). The hand is a flipper, with poorly defined digits and interdigital webbing. The humerus has a large deltoid tubercle, and the ulna a large olecranon. The hand is highly asymmetrical, and the thumb is longer and more robust than the other digits. A cartilaginous element attached to the distal phalanx lengthens each finger well beyond the small nail.

Otariid hindlimbs are not used in aquatic propulsion but serve a function in steering. Their femur is short and powerful. The foot is disproportionally long and symmetrical, with interdigital webbing, greatly enlarged first and fifth digits, and cartilage extensions to each distal phalanx (J. E. King 1983). Unlike phocids, otariids can put the soles of their feet on the ground, and the limbs can support their body while walking on land (Beentjes 1990). There are consistent differences between the terrestrial locomotor patterns among the two subfamilies of otariids (arctocephalines and otariines), but these are not correlated to morphological differences (Beentjes 1990).

Birds

Most birds are classified in the subclass Neornithes and differ from other amniotes, including early birds, such as *Ar-*

chaeopteryx. In Neornithes, the body is differentiated into three separate anatomical and functional subregions, each functioning as a unit called a "locomotor module" (Gatesy 2002; Gatesy and Dial 1996). The pectoral (wing) module operates in flight, in coordination with the caudal (tail) module (which helps to steer and brake the animal). The pelvic (hindlimb) module functions in terrestrial locomotion, with the main thrust provided by flexion of the knee by the hamstring muscles; the femur is held largely horizontally, to balance the center of gravity over the foot (Gatesy 1990). This is in contrast to the original (lizardlike) amniote design, where much of the hindlimb retraction force is produced by musculature running from the femur back to the base of the tail. As a result, birds are "mosaics of locomotor modules" (Gatesy and Dial 1996, 338), and flying, walking, and steering with the tail are mechanically independent.

Aquatic adaptations in birds therefore involve either the forelimbs or the hindlimbs, leaving the other pair of limbs relatively unmodified. In birds, the evolution of aquatic life has repeatedly occurred along both routes (figs. 18.4 and 18.5). Our hypothesis as to how these modes of aquatic life relate is presented in figure 18.6. Aquatic birds and their adaptations were discussed by Storer (1960), Feduccia (1980, 1996), Raikow (1985), and Raikow et al. (1988).

Pelvic Walking, Wading, and Flying

Shorebirds (Charadriiformes), wading birds (in the order Gruiformes), herons (Ardeidae), and storks (Ciconiiformes) are not committed swimmers even though they are often associated with water. Their wings are used to fly, their legs mostly to walk on shore or wade. The more specialized members of these groups have broadened toes, to distribute weight on soft substrates, and long legs and necks, to allow for feeding in deeper water.

Other birds fly over the water surface and feed from it without significant aquatic locomotion. The osprey *Pandion haliaeetus* grabs surface-swimming fishes in its claws when flying over the water. Some others such as gannets and boobies (Sulidae) plunge-dive: the northern gannet *Sula bassana* uses the height gained by flight to dive vertically onto prey, with the wings folded back to minimize resistance (Andrew C. Kitchener, pers. comm. 2003). Buoyancy pushes *Sula* to the surface, and the wings of gannets are not significantly modified (Pennycuick 1987).

Alternating Pelvic Surface Paddling

Pelvic paddling is a commonly used mode of aquatic locomotion, which allows the wings to retain their original function of flight in air. In ducks (Anseriformes) the hindlimbs

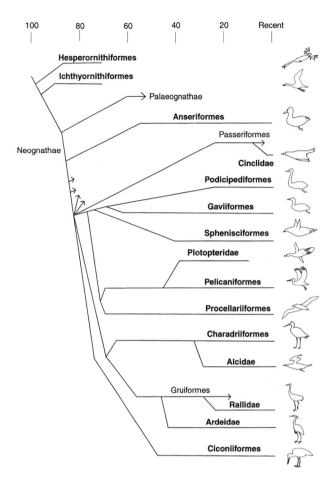

Figure 18-4 Phylogenetic tree showing the diversity of aquatic birds, with approximate relationships and divergence times in millions of years before present. Clades with largely or wholly aquatic members are shown in boldface (most terrestrial birds are not indicated). It is apparent that aquatic adaptations arose independently many times. (Compiled from Unwin 1993; Feduccia 1996; Benton 1997; and Padian and Chiappe 1998.)

provide thrust by drag-based locomotion with webbed feet (Baudinette and Gill 1985). The least modified swimming birds simply float on the surface and feed while floating and swimming—for instance, swans, geese, dabbling ducks (part of Anseriformes), some pelicans (Pelicaniformes), and gulls (part of Charadriiformes). The phalaropes (part of Charadriiformes) extend feeding range using their hindlimbs. They spin on one spot on the surface, using their limbs to create an upward suction of water to bring up prey from deeper water (Obst et al. 1996).

Other surface swimmers extend their feeding reach by actively diving using their hindlimbs for propulsion. These include the diving ducks and the cormorants. Raikow (1985) identifies some specific adaptations in diving birds. These include the short femur held laterally with limited protraction and retraction in the plane parallel to the water surface. Also, the tibiotarsus is long and extends parallel to the verte-

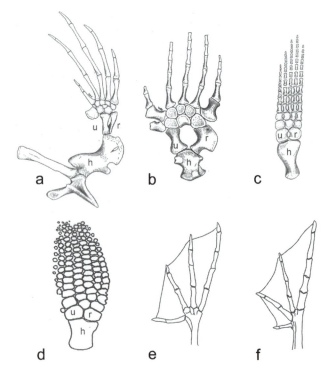

Figure 18-5 Aquatic adaptations of the limbs of reptiles and birds. Left forelimbs of the leatherback turtle *Dermochelys* (a; includes shoulder girdle), the mosasaur *Clidastes* (b), the plesiosaur *Elasmosaurus* (c; in life digits were probably less straight than in this reconstruction), and the ichthyosaur *Ichthyosaurus* (d) in dorsal view. Right feet of the modern cormorant *Phalacrocorax* (e) and the Cretaceous bird *Hesperornis* (f) in dorsal view. Humerus (h), radius (r), and ulna (u) are indicated. (a–c modified from Howell 1930; d from Motani 1999; e–f from Heilmann 1927.)

bral column, and is held close to the body, while extension at the knee is limited. The power stroke in diving birds involves mostly foot extension (between tibiotarsus and tarsometatarsus), and this joint is placed at the level of the tail.

Alternating Pelvic Submerged Paddling

Diving birds, which commonly paddle with their feet underwater, display anatomical modifications for this type of locomotion. The loons (Gaviiformes) and the unrelated and extinct Cretaceous Hesperornithiformes such as *Hesperornis* evolved these adaptations in parallel (Storer 1960; fig. 18.5e and f). In loons and *Hesperornis*, the hindlimbs are reoriented to the rear of the body. The tarsometatarsi are narrow mediolaterally during the recovery stroke when the foot is brought forward folded, in order to minimize resistance (Raikow 1985). The orientation of the legs in these birds makes them unable or nearly unable to walk on land, and instead they push themselves along on their bellies, similar to phocid seals.

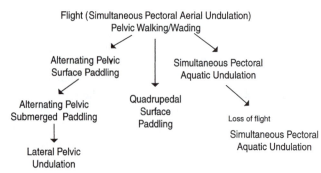

Figure 18-6 Evolutionary hypothesis for transitions between aquatic locomotor modes in birds.

Lateral Pelvic Undulation

Whereas loons are paddlers using drag-based means of swimming, grebes (Podicipediiformes) apparently use lift-based locomotion with the hindlimbs (lateral pelvic undulation). These birds dive with spread toes, which they move in a laterally undulating fashion rather than a backward paddle. The toes of grebes are asymmetric, and each toe cross section has the shape of a hydrofoil (Johansson and Lindhe Norberg 2000).

Quadrupedal Surface Paddling

Rarely do birds use both fore- and hindlimbs in aquatic locomotion. However, this occurs in *Tachyeres*, flightless steamer ducks, which combine simultaneous beats of the forelimbs with alternating beats of the hindlimbs in paddling locomotion on the surface (Livezey and Humphrey 1983).

Simultaneous Pectoral Aquatic Undulation

The hindlimb morphology of pelvic swimmers represents a compromise between the demands of land locomotion and those of swimming. Using the forelimbs for swimming frees the hindlimbs for walking on land. Patterns of movement of the wings are similar in air and in water, and there is no equivalent of the paddling stage that precedes other forms of lift-based locomotion. Some dippers (Cinclidae, Passeriformes) use their relatively short wings to fly underwater in streams, with an immediate transition between aerial and underwater flight, and they also walk underwater, gripping rocks with their claws, and surface-swim with their legs in a paddling motion (Goodge 1959; Raikow et al. 1988; S. J. Tyler and Ormerod 1994).

Locomotion in water does differ significantly from locomotion in air, and the design of the wing reflects these differences. For instance, in water the weight a bird has to support is far less than in air, but the fluid in which its wings work is denser and more viscous. The combination of aerial and aquatic flight reduces the efficiency of both. Diving petrels (in the order Procellariiformes) and some alcids (auks; in the order Charadriiformes), such as the murres (*Uria*) and puffins (*Fratercula*), fly in air and in water (Storer 1960; Pennycuick 1987). These alcids essentially combine an aerial flight apparatus of reduced wing area (the wings of these birds are short and are partly folded when swimming) with behavioral adaptations to underwater swimming (Raikow et al. 1988; Gaston and Jones 1998). The resulting high wing loadings during aerial flight in alcids mean increased wing beat frequency, flight speed, and power input, and in turn higher costs of locomotion (Pennycuick 1987). High wing loading also means that alcids cannot glide at low speeds. Larger alcids such as murres *Uria* may be unable to glide at all except in high winds, and their landing technique on ledges and cliffs uses steep climbs in a ballistic trajectory to reduce their airspeed to zero at the desired landing point (Pennycuick 1987; Gaston and Jones 1998). Underwater flight is also compromised. For these reasons, flying alcids are relatively short-ranged aerial foragers compared to similar seabirds. This may also limit their body size: the murres *Uria* and razorbills *Alca* are the largest alcids that can use their wings for flight in air and in water (Bannasch 1994).

However, loss of the requirement for aerial flight allows optimization for aquatic flight and, separately, the evolution of large size. The largest alcid was the recently extinct great auk *Pinguinus impennis*. It was unable to fly in air, as were certain other extinct auks such as the mancallines (Gaston and Jones 1998). *Pinguinus* was therefore presumably able to increase the optimization of its wing morphology for swimming and to evolve larger size. However, locomotion may not have been the only factor affecting body size, since larger size has thermoregulatory advantages in cold boreal waters. Larger size may also have enabled *Pinguinus* to dive deeper for food, judging from the correlation of size and diving depth reported in living forms by J. F. Platt and Nettleship (1985).

Flightlessness allowed penguins, also, to evolve large size (Simpson 1975; Pennycuick 1987; Feduccia 1996). Pennycuick (1987) calculated that a medium-sized penguin would need to beat its wings at the physically impossible hummingbird-like frequency of 50 Hz to fly in air. Flightlessness also enabled penguins to optimize their body form for underwater swimming, which is energetically efficient compared to surface locomotion (Kooyman 1989).

Penguins illustrate the differences between flight in water and in air, and the resulting modifications of their skeletons and musculature (B. D. Clark and Bemis 1979; Bannasch 1994; Raikow et al. 1988). The joints within each wing have a reduced range of movement, and the bones are shorter and stiffer. This needs less internal wing musculature (penguins do not need to fold their much shorter wings to move on land, or to make the rapid and variable control movements necessary for aerial flight). The wing bones are flattened (also seen in alcids; Gaston and Jones 1998) and covered in firm connective tissue, and the wing externally becomes a smooth flipper covered by many very short and stiff feathers. Such a smaller wing can still provide sufficient thrust because water is denser than air and thus less water has to be moved caudad to produce thrust. In addition, penguins are close to being weightless in water, and they therefore need to produce little or no upward lift force, compared to birds flying in air. Flying birds produce lift on both the upstroke and downstroke, and the same is probably true for penguins.

The wings of penguins are near the midbody, which compromises maneuverability because it is difficult to produce a turning moment, so penguins also steer with the head, tail, and feet, and the powered wings are also actively used to change direction (B. D. Clark and Bemis 1979; Hui 1985). The location of the feet near the posterior extremity of the streamlined body is helpful in steering and also in the extreme streamlining noted in penguins by Bannasch (1994). This leads to the characteristically upright stance on land compared to most other birds, in which the body is far more horizontal. This appears to be at the price of increased inefficiency of land locomotion (Baudinette and Gill 1985). *Pinguinus* also had an upright stance on land (Gaston and Jones 1998).

Plotopterids, members of an extinct family of pelicaniforms of the North Pacific, were heavily modified and flightless forms convergent on penguins in their appearance and lifestyle and include the largest diving birds known (length 1.8 m; S. L. Olson and Hasegawa 1979; Feduccia 1996).

Plotopterids and penguins represent a remarkable example of convergent evolution, although their respective ancestors show differences in detail in the actual pattern of underwater flight (Raikow et al. 1998). The flying alcids are the Northern Hemisphere analogues of the Southern Hemisphere fliers in air and water, the diving petrels. Likewise, the flightless alcid *Pinguinus* and the independently evolved plotopterids are the Northern analogues of the Southern Hemisphere penguins, themselves thought to have evolved from forms similar to diving petrels (Storer 1960; Simpson 1975; S. L. Olson and Hasegawa 1979; Feduccia 1996).

The water-repellent feathers and air sac system normal in birds can trap so much air that it is hard to remain underwater (M. A. Taylor 1993, 2000). Most diving birds (e.g., darters, anhingas, and cormorants) therefore have wettable feathers (Owre 1967). However, penguins have nonwettable feathers, presumably as a thermoregulatory measure, as well as fatty deposits for thermal insulation. Penguins counteract their positive buoyancy by increasing the density and solidity of their long bones of their limbs, swallowing rocks, and losing pneumatization (Meister 1962; M. A. Taylor 2000).

Reptiles

The nonmammalian, nonavian amniotes, informally termed "reptiles," comprise a heterogeneous assemblage that includes many aquatic groups (reviewed by Benton 1993, 1997; Callaway and Nicholls 1997; R. L. Carroll 1985, 1987, 1997a; Massare 1994, 1997; Mazin 2001; Mazin and de Buffrénil 2001; Seymour 1982; and Caldwell 2002).

Locomotion in living groups is relatively poorly known, and experimental studies are often based on young or juvenile individuals, rather than adults. Also, most aquatic reptiles are extinct, and it is difficult to be certain about their locomotor mode or modes. Therefore, we focus on selected major groups and on locomotor parallels with aquatic birds and mammals. Figure 18.7 presents a phylogeny of marine reptiles, figure 18.5 shows anatomical adaptations, and figure 18.8 shows proposed evolutionary transitions in locomotor modes.

It is likely that basal amniotes were facultatively aquatic (whether or not any lineages were actually always aquatic as Romer [1974] suggested). The original reptiles—that is, the basal amniotes—had a long body and tail and four sprawling legs, similar to lizards. This is the primitive body plan of all tetrapods, which were originally aquatic (Clack 2002b). On land and in water, fast locomotion in lizards involves lateral undulations of the vertebral column, which may be combined on land with the forward and backward movements of sprawling limbs (Guibe 1970). Reptile physiology also facilitates aquatic life (Seymour 1982). Thus many lizards, for example, are facultatively aquatic (Halliday and Adler 1986). Among living lizards, the Galápagos marine iguana *Amblyrhynchus* has a slightly flattened tail and some webbing at the base of the toes (Bartholomew 1987; W. R. Dawson et al. 1977; Tracy and Christian 1985).

Specialist aquatic forms are almost completely absent from two major groups of basal amniotes, the therapsids (T. S. Kemp 1982; G. M. King 1991) and the dinosaurs, although their respective descendants—mammals and birds— contain many aquatic forms. The reasons probably lie in their modifications of the original amniote body plan. Reduction of the tail musculature weakened the inherent swimming ability of at least the later therapsids. The dinosaurs

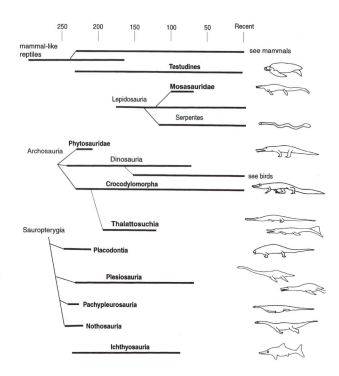

Figure 18-7 Phylogenetic tree showing the diversity of aquatic amniotes (exclusive of mammals and birds) with approximate relationships and divergence times in millions of years before present. Clades with largely or wholly aquatic members are shown in boldface. It is apparent that aquatic adaptations arose independently many times. Most terrestrial amniotes are not indicated. (Data mainly from R. L. Carroll 1987 and Benton 1993, 1997.)

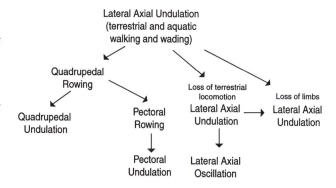

Figure 18-8 Evolutionary hypothesis for transitions between aquatic locomotor modes in nonavian, nonmammalian amniotes.

had an erect stance with the femur more or less vertical at rest (Gatesy 1990). However, as with most amniotes, all could probably swim briefly if they had to. Certain fossil trackways appear to have been left by swimming dinosaurs, floating with body and tail straight out on the surface and long hindlegs kicking out powerfully (Coombs 1980; Whyte and Romano 2002).

Wading, Flying, and Bottom Walking

The theropod dinosaur *Baryonyx walkeri* apparently fed on freshwater fish, not by swimming but by wading and gaffing with the huge claw on each hand, much as grizzly bears, *Ursus arctos*, do today, as well as snapping with its jaws (Charig and Milner 1997).

Some pterosaurs (a group of flying archosaurs) were wading filter feeders, while others fed on the wing, snapping up food from the sea surface. Apparently, as a group, pterosaurs lacked plunge divers, underwater fliers, or surface floaters. This may be because the forelimbs and hindlimbs were tied together by the flight membrane as a single locomotor module, in contrast to birds, where the forelimbs and hindlimbs can specialize separately (Wellnhofer 1991; David M. Unwin, pers. comm. 2003).

Some reptiles were bottom walkers: waders with bodies dense enough to remain on the sea bed even when fully submerged, for example when seeking food. Some freshwater chelonians walk underwater rather than swim as their main means of locomotion (Zug 1971; Renous et al. 2000). The extinct placodonts could apparently walk on, and grip, the sea floor when foraging for hard-shelled invertebrates. They had rigid, barrel-like bodies (probably for large lungs) and dense, heavy skeletons to provide bony ballast. Unarmored placodonts probably swam with their long tails, and perhaps also with their limbs, but the superficially turtlelike armored forms seem to have swum by paddling with the hindlimbs and possibly also forelimbs (Pinna and Nosotti 1989; Mazin and Pinna 1993; de Buffrénil and Mazin 1992; Storrs 1993; Renesto and Tintori 1995; Rieppel 1995; M. A. Taylor 2000, 2002). In contrast, the Galápagos marine iguana is buoyant but clings to rocks with clawed feet (Hobson 1965).

Lateral Axial Undulation and Oscillation

The primitive amniote means of swimming, as seen in modern crocodilians and lizards, uses lateral undulation of the tail and to some extent the body to generate thrust (Manter 1940; Hobson 1965; Frey 1982a; Seymour 1982; Fish 1984b; Bartholomew 1987; J. Davenport and Sayer 1989; J. Davenport et al. 1990). The limbs are normally held immobile close to the body to minimize drag. However, this applies primarily to faster straight-line swimming. In crocodilians, the limbs and the partly webbed hands and feet are actively used in water for directional control, braking, and minor corrective movements when the animal is stationary. Limbs are used independently of the tail with a drag-based paddling action in very slow swimming, at least in young crocodilians. Limbs are also used with the tail in swimming at intermediate speeds, while limbs also play roles in locomotion during

escaping and in attacking and processing prey (J. Davenport and Sayer 1989; J. Davenport et al. 1990).

Such limbs do not need to be highly adapted for underwater movement and are in any case needed for use on land. Nevertheless, modern crocodilian limbs are somewhat adapted for aquatic locomotion, being short and sprawling—except when moving quickly on land—with broad webbed hands and feet, and are therefore derived compared to those of the early crocodilians, which were terrestrial, like other early members of the Crocodylomorpha and basal archosaurs generally (Parrish 1987; Gatesy 1991).

The modern crocodilians are essentially amphibious. Another group of archosaurs, the phytosaurs, independently evolved a similar lifestyle and body form (R. L. Carroll 1987). Those crocodilians and phytosaurs form one end of a spectrum of increasingly aquatic axial swimmers.

Thalattosuchians (Mesozoic marine crocodilians) exemplify the next stages of aquatic adaptation (Hua 1994; Hua and Buffetaut 1997). The coastal and estuarine teleosaurids had forelimbs reduced to flattened mittens, possibly held close to the body in axillary depressions to reduce drag. The hindlimbs were larger and bore webbed feet. The open-sea metriorhynchids were more streamlined, with the development of a caudal fin in advanced forms. They retained hindlimbs, presumably for breeding on land.

Mosasaurs and ichthyosaurs show further development of aquatic adaptation. Mosasaurs were large Cretaceous marine reptiles, possibly varanoid lizards, and swam by lateral undulation (Bell 1997; R. L. Carroll 1997a, 1997b; Lingham-Soliar 1991; Massare 1994; Mulder 2001). The mosasaurian forelimb (fig. 18.5b), whose function is discussed below with that of ichthyosaurs, was transformed by shortening of the long bones, reduction of flexibility within the limb, loss of claws on the terminal phalanges, often—but not always—hyperdactyly (additional phalanges), and embedding of the digits in connective tissue to form an externally smooth fin or flipper (D. A. Russell 1967; Caldwell 1996; R. L. Carroll 1997b).

Ichthyosaurs were highly specialized for aquatic locomotion. Earlier ichthyosaurs somewhat resembled mosasaurs or even eels, but later ichthyosaurs evolved oscillatory swimming, with thrust produced by a specialized caudal fin, convergently with cetaceans, sharks, and tunas (Massare and Callaway 1990; McGowan 1991, 1992; Massare 1994; Lingham-Soliar and Reif 1998; Sander 2000; M. A. Taylor 2000, 2001; Buchholtz 2001; Motani et al. 1996). The forelimb is remarkably modified, among the most highly derived of any amniote; major changes obscure the homologies of distal elements (Johnson 1979; McGowan 1972, 1991; Caldwell 1997a, 1997c, 2002; R. L. Carroll 1997a, 1997b; Motani 1999; Sander 2000; Galis et al. 2001). The hindlimb is broadly similar but smaller than the forelimb. Specific fore- and hindlimb adaptations of ichthyosaurs (fig. 18.5d) include greatly shortened humerus, radius, ulna, femur, tibia, and fibula, and loss of morphological differentiation between carpals, tarsals, metacarptals, metatarsals, and phalanges (which all become rounded or polygonal discs). The ability to move on land is lost, and elbow, knee, wrist, and ankle joints are no longer distinct. Distally, the separation of digits is lost, often yielding an interlocking mosaic of fused digital elements. There may be hyperphalangy (the addition of phalanges within each digit), a reduced number of digits in some ichthyosaurs and an apparent increase in number of digits (polydactyly) in others, and loss of claws. The tail of ichthyosaurs evolved a finlike hydrofoil outline with connective tissue, as shown by exceptionally preserved fossils (Martill 1993; Bardet and Fernández 2000).

Ichthyosaurs, and perhaps mosasaurs, bred by giving live birth in the water (Böttcher 1990; Deeming et al. 1993; Dobie et al. 1986). Without any need to function on land, their limbs became specialized for use as control surfaces, highly functional hydrofoil fins pivoted at the base and controlling the animal's movements, sometimes (in ichthyosaurs) in conjunction with changes in the direction of thrust produced by the tail itself (McGowan 1991, 1992). Fin location, size, and shape were variable, no doubt reflecting different compromises between stability, energy efficiency, and maneuverability, as in cetaceans and sharks (Fish 2002). It has been suggested that certain ichthyosaurs and mosasaurs swam by using large forefins in a lift-based mode rather than by lateral undulation of the body and tail. This seems unlikely, as fins would probably only have been used at slow speeds, at which drag-based paddling is in any case more effective (Lingham-Soliar 1992, 1999; Riess 1986; Massare 1994; Nicholls and Godfrey 1994; Buchholtz 2001). Large foreflippers could very well be useful for paddling in slow swimming, and in tight turning, which is necessary in some habitats and ways of hunting prey (Fish and Battle 1995; Fish 2002).

The limb bones of ichthyosaurs show a relatively lightweight structure interpreted as a buoyancy control strategy (de Buffrénil and Mazin 1990; Massare 1994; M. A. Taylor 2000).

The evolution of anguilliform ("eel-like") swimming by lateral undulation of long, slender bodies, associated with reduction of limbs, happened in several lineages of marine squamates such as the dolichosaurs and other minor groups (not all shown in fig. 18.7; Caldwell 2000; Caldwell et al. 1995; Caldwell and Cooper 1999; R. L. Carroll 1985, 1997a; M. S. Y. Lee and Caldwell 2000). Several lineages of squamates (snakes, terrestrial legless lizards, and amphisbaenians) went on to lose limbs and limb girdles (Gans 1962; J. L. Edwards 1985). Limb reduction and loss are common adaptations for burrowing in earth and litter (Gans 1975) but are also useful adaptations to life in water, especially for small predators seeking prey in confined environments such as coral reef crevices. Indeed, some "terrestrial" snakes swim well, while several groups of sea snakes have adapted to marine life, often with a flattened paddlelike tail or keel (Dunson 1975; Gans 1975; Heatwole et al. 1978).

Furthermore, certain early snakes, such as *Pachyrhachis*, have pachyostotic ribs for ballast in water (M. A. Taylor 2000) and compressed tails for swimming; their fossils are found in marine deposits (Caldwell and Lee 1997; Scanlon et al. 1999). *Pachyrhachis* retained hindlimbs with a well-developed pelvic girdle and femora, a short tibia and fibula, and diminutive tarsals. More distal elements were not present (or are not preserved).

It has therefore been suggested that snakes are primitively semiaquatic or aquatic, and that they may furthermore be descended from the same lineage of marine squamates as the mosasaurs and dolichosaurs, but there is no consensus (e.g., Caldwell 1999; Scanlon et al. 1999; M. S. Y. Lee and Caldwell 2000; Tchernov et al. 2000).

Rowing

Rowing is seen in reptiles with relatively immobile vertebral columns.

Freshwater chelonians row quadrupedally with webbed feet in drag-based locomotion. The left forelimb and right hindlimb act together, and are followed by the other diagonal pair of limbs (J. Davenport et al. 1984; Zug 1971). Freshwater chelonians show compromises between land and water locomotion: they are markedly poorer swimmers than marine forms, but they are much better walkers and even climbers on land.

The extinct sauropterygians appear to show the evolution (probably with some parallelism) of various stages of locomotor mode from drag-based rowing in pachypleurosaurs, through an intermediate nothosaurian grade, to the plesiosaurian mode of primarily lift-based underwater flight (Sues 1987; Storrs 1993; Lin and Rieppel 1998; Rieppel 1994, 1997). Most pachypleurosaurs apparently combined tail undulation and rowing with their forelimbs, which were robust but otherwise relatively unmodified, except for some reduction of digits in later forms. They used hindlimbs for steering and perhaps also for land locomotion. The remaining nonplesiosaurian sauropterygians, the nothosaurs, tend to show stiffer bodies, weaker and shorter tails, and forelimbs in which the humerus is stout and distally flattened and the radius and ulna are wide and flattened. The limb is made even wider by a large space between each radius and ulna, and some stiffening and hyperphalangy. This suggests a sculling type of locomotion based on the forelimb, intermediate between rowing and flying, generating thrust through drag and lift.

Quadrupedal Undulation

The plesiosaurs were sauropterygians specialized for lift-based thrust, which is more efficient than drag-based rowing in terms of cost of locomotion (Fish 1996). Uniquely for tetrapods, they swam with four limbs of comparable size and essentially the same structure.

The plesiosaurian limb is as exceptionally derived as the ichthyosaurian limb (Robinson 1975; Caldwell 1997b, 1997c, 2002; Lin and Rieppel 1998). Specializations (fig. 18.5c) include a massive humerus or femur; reduction of lower limb and wrist/ankle elements to bony discs or polygons (not complete in early plesiosaurs); loss of functional differentiation within the limb, especially elimination of the elbow/knee and wrist/ankle; extreme elongation of the distal portion through hyperphalangy (an increase in the number of phalanges in a digit); possible addition of cartilaginous digits (Caldwell 1997b); no claws on terminal phalanges; and extremities covered by connective tissue to give the external appearance of a flipperlike limb, projecting laterally from the body.

The limb of plesiosaurs was a single tapering unit, which was fairly stiff, most stiff at the base, but gradually more flexible toward the tip, and from front edge to rear edge, forming a hydrofoil optimized for lift-based underwater flight. This basic structure and by implication its development (Caldwell 1997b, 2002) were highly conserved, although different plesiosaur genera do show slight variation in detail and proportions, presumably in varying tradeoffs between acceleration and maneuverability versus cruising efficiency (O'Keefe 2001).

Plesiosaurs lacked dedicated control surfaces, though could have used their heads and necks, as in otariids (Fish et al. 2003). The presence of two pairs of underwater "wings," one pair in front of and one pair behind the center of mass, suggests that, at least when plesiosaurs were swimming at moderate speeds, they steered partly by vectoring the thrust from fore- and hindlimbs differentially. The common structure of the four limbs is strong evidence that all worked primarily as hydrofoils (Robinson 1977; S. J. Godfrey 1984). This is contrary to suggestions that the hindlimb was primarily used in a rowing action and for steering and braking (Lingham-Soliar 2000)—though this was doubtless true for the hindlimbs, and also the forelimbs, during slow swimming, maneuvering, and backing up (Riess and Frey 1991).

Plesiosaurs beat their limbs symmetrically, though the two pairs would need to operate with a phase difference that optimized the interaction of the hindlimbs with the vortices coming off the forelimbs (M. A. Taylor cited in Halstead

1989; Fish 1999 and pers. comm. 2003). Beyond this, it has proved difficult to say more about how they swam, as they are very derived without close living relatives (Robinson 1975; S. J. Godfrey 1984; Alexander 1989; Halstead 1989; Nicholls and Russell 1991; Riess and Frey 1991; Lingham-Soliar 2000). Probably plesiosaurs swam with a sea lion–like action, the power stroke being backward, and first downward then upward, with the forward recovery stroke being essentially horizontal (S. J. Godfrey 1984). This action is somewhat intermediate between sea turtle– or penguinlike flying and freshwater turtle–like rowing, although the limb structure suggests that thrust production was primarily lift based.

Pectoral Undulation

Earlier marine chelonians were broadly similar to modern freshwater forms in locomotor habits. The most recent radiation, the Chelonioidea, includes the extinct Cretaceous Protostegidae and the living Dermochelyidae and Cheloniidae. All three families evolved underwater flight through lift-based pectoral undulation by the forelimbs, apparently partly in parallel (Nicholls 1997; Hirayama 1998). Adult sea turtles mainly swim by strong simultaneous and symmetrical beats of the forelimbs in underwater flight, while the hindlimbs retain a more primitive form to help in terrestrial locomotion and nesting (Zug 1971; Hendrickson 1980; J. Davenport et al. 1984; Renous et al. 2000). Each forelimb is externally a hypertrophied hydrofoil in external shape (fig. 18.5a). It operates in a largely vertical figure-eight motion, often producing lift on both the up and the down strokes, with the two forelimbs beating in phase.

The only living dermochelyid, the leatherback turtle *Dermochelys coriacea*, is highly optimized for pelagic cruising with long forelimbs, and the hindlimbs make no contribution to thrust. Even on land, the leatherback mainly uses simultaneous symmetrical movements of the forelimbs, and the hindlimbs provide only some assistance and dig the nest (J. Davenport and Pearson 1994; Hendrickson 1980; J. Davenport 1987; Renous et al. 1989; Renous and Bels 1991, 1993). In contrast, cheloniids are neritic, living in shallow water and often feeding benthically. It is clear that they can use limbs in various patterns depending on species, need, habitat, lifestyle, and age (Hendrickson 1980; J. Davenport and Pearson 1994; Nicholls 1997). On the whole, synchronous beats of the forelimbs are used for faster swimming, while the hindlimbs are used for steering and braking, and also for slow swimming, maneuvering, and backing up, for instance during benthic feeding (W. F. Walker 1971; J. Davenport et al. 1984; J. Davenport and Pearson 1994; J. Davenport and Clough 1986).

Discussion

M. A. Taylor (2000, 2002) summarized the major questions relating to the origin of aquatic adaptations in amniotes, noting that the interplay between phylogeny, ecology, habitat, and function shapes the transitional forms between aquatic amniotes and their terrestrial forebears. In this chapter we are concerned only with limb morphology, which is influenced mostly by its role in locomotion. Taylor identified several stages of increasing adaptations to the aquatic environment that are likely to occur or have been documented in most aquatic amniote lineages. These stages initially pass from an amphibious animal showing no aquatic adaptations to an amphibious animal counteracting its buoyancy with bone used as ballast. Two different tracks can be followed after that, one for slow-moving taxa (herbivores or benthic feeders) and another for fast-moving predators. The latter category contains limb-based locomotor patterns as well as tail-based locomotor patterns.

Our study here refines this system by hypothesizing specific swimming modes and their transitions. Our interest is in the similarities between the three amniote groups (mammals, birds, and "reptiles"), and we conclude that, although the ancestral locomotor patterns of these three amniote groups differ greatly, the most derived locomotor patterns and their associated morphologies are very similar, providing strong evidence that selection for locomotor efficiency has indeed dominated the limb morphologies of aquatic amniotes.

Acknowledgments

We thank M. J. Benton, A. R. I. Cruickshank, F. Fish, B. Garner, T. S. Kemp, A. Kitchener, L. S. Stevens, G. Swinney, D. M. Unwin, and M. A. Whyte for discussion and assistance, and the staff of the National Museums of Scotland Library for assistance.

Chapter 19 Sesamoids and Ossicles in the Appendicular Skeleton

Matthew K. Vickaryous and Wendy M. Olson

This bone can never be burned or corrupted in all eternity for its ground substance is of celestial origin and watered with heavenly dew.   —The Rabbi Ushaia, AD 210

THE SKELETAL ELEMENT described above—the Bone of Lus (Luz)—is presumed to be one of the ossa "sesamoidea" hallucis, positioned along the plantar surface of the first metatarsophalangeal joint (Inge and Ferguson 1933; Helal 1981; fig. 19.1). Although unconventional, the statement by Rabbi Ushaia illustrates the historica and confused fascination with various poorly understood skeletal masses. Among ancient philosophers, the Bone of Lus was of cultural importance and has even been described as the receptacle of the soul (Inge and Ferguson 1933; Helal 1981). In modern times the significance of this element has shifted to a more clinical nature—these ossicles are frequently the sites of inflammation, physical damage, and disease.

The development and evolutionary significance of diminutive and histologically complex elements such as sesamoids, lunulae (meniscal elements), mineralized tendons and ligaments, and others is typically downplayed during considerations of the appendicular skeleton. Yet whereas biologists and anatomists often overlook these skeletal outliers, they remain important topics for pathological and biomechanical consideration. As a group, sesamoids and their ilk represent something of an anatomical enigma, with an enormous degree of variability in size, shape, and position both within and between taxa. Consequently, most skeletal descriptions relegate these elements to passages that summarize "heterotopic" or "accessory" bones and cartilages, predisposing them to continued marginalization. By way of their very proximity to (and indeed often entrenchment within) the musculotendinous system, these elements may be strongly correlated with function. However, while mechanobiological stimuli clearly contribute to the onset, form, and histological composition of these skeletal masses, there exist several examples of development even in the absence of extrinsic factors. We admit that the subject is complex and confused. As we have approached it, our task is to analyze the evolution and development of these unwonted masses as regular (i.e., consistent, nonpathological) components of the skeletal system.

A review of the literature suggests that the term "sesamoid" is often used as a wastebasket for all manner of small and unusual skeletal elements. Beyond the superficial level, however, many of these components are quite dissimilar in terms of their development and skeletally mature state. Bones, cartilage nodules, and other connective tissue condensations that do not consistently contribute (either inter- or intrataxonomically) to the head, axial, or appendicular skeletons, are often referred to as "accessory" (supplementary yet nonessential), "ectopic" (out of place), "extraskeletal" (outside or beyond the skeleton), or "heterotopic" (different place). However, these terms are neither synonymous nor accurate. Furthermore, they often carry the implication of pathology or abnormality. We recommend that such descriptors not be used without additional clarification.

In order to draw attention to the distinctiveness of many of these forms we have restricted our use of the term "sesamoid" to a relatively narrow application—skeletal elements that develop within a continuous band of regular dense connective tissue (tendon or ligament) adjacent to an articula-

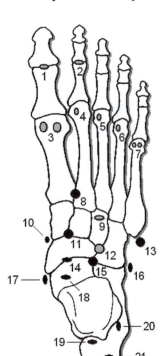

Figure 19-1 Schematic illustration of a (right) human foot skeleton with the putative location of various sesamoids and inconstant ossicles, many of which may be bi- or even tripartite (adapted from Burman and Lapidus 1931 and W. G. Hamilton 1985). The foot skeleton is a frequent site of ossicle development, although in most cases the etiology remains unclear. Ossicles positioned along the superior surface of the foot, black; ossicles situated along the plantar surface, white. Ossicles: (1) and (2) os interphalangeus; (3) ossa "sesamoidea" hallucis (medial and lateral); (4) metatarsophalangeal sesamoids of the second, (5) third, (6) fourth, and (7) fifth digits, plantar surface; (8) os intermetatarseum; (9) os unci; (10) os paracuneiforme; (11) os intercuneiforme; (12) cubois secundaris; (13) os Vesalianum (W. G. Hamilton 1985; alternatively, the os Vesalianum has been viewed as a fracture of the tuberosity of the fifth metatarsal [Burman and Lapidus 1931] and not a true element); (14) os talo-naviculare dorsale (= os supranaviculare; astragalo-scaphoid ossicle); (15) os calcaneus secundarius; (16) os peroneum; (17) os tibiale externum (= accessory navicular/scaphoid); (18) os supratalare; (19) os trigonum; (20) os sustentaculi (os proprium sustentaculi); (21) os supracalcaneum.

tion or joint. All overlooked skeletal elements of the appendicular system, including sesamoids, are herein considered to be appendicular ossicles, and fall into two broad categories: those that initially develop within a tendon or ligament (intratendinous elements) and those that develop adjacent to joints and articulations but not initially within a tendon or ligament (periarticular elements). The former category includes sesamoids as well as mineralized tendons and ligaments; the latter includes lunulae (meniscal ossicles) and a variety of curious elements most commonly associated with the autopodium. Two final caveats: although the term "ossicle" technically refers to a small bone, our adoption of this descriptor is based on its ubiquity within the literature and has nothing to do with histology: appendicular ossicles may consist of hyaline cartilage, fibrocartilage, bone, and/or any combination of transitional/intermediate skeletal tis-

sue(s). Furthermore our categorization of appendicular ossicles is not intended to infer any evolutionary relationships, but rather is merely presented as a convenient means of organizing this rather broad assemblage of unusual elements.

Although most appendicular ossicles receive little more than anecdotal reference in the literature, the vast majority of tetrapod lineages develop one or more such elements. Perhaps expectedly these elements have an imperfect fossil record, although a few remarkable examples predating the Jurassic period (200+ mya) are known. A comprehensive review of these intriguing skeletal elements is somewhat beyond the scope of a single chapter. In particular, we only occasionally draw on the extensive field of genetic and molecular determinants of bone and cartilage formation. To supplement our review of the existing data set, we offer new observations on development of both intratendinous and periarticular elements using the unusual pipid anuran *Hymenochirus boettgeri* as a case study of histogenesis, paralysis, and regeneration.

Intratendinous Elements

Sesamoids

We begin our discussion with the most familiar group of appendicular ossicles, the sesamoids. The stereotypical sesamoid is a relatively small concentration of skeletal tissue with an innocuous, generally ovoid (seedlike) morphology, shrouded within a band of regular dense connective tissue (tendon or ligament), and located proximate to a bony prominence over which the connective tissue wraps. Although sesamoids are often categorized on the basis of their histological composition (e.g., osseous orthosesamoids or cartilaginous hemisesamoids; see also Bizarro 1921; K. Pearson and Davin 1921a, 1921b), in most cases this characterization is somewhat arbitrary. Whereas skeletal preparations reveal a nucleus of resilient material, serial histology often demonstrates a suite of insensibly grading tissues including bone, osteoid, fibrocartilage, calcified cartilage, and hyaline cartilage (see below).

Vexingly, many sesamoids (and indeed other forms of intratendinous and periarticular elements) appear to demonstrate an inconstant distribution, both between and within taxa. One example of such variability is the os peroneum of the m. peroneus longus, a sesamoid that is unknown in pentadactylous nonprimate mammals. In pongid apes the os peroneum is rare (reportedly present in 2 out of 11 specimens), and in *Homo sapiens* this bony sesamoid is infrequently found laterally adjacent to the cuboid bone of the foot (14–20% of the population; Le Minor 1987; Sarin et al. 1999; fig. 19.1). When present in humans, this sesamoid may be bi- or

Table 19-1 Representative sesamoids found in tetrapods

Sesamoid	Region of body	Musculotendinous association	Taxa	Reference
fibrocartilago (os) humerocapsularis	shoulder	M. deltoideus major, pars cranalis	some avians	Baumel and Raikow 1993
patella ulnaris (=os sesamoideum m. scapulotriceps)	elbow	M. scapulotriceps	some pipid anurans (*Hymenochirus* spp., *Pipa pipa*), avians (e.g., hummingbirds, the swift *Micropus apus*, spheisciforms) and mammals and most nonophidian squamates (excluding teiioids and *Xantusia* spp.)	K. Pearson and Davin 1921b; Barnett and Lewis 1958; Hudson et al. 1969; Sokol 1962; Schreiweis 1982; Vanden Berge and Storer 1995; Baumel and Raikow 1993; Baumel and Witmer 1993; Trueb et al. 2000; W. M. Olson 2000; Maisano 2001, 2002a, 2002b
prepollex (= os radiale externum)	carpus	M. abductor pollicus longus	most nonungulate placentals	Le Minor 1994
os prominens	carpometacarpus	propatagial ligament	some falconiform and strigid avians	Baumel and Raikow 1993
palmar sesamoid	metacarpus / metatarsus	palmar aponeurosis	some placental mammals (e.g. *Chaetophracus*) and most nonophidian squamates (excluding gekkotans)	Romer 1956; Gabe et al. 1967; Maisano 2002a, 2002b
(tibial) patella	knee	Mm. Quadriceps / M. femorotibiales, M. iliotibiales	most nonophidian squamates, birds and placental mammals, monotremes, and peramelid marsupials	Schufeldt 1882; Gabe et al. 1967; Baumel and Raikow 1993; Maisano 2001, 2002a, 2002b; Reese et al. 2001
lateral and medial fabellae	knee	M. gastrocnemius, pars externus and pars internus	most nonungulate placentals	K. Pearson and Davin 1921b
cyamella	knee	M. popliteus	some nonungulate placentals	K. Pearson and Davin 1921b
os peroneum	tarsus	M. peroneus longus	cercopithecid primates	Le Minor 1987
ossa sesamoidea metacarpophalangealia volaris / dorsalis, ossa metatarsophalangealia plantaris / dorsalis	interphalangeal, metacarpophalangeal, metatarsophalangeal	M. extensor / flexor digiti	some nonophidian squamates, many mammals (e.g., chiropterans, rodents, tubulidentates, insectivores, carnivores, and at least one fossil cetacean), a †fossil turtle, †armoured dinosaur, some †pterosaurs	Kluge 1962; Gabe et al. 1967; Maryanska 1977; Gaffney 1990; S. C. Bennett 2001a; Gingerich et al. 2001; Maisano 2002b

multipartite (30% of all examples; Burman and Lapidus 1931) and may have a unilateral distribution (40% of cases). In contrast, among cercopithecid (old world) monkeys and hylobatid apes (gibbons) the os peroneum is virtually ubiquitous and always bilateral. Based on such observations sesamoids are often categorized as either constant (omnipresent) or inconstant (fluctuating; Burman and Lapidus 1931; W. G. Hamilton 1985).

Diversity and Distribution

The diversity of sesamoids, in terms of topology, morphology, and taxonomy, is immense, and an exhaustive listing is well beyond the scope of this chapter; a number of represen-

tative elements are presented in table 19.1. The oldest known examples date back to the Permian (295–250 mya) and are found articulated along the palmar (volar) surface of the manus of early lizardlike reptiles called †captorhinids († = extinct taxon; Holmes 1977). These sesamoids are positioned adjacent to the metacarpophalangeal and interphalangeal joints, and likely were associated with the digital flexor tendons (see fig. 19.2a). Whereas modern turtles are typically characterized as lacking appendicular ossicles (Haines 1942c; W. F. Walker 1973), with the possible exception of the patelloid (Haines 1969), the earliest known representative, †*Proganochelys quenstedi* (Late Triassic, 230–200 mya), demonstrates unpaired sesamoids associated with the dor-

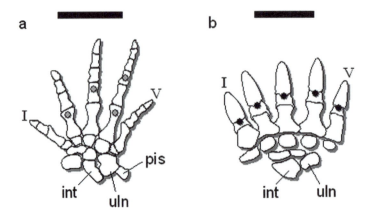

Figure 19-2 Schematic illustrations in dorsal view depicting the distribution of the earliest known appendicular sesamoids (circles). Sesamoids positioned along superior surface, black; sesamoids situated along palmar surface, white. (a) Manus of †*Captorhinus* sp., a Permian (295–250 mya) reptile (modified from Holmes 1977). Scale bar = 1 cm. (b) Manus of †*Proganochelys quenstedi*, a Late Triassic (230–200 mya) turtle (modified from Gaffney 1990). Abbreviations: I, digit I; V, digit V; int, intermedium; pis, pisiform; uln, ulnare. Scale bar = 2 cm.

sum of both the manus and pes, presumably entrenched within the digital extensor tendon (Gaffney 1990; fig. 19.2b). Unpaired sesamoids of the digital extensor tendons have also been identified in the manus of various †pterosaurs (Triassic-Cretaceous, 200–65 mya; Unwin 1988b; S. C. Bennett 2001a) and the armored dinosaur †*Saichania chulsanensis* (Late Cretaceous, 83–73 mya; Maryanska 1977). Comparable unpaired sesamoids in digital flexor/extensor tendons are known for lizards (Kluge 1962; Maisano 2002b), birds (Vanden Berge and Storer 1995), and mammals (Gabe et al. 1967; Romankowowa 1961), including the fossil (47 mya) cetacean †*Rodhocetus balochistanesis* (Gingerich et al. 2001; see table 19.1). In addition, some anurans (e.g., *Rana esculenta*, *Pipa pipa*) have a carpal sesamoid embedded within the tendon of the m. extensor carpi radialis (Gaupp 1896; Trueb et al. 2000).

Although the brachium/antebrachium regions of the forelimb are less commonly associated with sesamoids, a prominent ulnar patella (patella ulnaris) is known from numerous taxa, including some pipid anurans (fig. 19.3), avians, mammals, and most squamates (see table 19.1; fig. 19.4). In sphenisciforms (penguins) there are two ulnar patellae, one for each of the m. scapulotriceps and m. humerotriceps (Barnett and Lewis 1958; Schreiweis 1982). Barnett and Lewis (1958) observed a correspondence between the absence of the olecranon and presence of the ulnar patella in various avian species. Similarly, they noted in scandentians (tree shrews) and the chiropteran *Rousettus* that the presence of an ulnar patella appeared to replace the lateral epicondyle.

The oldest known sesamoid (or possibly lunula) from the knee region of the hindlimb is a small skeletal center observed between the femur and tibia of the bizarre reptilian †*Macrocnemius bassanii*, from the Middle Triassic (240–230

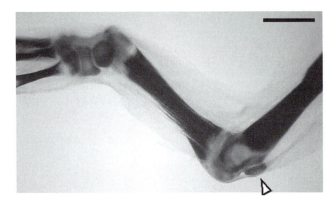

Figure 19-3 *Hymenochirus boettgeri* forearm whole mount double-stained for cartilage and bone (humerus to right of image), detailing the position of the ulnar patella (white arrowhead). Metamorphosed subadult, Nieuwkoop and Faber 1967 stage 66+. Scale bar = 1 mm.

mya; Rieppel 1989). In addition to the patella (detailed below), there are a variety of other sesamoids found adjacent to the knee joint of many tetrapods, including the lateral fabella (the anuran *Hymenochirus* spp., some lizards and mammals; fig. 19.4), medial fabella (mammals), and cyamella (or popliteal sesamoid, some birds and mammals; K. Pearson and Davin 1921a, 1921b; Vanden Berge and Storer 1995; W. M. Olson 2000; see table 19.1). Although both fabellae often co-occur, the lateral fabella is more likely to develop alone than the medial fabella (K. Pearson and Davin 1921b). In many nonplacental mammals (e.g., marsupials and †multituberculates), the proximal end of the fibula articulates with a sesamoid called the parafibula that ligamentously connects with the lateral condyle of the femur and provides the origin for the m. gastrocnemius lateralis (Barnett and Lewis 1958; O. J. Lewis 1989; Kielan-Jaworoska and Gambayan 1994).

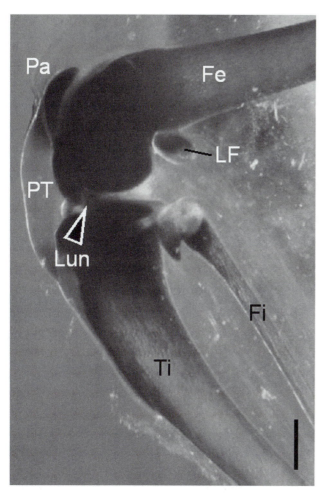

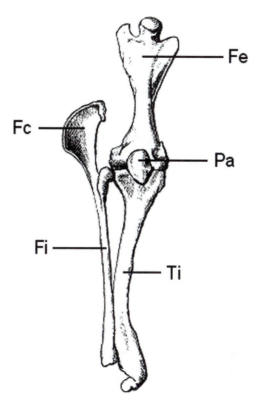

Figure 19-4 *Dipodomys* sp. (kangaroo rat) hindlimb whole mount double-stained for cartilage and bone, detailing the position of the patella (Pa) and other neighboring ossicles adjacent to the knee joint, including the anterolateral lunula (black arrowhead, Lun) and lateral fabella (LF). Other abbreviations: Fe, femur; Fi, fibula; PT, patellar tendon; Ti, tibia. Scale bar = 1 mm.

Figure 19-5 Schematic illustration of an *Ornithorhynchus anatinus* (platypus) hindlimb skeleton (right; adapted from Flower 1876). The massive fibular crest (Fc) reportedly represents a traction epiphysis and is homologous with the parafibula (K. Pearson and Davin 1921b; O. J. Lewis 1989; see text for details). Other abbreviations: Fe, femur; Fi, fibula; Pa, patella; Ti, tibia.

K. Pearson and Davin (1921b) demonstrated that the pronounced proximal fibula crest of the monotreme *Ornithorhychus anatinus* develops from an independent center of ossification, suggesting a consolidation of the parafibula with the fibula (see fig. 19.5; Barnett and Lewis 1958).

Among anurans, including various pipids, ranids, and sooglossids, sesamoids are most commonly associated with the tarsus (Nussbaum 1982; W. M. Olson 2000; Trueb et al. 2000). Such sesamoids include the os sesamoides tarsalia, the cartilagines plantares, and the cartilago sesamoides. The remaining two groups of amphibians, the gymnophiones and urodeles, do not appear to possess appendicular sesamoids. Sesamoids in the region of the tarsus have also been reported for some birds (Baumel and Raikow 1993; Vanden Berge and Storer 1995), lizards (Maisano 2001, 2002a, 2002b), and primate mammals (Le Minor 1987).

Patella and Patelloid

The predominant sesamoid for biomechanical, orthopaedic, pathologic, and evolutionary study is the tibial patella. The tibial patella (hereafter "the patella") is a large, generally well-ossified sesamoid positioned cranially adjacent to the distal end of the femur (figs. 19.4, 19.5). The patella has an extremely broad taxonomic distribution, and notwithstanding the obvious absence in those taxa not developing hindlimbs (e.g., snakes, cetaceans), it is constant among most modern lizards (Maisano 2001, 2002a, 2002b), birds (Schufeldt 1886), and various mammals including monotremes, placentals (Gabe et al. 1967), and at least one †multituberculate (Kielan-Jaworowska and Gambaryan 1994). Major lineages failing to develop the patella include lissamphibians, nonavian archosaurs (e.g., crocodylians, †pterosaurs, nonavian †dinosaurs), turtles, and marsupials, with the exception of peramelids (bandicoots; Reese et al. 2001) and the marsupial mole (*Notoryctes typhlops;* Barnett and Lewis 1958).

In humans, the patella is the only sesamoid to be counted regularly among the bony elements of the body (a common total for humans, including the paired patellae, is 206; T. D.

White 1991; Rosse and Gaddum-Rosse 1997). Although relatively simple in morphology, the adult (skeletally mature) patella has two important borders, a proximal pole and distal or inferior pole, and two main surfaces, a superficial external surface and a deep articular surface. In histological section the skeletally mature bony aspect of the patella is composed of a lamellar bone cortex with a trabeculated core (including marrow; J. Clark and Stechschulte 1998) and an articular cartilage-lined deep (articular) surface. Although some previous authors suggested that the patella is not a true sesamoid, claiming that it developed deep to (i.e., not within) the tendon complex of the knee (e.g., Brooke 1937), this notion has since become widely discounted. Indeed, the patella is always embedded within the tendon for the m. quadriceps femoris or its homologous counterpart (e.g., the m. femorotibiales and m. iliotibialis of reptilians including birds). Distal to the patella, the tendon continues across the knee to insert on the tibial tuberosity (fig. 19.4). The portion of the tendon proximal to the patella is the quadriceps tendon; the portion of the tendon distal to the patella is usually referred to as the "patellar ligament." However, Bland and Ashhurst (1997) have argued that the so-called patellar ligament more closely resembles a tendon than a ligament in histological structure (ligaments generally have rounder cells and more type III collagen than tendons). This notion is further supported by the developmental continuity of the quadriceps-patellar tendon complex (see below). Consequently, we have adopted the designation "patellar tendon" (*sensu* Bland and Ashhurst 1997). Tendon fibers from the quadriceps and patellar tendons may insert directly into the lamellar bone or invest between lamellar systems (J. Clark and Stechschulte 1998).

In addition to the bony patella, peramelid marsupials (Reese et al. 2001) and various placental mammals (e.g., rodents, lemurs, lagomorphs; Gabe et al. 1967; Jungers et al. 1980; Bland and Ashhurst 1997) develop an additional sesamoid nested deep within the quadriceps tendon. Characteristically this more proximally positioned element—the patelloid (= fibrocartilago sesamoidea parapatellaris, secondary patella, cartilago suprapatellaris, suprapatella)—is composed largely of fibrocartilage. A comparable fibrocartilaginous element is also common to nonperamelid marsupials (i.e., those lacking a bony patella). The distinct topographic position, characteristic histology, and co-occurrence of the patella and patelloid in peramelids and various placentals suggest that the patelloid is not the unossified homologue of patella. A patelloid-like structure has also been reported in a representative crocodylian (F. G. Parsons 1908) and turtle (*Terrapene carolina*; Haines 1969).

Development and Histology

By far the majority of data concerning development of the patella and patelloid are derived from placental mammals, in particular lagomorphs (*Oryctolagus*), rodents (*Rattus*), and hominids (*Homo*), as well as the domestic fowl (*Gallus*). Shortly after chondrification of the articular ends of the femur and tibia, the complex consisting of the presumptive quadriceps and patellar tendons and intervening patella is visible as a continuous band of fibrous connective tissue spanning the mesenchymal interzone along the anterior surface of the presumptive knee joint (in *Oryctolagus* 17d of gestation, *Homo* O'Rahilly stages 18–19 [weeks 6–7], *Gallus* HH stages 19–20; see Gardner and O'Rahilly 1968; Bland and Ashhurst 1997; Mérida-Velasco et al. 1997). The patella develops within this band as a longitudinally arranged cluster of cells adjacent to the distal end of the femur (Gardner and O'Rahilly 1968; Bland and Ashhurst 1997; Mérida-Velasco et al. 1997). Immunohistological studies on *Oryctolagus* indicate that most cells representing the future quadriceps-patellar tendon secrete collagen types I, III, and V. However, a small region localized deep within the tendon and representing the presumptive patella binds with antibodies to type II collagen (Bland and Ashhurst 1997). With subsequent development (*Oryctolagus* 25d of gestation, *Homo* O'Rahilly stage 22, *Gallus* HH stage 36) the region of the presumptive patella becomes chondrified, forming a hyaline cartilage mass that segments the formerly continuous tendinous insertion of the quadriceps, excluding a narrow communication of fibers across the external (superficial) patellar surface. Distally from the inferior pole, the hyaline cartilage patella grades through a fibrocartilaginous transition into the patellar tendon (see fig. 19.6; Bland and Ashhurst 1997). Prior to hatching (HH stage 42), the biconvex patella of *Gallus* develops two discrete concavities corresponding to the positions of the femoral condyles (Niven 1933).

In neonate *Oryctolagus*, J. Clark and Stechschulte (1998) found that the quadriceps tendon adjacent to and passing in advance of the patella had no cartilage-like characteristics. However, by two weeks postnatal development chondrocyte-like cells were observed in the tendinous tissue juxtaposed with the patella. By six weeks postnatal, this fibrocartilaginous zone, connecting the proximal pole of the patella with the quadriceps tendon, had become remodeled into osteoid and bony tissue. Based on these observations, J. Clark and Stechschulte (1998) hypothesized that the bony patella develops from two independent sources: a deeply nested hyaline cartilage mass and a superficial fibrocartilaginous transition. In humans the patella is typically associated with a single center of ossification; the presence of two or more centers that then fail to fuse is reported to give rise to a fragmented appearance (so-called bipartite or tripartite patellae) or even

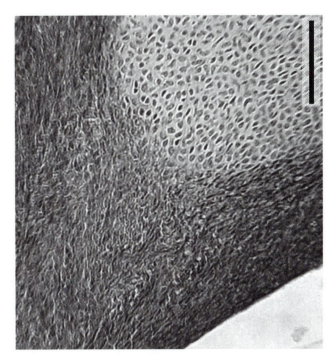

Figure 19-6 *Gallus gallus* serial section, stained with Masson's trichrome, detailing the distal pole of the developing patella. Late-stage (Hamilton and Hamburger stage 44; fully formed) embryo. Hyaline cartilage of the developing patella (upper right corner of image) merges through a fibrocartilaginous transition (lower left corner of image) into the patellar tendon. Scale bar = 100 μm.

patellar duplication (Tria and Alicea 1995). Unlike most other sesamoids, ossification of the patella (exclusive of the articular surface) occurs relatively early in ontogeny (*Oryctolagus*, 3–6 weeks postnatal; *Homo*, 14 weeks; *Gallus*, 11 weeks; see Gardner and O'Rahilly 1968; Bland and Ashhurst 1997; Mérida-Velasco et al. 1997).

Similar to the patella, the patelloid is nested deep within the quadriceps tendon, adjacent to the articular surface of the femur. Histologically, the irregular arrangement of the chondrocyte-like cells within the patelloid generally differs from the more orderly, longitudinal distribution of cartilage cells observed in the fibrocartilaginous transition from patella to patellar tendon. Starting at three weeks postnatal, and coinciding with the acquisition of ricochetal locomotory patterns, the patelloid of *Oryctolagus* begins to bind with antibodies to collagen type II (Bland and Ashhurst 1997). In contrast, the patelloid of the rodent *Rattus* does not demonstrate type II binding until nearly one year postnatal (Benjamin et al. 1991; Ralphs et al. 1991, 1992).

Details of the development and histology of the patella and patelloid in other most mammals is lacking. Observations made on the megachiropteran *Pteropus* ("flying fox") suggest that while an osseous patella is lacking, a narrow band of hyaline cartilage overlain by fibrocartilage is present deep within the quadriceps-patellar tendon (B. J. Smith et al. 1995). In most other chiropterans an osseous patella is present, with the proximal pole nested deep within a matrix of fibrocartilage that grades into the quadriceps tendon (B. J. Smith et al. 1995).

Reese et al. (2001) surveyed 30 marsupial species and found that whereas the patelloid is always present, the patella is almost always absent. Similar to that of lagomorphs, rodents, chiropterans, and nonanthropoid primates (e.g., tarsiers, galagids, lemurs; Gabe et al. 1967; Jungers et al. 1980), the marsupial patelloid consists largely of fibrocartilage, localized deep within the quadriceps tendon. Based on aspects of sectional area, collagen orientation, cell morphology, and histology, Reese et al. (2001) identified five characteristic patelloid "types" that appear to correlate with function, locomotory patterns, and taxonomy. Among terrestrial quadrupeds such as the carnivorous dasyurids (e.g., quolls), the patelloid is relatively large (two-thirds the sectional area of the quadriceps tendon) with interwoven collagen fibers and differentiated chondrocytes. In contrast, the patelloid of bipedal and ricochetal macropodoids (so-called real kangaroos plus rat-kangaroos) is relatively small (one-quarter to one-half the sectional area of the quadriceps tendon), with parallel collagen fibers and poorly differentiated chondrocytes. Peramelids (bandicoots), which have a hopping gait similar to that of lagomorphs, had both an osseous patella and a more proximal fibrocartilaginous patelloid. The patella may (e.g., *Oryctolagus*) or may not (*Rattus*, peramelids) be separated from the patelloid by a pad of adipose tissue (Bland and Ashhurst 1997).

Patellar Experimentation and Genetic Control

The patella is often treated as an exemplar of sesamoids, but it is quite atypical in two important respects: (1) it initially develops and begins to ossify relatively early in ontogeny, and (2) it will develop even in the absence of mechanobiological stimuli. As observed by Murray (1936, 4), "the first formation of the patella is a self-differentiation, and neither functional activity nor mechanical forces due to growth pressure, etc. are required." Experimental work on *Gallus* has demonstrated that the patella will develop in culture (Murray and Huxley 1925; Niven 1933) or under conditions of complete musculoskeletal paralysis (Drachman and Sokoloff 1966; Hosseini and Hogg 1991). Murray and Huxley (1925) grafted the limb bud of one four-day embryo onto the chorioallantois of a second. After five days the bud had skeletally differentiated to form a femur, with a head, shaft and trochanter, and a patella. Niven (1933) removed presumptive patellar mesenchyme from HH stage 31 to 35 *Gallus* embryos and cultivated them *in vitro*. Mesenchyme from the early stages did develop cartilage, although the overall form of the skeletal mass was irregular in morphology. Tissue from later-stage embryos (in particular HH stage 35) developed a cartilaginous mass that closely resembled the *in situ* developing patella.

Drachman and Sokoloff (1966) examined skeletal development in *Gallus* following artificially induced paralysis via neuromuscular blocking agents (decamethonium and botulinum toxin) and spinal cord extirpation. Whereas the patella continued to develop in paralyzed embryos (albeit somewhat reduced in size), a second sesamoid, the os sesamoideum intertarsale (plantar tarsal sesamoid), did not (Drachman and Sokoloff 1966). Extrinsic stimuli play an important role in shaping and determining the relative timing of patella skeletogenesis. Brunner (1891) observed that in subadult hominids (*Homo*) with extremely reduced or completely absent patella, development could be prompted with movements mimicking normal patterns of flexion and extension. As translated by Beresford (1981, 52), "the patella, which has no genetic relation with the tibia, develops after this as a sesamoid that owes its development to the rubbing of the well functioning quadriceps femoris tendon on the already established femoral epiphyseal cartilage." Carey et al. (1927) removed the patella from the quadriceps femoris tendon of subadult canids (*Canis familiaris*) and noted that if movement was permitted across the joint the patella would regenerate, forming both cartilage and bone. If movement was inhibited no regeneration occurred. Perhaps not surprisingly, even after seven weeks the *Gallus* patellae cultured *in vitro* by Niven (1933) never developed articular surfaces.

Details concerning the molecular and genetic control of development of the patella (and other appendicular ossicles) remain largely unexplored. One notable exception is the ongoing research on nail-patellar syndrome (NPS; onycho-osteodysplasia, MIM #161200), an autosomally inherited condition characterized by hypoplastic nails, renal insufficiencies, and skeletal dysplasia including hypoplastic or absent patellae (H. Chen et al. 1998; H. Chen and Johnson 1999; Clough et al. 1999; Dreyer et al. 2000). NPS results from mutations of the *LMX1B* gene (Dreyer et al. 1998a; 2000). Murine knockouts with homozygous mutations to the orthologue *Lmx1b* have since been created that demonstrate similar, albeit more severe, abnormalities. Recent work using *in situ* hybridization has demonstrated that *LMX1B* is a mesenchymal determinant of the dorsoventral limb axis (H. Chen et al. 1998; Clough et al. 1999) and that *Lmx1b* is strongly expressed in tissues on the anterior surface of the limb such as the presumptive quadriceps/patellar tendon (Dreyer et al. 2000). As a result of the reduced dorsalization, *Lmx1b*$^{-/-}$ mutants, similar to patients with NPS, exhibit a variety of skeletal and musculotendinous problems, including a reduced or absent patella and ectopic

muscle insertions. In *Lmx1b*$^{-/-}$ mutants, symmetrical dorsoventral patterning of tendons and autopodial sesamoids normally confined to the volar (palmar) surface have also been noted (H. Chen and Johnson 1999; Clough et al. 1999). Whereas an orthologous gene in *Gallus* (*Lmx-1*) is known to play a role in dorsoventral limb patterning (Riddle et al. 1995; H. Chen et al. 1998), it remains unclear if the patella and other appendicular ossicles are affected.

Targeted disruption of the murine HoxA and HoxD genes results in various skeletal dysplasias of the appendicular system, including the patella (Cruz et al. 1999; Branford et al. 2000). In most *Hoxd9/Hoxd10* double mutants the span of the interzone between the femur and tibia is reduced and the patella becomes displaced proximally (Cruz et al. 1999). Furthermore, the patella is often misshapen and demonstrates a fragmented appearance. From whole-mount specimens it has been observed that some *Hoxd9/Hoxd10* double mutants develop a small additional mineralization, positioned proximal to the patella along the femur. At present it is uncertain if this ossicle represents a homologue of the patelloid, a portion of the fragmentary patella, or an ectopic element. The patella of *Hoxd9/Hoxd10* double mutants is also fragmented in appearance (Cruz et al. 1999). Both *Hoxa9/Hoxd9* and *Hoxd9/Hoxd10* knockouts may also develop one or more additional, seemingly ectopic, sesamoids (or other appendicular ossicles) at the elbow joint (Fromental-Ramain et al. 1996a; Cruz et al. 1999). Disruption of *Hoxa10* results in the malformation and displacement of the lateral fabella and reduction or absence of the medial fabella (Branford et al. 2000).

Traction Epiphyses

Although perhaps it is not widely recognized, there is considerable evidence that supports the notion of sesamoids occasionally becoming morphologically integrated with other skeletal elements. F. G. Parsons (1904) observed that in some instances the bony projection of insertion for a tendon or ligament develops independent of the limb element proper. He referred to these secondarily derived projections as "traction epiphyses" and suggested that they represented sesamoids that became incorporated onto long bones of the appendicular skeleton. This hypothesis originally met with skepticism. K. Pearson and Davin (1921a, 1921b) suggested that the opposite is true, that sesamoids represented disarticulated traction epiphyses, and Haines (1940; 1942c) argued that traction epiphyses and sesamoids were completely independent of one another. However, subsequent work by O. J. Lewis (1958, 1989; Barnett and Lewis 1958) and others (Heilmann 1926; Broome and Houghton 1989) has bolstered Parsons's hypothesis. Developmental evidence from both a lizard

(*Uromastix spinifer*) and a marsupial (*Wallabia bicolor* [formerly *Macropus ualabatus*]) demonstrates that the tibial tuberosity is derived from an independent condensation that originates within the patellar ligament and subsequently conjoins with the proximocranial surface of the tibia (O. J. Lewis 1958). Broome and Houghton (1989) described a case of a congenital malformation in a human where a small accessory osseous sesamoid was found distal to the patella, in advance of the proximocranial end of the tibia; this individual lacked a protruding tibial tuberosity. As noted previously, marsupials and †multituberculates develop a parafibula proximally adjacent to the fibula. Barnett and Lewis (1958) observed that in at least one marsupial taxon (*Phascolomys*; wombat), the parafibula could either be an independent element or fused to the fibula. According to O. J. Lewis (1989), a mobile parafibula is the plesiomorphic condition for mammals, whereas an enlarged fibular crest (e.g., monotremes; fig. 19.5) is apomorphic.

Perhaps the most compelling evidence supporting the sesamoid–traction epiphysis hypothesis comes from various avians, where a separate ossification fuses with the proximal end of the tibia to from the cranial cnemial crest (= tibial tuberosity; Barnett and Lewis 1958; Hogg 1980; Hutchinson 2002). In addition, in some avians the patella also becomes incorporated onto the hypertrophic cnemial crest (a so-called double traction epiphysis; Hutchinson 2002). Whereas the fossil avian †*Hesperornis* has a lengthy patella articulating with proximocranial end of the tibiotarsus (figs. 19.7a, b), in several modern lineages (including gaviiforms, ciconiiforms, and podicipediforms) an unconstrained patella is absent, whereas a prominent patella-like cnemial crest develops (figs. 19.7c, d; D'A. W. Thompson 1890; Heilman 1926; Barnett and Lewis 1958; Baumel and Witmer 1993; Hutchinson 2002; see also K. Pearson and Davin 1921b).

Mineralized Tendons and Ligaments

Among various tetrapods—most notably modern birds—mineralized tendons and (to a lesser extent) ligaments are occasionally encountered as nonpathologic skeletal components. Frequently these elements are referred to as "ossified," although rarely is there an unequivocal demonstration of osteocytes (Rooney 1994). Ostensibly, mineralized tendons (and ligaments) may be likened to sesamoids in that each develops within a collagen-rich band of dense regular connective tissue. However numerous differences qualify mineralized tendons as developmentally, evolutionarily, and functionally distinct from sesamoids. Whereas sesamoids tend to be localized adjacent to bony articulations and have a compact morphology, mineralized tendons are slender and elon-

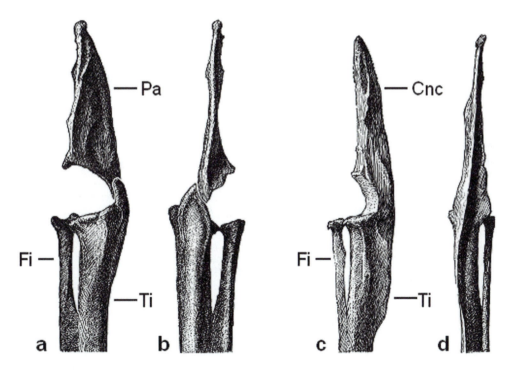

Figure 19-7 Schematic illustration of the left knees of a fossil and a modern avian, in two views. (a) Medial and (b) cranial views of the unconstrained yet elongate patella of †*Hesperornis regalis*. (c) Medial and (d) cranial views of the hypertrophic cranial cnemial crest of *Colymbus glacialis*. The obvious similarities between these morphologies have suggested that in some avians the patella may become assimilated with the tibia as a traction epiphysis (modified from Heilmann 1927). Abbreviations: Cnc, hypertrophic cranial cnemial crest; Fi, fibula; Pa, patella; Ti, tibia.

gate and terminate in advance of the joint. Proximal and distal tendinous connections with muscle and bone, and any spans that traverse joints, remain unmineralized. Furthermore, mineralized tendons appear to lack a discrete cartilaginous phase of development (see below).

Vertebrate tendon is a dense, regular fibrous connective tissue that links contractile muscle to bone or (less frequently) other muscles (e.g., the digastric tendon). The majority of the tendon mass is extracellular matrix (ECM), principally type I collagen (approaching 95% of the matrix; Rooney 1994; Landis and Silver 2002). In addition, the ECM may include limited amounts of collagen types III, V, and VI (Rooney 1994) and various proteoglycans (e.g., decorin and aggrecan; Koob and Summers 2002). The cellular content is mostly fibroblasts, so-called tenocytes.

In avians, the tendon of insertion for the m. gastrocnemius (the "Achilles tendon") of the domestic turkey (*Meleagris gallopavo*) provides the primary model for mineralized tendon study. Morphologically the tendon resembles an upright Y, with a relatively thick, cylindrical segment distally and a proximal bifurcation leading to smaller fan-shaped segments that merge into the muscle bellies (Landis and Silver 2002). Beginning around 12 weeks posthatching (M. A. Bennett and Stafford 1988), the tendon initiates a histological transformation that superficially resembles the transition from cartilage to bone at the epiphyseal plate (Landis and Silver 2002). Tenocytes proximal to the m. gastrocnemius are characteristically fusiform in shape. Distal to the muscle, approaching the bifurcation, the tenocytes undergo hypertrophy and adopt a stellate morphology with a complex series of interconnections. Proximal to the bifurcation, the hypertrophic tenocytes begin to produce vesicles containing calcium and phosphorous. The earliest appearance of mineralization is correlated with the presence of these vesicles (Landis and Silver 2002). Eventually the vesicles diffuse their contents into the ECM and the crystals of apatite formed become organized on, and within the spaces between, the collagen fibrils (Rooney 1994; Landis and Silver 2002).

Mineralized tendons have been reported in both the fore- and hindlimbs of many avian groups including galliforms, tinamids, podicipedids, falconids, larids, strigids (fig. 19.8), the hoatzin (*Opisthocomus hoazin*), and dendrocolaptine passeriforms (woodcreepers; Hudson et al. 1959, 1965; M. A. Bennett and Stafford 1988; Bledsoe et al. 1993; Vanden Berge and Storer 1995). Outside of crown-group avians, mineralized tendons have been identified in the fossil ornithurine †*Apsaravis ukhaana* (Late Cretaceous, 80–70 mya) adjacent to the caudal surface of the tarsometatarsus (Norell and Clarke 2001; Clarke and Norell 2002; Hutchinson 2002). Hutchinson (2002) phylogenetically mapped the presence of mineralized tendons and concluded that these skeletal ele-

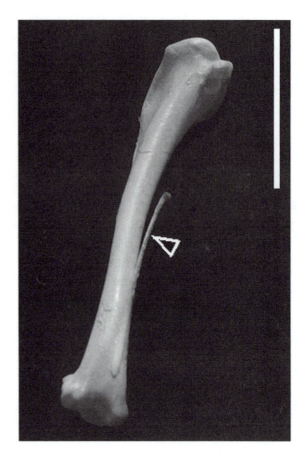

Figure 19-8 The caudal surface of the right humerus of *Asio otus* (long-eared owl) (proximal toward top of image), with a mineralized tendon of the complexus m. deltoideus (black arrowhead). Scale bar = 4 cm.

ments may be plesiomorphic for modern birds and have likely been overlooked or unpreserved in a variety of fossil taxa. Mineralized tendons have also been identified in the anuran *Hymenochirus* (W. M. Olson 2000; plate 19.1) and a †pterosaur specimen (S. C. Bennett 2001a). Regrettably, the taxonomic distribution of mineralized tendons is far from clear, and Bledsoe et al. (1993, 289) caution that "the absence of reported instances in descriptive studies should not be assumed to mean absence of the condition."

Periarticular Elements

Lunulae

Associated immediately adjacent to or deep within joints and joint cavities are a variety of constant forming, non-pathological ossicles. These elements, herein designated as "periarticular," differ from sesamoids and mineralized tendons in aspects of position (they are often found inside the joint capsule) and development (see below). Although both intratendinous and periarticular elements are distinct from one another, they often co-occur in the same individual. The best-known periarticular elements are the meniscal ossicles or lunulae of the knee and other synovial joints. Characteristically, skeletally mature lunulae form osseous wedges with a crescentic morphology, nested within fibrocartilaginous menisci (Barnett and Lewis 1958; O. J. Lewis 1989). Those at the knee reside between the distal femoral condyles and the tibial plateau (Haines 1969; McDevitt and Webber 1990). K. Pearson and Davin (1921a) offer several possible locations for femorotibial lunulae, including antero-lateral, antero-mesial, postero-lateral, postero-mesial and possibly antero-/postero-sagittal. Lunulae may also occur in menisci of other joints, including the hand or wrist (see below).

Diversity and Distribution

Lunulae are reported in lissamphibians (viz., anurans), squamates, birds, and a wide variety of mammals. K. Pearson and Davin (1921b) note that whereas anurans do not have sesamoids at the knee joint, some taxa develop lunulae (e.g., *Leptodactylus pentadactylus, Docidophryne gigantean;* also *Hymenochirus,* see below).

Lunulae are extremely common within nonophidian squamates (K. Pearson and Davin 1921b; Haines 1942c, 1969; Maisano 2001, 2002a, 2002b) with some "lizards" (e.g., iguanids, *Varanus*) demonstrating as many as five different elements. All exist as cartilaginous precursors in neonates and ossify either at hatching/birth or later in ontogeny (Maisano 2001, 2002a). Lunulae also occur in *Sphenodon* and possibly the *Varanus*-like fossils forms †*Pontosaurus* and †*Opetiosaurus* (= †*Carsosaurus;* Barnett 1954; Haines 1969). In Aves, they are reportedly present in *Syrnium* (wood owl) and *Dendrocygna* (tree duck; K. Pearson and Davin 1921b).

Lunulae of the knee joint have also been reported in many groups of mammals, including edentates, insectivores, lagomorphs, carnivores, and primates, and they are extremely common in rodents (K. Pearson and Davin 1921b; Haines 1942c; Pedersen 1949; Barnett 1954; M. Walker et al. 2002; fig. 19.4). In felids, lunulae are present in the cranial end of the medial meniscus, arising from a single ossification center; in many large taxa (e.g., *Panthera* spp.) lunulae, along with patellae, cyamellae, and lateral fabellae, are all present at skeletal maturity (M. Walker et al. 2002). Femorotibial lunulae are reported to be absent in monotremes, marsupials (although see below), and ungulates (K. Pearson and Davin 1921b), and possibly chiropterans (the megachiropteran *Pteropus* is exceptional among mammals in that it completely lacks menisci, and hence lunulae; Barnett 1954). Among primates anterior lunulae are present in several nonanthropoids (e.g., *Lemur, Galago, Nycticebus;* Barnett 1954).

Various other examples of lunulae exist outside the femorotibial joint. All marsupials, with the exception of peramelids, possess a prominent meniscus between the fibula and the talus (the so-called intermedium), which may contain a large pyramidal lunula (O. J. Lewis 1989; *contra* K. Pearson and Davin 1921b). In *Phalanger,* this meniscus extends proximally to separate the tibia and the fibula; the posterolateral and posteromedial corners of the lunula (which may be cartilaginous or ossified) are fixed to the limb elements via ligaments (O. J. Lewis 1989). Similarly, in hominids there is (pleisomorphically) a fibrocartilaginous meniscus in the antebrachiocarpal joint, separating the ulna from the carpus. In *Hylobates* (gibbons) and occasionally in *Gorilla,* this meniscus contains a lunula, the os Daubentonii, which separates the ulnar styloid process and the triquetral and articulates with the pisiform (O. J. Lewis 1989). The os Daubentonii is preformed in hyaline cartilage and may comprise the entire meniscus in adults. Based on homology, small ossifications that occasionally appear in the human wrist (in the region of the triangular fibrocartilage complex) are thus most appropriately referred to as lunulae.

Development and Histology

Joints in the appendicular skeleton form via the segmentation of initially continuous precartilaginous condensations. Development of a diarthrodal joint can be divided into two main phases: formation of the interzone, which determines the location of the articulation, followed by subsequent cavitation, development, and maintenance of the joint cavity (Mikić et al. 2000a, 2000b). Morphologically, the interzone is first visible as a region of high cell density, presaged by *Gdf5* expression (Storm and Kingsley 1996). Selective cell death (apoptosis) and changes in the composition of the ECM give rise to three layers, two chondrogenic layers that form the articular surfaces of the long bones and a more diffuse intermediate layer (Mikić et al. 2000a, 2000b). The menisci, synovial membranes, and the ligaments surrounding the joint arise from this intermediate layer and from mesenchymal condensations lateral to the joint (Storm and Kingsley 1996; Wolfman et al. 1997; Mikić et al. 2000b). *Gdf5* expression eventually becomes restricted to the articular surfaces of the joint (Wolfman et al. 1997). *Wnt14* has also been implicated in the early development of joints. *Wnt14* induces ectopic interzones, *in vitro;* it also induces *Gdf5* expression and the production of hyaluronan receptors, which are involved in the process of joint cavitation (Hartmann and Tabin 2001). *Wnt14* appears to block cartilage differentiation, and may be involved in diverting cells to a different morphogenetic pathway; interestingly, *Wnt14* cannot elicit this response from nonchondrogenic cell types (Hartmann and Tabin 2001).

In modern hominids, the interzone of the knee joint begins to differentiate morphologically at O'Rahilly stage 18–19 (Mérida-Velasco et al. 1997). By O'Rahilly stage 22, there are two bands of articular cartilage covering the femoral condyles and the upper surface of the tibia, separated by a more diffuse medial band of mesenchyme. The menisci begin to organize from the lateral portions of the interzone, and joint cavitation begins; by week nine, the meniscal cavities are developed and the menisci are attached peripherally to the joint capsule (via coronary ligaments) and to the anterior and posterior aspects of the upper surface of the tibia. In humans the menisci remain cartilaginous and lunulae are rare (Haines 1969); when they do occur, most commonly (but not exclusively) because of pathology or injury, they form in the posterior portion of the medial meniscus (M. Walker et al. 2002). Pedersen (1949) cites several reports of lunulae in humans, with no evidence of trauma or degeneration, and argues that in such cases the lunulae may be considered vestigial.

Formation of the interzone is primarily under genetic control. The second (cavitation and maintenance) phase of joint development, however, is highly dependent on movement of the limbs (Storm and Kingsley 1996; Mikić et al. 2000a; Hartmann and Tabin 2001). In subadult *Gallus* paralyzed early during development, the tibiofemoral menisci begin to develop but then degenerate in the absence of biomechanical stimuli; the os sesamoideum intertarsale also fails to develop (Mikić et al. 2000a; see also Drachman and Sokoloff 1966). Degeneration or loss of joint structures (e.g., menisci, sesamoids) is correlated with changes in the expression of tenascin-C and collagens I and XII, suggesting that these molecules are sensitive to changes in mechanical environment (Mikić et al. 2000b). Collagen XII shows normal expression in tissues that persist, including the patellar tendon.

Mature menisci consist of an avascular, aneural fibrocartilaginous matrix surrounded by a layer of connective tissue. The composition of the ECM varies by species but is distinct from both articular and hyaline cartilage in containing high levels of adhesive glycoproteins and fibrillar collagens (McDevitt and Webber 1990). Lunulae tend to form in steeply wedged or biconcave menisci (Barnett 1954). The outer layer consists of lamellar bone, the inner of cancellous bone (which may or may not contain marrow); the proximal and distal articular surfaces are covered with hyaline cartilage (M. Walker et al. 2002). Lunulae are always prefigured by a nucleus of hyaline cartilage (K. Pearson and Davin 1921a; Haines 1969).

Menisci are a normal component of synovial joints and may function to enlarge the functional surface of the joint, as well as contributing to the stability of the articulation between the limb elements; menisci are adapted to the morphological incongruencies between the femoral condyles and the

tibial plateau (Bizarro 1921; Haines 1969). Bizarro (1921) has even suggested that all condylarthroses require the presence of such periarticular elements. In addition, lunulae may serve both protective and biomechanical roles—for example, resisting compressive forces or acting as a fulcrum (M. Walker et al. 2002). Haines (1942c) hypothesized that lunulae develop whenever menisci reach a critical size threshold. As they are typically embedded within the thickest region of the meniscus (O. J. Lewis 1989), they may, at least in part (like most sesamoids), be the result of biomechanical induction, regardless of any resulting biomechanical advantage. In *Rattus*, it was found that with increased exercise there is a corresponding, albeit reversible, increase in collagen and proteoglycan production in the posterior portion of the lateral meniscus (McDevitt and Webber 1990). Pedersen (1949) noted that lunulae are occasionally found in this posterior lateral position, but only in aged rodents.

Problematic "Sesamoids" and Enigmatic Appendicular Ossicles

As mentioned previously, many appendicular ossicles defy discrete characterization. By constraining our definition of sesamoids and lunulae to require knowledge of positional context during skeletogenesis, various elements known best from skeletally mature individuals cannot be accurately categorized. In *Ailuropoda* (giant panda), the famous "thumb" (radial sesamoid; os sesamoideum radiale) is a large digitlike preaxial element that articulates with the scapholunatum (radial carpal; intermedioradial carpal) and the medial border of the first metacarpal (D. D. Davis 1964; H. Endo et al. 1999). It is far from unique, however, as similar albeit significantly smaller elements (variably identified as the prepollex, os radiale externum, or os falciforme) are found in various other carnivores, as well as marsupials, insectivores, chiropterans, nonhominid primates, and rodents (Romankowowa 1961; Le Minor 1994; Thorington and Darrow 2000; Prochel and Sánchez-Villagra 2003; see below). However, whereas in most noncarnivores the radial sesamoid is entrenched within the tendon for the m. abductor pollicis longus (Le Minor 1994), in *Ailuropoda* and other ursids this element independently attaches to two separate muscles (m. abductor longus and the m. opponens pollicis). In the absence of developmental data it remains unclear if the radial "sesamoid" of *Ailuropoda* is indeed a true hypertrophic sesamoid (forming within the tendon for either muscle or an adjacent ligament), a remodeled periarticular ossicle (e.g., lunula), or a neomorphic derivative of the carpal series.

In addition to the preaxial "thumb," the manus of *Ailuro-*

poda also demonstrates a pronounced postaxial pisiform (the largest bone of the carpus), and the pes has a corresponding (preaxial) enlarged tibial sesamoid ("panda's toe"). Similar to the "thumb," neither of these elements are unique to *Ailuropoda* among carnivores (although they are most pronounced in the former), each has multiple musculotendinous attachments, and none is well described developmentally (D. D. Davis 1964).

The so-called sesamoids at metacarpophalangeal and interphalangeal joints in *Homo* appear to form from thickened fibrocartilaginous "palmar plates" of the capsule (O. J. Lewis 1989). Flexor tendons are channeled within a median depression along the "palmar plate" coursed between wedge-shaped margins. Frequently, these fibrocartilaginous margins develop centers of mineralization, resulting in each joint capsule being associated with one or even a bilateral pair of ossicles. Tendons and ligaments then secondarily attach. Accordingly, O. J. Lewis (1989) suggests that these elements more closely resemble modified lunulae than sesamoids. A similar argument is conceivable for the metatarsophalangeal "sesamoids" at the base of the hallux (the bone of Lus or ossa "sesamoidea" hallucis) and the other pedal phalanges (fig. 19.1).

Among many adhesive pad-bearing gekkonid squamates, a comparable situation exists. In various taxa, the interphalangeal joints are associated with unique bilaterally arranged cartilaginous elements—so-called paraphalanges (Wellborn 1933/1997; Russell and Bauer 1988). The developmental context of paraphalanges is unknown, although in at least one genus (*Hemidactylus*) they appear to reside within lateral digital tendons and thus may arguably represent sesamoids *sensu stricto* (Russell and Bauer 1988). Interestingly, the presence of paraphalanges is not correlated with phylogeny but instead appears to be related to functional and structural properties of the autopodium (Russell and Bauer 1988), a situation comparable with the patelloid of marsupials (see above).

In gliding and flying taxa there is often an assortment of unusual skeletal rods and extensions that support the wing and gliding membranes (patagia). Such patagial supports as the styliform cartilage of glissant mammals, the calcar of chiropterans, and the pteroid elements of †pterosaurs represent uniquely derived skeletal modifications that are common if not ubiquitous within respective lineages, and which are often cited as support for monophyly, although little is known concerning their development and histological composition.

Several mammalian lineages (including both marsupials and placentals) have independently evolved the capacity to glide. In the rodent clade Pteromyinae (flying squirrels), addi-

tional patagial support at the forelimb is provided by a shaft of skeletal tissue that articulates with the lateral border of the carpus. This novel element, the styliform cartilage, extends the patagium dorsolaterally and appears to act similar to an airplane winglet, functioning to reduce drag (Thorington et al. 1998). Extension of the pteromyine winglet is achieved by an unique system of connective tissue linkages wherein an abductor of the thumb (the m. abductor pollicis longus) inserts on the prepollex (os falciforme; a putative sesamoid), which is connected across the palm via a ligament to the styliform cartilage (Thorington et al. 1998). The styliform cartilage has variably been interpreted as a modified "ulnar" sesamoid (Gupta 1966), pisiform (Oshida et al. 2000), or hypothenar cartilage (Thorington et al. 1998; Thorington and Stafford 2001), and while its homology remains muddled, it does not reside within a tendon or ligament, which precludes it from our definition of a sesamoid. In other glissant mammals (e.g., anomalurid rodents, the marsupial *Petauroides;* Johnson-Murray 1987) a comparable patagial support element may be found articulating with the olecranon process of the ulna.

Members of Chiroptera (bats) are characterized by numerous skeletal modifications to both the fore- and hindlimbs. Included among these modifications is a variably sized splint of hyaline cartilage medially adjacent to or articulating with the tarsus. In contrast to the styliform cartilage, this novel skeletal mass, the "calcar," acts to support the caudal-most component of the wing membrane (the uropatagium). A review of hindlimb morphology in various representative chiropterans (Schutt and Simmons 1998) noted that while most (but not all) forms develop a calcarlike splint juxtaposed with the tarsus, two discrete morphologies exist, divided along lines of phylogenetic descent. In megachiropterans (fruit bats) the element emerges as a laterally directed spur from the distal tendon of the m. gastrocnemius, proximal to the insertion on the calcaneum; in microchiropterans this element articulates directly with the calcaneum. Schutt and Simmons (1998) applied the term "uropatagial spur" to the element of megachiropterans and noted that is also different from the true calcar of microchiropterans in details of associated musculature. Furthermore, whereas the microchiropteran calcar may become calcified, the megachiropteran uropatagial spur does not. Limited developmental work indicates that the calcar of the microchiropteran *Myotis lucifugus* (little brown bat) is one of the last skeletal elements to form, although adult proportions are attained prior to volancy (R. A. Adams and Thibault 1999). During postnatal ontogeny, the hyaline cartilage composing the proximal portion of the calcar becomes invested with calcium salts. The presence of a calcar has also been affirmed in the fossil chiropterans dating back to the Oligocene (33–24 myo; †*Archaeopteropus;* Schutt and Simmons 1998). An additional skeletal orphan supporting the wing membrane was reported along the leading edge of manual digit I of *Barbastella barbastellus* (Romankowowa 1961).

The extinct archosaurian group †Pterosauria has long, narrow wings supported by the bones of the brachium and antebrachium, an incredibly lengthy fourth digit, and one or two pteroid elements. Similar to the styliform cartilage, the †pterosaurian pteroid articulates with the carpus along the preaxial border, supporting the small leading-edge flap of wing membrane. In many of the earliest known, smaller forms (e.g., basal †pterosaurs or †"rhamphorhynchoids") the pteroid consists of a bipartite articulating set of elements; in larger, more recently derived members (e.g., †pterodactyloids) it is a single elongate element (S. C. Bennett 2001a). Unique among patagial supports (and skeletal orphans in general), the pteroid element of some †pterodactyloids is pneumatized (e.g., †*Anhanguera,* †*Pteranodon;* Unwin et al. 1996; S. C. Bennett 2001a). Rare subadult pterosaurs demonstrate that the pteroid is present relatively early in (presumably postnatal) ontogeny, preceding the fusion of the epiphyses and composite elements (e.g., the scapulocoracoid; Unwin et al. 1996). At least as preserved by the fossil record, the pteroid exhibits a similar superficial appearance to other elements of the wing, with a finely foveated or grainy texture in subadults and a smooth unmarked surface in adult material (S. C. Bennett 1993; Unwin et al. 1996). Histologically, the pteroid is composed of fibrolamellar bone with many spindle-shaped lacunae (Unwin et al. 1996).

Hymenochirus: A Case Study

The anuran genus *Hymenochirus* (Pipidae) is presented as a particularly intriguing model for the study of appendicular ossicles, in that up to 10 different sesamoids and lunulae are consistently present (W. M. Olson 2000). Of these, the radial sesamoid and the ulnar patella (fig. 19.4), are located in the forelimb will not be discussed further. The remaining 7 are found in the hindlimb. The knee joint includes a lateral fabella and two lunulae (the antero- and posterolateral lunulae [= "posterior lunula"; W. M. Olson 2000]). To either side of the tibiofibular-tarsal joint there are two sesamoids embedded within the tendon for the m. plantaris longus, the proximal and distal os sesamoides tarsalia (OST). In the distal region of this joint, the os tibialis anticus (intratendinous) and cartilago sesamoides (within the calcaneal ligament) occur much more rarely and are typically confined to larger or older individuals. And finally, a series of four elements, the cartilagines plantares, are located with the digital flexor ten-

dons on the plantar surface of the foot. Within the Pipidae many (but not all) of these sesamoids appear to be unique to *Hymenochirus;* their presence and distribution may be correlated with the highly specialized, unique suction-feeding mode of *Hymenochirus* (W. M. Olson 2000; Deban and Olson 2002; plates 19.1–19.4).

Additional appendicular ossicles are not uncommon in *Hymenochirus,* particularly in older individuals. Small mineralized, intratendinous nodules often appear in both the knee and ankle joint regions; entire tendons or, more rarely, muscle sheaths may also become completely mineralized (W. M. Olson 2000; plate 19.1). Such additional elements are typically not preformed in cartilage, and their presence and location vary considerably. These inconstant ossicles may represent senescence or pathological effects of hyperossification.

Development and Histology

This study includes a complete developmental series of *Hymenochirus,* ranging from tadpoles at early limb bud stages (Nieuwkoop and Faber 1967 [NF] stage 50) to skeletally mature adults. Following the methods outlined in W. M. Olson (2000), specimens were euthanized, fixed, and then either cleared and double-stained for bone and cartilage (see W. M. Olson 2000 for protocol) or serially sectioned (6 μm parasagittal sections) after decalcification (Cal-EX decalcifying solution, Fisher Scientific), dehydration through a graded ethanol series, and embedding in Paraplast. Sections were stained with a modified Masson's trichrome (Witten and Hall 2003).

In whole mount, all sesamoids and lunulae are Alcian-positive (cartilaginous) when they first appear (e.g., plates 19.1, 19.2), with the mineralization of most not occurring until after metamorphosis is complete (table 19.2). Four, including both lunulae, distal OST, and the plantares, are present early during metamorphosis (NF stage 56) as cell-rich precartilaginous condensations (plates 19.3a, 19.4a), concomitant with the initial phases of ossification of the hindlimb diaphyses (W. M. Olson 1998). Perhaps not surprisingly, development of these presumptive ossicles lags behind that of the articular cartilages bounding the interzone. Subsequent development (histogenesis) of these elements is also synchronous, and by NF stage 58 the chondroblasts composing each have begun to hypertrophy (plates 19.3b, 19.4b). Shortly thereafter (NF stage 62; plates 19.3c, 19.4c) cells within the center of each element have secreted enough matrix to become widely separated and may now be termed chondrocytes; cells interconnecting the presumptive lunulae (plate 19.4c) remain as chondroblasts. The developing hyaline cartilage nodules (plates 19.3c, 19.4c) demonstrate both

Table 19-2 Skeletogenesis sequence for hindlimb appendicular ossicles in *Hymenochirus boettgeri* (based on 42 specimens)

Stage	lunulae	fabella	proximal OST	distal OST	os tibialis anticus	cartilago plantares
				cart		cart
57	cart			cart		cart
58	cart			cart		cart
58–60	cart			cart		cart
61	cart			cart		cart
62	cart	?cart	?cart	cart	?cart	cart
63	cart	cart		cart		cart
64	cart	cart		cart	?cart	cart
65	cart	cart		cart	?cart	cart
66	cart	cart		cart	?cart	cart
66+1 month	cart	cart		*min		cart
66+2 months	min	cart		min	cart	*min
66+3 months	min	min		min	min	min
66+4 months	min	min		min	min	min
adult	min	min	min	min	min	min

OST = os sesamoides tarsale
?cart = cartilaginous, very faintly visible in whole-mount preparations
cart = cartilaginous
*min = starting to mineralize
min = mineralized

isogenous groups of chondrocytes (indicative of interstitial growth) within the center of the element, and appositional growth along their periphery. When present, the fabella, proximal OST, and os tibialis anticus also begin to appear in whole mount as skeletal elements at this time.

Immediately prior to metamorphic climax (NF stages 62–65), the fabella appears in nearly 50% of individuals; however, once metamorphosis is complete (NF stage 66+) it is virtually ubiquitous (W. M. Olson 2000; plate 19.2a). The presence of the os tibialis anticus is much less consistent. It occurs rarely in premetamorphic and young postmetamorphic individuals, and is present in only 40% of mature adults (W. M. Olson 2000; plate 19.1b, c). The distal OST begins to mineralize within the first month following completion of metamorphosis (NF stage 66+1), whereas the lunulae and the cartilagines plantares initiate mineralization in the second month. In all three ossicles the hyaline cartilage gradually begins to mineralize near the center of the element.

In the adult the histology of these elements is complex. The distal OST (plate 19.3d–f) includes tendinous attachment sites consisting of fibrocartilage with hyaline cells (fibrohyaline-cell cartilage; Benjamin 1990) that grade insensibly into a hyaline/calcified hyaline cartilage cortex, surrounding a bony center and marrow cavity. The lunulae also become calcified, with the antereolateral element becoming more heavily mineralized than its posterolateral counterpart (compare plates 19.4d, f). By this time the tissue inter-connecting the lunulae (plate 19.4e) consists of fibrocartilage with isogenous groups of cells arranged in rows between densely packed type I collagen bundles. The fabella (plate 19.4f) and the os tibialis anticus (when present) mineralize during the third and fourth postmetamorphic months, respectively. The proximal OST, with one possible exception, does not appear until sexual maturity.

Xenopus laevis and *Pipa pipa*, close relatives of *Hymenochirus*, possess fewer appendicular sesamoids (one and five, respectively). These sesamoids also do not appear in whole-mount preparations (as centers of chondrification) until after the long bones have begun to ossify (Trueb and Hanken 1992; Trueb et al. 2000).

Experimental Analyses

Many of the hindlimb sesamoids in *Hymenochirus* develop during metamorphosis, prior to the limbs becoming fully functional. As with the other limb elements, they are also capable of regeneration following amputation. Girvan et al. (2002) examined the extent of hindlimb regeneration following amputation at two different planes (ankle vs. knee joint) and four different developmental stages (NF stages 52, 54, 56, 59) in *Hymenochirus*. Like many anurans, *Hymenochirus* gradually loses the ability to regenerate as it progresses through metamorphosis, and this loss appears to be correlated with onset of ossification (Girvan et al. 2002). Access to the specimens of Girvan et al. permits us to reanalyze the data, with a particular focus on ossicle regeneration (i.e., the lunulae, fabella, distal OST, and cartilagines plantares).

In general, the extent of sesamoid regeneration matched the extent of limb regeneration. Following amputation at the knee joint, the ability to regenerate the tibiofibula was greatest at the earliest staged amputations (NF stage 52) and was lost by the latest stages (NF stages 56 and 59). In all cases, the tarsus and pes were more severely affected than the rest of the limb, and regeneration was incomplete (Girvan et al. 2002; plate 19.1d); the fabella was invariably absent in regenerates (plate 19.2d). The distal OST was always present in the stage 52 group, although it was slightly displaced proximally (cf. plate 19.1). Its presence was more variable in the stage 54 group; it was also more proximally displaced. The cartilagines plantares were only present in 30% of the stage 52 and 54 regenerates and were always abnormal. The variable presence, displacement, and abnormality of the distal sesamoids appeared to be directly correlated with abnormal, incomplete regeneration of the tarsus and pes. Significantly, the lunulae were always present in regenerates, comparably developed to the contralateral controls (plate 19.2), as long as the joint capsule was retained, regardless of the condition of the tibiofibula.

For specimens amputated at the ankle joint, the lunulae were not affected, but the development of the fabella (when present on the contralateral side) was slightly delayed. The distal OST and cartilagines plantares were present when their respective limb segments were present; any abnormality could be correlated with structural abnormality of the regenerate limb. As with the knee amputations, the distal OST was often displaced proximally. The cartilagines plantares were present in approximately 50% of stage 52 and 54 regenerates and again were abnormal. None of the tadpoles in the original study showed normal or complete regeneration of the digits (Girvan et al. 2002); it is not surprising that the cartilagines plantares (located within the digital flexor tendons) were the most severely affected of the sesamoids.

Of the four sesamoids examined in this study, three appear as early as stage 56, when the hindlimbs are small but well differentiated and the forelimbs have not yet erupted. The fourth, the fabella, does not appear until much later in ontogeny (stage 63), when the tail has already started to regress. The fabella does not regenerate following amputation at the knee joint and is delayed after amputation at the ankle joint. Amputation appears to impact the late-developing fabella much more strongly than the other sesamoids (see plate 19.2).

Denervation (by transection of the sciatic nerve) in *Hymenochirus* at stages 63 and 66, leading to complete (if temporary) paralysis, had mixed effects on the sesamoids of the hindlimb. Similar to the regeneration experiments, induced paralysis may have delayed the appearance of the fabella in some individuals and slowed the rate of mineralization of the fabella, the distal OST, and the cartilagines plantares, but the lunulae appeared to be completely unaffected (H. T. Kim et al. 2002). Given the timing of the denervation relative to the normal ontogeny of the sesamoids (see table 19.2), these results are perhaps not surprising. However, they do point to a role for movement and biomechanics in the later stages of sesamoid development (mineralization and maintenance). It is also interesting to note that the intratendinous sesamoids were more strongly affected by paralysis than were the periarticular lunulae.

Biomechanics

Whereas anatomists frequently overlook appendicular ossicles, these elements have received considerable attention by developmental biologists and biomechanists and have proven a useful system for the elucidation of skeletogenic mechanisms, in particular the exploration of chondrogenic or osteogenic differentiation pathways and signal transduction systems. Similarly, sesamoids have been employed in the study of metaplasia and ectopic ossification, both issues of profound medical interest. Paralysis, surgery, or trauma can all lead to the production of ectopic bone (Rooney 1994), and there is a correlation among increased number of sesamoids, increased incidence and severity of pathological mineralization (such as osteoarthritis), and increased age in both humans (Sarin et al. 1999) and frogs (W. M. Olson 2000).

The traditionally strong link between sesamoid biology and biomechanics has yielded some of the most common explanations for the presence of sesamoids—protection, increased leverage, even phylogenetic remnance—emphasizing the importance of past or present biomechanical function. Sesamoids and other appendicular ossicles are most common in musculoskeletal regions subject to friction, compression, or torsion (Benjamin and Ralphs 1997b). Furthermore, the formation of fibrocartilage, cartilage, and/or bone is a natural response of connective tissue to increased loading (Scapinelli and Little 1970). Such tissues also occur pathologically in response to degeneration or damage (see, e.g., McClure 1983; Rooney 1994). The source of the presumptive chondrocytes may be progenitor cells (reservoirs of undifferentiated cells present in connective, subcutaneous, or muscular tissue; Urist 1980; Wolfman et al. 1997; Bosch et al. 2000) or, as is commonly suggested, the result of chondrometaplasia. Fibrocartilage may differentiate into true cartilage, which may then undergo endochondral ossification, resulting in an ossicle.

Biomechanical stimuli, which play such a critical role in both skeletogenesis proper and the physiological adaptation of bone, may have an even more significant role in the development of sesamoids. There is an ever-growing body of evidence documenting the capacity of connective tissue to alter its histological profile, cellular structure, and genetic expression pattern in direct response to changes in mechanical loading environment. The response of connective tissue to mechanical loading varies greatly. Increased loading may lead to the formation of fibrocartilage, cartilage, or even ossicles in some cases, but may have no effect in others (see below). The effects of decreased mechanical loading, such as occurs during trauma, paralysis, or immobilization, were discussed earlier in the sections on patellar experimentation and development of periarticular elements. Decreased loading can prevent sesamoid formation, cause existing sesamoids to degenerate, or, again, have no effect. Perhaps nowhere do we find a more perfect balance between genetic and epigenetic control of morphogenesis.

Response to Loading

Connective tissue cells (such as tenocytes and other fibroblasts) respond to increased compressive load by changing the composition of the surrounding extracellular matrix (ECM; Scapinelli and Little 1970; Benjamin and Ralphs 1997a). The rat Achilles tendon, for example, contains three phenotypically and developmentally distinct fibrocartilages (Rufai et al. 1992). Two, calcaneal and compressive fibrocartilage, are not present at birth but appear with the onset of locomotion; tenocytes respond to increased loading by undergoing chondrometaplasia and synthesizing and secreting ECM more typically associated with cartilage. The same functionally mediated structural gradation is found in enthesial fibrocartilage (Benjamin et al. 1991; Benjamin and Ralphs 1997a, 1997b). The flexor digitorum profundus tendon (of oxen, rabbits, and dogs) also has distinct histological properties, depending on whether it is primarily subject to tensional or compressive forces (K. G. Vogel and Koob 1989). In tensional regions, cells are elongate and widely dispersed, and the ECM contains parallel fibers, low levels of glycosaminoglycans (GAGs), lots of collagen (but no collagen II), and primarily small proteoglycans. Under compression, the cells are rounded and arranged in columns, fibers are interwoven, and the ECM has high levels of GAGs, a higher percentage of water, collagen II, and primarily large proteoglycans. In short, the compressional regions are more like cartilage.

Biomechanical loading can also alter the three-dimensional configuration of cartilage; changes in cartilage fiber orientation and cell shape can lead to a coordinated cellular and molecular response to mechanical environment

(reviewed in D. R. Carter and Beaupré 2001). Chondrocytes and tenocytes often change orientation in the direction of the loading force, with corresponding deformation of nuclear or cell shape (I. Takahashi et al. 1998; Arnoczky et al. 2002). Integrins, transmembrane receptors that link the intracellular cytoskeleton to ECM and to other cells, may act as bidirectional mechanotransducers (Ingber 1991). Through integrins, force on the ECM can act directly on the cytoskeleton, leading to a cascade of intracellular responses including activation or suppression of gene expression. Similarly, any change in cytoskeletal tension can lead to changes in cell shape, remodeling of ECM, and ultimately changes in tissue organization and function. The bidirectionality of the transduction mechanism reinforces and coordinates the collective response.

In a 3-D gel culture system of embryonic mesenchymal cells, I. Takahashi et al. (1998) found that static compressive force promoted chondrogenesis, in part via increased proliferation and recruitment of chondrocytes. Compression led to an increased production of collagen II and aggrecan, as well as upregulation of Sox9 (a transcriptional activator of collagen II) and downregulation of IL-1beta (a repressor of collagen II and aggrecan). In another study, chondrocytes under continuous hydrostatic pressure responded by changing gene expression of 51/588 genes examined by cDNA microarray analysis (Sironen et al. 2002). Many of the activated genes were transcription factors and genes involved with heat shock response, growth arrest, and differentiation pathways; differentiation inhibitors were down-regulated. Clearly, external mechanical force can directly impact gene expression.

Osteogenic Index

Most true sesamoids develop by endochondral ossification, in a manner similar to long bone and following the same chondrogenic and osteogenic pathways (Sarin and Carter 2000). Sesamoid cartilage forms early and then becomes mineralized or ossified; mature sesamoids may even contain hematopoetic marrow. Ossification is characterized by the formation of mutliple ossific nuclei, and both the initiation and location of these nuclei appear to be directly correlated with mechanical stimulus. Using finite element analysis, D. R. Carter and colleagues have calculated an "osteogenic index" parameter, which can be used to predict the likelihood and pattern of ossification of sesamoids under different stress regimes (D. R. Carter et al. 1991, 1998b; Sarin and Carter 2000). The osteogenic index predicts when and where ossification nuclei will develop, given the mechanical stress history and the morphology or structure of the joint (for cartilaginous sesamoids embedded in tendons wrapping around

joints). For nonconforming joints, the model closely mim-
icked actual developmental data, in that ossification started
in predicted areas of highest mechanical stress (Sarin and
Carter 2000).

Summary and Conclusions

The utility of appendicular ossicles as phylogenetic characters has only recently become established, and regrettably
only few detailed examples exist. In their monumental study
of ossicles of the tetrapod knee, K. Pearson and Davin
(1921b) clearly became discouraged with the nonconstant
nature of ossicle distribution (e.g., abundant in rodents, entirely absent in some chiropterans, reportedly absent in anurans, though see below). However, they did acknowledge
that with further study these elements would undoubtedly
provide important evolutionary information. As discussed
by W. M. Olson (2000), in order to be informative, elements
such as sesamoids, mineralized tendons, and lunulae must
be ontogenetically stable and relatively constant within a
population. Furthermore, if multiple elements share a common developmental mechanism or trigger, then these ossicles cannot be treated as independent variables.

The anuran *Hymenochirus* exhibits both hyperossification and a large number of appendicular ossicles (W. M. Olson 2000). However, not all of these elements are unique to
this taxon. Indeed, all members of the clade Pipidae appear
to possess a radial sesamoid, and the patella ulnaris and at
least one of the os sesamoides tarsalia (proximal and/or distal) appears to be a synapomorphy of an inclusive group
(Pipinae; see fig. 19.9). Future studies examining in detail the
skeletal morphology of *Pseudhymenochirus* or other species
of *Pipa* may determine whether any of the various ossicles
present in *Hymenochirus* is actually an autapomorphy or
one of several shared characters of a larger lineage (e.g., hy-
menochirine pipids, or even Pipinae).

Nonophidian squamates (and perhaps *Sphenodon*) have
also been characterized as demonstrating a large number of
ossicles, mostly sesamoids (Maisano 2002b). Maisano observed that the distribution of such elements was highly conserved at the species level, and thus they offered phylogenetically informative information (fig. 19.10). Hutchinson (2002)
included characters such as the presence of an ossified patella
and mineralized tendons for the digital flexor musculature
to evaluate the phylogeny of crown-group avians and their
out-groups.

Because sesamoids and other ossicles are too often
viewed as being "of more functional than evolutionary significance" (Hildebrand and Goslow 2001, 166), they are of-

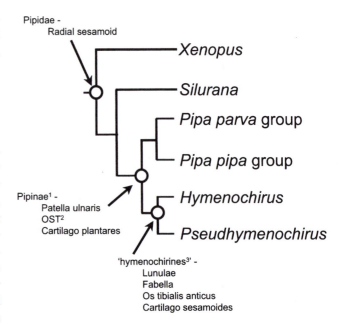

Figure 19-9 Pipidae phylogeny (modified from Cannatella and Trueb 1988a,
1988b) illustrating the distribution of appendicular ossicles (modified from
W. M. Olson 2000). Footnotes: 1, the patella ulnaris and os sesamoides tarsale
(OST) have only been described for *Hymenochirus* spp. and *Pipa pipa*, and may
not be a synapomorphy of Pipinae; 2, it is not clear if *Pipa pipa* has the distal or
proximal OST, or some combination thereof; 3, although details on its skeleton
are lacking, *Pseudhymenochirus* is widely assumed to be similar osteologically
with *Hymenochirus* (see W. M. Olson 2000 for details). The *Pipa parva* group
includes two species (*P. parva* and *P. myersi*), and the *Pipa pipa* group includes
four (*P. pipa, P. carvalhoi, P. arrabali, P. snethlageae*).

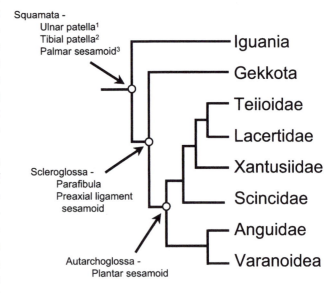

Figure 19-10 Squamate phylogeny (modified from Frost and Etheridge 1989)
illustrating the distribution of appendicular sesamoids (data from Maisano
2002b). Footnotes: 1, the presence of an ulnar patella is reversed in teiioids,
Xantusia riversiana and *X. vigilis*; 2, the presence of a tibial patella is reversed in
Agama and *Basiliscus*; 3, the presence of the palmar sesamoid is reversed in
gekkotans.

ten neglected in anatomical descriptions. We argue that sesamoids and other ossicles are important aspects of skeletal morphology and that they can be phylogenetically informative. As with any phylogenetic character, they must be analyzed and interpreted with caution. Careful analysis can determine the constancy, variability, or developmental stability of individual ossicles, and whether or not they are correlated with pathology or senescence. Similarly, by analyzing their phylogenetic distribution, one can form different testable hypotheses concerning the presence of ossicles in a particular taxon—for example, whether the ossicles are phylogenetic remnants or new structures, or whether the taxon may be predisposed to form sesamoids under certain boundary conditions (such as a global lowered ossification threshold). Diogo et al. (2001), when confronted with unusual skeletal structures in the catfish suspensorium, used a combination of phylogenetic and developmental analyses to determine that the structures were most likely sesamoids.

Whereas we have confined our efforts to the appendicular skeletal system, a wide variety of similar ossicles may be found throughout other regions of the body, including the head (e.g., sclerotic rings), heart (e.g., cardiac ossicles), genitalia (e.g., baculum), and along the vertebral column (e.g., ossified tendons). All such ossicles, commonly grouped under the term "sesamoid," show a perhaps surprising degree of developmental and histological complexity. As stated in the introduction, sesamoids may develop in the complete absence of extrinsic factors. They may also fail to appear in locations where we might otherwise expect them, based on either functional or phylogenetic grounds. Sesamoids may represent a physiological (and evolutionary) adaptation of connective tissue to compression (Summers and Koob 2002), or they may represent the "selective incorporation of acquired characters into the genetical make-up of a species" (Barnett and Lewis 1958, 600).

Sesamoids are a class of musculoskeletal structures born of the principle of self-regulation. Presence or absence of sesamoids and other appendicular ossicles can be attributed to a number of factors acting in combination. Biomechanical loading environment (current and historical), access to progenitor cells, histological profile (structure, complexity) of the connective tissue in which the sesamoid forms, genetic profile, developmental history, and evolutionary or epigenetic history (including the possibility of genetic assimilation) must all enter into a complete analysis of sesamoids. Given this range of factors, sesamoids provide a perfect system in which to study the interaction of genetics and epigenetics, or the modulation of genetic and developmental thresholds by the environment (D. R. Carter et al. 1991, 1998b; Sarin et al. 1999; G. B. Müller 2003). We find it ironic that two of the most intriguing aspects of their biology, their variability and their strong correlation with function, have led to their offhanded dismissal by many anatomists and evolutionary systematists.

References

A

Abad, V., J. A. Uyeda, H. T. Temple, F. De Luca, and J. Baron. 1999. Determinants of spatial polarity in the growth plate. *Endocrinology* 140:958–962.

Abercromby, R. F. 1922. Crocodile (*C. palustris*) burying its food. *Journal of the Bombay Natural History Society* 28:533.

Abeysinghe, D.C., S. Dasgupta, H. E. Jackson, and J. T. Boyd. 2002. Novel MEMS pressure and temperature sensors fabricated on optical fibers. *Journal of Micromechanics and Microengineering* 12:229–235.

Acampora, D., G. R. Merlo, L. Paleari, B. Zerega, M. P. Postiglione, S. Mantero, E. Bober, O. Barbieri, A. Simeone, and G. Levi. 1999. Craniofacial, vestibular and bone defects in mice lacking the Distal-less-related gene *Dlx5. Development* 126:3795–3809.

Adams, C. S., and I. M. Shapiro. 2002. The fate of the terminally differentiated chondrocyte: Evidence for microenvironmental regulation of chondrocyte apoptosis. *Critical Reviews in Oral Biology Medicine* 13:465–473.

Adams, R. A., and K. M. Thibault. 1999. Growth, development, and histology of the calcar in the little brown bat, *Myotis lucifugus* (Vespertilionidae). *Acta Chiropterologica* 1:215–221.

Adriaens, D., S. Devaere, G. G. Teugels, B. Dekegel, and W. Verraes. 2002. Intraspecific variation in limblessness in vertebrates: A unique example of microevolution. *Biological Journal of the Linnaean Society* 75:367–377.

Agarwal, P., J. N. Wylie, J. Galceran, O. Arkhitko, C. Li, C. Deng, R. Grosschedl, and B. G. Bruneau. 2003. Tbx5 is essential for forelimb bud initiation following patterning of the limb field in the mouse embryo. *Development* 130:623–633.

Agata, K., T. Tanaka, C. Kobayashi, K. Kato, and Y. Saitoh. 2003. Intercalary regeneration in planarians. *Developmental Dynamics* 226:308–316.

Agrawal, V. C. 1967. Skull adaptations in fossorial rodents. *Mammalia* 31:300–312.

Ahlberg, P. E. 1989. Paired fin skeletons and relationships of the fossil group Porolepiformes (Osteichthyes: Sarcopterygii). *Zoological Journal of the Linnean Society* 96:119–166.

Ahlberg, P. E. 1991. A re-examination of sarcopterygian interrelationships, with special reference to the Porolepiformes. *Zoological Journal of the Linnean Society* 103:241–287.

Ahlberg, P. E. 1995. *Elginerpeton pancheni* and the earliest tetrapod clade. *Nature* 373:420–425.

Ahlberg, P. E. 1998. Postcranial stem tetrapod remains from the Devonian of Scat Craig, Morayshire, Scotland. *Zoological Journal of the Linnean Society* 122:99–141.

Ahlberg, P. E., and J. A. Clack. 1998. Lower jaws, lower tetrapods: A review based on the Devonian genus *Acanthostega. Transactions of the Royal Society of Edinburgh, Earth Sciences* 89:11–46.

Ahlberg, P. E., and Z. Johanson. 1998. Osteolepiforms and the ancestry of tetrapods. *Nature* 395:792–794.

Ahlberg, P. E., E. Luksevics, and O. Lebedev. 1994. The first tetrapod finds from the Devonian (Upper Famennian) of Latvia. *Philosophical Transactions of the Royal Society of London B* 343:303–328.

Ahlberg, P. E., E. Luksevics, and O. Mark-Kurik. 2000. A near-tetrapod from the Baltic Middle Devonian. *Palaeontology* 43:533–548.

Ahlberg, P. E., and A. R. Milner. 1994. The origin and early diversification of tetrapods. *Nature* 368:507–514.

Ahlberg, P. E., and N. H. Trewin. 1995. The postcranial skeleton of the Middle Devonian lungfish *Dipterus valenciennesi. Transactions of the Royal Society of Edinburgh, Earth Sciences* 85:159–175.

Ahn, D. G., M. J. Kourakis, L. A. Rohde, L. M. Silver, and R. K. Ho. 2002. T-box gene *tbx5* is essential for formation of the pectoral limb bud. *Nature* 417:754–758.

Ahn, K., Y. Mishina, M. C. Hanks, R. R. Behringer, and E. B. Crenshaw III. 2001. BMPR-IA signaling is required for the formation of the apical ectodermal ridge and dorsal-ventral patterning of the limb. *Development* 128:4449–4461.

Ahn, S., and A. L. Joyner. 2004. Dynamic changes in the response of cells to positive hedgehog signaling during mouse limb patterning. *Cell* 118:505–516.

Ahuja, H. S., W. James, and Z. Zakeri. 1997. Rescue of the limb deformity in hammertoe mutant mice by retinoic acid-induced cell death. *Developmental Dynamics* 208:466–481.

Akimenko, M. A., and M. Ekker. 1995. Anterior duplication of the *Sonic hedgehog* expression pattern in the pectoral fin buds of zebrafish treated with retinoic acid. *Developmental Biology* 170:243–247.

Akimenko, M. A., M. Ekker, J. Wegner, W. Lin, and M. Westerfield. 1994. Combinatorial expression of three zebrafish genes related to Distal-less: Part of a homeobox gene code for the head. *Journal of Neuroscience* 14:3475–3486.

Akimenko, M.-A., M. Ekker, and M. Westerfield. 1991. Characterization of three zebrafish genes related to Hox-7. In *Developmental patterning of the vertebrate limb*, ed. J. R. Hinchliffe, 61–63. New York: Plenum.

Akimenko, M. A., S. L. Johnson, M. Westerfield, and M. Ekker. 1995. Differential induction of four *msx* homeobox genes during fin development and regeneration in zebrafish. *Development* 121:347–357.

Akimenko, M. A., M. Marí-Beffa, J. Becerra, and J. Géraudie. 2003. Old questions, new tools, and some answers to the mystery of fin regeneration. *Developmental Dynamics* 226:190–201.

Akita, K. 1996. The effect of the ectoderm on the dorsoventral pattern of epidermis, muscles and joints in the developing chick leg: A new model. *Anatomy and Embryology* (Berlin) 193:377–386.

Akiyama, H., M. C. Chaboissier, J. F. Martin, A. Schedl, and B. de Crombrugghe. 2002. The transcription factor Sox9 has essential roles in successive steps of the chondrocyte differentiation pathway and is required for expression of Sox5 and Sox6. *Genes and Development* 16:2813–2828.

Alappat, S., Z. Y. Zhang, and Y. P. Chen. 2003. Msx homeobox gene family and craniofacial development. *Cell Research* 13:429–442.

Alberch, P. 1985a. Developmental constraints: Why St. Bernards often have an extra digit and poodles never do. *American Naturalist* 126:430–433.

Alberch, P. 1985b. Problems with interpretation of developmental sequences. *Systematic Zoology* 34:46–58.

Alberch, P., and E. A. Gale. 1983. Size dependence during the development of the amphibian foot: Colchicine-induced digital loss and reduction. *Journal of Embryology and Experimental Morphology* 76:177–197.

Alberch, P., and E. A. Gale. 1985. A developmental analysis of an evolutionary trend: Digital reduction in amphibians. *Evolution* 39:8–23.

Alberch, P., S. J. Gould, G. F. Oster, and D. B. Wake. 1979. Size and shape in ontogeny and phylogeny. *Paleobiology* 5:296–317.

Alberty, A., and J. Peltonen. 1993. Proliferation of the hypertrophic chondrocytes of the growth plate after physeal distraction: An experimental study in rabbits. *Clinical Orthopaedics* 297:7–11.

Alexander, R. M. 1983a. *Animal mechanics*. Oxford: Blackwell Scientific.

Alexander, R. M. 1983b. The history of fish mechanics. In *Fish biomechanics*, ed. D. Weihs, 1–35. New York: Praeger.

Alexander, R. M. 1984. *Elastic mechanisms in animal movement*. Cambridge: Cambridge University Press.

Alexander, R. M. 1989. *Dynamics of dinosaurs and other extinct giants*. New York: Columbia University Press.

Alexander, R. M. 1994. The flight of the pterosaur. *Nature* 371:12–13.

Alexander, R. M. 2002. *Principles of animal locomotion*. Princeton, NJ: Princeton University Press.

Alexander, R. M., and A. S. Jayes. 1983. A dynamic similarity hypothesis for the gaits of quadrupedal mammals. *Journal of Zoology, London* 201:135–152.

Alho, C. J. R., and L. F. M. Pádua. 1982. Reproductive parameters and nesting behavior of the Amazon turtle *Podocnemis expansa* (Testudinata: Pelomedusidae) in Brazil. *Canadian Journal of Zoology* 60:97–103.

Allard, H. A. 1948. The eastern box-turtle and its behavior. *Journal of the Tennessee Academy of Sciences* 23:307–321.

Alonso, M., Y. A. Tabata, M. G. Rigolino, and R. Y. Tsukamoto. 2000. Effect of induced triploidy on fin regeneration of juvenile rainbow trout, *Oncorhynchus mykiss*. *Journal of Experimental Zoology* 287:493–502.

Altabef, M., J. D. Clarke, and C. Tickle. 1997. Dorso-ventral ectodermal compartments and origin of apical ectodermal ridge in developing chick limb. *Development* 124:4547–4556.

Altabef, M., C. Logan, C. Tickle, and A. Lumsden. 2000. Engrailed-1 misexpression in chick embryos prevents apical ridge formation but preserves segregation of dorsal and ventral ectodermal compartments. *Developmental Biology* 222:307–316.

Altabef, M., and C. Tickle. 2002. Initiation of dorso-ventral axis during chick limb development. *Mechanisms of Development* 116:19–27.

Altenbach, J. S., and J. W. Hermanson. 1987. Bat flight muscle function and the scapulohumeral lock. In *Recent advances in the study of bats*, ed. M. B. Fenton, P. A. Racey, and J. M. V. Rayner, 100–118. Cambridge: Cambridge University Press.

Altringham, J. D. 1996. *Bats: Biology and behaviour*. Oxford: Oxford University Press.

Amaya, E., T. J. Musci, and M. W. Kirschner. 1991. Expression of a dominant negative mutant of the FGF receptor disrupts mesoderm formation in Xenopus embryos. *Cell* 66:257–270.

Ambler, C. A., J. L. Nowicki, A. C. Burke, and V. L. Bautch. 2001. Assembly of trunk and limb blood vessels involves extensive migration and vasculogenesis of somite-derived angioblasts. *Developmental Biology* 234:352–364.

Amundson, R. 1989. The trials and tribulations of selectionist explanations. In *Issues in evolutionary epistemology*, ed. K. Hahlweg and C. A. Hooker, 413–432. New York: State University of New York Press.

Amundson, R. 2007. Richard Owen and animal form. In *The nature of limbs* by Richard Owen, ed. R. Amundson. Chicago: University of Chicago Press, forthcoming.

Anderson, J. S. 2001. The phylogenetic trunk: Maximal inclusion of taxa with missing data in an analysis of the Lepospondyli (Vertebrata, Tetrapoda). *Systematic Biology* 50:170–193.

Anderson, J. S. 2002. A revision of Phlegethontiidae (Tetrapoda, Lepospondyli, Aïstopoda). *Journal of Paleontology* 76:1029–1046.

Anderson, J. S. 2003. A new aïstopod (Tetrapoda: Lepospondyli) from Mazon Creek, Illinois. *Journal of Vertebrate Paleontology* 23:79–88.

Anderson, J. S., R. L. Carroll, and T. Rowe. 2003. New information on *Lethiscus stocki* (Tetrapoda, Lepospondyli, Aïstopoda) from high resolution computerized tomography and a phylogenetic analysis of Aïstopoda. *Canadian Journal of Earth Sciences* 40:1071–1083.

Andersson, K., and L. Werdelin. 2003. The evolution of cursorial carnivores in the Tertiary: Implications of the elbow joint morphology. *Biology Letters* 270:S163–S165.

Andersson, L. G. 1908. A remarkable new gecko from South Africa and a new *Stenocercus* species from South-America in the Natural Museum in Wiesbaden. *Jahrbücher des Nassauischen Vereins für Naturkunde* 61:299–306.

Andrews, S. M. 1985. Rhizodont crossopterygian fish from the Dinantian of Foulden Berwickshire, Scotland, with a reevaluation of this group. *Transactions of the Royal Society of Edinburgh, Earth Sciences* 76:67–95.

Andrews, S. M., and R. L. Carroll. 1991. The order Adelospondyli: Carboniferous lepospondyl amphibians. *Transactions of the Royal Society of Edinburgh, Earth Sciences* 82:239–275.

Andrews, S. M., and T. S. Westoll. 1970a. The postcranial skeleton of *Eusthenopteron foordi* Whiteaves. *Transactions of the Royal Society of Edinburgh* 68:207–329.

Andrews, S. M., and T. S. Westoll. 1970b. The postcranial skeleton of rhipidistian fishes excluding Eustenopteron. *Transactions of the Royal Society of Edinburgh* 68:391–489.

Aoto, K., T. Nishimura, K. Eto, and J. Motoyama. 2002. Mouse GLI3 regulates Fgf8 expression and apoptosis in the developing neural tube, face, and limb bud. *Developmental Biology* 251:320–332.

Archer, C. W., H. Morrison, and A. A. Pitsillides. 1994. Cellular aspects of the development of diarthrodial joints and articular cartilage. *Journal of Anatomy* 184:447–456.

Archer, S. D., and I. A. Johnston. 1989. Kinematics of labriform and subcarangiform swimming in the antarctic fish *Notothenia neglecta*. *Journal of Experimental Biology* 143:195–210.

Arlton, A. V. 1936. An ecological study of the mole. *Journal of Mammalogy* 17:349–371.

Arnason U., A. Gullberg, and A. Janke. 2001. Molecular phylogenetics of gnathostomous (jawed) fishes: Old bones, new cartilage. *Zoologica Scripta* 30:249–255.

Arnoczky, S. P., M. Lavagnino, J. H. Whallon, and A. Hoonjan. 2002. In situ cell nucleus deformation in tendons under tensile load: A morphological analysis using confocal laser microscopy. *Journal of Orthopaedic Research* 20:29–35.

Arnold, E. N. 1995. Identifying the effects of history on adaptation: Origins of different sand-diving techniques in lizards. *Journal of Zoology, London* 235:351–388.

Arnold, G. P., P. W. Webb, and B. H. Holford. 1991. The role of the pectoral fins in station-holding of Atlantic salmon parr (*Salmo salar* L.). *Journal of Experimental Biology* 156:625–629.

Arratia, G., H.-P. Schultze, and J. Casciotta. 2001. Vertebral column and associated elements in dipnoans and comparison with other fishes: Development and homology. *Journal of Morphology* 250:101–172.

Arreola, V. I., and M. W. Westneat. 1996. Mechanics of propulsion by multiple fins: Kinematics of aquatic locomotion in the burrfish (*Chilomycterus schoepfi*). *Proceedings of the Royal Society of London B* 263:1689–1696.

Asana, J. J. 1941. Further observations on the egg-laying habits of the lizard, *Calotes versicolor* (Boulenger). *Journal of the Bombay Natural History Society* 42:937–940.

Atkinson, R. J. A., and A. C. Taylor. 1991. Burrows and burrowing behaviour of fish. In *The environmental impact of burrowing animals and animal burrows*, ed. P. S. Meadows and A. Meadows, 133–155. Symposia of the Zoological Society of London no. 63. Oxford: Clarendon Press.

Aubin, J., M. Lemieux, M. Tremblay, R. Behringer, and L. Jeannotte. 2002. Transcriptional interferences at the Hoxa4/Hoxa5 locus: Importance for correct Hoxa5 expression for the proper development of the axial skeleton. *Developmental Dynamics* 212:141–156.

Auffenberg, W. 1966. The carpus of land tortoises (Testudininae). *Bulletin of the Florida State Museum, Biological Sciences* 10:159–191.

Auffenberg, W. 1976. The genus *Gopherus* (Testudinidae). Pt. 1: Osteology and relationships of extant species. *Bulletin of the Florida State Museum, Biological Sciences* 20:47–110.

Auffenberg, W. 1981. *The behavioral ecology of the Komodo monitor*. Gainesville: University Presses of Florida.

Auffenberg, W., and W. G. Weaver Jr. 1969. *Gopherus berlandieri* in southeastern Texas. *Bulletin of the Florida State Museum, Biological Sciences* 13:141–203.

Aulehla, A., C. Wehrle, B. Brand-Saberi, R. Kemler, A. Gossler, B. Kanzler, and B. G. Herrmann. 2003. Wnt3a plays a major role in the segmentation clock controlling somitogenesis. *Developmental Cell* 4:395–406.

Ayres, J. A., L. Shum, A. N. Akarsu, R. Dashner, K. Takahashi, T. Ikura, H. C. Slavkin, and G. H. Nuckolls. 2001. DACH: Genomic characterization, evaluation as a candidate for postaxial polydactyly type A2, and developmental expression pattern of the mouse homologue. *Genomics* 77:18–26.

B

Backhouse, K. M. 1961. Locomotion of seals with particular reference to the forelimb. *Symposia of the Zoological Society of London* 5:59–75.

Báez, A. M., and N. G. Basso. 1996. The earliest known frogs of the Jurassic of South America: Review and cladistic appraisal of their relationships. *Münchner Geowissichaftliche Abhandlungen A* 30:131–158.

Bailey, W. J., J. L. Slighton, and M. Goodman. 1992. Rejection of the flying primate hypothesis by phylogenetic evidence from the E-globin gene. *Science* 256:86–89.

Bailòn-Plaza, A., A. O. Lee, E. C. Veson, C. E. Farnum, and M. C. H. van der Meulen. 1999. BMP-5 deficiency alters chondrocytic activity in the mouse proximal tibial growth plate. *Bone* 24:211–216.

Bailòn-Plaza, A., and M. C. H. van der Meulen. 2001. A mathematical framework to study the efects of growth factor influences on fracture healing. *Journal of Theoretical Biology* 212:191–209.

Bajpai, S., and J. G. M. Thewissen. 2000. A new, diminutive Eocene whale from Kachchh (Gujarat, India) and its implications for locomotor evolution of cetaceans. *Current Science* 79:1478–1482.

Bakker, R. T., and P. M. Galton. 1974. Dinosaurian monophyly and a new class of vertebrates. *Nature* 248:168–172.

Balart, E. F. 1995. Development of the vertebral column, fins and fin supports in the Japanese anchovy, *Engraulis japonicus* (Clupeiformes: Engraulididae). *Bulletin of Marine Science* 56:495–522.

Balfour, F. M. 1881. On the development of the skeleton of the paired fins of Elasmobranchii, considered in relation to its bearings on the nature of the limbs of the vertebrata. *Proceedings of the Zoological Society of London* 1881:656–671.

Balfour, F. M. 1885. *The works of Francis Maitland Balfour*. Ed. M. Foster and A. Sedgwick. 4 vols. London: Macmillan.

Balinsky, B. I. 1933. Das Extremitätenseitenfeld, seine Ausdehnung und Beschalfenheit. *Tonx Arch.* 130:704–747.

Balinsky, B. I. 1975. *An introduction into embryology*. Philadelphia: W. B. Saunders.

Ball, S. C. 1936. The distribution and behavior of the spadefoot toad in Connecticut. *Transactions of the Connecticut Academy of Arts and Sciences* 32:351–379.

Ballanti, P., S. Minisola, M. T. Pacitti, L. Scarnecchia, R. Rosso, G. F. Mazzuoli, and E. Bonucci. 1997. Tartrate-resistant acid phosphate activity as osteoclastic marker: Sensitivity of cytochemical assessment and serum assay in comparison with standardized osteoclast histomorphometry. *Osteoporosis International* 7:39–43.

Balmain, N., A. Moscofian, and P. Cuisinier-Gleizes. 1983. The mineralized ring, a single structure peculiar to long bone growth. *Calcified Tissue International* 35:232–236.

Bang, A. G., N. Papalopulu, M. D. Goulding, and C. Kintner. 1999. Expression of Pax-3 in the lateral neural plate is dependent on a Wnt-mediated signal from posterior nonaxial mesoderm. *Developmental Biology* 212:366–380.

Bannasch, R. 1994. Functional anatomy of the "flight" apparatus in penguins. In *Mechanics and physiology of animal swimming*, ed. L. Maddock, Q. Bone, and J. M. V. Rayner, 163–192, 205–229. Cambridge: Cambridge University Press.

Barbour, T. 1926. *Reptiles and amphibians: Their habits and adaptations*. Boston: Houghton Mifflin.

Bardet, N., and M. Fernández. 2000. A new ichthyosaur from the Upper Jurassic lithographic limestones of Bavaria. *Journal of Paleontology* 74:503–511.

Barnett, C. H. 1954. The structure and functions of fibrocartilages within vertebrate joints. *Journal of Anatomy* 88:363–368.

Barnett, C. H., and O. J. Lewis. 1958. The evolution of some traction epiphyses in birds and mammals. *Journal of Anatomy* 92:593–601.

Baron, J., K. O. Klein, M. J. Colli, J. A. Yanovski, J. A. Novosad, J. D. Bacher, and G. B. Cutler. 1994. Catch-up growth after glucocorticoid excess: A mechanism intrinsic to the growth plate. *Endocrinology* 135:1367–1371.

Barrell, J. C. 1916. Influences of Silurian-Devonian climates on the rise of air-breathing vertebrates. *Bulletin of the Geological Society of America* 27:387–436.

Barreto, C., R. M. Albrecht, D. E. Djorling, J. R. Horner, and N. J. Wilsman. 1993. Evidence of the growth plate and the growth of long bones in juvenile dinosaurs. *Science* 262:2020–2023.

Barreto, C., and N. J. Wilsman. 1994. Hypertrophic chondrocyte volume and growth rates in avian growth plates. *Research in Veterinary Science* 56:43–61.

Barrow, J. R., K. R. Thomas, O. Boussadia-Zahui, R. Moore, R. Kemler, M. R. Capecchi, and A. P. McMahon. 2003. Ectodermal Wnt3/beta-catenin signaling is required for the establishment and maintenance of the apical ectodermal ridge. *Genes and Development* 17:394–409.

Barsbold, R., and H. Osmólska. 1990. Ornithomimosauria. In *The Dinosauria*, ed. D. B. Weishampel, P. Dodson, and H. Osmólska, H., 225–248. Berkeley: University of California Press.

Bartholomew, G. A. 1987. Living in two worlds: The marine iguana, *Amblyrhynchus cristatus*. In *Comparative physiology: Life in water and on land*, ed. P. Dejours, L. Bolis, C. R. Taylor, and E. R. Weibel, 389–400. Padua, Italy: Liviana Press.

Bartsch, P., S. Gemballa, and T. Piotrowski. 1997. The embryonic and larval development of *Polypterus senegalus* Cuvier, 1829: Its staging with reference to external and skeletal features, behaviour and locomotory habits. *Acta Zoologica* (Stockholm) 78:309–328.

Başoğlu, M., and S. Zaloğlu. 1964. Morphological and osteological studies in *Pelobates syriacus* from Izmir Region, Western Anatolia (Amphibia, Pelobatidae). *Senckenbergiana Biologica* 45:233–242.

Bataille, G. 1955. *Lascaux; or, The birth of art: Prehistoric painting*. Lausanne, Switzerland: Skira.

Bateman, J. A. 1959. Laboratory studies of the golden mole and mole-rat. *African Wild Life* 13:65–71.

Baudinette, R. V., and P. Gill. 1985. The energetics of "flying" and "paddling" in water: Locomotion in penguins and ducks. *Journal of Comparative Physiology B* 155:373–380.

Bauer, A. M., and A. P. Russell. 1991. Pedal specialisations in dune-dwelling geckos. *Journal of Arid Environments* 20:43–62.

Bauer, H., A. Meier, M. Hild, S. Stachel, A. Economides, D. Hazelett, R. M. Harland, and M. Hammerschmidt. 1998. Follistatin and Noggin are excluded from the zebrafish organizer. *Developmental Biology* 204:488–507.

Baum, D. A., and A. Larson. 1991. Adaptation reviewed: a phylogenetic methodology for studying character macroevolution. *Systematic Zoology* 40:1–18.

Baumel, J. J., A. S. King, J. E. Breazile, H. E. Evans, and J. C. Vanden Berge. 1993. *Handbook of avian anatomy: Nomina anatomica avium*. 2nd ed. Cambridge, MA: Nuttall Ornithological Club.

Baumel, J. J., and R. J. Raikow. 1993. Arthrologia. In *Handbook of avian anatomy: Nomina Anatomica Avium*, 2nd ed., ed. J. J. Baumel, 133–187. Cambridge, MA: Nuttal Ornithological Club.

Baumel, J. J., and L. M. Witmer. 1993. Osteologia. In *Handbook of avian anatomy: Nomina anatomica avium*, 2nd ed., ed. J. J. Baumel, 45–132. Cambridge, MA: Nuttal Ornithological Club.

Baur, G. 1894. The relationship of the lacertilian genus *Anniella*, Gray. *Proceedings of the United States National Museum* 17:345–351.

Baur, G. 1896. The Stegocephali: A phylogenetic study. *Anatomischer Anzeiger* 11:657–673.

Baur, S. T., J. J. Mai, and S. M. Dymecki. 2000. Combinatorial signaling through BMP receptor IB and GDF5: Shaping of the distal mouse limb and the genetics of distal limb diversity. *Development* 127:605–619.

Beaumont, E. I., and T. R. Smithson. 1998. The cranial morphology and relationships of the aberrant Carboniferous amphibian *Spathicephalis mirus* Watson. *Zoological Journal of the Linnean Society* 122:187–209.

Becerra, J., L. C. U. Junqueira, I. J. Bechara, and G. S. Montes. 1996. Regeneration of fin rays in teleosts: A histochemical, radioautographic, and ultrastructural study. *Archives of Histology and Cytology* 59:15–35.

Becerra, J., G. S. Montes, S. R. R. Bexiga, and L. C. U. Junquiera. 1983. Structure of the tail fin in teleosts. *Cell and Tissue Research* 230:127–137.

Bechara, I. J., P. P. Joazeiro, M. Marí-Beffa, J. Becerra, and G. S. Montes. 2000. Collagen-affecting drugs impair regeneration of teleost tail fins. *Journal of Submicroscopy and Cytology and Pathology* 32:273–280.

Beckers, J., M. Gérard, and D. Duboule. 1996. Transgenic analysis of a potential Hoxd-11 limb regulatory element present in tetrapods and fish. *Developmental Biology* 180:543–553.

Beddard, F. E. 1905. The rudimentary hind limbs of the boine snakes. *Nature* 72:630.

Beentjes, M. P. 1990. Comparative terrestrial locomotion of the Hooker's sea lion (*Phocarctos hookeri*) and the New Zealand fur seal (*Arctocephalus forsteri*): Evolutionary and ecological implications. *Zoological Journal of the Linnean Society* 98:307–325.

Belkovich, V. M., and A. V. Yablokov. 1965. O strukture stada zubatykh kitobraznykh. *Morskie Mlekopitayushchie* 1965:25–29.

Bell, G. L. 1997. Introduction to part IV: Mosasauridae. In *Ancient marine reptiles*, ed. J. M. Callaway and E. L. Nicholls, 281–292. San Diego: Academic Press.

Bellairs, A. D'A. 1950. The limbs of snakes with special reference to the hind limb rudiments of *Trachyboa boulengeri*. *British Journal of Herpetology* 1:73–83.

Bellairs, A. D'A. 1969. *The life of reptiles*. Vol. 1. London: Weidenfeld and Nicolson.

Bellus, G. A., I. McIntosh, E. A. Smith, A. S. Aylsworth, I. Kaitila, W. A. Horton, G. A. Greenhaw, J. T. Hecht, and C. A. Francomano. 1995. A recurrent mutation in the tyrosine kinase domain of fibroblast growth factor receptor 3 causes hypochondroplasia. *Nature: Genetics* 10:357–359.

Bemis, W. E., and L. Grande. 1999. Development of the median fins of the North American paddlefish (*Polyodon spathula*), and a reevaluation of the lateral fin-fold hypothesis. In *Mesozoic Fishes 2,* ed. G. Aratia and H.-P. Schultze, 41–68. Munich: Pfeil.

Bendall, A. J., and C. Abate-Shen. 2000. Roles for Msx and Dlx homeoproteins in vertebrate development. *Gene* 247:17–31.

Bendall, A. J., G. Hu, G. Levi, and C. Abate-Shen. 2003. Dlx5 regulates chondrocyte differentiation at multiple stages. *International Journal of Developmental Biology* 47:335–344.

Benjamin, M. 1990. The cranial cartilages of teleosts and their classification. *Journal of Anatomy* 169:153–172.

Benjamin, M., and J. R. Ralphs. 1997a. Cells, tissues, and structures of the musculoskeletal system. *Current Opinion in Orthopedics* 8:29–33.

Benjamin, M., and J. R. Ralphs. 1997b. Tendons and ligaments: An overview. *Histology and Histopathology* 12:1135–1144.

Benjamin, M., J. R. Ralphs, and O. S. Eberewariye. 1992. Cartilage and related tissues in the trunk and fins of teleosts. *Journal of Anatomy* 181:113–118.

Benjamin, M., R. N. S. Tyers, and J. R. Ralphs. 1991. Age related changes in tendon fibrocartilage. *Journal of Anatomy* 179:127–136.

Benke, H. 1993. Investigations on the osteology and the functional morphology of the flipper of whales and dolphins (Cetacea). *Investigations on Cetacea* 24:9–252.

Bennett, D. H. 1972. Notes on the terrestrial wintering of mud turtles (*Kinosternon subrubrum*). *Herpetologica* 28:245–247.

Bennett, D. H., J. W. Gibbons, and J. C. Franson. 1970. Terrestrial activity in aquatic turtles. *Ecology* 51:738–740.

Bennett, M. A., and J. A. Stafford. 1988. Tensile properties of calcified and uncalcified avian tendons. *Journal of Zoology* 214:343–351.

Bennett, S. C. 1990. A pterodactyloid pterosaur pelvis from the Santana formation of Brazil: Implications for terrestrial locomotion. *Journal of Vertebrate Paleontology* 10:80–85.

Bennett, S. C. 1993. The ontogeny of *Pteranodon* and other pterosaurs. *Paleobiology* 19:92–106.

Bennett, S. C. 1996. The phylogenetic position of the Pterosauria within the Archosauromorpha. *Zoological Journal of the Linnean Society* 118:261–308.

Bennett, S. C. 1997. The arboreal leaping theory of the origin of pterosaur flight. *Historical Biology* 12:265–290.

Bennett, S. C. 2001a. The osteology and functional morphology of the Late Cretaceous pterosaur *Pteranodon*. Part 1: General description of osteology. *Palaeontographica Abteilung A* 260:1–112.

Bennett, S. C. 2001b. The osteology and functional morphology of the Late Cretaceous pterosaur *Pteranodon*. Part 2: Size and functional morphology. *Palaeontographica Abteilung A* 260:113–153.

Bent, A. C. 1938. Life histories of North American birds of prey. Part 2: Orders Falconiformes and Strigiformes. *United States National Museum Bulletin* 170:1–482.

Benton, M. J. 1993. Reptilia. In *The fossil record 2,* ed. M. J. Benton, 681–716. London: Chapman and Hall.

Benton, M. J. 1997. *Vertebrate palaeontology.* 2nd ed. London: Chapman and Hall.

Benton, M. J. 1999. *Scleromochlus taylori* and the origin of dino-

saurs and pterosaurs. *Philosophical Transactions of the Royal Society of London B* 354:1423–1446.

Beresford, W. A. 1981. *Chondroid bone, secondary cartilage and metaplasia.* Baltimore: Urban and Schwarzenberg.

Beresford, W. A. 1993. Cranial skeletal tissues: Diversity and evolutionary trends. In *The skull,* vol. 2, *Patterns of structural and systematic diversity,* ed. J. Hanken and B. K. Hall, 69–130. Chicago: University of Chicago Press.

Bergwitz, C., P. Klein, H. Kohno, S. A. Forman, K. Lee, D. Rubin, and H. Jüppner. 1998. Identification, functional characterization, and developmental expression of two nonallelid parathyroid Hormone (PTH)/PTH-related peptide receptor isoforms in *Xenopus laevis* (Daudin). *Endocrinology* 139:723–732.

Berman, D. S., and A. C. Henrici. 2003. Homology of the astragalus and structure and function of the tarsus of Diadectidae. *Journal of Paleontology* 77:172–188.

Berman, D. S., A. C. Henrici, S. S. Sumida, and T. Martens. 2000. Redescription of *Seymouria sanjuanensis* (Seymouriamorpha) from the Lower Permian of Germany based on complete, mature specimens with a discussion of paleoecology of the Bromacker locality assemblage. *Journal of Vertebrate Paleontology* 20:253–268.

Berman, D. S., R. A. Kissel, A. C. Henrici, S. S. Sumida, and T. Martens. 2004. A new diadectid (Diadectomorpha), *Orobates pabsti* from the Early Permian of central Germany. *Annals of the Carnegie Museum* 35:1–35.

Bernays, A. 1878. Die entwicklungsgeschichte des kniegelenkes des menschen mit bemerkugen uber die gelenke im allgemeinen. *Morphologisches Jahrbuch* 4.

Berta, A. 1994. What is a whale? *Science* 263:180–181.

Bertin, L. 1958. Squelette appendiculaire. In *Traité de zoologie: Anatomie, systématique, biologie,* vol. 13, pt. 1, ed. P. P. Grassé, 710–747. Paris: Masson.

Bertram, J. E. A., and A. A. Biewener. 1990. Differential scaling of the long bones in the terrestrial Carnivora and other mammals. *Journal of Morphology* 204:157–169.

Bertram, J. E. A., L. S. Greenberg, T. Miyake, and B. K. Hall. 1997. Paralysis and long bone growth in the chick: Growth shape trajectories of the pelvic limb. *Growth Development and Aging* 61:51–60.

Bertram, J. E. A., and S. M. Swartz. 1991. The "law of bone transformation": A case of crying Wolff? *Biological Reviews* 66:245–273.

Bezuidenhout, A. J., and C. D. Seegers. 1996. The osteology of the African elephant (*Loxodonta Africana*): Vertebral column, ribs and sternum. *Onderstepoort Journal of Veterinary Research* 63:131–147.

Bi, W., W. Huang, D. J. Whitworth, J. M. Deng, Z. Zhang, R. R. Behringer, and B. de Crombrugghe. 2001. Haploinsufficiency of Sox9 results in defective cartilage primordia and premature skeletal mineralization. *Proceedings of the National Academy of Sciences, USA* 98:6698–6703.

Bianco, P., F. D. Cancedda, M. Riminucci, and R. Cancedda. 1998. Bone formation via cartilage models: The "borderline" chondrocyte. *Matrix Biology* 17:185–192.

Biewener, A. A. 1982. Bone strength in small mammals and bipedal birds: Do safety factors change with body size? *Journal of Experimental Biology* 98:289–301.

Biewener, A. A., and J. E. Bertram. 1993. Mechanical loading and bone growth *in vivo.* In *Bone,* vol. 7, *Bone growth—B,* ed. B. K. Hall, 1–36. Boca Raton, FL: CRC Press.

Biewener, A. A., K. P. Dial, and G. E. Goslow Jr. 1992. Pectoralis muscle force and power output during flight in the starling. *Journal of Experimental Biology* 164:1–18.

Biknevicius, A. R. 1993. Biomechanical scaling of limb bones and differential limb use in caviomorph rodents. *Journal of Mammalogy* 74:95–107.

Bilezikian, J. P. 2002. Sex steroids, mice, and men: When androgens and estrogens get very close together. *Journal of Bone and Mineral Research* 17:563–566.

Birnie, J. H. 1934. Regeneration of the tail-fins of Fundulus embryos. *Biological Bulletin* 66:316–325.

Birnie, J. H. 1947. Regeneration and transplantation of fin rays in the goldfish. *Anatomical Record* 99:648.

Bishop, S. C. 1941. The salamanders of New York. *New York State Museum Bulletin* 324:1–365.

Bishop, S. C. 1943. *Handbook of salamanders: The salamanders of the United States, of Canada, and of lower California.* Ithaca, NY: Comstock.

Bitgood, M. J., and A. P. McMahon. 1995. Hedgehog and Bmp genes are coexpressed at many diverse sites of cell-cell interaction in the mouse embryo. *Developmental Biology* 172:126–138.

Bizarro, A. H. 1921. On sesamoid and supernumerary bones of the limbs. *Journal of Anatomy* 55:256–268.

Blair, W. F. 1960. *The rusty lizard: A population study.* Austin: University of Texas Press.

Blake, R. W. 1976. On seahorse locomotion. *Journal of the Marine Biological Association, UK* 56:939–949.

Blake, R. W. 1977. On ostraciiform locomotion. *Journal of the Marine Biological Association, UK* 57:1047–1055.

Blake, R. W. 1979a. The energetics of hovering in the mandarin fish (*Synchropus picturatus*). *Journal of Experimental Biology* 82:25–33.

Blake, R. W. 1979b. The mechanics of labriform locomotion. I: Labriform locomotion in the angelfish (*Pterophyllum eimekei*): An analysis of the power stroke. *Journal of Experimental Biology* 82:255–271.

Blake, R. W. 1979c. The swimming of the mandarin fish *Synchropus picturatus* (Callionyiidae: Teleostei). *Journal of the Marine Biological Association, UK* 59:421–428.

Blake, R. W. 1980. The mechanics of labriform locomotion. II: An analysis of the recovery stroke and the overall fin-beat cycle propulsive efficiency in the angelfish. *Journal of Experimental Biology* 85:337–342.

Blake, R. W. 1981. Influence of pectoral fin shape on thrust and drag in labriform locomotion. *Journal of Zoology, London* 194:43–66.

Blake, R. W. 1983a. *Fish locomotion.* Cambridge: Cambridge University Press.

Blake, R. W. 1983b. Hovering performance of a negatively buoyant fish. *Canadian Journal of Zoology* 61:2629–2630.

Blanc, M. 1953. Contribution à l'étude de l'ostéogénèse chez les poissons téléostéens. *Mémoires du Muséum National d'Histoire Naturelle*, series A, 7:1–146.

Blanco, M. J., and P. Alberch. 1992. Caenogenesis, developmental variability, and evolution in the carpus and tarsus of the marbled newt *Triturus marmoratus*. *Evolution* 46:677–687.

Blanco, M. J., B. Y. Misof, and G. P. Wagner. 1998. Heterochronic differences of Hoxa-11 expression in Xenopus fore- and hind limb development: Evidence for a lower limb identity of the anuran ankle bones. *Development Genes and Evolution* 208:175–187.

Bland, Y. S., and Ashhurst, D. E. 1997. Fetal and postnatal development of the patella, patellar tendon and suprapatella in the rabbit: Changes in the distribution of the fibrillar collagens. *Journal of Anatomy* 190:327–342.

Blau, H. M. 2002. A twist of fate. *Nature* 419:437.

Blau, H. M., T. Brazelton, G. Keshet, and F. Rossi. 2002. Something in the eye of the beholder. *Science* 298:361–362.

Blau, H. M., T. R. Brazelton, and J. M. Weimann. 2001. The evolving concept of a stem cell: Entity or function? *Cell* 105:829–841.

Bledsoe, A. H., R. Raikow, and A. G. Glasgow. 1993. Evolution and functional significance of tendon ossification in woodcreepers (Aves: Passiformes: Dendrocolaptinae). *Journal of Morphology* 215:289–300.

Bloch, J. I., and D. M. Boyer. 2002. Grasping primate origins. *Science* 298:1606–1610.

Bloch, J. I., and D. M. Boyer. 2003. Response to comment on "Grasping primate origins." *Science* 300:741.

Bodemer, C. W. 1958. The development of nerve-induced supernumerary limbs in the adult newt, *Triturus viridescens*. *Journal of Morphology* 102:555–581.

Bodemer, C. W. 1959. Observations on the mechanism of induction of supernumerary limbs in adult *Triturus viridescens*. *Journal of Experimental Zoology* 140:79–99.

Boersma, B., and J. M. Wit. 1997. Catch-up growth. *Endocrine Reviews* 18:646–661.

Bogert, C. M. 1964. Amphisbaenids are a taxonomic enigma. *Natural History* 73 (7): 17–25.

Boilly, B., K. P. Cavanaugh, D. Thomas, H. Hondermarck, S. V. Bryant, and R. A. Bradshaw. 1991. Acidic fibroblast growth factor is present in regenerating limb blastemas of axolotls and binds specifically to blastema tissues. *Developmental Biology* 145:302–310.

Boisvert, C. A. 2005. The pelvic fin and girdle of *Panderichthys* and the origin of tetrapod locomotion. *Nature* 438:1145–1147.

Bolt, J. R. 1969. Lissamphibian origins: Possible protolissamphibian from the Lower Permian of Oklahoma. *Science* 166:888–891.

Bolt, J. R. 1991. Lissamphibian origins. In *Origins of the higher groups of tetrapods*, ed. H.-P. Schultze and L. Trueb, 194–222. Ithaca, NY: Cornell University Press.

Bolt, J. R., and R. E. Lombard. 2000. Palaeobiology of *Whatcheeria deltae*, a primitive Mississippian tetrapods. In *Amphibian biology*, vol. 4, *Palaeontology: The evolutionary history of amphibians*, ed. H. Heatwole and R. L. Carroll, 1044–1052. Chipping Norton, Australia: Surrey Beatty and Sons.

Bordat, C. 1987. Etude ultrastructurale de l'os des vertèbres du sélacien *Scyliorhinus canicula* L. *Canadian Journal of Zoology* 65:1435–1444.

Borday, V., C. Thaëron, F. Avaron, A. Brulfert, D. Casane, P. Laurenti, and J. Géraudie. 2001. *evx1* transcription in bony fin rays segment boundaries leads to a reiterated pattern during zebrafish fin development and regeneration. *Developmental Dynamics* 220:91–98.

Borelli, G. A. 1680. *De Motu Animalium.* Trans. P. Maquet. 1989. Berlin: Springer-Verlag.

Borsuk-Białynicka, M., and S. E. Evans. 2002. The scapulocoracoid of an Early Triassic stem-frog from Poland. *Acta Palaeontologica Polonica* 47:79–96.

Bosch, P., D. S. Musgrave, J. Y. Lee, J. Cummins, F. Shuler, S. C. Ghivizzani, C. Evans, P. D. Robbins, J. Huard. 2000. Osteoprogenitor cells within skeletal muscle. *Journal of Orthopaedic Research* 18:933–944.

Bossy, K. A., and A. C. Milner. 1998. Order Nectridea. In *Encyclopedia of paleoherpetology,* vol. 1, *Lepospondyli,* ed. P. Wellnhofer, 73–108. Munich: Pfeil.

Böttcher, R. 1990. Neue Erkenntnisse über die Fortpflanzungsbiologie der Ichthyosaurier (Reptilia). *Stuttgarter Beiträge zur Naturkunde B* 164:1–51.

Bou, J., A. Casinos, J. Ocaña. 1987. Allometry of the limb long bones of insectivores and rodents. *Journal of Morphology* 192:113–123.

Boulenger, G. A. 1896. *The tailless batrachians of Europe.* Part 1. London: Ray Society.

Boulenger, G. A. 1897. *The tailless batrachians of Europe.* Part 2. London: Ray Society.

Boulet, A. M., and M. R. Capecchi. 2004. Multiple roles of Hoxa11 and Hoxd11 in the formation of the mammalian forelimb zeugopod. *Development* 131:299–309.

Boulet, A. M., A. M. Moon, B. R. Arenkiel, and M. R. Capecchi. 2004. The roles of Fgf4 and Fgf8 in limb bud initiation and outgrowth. *Developmental Biology* 273:361–372.

Bouvet, J. 1974a. Différenciation et ultrastructure du squelette distal de la nageoire pectorale chez la truite indigène (*Salmo trutta fario* L.). I: Différenciation et ultrastructure des actinotriches. *Archives d'Anatomie Microscopique* 63:79–96.

Bouvet, J. 1974b. Différenciation et ultrastructure du squelette distal de la nageoire pectorale chez la truite indigène (*Salmo trutta fario* L.). II: Différenciation et ultrastructure des lépidotriches. *Archives d'Anatomie Microscopique* 63:323–335.

Bowler, P. J. 1996. *Life's splendid drama: Evolutionary biology and the reconstruction of life's ancestry, 1860–1940.* Chicago: University of Chicago Press.

Bowen, J., J. R. Hinchliffe, T. J. Horder, and A. M. Reeve. 1989. The fate map of the chick forelimb-bud and its bearing on hypothesized developmental control mechanisms. *Anatomy and Embryology* (Berlin) 179:269–283.

Boy, J. A. 1996. Ein neuer Eryopoids (Amphibia: Temnospondyli) aus dem saarpfalzischen Rotliegend (Unter-Perm; Sudwest-Deutschland). *Mainzer Geowissenschaftliche Mitteilungen* 25:7–26.

Boy, J. A., and H.-D. Sues. 2000. Branchiosaurs: Larvae, metamorphosis and heterochrony in Temnospondyls and Seymouriamorphs. In *Amphibian biology,* vol. 4, *Palaeontology: The evolutionary history of amphibians,* ed. H. Heatwole and R. L. Carroll, 1150–1197. Chipping Norton, Australia: Surrey Beatty and Sons.

Boyden, L. M., J. Mao, J. Belsky, L. Mitzner, A. Farhi, M. A. Mitnick, D. Wu, K. Insogna, and P. P. Lifton. 2002. High bone density due to a mutation in LDL-receptor-related protein 5. *New England Journal of Medicine* 346:1513–1521.

Boyer, M. I., P. W. Bray, C. V. A. Bowen. 1994. Epiphyseal plate transplantation: An historical overview. *British Journal of Plastic Surgery* 47:663–669.

Boyle, W. J., W. S. Simonet, and D. L. Lacey. 2003. Osteoclast differentiation and activation. *Nature* 423:337–342.

Bragg, A. N. 1965. *Gnomes of the night: The spadefoot toads.* Philadelphia: University of Pennsylvania Press.

Brain, C. K. 1962. A review of the gecko genus *Ptenopus* with the description of a new species. *Cimbebasia* 1:1–18.

Bramble, D. M. 1982. *Scaptochelys:* Generic revision and evolution of gopher tortoises. *Copeia* 1982:852–867.

Bramwell, C. D., and G. R. Whitfield. 1974. Biomechanics of *Pteranodon. Philosophical Transactions of the Royal Society of London B* 267:503–581.

Brandon, R. A. 1965. Morphological variation and ecology of the salamander *Phaeognathus hubrichti. Copeia* 1965:67–71.

Brandon, R. N. 1990. *Adaptation and environment.* Princeton, NJ: Princeton University Press.

Brandstätter, R., B. Misof, C. Pazmandi, and G. P. Wagner. 1990. Micro-anatomy of the pectoral fin in blennies (Blenniini, Blennioidea, Teleostei). *Journal of Fish Biology* 37:729–743.

Branford, W. W., G. V. Benson, L. Ma, R. L. Maas, and S. S. Potter. 2000. Characterization of *Hoxa-10/Hoxa-11* trans heterozygotes reveals functional redundancy and regulatory interactions. *Developmental Biology* 224:373–387.

Braus, H. 1906a. Die Entwicklung der Form der Extremitäten und des Extremitätenskeletts. In *Handbuch der vergleichenden und experimentellen Entwicklungslehre der Wirbeltiere,* ed. O. Hertwig, 167–338. Jena, Germany: Gustav Fisher.

Braus, H. 1906b. Ist die Bildung des Skeletes von den Muskelanlagen abhängig? Eine experimentelle Untersuchung an der Brustflosse von Haiembryonen. *Morphologie Jahrbuch* 35:240–321.

Braus, H. 1919. Der Brustschulterapparat der Froschlurche. *Sitzungsberichte der Heidelberger Akademie der Wissenschaften. Abteilung B. Abhandlungen* 15.

Breder, C. M., Jr. 1926. The locomotion of fishes. *Zoologica (New York)* 4:159–296.

Breder, C. M., Jr., and H. E. Edgerton. 1942. An analysis of the locomotion of the seahorse, *Hippocampus,* by means of high speed cinematography. *Annals of the New York Academy of Sciences* 43:145–172.

Breur, G. J., B. A. VanEnkevort, C. E. Farnum, and N. J. Wilsman. 1991. Linear relationship between the volume of hypertrophic chondrocytes and the rate of longitudinal growth in growth plates. *Journal of Orthopaedic Research* 9:348–359.

Breur, G. J., C. E. Farnum, G. A. Padgett, and N. J. Wilsman. 1992. Cellular basis of decreased rate of longitudinal growth in pseudoachondroplastic dogs. *Journal of Bone and Joint Surgery* 74:516–528.

Brimley, C. S. 1920. Notes on *Amphiuma* and *Necturus. Copeia* 1920:5–7.

Brinkman, D. 1980. The hindlimb step cycle of *Caiman sclerops* and the mechanics of the crocodile tarsus and metatarsus. *Canadian Journal of Zoology* 58:2187–2200.

Brockes, J. P. 1997. Amphibian limb regeneration: Rebuilding a complex structure. *Science* 276:81–87.

Brockes, J. P. 1998. Progenitor cells for regeneration: Origin by reversal of the differentiated state. In *Cellular and molecular basis of regeneration,* ed. P. Ferretti and J. Géraudie, 63–77. Chichester, UK: Wiley.

Brockes, J. P., and A. Kumar. 2002. Plasticity and reprogramming of differentiated cells in amphibian regeneration. *Nature Reviews Molecular Cellular Biology* 3:566–574.

Broglio, F., E. Arvat, A. Benso, M. Papotti, G. Muccioli, R. Deghenghi, and E. Ghigo. 2002. Ghrelin: Endocrine and non-endocrine actions. *Journal of Pediatric Endocrinology and Metabolism* 15:1219–1227.

Broili, F., and J. Schröder. 1937. Beobachtungen an Wirbeltieren der Karooformation. XXV: Über *Micropholis* Huxley. *Sitzungsberichte Akademie der Wissenschaften München* 1937:19–38.

Brongersma, L. D., and W. Helle. 1951. Notes on Indo-Australian snakes. I: Koninklijke Nederlandse Akademie van Wetenschappen. *Proceedings of the Section of Sciences,* series C, *Biological and Medical Sciences* 54:1–8.

Brooke, R. 1937. The treatment of fractured patella by excision: A study of morphology and function. *British Journal of Surgery* 24:733–747.

Broom, R. 1913. On the origin of the Cheiropterygium. *Bulletin of the American Museum of Natural History* 32:265–272.

Broome, G. H. H., and G. R. Houghton. 1989. A congenital abnormality of the tibial tuberosity representing the evolution of traction epiphyses. *Journal of Anatomy* 165:275–278.

Brower, J. C. 1982. The aerodynamics of an ancient flying reptile. *Syracuse Scholar* 1982:45–57.

Brown, B. 1914. *Leptoceratops,* a new genus of Ceratopsia from the Edmonton Cretaceous of Alberta. *Bulletin of the American Museum of Natural History* 33:539–548.

Brown, B., and E. M. Schlaikjer. 1940. The structure and relationships of *Protoceratops. Annals of the New York Academy of Sciences* 40:133–266.

Brown, L. E. 1996. Trend toward extinction of the unusual forward burrowing Illinois chorus frog. *Reptile and Amphibian Magazine* 1996 (November/December): 70–73.

Brown, L. E., H. O. Jackson, and J. R. Brown, J. R. 1972. Burrowing behavior of the chorus frog *Pseudacris streckeri. Herpetologica* 28:325–328.

Brown, L. E., and D. B. Means. 1984. Fossorial behavior and ecology of the chorus frog *Pseudacris ornata. Amphibia-Reptilia* 5:261–273.

Brown, S., J. Fellers, T. Shippy, R. Denell, M. Stauber et al. 2001. A strategy for mapping bicoid on the phylogenetic tree. *Current Biology* 11:R43–R44.

Brulfert, A., M. J. Monnot, and J. Geraudie. 1998. Expression of two even-skipped genes eve1 and evx2 during zebrafish fin morphogenesis and their regulation by retinoic acid. *International Journal of Developmental Biology* 42:1117–1124.

Bruneau, S., P. Mourrain, and F. M. Rosa. 1997. Expression of *contact,* a new zebrafish DVR member, marks mesenchymal cell lineages in the developing pectoral fins and head and is regulated by retinoic acid. *Mechanisms of Development* 65:163–173.

Brunet, L. J., J. A. McMahon, A. P. McMahon, and R. M. Harland. 1998. Noggin, cartilage morphogenesis, and joint formation in the mammalian skeleton. *Science* 280:1455–1457.

Brunner, C. 1891. Ueber Genese, congenitalen Mangel und rudimentäre Bildung der Patella. *Virchows Archiv Abteilung a Pathologische Anatomie* 124:358–372.

Bryant, S. V., T. Endo, and D. M. Gardiner. 2002. Vertebrate limb regeneration and the origin of limb stem cells. *International Journal of Developmental Biology* 46:887–896.

Bryant, S. V., V. French, and P. J. Bryant. 1981. Distal regeneration and symmetry. *Science* 212:993–1002.

Bryant, S. V., D. M. Gardiner, and K. Muneoka. 1987. Limb development and regeneration. *American Zoologist* 27:675–696.

Bryant, S. V., and L. Iten. 1974. The regulative ability of the limb regeneration blastema of *Notophthalmus viridescens:* Experiments *in situ. Wilhelm Roux's Archives of Developmental Biology* 174:90–101.

Bryant, S. V., and K. Muneoka. 1986. Views of limb development and regeneration. *Trends in Genetics* 2:153–159.

Buchholtz, E. A. 1998. Implications of vertebral morphology for

locomotor evolution in early Cetacea. In *The emergence of whales: Evolutionary patterns in the origin of Cetacea,* ed. J. G. M. Thewissen, 325–351. New York: Plenum Press.

Buchholtz, E. A. 2001. Swimming styles in Jurassic ichthyosaurs. *Journal of Vertebrate Paleontology* 21:61–73.

Buckwalter, J. A., and R. D. Sjolund. 1989. Growth of corn stalks and long bones: Does longitudinal bone growth depend upon a matrix-directed hydraulic mechanism? *Iowa Orthopaedic Journal* 9:25–31.

Buffrénil, V. de, and J.-M. Mazin. 1990. Bone histology of the ichthyosaurs: Comparative data and functional interpretation. *Paleobiology* 16:435–447.

Buffrénil, V. de, and J.-M. Mazin. 1992. Contribution de l'histologie osseuse à l'interprétation paléobiologique du genre *Placodus* Agassiz, 1833 (Reptilia, Placodontia). *Revue de Paléobiologie* 11:397–407.

Bullock, A., R. Marks, and R. Roberts. 1978. The cell kinetics of teleost fish epidermis: Epidermal mitotic activity in relation to wound healing at varying temperatures in plaice (*Pleuronectes platessa*). *Journal of Zoology* 185:197–204.

Burger, J. 1977. Determinants of hatching success in diamondback terrapin, *Malaclemys terrapin. American Midland Naturalist* 97:444–464.

Burgess, T. L., Y. Qian, S. Kaufman, B. D. Ring, G. Van, C. Capparelli, M. Kelley, et al. 1999. The ligand for osteoprotegerin (OPGL) directly activates mature osteoclasts. *Journal of Cell Biology* 145:527–538.

Burke, A. C. 1989. Development of the turtle carapace: Implications for the evolution of a novel Bauplan. *Journal of Morphology* 199:363–378.

Burke, A. C., and P. Alberch. 1985. The development and homology of the chelonian carpus and tarsus. *Journal of Morphology* 186:119–131.

Burke, A. C., and A. Feduccia. 1997. Developmental patterns and the identification of homologies in the avian hand. *Science* 278:666–668.

Burman, M. S., and P. W. Lapidus. 1931. The functional disturbances caused by the inconstant bones and sesamoids of the foot. *Archives of Surgery* 22:936–975.

Burnham, D. A., K. L. Derstler, P. J. Currie, R. T. Bakker, Z. Zhou, and J. H. Ostrom. 2000. Remarkable new birdlike dinosaur (Theropoda: Maniraptora) from the Upper Cretaceous of Montana. *University of Kansas Paleontological Contributions* 13:1–14.

Burton, T. C. 2001. Variation in the foot muscles of frogs of the family Myobatrachidae. *Australian Journal of Zoology* 49:539–559.

Bushdid, P. B., D. M. Brantley, F. E. Yull, G. L. Blaeuer, L. H. Hoffman, L. Niswander, and L. D. Kerr. 1998. Inhibition of NF-kappaB activity results in disruption of the apical ectodermal ridge and aberrant limb morphogenesis. *Nature* 392:615–618.

Bushdid, P. B., C.-L. Chen, D. M. Brantley, F. Yull, R. Raghow, L. D. Kerr, and J. V. Barnett. 2001. NF-kappaB mediates FGF signal regulation of msx-1 expression. *Developmental Biology* 237:107–115.

Bustard, H. R., P. Greenham, and C. Limpus. 1975. Nesting behaviour of loggerhead and flatback turtles in Queensland, Australia. *Proceedings of the Koninklijke Nederlandse Akademie van Wetenschappen,* series C, 78:111–122.

Buxton, P., B. K. Hall, C. W. Archer, and P. Francis-West. 2003.

Secondary chondrocyte-derived Ihh stimulates proliferation of periosteal cells during chick development. *Development* 130:4729–4739.

Buxton, P. G., K. Kostakopoulou, P. Brickell, P. Thorogood, and P. Ferretti. 1997. Expression of the transcription factor slug correlates with growth of the limb bud and is regulated by FGF-4 and retinoic acid. *International Journal of Developmental Biology* 41:559–568.

Bystrow, A. P. 1938. *Divinosaurus* als neotenische Form der Stegocephalen. *Acta Zoologica* 19:209–295.

C

Cagle, F. R. 1948. Observations on a population of the salamander, *Amphiuma tridactylum* Cuvier. *Ecology* 29:479–491.

Caldwell, M. W. 1996. Ontogeny and phylogeny of the mesopodial skeleton in mosasauroid reptiles. *Zoological Journal of the Linnean Society* 116:407–436.

Caldwell, M. W. 1997a. Limb ossification patterns of the ichthyosaur *Stenopterygius*, and a discussion of the proximal tarsal row of ichthyosaurs and other neodiapsid reptiles. *Zoological Journal of the Linnean Society* 120:1–25.

Caldwell, M. W. 1997b. Limb osteology and ossification patterns in *Cryptoclidus* (Reptilia: Pleiosauroidea) with a review of sauropterygian limbs. *Journal of Vertebrate Palentology* 17:295–307.

Caldwell, M. W. 1997c. Modified perichondral ossification and the evolution of paddle-like limbs in ichthyosaurs and plesiosaurs. *Journal of Vertebrate Paleontology* 17:534–547.

Caldwell, M. W. 1999. Squamate phylogeny and the relationships of snakes and mosasauroids. *Zoological Journal of the Linnean Society* 125:115–147.

Caldwell, M. W. 2000. On the aquatic squamate *Dolichosaurus longicollis* Owen, 1850 (Cenomanian, Upper Cretaceous) and the evolution of elongate necks in squamates. *Journal of Vertebrate Paleontology* 20:720–735.

Caldwell, M. W. 2002. From fins to limbs to fins: Limb evolution in fossil marine reptiles. *American Journal of Medical Genetics* 112:236–249.

Caldwell, M. W., and J. A. Cooper. 1999. Redescription, palaeobiogeography and palaeoecology of *Coniasaurus crassidens* Owen, 1850 (Squamata) from the Lower Chalk (Cretaceous, Cenomanian) of SE England. *Zoological Journal of the Linnean Society* 127:423–452.

Caldwell, M. W., R. L. Carroll, and H. Kaiser. 1995. The pectoral girdle and forelimb of *Carsosaurus marchesetti* (Aigialosauridae), with a preliminary phylogenetic analysis of mosasauroids and varanoids. *Journal of Vertebrate Paleontology* 15:516–531.

Caldwell, M. W., and M. S. Y. Lee. 1997. A snake with legs from the marine Cretaceous of the Middle East. *Nature* 386:705–709.

Callaway, J. M., and E. L. Nicholls, eds. 1997. *Ancient marine reptiles.* San Diego, CA: Academic Press.

Cameron, J., and J. F. Fallon. 1977. Absence of cell death during development of free digits in amphibians. *Developmental Biology* 55:331–338.

Cameron, J. A., A. R. Hilgers, and T. J. Hinterberger. 1986. Evidence that reserve cells are a source of regenerated adult newt muscle in-vitro. *Nature* 321:607–610.

Camin, S., L. Madoery, and V. Roig. 1995. The burrowing behavior of *Ctenomys mendocinus* (Rodentia). *Mammalia* 59:9–17.

Camp, C. L. 1923. Classification of the lizards. *Bulletin of the American Museum of Natural History* 48:289–481.

Camp, C. L., and N. Smith. 1942. Phylogeny and functions of the distal ligaments of the horse. *Memoirs of the University of California, Berkeley* 13:1–73.

Campbell, B. 1938. A reconsideration of the shoulder musculature of the Cape golden mole. *Journal of Mammalogy* 19:234–240.

Campbell, B. 1939. The shoulder anatomy of the moles: A study in phylogeny and adaptation. *American Journal of Anatomy* 64:1–39.

Campbell, G. 2002. Distalization of the *Drosophila* leg by graded EGF-receptor activity. *Nature* 418:781–785.

Campbell, H. W. 1972. Ecological or phylogenetic interpretations of crocodilian nesting habits. *Nature* 238:404–405.

Campbell, K. E., Jr. 1992. *Papers in avian paleontology, honoring Pierce Brodkorb.* Los Angeles, CA: Natural History Museum of Los Angeles County.

Campisi, J. 1996. Replicative senescence: An old lives' tale? *Cell* 84:497–500.

Cancedda, R., F. D. Cancedda, and P. Castagnola. 1995. Chondrocyte differentiation. *International Review of Cytology* 159:265–358.

Cannatella, D. C., and L. Trueb. 1988a. Evolution of pipoid frogs: Intergeneric relationships of the aquatic frog family Pipidae (Anura). *Zoological Journal of the Linnean Society* 94:1–38.

Cannatella, D. C., and L. Trueb. 1988b. Evolution of pipoid frogs: Morphology and phylogenetic relationships of *Pseudhymenochirus*. *Journal of Herpetology* 22:439–456.

Capdevila, J., and J. C. Izpisúa Belmonte. 2001. Perspectives on the evolutionary origin of tetrapod limbs. In *The character concept in evolutionary biology,* ed. G. P. Wagner, 531–558. San Diego, CA: Academic Press.

Capdevila, J., and R. L. Johnson. 1998. Endogenous and ectopic expression of noggin suggests a conserved mechanism for regulation of BMP function during limb and somite patterning. *Developmental Biology* 197:205–217.

Capdevila, J., T. Tsukui, C. Rodriguez-Esteban, V. Zappavigna, and J. C. Izpisúa Belmonte. 1999. Control of vertebrate limb outgrowth by the proximal factor Meis2 and distal antagonism of BMPs by Gremlin. *Molecular Cell* 4:839–849.

Caputo, V., B. Lanza, and R. Palmieri. 1995. Body elongation and limb reduction in the genus *Chalcides* Laurenti 1768 (Squamata Scincidae): A comparative study. *Tropical Zoology* 8:95–152.

Carey, E. J., W. Zeit, and B. F. McGrath. 1927. Studies in the dynamics of histogenesis. *Journal of Anatomy* 40:127–158.

Carlevaro, M. F., S. Cermelli, R. Cancedda, and F. D. Cancedda. 2000. Vascular endothelia growth factor (VEGF) in cartilage neovascularization and chondrocyte differentiation: Auto-paracrine role during endochondral bone formation. *Journal of Cell Science* 113:49–69.

Carling, J., T. L. Williams, and G. Bowtell. 1998. Self-propelled anguilliform swimming: simultaneous solution of the two-dimensional Navier-Stokes equations and Newton's laws of motion. *Journal of Experimental Biology* 201:3143–3166.

Carlson, B. M. 1998. Development and regeneration, with special emphasis on the amphibian limb. In *Cellular and molecular basis of regeneration,* ed. P. Ferretti and J. Géraudie, 45–61. Chichester, UK: Wiley.

Carlson, B. M. 2003. Muscle regeneration in amphibians and

mammals: Passing the torch. *Developmental Dynamics* 226:167–181.

Carlson, M. R. J., S. V. Bryant, and D. M. Gardiner. 1998. Expression of Msx-2 during development, regeneration, and wound healing in axolotl limbs. *Journal of Experimental Zoology* 282:715–723.

Carlson, M. R. J., Y. Komine, S. V. Bryant, and D. M. Gardiner. 2001. Expression of *Hoxb13* and *Hoxc10* in developing and regenerating axolotl limbs and tails. *Developmental Biology* 229:396–406.

Carlsson, A. 1904. Zur Anatomie des *Notoryctes typhlops*. *Zoologische Jahrbücher. Abteilung für Anatomie und Ontogenie der Tiere* 20:81–122.

Caronia, G., F. R. Goodman, C. M. McKeown, P. J. Scambler, and V. Zappavigna. 2003. An I47L substitution in the HOXD13 homeodomain causes a novel human limb malformation by producing a selective loss of function. *Development* 130:1701–1712.

Carothers, J. H. 1986. An experimental confirmation of morphological adaptation: Toe fringes in the sand-dwelling lizard *Uma scoparia. Evolution* 40:871–874.

Carpenter, C. C. 1966. The marine iguana of the Galápagos Islands, its behavior and ecology. *Proceedings of the California Academy of Sciences,* series 4, 34:329–376.

Carpenter, C. C. 1982. The bullsnake as an excavator. *Journal of Herpetology* 16:394–401.

Carpenter, E. M., J. M. Goddard, A. P. Davis, T. P. Nguyen, and M. R. Capecchi. 1997. Targeted disruption of Hoxd-10 affects mouse hindlimb development. *Development* 124:4505–4514.

Carr, A. 1952. *Handbook of turtles: The turtles of the United States, Canada, and Baja California.* Ithaca, NY: Cornell University Press.

Carr, A., H. Hirth, and L. Ogren. 1966. The ecology and migrations of sea turtles. 6: The hawksbill turtle in the Caribbean Sea. *American Museum Novitates* 2248:1–29.

Carr, A., and L. Ogren. 1959. The ecology and migrations of sea turtles. 3: *Dermochelys* in Costa Rica. *American Museum Novitates* 1958:1–29.

Carr, A., L. Ogren, and C. McVea. 1980. Apparent hibernation by the Atlantic loggerhead turtle *Caretta caretta* off Cape Canaveral, Florida. *Biological Conservation* 19:7–14.

Carrano, M. T. 1999. What, if anything, is a cursor? Categories versus continua for determining locomotor habit in mammals and dinosaurs. *Journal of Zoology, London* 247:29–42.

Carrier, D. R., and C. G. Farmer. 2000. The evolution of pelvic aspiration in archosaurs. *Paleobiology* 26:271–293.

Carrig, C. B., and J. A. Wortman. 1981. Acquired dysplasias of the canine radius and ulna. *Compendium on Continuing Education* 3:557–565.

Carrio, R., M. Lopez-Hoyos, J. Jimeno, M. A. Benedict, R. Merino, A. Benito, J. L. Fernandez-Luna, G. Nunez, J. A. Garcia-Porrero, and J. Merino. 1996. A1 demonstrates restricted tissue distribution during embryonic development and functions to protect against cell death. *American Journal of Pathology* 149:2133–2142.

Carroll, R. L. 1964a. The earliest reptiles. *Journal of the Linnean Society (Zoology)* 45:61–83.

Carroll, R. L. 1964b. Early evolution of the dissorophid amphibians. *Bulletin of the Museum of Comparative Zoology, Harvard* 131:163–250.

Carroll, R. L. 1964c. The relationship of the rhachitomous amphibian *Parioxys. American Museum Novitates* 2167:1–11.

Carroll, R. L. 1967a. Labyrinthodonts from the Joggins Formation. *Journal of Paleontology* 41:111–142.

Carroll, R. L. 1967b. A limnoscelid reptile from the Middle Pennsylvanian. *Journal of Paleontology* 41:1256–1261.

Carroll, R. L. 1968. The postcranial skeleton of the Permian microsaur *Pantylus. Canadian Journal of Zoology* 46:1175–1192.

Carroll, R. L. 1969a. A Middle Pennsylvanian captorhinomorph, and the interrelationships of primitive reptiles. *Journal of Paleontology* 43:151–170.

Carroll, R. L. 1969b. A new family of Carboniferous amphibians. *Palaeontology* 12:537–548.

Carroll, R. L. 1970. The ancestry of reptiles. *Philosophical Transactions of the Royal Society of London B* 257:267–308.

Carroll, R. L. 1985. Evolutionary constraints in aquatic diapsid reptiles. *Special Papers in Palaeontology* 33:145–155.

Carroll, R. L. 1987. *Vertebrate paleontology and evolution.* New York: Freeman.

Carroll, R. L. 1988. An articulated gymnarthrid microsaurs (Amphibia) from the Upper Carboniferous of Czechoslovakia. *Acata Zoologica Cracovi* 31:441–450.

Carroll, R. L. 1991a. *Batropetes* from the Lower Permian of Europe: A microsaur, not a reptile. *Journal of Vertebrate Paleontology* 11:229–242.

Carroll, R. L. 1991b. The origin of reptiles. In *Origins of the higher groups of tetrapods,* ed. H.-P. Schultze and L. Trueb, 331–353. Ithaca, NY: Cornell University Press.

Carroll, R. L. 1997a. Mesozoic marine reptiles as models of long-term, large-scale evolutionary phenomena. In *Ancient marine reptiles,* ed. J. M. Callaway and E. L. Nicholls, 467–489. San Diego, CA: Academic Press.

Carroll, R. L. 1997b. *Patterns and processes in vertebrate evolution.* Cambridge: Cambridge University Press.

Carroll, R. L. 1999. Homology among divergent Paleozoic tetrapod clades. In *Homology,* ed. G. R. Bock and B. Cardew, 47–64. Chichester: Wiley.

Carroll, R. L. 2000a. *Eocaecilia* and the origin of caecilians. In *Amphibian biology,* vol. 4, *Palaeontology: The evolutionary history of amphibians,* ed H. Heatwole and R. L. Carroll, 1402–1411. Chipping Norton, Australia: Surrey Beatty and Sons.

Carroll, R. L. 2000b. Lepospondyls. In *Amphibian biology,* vol. 4, *Palaeontology: The evolutionary history of amphibians,* ed. H. Heatwole and R. L. Carroll, 1198–1269. Chipping Norton, Australia: Surrey Beatty and Sons.

Carroll, R. L. 2000c. The lissamphibian enigma. In *Amphibian biology,* vol. 4, *Palaeontology: The evolutionary history of amphibians,* ed. H. Heatwole and R. L. Carroll, 1270–1273. Chipping Norton, Australia: Surrey Beatty and Sons.

Carroll, R. L. 2001a. Chinese salamanders tell tales. *Nature* 410:534–536.

Carroll, R. L. 2001b. The origin and early radiation of terrestrial vertebrates. *Journal of Paleontology* 75:1202–1213.

Carroll, R. L. 2002. Early land vertebrates. *Nature* 408:35–36.

Carroll, R. L. 2004. The importance of branchiosaurs in determining the ancestry of the modern amphibian orders. *Neues Jahrbuch für Geologie und Paläontologie, Abhandlungen* 232:157–180.

Carroll, R. L., and D. Baird. 1968. The Carboniferous amphibian

Tuditanus [*Eosauravus*] and the distinction between microsaurs and reptiles. *American Museum Novitates* 2337:1–50.

Carroll, R. L., C. Boisvert, J. Bolt, D. M. Green, N. Philip, C. Rolian, R. Schoch, and A. Tarenko. 2004. Changing patterns of ontology from osteolepiform fish through Permian tetrapods as a guide to the early evolution of land vertebrates. In *Recent advances in the origin and early radiation of vertebrates*, ed. G. Arratia, R. Cloutier, and M. V. H. Wilson, 321–343. Munich: Pfeil.

Carroll, R. L., K. A. Bossy, A. C. Milner, S. M. Andrews, and C. F. Wellstead. 1998. Lepospondyli. In *Encyclopedia of paleoherpetology*, vol. 1, *Lepospondyli*, ed. P. Wellnhofer, xii–216. Munich: Pfeil.

Carroll, R. L., P. Bybee, and W. Tidwell. 1991. The oldest microsaur (Amphibia). *Journal of Paleontology* 65:314–322.

Carroll, R. L., and J. Chorn. 1995. Vertebral development in the oldest microsaur and the problem of "Lepospondyl" relationships. *Journal of Vertebrate Paleontology* 15:37–56.

Carroll, R. L., and P. J. Currie. 1975. Microsaurs as possible apodan ancestors. *Zoological Journal of the Linnean Society* 57:229–247.

Carroll, R. L., and P. Gaskill. 1978. The order Microsauria. *Memoirs of the American Philosophical Society* 126:1–211.

Carroll, R. L., J. Irwin, and D. M. Green. 2005. Thermal physiology and the origin of terrestriality in vertebrates. *Zoological Journal of the Linnean Society* 143:345–346.

Carroll, S. B., J. K. Grenier, and S. D. Weatherbee. 2001. *From DNA to diversity*. Malden, MA: Blackwell Science.

Carter, D. R., T. E. Orr, D. P. Fyhrie, and D. J. Schurman. 1987. Influences of mechanical stress on prenatal and postnatal development. *Clinical Orthopaedics* 219:237–250.

Carter, D. R., and G. S. Beaupré. 2001. *Skeletal function and form: Mechanobiology of skeletal development, aging, and regeneration*. Cambridge: Cambridge University Press.

Carter, D. R., G. S. Beaupré, N. J. Giori, and J. A. Helms. 1998a. Mechanobiology of skeletal regeneration. *Clinical Orthopaedics and Related Research* 355S:S41–S55.

Carter, D. R., B. Mikíc, and K. Padian. 1998b. Epigenetic mechanical factors in the evolution of long bone epiphyses. *Zoological Journal of the Linnean Society* 123:163–178.

Carter, D. R., M. C. H. van der Meulen, and G. S. Beaupré. 1998c. Mechanobiologic regulation of osteogenesis and arthrogenesis. In *Skeletal growth and development: Clinical issues and basic science advances*, ed. J. Buckwalter, M. Ehrlich, L. Sandell, and S. Trippel, 99–130. Rosemont, IL: American Academy of Orthopaedic Surgeons.

Carter, D. R., and M. Wong. 1988. Mechanical tresses and endochondral ossification in the chondroepiphysis. *Journal of Orthopaedic Research* 6:148–154.

Carter, D. R., M. Wong, and T. E. Orr. 1991. Musculoskeletal ontogeny, phylogeny, and functional adaptation. *Journal of Biomechanics* 24 (suppl. 1): 3–16.

Carter, T. C. 1954. The genetics of luxate mice. IV: Embryology. *Journal of Genetics* 52:1–35.

Cartmill, M. 1974. Pads and claws in arboreal locomotion. In *Primate locomotion*, ed. F. A. Jenkins Jr., 45–83. New York: Academic Press.

Caruccio, N. C., A. Martinez-Lopex, M. Harris, L. Dvorak, J. Bitgood, B. K. Simandl, and J. F. Fallon. 1999. Constitutive activation of sonic hedgehog signalling in the chicken mutant

talpid2: Shh-independent outgrowth and polarizing activity. *Developmental Biology* 212:137–149.

Case, E. C. 1910. New or little known reptiles from the Permian (?) of Texas. *Bulletin of the American Museum of Natural History* 28:163–181.

Case, E. C. 1911. Revision of the Amphibia and Pisces of the Permian of North America. *Publications of the Carnegie Institute, Wash.* 146:1–148.

Case, E. C. 1935. Description of a collection of associated skeletons of *Trimerorhachis*. *Contributions of the Museum of Paleontology, University of Michigan* 4:227–274.

Cash, D. E., P. B. Gates, Y. Imokawa, and J. P. Brockes. 1998. Identification of newt connective tissue growth factor as a target of retinoid regulation in limb blastemal cells. *Gene* 222:119–124.

Casinos, A., J.-P. Gasc, S. Renous, and J. Bou. 1983. Les modalités de fouissage de *Pitymys duodecimcostatus* (Mammalia, Arvicolidae). *Mammalia* 47:27–36.

Casinos, A., C. Quintana, and C. Viladiu. 1993. Allometry and adaptation in the long bones of a digging group of rodents (Ctenomyinae). *Zoological Journal of the Linnean Society* 107:107–115.

Castañeda, M. R., and T. Alvarez. 1968. Contribución al conocimiento de la Osteología apendicular de *Bipes* (Reptilia: Amphisbaenia). *Anales de la Escuela Nacional de Ciencias Biológicas* (Mexico) 17:189–223.

Castanet, J., H. Francillon-Vieillot, F. J. Meunier, and A. de Ricqlès. 1993. Bone and individual aging. In *Bone*, vol. 7, *Bone growth—B*, ed. B. K. Hall, 245–283. Boca Raton, FL: CRC Press.

Castanet, J., and E. Smirina. 1990. Introduction to the skeletochronological method in amphibians and reptiles. *Annales des Sciences Naturelles, Zoologie* 11:191–196.

Caubit, X., R. Thangarajah, T. Theil, J. Wirth, H. G. Nothwang, U. Ruther, and S. Krauss. 1999. Mouse Dac, a novel nuclear factor with homology to Drosophila dachshund shows a dynamic expression in the neural crest, the eye, the neocortex, and the limb bud. *Developmental Dynamics* 214:66–80.

Cecconi, F., G. Alvarez-Bolado, B. I. Meyer, K. A. Roth, and P. Gruss. 1998. Apaf1 (CED-4 homolog) regulates programmed cell death in mammalian development. *Cell* 94:727–737.

Chadwick, C. S. 1940. Some notes on the burrows of *Plethodon metcalfi*. *Copeia* 1940:50.

Chalkley, D. T. 1954. A quantitative histological analysis of forelimb regeneration in *Triturus viridescens*. *Journal of Morphology* 94:21–70.

Chalkley, D. T. 1959. The cellular basis of limb regeneration. In *Regeneration in vertebrates*, ed. C. S. Thornton, 34–58. Chicago: University of Chicago Press.

Chambers, T. J. 2000. Regulation of the differentiation and function of osteoclasts. *Journal of Pathology* 192:4–13.

Chan, W. Y., K. K. H. Lee, and P. P. L. Tam. 1991. Regenerative capacity of forelimb buds after amputation in mouse embryos at the early-organogenesis stage. *Journal of Experimental Zoology* 260:74–83.

Chandross, R. J., and R. S. Bear. 1979. Comparison of mammalian collagen and elasmobranch elastoidin fiber structures, based on electron density profiles. *Journal of Molecular Biology* 130:215–229.

Chang, H. Y., J. T. Chi, S. Dudoit, C. Bondre, M. van de Rijn, D. Botstein, and P. O. Brown. 2002. Diversity, topographic differentiation, and positional memory in human fibroblasts.

Proceedings of the National Academy of Sciences, USA 99:12877–12882.

Chang, M. M., and M. Zhu. 1993. A new Middle Devonian osteolepidid from Quijing, Yunnan. *Memoirs of the Association of Australasian Palaeontologists* 15:183–198.

Chang, Y. H., H. W. C. Huang, C. M. Hamerski, and R. Kram. 2000. The independent effects of gravity and inertia on running mechanics. *Journal of Experimental Biology* 203:229–238.

Chantell, C. J. 1968. The osteology of *Pseudacris* (Amphibia: Hylidae). *American Midland Naturalist* 80:381–391.

Chapman, R. N. 1919. A study of the correlation of the pelvic structure and the habits of certain burrowing mammals. *American Journal of Anatomy* 25:185–219.

Charig, A. J., and A. C. Milner. 1997. *Baryonyx walkeri,* a fish-eating dinosaur from the Wealden of Surrey. *Bulletin of the Natural History Museum London (Geology)* 53:11–70.

Charité, J., W. de Graaff, S. Shen, and J. Deschamps. 1994. Ectopic expression of Hoxb-8 causes duplication of the ZPA in the forelimb and homeotic transformation of axial structures. *Cell* 78:589–601.

Charité, J., D. G. McFadden, and E. N. Olson. 2000. The bHLH transcription factor dHAND controls Sonic hedgehog expression and establishment of the zone of polarizing activity during limb development. *Development* 127:2461–2470.

Chautan, M., G. Chazal, F. Cecconi, P. Gruss, and P. Golstein. 1999. Interdigital cell death can occur through a necrotic and caspase-independent pathway. *Current Biology* 9:967–970.

Chellaiah, M., C. Fitzgerald, U. Alvarez, and K. Hruska. 1998. c-Src is required for stimulation of gelsolin-associated phosphatidylinositol 3-kinase. *Journal of Biological Chemistry* 273:11908–11916.

Chen, D., X. Ji, M. A. Harris, J. Q. Feng, G. Karsenty, A. J. Celeste, V. Rosen, G. R. Mundy, and S. E. Harris. 1998. Differential roles for bone morphogenetic protein (BMP) receptor type IB and IA in differentiation and specification of mesenchymal precursor cells to osteoblast and adipocyte lineages. *Journal of Cell Biology* 142:295–305.

Chen, H., Y. Lun, D. Ovchinnokov, H. Kokubo, K. C. Oberg, C. V. Pepicelli, L. Gan, R. Lee, and R. L. Johnson. 1998. Limb and kidney defects in Lmx1b mutant mice suggest an involvement of LMX1B in human nail-patella syndrome. *Nature Genetics* 19:51–55.

Chen, H., and R. L. Johnson. 1999. Dorsoventral patterning of the vertebrate limb: A process governed by multiple events. *Cell and Tissue Research* 296:67–73.

Chen, L., R. Adar, X. Yang, E. O. Monsonego, C. Li, P. V. Hauschka, A. Yayon, and C. X. Deng. 1999. Gly369Cys mutation in mouse FGFR3 causes achondroplasia by affecting both chondrogenesis and osteogenesis. *Journal of Clinical Investigation* 104:1517–1525.

Chen, P.-J., Z.-M. Dong, and S.-N. Zhen. 1998. An exceptionally well-preserved theropod dinosaur from the Yixian Formation of China. *Nature* 391:147–152.

Chen, Y., V. Knezevic, V. Ervin, R. Hutson, Y. Ward, et al. 2004. Direct interaction with Hoxd proteins reverses Gli3-repressor function to promote digit formation downstream of Shh. *Development* 131:2339–2347.

Chen, Z. F., and R. R. Behringer. 1995. Twist is required in head mesenchyme for cranial neural tube morphogenesis. *Genes and Development* 9:686–699.

Chevallier, A., M. Kieny, and A. Mauger. 1977. Limb-somite relationship: Origin of the limb musculature. *Journal of Embryology and Experimental Morphology* 41:245–258.

Cheverud, J. M., E. J. Routman, F. A. M. Duarte, B. vanSwinderen, K. Cothran, and C. Perel. 1996. Quantitative trait loci for murine growth. *Genetics* 142:1305–1319.

Chiang, C., Y. Litingtung, M. P. Harris, B. K. Simandl, Y. Li, et al. 2001. Manifestation of the limb prepattern: Limb development in the absence of Sonic Hedgehog function. *Developmental Biology* 236:421–435.

Chiappe, L. M., and G. J. Dyke. 2003. The Mesozoic radiation of birds. *Annual Review of Ecology and Systematics* 33:91–124.

Chiappe, L. M., and L. M. Witmer. 2002. *Mesozoic birds: Above the heads of dinosaurs.* Berkeley: University of California Press.

Chiu, C., and G. P. Wagner. 2000. Molecular evolution across the fin-limb transition. *American Zoologist* 40:973.

Chiu, C., D. Nonaka, L. Xue, C. T. Amemiya, and G. P. Wagner. 2000. Evolution of *Hoxa-11* in lineages phylogenetically positioned along the fin-limb transition. *Molecular Phylogenetics and Evolution* 17:305–316.

Chiu, C.-H., and M. W. Hamrick. 2002. Evolution and development of the primate limb skeleton. *Evolutionary Anthropology* 11:94–107.

Choquenot, D., and A. E. Greer. 1989. Intrapopulational and interspecific variation in digital limb bones and presacral vertebrae of the genus *Hemiergis* (Lacertilia, Scincidae). *Journal of Herpetology* 23:274–281.

Christ, B., H. J. Jacob, and M. Jacob. 1977. Experimental analysis of the origin of the wing musculature in avian embryos. *Anatomy and Embryology* (Berlin) 150:171–186.

Christen, B., and J. M. W. Slack. 1997. FGF-8 is associated with anteroposterior patterning and limb regeneration in Xenopus. *Developmental Biology* 192:455–466.

Christensen, R. N., M. Weinstein, and R. A. Tassava. 2002. Expression of fibroblast growth factors 4, 8, and 10 in limbs, flanks, and blastemas of *Ambystoma. Developmental Dynamics* 223:193–203.

Christiansen, P. 2002. Locomotion in terrestrial mammals: The influence of body mass, limb length, and bone proportions on speed. *Zoological Journal of the Linnean Society* 136:685–714.

Chuang, P. T., and A. P. McMahon. 1999. Vertebrate Hedgehog signalling modulated by induction of a Hedgehog-binding protein. *Nature* 397:617–621.

Chubb, S. H. 1932. Vestigial clavicles and rudimentary sesamoids. *American Naturalist* 66:376–381.

Chu-chien, H. 1982. The ecology of the Chinese alligator and changes in its geographical distribution. In *Proceedings of the fifth working meeting of the Crocodile Specialist Group,* 54–62. Gland, Switzerland: International Union for Conservation of Nature and Natural Resources.

Church, V., T. Nohno, C. Linker, C. Marcelle, and P. Francis-West. 2002. Wnt regulation of chondrocyte differentiation. *Journal of Cell Science* 115:4809–4818.

Cianfarini, S., C. Geremia, C. D. Scott, and D. Germani. 2002. Growth, IGF system, and cortisol in children with intrauterine growth retardation: Is catch-up growth affected by reprogramming of the hypothalamic-pituitary-adrenal axis? *Pediatric Research* 51:94–99.

Cihak, R. 1972. Ontogenesis of the skeleton and intrinsic muscles of the human hand and foot. *Ergebnisse der Anatomie und Entwicklungsgeschichte* 46:1–194.

Clack, J. A. 1987. *Pholiderpeton scutigerum* Huxley, an amphibian from the Yorkshire coal measures. *Philosophical Transactions of the Royal Society of London B* 318:1–107.

Clack, J. A. 1994. *Silvanerpeton miripeds*, a new anthracosauroid from the Viséan of East Kirkton, West Lothian, Scotland. *Transactions of the Royal Society of Edinburgh, Earth Sciences* 84:369–376.

Clack, J. A. 1997. Devonian tetrapod trackways and trackmakers: A review of the fossils and footprints. *Palaeogeography, Paleoclimatology, and Palaeoecology* 130:227–250.

Clack, J. A. 1998a. The British Carboniferous tetrapod *Crassigyrinus scoticus* (Lydekker): Cranial anatomy and relationships. *Transactions of the Royal Society of Edinburgh, Earth Sciences* 88:127–142.

Clack, J. A. 1998b. A new Early Carboniferous tetrapod with a mélange of crown-group characters. *Nature* 394:66–69.

Clack, J. A. 2000. The origin of tetrapods. In *Amphibian biology*, vol. 4, *Palaeontology: The evolutionary history of amphibians*, ed. H. Heatwole and R. L. Carroll, 979–1029. Chipping Norton, Australia: Surrey Beatty and Sons.

Clack, J. A. 2001. *Eucritta melanolimnetes* from the Early Carboniferous of Scotland, a stem tetrapod showing a mosaic of characteristics. *Transactions of the Royal Society of Edinburgh, Earth Sciences* 92:75–95.

Clack, J. A. 2002a. An early tetrapod from "Romer's Gap." *Nature* 418:72–76.

Clack, J. A. 2002b. *Gaining ground: The origin and evolution of tetrapods.* Bloomington: Indiana University Press.

Clack, J. A., and R. L. Carroll. 2000. Early Carboniferous tetrapods. In *Amphibian biology*, vol. 4, *Palaeontology: The evolutionary history of amphibians*, ed. H. Heatwole and R. L. Carroll, 1030–1043. Chipping Norton, Australia: Surrey Beatty and Sons.

Clack, J. A., and M. I. Coates. 1995. *Acanthostega*: A primitive aquatic tetrapod? *Bulletin du Muséum National d'Histoire Naturelle, Paris* 17:359–373.

Clack, J. A., and S. M. Finney. 2005. *Pederpes finneyae*, an articulated tetrapod from the Tournaisian of western Scotland. *Journal of Systematic Palaeontology* 2:311–346.

Clark, B. D., and W. Bemis. 1979. Kinematics of swimming of penguins at the Detroit Zoo. *Journal of Zoology, London* 188:411–428.

Clark, J., and D. J. Stechschulte Jr. 1998. The interface between bone and tendon at an insertion site: A study of the quadriceps tendon insertion. *Journal of Anatomy* 193:605–616.

Clark, J. M., J. A. Hopson, R. Hernández, D. E. Fastovsky, and M. Montellano. 1998. Foot posture in a primitive pterosaur. *Nature* 391:886–889.

Clark, J. M., M. A. Norell, and P. J. Makovicky. 2002. Cladistic approaches to the relationships of birds to other theropod dinosaurs. In *Mesozoic birds: Above the heads of dinosaurs*, ed. L. M. Chiappe and L. M. Witmer, 31–61. Berkeley: University of California Press.

Clarke, J. A., and M. A. Norell. 2002. The morphology and phylogenetic position of *Apsaravis ukhaana* from the Late Cretaceous of Mongolia. *American Museum Novitates* 3387:1–47.

Clarke, P. G., and S. Clarke. 1995. Historic apoptosis. *Nature* 378:230.

Claussen, D. L., P. M. Daniel, S. Jiang, and N. A. Adams. 1991. Hibernation in the eastern box turtle, *Terrapene c. carolina*. *Journal of Herpetology* 25:334–341.

Clevedon Brown, J., and D. W. Yalden. 1973. The description of

mammals. 2: Limbs and locomotion of terrestrial mammals. *Mammal Review* 3:107–135.

Clough, M. V., J. D. Hamlington, and I. McIntosh. 1999. Restricted distribution of loss-of-function mutations within the *LMX1B* genes of nail-patella syndrome patients. *Human Mutation* 14:459–465.

Cloutier, R. 1996. The primitive actinistian *Miguashaia bureaui* Schultze (Sarcopterygii). In *Devonian fishes and plants of Miguasha, Quebec, Canada*, ed. H.-P. Schultze and R. Cloutier, 227–247. Munich: Pfeil.

Cloutier, R., and P. E. Ahlberg. 1996. Morphology, characters, and the interrelationships of basal sarcopterygians. In *Interrelationships of fishes*, ed. M. L. J. Stiassny, L. R. Parenti, and G. D. Johnson, 325–337. London: Academic Press.

Cloutier, R., and H.-P. Schultze. 1996. Porolepiform fishes (Sarcopterygii). In *Devonian fishes and plants of Miguasha, Quebec, Canada*, ed. H.-P. Schultze and R. Cloutier, 248–270. Munich: Pfeil.

Coates, M. I. 1991. New palaeontological contributions to limb ontogeny and phylogeny. In *Developmental patterning of the vertebrate limb*, ed. J. R. Hinchliffe, J. M. Hurlé, and D. Summerbell, 325–338. New York: Plenum Press.

Coates, M. I. 1994. The origin of vertebrate limbs. In *The evolution of developmental mechanisms*, ed. M. Akam, P. Holland, P. Ingham, and W. Wray, 169–180. London: Academic Press.

Coates, M. I. 1995. Limb evolution: Fish fins or tetrapod limbs; A simple twist of fate? *Current Biology* 5:844–848.

Coates, M. I. 1996. The Devonian tetrapod *Acanthostega gunnari* Jarvik: Postcranial anatomy, basal tetrapod interrelationships and patterns of skeletal evolution. *Transactions of the Royal Society of Edinburgh, Earth Sciences* 87:363–421.

Coates, M. I. 2001. Origin of tetrapods. In *Palaeobiology II*, ed. D. E. G. Briggs and P. R. Crowther, 74–79. London: Blackwell Science.

Coates, M. I. 2003. The evolution of paired fins. *Theory in Biosciences* 122:266–287.

Coates, M. I., and J. A. Clack. 1990. Polydactyly in the earliest known tetrapod limbs. *Nature* 347:66–69.

Coates, M. I., and J. A. Clack. 1995. Romer's Gap: Tetrapod origins and terrestriality. *Bulletin du Muséum National d'Histoire Naturelle, Paris*, series 4, 17 (C): 373–388.

Coates, M. I., and M. J. Cohn. 1998. Fins, limbs, and tails: Outgrowths and axial patterning in vertebrate evolution. *BioEssays* 20:371–381.

Coates, M. I., and M. J. Cohn. 1999. Vertebrate axial and appendicular patterning: The early development of paired appendages. *American Zoologist* 39:676–685.

Coates, M. I., J. E. Jeffery, and M. Ruta. 2002. Fins to limbs: What the fossils say. *Evolution and Development* 4:390–401.

Cock, A. G. 1966. Genetical aspects of metrical growth and form in animals. *Quarterly Review of Biology* 41:131–190.

Cockrum, L. 1941. Notes on *Siren intermedia*. *Copeia* 1941:265.

Coddington, J. A. 1988. Cladistic tests of adaptational hypotheses. *Cladistics* 4:3–22.

Coe, W. R., and B. W. Kunkel. 1906. Studies on the California limbless lizard, *Anniella*. *Transactions of the Connecticut Academy of Arts and Sciences* 12:1–55.

Cohn, M. J. 2000. Giving limbs a hand. *Nature* 406:953–954.

Cohn, M. J. 2001. Developmental mechanisms of vertebrate limb evolution. *Novartis Foundation Symposium* 232:47–57.

Cohn, M. J., and P. E. Bright. 1999. Molecular control of vertebrate limb development, evolution and congenital malformations. *Cell and Tissue Research* 296:3–17.

Cohn, M. J., J. C. Izpisúa Belmonte, H. Abu, J. K. Heath, and C. Tickle. 1995. Fibroblast growth factors induce additional limb development from the flank of chick embryos. *Cell* 80:739–746.

Cohn, M. J., C. O. Lovejoy, L. Wolpert, and M. I. Coates. 2002. Branching, segmentation and the metapterygial axis: Pattern versus process in the vertebrate limb. *BioEssays* 24:460–465.

Cohn, M. J., K. Patel, R. Krumlauf, D. G. Wilkinson, J. D. Clarke, and C. Tickle. 1997. Hox9 genes and vertebrate limb specification. *Nature* 387:97–101.

Cohn, M. J., and C. Tickle. 1996. Limbs: A model for pattern formation within the vertebrate body plan. *Trends in Genetics* 12:253–256.

Cohn, M. J., and C. Tickle. 1999. Developmental basis of limblessness and axial patterning in snakes. *Nature* 399:474–479.

Colbert, E. H., and J. W. Kitching. 1975. The Triassic reptile *Procolophon* in Antarctica. *American Museum Novitates* 2566:1–23.

Cole, L. K., and L. S. Ross. 2001. Apoptosis in the developing zebrafish embryo. *Developmental Biology* 240:123–142.

Cole, N. J., M. Tanaka, A. Prescott, and C. Tickle. 2003. Expression of limb initiation genes and clues to the morphological diversification of threespine stickleback. *Current Biology* 13:R951–R952.

Colton, H. S. 1929. How bipedal habit affects the bones of the hind legs of the albino rat. *Journal of Experimental Zoology* 53:1–11.

Colvin, J. S., B. A. Bohne, G. W. Harding, D. G. McEwen, and D. M. Ornitz. 1996. Skeletal overgrowth and deafness in mice lacking fibroblast growth factor receptor 3. *Nature: Genetics* 12:390–397.

Combes, S. A., and T. L. Daniel. 2001. Shape, flapping and flexion: Wing and fin design for forward flight. *Journal of Experimental Biology* 204:2073–2085.

Compagno, L. J. V. 1973. Interrelationships of living elasmobranchs. *Journal of the Linnean Society (Zoology)* 53:15–61.

Compton, A. W. 1981. Courtship and nesting behavior in the freshwater crocodile, *Crocodylus johnstoni,* under controlled conditions. *Australian Wildlife Research* 8:443–450.

Conant, E. B. 1970. Regeneration in the African lungfish, Protopterus. I: Gross aspects. *Journal of Experimental Zoology* 174:15–32.

Congello, K. 1978. Nesting and egg laying behavior in *Terrapene carolina. Proceedings of the Pennsylvania Academy of Science* 52:51–56.

Connelly, T. G., and F. L. Bookstein. 1983. Method for 3-dimensional analysis of patterns of thymidine labeling in regenerating and developing limbs. *Progress in Clinical Biological Research* 110 (part A): 525–536.

Connor, J. R., R. A. Dodds, I. E. James, and M. Gowen. 1995. Human osteoclast and giant cell differentiation: The apparent switch from nonspecific esterase to tartrate resistant acid phosphatase activity coincides with the *in situ* expression of osteopontin mRNA. *Journal of Histochemistry and Cytochemistry* 43:1193–1201.

Coombs, W. P. 1980. Swimming ability of carnivorous dinosaurs. *Science* 207:1198–1200.

Coombs, W. P., and T. Maryanska. 1990. Ankylosauria. In *The Di-*

nosauria, ed. D. B. Weishampel, P. Dodson, and H. Osmólska, 456–483. Berkeley: University of California Press.

Cooper, C., M. K. Javaid, P. Taylor, K. Walker-Bone, E. Dennison, and N. Arden. 2002. The fetal origins of osteoporotic fracture. *Calcified Tissue International* 70:391–394.

Cope, E. D. 1871. Contribution to the ichthyology of the Lesser Antilles. *Transactions of the American Philosophical Society* 14:445–483.

Cope, E. D. 1889. The Batrachia of North America. *Bulletin of the United States National Museum* 34:1–525.

Cope, E. D. 1892a. On degenerate types of scapular and pelvic arches in the Lacertilia. *Journal of Morphology* 7:223–244.

Cope, E. D. 1892b. On the phylogeny of the Vertebrata. *Proceedings of the American Philosophical Society* 30:278–281.

Cope, E. D. 1896. *The primary factors of organic evolution.* Chicago: Open Court.

Cope, E. D. 1900. The crocodilians, lizards, and snakes of North America. *Report of the United States National Museum* 1898:153–1270.

Cornwall, I. W. 1956. *Bones for the archaeologist.* New York: Macmillan.

Cote, S., R. Carroll, R. Cloutier, and L. Bar-Sagi. 2002. Vertebral development in the Devonian sarcopterygian fish *Eusthenopteron foordi* and the polarity of vertebral evolution in non-amniote tetrapods. *Journal of Vertebrate Paleontology* 22:487–502.

Cott, H. B. 1961. Scientific results of an inquiry into the ecology and economic status of the Nile crocodile (*Crocodylus niloticus*) in Uganda and Northern Rhodesia. *Transactions of the Zoological Society of London* 29:211–356.

Coulier, F., P. Pontarotti, R. Roubin, H. Hartung, M. Goldfarb, and D. Birnbaum. 1997. Of worms and men: An evolutionary perspective on the fibroblast growth factor (FGF) and FGF receptor families. *Journal of Molecular Evolution* 44:43–56.

Cracraft, J. 1986. The origin and early diversification of birds. *Paleobiology* 12:383–399.

Cracraft, J., and J. A. Clarke. 2001. The basal clades of modern birds. In *New perspectives on the origin and early evolution of birds: Proceedings of the International Symposium in Honor of John H. Ostrom,* ed. J. A. Gauthier and L. F. Gall, 143–156. New Haven, CT: Peabody Museum of Natural History, Yale University.

Craig, F. M. 1987. The development of the chick limb. PhD diss., University of London.

Craig, F. M., G. Bentley, and C. W. Archer. 1987. The spatial and temporal pattern of collagens I and II and keratan sulphate in the developing chick metatarsophalangeal joint. *Development* 99:383–391.

Craig, F. M., M. T. Bayliss, G. Bentley, and C. W. Archer. 1990. A role for hyaluronan in joint development. *Journal of Anatomy* 171:17–23.

Crawshaw, P. G., Jr., and G. B. Schaller. 1980. Nesting of Paraguayan caiman (*Caiman yacare*) in Brazil. *Papéis Avulsos de Zoologia* 33:283–292.

Cresko, W. A., A. Amores, C. Wilson, J. Murphy, M. Currey, P. Phillips, M. A. Bell, C. B. Kimmel., and J. H. Postlethwait. 2004. Parallel genetic basis for repeated evolution of armor loss in Alaskan threespine stickleback populations. *Proceedings of the National Academy of Sciences, USA* 101:6050–5.

Crews, L., P. B. Gates, R. Brown, A. Joliot, C. Foley, J. P. Brockes, and A. Gann. 1995. Expression and activity of the newt Msx-1 gene in relation to limb regeneration. *Proceedings of the Royal Society of London B* 259:161–171.

Crocoll, A., U. Herzer, N. B. Ghyselinck, P. Chambon, and A. C. B. Cato. 2002. Interdigital apoptosis and down regulation of BAG-1 expression in mouse autopods. *Mechanisms of Development* 111:149–152.

Crossley, P. H., G. Minowada, C. A. McArthur, and G. R. Martin. 1996. Roles for FGF8 in the induction, initiation and maintenance of chick limb development. *Cell* 84:127–136.

Crotwell, P. L., T. G. Clark, and P. M. Mabee. 2001. *Gdf5* is expressed in the developing skeleton of median fins of late-stage zebrafish, *Danio rerio*. *Development Genes and Evolution* 211:555–558.

Crotwell, P. L., A. R. Sommervold, and P. M. Mabee. 2004. Expression of *bmp2a* and *bmp2b* in late-stage zebrafish median fin development. *Gene Expression Patterns* 5:291–296.

Crowe, R., J. Zikherman, and L. Niswander. 1999. Delta-1 negatively regulates the transition from prehypertrophic to hypertrophic chondrocytes during cartilage formation. *Development* 126:987–998.

Crumly, C. R. 1990. The case of the two-legged "lizard." *Environment West Magazine* 1990 (Fall): 20–24.

Crumly, C. R. 1994. Phylogenetic systematics of North American tortoises (Genus *Gopherus*): evidence for their classification. In *Biology of North American tortoises*, ed. R. B. Bury and D. J. Germano, 7–32. Washington, DC: U.S. Department of the Interior National Biological Survey.

Crumly, C. R., and M. R. Sánchez-Villagra. 2004. Patterns of variation in the phalangeal formulae of land tortoises (Testudinidae): Developmental constraint, size, and phylogenetic history. *Journal of Experimental Zoology (Molecular and Developmental Evolution)* 302B:134–146.

Cruz, C. C., A. Der-Avakian, D. D. Spyropoulos, D. D. Tieu, and E. M. Carpenter. 1999. Targeted disruption of *Hoxd9* and *Hoxd10* alters locomotor behavior, vertebral identity, and peripheral nervous development. *Developmental Biology* 216:595–610.

Cruz-Orivé, L.-M., and E. B. Hunziker. 1986. Stereology for anisotropic cells: Application to growth cartilage. *Journal of Microscopy* 143:57–80.

Cubbage, C. C., and P. M. Mabee. 1996. Development of the cranium and paired fins in the zebrafish *Danio rerio* (Ostariophysi, Cyprinidae). *Journal of Morphology* 229:121–160.

Currey, J. D. 1987. The evolution of the mechanical properties of bone. *Journal of Biomechanics* 20:1035–1044.

Currey, J. D., and R. M. Alexander. 1985. The thickness of the walls of tubular bones. *Journal of Zoology, London* 206:453–468.

Cuvier, G. 1805. *Leçons d'anatomie comparée*. Paris: Baudouin.

D

Daeschler, E. B., and N. H. Shubin. 1998. Fish with fingers? *Nature* 391:133.

Daeschler, E. B., N. H. Shubin, K. S. Thomson, and W. W. Amaral. 1994. A Devonian tetrapod from North America. *Science* 265:639–642.

Dahn, R. D., and J. F. Fallon. 2000. Interdigital regulation of digit identity and homeotic transformation by modulated BMP signaling. *Science* 289:438–441.

Dalla Vecchia, F. M. 1998. New observations on the osteology and taxonomic status of *Preondactylus buffarinii* Wild, 1984 (Reptilia, Pterosauria). *Bollettino della Società Paleontologica Italiana* 36:355–366.

Dal Sasso, C., and M. Signore. 1998. Exceptional soft-tissue preservation in a theropod dinosaur from Italy. *Nature* 392:383–387.

D'Alton, E. 1836. *Pythonis ac Boarum Ossibus Commentatio*. Halle, Germany.

Daluiski, A., T. Engstrand, M. E. Bahamonde, L. W. Gamer, E. Agius, S. L. Stevenson, K. Cox, V. Rosen, and K. M. Lyons. 2001. Bone morphogenetic protein-3 is a negative regulator of bone density. *Nature: Genetics* 27:84–88.

Daly, E. 1994. The Amphibamidae (Amphibia: Temnospondyli), with a description of a new genus from the Upper Pennsylvanian of Kansas. *Miscellaneous Publications of the University of Kansas Museum of Natural History* no. 85:1–59.

Danforth, C. H. 1913. The myology of *Polyodon*. *Journal of Morphology* 24:107–146.

Daniel, T. L. 1988. Forward flapping flight from flexible fins. *Canadian Journal of Zoology* 66:630–638.

Darwin, C. R. 1859. *On the origin of species*. Facsimile of the 1st ed., with introd. by Ernest Mayr. Cambridge, MA: Harvard University Press, 1964.

Das, P., and P. Mohanty-Hejmadi. 2000. Vitamin A mediated limb deformities in the common Indian toad, *Bufo melanostictus* (Schneider). *Indian Journal of Experimental Biology* 38:258–264.

da Silva, S. M., P. B. Gates, and J. P. Brockes. 2002. The newt ortholog of CD59 is implicated in proximodistal identity during amphibian limb regeneration. *Developmental Cell* 3:547–555.

Davenport, J. 1987. Locomotion in hatchling leatherback turtles *Dermochelys coriacea*. *Journal of Zoology, London* 212:85–101.

Davenport, J., and W. Clough. 1986. Swimming and diving in young loggerhead sea turtles (*Caretta caretta* L.). *Copeia* 1986:53–57.

Davenport, J., and G. A. Pearson. 1994. Observations on the swimming of the Pacific Ridley, *Lepidochelys olivacea* (Eschscholtz, 1829): Comparisons with other sea-turtles. *Herpetological Journal* 4:60–63.

Davenport, J., and M. D. J. Sayer. 1989. Observations on the aquatic locomotion of young salt-water crocodiles (*Crocodylus porosus* Schneider). *Herpetological Journal* 1:356–361.

Davenport, J., D. J. Grove, J. Cannon, T. R. Ellis, and R. Stables. 1990. Food capture, appetite, digestion rate and efficiency in hatchling and juvenile *Crocodylus porosus*. *Journal of Zoology, London* 220:569–592.

Davenport, J., S. A. Munks, and P. J. Oxford. 1984. A comparison of the swimming of marine and freshwater turtles. *Proceedings of the Royal Society of London B* 220:447–475.

Davenport, T. G., L. A. Jerome-Majewska, and V. E. Papaioannou. 2003. Mammary gland, limb and yolk sac defects in mice lacking *Tbx3*, the gene mutated in human ulnar mammary syndrome. *Development* 130:2263–2273.

Davidson, D. R., A. Crawley, R. E. Hill, and C. Tickle. 1991. Position-dependent expression of two related homeobox genes in developing vertebrate limbs. *Nature* 352:429–431.

Davidson, E. 2001. *Genomic regulatory systems*. San Diego, CA: Academic Press.

Davies, M. 1984. Osteology of the myobatrachine frog *Arenophryne rotunda* Tyler (Anura: Leptodactylidae) and comparisons with other myobatrachine genera. *Australian Journal of Zoology* 32:789–802.

Davis, A. P., and M. R. Capecchi. 1994. Axial homeosis and appendicular skeleton defects in mice with a targeted disruption of hoxd-11. *Development* 120:2187–2198.

Davis, A. P., and M. R. Capecchi. 1996. A mutational analysis of the 5′ HoxD genes: Dissection of genetic interactions during limb development in the mouse. *Development* 122:1175–1185.

Davis, A. P., D. P. Witte, H. M. Hsieh-Li, S. S. Potter, and M. R. Capecchi. 1995. Absence of radius and ulna in mice lacking hoxa-11 and hoxd-11. *Nature* 375:791–795.

Davis, D. D. 1946. Observations on the burrowing behavior of the hog-nosed snake. *Copeia* 1946:75–78.

Davis, D. D. 1964. The giant panda, a morphological study of evolutionary mechanisms. *Fieldiana: Zoological Memoirs* 3:1–339.

Davis, M. C., N. H. Shubin, and E. B. Daeschler. 2001. Immature rhizodontids from the Devonian of North America. *Bulletin of the Museum of Comparative Zoology, Harvard University* 156:171–187.

Davis, M. C., N. H. Shubin, and A. Force. 2004. Pectoral fin and girdle development in the basal actinopterygians *Polyodon spathula* and *Acipenser transmontanus. Journal of Morphology* 262:608–628.

Davison, W. 1988. The myotomal muscle of labriform swimming fish is not designed for high speed sustained swimming. *New Zealand Natural Sciences* 15:37–42.

Dawson, A. B. 1929. A histological study of the persisting cartilage plates in retarded or lapsed endochondral union in the albino rat. *Anatomical Record* 43:109–129.

Dawson, A. B. 1935. The sequence of epiphyseal union in the skeleton of the mouse with special reference to the phenomenon of "lapsed" union. *Anatomical Record* 63:93–99.

Dawson, W. R., G. A. Bartholomew, and A. F. Bennett. 1977. A reappraisal of the aquatic specializations of the Galápagos marine iguana (*Amblyrhynchus cristatus*). *Evolution* 31:891–897.

Dealy, C. N., A. Roth, D. Ferrari, A. M. Brown, and R. A. Kosher. 1993. Wnt-5a and Wnt-7a are expressed in the developing chick limb bud in a manner suggesting roles in pattern formation along the proximodistal and dorsoventral axes. *Mechanisms of Development* 43:175–186.

Deban, S. M., and W. Olson. 2002. Suction feeding by a tiny predatory tadpole. *Nature* 420:41–42.

deCrombrugghe, B. B., V. Lefebvre, K. Nakashima. 2001. Regulatory mechanisms in the pathways of cartilage and bone formation. *Current Opinion in Cell Biology* 13:721–727.

deCrombrugghe, B. B., V. Lefebvre, and K. Nakashima. 2002. Deconstructing the molecular biology of cartilage and bone formation. In *Mouse development: Patterning, morphogenesis, and organogenesis*, ed. J. Rossant and P. P. L. Tam, 279–295. San Diego, CA: Academic Press.

Deeming, D.C., L. B. Halstead, M. Manabe, and D. M. Unwin. 1993. An ichthyosaurian embryo from the Lower Lias (Jurassic: Hettangian) of Somerset, England, with comments on the reproductive biology of ichthyosaurs. *Modern Geology* 18:423–442.

DeHaan, R. L., and H. Ursprung. 1965. *Organogenesis*. New York: Holt, Rinehart and Winston.

Deitz, D. C., and T. C. Hines. 1980. Alligator nesting in north-central Florida. *Copeia* 1980:249–258.

Dell'Orbo, C., L. Gioglio, and D. Quacci. 1992. Morphology of epiphyseal apparatus of a ranad frog (*Rana esculenta*). *Histology and Histopathology* 7:267–273.

Deng, C., A. Wynshaw-Boris, F. Zhou, A. Kuo, and P. Leder. 1996. Fibroblast growth factor receptor 3 is a negative regulator of bone growth. *Cell* 84:911–921.

Denny, M. W. 1993. *Air and water: The biology and physics of life's media*. Princeton, NJ: Princeton University Press.

Dent, J. N. 1962. Limb regeneration in larvae and metamorphosing individuals of the South African clawed toad. *Journal of Morphology* 110:61–77.

de Queiroz, K., and J. Gauthier. 1990. Phylogeny as a central principle in taxonomy: Phylogenetic definitions of taxon names. *Systematic Zoology* 39:449–480.

de Queiroz, K., and J. Gauthier. 1992. Phylogenetic taxonomy. *Annual Review of Ecology and Systematics* 23:449–480.

de Queiroz, K., and J. Gauthier. 1994. Toward a phylogenetic system of biological nomenclature. *Trends in Ecology and Evolution* 9:27–31.

Deraniyagala, P. E. P. 1958. Reproduction in the monitor lizard *Varanus bengalensis* (Daudin). *Spolia Zeylanica* 28:161–166.

De Ricqlès, A. J., F. J. Meunier, J. Castenet, and H. Francillon-Vieillot. 1991. Comparative microstructure of bone. In *Bone*, vol. 3, *Bone matrix and bone specific products*, ed. B. K. Hall, 1–77. Boca Raton, FL: CRC Press.

Deuchar, E. M. 1976. Regeneration of amputated limb-buds in early rat embryos. *Journal of Embryology and Experimental Morphology* 35:345–354.

Dewulf, N., K. Verschueren, O. Lonnoy, A. Moren, S. Grimsby, K. Vande Spiegle, K. Miyazono, D. Huylebroeck, and P. Ten Dijke. 1995. Distinct spatial and temporal expression patterns of two type I receptors for bone morphogenetic proteins during mouse embryogenesis. *Endocrinology* 136:2652–2663.

Dial, K. P. 1992a. Activity patterns of the wing muscles of the pigeon (*Columba livia*) during different modes of flight. *Journal of Experimental Zoology* 262:357–373.

Dial, K. P. 1992b. Avian forelimb muscles and nonsteady flight: Can birds fly without using the muscles in their wings? *Auk* 109:874–885.

Dial, K. P., G. E. Goslow Jr., and F. A. Jenkins Jr. 1991. The functional anatomy of the shoulder in the European Starling (*Sturnus vulgaris*). *Journal of Morphology* 207:327–344.

Dial, K. P., S. R. Kaplan, G. E. Goslow Jr., and F. A. Jenkins Jr. 1987. Structure and neural control of the pectoralis in pigeons: Implications for flight muscles. *Anatomical Record* 218:284–287.

Dial, K. P., S. R. Kaplan, G. E. Goslow Jr., and F. A. Jenkins Jr. 1988. A functional analysis of the primary upstroke and downstroke muscles in the domestic pigeon (*Columba livia*) during flight. *Journal of Experimental Biology* 134:1–16.

Dickinson, M. H. 1996. Unsteady mechanisms of force generation in aquatic and aerial locomotion. *American Zoologist* 36:537–554.

Dickinson, M. H., F. O. Lehmann, and S. P. Sane. 1999. Wing rotation and the aerodynamic basis of insect flight. *Science* 284:1954–1960.

Dickson, G. R. 1982. Ultrastructure of growth cartilage in the proximal femur of the frog, *Rana temporaria. Journal of Anatomy* 135:549–564.

Diekwisch, T. G. H., B. J. Berman, X. Anderton, B. Gurinsky, A. J. Ortega, P. G. Satchell, M. Williams, et al. 2002. Membranes, minerals, and proteins of developing vertebrate enamel. *Microscopy Research and Technique* 59:373–395.

Diogo, R., C. Oliveira, and M. Chardon, M. 2001. On the homologies of the skeletal components of catfish (Teleostei: Siluriformes) suspensorium. *Belgian Journal of Zoology* 131:93–109.

D'Jamoos, C. A., G. McMahon, and P. A. Tsonis. 1998. Fibroblast growth factor receptors regulate the ability for hindlimb regeneration in Xenopus laevis. *Wound Repair Regeneration* 6:388–397.

Dobie, J. L., D. R. Womochel, and G. L. Bell. 1986. A unique sacroiliac contact in mosasaurs (Sauria, Varanoidea, Mosasauridae). *Journal of Vertebrate Paleontology* 6:197–199.

Dobrowolska, H. 1973. Body-part proportions in relation to the mode of locomotion in anurans. *Zoologica Poloniae* 23:59–108.

Dodson, P. 1975. Functional and ecological significance of relative growth in *Alligator. Journal of Zoology, London* 175:315–355.

Dollé, P., A. Dierich, M. LeMeur, T. Schimmang, B. Schubaur, P. Chambon, and D. Duboule. 1993. Disruption of the *Hoxd-13* gene induces localized heterochrony leading to mice with neotenic limbs. *Cell* 75:431–441.

Dollé, P., J. C. Izpisúa Belmonte, H. Falkenstein, A. Renucci, and D. Duboule. 1989. Coordinate expression of the murine Hox-5 complex homoeobox-containing genes during limb pattern formation. *Nature* 342:767–772.

Dollo, L. 1895. Sur la phylogenie des Dipneustes. *Bulletin de la Société Belgique de Geologie, de Paleontologie et d'Hydrogeologie* 9:79–128.

Dollo, L. 1899. Les ancêtres de marsupiaux étaient-ils arboricoles? *Travail de la Station de Zoologie du Wimereux* 7:188–203.

Dong, Z.-M. 1993. A Lower Cretaceous enantiornithine bird from the Ordos Basin of Inner Mongolia, People's Republic of China. *Canadian Journal of Earth Sciences* 30:2177–2179.

Donley, J. M., and R. E. Shadwick. 2003. Steady swimming muscle dynamics in the leopard shark *Triakis semifasciata. Journal of Experimental Biology* 206:1117–1126.

Dowthwaite, G. P., J. C. W. Edwards, and A. A. Pitsillides. 1998. An essential role for the interaction between hyaluronan and hyaluronan binding proteins during joint development. *Journal of Histochemistry and Cytochemistry* 46:641–651.

Dowthwaite, G. P., A. C. Ward, J. Flannely, R. F. L. Suswillo, C. R. Flannery, C. W. Archer, and A. A. Pitsillides. 1999. The effect of mechanical strain on hyaluronan metabolism in embryonic fibrocartilage cells. *Matrix Biology* 18:523–532.

Drachman, D. B., and L. Sokoloff. 1966. The role of movement in embryonic joint development. *Developmental Biology* 14:401–420.

Dresden, M. H., and J. Gross. 1970. The collagenolytic enzyme of the regenerating limb of the newt *Triturus viridescens. Developmental Biology* 22:129–137.

Dreyer, S. D., R. Morella, M. S. German, B. Zabel, A. Winterpacht, G. P. Lunstrum, W. A. Horton, K. C. Oberg, and B. Lee. 2000. *LMX1B* transactivation and expression in nail-patella syndrome. *Human Molecular Genetics* 9:1067–1074.

Dreyer, S. D., G. Zhou, A. Baldini, A. Winterpacht, B. Zabel, W. Cole, R. Johnson, and B. Lee. 1998a. Mutations in *LMX1B* cause abnormal skeletal patterning and renal dysplasia in nail patella syndrome. *Nature Genetics* 19:47–51.

Dreyer, S. D., G. Zhou, and B. Lee. 1998b. The long and the short of it: Developmental genetics of the skeletal dysplasias. *Clinical Genetics* 54:464–473.

Drossopoulou, G., K. E. Lewis, J. J. Sanz-Ezquerro, N. Nikbakht, A. P. McMahon, et al. 2000. A model for anterioposterior patterning of the vertebrate limb based on sequential long- and short-range Shh signalling and Bmp signalling. *Development* 127:1337–1348.

Drucker, E. G. 1996. The use of gait transition speed in comparative studies of fish locomotion. *American Zoologist* 36:555–566.

Drucker, E. G., and J. S. Jensen. 1996a. Pectoral fin locomotion in the striped surfperch. I: Kinematic effects of swimming speed and body size. *Journal of Experimental Biology* 199:2235–2242.

Drucker, E. G., and J. S. Jensen. 1996b. Pectoral fin locomotion in the striped surfperch. II: Scaling swimming kinematics and performance at a gait transition. *Journal of Experimental Biology* 199:2243–2252.

Drucker, E. G., and J. S. Jensen. 1997. Kinematic and electromyographic analysis of steady pectoral fin swimming in the surfperches. *Journal of Experimental Biology* 200:1709–1723.

Drucker, E. G., and G. V. Lauder. 1999. Locomotor forces on a swimming fish: Three-dimensional vortex wake dynamics quantified using digital particle image velocimetry. *Journal of Experimental Biology* 202:2393–2412.

Drucker, E. G., and G. V. Lauder. 2000. A hydrodynamic analysis of fish swimming speed: Wake structure and locomotor force in slow and fast labriform swimmers. *Journal of Experimental Biology* 203:2379–2393.

Drucker, E. G., and G. V. Lauder. 2001a. Locomotor function of the dorsal fin in teleost fishes: Experimental analysis of wake forces in sunfish. *Journal of Experimental Biology* 204:2943–2958.

Drucker, E. G., and G. V. Lauder. 2001b. Wake dynamics and fluid forces of turning maneuvers in sunfish. *Journal of Experimental Biology* 204:431–442.

Drucker, E. G., and G. V. Lauder. 2002a. Experimental hydrodynamics of fish locomotion: Functional insights from wake visualization. *Integrative and Comparative Biology* 42:243–257.

Drucker, E. G., and G. V. Lauder. 2002b. Wake dynamics and locomotor function in fishes: Interpreting evolutionary patterns in pectoral fin design. *Integrative and Comparative Biology* 42:997–1008.

Drucker, E. G., and G. V. Lauder. 2003. Function of pectoral fins in rainbow trout: Behavioral repertoire and hydrodynamic forces. *Journal of Experimental Biology* 206:813–826.

Dublin, L. I. 1903. Adaptations to aquatic, arboreal, fossorial, and cursorial habits in mammals. II: Arboreal adaptations. *American Naturalist* 37:731–736.

Dubost, G. 1968. Les mammifères souterrains. *Revue d'Écologie et de Biologie du Sol* 5:99–197.

Dubrulle, J., M. J. McGrew, and O. Pourquie. 2001. FGF signaling controls somite boundary position and regulates segmentation clock control of spatiotemporal *Hox* gene activation. *Cell* 106:219–232.

Ducey, P. K., D. R. Formanowicz Jr., L. Boyet, J. Mailloux, and R. A. Nussbaum. 1993. Experimental examination of burrowing behavior in caecilians (Amphibia: Gymnophiona): Effects of soil compaction on burrowing ability of four species. *Herpetologica* 49:450–457.

Ducy, P., R. Zhang, V. Geoffroy, A. L. Ridall, and G. Karsenty.

1997. Osf2/Cbfa1: A transcriptional activator of osteoblast differentiation. *Cell* 89:747–754.

Dudley, A. T., M. A. Ros, and C. J. Tabin. 2002. A re-examination of proximodistal patterning during vertebrate limb development. *Nature* 418:539–544.

Duellman, W. E., and A. Schwartz. 1958. Amphibians and reptiles of southern Florida. *Bulletin of the Florida State Museum, Biological Sciences* 3:181–324.

Duellman, W. E., and L. Trueb. 1986. *Biology of amphibians.* New York: McGraw-Hill.

Duerden, J. E. 1922. Degeneration in the limbs of South African serpentiform lizards (*Chamaesaura*). *South African Journal of Science* 19:269–275.

Duerden, J. E., and R. Essex. 1923a. The pelvic girdle in the snake *Glauconia*. *South African Journal of Science* 20:354–356.

Duerden, J. E., and R. Essex. 1923b. The pelvic girdle of the snake *Typhlops*. *South African Journal of Natural History* 4:173–186.

Duncker, G. 1905. Über Regeneration des Schwanzendesbei Syngnathiden. *Wilhem Roux's Archives of Developmental Biology* 20:30–37.

Dunis, D. A., and M. Namenwirth. 1977. The role of grafted skin in the regeneration of X-irradiated axolotl limbs. *Developmental Biology* 56:97–109.

Dunn, E. R., and J. A. Tihen. 1944. The skeletal anatomy of *Liotyphlops albirostris*. *Journal of Morphology* 74:287–295.

Dunson, W. A., ed. 1975. *The biology of sea snakes.* Baltimore: University Park Press.

Dupe, V., N. B. Ghyselinck, V. Thomazy, L. Nagy, P. J. Davies, P. Chambon, and M. Mark. 1999. Essential roles of retinoic acid signalling in interdigital apoptosis and control of BMP-7 expression in mouse autopods. *Developmental Biology* 208:30–43.

Duprez, D. M., E. J. Bell, M. K. Richardson, C. W. Archer, L. Wolpert, P. M. Brickell, and P. H. Francis-West. 1996a. Overexpression of BMP-2 and BMP-4 alters the size and shape of developing skeletal elements in the chick limb. *Mechanisms in Development* 57:145–157.

Duprez, D. M., M. Coltey, H. Amthor, P. M. Brickell, and C. Tickle. 1996b. Bone morphogenetic protein-2 (BMP-2) inhibits muscle development and promotes cartilage formation in chick limb bud cultures. *Developmental Biology* 174:448–452.

Duprez, D. M., K. Kostakopoulou, P. H. Francis-West, C. Tickle, and P. M. Brickell. 1996c. Activation of Fgf-4 and *HoxD* gene expression by BMP-2 expressing cells in the developing chick limb. *Development* 122:1821–1828.

Du Shane, G. P. 1935. An experimental study of the origin of pigment cells in amphibia. *Journal of Experimental Zoology* 72:1–31.

Dutta, S. K., and B. Pradhan. 1985. Burrowing specializations in anurans. *Pranikee* 6:41–52.

Dyke, G. J., B. E. Gulas, and T. M. Crowe. 2003. Suprageneric relationships of galliform birds (Aves, Galliformes): A cladistic analysis of morphological characters. *Zoological Journal of the Linnean Society* 137:227–244.

Dyke, G. J., and J. M. V. Rayner. 2001. Forelimb shape and the evolution of birds. In *New perspectives on the origin and early evolution of birds: Proceedings of the International Symposium in Honor of John H. Ostrom*, ed. J. A. Gauthier and L. F. Gall, 276–282. New Haven, CT: Peabody Museum of Natural History, Yale University.

E

Eaton, G. F. 1910. Osteology of *Pteranodon. Memoirs of the Connecticut Academy of Arts and Sciences* 2:1–38.

Ebensperger, L. A., and F. Bozinovic. 2000. Energetics and burrowing behaviour in the semifossorial degu *Octodon degus* (Rodentia: Octodontidae). *Journal of Zoology, London* 252:179–186.

Eberhard, W. G. 2001. Multiple origins of a major novelty: Moveable abdominal lobes in male sepsid flies (Diptera: [S]epsidae), and the question of developmental constraints. *Evolution and Development* 3:206–222.

Eblaghie, M. C., S. J. Lunn, R. J. Dickinson, A. E. Munsterberg, J. J. Sanz-Ezquerro, E. R. Farrell, J. Mathers, S. M. Keyse, K. Storey, and C. Tickle. 2003. Negative feedback regulation of FGF signaling levels by Pyst1/MKP3 in chick embryos. *Current Biology* 13:1009–1018.

Echelard, Y., D. J. Epstein, B. St-Jacques, L. Shen, J. Mohler, J. A. McMahon, and A. P. McMahon. 1993. Sonic hedgehog, a member of a family of putative signaling molecules, is implicated in the regulation of CNS polarity. *Cell* 75:1417–1430.

Echeverri, K., J. D. W. Clarke, and E. M. Tanaka. 2001. *In vivo* imaging indicates muscle fiber dedifferentiation is a major contributor to the regenerating tail blastema. *Developmental Biology* 236:151–164.

Ede, D. A. 1971. Control of form and pattern in the vertebrate limb. *Symposium of the Society of Experimental Biology* 25:235–254.

Edwards, J. C., L. S. Wilkinson, H. M. Jones, P. Soothill, K. J. Henderson, J. G. Worrall, and A. A. Pitsillides. 1994. The formation of human synovial joint cavities: A possible role for hyaluronan and CD44 in altered interzone cohesion. *Journal of Anatomy* 185:355–367.

Edwards, J. L. 1985. Terrestrial locomotion without appendages. In *Functional vertebrate morphology*, ed. M. Hildebrand, D. M. Bramble, K. F. Liem, and D. B. Wake, 159–172. Cambridge, MA: Harvard University Press.

Edwards, J. L. 1989. Two perspectives on the evolution of the tetrapod limb. *American Zoologist* 29:235–254.

Egar, M. W. 1988. Accessory limb production by nerve-induced cell proliferation. *Anatomical Record* 221:550–564.

Eggli, P. S., W. Hermann, E. R. Hunziker, and R. K. Schenk. 1985. Matrix compartments in the growth plate of the proximal tibia of the rat. *Anatomical Record* 211:246–257.

Ekanayake, S., and B. K. Hall. 1987. The development of acellularity of the vertebral bone of the Japanese medaka, *Oryzias latipes* (Teleostei; Cyprinidontidae). *Journal of Morphology* 193:253–261.

Ekanayake, S., and B. K. Hall. 1988. Ultrastructure of the osteogenesis of acellular vertebral bone in the Japanese medaka, *Oryzias latipes* (Teleostei, Cyprinidontidae). *American Journal of Anatomy* 182:241–249.

Ellington, C. P., C. Van den Berg, A. P. Willmott, and A. L. R. Thomas. 1996. Leading-edge vortices in insect flight. *Nature* 384:626–630.

Elsner, R. 1999. Living in water: Solutions to physiological problems. In *Biology of marine mammals*, ed. J. E. Reynolds III and S. A. Rommel, 73–116. Washington, DC: Smithsonian Institution Press.

Eltringham, S. K. 1999. *The hippos: Natural history and conservation.* London: Academic Press.

Emerson, S. B. 1971. The fossorial frog adaptive zone: A study of convergence and parallelism in the Anura. PhD diss., University of Southern California.

Emerson, S. B. 1976. Burrowing in frogs. *Journal of Morphology* 149:437–458.

Emerson, S. B. 1979. The ilio-sacral articulation in frogs: Form and function. *Biological Journal of the Linnean Society* 11:153–168.

Emerson, S. B. 1982. Frog postcranial morphology: Identification of a functional complex. *Copeia* 1982:603–613.

Emerson, S. B. 1988a. Convergence and morphological constraint in frogs: Variation in post-cranial morphology. *Fieldiana: Zoology*, n.s., 43:1–19.

Emerson, S. B. 1988b. Testing for historical patterns of change: A case study with frog pectoral girdles. *Paleobiology* 14:174–186.

Emerson, S. B., and D. Diehl. 1980. Toe pad morphology and mechanisms of sticking in frogs. *Biological Journal of the Linnean Society* 13:199–216.

Endo, H., T. Makita, M. Sasaki, K. Arishima, M. Yamamoto, and Y. Hayashi. 1999. Comparative anatomy of the radial sesamoid in the Polar Bear (*Ursus maritimus*), the Brown Bear (*Ursus arctos*) and the Giant Panda (*Ailuropoda melanoleuca*). *Journal of Veterinary Medical Science* 61:903–907.

Endo, T., H. Yokoyama, K. Tamura, and H. Ide. 1997. Shh expression in developing and regenerating limb buds of Xenopus laevis. *Developmental Dynamics* 209:227–232.

Engleman, V. W., G. A. Nickols, F. P. Ross, M. A. Horton, D. W. Griggs, S. L. Settle, P. G. Ruminski, and S. L. Teitelbaum. 1997. A peptidomimetic antagonist of the alpha(v)beta3 integrin inhibits bone resorption in vitro and prevents osteoporosis in vivo. *Journal of Clinical Investigation* 99:2284–2292.

Engler, E. 1929. Untersuchungen zur Anatomie und Entwicklungsgeschichte des Brustschulterapparates der Urodelen. *Acta Zoologica* 10:143–229.

English, A. W. M. 1976a. Limb movements and locomotor function in the California sea lion (*Zalophus californianus*). *Journal of Zoology, London* 178:341–364.

English, A. W. M. 1976b. Structural correlates of forelimb function in fur seals and sea lions. *Journal of Morphology* 151:325–352.

Engsig, M. T., Q. J. Chen, T. H. Vu, A. C. Pedersen, B. Therkidsen, L. R. Lund, K. Henriksen, et al. 2000. Matrix metalloproteinase 9 and vascular endothelial growth factor are essential for osteoclast recruitment into developing long bones. *Journal of Cell Biology* 151:879–889.

Enlow, D. H. 1962. A study of the post-natal growth and remodeling of bone. *American Journal of Anatomy* 110:79–101.

Enlow, D. H. 1969. The bone of reptiles. In *Biology of the reptilia*, vol. 1, ed. C. Gans, 45–80. New York: Academic Press.

Enomoto-Iwamoto, M., M. Iwamoto, Y. Mukudai, Y. Kawakami, T. Nohno, Y. Higuchi, S. Takemoto, H. Ohuchi, S. Noji, and K. Kurisu. 1998. Bone morphogenetic protein signaling is required for maintenance of differentiated phenotype, control of proliferation, and hypertrophy in chondrocytes. *Journal of Cell Biology* 140:409–418.

Enomoto-Iwamoto, M., J. Kitagaki, E. Koyama, Y. Tamamura, C. Wu, N. Kanatani, T. Koike, et al. 2002. The Wnt antagonist Frzb-1 regulates chondrocyte maturation and long bone development during limb skeletogenesis. *Developmental Biology* 251:142–156.

Erdmann, K. 1933. Zur Entwicklung des knöchernen Skelets von Triton and Rana unterbesonderer Berücksichtigung der Zeitfolge der Ossificiationen. *Zeitschrift für Anatomie und Entwicklungseschichte* 101:566–651.

Erickson, G. M., and C. A. Brochu. 1999. How the "terror crocodile" grew so big. *Nature* 398:205–206.

Erickson, G. M., J. Catanese III, and T. M. Keaveny. 2002. Evolution of the biomechanical material properties of the femur. *Anatomical Record* 268:115–124.

Erickson, G. M., K. C. Rogers, and S. A. Yerby. 2001. Dinosaurian growth patterns and rapid avian growth rates. *Nature* 412:429–433.

Ericson, P. G. P. 1997. Systematic relationships of the palaeogene family Presbyornithidae (Aves: Anseriformes). *Zoological Journal of the Linnean Society* 121:429–483.

Ericson, P. G. P., T. J. Parsons, and U.S. Johansson. 2001. Morphological and molecular support for nonmonophyly of the Galloanserae. In *New perspectives on the origin and early evolution of birds: Proceedings of the International Symposium in Honor of John H. Ostrom*, ed. J. A. Gauthier and L. F. Gall, 157–168. New Haven, CT: Peabody Museum of Natural History, Yale University.

Erikson, G. E. 1963. Brachiation in New World monkeys and in anthropoid apes. *Zoological Society of London, Symposium* 10:135–164.

Erlebacher, A., E. H. Filvaroff, S. E. Gitelman, and R. Derynck. 1995. Toward a molecular understanding of skeletal development. *Cell* 80:371–378.

Ernst, C. H., and R. W. Barbour. 1989. *Turtles of the world*. Washington, DC: Smithsonian Institution Press.

Essex, R. 1927. Studies in reptilian degeneration. *Proceedings of the Zoological Society of London* 1927:879–945.

Esteban, M., M. Garcia-Paris, D. Buckley, and J. Castanet. 1999. Bone growth and age in *Rana saharica*, a water frog living in a desert environment. *Annales Zoologici Fennici* 36:43–62.

Esteban, M., M. Garcia-Paris, and J. Castanet. 1996. Bone growth and life history traits of frogs (*Rana perezi*) from a warm temperate climate area. *Canadian Journal of Zoology* 74:1914–1921.

Evans, H. E. 1955. The osteology of a worm snake, *Typhlops jamaicensis* (Shaw). *Anatomical Record* 122:381–396.

Evans, H. E. 1993. *Miller's anatomy of the dog*. 3rd ed. Philadelphia: W. B. Saunders.

Evans, S., and D. Sigogneau-Russell. 2001. A stem-group caecilian (Lissamphibia: Gymnophiona) from the Lower Cretaceous of North Africa. *Palaeontology* 44:259–273.

Evans, S. E., and A. R. Milner. 1996. A metamorphosed salamander from the Early Cretaceous of Las Hoyas, Spain. *Philosophical Transactions of the Royal Society of London B* 351:627–646.

Everly, A. W. 2002. Stages of development of the goosefish, *Lophius americanus*, and comments on the phylogenetic significance of the development of the luring apparatus in Lophiiformes. *Environmental Biology of Fishes* 64:393–417.

Ewart, J. C. 1894. The development of the skeleton of the limbs of the horse, with observations on polydactyly. *Journal of Anatomy and Physiology* 28:236–256, 342–369.

Ewer, R. F. 1973. *The carnivores*. Ithaca, NY: Comstock.

Exposito, J.-Y., C. Cluzel, R. Garrone, and C. Lethias. 2002. Evolution of collagens. *Anatomical Record* 268:302–316.

F

Fabrezi, M. 1993. The anuran tarsus. *Alytes* 11:47–63.

Fabrezi, M. 2001. A survey of prepollex and prehallux variation in anuran limbs. *Zoological Journal of the Linnean Society* 131:227–248.

Fabrezi, M., A. Marcus, and G. Scrocchi. 1985. Contribución al conocimiento de los Leptotyphlopidae de Argentina. I: *Leptotyphlops weyrauchi* y *Leptotyphlops albipuncta. Cuadernos de Herpetología, Asociación Herpetológica Argentina* 1:1–20.

Falconer, D. S., and T. F. C. Mackay. 1996. *Introduction to quantitative genetics.* Essex, UK: Longman.

Fallon, J. F., and J. A. Cameron. 1977. Interdigital cell death during limb development of turtle and lizard with an interpretation of evolutionary significance. *Journal of Embryology and Experimental Morphology* 40:285–289.

Fallon, J. F., and A. I. Caplan. 1983. *Limb development and regeneration.* Part A. New York: Liss.

Fallon, J. F., A. Lopez, M. A. Ros, M. P. Savage, B. B. Olwin, and B. K. Simandl. 1994. FGF-2: Apical ectodermal ridge growth signal for chick limb development. *Science* 264:104–107.

Fariña, R. A., and S. F. Vizcaíno. 1997. Allometry of the bones of living and extinct armadillos (Xenarthra, Dasypoda). *Zeitschrift für Säugetierkunde* 62:65–70.

Farmer, C. G., and D. R. Carrier. 2000. Pelvic aspiration in the American alligator (*Alligator mississippiensis*). *Journal of Experimental Biology* 203:1679–1687.

Farnum, C. 1993. Differential growth rates of long bones. In *Bone,* vol. 8, *Mechanisms of bone development and growth,* ed. B. K. Hall, 193–222. Boca Raton, FL: CRC Press.

Farnum, C. E., R. Lee, K. O'Hara, and J. P. G. Urban. 2002. Volume increase in growth plate chondrocytes during hypertrophy: The contribution of organic osmolytes. *Bone* 30:574–581.

Farnum, C. E., A. O. Lee, K. O'Hara, and N. J. Wilsman. 2003. Effect of short-term fasting on bone elongation rates: An analysis of catch-up growth in young male rats. *Pediatric Research* 53:33–41.

Farnum, C. E., A. Nixon, A. O. Lee, D. T. Kwan, L. Belanger, and N. J. Wilsman. 2000. Quantitative three-dimensional analysis of chondrocytic kinetic responses to short-term stapling of the rat proximal tibial growth plate. *Cells Tissues Organs* 167:247–258.

Farnum, C. E., J. Turgai, and N. J. Wilsman. 1990. Visualization of living termnal hypertrophic chondrocytes of growth plate cartilage *in situ* by differential interference conrast microscopy and time-lapse cinematography. *Journal of Orthopaedic Research* 8:750–763.

Farnum, C. E., and N. J. Wilsman. 1986. *In situ* localization of lectin-binding glycoconjugates in the matrix of growth-plate cartilage. *American Journal of Anatomy* 176:65–82.

Farnum, C. E., and N. J. Wilsman. 1987. Morphologic stages of the terminal hypertrophic chondrocyte of growth plate cartilage. *Anatomical Record* 29:221–232.

Farnum, C. E., and N. J. Wilsman. 1988. Lectin-binding histochemistry of intracellular and extracellular glycoconjugates of the resesrve cell zone of growth plate cartilage. *Journal of Orthopaedic Research* 6:166–179.

Farnum, C. E., and N. J. Wilsman. 1989. Cellular turnover at the chondroosseous junction of growth plate cartilage: Analysis by serial section s at the light microscopical level. *Journal of Orthopaedic Research* 7:654–666.

Farnum, C. E., and N. J. Wilsman. 1993. Determination of proliferative characteristics of growth plate chondrocytes by labeling with bromodeoxyuridine. *Calcified Tissue International* 52:110–119.

Farnum, C. E., and N. J. Wilsman. 1998a. Effects of distraction and compression on growth plate function. In *Skeletal growth and development: Clinical issues and basic science advances,* ed. J. Buckwalter, M. Ehrlich, L. Sandell, and S. Trippel, 517–531. Rosemont, IL: American Academy of Orthopaedic Surgeons.

Farnum, C. E., and N. J. Wilsman. 1998b. Growth plate cellular function. In *Skeletal growth and development: Clinical issues and basic science advances,* ed. J. Buckwalter, M. Ehrlich, L. Sandell, and S. Trippel, 187–223. Rosemont, IL: American Academy of Orthopaedic Surgeons.

Farnum, C. E., and N. J. Wilsman. 2001. Converting a differentiation cascade into longitudinal growth: Stereology and analysis of transgenic animals as tools for understanding growth plate function. *Current Opinion in Orthopaedics* 12:428–432.

Farnum, C. E., and N. J. Wilsman. 2002. Chondrocyte kinetics in the growth plate. In *The growth plate,* ed. I. M. Shapiro, B. Boyan, and H. C. Anderson, 245–257. Washington, DC: IOS Press.

Farquharson, C., D. Lester, E. Seawright, D. Jeffries, and B. Houston. 1999. Microtubules are potential regulators of growth-plate chondrocyte differentiation and hypertrophy. *Bone* 25:405–412.

Farrell, E. R., and A. E. Munsterberg. 2000. csal1 is controlled by a combination of FGF and Wnt signals in developing limb buds. *Developmental Biology* 225:447–458.

Farrell, E. R., G. Tosh, E. Church, and A. E. Munsterberg. 2001. Cloning and expression of CSAL2, a new member of the spalt gene family in chick. *Mechanisms of Development* 102:227–230.

Faustini-Fustini, M., V. Rochira, and C. Carani. 1999. Oestrogen deficiency in men: Where are we today? *European Journal of Endocrinology* 140:111–129.

Faustino, M., and D. M. Power. 1999. Development of the pectoral, pelvic, dorsal and anal fins in cultured sea bream. *Journal of Fish Biology* 54:1094–1110.

Fedak, T. J., and B. K. Hall. 2004. Perspectives on hyperphalangy: Patterns and processes. *Journal of Anatomy* 204:151–163.

Feduccia, A. 1980. *The age of birds.* Cambridge, MA: Harvard University Press.

Feduccia, A. 1996. *The origin and evolution of birds.* New Haven, CT: Yale University Press.

Feduccia, A., and J. Nowicki. 2002. The hand of birds revealed by early ostrich embryos. *Naturwissenschaften* 89:391–393.

Fekete, D. M., and J. P. Brockes. 1987. A monoclonal antibody detects a difference in the cellular composition of developing and regenerating limbs of newts. *Development* 99:589–602.

Feldkamp, S. D. 1987. Swimming in the California sea lion: Morphometrics, drag and energetics. *Journal of Experimental Biology* 131:117–135.

Felger, R. S., K. Cliffton, and P. J. Regal. 1976. Winter dormancy in sea turtles: Independent discovery and exploitation in the Gulf of California by two local cultures. *Science* 191:283–285.

Felisbino, S. L., and H. F. Carvalho. 1999. The epiphseal cartilage and growth of long bones in *Rana catesbeiana. Tissue and Cell* 31:301–307.

Felisbino, S. L., and H. F. Carvalho. 2000. The osteochondral ligament: A fibrous attachment between bone and articular cartilage in *Rana catesbeiana*. *Tissue and Cell* 32:527–536.

Felisbino, S. L., and H. S. Carvalho. 2001. Growth cartilage calcification and formation of bone trabeculae are late and dissociated events in the endochondral ossification of *Rana catesbeiana*. *Cell and Tissue Research* 306:319–323.

Felisbino, S. L., and H. S. Carvalho. 2002. Ectopic mineralization of articular cartilage in the bullfrog *Rana catesbeiana* and its possible involvement in bone closure. *Cell and Tissue Research* 307:357–365.

Fell, H. B. 1925. The histogenesis of cartilage and bone in the long bones of the embryonic fowl. *Journal of Morphology and Physiology* 40:417–459.

Fell, H. B., and R. G. Canti. 1934. Experiments on the development *in vitro* of the avian knee joint. *Proceedings of the Royal Society of London B* 116:316–351.

Felts, W. J., and F. A. Spurrell. 1966. Some structural and developmental characteristics of cetacean (odontocete) radii: A study of adaptive osteogenesis. *American Journal of Anatomy* 118:103–134.

Ferguson, C., T. Miclau, D. Hu, E. Alpern, and J. A. Helms. 1998. Common molecular pathways in skeletal morphogenesis and repair. *Annals of the New York Academy of Science* 857:33–42.

Ferguson, C., E. Alpern, T. Miclau, and J. A. Helms. 1999. Does adult fracture repair recapitulate embryonic skeletal formation? *Mechanisms of Development* 87:47–66.

Fernandez-Teran, M., M. E. Piedra, I. S. Kathiriya, D. Srivastava, J. C. Rodriguez-Rey, and M. A. Ros. 2000. Role of dHAND in the anterior-posterior polarization of the limb bud: Implications for the Sonic hedgehog pathway. *Development* 127:2133–2142.

Ferrari, D., and R. A. Kosher. 2002. Dl55 is a positive regulator of chondrocyte differentiation during endochondral ossification. *Developmental Biology* 252:257–270.

Ferreira, L. C. G., R. J. Beamish, and J. H. Youson. 1999. Macroscopic structure of the fin-rays and their annuli in pectoral and pelvic fins of chinook salmon, *Oncorhynchus tshawytscha*. *Journal of Morphology* 239:297–320.

Ferretti, P., and J. P. Brockes. 1991. Cell origin and identity in limb regeneration and development. *Glia* 4:214–224.

Ferretti, P., and C. Tickle. 2006. The limbs. In *Embryos, genes, and birth defects*, ed. P. Ferretti, A. Copp, C. Tickle, and G. Moore. 2nd ed. New York: Wiley. In press.

Fischer, S., B. W. Draper, and C. J. Neumann. 2003. The zebrafish fgf24 mutant identifies an additional level of Fgf signaling involved in vertebrate forelimb initiation. *Development* 130:3515–3524.

Fish, F. E. 1982. Function of the compressed tail of surface swimming muskrats (*Ondatra zibethicus*). *Journal of Mammalogy* 63:591–597.

Fish, F. E. 1984a. Kinematics of undulatory swimming in the American alligator. *Copeia* 1984:839–843.

Fish, F. E. 1984b. Mechanics, power output and efficiency of the swimming muskrat (*Ondatra zibethicus*). *Journal of Experimental Biology* 110:183–201.

Fish, F. E. 1993. Influence of hydrodynamic design and propulsive mode on mammalian swimming energetics. *Australian Journal of Zoology* 42:79–101.

Fish, F. E. 1994. Association of propulsive swimming mode with behavior in river otters (*Lutra canadensis*). *Journal of Mammalogy* 75:989–997.

Fish, F. E. 1996. Transitions from drag-based to lift-based propulsion in mammalian swimming. *American Zoologist* 36:628–641.

Fish, F. E. 1998a. Comparative kinematics and hydrodynamics of odontocete cetaceans: Morphological and ecological correlates with swimming performance. *Journal of Experimental Biology* 201:2867–2877.

Fish, F. E. 1998b. Imaginative solutions by marine organisms for drag reduction. *Proceedings of the International Symposium on Seawater Drag Reduction*, ed. J. C. S. Meng, 443–450. Newport, RI: Naval Undersea Warfare Center.

Fish, F. E. 1999. Energetics of swimming and flying in formation. *Comments on Theoretical Biology* 5:283–304.

Fish, F. E. 2001. A mechanism for evolutionary transition in swimming mode by mammals. In *Secondary adaptation of tetrapods to life in water: Proceedings of the International Meeting, Poitiers, 1996*, ed. J.-M. Mazin and V. de Buffrénil, 261–282. Munich: Pfeil.

Fish, F. E. 2002. Balancing requirements for stability and maneuverability in cetaceans. *Integrative and Comparative Biology* 42:85–93.

Fish, F. E., and J. M. Battle. 1995. Hydrodynamic design of the humpback whale flipper. *Journal of Morphology* 225:51–60.

Fish, F. E., and R. V. Baudinette. 1999. Energetics of locomotion by the Australian water rat (*Hydromys chrysogaster*): A comparison of swimming and running in a semi-aquatic mammal. *Journal of Experimental Biology* 202:353–363.

Fish, F. E., P. B. Frappell, and M. P. Sarre. 1997. Energetics of swimming by the platypus *Ornithorhynchus anatinus*: Metabolic effort associated with rowing. *Journal of Experimental Biology* 200:2647–2652.

Fish, F. E., P. B. Frappell, R. V. Baudinette, and P. M. Macfarlane. 2001. Energetics of terrestrial locomotion of the platypus *Ornithorhynchus anatinus*. *Journal of Experimental Biology* 204:797–803.

Fish, F. E., and C. A. Hui. 1991. Dolphin swimming: A review. *Mammal Review* 4:181–195.

Fish, F. E., J. Hurley, and D. P. Costa. 2003. Maneuverability by the sea lion *Zalophus californianus*: Turning performance of an unstable body design. *Journal of Experimental Biology* 206:667–674.

Fish, F. E., S. Innes, and K. Ronald. 1988. Kinematics and estimated thrust production of swimming harp and ringed seals. *Journal of Experimental Biology* 137:157–173.

Fish, F. E., J. Smelstoys, R. V. Baudinette, and P. S. Reynolds. 2002. Fur does not fly, it floats: Buoyancy of pelage in semiaquatic mammals. *Aquatic Mammals* 28:103–112.

Fish, F. E., and B. R. Stein. 1991. Functional correlates of differences in bone density among terrestrial and aquatic genera in the family Mustelidae (Mammalia). *Zoomorphology* 110:339–345.

Fisher, H. I. 1940. The occurrence of vestigial claws on the wings of birds. *American Midland Naturalist* 23:234–243.

Fisher, H. I. 1946. Adaptations and comparative anatomy of the locomotor apparatus of new world vultures. *American Midland Naturalist* 35:545–727.

Fitch, H. S. 1989. A field study of the slender glass lizard, *Ophisaurus attenuatus*, in northeastern Kansas. *Occasional Papers of the Museum of Natural History, The University of Kansas* 125:1–50.

Fjeld, T. O., and H. Steen. 1990. Growth retardation after experimental limb lengthening by epiphyseal distraction. *Journal of Pediatric Orthopedics* 10:463–466.

Flake, A. W. 2001. Fate mapping of stem cells. In *Stem cell biology,* ed. D. R. Marshak, D. Gottlieb, and R. L. Gardner, 375–397. New York: Cold Spring Harbor Laboratory Press.

Fleming, M. W., and R. A. Tassava. 1981. Preamputation and postamputation histology of the neonatal opossum hindlimb: Implications for regeneration experiments. *Journal of Experimental Zoology* 215:143–149.

Flower, W. H. 1870. *An introduction to the osteology of the Mammalia.* London: Macmillan.

Flower, W. H. 1876. *An introduction to the osteology of the Mammalia.* 2nd ed. London: Macmillan.

Flyger, V., and M. R. Townsend. 1968. The migration of polar bears. *Scientific American* 218:108–116.

Forey, P. L. 1998. *History of the coelacanth fishes.* London: Chapman and Hall.

Foster, J. W., M. A. Dominguez-Steglich, S. Guioli, G. Kowk, P. A. Weller, M. Stevanovic, J. Weissenbach, et al. 1994. Campomelic dysplasia and autosomal sex reversal caused by mutations in an SRY-related gene. *Nature* 372:525–530.

Foster, M. S., and R. W. McDiarmid. 1983. *Rhinophrynus dorsalis.* In *Costa Rican natural history,* ed. D. H. Janzen, 419–421. Chicago: University of Chicago Press.

Fouquette, M. J., Jr. 1969. Rhinophrynidae, *Rhinophrynus, R. dorsalis. Catalogue of American Amphibians and Reptiles* 78:1–2.

Fouquette, M. J., Jr., and D. A. Rossman. 1963. Noteworthy records of Mexican amphibians and reptiles in the Florida State Museum and the Texas Natural History Collection. *Herpetologica* 19:185–201.

Fraas, E. 1889. Die Labyrinthodonten der schwäbischen Trias. *Palaeontographica* 36:1–158.

Francillon-Vieillot, H., V. de Buffrénil, J. Castanet, J. Géraudie, F. J. Meunier, J. Y. Sire, L. Zylberberg, and A. de Ricqlès. 1990. Microstructure and mineralization of vertebrate skeletal tissues. In *Skeletal biomineralization: Patterns, processes, and evolutionary trends,* ed. J. G. Carter, 471–530. New York: Van Nostrand Reinhold.

Francis, E. T. B. 1934. *The anatomy of the salamander.* Oxford: Clarendon Press.

Francis, P. H., M. K. Richardson, P. M. Brickell, and C. Tickle. 1994. Bone morphogenetic proteins and a signalling pathway that controls patterning in the developing chick limb. *Development* 120:209–218.

Francis-West, P. H., A. Abdelfattah, P. Chen, C. Allen, J. Parish, R. Ladher, S. Allen, S. MacPherson, F. P. Luyten, and C. W. Archer. 1999. Mechanisms of GDF-5 action during skeletal development. *Development* 126:1305–1315.

Francis-West, P. H., J. Parish, K. Lee, and C. W. Archer. 1999. BMP/GDF-signalling interactions during synovial joint development. *Cell and Tissue Research* 296:111–119.

Francis-West, P. H., K. E. Robertson, D. A. Ede, C. Rodriguez, J. C. Izpisúa Belmonte, B. Houston, D. W. Burt, C. Gribbin, P. M. Brickell, and C. Tickle. 1995. Expression of genes encoding bone morphogenetic proteins and sonic hedgehog in talpid (ta³) limb buds: Their relationships in the signalling cascade involved in limb patterning. *Developmental Dynamics* 203:187–197.

Franssen, R. A., S. Marks, D. Wake, and N. Shubin. 2005. Limb chondrogenesis of the seepage salamander, *Desmognathus aeneus* (Amphibia: Plethodontidae). *Journal of Morphology* 265:87–101.

Franzoso, G., L. Carlson, L. Xing, L. Poljak, E. W. Shores, K. D. Brown, A. Leonardi, T. Tran, B. F. Boyce, and U. Siebenlist. 1997. Requirement for NF-kappaB in osteoclast and B-cell development. *Genes and Development* 11:3482–3496.

Freeman, J. R. 1958. Burrowing in the salamanders *Pseudobranchus striatus* and *Siren lacertina. Herpetologica* 14:130.

Freeman, R. A. 1886. The anatomy of the shoulder and upper arm of the mole (*Talpa europaea*). *Journal of Anatomy and Physiology* 20:201–219.

French, V., P. J. Bryant, and S. V. Bryant. 1976. Pattern regulation in epimorphic fields. *Science* 193:969–981.

Frey, E. 1982a. Ecology, locomotion and tail muscle anatomy of crocodiles. *Neues Jahrbuch für Geologie und Paläontologie, Monatshefte* 1982:194–199.

Frey, E. 1982b. *Ophisaurus apodus* (Lacertilia, Anguidae): A stemming digger? *Neues Jahrbuch für Geologie und Paläontologie, Abhandlungen* 164:217–221.

Frey, R. W. 1975. *The study of trace fossils: A synthesis of principles, problems, and procedures in ichnology.* New York: Springer-Verlag.

Fricke, H., and K. Hissmann. 1992. Locomotion, fin coordination and body form of the living coelacanth *Latimeria chalumnae. Environmental Biology of Fishes* 34:329–356.

Fröhlich, M. W. 2003. An evolutionary scenario for the origin of flowers. *Nature Reviews Genetics* 4:559–566.

Fromental-Ramain, C., X. Warot, S. Lakkaraju, B. Favier, H. Haack, C. Birling, A. Dierich, P. Dollé, and P. Chambon. 1996a. Specific and redundant functions of the paralogous Hoxa-9 and Hoxd-9 genes in forelimb and axial skeleton patterning. *Development* 122:461–472.

Fromental-Ramain, C., X. Warot, N. Messadecq, M. LeMeur, P. Dollé, and P. Chambon. 1996b. Hoxa-13 and Hoxd-13 play a crucial role in the patterning of the limb autopod. *Development* 122:2997–3011.

Frost, D. R., and V. Etheridge. 1989. A phylogenetic analysis and taxonomy of iguanian lizards (Reptilia: Squamata). *University of Kansas Museum of Natural History Miscellaneous Publications* 81:1–65.

Fürbringer, M. 1870. *Die Knochen und Muskeln der Extremitäten bei den schlangenähnlichen Sauriern. Vergleichend-anatomische Abhandlung.* Leipzig: Wilhelm Engelmann.

Fürbringer, M. 1876. Zur vergleichenden Anatomie der Schultermuskeln: III. *Gegenbaurs Morphologisches Jahrbuch* 1:636–816.

Fürbringer, M. 1888. *Untersuchungen zur Morphologie und Systematik der Vögel.* I: *Specieller Theil.* Amsterdam: T. J. Van Holkema.

Fürbringer, M. 1900. Zur vergleichenden Anatomie des Brustschulterapparates und der Schultermuskeln. *Jenaische Zeitschrift für Naturwissenschaft* 34:215–718.

Furness, R. W. 1991. The occurrence of burrow-nesting among birds and its influence on soil fertility and stability. In *The environmental impact of burrowing animals and animal burrows,* ed. P. S. Meadows and A. Meadows, 53–67. Symposia of the Zoological Society of London no. 63. Oxford: Clarendon Press.

Fuse, N., S. Hirose, and S. Hayashi. 1996. Determination of wing cell fate by the escargot and snail genes in Drosophila. *Development* 122:1059–1067.

Futuyma, D. J. 1998. *Evolutionary biology.* Sunderland, MA: Sinauer Associates.

G

Gabe, M., J.-P. Gasc, J. Lessertisseur, R. Saban, and D. Starck. 1967. *Traité de zoologie: Anatomie, systématique, biologie,* vol. 16, *Mammiferes teguments et squelette.* Paris: Masson.

Gadow, H. 1882. Beiträge zur Myologie der hinteren Extremität der Reptilien. *Gegenbaurs Morphologisches Jahrbuch* 7:329–466.

Gadow, H. 1901. *Amphibia and reptiles.* Cambridge Natural History, vol. 8. Cambridge, UK: Macmillan.

Gaffney, E. S. 1979. Tetrapod monophyly: A phylogenetic analysis. *Bulletin of the Carnegie Museum of Natural History* 13:92–105.

Gaffney, E. S. 1990. The comparative osteology of the Triassic turtle *Proganochelys. Bulletin of the American Museum of Natural History* 194:1–263.

Gafni, R. I., and J. Baron. 2000. Catch-up growth: Possible mechanisms. *Pediatric Nephrology* 14:616–619.

Gafni, R. I., M. R. Weise, T. Daniel, J. L. Meyers, K. M. Barnes, S. De-Levi, and J. Baron. 2001. Catch-up growth is associated with delayed senescence of the growth plate in rabbits. *Pediatric Research* 50:618–623.

Galceran, J., I. Farinas, M. J. Depew, H. Clevers, and R. Grosschedl. 1999. *Wnt3a*$^{-/-}$-like phenotype and limb deficiency in *Lef1*$^{-/-}$*Tcf1*$^{-/-}$ mice. *Genes and Development* 13:709–717.

Galilei, G. 1914. *Dialogues concerning two new sciences.* Trans. H. Crew and A. De Salvio. New York: Macmillan. (Orig. pub. 1637.)

Galindo, M. I., S. A. Bishop, S. Greig, and J. P. Couso. 2002. Leg patterning driven by proximal-distal interactions and EGFR signaling. *Science* 297:256–259.

Galis, F., J. J. M. v. Alphen, and J. A. J. Metz. 2001. Why five fingers? Evolutionary constraints on digit numbers. *Trends in Ecology and Evolution* 16:637–646.

Galton, P. M. 1990a. Basal Sauropodomorpha: Prosauropoda. In *The Dinosauria,* ed. D. B. Weishampel, P. Dodson, and H. Osmólska, 320–344. Berkeley: University of California Press.

Galton, P. M. 1990b. Stegosauria. In *The Dinosauria,* ed. D. B. Weishampel, P. Dodson, and H. Osmólska, 435–455. Berkeley: University of California Press.

Gambaryan, P. P. 1960. Adaptive features of locomotor organs in burrowing mammals. [In Russian.] Yerevan: Akademiya Nauk Armyanskoy SSR.

Gambaryan, P. P. 1974. *How mammals run: Anatomical adaptations.* New York: Wiley.

Gambaryan, P. P. 2002. Ways of adaptive changes in claws of digging mammals. [In Russian.] *Zoologicheskii Zhurnal* 81:978–990.

Gambaryan, P. P., and J.-P. Gasc. 1993. Adaptive properties of the musculoskeletal system in the mole-rat *Myospalax myospalax* (Mammalia, Rodentia), cinefluorographical, anatomical, and biomechanical analyses of burrowing. *Zoologische Jahrbücher. Abteilung für Anatomie und Ontogenie der Tiere* 123:363–401.

Gambaryan, P. P., and Z. Kielan-Jaworowska. 1997. Sprawling versus parasagittal stance in multituberculate mammals. *Acta Palaeontologica Polonica* 42:13–44.

Gañan, Y., D. Macias, R. D. Basco, R. Merino, and J. M. Hurlé. 1998. Morphological diversity of the avian foot is related with the pattern of msx gene expression in the developing autopod. *Developmental Biology* 196:33–41.

Gañan, Y., D. Macias, M. Duterque-Coquillaud, M. A. Ros, and J. M. Hurlé. 1996. Role of TGFβs and BMPs as signals controlling the position of the digits and the areas of interdigital cell death in the developing chick limb autopod. *Development* 122:2349–2357.

Ganguly, D. N., and A. Nag. 1964. On the functional morphology of the pectoral girdle and the acranial myomeric musculature of a benthonic teleostean fish *Ophichthys boro* (Ham. Buch.). *Anatomischer Anzeiger* 115:405–417.

Gans, C. 1958. Modifications of the head joint in acrodont amphisbaenids and their functional implication. *Anatomical Record* 132:441.

Gans, C. 1960. Studies on amphisbaenids (Amphisbaenia, Reptilia). 1: A taxonomic revision of the Trogonophinae, and a functional interpretation of the amphisbaenid adaptive pattern. *Bulletin of the American Museum of Natural History* 119:129–204.

Gans, C. 1962. Terrestrial locomotion without limbs. *American Zoologist* 2:167–182.

Gans, C. 1968. Relative success of divergent pathways in amphisbaenian specialization. *American Naturalist* 102:345–362.

Gans, C. 1969. Amphisbaenians: Reptiles specialized for a burrowing existence. *Endeavour* 28:146–151.

Gans, C. 1973a. Locomotion and burrowing in limbless vertebrates. *Nature* 242:414–415.

Gans, C. 1973b. Uropeltid snakes: Survivors in a changing world. *Endeavour* 32:60–65.

Gans, C. 1974. *Biomechanics: An approach to vertebrate biology.* Ann Arbor: University of Michigan Press.

Gans, C. 1975. Tetrapod limblessness: Evolution and functional corollaries. *American Zoologist* 15:455–467.

Gans, C. 1976. Aspects of the biology of uropeltid snakes. In *Morphology and biology of reptiles,* ed. A. D.'A. Bellairs and C. B. Cox, 191–204. London: Linnean Society.

Gans, C. 1978. The characteristics and affinities of the Amphisbaenia. *Transactions of the Zoological Society of London* 34:347–416.

Gans, C., and D. Baic. 1977. Regional specialization of reptilian scale surfaces: Relation of texture and biologic role. *Science* 195:1348–1350.

Gans, C., H. C. Dessauer, and D. Baic. 1978. Axial differences in the musculature of uropeltid snakes: The freight-train approach to burrowing. *Science* 199:189–192.

Gans, C., and J.-P. Gasc. 1990. Tests on the locomotion of the elongate and limbless reptile *Ophisaurus apodus* (Sauria: Anguidae). *Journal of Zoology, London* 220:517–536.

Gans, C., and T. S. Parsons. 1966. On the origin of the jumping mechanism in frogs. *Evolution* 20:92–99.

Gao, K.-Q., and N. H. Shubin. 2001. Late Jurassic salamanders from northern China. *Nature* 410:574–577.

Gao, K.-Q., and N. H. Shubin. 2003. Earliest known crown-group salamanders. *Nature* 422:424–428.

Gao, K.-Q., and Y. Wang. 2001. Mesozoic anurans from Liaoning Province, China, and phylogenetic relationships of archaeobatrachian anuran clade. *Journal of Vertebrate Paleontology* 21:574–577.

Garcia-Domingo, D., E. Leonardo, A. Grandien, P. Martinez, J. P. Albar, J. C. Izpisúa Belmonte, and C. Martinez. 1999. DIO-1 is a gene involved in onset of apoptosis in vitro, whose misexpression disrupts limb development. *Proceedings of the National Academy of Sciences, USA* 96:7992–7997.

Gardiner, D. M., B. Blumberg, Y. Komine, and S. V. Bryant. 1995. Regulation of *HoxA* expression in developing and regenerating axolotl limbs. *Development* 121:1731–1741.

Gardiner, D. M., and S. V. Bryant. 1996. Molecular mechanisms in the control of limb regeneration: The role of homeobox genes. *International Journal of Developmental Biology* 40:797–805.

Gardiner, D. M., and S. V. Bryant. 1998. The tetrapod limb. In *Cellular and molecular basis of regeneration: From invertebrates to humans*, ed. P. Ferretti and J. Géraudie, 187–205. New York: Wiley.

Gardiner, D. M., M. R. J. Carlson, and S. Roy. 1999. Towards a functional analysis of limb regeneration. *Cell and Developmental Biology* 10:385–393.

Gardiner, D. M., T. Endo, and S. V. Bryant. 2002. The molecular basis of amphibian limb regeneration: Integrating the old with the new. *Seminars in Cell and Developmental Biology* 13:345–352.

Gardiner, D. M., K. Muneoka, and S. V. Bryant. 1986. The migration of dermal cells during blastema formation in axolotls. *Developmental Biology* 118:488–493.

Gardiner, D. M., M. A. Torok, L. M. Mullen, and S. V. Bryant. 1998. Evolution of vertebrate limbs: Robust morphology and flexible development. *American Zoologist* 38:659–671.

Gardner, E., and R. O'Rahilly. 1968. The early development of the knee joint in staged human embryos. *Journal of Anatomy* 102:289–299.

Garland, T., Jr., and C. M. Janis. 1993. Does metatarsal/femur ratio predict the maximum running speed in cursorial mammals? *Journal of Zoology, London* 229:133–151.

Garrault, H. 1936. Developpement des fibres d'elastoidine (actinotrichia) chez les Salmonides. *Archives d'Anatomie Microscopique* 32:105–137.

Gasc, J.-P. 1966. Les rapports anatomiques du membre pelvien vestigial chez les squamates serpentiformes. B: *Python sebae* (Seba). *Bulletin du Muséum National d'Histoire Naturelle, Paris*, series 2, 38:99–110.

Gasc, J.-P. 1968. Contribution a l'ostéologie et a la myologie de *Dibamus novaeguineae* Gray (Sauria, Reptilia): Discussion systématique. *Annales des Sciences Naturelles, Zoologie, Paris*, series 12, 10:127–150.

Gasc, J.-P. 1982. Le mécanisme du fouissage chez *Amphisbaena alba* (Amphisbaenidae, Squamata). *Vertebrata Hungarica* 21:147–155.

Gasc, J.-P., F. K. Jouffroy, S. Renous, and F. von Blottnitz. 1986. Morphofunctional study of the digging system of the Namib Desert golden mole (*Eremitalpa granti namibensis*): Cinefluorographical and anatomical analysis. *Journal of Zoology, London A* 208:9–35.

Gasc, J.-P., and S. Renous. 1974. Les rapports anatomiques du membre pelvien vestigial chez les Squamates serpentiformes. II: *Scelotes brevipes* et *Scelotes inornatus* (Scincidae, Sauria). *Bulletin du Muséum National d'Histoire Naturelle, Paris*, series 3, *Zoologie* 186:1701–1712.

Gasc, J.-P., and S. Renous. 1979. La région pelvi-cloacale de *Dibamus* (Squamata, Reptilia). Nouvelle contribution à sa position systématique. *Bulletin du Muséum National d'Histoire Naturelle, Paris*, series 4, sec. A, 1:659–684.

Gasc, J.-P., S. Renous, A. Casinos, E. Laville, and J. Bou. 1985. Comparison of diverse digging patterns in some small mammals. In *Vertebrate morphology*, ed. H. R. Duncker and G. Fleischer, 35–38. Stuttgart: Gustav Fischer Verlag.

Gaston, A. J., and I. L. Jones. 1998. *The auks: Alcidae.* Oxford: Oxford University Press.

Gatesy, S. M. 1990. Caudofemoralis musculature and the evolution of theropod locomotion. *Paleobiology* 16:170–186.

Gatesy, S. M. 1991. Hind limb movements of the American alligator (*Alligator mississippiensis*) and postural grades. *Journal of Zoology, London* 224:577–588.

Gatesy, S. M. 2002. Locomotor evolution on the line to modern birds. In *Mesozoic birds: Above the heads of dinosaurs*, ed. L. M. Chiappe and L. M. Witmer, 432–447. Berkeley: University of California Press.

Gatesy, S. M., and K. P. Dial. 1996. Locomotor modules and the evolution of avian flight. *Evolution* 50:331–340.

Gatesy, S. M., and K. M. Middleton. 1997. Bipedalism, flight, and the evolution of theropod locomotor diversity. *Journal of Vertebrate Paleontology* 17:308–329.

Gaudin, T. J., and D. G. Branham. 1998. The phylogeny of the Myrmecophagidae (Mammalia, Xenarthra, Vermilingua) and the relationship of *Eurotamandua* to the Vermilingua. *Journal of Mammalian Evolution* 5:237–265.

Gaupp, E. 1896. *A. Ecker's und R. Wiedersheim's Anatomie des Frosches.* Brunswick, Germany: Friedrich Viewig und Sohn.

Gauthier, J. A. 1984. A cladistic analysis of the higher systematic categories of the Diapsida. PhD diss., University of California, Berkeley.

Gauthier, J. A. 1986. Saurischian monophyly and the origin of birds. In *The origin of birds and the evolution of flight: Memoirs of the California Academy of Sciences*, vol. 8, ed. K. Padian, 1–55. San Francisco: California Academy of Sciences.

Gauthier, J. A., R. Estes, and K. de Queiroz. 1988a. A phylogenetic analysis of Lepidosauromorpha. In *Phylogenetic relationships of the lizard families*, ed. R. Estes and G. Pregill, 15–98. Stanford, CA: Stanford University Press.

Gauthier, J. A., and L. F. Gall. 2001. *New perspectives on the origin and early evolution of birds: Proceedings of the International Symposium in Honor of John H. Ostrom.* New Haven, CT: Peabody Museum of Natural History, Yale University.

Gauthier, J. A., A. G. Kluge, and T. Rowe. 1988b. Amniote phylogeny and the importance of fossils. *Cladistics* 4:105–209.

Gauthier, J. A., A. G. Kluge, and T. Rowe. 1988c. The early evolution of the Amniota. In *The phylogeny and classification of the tetrapods*, vol. 1, *Amphibians, reptiles, birds*, ed. M. J. Benton, 103–155. Oxford: Clarendon Press.

Gavaia, P. J., M. T. Dinis, and M. L. Cancella. 2002. Osteological development and abnormalities of the vertebral column and caudal skeleton in larval and juvenile stages of hatchery-reared Senegal sole (*Solea senegalensis*). *Aquaculture* 211:305–323.

Gaymer, R. 1971. New method of locomotion in limbless terrestrial vertebrates. *Nature* 234:150–151.

Geerlink, P. J. 1983. Pectoral fin kinematics of *Coris formosa* (Teleostei, Labridae). *Netherlands Journal of Zoology* 33:515–531.

Geerlink, P. J. 1987. The role of the pectoral fins in braking of mackerel, cod and saithe. *Netherlands Journal of Zoology* 37:81–104.

Geerlink, P. J. 1989. Pectoral fin morphology: A simple relation with movement pattern? *Netherlands Journal of Zoology* 39:166–193.

Gegenbaur, C. 1874. *Grundriss der vergleichenden Anatomie.* Leipzig: Wilhelm Engelmann.

Gegenbaur, C. 1878. Elements of comparative anatomy. Transl. F. Jeffrey Bell. London: Macmillan.

Gehlbach, F. R., R. Gordon, and J. B. Jordan. 1973. Aestivation of the salamander, *Siren intermedia. American Midland Naturalist* 89:455–463.

Georgopoulos, N. A., K. B. Markou, A. Theodoropoulou, G. A. Vagenakis, D. Benardot, M. Leglise, J. C. A. Dimopoulos, and A. G. Vagenkis. 2001. Height velocity and skeletal maturation in elite female rhythmic gymnasts. *Journal of Clinical Endocrinology and Metabolism* 86:5159–5164.

Georgopoulos, N. A., K. B. Markou, A. Theodoropoulou, D. Benardot, M. Leglise, and A. G. Vagenkis. 2002. Growth retardation in artistic compared with rhythmic elite female gymnasts. *Journal of Clinical Endocrinology and Metabolism* 87:3169–3173.

Géraudie, J. 1977. Initiation of the actinotrichial development in the early fin bud of the fish, *Salmo. Journal of Morphology* 151:353–362.

Géraudie, J. 1980. Mitotic activity in the pseudoapical ridge of the trout pelvic fin bud, *Salmo gairdneri. Journal of Experimental Zoology* 214:311–316.

Géraudie, J. 1981. Consequences of cell death after nitrogen mustard treatment on skeletal pelvic fin morphogenesis in the trout, *Salmo gairdneri* (Pisces, Teleostei). *Journal of Morphology* 170:181–194.

Géraudie, J. 1988. Fine structural peculiarities of the pectoral fin dermoskeleton of two Brachiopterygii, *Polypterus senegalus* and *Calamoichthys calabaricus* (Pisces, Osteichthyes). *Anatomical Record* 221:455–468.

Géraudie, J., M. A. Akimenko, and M. M. Smith. 1998. The dermal skeleton. In *Cellular and molecular basis of regeneration: From invertebrates to humans,* ed. P. Ferretti and J. Géraudie, 167–185. Chichester: Wiley.

Géraudie, J., and V. Borday Birraux. 2003. Posterior hoxa genes expression during zebrafish bony ray development and regeneration suggests their involvement in scleroblast differentiation. *Development Genes and Evolution* 213:182–186.

Géraudie, J., A. Brulfert, M.-J. Monnot, and P. Ferretti. 1994. Teratogenic and morphogenetic effects of retinoic acid on the regenerating pectoral fin in zebrafish. *Journal of Experimental Zoology* 269:12–22.

Géraudie, J., and P. Ferretti. 1997. Correlation between RA-induced apoptosis and patterning defects in regenerating fins and limbs. *International Journal of Developmental Biology* 41:29–532.

Géraudie, J., and P. Ferretti. 1998. Gene expression during amphibian limb regeneration. *International Reviews of Cytology* 180:1–50.

Géraudie, J., and Y. François. 1973. Les premiers stades de la formation de l'ébauche de nageoire pelvienne de truite (*Salmo fario* et *Salmo gairdneri*). I: Étude anatomique. *Journal of Embryology and Experimental Morphology* 29:221–237.

Géraudie, J., and W. J. Landis. 1982. The fine structure of the developing pelvic fin dermal skeleton in the trout *Salmo gairdneri. American Journal of Anatomy* 163:141–156.

Géraudie, J., and F. J. Meunier. 1980. Elastoidin actinotrichia in coelacanth fins: A comparison with teleosts. *Tissue and Cell* 12:637–645.

Géraudie, J., and F. J. Meunier. 1982. Comparative fine structure of the osteichthyan dermotrichia. *Anatomical Record* 202:325–328.

Géraudie, J., and F. J. Meunier. 1984. Structure and comparative morphology of camptotrichia of lungfish fins. *Tissue and Cell* 16:217–236.

Géraudie, J., M. J. Monnot, A. Brulfert, and P. Ferretti. 1995. Caudal fin regeneration in wild type and *long-fin* mutant zebrafish is affected by retinoic acid. *International Journal of Developmental Biology* 39:373–381.

Géraudie, J., and M. Singer. 1977. Relation between nerve fiber number and pectoral fin regeneration in the teleost. *Journal of Experimental Zoology* 199:1–8.

Géraudie, J., and M. Singer. 1985. Necessity of an adequate nerve supply for regeneration of the amputated pectoral fin in the teleost Fundulus. *Journal of Experimental Zoology* 234:367–374.

Géraudie, J., and M. Singer. 1992. The fish fin regenerate. In *Fishes and amphibians,* ed. C. H. Taban and B. Boilly, 62–72. Basel, Switzerland: S. Karger.

Gerhard, G. S., E. J. Kauffman, X. Wang, R. Stewart, J. L. Moore, C. J. Kasales, E. Demidenko, and K. C. Cheng. 2002. Life spans and senescent phenotypes in two strains of Zebrafish (*Danio rerio*). *Experimental Gerontology* 37:1055–1068.

Gerstenfeld, L. C., D. M. Cullinane, G. L. Barnes, D. T. Graves, and T. A. Einhorn. 2003. Fracture healing as a post-natal developmental process: Molecular, spatial, and temporal aspects of its regulation. *Journal of Cellular Biochemistry* 88:873–884.

Gerstner, C. L. 1999. Maneuverability of four species of coral-reef fish that differ in body and pectoral-fin morphology. *Canadian Journal of Zoology* 77:1102–1110.

Gharib, M., F. Pereira, D. Dabiri, J. R. Hove, and D. Modarress. 2002. Quantitative flow visualization: Toward a comprehensive flow diagnostic tool. *Integrative and Comparative Biology* 42:964–970.

Ghosh, S., S. Roy, A. Pfeifer, M. Schmitt, I. M. Verma, S. V. Bryant, and D. M. Gardiner. 2001. Advances in functional analysis in urodele amphibians. *Axolotl News* 29:4–8.

Giampaoli, S., S. Bucci, M. Ragghianti, G. Mancino, F. Zhang, and P. Ferretti. 2003. Expression of FGF2 in the limb blastema of two Salamandriddae correlates with their regenerative capability. *Proceedings of the Royal Society of London B* 270:2197–2205.

Gibb, A. C., B. C. Jayne, and G. V. Lauder. 1994. Kinematics of pectoral fin locomotion in the bluegill sunfish *Lepomis macrochirus. Journal of Experimental Biology* 189:133–161.

Gibbs, P. D., and M. C. Schmale. 2000. GFP as a genetic marker scorable throughout the life cycle of transgenic zebrafish. *Marine Biotechnology* (New York) 2:107–125.

Gibson, G. 1998. Active role of chondrocyte apoptosis in endochondral ossification. *Microscopy Research and Technique* 43:191–204.

Gibson, G. 1999. Developmental evolution: going beyond the "just so." *Current Biology* 9:942–945.

Gibson-Brown, J. J., S. I. Agulnik, D. L. Chapman, M. Alexiou, N. Garvey, L. M. Silver, and V. E. Papaioannou. 1996. Evidence of a role for T-box genes in the evolution of limb morphogenesis and the specification of forelimb/hindlimb identity. *Mechanisms of Development* 56:93–101.

Gilbert, B. M. 1973. *Mammalian osteo-archaeology: North America*. Columbia, MO: Missouri Archaeological Society.

Gilbert, S. F., J. M. Opitz, and R. A. Raff. 1996. Resynthesizing evolutionary and developmental biology. *Developmental Biology* 173:357–372.

Gille, U., and F. V. Salomon. 1995. Bone growth in ducks through mathematical models with special reference to the Janoschek growth curve. *Growth, Development and Aging* 59:207–214.

Gillis, G. B. 1998. Environmental effects on undulatory locomotion in the American eel Anguilla rostrata: Kinematics in water and on land. *Journal of Experimental Biology* 201:949–961.

Gilmore, C. W. 1936. Osteology of *Apatosaurus*, with special reference to specimens in the Carnegie Museum. *Memoirs of the Carnegie Museum* 11:1–300.

Gingerich, P. D. 1987. Early Eocene bats (Mammalia, Chiroptera) and other vertebrates in fresh water limestones of the Willwood Formation Clark's Fork Basin, Wyoming. *Contributions from the Museum of Paleontology, University of Michigan* 27:275–320.

Gingerich, P. D., M. ul Haq, I. S. Zalmout, I. Hussain Khan, and M. Sadiq Malkani. 2001. Origin of whales from early artiodactyls: Hands and feet of Eocene Protocetidae from Pakistan. *Science* 293:2239–2242.

Girard, I., M. W. McAleer, J. S. Rhodes, and T. Garland. 2001. Selection for high voluntary wheel-running increases speed and intermittency in house mice (Mus domesticus). *Journal of Experimental Biology* 204:4311–4320.

Girvan, J. E., W. M. Olson, and B. K. Hall. 2002. Hind-limb regeneration in the dwarf African clawed frog, *Hymenochirus boettgeri* (Anura: Pipidae). *Journal of Herpetology* 36:537–543.

Gleeson, T. T., and T. V. Hancock. 2001. Modeling the metabolic energetics of brief and intermittent locomotion in lizards and rodents. *American Zoologist* 41:211–218.

Glücksmann, A. 1951. Cell death in normal development. *Biological Review* 26:59–86.

Godfrey, G., and P. Crowcroft. 1960. *The life of the mole*. London: Museum Press.

Godfrey, S. J. 1984. Plesiosaur subaqueous locomotion: A reappraisal. *Neues Jahrbuch für Geologie und Paläontologie, Monatshefte* 1984:661–672.

Godfrey, S. J. 1989. The postcranial skeletal anatomy of the Carboniferous tetrapod *Greererpeton burkemorani*. *Philosophical Transactions of the Royal Society of London B* 323:75–133.

Goldstein, B. 1972. Allometric analysis of relative humerus width and olecranon length in some unspecialized burrowing mammals. *Journal of Mammalogy* 53:148–156.

Gong, Y., D. Krakow, J. Marcelino, D. Wilkin, D. Chitayat, R. Babul-Hirji, L. Hudgins, et al. 1999. Heterozygous mutations in the gene encoding noggin affect human joint morphogenesis. *Nature: Genetics* 21:302–304.

Gong, Y., R. B. Slee, N. Fukai, G. Rawadi, S. Roman-Roman, A. M. Reginato, H. Wang, et al. 2001. LDL receptor-related protein 5 (LRP5) affects bone accrual and eye development. *Cell* 107:513–523.

Gong, Z., B. Ju, and H. Wan. 2001. Green fluorescent protein (GFP) transgenic fish and their applications. *Genetica* 111:213–225.

Gonyea, W. J. 1976. Adaptive differences in the body proportions of large felids. *Acta Anatomica* 96:81–96.

Goodge, W. R. 1959. Locomotion and other behavior of the Dipper. *Condor* 61:4–17.

Goodrich, E. S. 1904. On the dermal fin-rays of fishes—living and extinct. *Quarterly Journal of Microscopical Science* 47:565–522.

Goodrich, E. S. 1930. *Studies on the structure and development of vertebrates*. London: Macmillan. Repr. with introd. by Keith S. Thomson. Chicago: University of Chicago Press, 1986.

Gordon, M. S., H. G. Chin, and M. Vojkovich. 1989. Energetics of swimming in fishes using different modes of locomotion, I: Labriform swimmers. *Fish Physiology and Biochemistry* 6:341–352.

Gorman, M. L., and R. D. Stone. 1990. *The natural history of moles*. Ithaca, NY: Cornell University Press.

Goss, R. J. 1956. Regenerative inhibition following amputation and immediate insertion into the body cavity. *Anatomical Record* 126:15–27.

Goss, R. J. 1969a. *Principles of regeneration*. New York: Academic Press.

Goss, R. J. 1969b. Regeneration in fishes. In *Principles of regeneration*, ed. R. J. Goss, 113–139. New York: Academic Press.

Goss, R. J., and M. W. Stagg. 1957. The regeneration of fins and fin rays in *Fundulus heteroclitus*. *Journal of Experimental Zoology* 136:487–508.

Goto, T., K. Nishida, and K. Nakaya. 1999. Internal morphology and function of paired fins in the epaulette shark, Hemiscyllium ocellatum. *Ichthyological Research* 46:281–287.

Gould, S. J. 1991. The disparity of the Burgess Shale arthropod fauna and the limits of cladistic analysis: Why we must strive to quantify morphospace. *Paleobiology* 17:411–423.

Gould, S. J., and R. Lewontin. 1978. The spandrels of San Marco and the Panglossian paradigm: A critique of the adaptationist programme. *Proceedings of the Royal Society of London* 205:581–598.

Gould, S. J., and E. S. Vrba. 1982. Exaptation: A missing term in the science of form. *Paleobiology* 8:4–15.

Graham, T. E., and A. A. Graham. 1991. *Trionyx spiniferus spiniferus* (Eastern spiny softshell): Burying behavior. *Herpetological Review* 22:56–57.

Grande, L., and W. E. Bemis. 1991. Osteology and phylogenetic relationships of fossil and recent paddlefishes (Polyodontidae) with comments on the interrelationships of Acipenseriformes. *Journal of Vertebrate Paleontology* 11:1–21.

Grandel, H., B. W. Draper, and S. Schulte-Merker. 2000. dackel acts in the ectoderm of the zebrafish pectoral fin bud to maintain AER signaling. *Development* 127:4169–4178.

Grandel, H., K. Lun, G. J. Rauch, M. Rhinn, T. Piotrowski, C. Houart, P. Sordino, et al. 2002. Retinoic acid signalling in the zebrafish embryo is necessary during pre-segmentation stages to pattern the anterior-posterior axis of the CNS and to induce a pectoral fin bud. *Development* 129:2851–2865.

Grandel, H., and S. Schulte-Merker. 1998. The development of the paired fins in the zebrafish (*Danio rerio*). *Mechanisms of Development* 79:99–120.

Grassé, P.-P. 1955a. Anatomie. In *Traité de zoologie*, ed. P.-P. Grassé, 1729–1773. Paris: Masson.

Grassé, P.-P. 1955b. Ordre des Édentés: Formes actuelles. In *Traité de zoologie: Anatomie, systématique, biologie*, vol. 17, *Mammifères: Les ordres; Anatomie, éthologie, systématique*, pt. 2, ed. P.-P. Grassé, 1182–1246. Paris: Masson.

Gray, H. 1988. *Gray's anatomy*. Ed. T. P. Pick and R. Howden. London: Galley-Press.

Gray, J. 1933. Studies in animal locomotion. I: The movement of fish with special reference to the eel. *Journal of Experimental Biology* 10:88–104.

Gray, J. 1968. *Animal locomotion.* London: Weidenfeld and Nicolson.

Green, D. M. 1981. Adhesion and the toe-pads of treefrogs. *Copeia* 1981:790–796.

Greene, E. C. 1935. *Anatomy of the rat.* Philadelphia: American Philosophical Society.

Greene, H. W. 1986. Diet and arboreality in the emerald monitor, *Varanus prasinus,* with comments on the study of adaptation. *Fieldiana: Zoology* 31:1–12.

Greer, A. E. 1970. A subfamilial classification of scincid lizards. *Bulletin of the Museum of Comparative Zoology* 139:151–184.

Greer, A. E. 1971. Evolutionary and systematic significance of crocodilian nesting habits. *Nature* 227:523–524.

Greer, A. E. 1985. The relationships of the lizard genera *Anelytropsis* and *Dibamus. Journal of Herpetology* 19:116–156.

Greer, A. E. 1987. Limb reduction in the lizard genus *Lerista.* 1: Variation in the number of phalanges and presacral vertebrae. *Journal of Herpetology* 21:267–276.

Greer, A. E. 1989. *The biology and evolution of Australian lizards.* Chipping Norton, Australia: Surrey Beatty and Sons.

Greer, A. E. 1990. Limb reduction in the scincid lizard genus *Lerista.* 2: Variation in the bone complements of the front and rear limbs and the number of postsacral vertebrae. *Journal of Herpetology* 24:142–150.

Greer, A. E. 1991. Limb reduction in squamates: Identification of the lineages and discussion of the trends. *Journal of Herpetology* 25:166–173.

Greer, A. E. 1997. Does the limbless lygosomine skink *Isopachys borealis* really lack pectoral and pelvic girdles? *Journal of Herpetology* 31:461–462.

Greer, A. E., V. Caputo, B. Lanza, and R. Palmieri. 1998. Observations on limb reduction in the scincid lizard genus *Chalcides. Journal of Herpetology* 32:244–252.

Greer, A. E., and H. G. Cogger. 1985. Systematics of the reduced-limbed and limbless skinks currently assigned to the genus *Anomalopus* (Lacertilia: Scincidae). *Records of the Australian Museum* 37:11–54.

Greer, A. E., and C. Gans. 1983. The amphisbaenian carpus: How primitive is it? *Journal of Herpetology* 17:406.

Gregory, W. K. 1912. Notes on the principles of quadrupedal locomotion and on the mechanism of the limbs of hoofed animals. *Annals of the New York Academy of Science* 22:267–294.

Gregory, W. K. 1915. Present status of the problem of the origin of the tetrapoda, with special reference to the skull and paired limbs. *Annals of the New York Academy of Sciences* 26:317–383.

Gregory, W. K. 1929. *Our face from fish to man: A portrait gallery of our ancient ancestors and kinsfolk, together with a concise history of our best features.* New York: G. P. Putnam's Sons.

Gregory, W. K. 1951. *Evolution emerging.* New York: American Museum of Natural History and Columbia University Press.

Gregory, W. K., R. W. Miner, and G. K. Noble. 1923. The carpus of Eryops and the structure of the primitive chiropterygium. *Bulletin of the American Museum of Natural History* 48:279–288.

Gregory, W. K., and H. C. Raven. 1941. Studies on the origin and early evolution of paired fins and limbs. *Annals of the New York Academy of Sciences* 42:273–360.

Grigoriadis, A. E., Z. Q. Wang, M. G. Cecchini, W. Hofstetter,

R. Felix, H. A. Fleisch, and E. F. Wagner. 1994. c-Fos: A key regulator of osteoclast-macrophage lineage determination and bone remodeling. *Science* 266:443–448.

Grillo, H. C., C. M. Lapière, M. H. Dresden, and J. Gross. 1968. Collagenolytic activity in regenerating forelimbs of the adult newt (*Triturus viridescens*). *Developmental Biology* 17:571–583.

Grimsrud, C. D., P. R. Romano, M. D'Souza, J. E. Puzas, P. R. Reynolds, R. N. Rosier, and R. J. O'Keefe. 1999. BMP-6 is an autocrine stimulator of chondrocyte differentiation. *Journal of Bone and Mineral Research* 14:475–482.

Gritli-Linde, A., P. Lewis, A. P. McMahon, and A. Linde. 2001. The whereabouts of a morphogen: Direct evidence for short- and graded long-range activity of hedgehog signaling peptides. *Developmental Biology* 236:364–386.

Gross, J., and B. Dumsha. 1958. Elastoidin: A two component member of the collagen class. *Biochimica et Biophysica Acta* 28:268–270.

Grossmann, M., M. R. Sánchez-Villagra, and W. Maier. 2002. On the development of the shoulder girdle in Crocidura russula (Soricidae) and other placental mammals: Evolutionary and functional aspects. *Journal of Anatomy* 201:371–381.

Grotewold, L., and U. Ruther. 2002. Bmp, Fgf and Wnt signalling in programmed cell death and chondrogenesis during vertebrate limb development: The role of Dickkopf-1. *International Journal of Developmental Biology* 46:943–947.

Groth, J. G., and G. F. Barrowclough. 1999. Basal divergences in birds and the phylogenetic utility of the nuclear RAG-1 gene. *Molecular Phylogenetics and Evolution* 12:115–123.

Gruberg, E. R., and R. V. Stirling. 1972. Observations on the burrowing habits of the tiger salamander (*Ambystoma tigrinum*). *Herpetological Review* 4:85–89.

Grüneberg, H. 1952. *The genetics of the mouse.* The Hague: Martinus Nijhoff.

Gruss, P., and M. Kessel. 1991. Axial specification in higher vertebrates. *Current Opinions in Genetics and Development* 1:204–210.

Guha, U., W. A. Gomes, T. Kobayashi, R. G. Pestell, and J. A. Kessler. 2002. *In vivo* evidence that BMP signalling is necessary for apoptosis in the mouse limb. *Developmental Biology* 249:108–120.

Guibe, J. 1970. La musculature. In *Traité de zoologie: Anatomie, systématique, biologie,* ed. P. Grassé, 144–180. Paris: Masson.

Günther, A. 1858a. *Catalogue of the Batrachia Salientia in the collection of the British Museum.* London: Trustees of the British Museum.

Günther, A. 1858b. On the systematic arrangement of the tailless batrachians and the structure of *Rhinophrynus dorsalis. Proceedings of the Royal Society of London* 1858:339–352.

Gupta, B. B. 1966. Notes on the gliding mechanism in the flying squirrel. *Occasional Papers of the Museum of Zoology, University of Michigan* 645:1–7.

H

Haack, H., and P. Gruss. 1993. The establishment of murine Hox-1 expression domains during patterning of the limb. *Developmental Biology* 157:410–422.

Haacke, W. D. 1975. The burrowing geckos of southern Africa, 1 (Reptilia: Gekkonidae). *Annals of the Transvaal Museum* 29:197–243.

Haacke, W. D. 1976a. The burrowing geckos of southern Africa, 2 (Reptilia: Gekkonidae). *Annals of the Transvaal Museum* 30:13–28.

Haacke, W. D. 1976b. The burrowing geckos of southern Africa, 3 (Reptilia: Gekkonidae). *Annals of the Transvaal Museum* 30:29–39.

Haacke, W. D. 1976c. The burrowing geckos of southern Africa, 4 (Reptilia: Gekkonidae). *Annals of the Transvaal Museum* 30:53–70.

Haacke, W. D. 1976d. The burrowing geckos of southern Africa, 5 (Reptilia: Gekkonidae). *Annals of the Transvaal Museum* 30:71–89.

Haaijman, A., E. H. Burger, S. W. Goei, L. Nelles, P. ten Dijke, D. Huylebroeck, and A. L. Bronckers. 2000. Correlation between ALK-6 (BMPR-IB) distribution and responsiveness to osteogenic protein-1 (BMP-7) in embryonic mouse bone rudiments. *Growth Factors* 17:177–192.

Habersetzer, J. G., and G. Storch. 1987. Klassifikation und funktionelle Flügelmorphologie paläogener Fledermäuse (Mammalia, Chiroptera). *Courier Forschungsinstitut Senckenberg* 91:117–150.

Haeckel, E. 1876. *The history of creation; or, The development of the Earth and its inhabitants by the action of natural causes.* Transl. L. D. Schmidtz. New York: Appleton.

Haigh, J. J., H. P. Gerber, N. Ferrara, and E. F. Wagner. 2000. Conditional inactivation of VEGF-A in areas of collagen2a1 expression results in embryonic lethality in the heterozygous state. *Development* 127:1445–1453.

Hailman, J. P., and A. M. Elowson. 1992. Ethogram of the nesting female loggerhead (*Caretta caretta*). *Herpetologica* 48:1–30.

Haines, R. W. 1934. Epiphyseal growth in the branchial skeleton of fishes. *Quarterly Journal of Microscopical Science* 77:77–97.

Haines, R. W. 1937. Posterior end of Meckel's cartilage and related ossifications in bony fishes. *Quarterly Journal of Microscopical Science* 80:1–38.

Haines, R. W. 1938. The primitive form of epiphyses in the long bones of tetrapods. *Journal of Anatomy* 72:323–343.

Haines, R. W. 1939. The structure of the epiphyses in sphenodon and the primitive form of secondary centre. *Journal of Anatomy* 74:80–90.

Haines, R. W. 1940. Note on the independence of sesamoids and epiphysial centres of ossification. *Journal of Anatomy* 75:101–105.

Haines, R. W. 1941. Epiphyseal structure in lizards and marsupials. *Journal of Anatomy* 74:282–294.

Haines, R. W. 1942a. Eudiarthrodial joints in fishes. *Journal of Anatomy* 77:12–19.

Haines, R. W. 1942b. The evolution of epiphyses and of endochondral bone. *Biological Reviews* 17:267–292.

Haines, R. W. 1942c. The tetrapod knee joint. *Journal of Anatomy* 76:270–301.

Haines, R. W. 1958. Arboreal or terrestrial ancestry of placental mammals? *Quarterly Review of Biology* 33:1–23.

Haines, R. W. 1969. Epiphyses and sesamoids. In *Biology of the reptilia*, ed. C. Gans, 81–115. New York: Academic Press.

Haines, R. W. 1974. The pseudoepipysis of the first metacarpal of man. *Journal of Anatomy* 117:145–158.

Haines, R. W. 1975. The histology of eiphyseal union in mammals. *Journal of Anatomy* 120:1–25.

Hall, B. K. 1975. Evolutionary consequences of skeletal differentiation. *American Zoologist* 15:329–350.

Hall, B. K. 1978. *Developmental and cellular skeletal biology.* New York: Academic Press.

Hall, B. K. 1991. The evolution of connective tissue and skeletal tissues. In *Develomental patterning of the vertebrate limb*, ed. J. R. Hinchliffe, J. M., Hurlé, and D. Summerbell, 303–311. New York: Plenum Press.

Hall, B. K. 1998. *Evolutionary developmental biology.* 2nd ed. London: Chapman and Hall.

Hall, B. K. 2000. The neural crest as a fourth germ layer and vertebrates as quadroblastic not tribloblastic. *Evolution and Development* 2:3–5.

Hall, B. K. 2001. John Samuel Budgett (1872–1904): In pursuit of *Polypterus. BioScience* 51:399–407.

Hall, B. K. 2002. Palaeontology and evolutionary developmental biology: A science of the 19th and 21st centuries. *Palaeontology* 45:647–669.

Hall, B. K. 2005a. *Bones and cartilage: Developmental and evolutionary skeletal biology.* London: Elsevier/Academic Press.

Hall, B. K. 2005b. In Goethe's wake: Marvalee Wake's contribution to the development and evolution of a science of morphology. *Zoology* (special issue) 108:269–275.

Hall, B. K. 2007. Foreword. In *The nature of limbs* by Richard Owen. Chicago: University of Chicago Press, forthcoming.

Hall, B. K., and T. Miyake. 1995. Divide, accumulate, differentiate: Cell condensation in skeletal development revisited. *International Journal of Developmental Biology* 39:881–893.

Hall, B. K., and T. Miyake. 2000. All for one and one for all: Condensations and the initiation of skeletal development. *BioEssays* 22:138–147.

Hall, B. K., and P. E. Witten. Forthcoming. The origin and plasticity of skeletal tissues in vertebrate evolution and development. In *Major transitions in vertebrate evolution*, ed. S. Andeson and H.-D. Sues. Bloomington: Indiana University Press.

Hall, L. S., and R. L. Hughes. 1987. An evolutionary perspective of structural adaptations for environmental perceptions and utilization by the neonatal marsupials Trichosurus vulpecula (Phalangeridae) and Didelphis virginiana (Didelphidae). In *Possums and opossums: Studies in evolution*, vol. 1, ed. M. Archer, 251–271. Sydney, Australia: Surrey Beatty and Sons.

Hall-Craggs, E. C. B. 1965. An analysis of the jump of the lesser galago (Galago senegalensis). *Proceedings of the Zoological Society of London* 147:20–29.

Halleen, J. M., and R. Ranta. 2001. Tartrate-resistant acid phosphatase as a serum marker of bone resorption. *European Clinical Laboratory* 20:12–14.

Halliday, T. R., and K. Adler, eds. 1986. *The encyclopaedia of reptiles and amphibians.* London: Unwin Hyman.

Halloy, M., R. Etheridge, and G. M. Burghardt. 1998. To bury in sand: Phylogenetic relationships among lizard species of the *boulengeri* group, *Liolaemus* (Reptilia: Squamata: Tropiduridae), based on behavioral characters. *Herpetological Monographs* 12:1–37.

Halstead, L. B. 1989. Plesiosaur locomotion. *Journal of the Geological Society, London* 146:37–40.

Hamada, M., T. Nagai, N. Kai, Y. Tanoue, H. Mae, M. Hashimoto, K. Miyoshi, H. Kumagai, and K. Saeki. 1995. Inorganic constituents of bone of fish. *Fisheries Science* 61:517–520.

Hamilton, W. G. 1985. Surgical anatomy of the foot and ankle. *Clinical Symposia* 37:2–32.

Hammond, K. L., R. E. Hill, T. T. Whitfield, and P. D. Currie. 2002. Isolation of three zebrafish dachshund homologues and their expression in sensory organs, the central nervous system and pectoral fin buds. *Mechanisms of Development* 112:183–189.

Hamrick, M. W. 1999. Development of epiphyseal structure and function in *Didelphis virginiana* (Marsupiala, Didelphidae). *Journal of Morphology* 239:283–296.

Hamrick, M. W. 2001. Development and evolution of the mammalian limb: Adaptive diversification of nails, hooves, and claws. *Evolution and Development* 3:355–363.

Han, M., X. Yang, J. E. Farrington, and K. Muneoka. 2003. Digit regeneration is regulated by Msx1 and BMP4 in fetal mice. *Development* 130:5123–5132.

Han, M. J., J. Y. An, and W. S. Kim. 2001. Expression patterns of *Fgf8* during development and limb regeneration in the axolotl. *Developmental Dynamics* 220:40–48.

Hanken, J. 1982. Appendicular skeletal morphology in minute salamanders, genus *Thorius* (Amphibia: Plethodontidae): Growth regulation, adult size determination, and natural variation. *Journal of Morphology* 174:47–77.

Hanken, J. 1985. Morphological novelty in the limb skeleton accompanies miniaturization in salamanders. *Science* 229:871–874.

Hanken, J. 1993. Adaptation of bone growth to miniaturization of body size. In *Bone*, vol. 7, *Bone growth—B*, ed. B. K. Hall, 79–104. Boca Raton, FL: CRC Press.

Hanken, J., T. F. Carl, M. K. Richardson, L. Olsson, G. Schlosser, C. K. Osabutey, and M. W. Klymkowsky. 2001. Limb development in a "nonmodel" vertebrate, the direct-developing frog *Eleutherodactylus coqui. Journal of Experimental Zoology* 291:375–388.

Harder, W. 1964. *Anatomie der Fische: Sonderabdruck aus dem Handbuch der Binnenfischerei Mitteleuropas.* Vol. 2A. Stuttgart: E. Schweizerbart'sche Verlagsbuchhandlung.

Hardy, A., M. K. Richardson, P. H. Francis-West, C. Rodriguez, J. C. Izpisúa Belmonte, and D. Duprez. 1995. Gene expression, polarizing activity and skeletal patterning in reaggregated hind limb mesenchyme. *Development* 121:4329–4337.

Hare, W. C. D. 1961. The ages at which the centers of ossification appear roentgenographically in the limb bones of the dog. *American Journal of Veterinary Research* 22:825–835.

Harel, Z., and G. S. Tannenbaum. 1995. Long-term alterations in growth hormone and insulin secretion after temporary dietary protein restriction in early life in the rat. *Pediatric Research* 38:747–753.

Harfe, B. D., P. J. Scherz, S. Nissim, H. Tian, A. P. McMahon, and C. J. Tabin. 2004. Evidence for an expansion-based temporal Shh gradient in specifying vertebrate digit identities. *Cell* 118:517–528.

Harris, J. E. 1936. The role of the fins in the equilibrium of the swimming fish. I: Wind tunnel tests on a model of *Mustelus canis* (Mitchell). *Journal of Experimental Biology* 13:476–493.

Harris, J. E. 1937. The mechanical significance of the position and movements of the paired fins in the Teleostei. *Papers from Tortugas Laboratory* 31:173–189.

Harris, J. E. 1938. The role of the fins in the equilibrium of the swimming fish. II: The role of the pelvic fins. *Journal of Experimental Biology* 16:32–47.

Harris, M. P., J. F. Fallon, and R. O. Prum. 2002. Shh-Bmp2 signaling module and the evolutionary origin and diversification of feathers. *Journal of Experimental Zoology* 294 (2): 160–176.

Harris, V. A. 1964. *The life of the rainbow lizard.* London: Hutchinson.

Harrisson, B. 1961. *Lanthanotus borneensis:* Habits and observations. *Sarawak Museum Journal* 10:286–292.

Harrisson, T., and N. S. Haile. 1961. A rare earless monitor lizard from Borneo. *Nature* 190:12–13.

Hartmann, C., and C. J. Tabin. 2000. Dual roles of Wnt signaling during chondrogenesis in the chicken limb. *Development* 127:3141–3159.

Hartmann, C., and C. J. Tabin. 2001. Wnt-14 plays a pivotal role in inducing synovial joint formation in the developing appendicular skeleton. *Cell* 104:341–351.

Harvey, C. B., P. J. O'Shea, A. J. Scott, H. Robson, T. Siebler, S. M. Shalet, J. Samarut, O. Chassande, and G. R. Williams. 2002. Molecular mechanisms of thyroid hormone effects on bone growth and function. *Molecular Genetics and Metbolism* 75:17–30.

Hashimoto, K., Y. Yokouchi, M. Yamamoto, and A. Kuroiwa. 1999. Distinct signaling molecules control Hoxa-11 and Hoxa-13 expression in the muscle precursor and mesenchyme of the chick limb bud. *Development* 126:2771–2783.

Haswell, W. A. 1882. On the structure of the paired fins of *Ceratodus*, with remarks on the general theory of the vertebrate limb. *Proceedings of the Linnean Society of New South Wales* 7:2–11.

Hatori, M., K. J. Klatte, C. C. Teixeira, and I. M. Shapiro. 1995. End labeling studies of fragmented DNA in the avian growth plate: Evidence of apoptosis of terminally differentiated chondrocytes. *Journal of Bone and Mineral Research* 10:1960–1968.

Hay, E. D., and D. A. Fischman. 1961. Origin of the blastema in regenerating limbs of the newt *Triturus viridescens:* An autoradiographic study using tritiated thymidine to follow cell proliferation and migration. *Developmental Biology* 3:26–59.

Hayamizu, T. F., N. Wanek, G. Taylor, C. Trevino, R. Shi, R. Anderson, D. M. Gardiner, K. Muneoka, and S. V. Bryant. 1994. Regeneration of *HoxD* expression domains during pattern regulation in chick wing buds. *Developmental Biology* 161:504–512.

Hazlehurst, G. A., and J. M. V. Rayner. 1992. Flight characteristics of Triassic and Jurassic Pterosauria: An appraisal based on wing shape. *Paleobiology* 18:447–463.

Heaton, M. J., and R. R. Reisz. 1980. A skeletal reconstruction of the Early Permian captorhinid reptile *Eocaptorhinus laticeps* (Williston). *Journal of Paleontology* 54:136–143.

Heatwole, H. 1960. Burrowing ability and behavioral responses to desiccation of the salamander, *Plethodon cinereus. Ecology* 41:661–668.

Heatwole, H., and R. L. Carroll, eds. 2000. *Palaeontology: The evolutionary history of amphibians.* Vol. 4 of *Amphibian biology.* Chipping Norton, Australia: Surrey Beatty and Sons.

Heatwole, H., S. A. Minton, R. Taylor, and V. Taylor. 1978. Underwater observations on sea snake behaviour. *Records of the Australian Museum* 31:737–761.

Hedrick, T. L., B. W. Tobalske, and A. A. Biewener. 2002. Estimates of circulation and gait change based on a three-dimensional kinematic analysis of flight in cockatiels (*Nymphicus hollandicus*) and ringed turtle-doves (*Streptopelia risoria*). *Journal of Experimental Biology* 205:1389–1409.

Hedrick, T. L., B. W. Tobalske, and A. A. Biewener. 2003. How cockatiels (*Nymphicus hollandicus*) modulate pectoralis power output across light speeds. *Journal of Experimental Biology* 206:1363–1378.

Heglund, N. C., C. R. Taylor, and T. A. McMahon. 1974. Scaling stride frequency and gait to animal size: Mice to horses. *Science* 186:1112–1113.

Heilmann, G. A. 1926. *The origin of birds*. London: Witherby.

Heilmann, G. A. 1927. *The origin of birds*. New York: Appleton.

Heinrich, R. E., and A. R. Biknevicius. 1998. Skeletal allometry and interlimb scaling patterns in mustelid carnivorans. *Journal of Morphology* 235:121–134.

Helal, B. 1981. The great toe sesamoid bones: The Lus or Lost Souls of Ushuaia. *Clinical Orthopaedics and Related Research* 157:82–87.

Hellrung, H. 2003. Gerrothorax pustuloglomeratus, ein Temnospondyle (Amphibia) mit knöcherner Branchialkammer aus dem Unteren Keuper von Kupferzell (Süddeutschland). *Stuttgarter Baiträge zur Naturkunde Serie B (Geologie und Paläontologie)* 330:1–130.

Helms, J. A., C. H. Kim, G. Eichele, and C. Thaller, C. 1996. Retinoic acid signaling is required during early chick limb development. *Development* 122:1385–1394.

Henderson, J. H., and D. R. Carter. 2002. Mechanical induction in limb morphogenesis: Growth-generated strains and pressures. *Bone* 31:645–653.

Hendrickson, J. R. 1958. The green sea turtle, *Chelonia mydas* (Linn.) in Malaya and Sarawak. *Proceedings of the Zoological Society of London* 130:455–535.

Hendrickson, J. R. 1980. The ecological strategies of sea turtles. *American Zoologist* 20:597–608.

Hennig, W. 1966. *Phylogenetic systematics*. Urbana: University of Illinois Press.

Hérault, Y., J. Beckers, M. Gérard, and D. Duboule. 1999. Hox gene expression in limbs: Colinearity by opposite regulatory controls. *Developmental Biology* 208:157–165.

Hérault, Y., J. Beckers, T. Kondo, N. Fraudeau, and D. Duboule. 1998. Genetic analysis of a Hoxd-12 regulatory element reveals global versus local modes of controls in the HoxD complex. *Development* 125:1669–1677.

Hermanson, J. W., and J. S. Altenbach. 1983. The functional anatomy of the shoulder of the pallid bat, *Antrozous pallidus*. *Journal of Mammalogy* 64:62–75.

Herreid, C. F., II. 1958. Four-thumbed free-tailed bat. *Journal of Mammalogy* 39:481.

Herrmann, S., H. Hinrichs, K. D. Hinsch, and C. Surmann. 2000. Coherence concepts in holographic particle image velocimetry. *Experiments in Fluids* 29:S108–S116.

Herzog, H. A., Jr. 1975. An observation of nest opening by an American alligator *Alligator mississippiensis*. *Herpetologica* 31:446–447.

Heusinger, C. F. 1833. Untersuchungen über die Extremitäten der Ophidier, nebst Bemerkungen über die Extremitätenentwickelung im Allgemeinen. *Zeitschrift für die Organische Physik* 3:481–523, 653–654.

Heyer, W. R. 1972. A new limbless skink (Reptilia: Scincidae) from Thailand with comments on the generic status of the limbless skinks of Southeast Asia. *Fieldiana: Zoology* 58:109–129.

Hickman, G. C. 1983. Burrow structure of the talpid mole *Parascalops breweri* from Oswego County, New York State. *Zeitschrift für Säugetierkunde* 48:265–269.

Hickman, G. C. 1984. An excavated burrow of *Scalopus aquaticus* from Florida, with comments on Nearctic talpid/geomyid burrow structure. *Säugetierkundliche Mitteilungen* 31:243–249.

Hickman, G. C. 1985. Surface-mound formation by the Tuco-tuco, *Ctenomys fulvus* (Rodentia: Ctenomyidae), with comments on earth-pushing in other fossorial mammals. *Journal of Zoology, London A* 205:385–390.

Higashijima, S.-I., H. Okamoto, N. Ueno, Y. Hotta, and G. Eguchi. 1997. High-frequency generation of transgenic zebrafish which reliably express GFP in whole muscles or the whole body by using promoters of zebrafish origin. *Developmental Biology* 192:289–299.

Hildebrand, M. 1980. The adaptive significance of tetrapod gait selection. *American Zoologist* 20:255–267.

Hildebrand, M. 1982. *Analysis of vertebrate structure*. 2nd ed. New York: Wiley.

Hildebrand, M. 1985a. Digging of quadrupeds. In *Functional vertebrate morphology*, ed. M. Hildebrand, D. M. Bramble, K. F. Liem, and D. B. Wake, 89–109. Cambridge, MA: Belknap Press.

Hildebrand, M. 1985b. Walking and running. In *Functional vertebrate morphology*, ed. M. Hildebrand, D. M. Bramble, K. F. Liem, and D. B. Wake, 38–57. Cambridge, MA: Harvard University Press.

Hildebrand, M. 1988. *Analysis of vertebrate structure*. New York: Wiley.

Hildebrand, M., and G. E. Goslow. 2001. *Analysis of vertebrate structure*. 5th ed. New York: Wiley.

Hill, J. E. 1937. Morphology of the pocket gopher mammalian genus *Thomomys*. *University of California Publications in Zoology* 42:81–172.

Hill, J. E., and J. D. Smith. 1984. *Bats: A natural history*. Austin: University of Texas Press.

Himes, J. G. 2000. Burrowing ecology of the rare and elusive Louisiana pine snake, *Pituophis ruthveni* (Serpentes: Colubridae). *Amphibia-Reptilia* 22:91–101.

Himstedt, W. 1996. *Die Blindwühlen*. Magdeburg, Germany: Westarp Wissenschaften.

Hinchliffe, J. R. 1977. The chondrogenic pattern in chick limb morphogenesis: A problem of development *and* evolution. In *Vertebrate limb and somite morphogenesis*, ed. D. A. Ede, J. R. Hinchliffe, and M. Balls, 293–309. Cambridge: Cambridge University Press.

Hinchliffe, J. R. 1981. Cell death in embryogenesis. In *Cell death*, ed. I. D. Bowen and P. A. Lockshin, 35–46. London: Chapman and Hall.

Hinchliffe, J. R. 2002. Developmental basis of limb evolution. *International Journal of Developmental Biology* 46:835–845.

Hinchliffe, J. R., and D. Gumpel-Pinot. 1981. Control of maintenance and anterior-posterior patterning of the wing bud by its posterior margin (ZPA). *Journal of Embryology and Experimental Morphology* 62:63–82.

Hinchliffe, J. R., and M. K. Hecht. 1984. Homology of the bird wing skeleton. *Evolutionary Biology* 18:21–39.

Hinchliffe, J. R., J. Hurlé, and D. Summerbell, eds. 1991. *Developmental Patterning of the Vertebrate Limb*. NATO Advanced Science Institutes series A: Life Science. New York: Plenum.

Hinchliffe, J. R., and D. R. Johnson. 1980. *The development of the vertebrate limb*. New York: Oxford University Press.

Hinchliffe, J. R., and P. V. Thorogood. 1974. Genetic inhibition of mesenchymal cell death and the development of form and skeletal pattern in the limbs of talpid[3] (ta[3]) mutant chick embryos. *Journal of Embryology and Experimental Morphology* 31:747–760.

Hinchliffe, J. R., and E. I. Vorobyeva. 1999. Developmental basis of limb homology in urodeles: Heterochronic evidence from the primitive hynobiid family. In *Homology*, ed. G. R. Bock and B. Cardew, 95–105. Chichester, UK: Wiley.

Hinchliffe, R. J., E. I. Vorobyeva, and J. Géraudie. 2000. Evolution and the development of skeletal pattern in the limbs of amphibia: Is there a tetrapod developmental bauplan? *Developmental Dynamics* 219:141.

Hirayama, R. 1998. Oldest known sea turtle. *Nature* 392:705–708.

Hisaw, F. L. 1923. Observations on the burrowing habits of moles (*Scalopus aquaticus machrinoides*). *Journal of Mammalogy* 4:79–88.

Hoang, B. H., M. Moos Jr., S. Vukicevic, and F. P. Luyten. 1996. Primary structure and tissue distribution of FRZB, a novel protein related to *Drosophila* frizzled, suggest a role in skeletal morphogenesis. *Journal of Biological Chemistry* 271:26131–26137.

Hoang, B. H., J. T. Thomas, F. W. Abdul-Karim, K. M. Correia, R. A. Conlon, F. P. Luyten, and R. T. Ballock. 1998. Expression pattern of two Frizzled-related genes, Frzb-1 and Sfrp-1, during mouse embryogenesis suggests a role for modulating action of Wnt family members. *Developmental Dynamics* 212:364–372.

Hobson, E. S. 1965. Observations on diving in the Galápagos Marine Iguana, *Amblyrhynchus cristatus* (Bell). *Copeia* 1965:249–250.

Hochberg, Z. 2002a. Clinical physiology and pathology of the growth plate. *Best Practice and Research Clinical Endocrinology and Metabolism* 16:399–419.

Hochberg, Z. 2002b. *Endocrine control of skeletal maturation.* Basel, Switzerland: Kargers.

Hoeven, F. v. d., T. Schimmang, A. Volkmann, M. G. Mattei, B. Kyewski, and U. Ruther. 1994. Programmed cell death is affected in the novel mouse mutant Fused toes (Ft). *Development* 120:2601–2607.

Hoeven, F. v. d., J. Zákány, and D. Duboule. 1996. Gene transposition in the HoxD complex reveal a hierarchy of regulatory controls. *Cell* 85:1025–1035.

Hofbauer, L. C., D. L. Lacey, C. R. Dunstan, T. C. Spelsberg, B. L. Riggs, and S. Khosla. 1999. Interleukin-1beta and tumor necrosis factor-alpha, but not interleukin-6, stimulate osteoprotegerin ligand gene expression in human osteoblastic cells. *Bone* 25:255–259.

Hofmann, C., G. Luo, R. Balling, and G. Karsenty. 1996. Analysis of limb patterning in BMP-7-deficient mice. *Developmental Genetics* 19:43–50.

Hoffman, L., J. Miles, F. Avaron, L. Laforest, and M. A. Akimenko. 2002. Exogenous retinoic acid induces a stage-specific, transient and progressive extension of Sonic hedgehog expression across the pectoral fin bud of zebrafish. *International Journal of Developmental Biology* 46:949–956.

Hoge, A. 1947. Dimorfismo sexual nos Boídeos. *Memórias do Instituto Butantan* 20:181–188.

Hogg, D. A. 1980. A re-investigation of the centres of ossification in the avian skeleton at and after hatching. *Journal of Anatomy* 130:725–743.

Hoggard, N., L. Hunter, J. S. Duncan, L. M. Williams, P. Trayhurn, and J. G. Mercer. 1997. Leptin and leptin receptor mRNA and protein expression in the murine fetus and placenta. *Proceedings of the National Academy of Sciences, USA* 94:11073–11078.

Holbrook, J. E. 1842. *North American herpetology.* Vol. 5. Philadelphia: J. Dobson.

Holder, N. 1983. Developmental constraints and the evolution of vertebrate digit patterns. *Journal of Theoretical Biology* 104:451–471.

Holder, N. 1989. Organization of connective tissue patterns by dermal fibroblasts in the regenerating axolotl limb. *Development* 105:585–594.

Holliger, C. D. 1916. Anatomical adaptations in the thoracic limb of the California pocket gopher and other rodents. *University of California Publications in Zoology* 13:447–494.

Holmes, R. B. 1977. The osteology and musculature of the pectoral limb of small captorhinids. *Journal of Morphology* 152:101–140.

Holmes, R. B. 1980. Proterogyrinus scheelei and the early evolution of the labyrinthodont pectoral limb. In *The terrestrial environment and the origin of land vertebrates,* ed. A. L. Panchen, 351–376. London: Academic Press.

Holmes, R. B. 1984. The Carboniferous amphibian *Proterogyrinus scheelei* Romer, and the early evolution of tetrapods. *Philosophical Transactions of the Royal Society of London B* 306:531–527.

Holmes, R. B. 2000. Palaeozoic temnospondyls. In *Amphibian biology,* vol. 4, *Palaeontology: The evolutionary history of amphibians,* ed. H. Heatwole and R. L. Carroll, 1081–1120. Chipping Norton, Australia: Surrey Beatty and Sons.

Holmes, R. B., and R. L. Carroll. 1977. A temnospondyl amphibian from the Mississippian of Scotland. *Bulletin of the Museum of Comparative Zoology* 147:489–511.

Holmes, R. B., R. L. Carroll, and R. R. Reisz. 1998. The first articulated skeleton of *Dendrerpeton acadianum* (Temnospondyli, Dendrerpetontidae) from the Lower Pennsylvanian locality of Joggins, Nova Scotia, and a review of its relationships. *Journal of Vertebrate Paleontology* 18:64–79.

Holmgren, N. 1933. On the origin of the tetrapod limb. *Acta Zoologica, Stockholm* 14:185–295.

Holt, R. I. G. 2002. Fetal programming of the growth hormone-insulin-like growth factor axis. *Trends in Endocrinology and Metabolism* 13:392–397.

Honeycutt, R. L., and R. M. Adkins. 1993. Higher-level systematics of eutherian mammals: An assessment characters and phylogenetic hypotheses. *Annual Review of Ecology and Systematics* 24:279–306.

Hook, R. 1983. *Colosteus scutellatus* (Newberry), a primitive temnospondyl amphibian from the Middle Pennsylvanian of Linton, Ohio. *American Museum Novitates* 2770:1–41.

Hopson, J. A. 1995. Patterns of evolution in the manus and pes of non-mammalian therapsids. *Journal of Vertebrate Paleontology* 15:615–639.

Hornaday, W. T. 1904. *The American natural history.* New York: Charles Scribner's Sons.

Horner, J. R., A. De Ricqlès, and K. Padian. 1999. Variation in dinosaur skeletochonology indicators: Implications for age assessment and physiology. *Paleobiology* 25:295–304.

Horner, J. R., A. De Ricqlès, and K. Padian. 2000. Long bone histology of the hadrosaurid dinosaur *Maiasaura peeblesorum:* Growth dynamics and physiology based on an ontogenetic series of skeletal elements. *Journal of Vertebrate Paleontology* 20:115–129.

Horner, J. R., K. Padian, and A. De Ricqlès. 2001. Comparative osteohistology of some embryonic and perinatal archosaurs: Developmental and behavioral implications for dinosaurs. *Paleobiology* 27:39–58.

Horovitz, I. 2000. The tarsus of Ukhaatherium nessovi (Eutheria, Mammalia) from the Late Cretaceous of Mongolia: An appraisal of the evolution of the ankle in basal therians. *Journal of Vertebrate Paleontology* 20:547–560.

Horton, W. E., L. Feng, and C. Adams. 1998. Chondrocyte apoptosis in development, aging and disease. *Matrix Biology* 17:107–115.

Hosseini, A., and D. A. Hogg. 1991. The effects of paralysis on skeletal development in the chick embryo. I: General effects. *Journal of Anatomy* 177:159–168.

Houssay, F. 1912. *Forme, puissance et stabilité des poissons.* Paris: Hermann.

Hove, J. R., L. M. O'Bryan, M. S. Gordon, P. W. Webb, and D. Weihs. 2001. Boxfishes (Teleostei: Ostraciidae) as a model system for fishes swimming with many fins: Kinematics. *Journal of Experimental Biology* 204:1459–1471.

Howe, D. 1975. Observations on a captive marsupial mole, *Notoryctes typhlops. Australian Mammalogy* 1:361–365.

Howell, A. B. 1923. Mole notes. *Journal of Mammalogy* 4:253.

Howell, A. B. 1930. *Aquatic mammals: Their adaptations to life in the water.* Springfield, IL: Charles C. Thomas.

Howell, A. B. 1937a. Morphogenesis of the shoulder architecture. Part 5: Monotremata. *Quarterly Review of Biology* 12:191–205.

Howell, A. B. 1937b. Morphogenesis of the shoulder architecture. Part 6: Therian Mammalia. *Quarterly Review of Biology* 12:440–463.

Howell, A. B. 1937c. The swimming mechanism of the platypus. *Journal of Mammalogy* 18:217–222.

Howell, A. B. 1944. *Speed in animals: Their specialization for running and leaping.* Chicago: University of Chicago Press.

Howell, A. B. 1970. *Aquatic mammals.* New York: Dover.

Howes, G. B., and W. Ridewood. 1888. On the carpus and tarsus of the Anura. *Proceedings of the Zoological Society of London* 1888:141–182.

Howlett, C. R. 1979. The fine structure of the proximal growth plate of the avian tibia. *Journal of Anatomy* 128:377–399.

Howlett, C. R. 1980. The fine structure of the proximal growth late and metaphysis of the avian tibia: Endochondral osteogenesis. *Journal of Anatomy* 130:745–768.

Hoyt, D. F., and C. R. Taylor. 1981. Gait and the energetics of locomotion in horses. *Nature* 292:239–240.

Hsu, H., D. L. Lacey, C. R. Dunstan, I. Solovyev, A. Colombero, E. Timms, H. L. Tan, et al. 1999. Tumor necrosis factor receptor family member RANK mediates osteoclast differentiation and activation induced by osteoprotegerin ligand. *Proceedings of the National Academy of Sciences, USA* 96:3540–3545.

Hu, Y., V. Baud, M. Delhase, P. Zhang, T. Deerinck, M. Ellisman, R. Johnson, and M. Karin. 1999. Abnormal morphogenesis but intact IKK activation in mice lacking the IKKalpha subunit of IkappaB kinase. *Science* 284:316–320.

Hu, Y., Y. Wang, Z. Luo, and C. Li. 1997. A new symmetrodont mammal from China and its implications for mammalian evolution. *Nature* 390:137–142.

Hua, S. 1994. Hydrodynamique et modalités d'allégement chez *Metriorhynchus superciliosus* (Crocodylia, Thalattosuchia: implications paléoécologiques). *Neues Jahrbuch für Geologie und Paläontologie, Abhandlungen* 193:1–19.

Hua, S., and E. Buffetaut. 1997. Introduction to part V: Crocodylia. In *Ancient marine reptiles,* ed. J. M. Callaway and E. L. Nicholls, 357–374. San Diego, CA: Academic Press.

Huang, C. C., N. D. Lawson, B. M. Weinstein, and S. L. Johnson. 2003. reg6 is required for branching morphogenesis during blood vessel regeneration in zebrafish caudal fins. *Developmental Biology* 264:263–274.

Huang, R. J., Q. X. Zhi, K. Patel, J. Wilting, and B. Christ. 2000. Dual origin and segmental organization of the avian scapula. *Development* 127:3789–3794.

Huang, W., U. I. Chung, H. M. Kronenberg, and B. de Crombrugghe. 2001. The chondrogenic transcription factor Sox9 is a target of signaling by the parathyroid hormone-related peptide in the growth plate of endochondral bones. *Proceedings of the National Academy of Sciences, USA* 98:160–165.

Hudson, G. E., K. M. Hoff, J. Vanden Berge, and E. C. Trivette. 1969. A numerical study of the wing and leg muscle of Lari and Alcane. *Ibis* 111:459–529.

Hudson, G. E., P. J. Lanzillotti, and G. D. Edwards. 1959. Muscles of the pelvic limb in galliform birds. *American Midland Naturalist* 61:2–67.

Hudson, G. E., S. Y. C. Wang, and E. E. Provost. 1965. Ontogeny of the supranumerary sesamoids in the leg muscles of the ringnecked pheasant. *Auk* 82:427–437.

Hughes, P. C. R., and J. M. Tanner. 1970. The assessment of skeletal maturity in the growing rat. *Journal of Anatomy* 106:371–402.

Hui, C. 1985. Maneuverability of the Humboldt Penguin (*Spheniscus humboldti*) during swimming. *Canadian Journal of Zoology* 63:2165–2167.

Hunt, C. D., D. A. Ollerich, and F. H. Nielson. 1979. Morphology of the perforating cartilage canals in the proximal tibial growth plate of the chick. *Anatomical Record* 194:143–157.

Hunziker, E. B., and W. Herrmann. 1990. Ultrastructure of cartilage. In *Ultrastructure of skeletal tissues: Bone and cartilage in health and disease,* ed. E. Bonucci and P. M. Motta, 79–109. Boston: Kluwer Academic.

Hunziker, E. B., E. Kapfinger, and C. Saager. 1999. Hypertrophy of growth plate chondrocytes in vivo is accompanied by modulations in the activity state and surface area of their cytoplasmic organelles. *Histochemical Cell Biology* 112:115–123.

Hunziker, E. B., and R. K. Schenk. 1989. Physiological mechanisms adopted by chondrocytes in regulating longitudinal bone growth in rats. *Journal of Physiology* 414:45–71.

Hunziker, E. B., R. K. Schenk, and L.-M. Cruz-Orivé. 1987. Quantitation of chondrocyte performance in epiphyseal plates during longitudinal growth. *Journal of Bone and Joint Surgery* 69A:162–179.

Hurlé, J. M., and V. Climent. 1987. The regeneration of the interdigital tissue in Rallidae avian embryos (*Fulika atra* and *Gallinula chloropus*). *Archives of Biology* (Brussels) 98:299–316.

Hurlé, J. M., and E. Colvee. 1982. Surface changes in the embryonic interdigital epithelium during the formation of the free digits: A comparative study in the chick and duck foot. *Journal of Embryology and Experimental Morphology* 69:251–263.

Hurlé, J. M., V. Garcia-Martinez, Y. Ganan, V. Climent, and M. Blasco. 1987. Morphogenesis of the prehensile autopodium in the common chameleon (*Chamaeleo chamaeleo*). *Journal of Morphogenesis* 194:187–194.

Hurlé, J. M., and R. Merino. 2002. Apoptosis. In *Encyclopedia of evolution,* ed. M. Pagel, 61–64. New York: Oxford University Press.

Hurov, J. R. 1986. Soft-tissue interface: How do attachments of muscles, tendons, and ligaments change during growth? A light microscopic study. *Journal of Morphology* 189:313–325.

Hutchinson, J. R. 2002. The evolution of hindlimb tendons and muscles on the line leading to crown-group birds. *Comparative Biochemistry and Physiology Part A* 133:1051–1086.

Hutchinson, J. R., D. Famini, R. Lair, and R. Kram. 2003. Are fast-moving elephants really running? *Nature* 422:493–494.

Huxley, T. H. 1861. Preliminary essay upon the systematic arrangement of the fishes of the Devonian Epoch. In *Scientific memoirs* 2:417–460.

Huxley, T. H. 1868. On the animals which are most nearly intermediate between birds and reptiles. *Annals and Magazine of Natural History* 2 (4): 66–75.

Huxley, T. H. 1870a. Further evidence of the affinity between the dinosaurian reptiles and birds. *Quarterly Journal of the Geological Society of London* 26:12–31.

Huxley, T. H. 1870b. On the classification of the Dinosauria with observations on the Dinosauria of the Trias. I: The classification and affinities of the Dinosauria. *Quarterly Journal of the Geological Society of London* 26:32–38.

Huxley, T. H. 1876. Contributions to morphology: Ichthyopsida; On Ceratodus forsteri, with observations on the classification of fishes. In *Scientific memoirs* 4:84–124.

Huxley, T. H. 1880. On the application of the laws of evolution to the arrangement of the vertebrata, and more particularly of the Mammalia. *Proceedings of the Zoological Society of London* 43:649–662.

Huysseune, A. 1989. Morphogenetic aspects of the pharyngeal jaws and neurocranial apophysis in postembryonic *Astatotilapia elegans* (Trewavas, 1933) (Teleostei: Cichlidae). *Academiae Analecta* (Brussels) 51:11–35.

Huysseune, A. 2000. Skeletal system. In *The laboratory fish,* part 4: *Microscopic functional anatomy,* ed. G. K. Ostrander, 307–317. San Diego: Academic Press.

Huysseune, A., and J. Y. Sire. 1990. Ultrastructural observations on chondroid bone in the teleost fish *Hemichromis bimaculatus. Tissue and Cell* 22:371–383.

Huysseune, A., and J. Y. Sire. 1992. Development of cartilage and bone tissues of the anterior part of the mandible in cichlid fish: A light and TEM study. *Anatomical Record* 233:357–375.

Huysseune, A., and J. Y. Sire. 1998. Evolution of patterns and processes in teeth and tooth-related tissues in non-mammalian vertebrates. *European Journal of Oral Sciences* 106 (suppl. 1): 437–481.

Huysseune, A., and W. Verraes. 1990. Carbohydrate histochemistry of mature chondroid bone in *Astatotilapia elegans* (Teleostei: Cichlidae) with a comparison to acellular bone and cartilage. *Annales des Sciences Naturelles, Zoologie, Paris,* series 13, 11:29–43.

Hwang, S. H., M. A. Norell, Q. Ji, and K. Gao. 2002. New specimens of *Microraptor zhouianus* (Theropoda: Dromaeosauridae) from northeastern China. *American Museum Novitates* 3381:1–44.

I

Ihde, T. F., and M. E. Chittenden. 2002. Comparison of calcified structures for aging spotted sea trout. *Transactions of the American Fisheries Society* 131:634–642.

Ingber, D. 1991. Integrins as mechanochemical transducers. *Current Opinion in Cell Biology* 3:841–848.

Inge, G. A. L., and A. R. Ferguson. 1933. Surgery of the sesamoid bones of the great toe. *Archives of Surgery* 27:466–489.

Inohaya, K., and A. Kudo. 2000. Temporal and spatial patterns of cbfa1 expression during embryonic development in the teleost, *Oryzias latipes. Development Genes and Evolution* 210:570–574.

Inoue, J. G., M. Miya, K. Tsukamoto, and M. Nishida. 2003. Basal actinopterygian realtionships: A mitogenomic perspective on the phylogeny of the "ancient fish." *Molecular Phylogenetics and Evolution* 26:110–120.

Inoue, Y., T. Mito, K. Miyawaki, K. Matsushima, Y. Shinmyo, T. Heanue, G. Mardon, H. Ohuchi, and S. Noji. 2002. Correlation of expression patterns of homothorax, dachshund and Distal-less with the proximodistal segmentation of the cricket leg bud. *Mechanisms of Development* 113:141–148.

International working group on the constitutional diseases of bone (D. L. Rimoin, chair). 1998. International nomenclature and classification of the osteochondroysplasias 1997. *American Journal of Medical Genetics* 79:376–382.

Inuzuka, N. 1984. Skeletal restoration of the desmostylians: Herpetiform mammals. *Memoirs of the Faculty of Science, Kyoto University, Series of Biology* 9:157–253.

Ioan, D. M., D. Dagomiz, J. P. Fryns. 2002. Oculo-Dento-Digital dysplasia (OMIM* 164200): Full manifestation of the syndrome in a 9.5-year-old girl and type III syndactyly in the father. *Genetic Counseling* 13:187–189.

Iotsova, V., J. Caamano, J. Loy, Y. Yang, A. Lewin, and R. Bravo. 1997. Osteopetrosis in mice lacking NF-kappaB1 and NF-kappaB2. *Nature Medicine* 3:1285–1289.

Iovine, M. K., and S. L. Johnson. 2000. Genetic analysis of isometric growth control mechanisms in the zebrafish caudal fin. *Genetics* 155:1321–1329.

Irschick, D. J., and T. Garland. 2001. Integrating function and ecology in studies of adaptation: Investigations of locomotor capacity as a model system. *Annual Review of Ecology and Systematics* 32:367–396.

Irvin, B. C., and R. A. Tassava. 1998. Effects of peripheral nerve implants on the regeneration of partially and fully innervated urodele forelimbs. *Wound Repair and Regeneration* 6:382–387.

Isaac, A., M. J. Cohn, P. Ashby, P. Ataliotis, D. B. Spicer, J. Cooke, and C. Tickle. 2000. FGF and genes encoding transcription factors in early limb specification. *Mechanisms of Development* 93:41–48.

Ishikawa, Y., B. R. Genge, R. E. Wuthier, and L. N. Y. Wu. 1998. Thyroid hormone inibits growth and stimulates terminal differentiation of epiphyseal growth plate chondrocytes. *Journal of Bone and Mineral Research* 13:1398–1411.

Iten, L., and S. V. Bryant. 1973. Forelimb regeneration from different levels of amputation in the newt, *Notophthalmus viridescens. Wilhelm Roux's Archives of Developmental Biology* 173:263–282.

Ivachnenko, M. F. 1978. Urodelans from the Triassic and Jurassic of Soviet Central Asia. *Paleontological Journal* 12:362–368.

Iwamatsu, T. 1994. Stages of normal development in the medaka *Oryzias latipes. Zoological Science* 11:825–839.

Iwamoto, M., Y. Yoshinobu, E. Koyama, M. Enomoto-Iwamoto, K. Kurisu, H. Yeh, W. R. Abrams, J. Rosenbloom, and M. Pacifici. 2000. Transcription factor ERG variants and functional diversification of chondrocytes during limb long bone development. *Journal of Cell Biology* 150:27–39.

Iwamoto, M., J. Kitagaki, Y. Tamamura, C. Gentili, E. Koyama, H. Enomoto, T. Komori, M. Pacifici, and M. Enomoto-

Iwamoto. 2003. Runx2 expression and action in chondrocytes are regulated by retinoid signaling and parathyroid hormone-related peptide (PTHrP). *Osteoarthritis and Cartilage* 11:6–15.

Iwata, T., L. Chen, C. Li, D. A. Ovchinnikov, R. R. Behringer, C. A. Francomano, and C. X. Deng. 2000. A neonatal lethal mutation in FGFR3 uncouples proliferation and differentiation of growth plate chondrocytes in embryos. *Human Molecular Genetics* 9:1603–1613.

Iwata, T., C. L. Li, C. X. Deng, and C. A. Francomano. 2001. Highly activated Fgfr3 with the K644M mutation causes prolonged survival in severe dwarf mice. *Human Molecular Genetics* 10:1255–1264.

Izpisúa Belmonte, J. C., J. M. Brown, D. Duboule, and C. Tickle. 1992a. Expression of Hox-4 genes in the chick wing links pattern formation to the epithelial-mesenchymal interactions that mediate growth. *European Molecular Biology Organization Journal* 11:1451–1457.

Izpisúa Belmonte, J. C., D. A. Ede, C. Tickle, and D. Duboule. 1992b. The mis-expression of posterior Hox-4 genes in talpid (ta³) mutant wings correlates with the absence of antero-posterior polarity. *Development* 114:959–963.

Izpisúa Belmonte, J. C., C. Tickle, P. Dollé, L. Wolpert, and D. Duboule. 1991. Expression of the homeobox Hox-4 genes and the specification of position in chick wing development. *Nature* 350:585–589.

J

Jacenko, O., D. W. Roberts, M. R. Campbell, P. M. McManus, C. J. Gress, and Z. Tao. 2002. Linking hematopoiesis to endochondral skeletogenesis through analysis of mice transgenic for collagen X. *American Journal of Pathology* 160:2019–2034.

Jacobson, M. D., M. Weil, and M. C. Raff, M. 1996. Role of Ced-3/ICE-family proteases in staurosporine-induced programmed cell death. *Journal of Cell Biology* 133:1041–1051.

Jacobson, M. D., M. Weil, and M. C. Raff. 1997. Programmed cell death in animal development. *Cell* 88:347–354.

Jansen, K. P., A. P. Summers, and P. R. Delis. 2001. Spadefoot toads (*Scaphiopus holbrookii holbrookii*) in an urban landscape: Effects of nonnatural substrates on burrowing in adults and juveniles. *Journal of Herpetology* 35:141–145.

Janvier, P. 1996. *Early vertebrates*. Oxford: Clarendon Press.

Janvier, P., M. Arsenaul, and S. Desbiens. 2004. Calcified cartilage in the paired fins of the osteostracan *Escuminaspis laticeps* (Traquair 1880), from the late Devonian of Miguasha (Québec, Canada), with a consideration of the early evolution of the pectoral fin endoskeleton in vertebrates. *Journal of Vertebrate Paleontology* 24:773–779.

Jarvik, E. 1942. On the structure of the snout in crossopterygians and in lower gnathostomes in general. *Zoologica Bidrag* (Uppsala) 21:235–675.

Jarvik, E. 1959. Dermal fin-rays and Holmgren's principle of delamination. *Kungliga Svenska Vetenskaps-Akademiens Handlingar*, series 4, 6:1–51.

Jarvik, E. 1980. *Basic structure and evolution of vertebrates*. New York: Academic Press.

Jarvik, E. 1996. The Devonian tetrapod *Ichthyostega*. *Fossils and Strata* 40:1–213.

Jarvis, J. U. M., and J. B. Sale. 1971. Burrowing and burrow patterns of east African mole-rats *Tachyoryctes*, *Heliophobius* and *Heterocephalus*. *Journal of Zoology, London* 163:451–479.

Javaid, M. K., and C. Cooper. 2002. Prenatal and childhood influences on osteoporosis. *Best Practice and Research Clinical Endocrinology and Metabolism* 16:349–367.

Jayne, B. C. 1988. Muscular mechanisms of snake locomotion: An electromyographic study of lateral undulation of the Florida banded water snake (*Nerodia fasciata*) and the yellow rat snake (*Elaphe obsoleta*). *Journal of Morphology* 197:159–181.

Jayne, B. C., and D. J. Irschick. 2000. A field study of incline use and preferred speeds for the locomotion of lizards. *Ecology* 81:2969–2983.

Jarvik, E. 1980. *Basic structure and evolution of vertebrates*. London: Academic Press.

Jarvik, E. 1996. The Devonian tetrapod *Ichthyostega*. *Fossils and Strata* 40:1–206.

Jeffery, J. E. 2001. Pectoral fins of rhizodontids and the evolution of pectoral appendages in the tetrapod stem-group. *Biological Journal of the Linnean Society* 74:217–236.

Jena, N., C. Martin-Seisdedos, P. McCue, and C. M. Croce. 1997. BMP7 null mutation in mice: Developmental defects in skeleton, kidney, and eye. *Experimental Cell Research* 230:28–37.

Jenkins, F. A., Jr. 1970. Limb movements in a monotreme (*Tachyglossus aculeatus*): A cineradiographic analysis. *Science* 168:1473–1475.

Jenkins, F. A., Jr. 1971. The postcranial skeleton of African cynodonts: Problems in the early evolution of the mammalian postcranial skeleton. *Peabody Museum of Natural History Bulletin* 36:1–216.

Jenkins, F. A., Jr. 1973. The functional anatomy and evolution of the mammalian humero-ulnar articulation. *American Journal of Anatomy* 137:281–298.

Jenkins, F. A., Jr. 1974. Tree shrew locomotion and the origins of Primate arborealism. In *Primate locomotion*, ed. F. A. Jenkins Jr., 85–115. New York: Academic Press.

Jenkins, F. A., Jr. 1993. The evolution of the avian shoulder joint. *American Journal of Science* 293A:253–267.

Jenkins, F. A., Jr., and S. M. Camazine. 1977. Hip structure and locomotion in ambulatory and cursorial carnivores. *Journal of Zoology, London* 181:351–370.

Jenkins, F. A., Jr., K. P. Dial, and G. E. Goslow Jr. 1988. A cineradiographic analysis of bird flight: The wishbone in starlings is a spring. *Science* 241:1495–1498.

Jenkins, F. A., Jr., S. M. Gatesy, N. H. Shubin, and W. W. Amaral. 1997. Haramiyids and Triassic mammalian evolution. *Nature* 385:715–718.

Jenkins, F. A., Jr., and D. W. Krause. 1983. Adaptations for climbing in North American multituberculates (Mammalia). *Science* 220:712–715.

Jenkins, F. A., Jr., and D. McClearn. 1984. Mechanisms of hind foot reversal in climbing mammals. *Journal of Morphology* 182:197–219.

Jenkins, F. A., Jr., and F. R. Parrington. 1976. The postcranial skeletons of the Triassic mammals Eozostrodon, Megazostrodon, and Erythrotherium. *Philosophical Transactions of the Royal Society of London B* 273:387–431.

Jenkins, F. A., Jr., and C. R. Schaff. 1988. The Early Cretaceous mammal Gobiconodon (Mammalia, Triconodonta) from the Cloverly Formation in Montana. *Journal of Vertebrate Paleontology* 8:1–24.

Jenkins, F. A., Jr., and N. H. Shubin. 1998. *Prosalirus bitis* and the anuran caudopelvic mechanism. *Journal of Vertebrate Paleontology* 18:495–510.

Jenkins, F. A., Jr., N. H. Shubin, S. M. Gatesy, and K. Padian. 2001. A diminutive pterosaur (Pterosauria: Eudimorphodontidae) from the Greenlandic Triassic. *Bulletin of the Museum of Comparative Zoology* 156:151–170.

Jenkins, F. A., Jr., and D. M. Walsh. 1993. An early Jurassic caecilian with limbs. *Nature* 365:246–250.

Jenkins, F. A., Jr., D. M. Walsh, and R. L. Carroll. 2006. Anatomy of *Eocaecilia micropodia*, a limbed gymnophionan of the Early Jurassic. *Bulletin of the Museum of Comparative Zoology, Harvard*, in press.

Jennings, M. R. 1996. *Ambystoma californiense* (California tiger salamander): Burrowing ability. *Herpetological Review* 27:194.

Jepsen, G. L. 1970. Bat origins and evolution. In *Biology of bats*, vol. 1, ed. W. A. Wimsatt, 1–64. New York: Academic Press.

Jessen, H. 1966. Die Crossopterygier des Oberen Plattenkalkes (Devon) der Bergisch-Gladbach-Paffrather Mulde (Rheinisches Schiefergebirge) unter Berücksichtigung von amerikanischem und europäischem *Onychodus*-Material. *Arkiv för Zoologi* 18:305–389.

Jessen, H. 1972. Schultergürtel und Pectoralflosse bei Actinopterygiern. *Fossils and Strata* 1:1–101.

Ji, Q., P. J. Currie, M. A. Norell, and S. Ji. 1998. Two feathered dinosaurs from northeastern China. *Nature* 393:753–761.

Ji, Q., Z.-X. Luo, and S. Ji. 1999. A Chinese triconodont mammal and mosaic evolution of the mammalian skeleton. *Nature* 398:326–330.

Ji, Q., Z.-X. Luo, C.-X. Yuan, J. R. Wible, J.-P. Zhang, and J. A. Georgi. 2002. The earliest known eutherian mammal. *Nature* 416:816–822.

Jiang, R., Y. Lan, H. D. Chapman, C. Shawber, C. R. Norton, D. V. Serreze, G. Weinmaster, and T. Gridley. 1998. Defects in limb, craniofacial, and thymic development in Jagged2 mutant mice. *Genes and Development* 12:1046–1057.

Johanson, Z., and P. E. Ahlberg. 2001. Devonian rhizodontids and tristichopterids (Sarcopterygii; Tetrapodomorpha) from East Gondwana. *Transactions of the Royal Society of Edinburgh, Earth Sciences* 92:43–74.

Johanson, Z., P. E. Ahlberg, and A. Ritchie. 2003. The braincase and palate of the tetrapodomorph sarcopterygian *Mandageria fairfaxi*: Morphological variability near the fish-tetrapod transition. *Palaeontology* 46:212–234.

Johansson, L. C., and U. M. Lindhe Norberg. 2000. Asymmetric toes aid underwater swimming. *Nature* 407:582–583.

Johnson, R. 1979. The osteology of the pectoral complex of *Stenopterygius* Jaekel (Reptilia, Ichthyosauria). *Neues Jahrbuch für Geologie und Paläontologie, Abhandlungen* 159:41–86.

Johnson, R. L., and C. J. Tabin. 1997. Molecular models for vertebrate limb development. *Cell* 90:979–990.

Johnson, S. L., and P. Bennett. 1999. Growth control in the ontogenetic and regenerating zebrafish fin. *Methods in Cell Biology* 59:301–311.

Johnson, S. L., and J. A. Weston. 1995. Temperature-sensitive mutations that cause stage-specific defects in zebrafish fin regeneration. *Genetics* 141:1583–1595.

Johnson-Murray, J. L. 1987. The comparative myology of the gliding membranes in *Acrobates*, *Petauroides* and *Petaurus* contrasted with the cutaneous myology of *Hemibelideus* and *Pseudocheirus* (Marsupialia: Phalangeridae) and selected gliding Rodentia (Sciuridae and Anamoluridae). *Australian Journal of Zoology* 35:101–113.

Johnston, I. A., and J.-P. Camm. 1987. Muscle structure and differentiation in pelagic and demersal stages of the Antarctic teleost *Notothenia neglecta*. *Marine Biology* (Berlin) 94:183–190.

Jones, K. E., A. Purvis, A. MacLarnon, O. R. P. Bininda-Emonds, and N. B. Simmons. 2002. A phylogenetic supertree of the bats (Mammalia: Chiroptera). *Biological Reviews* 77:223–257.

Jones, T. R., A. G. Kluge, and A. J. Wolf. 1993. When theories and methodologies clash: A phylogenetic reanalysis of the North American ambystomatid salamanders (Caudata: Ambystomatidae). *Systematic Biology* 42:92–102.

Josephson, R. K. 1985. Mechanical power output from striated muscle during cyclic contraction. *Journal of Experimental Biology* 114:493–512.

Joss, J., and T. Longhurst. 2001. Lungfish paired fins. In *Major events in early vertebrate evolution*, ed. P. E. Ahlberg, 370–376. London: Taylor and Francis.

Joyce, W. G. 2000. The first complete skeleton of *Solnhofia parsoni* (Cryptodira, Erysternidae) from the Upper Jurassic of Germany and its taxonomic implications. *Journal of Paleontology* 74:684–700.

Joyce, W. G., J. F. Parham, and J. A. Gauthier. 2004. Developing a protocol for the conversion of rank-based taxon names to phylogenetically defined clade names, as exemplified by turtles. *Journal of Paleontology* 78:989–1013.

Ju, B., Y. Xu, J. He, J. Liao, T. Yan, C. L. Hew, T. J. Lam, and Z. Gong. 1999. Faithful expression of green fluorescent protein (GFP) in transgenic zebrafish embryos under control of zebrafish gene promoters. *Developmental Genetics* 25:158–167.

Jungers, W. L., F. K. Jouffroy, and J. T. Stern Jr. 1980. Gross structure and function of the quadriceps femoris in *Lemur fulvus*: An analysis based on telemetered electromyography. *Journal of Morphology* 164:287–299.

K

Kacem, A., F. J. Meunier, and J. L. Baglinière. 1998. A quantitative study of morphological and histological changes in the skeleton of *Salmo salar* during its anadromous migration. *Journal of Fish Biology* 53:1096–1109.

Kacem, A., S. Gustafsson, and F. J. Meunier. 2000. Demineralization of the vertebral skeleton in Atlantic salmon *Salmo salar* L. during spawning migration. *Comparative Biochemistry and Physiology A* 125:479–484.

Kähler, C. J., and J. Kompenhans. 2000. Fundamentals of multiple plane stereo particle image velocimetry. *Experiments in Fluids* 29:S70–S77.

Kanegae, Y., A. T. Tavares, J. C. Izpisúa Belmonte, and I. M. Verma. 1998. Role of Rel/NF-kappaB transcription factors during the outgrowth of the vertebrate limb. *Nature* 392:611–614.

Karaplis, A. C., A. Luz, J. Glowacki, R. T. Bronson, V. L. Tybulewicz, H. M. Kronenberg, and R. C. Mulligan. 1994. Lethal skeletal dysplasia from targeted disruption of the parathyroid hormone-related peptide gene. *Genes and Development* 8:277–289.

Kardon, G., J. K. Campbell, and C. J. Tabin. 2002. Local extrinsic signals determine muscle and endothelial cell fate and patterning in the vertebrate limb. *Developmental Cell* 3:533–545.

Karsenty, G. 1999. The genetic transformation of bone biology. *Genes and Development* 13:3037–3051.

Karsenty, G. 2003. The complexities of skeletal biology. *Nature* 423:316–318.

Karsenty, G., and E. F. Wagner. 2002. Reaching a genetic and molecular understanding of skeletal development. *Developmental Cell* 2:389–406.

Katagiri, T., S. Boorla, J. L. Frendo, B. L. Hogan, and G. Karsenty. 1998. Skeletal abnormalities in doubly heterozygous Bmp4 and Bmp7 mice. *Developmental Genetics* 22:340–348.

Kato, M., M. S. Patel, R. Levasseur, I. Lobov, B. H. Chang, D. A. Glass II, C. Hartmann, et al. 2002. Cbfa1-independent decrease in osteoblast proliferation, osteopenia, and persistent embryonic eye vascularization in mice deficient in Lrp5, a Wnt coreceptor. *Journal of Cell Biology* 157:303–314.

Kawakami, Y., T. Ishikawa, M. Shimabara, N. Tanda, M. Enomoto-Iwamoto, M. Iwamoto, T. Kuwana, A. Ueki, S. Noji, and T. Nohno. 1996. BMP signaling during bone pattern determination in the developing limb. *Development* 122:3557–3566.

Kawakami, Y., J. Capdevila, D. Buscher, T. Itoh, C. Rodriguez Esteban, and J. C. Izpisúa Belmonte. 2001. WNT signals control FGF-dependent limb initiation and AER induction in the chick embryo. *Cell* 104:891–900.

Kawakami, Y., J. Rodriguez-Leon, C. M. Koth, D. Buscher, T. Itoh, A. Raya, J. K. Ng, et al. 2003. MKP3 mediates the cellular response to FGF8 signalling in the vertebrate limb. *Nature: Cell Biology* 5:513–519.

Kawakami, Y., N. Wada, S. L. Nishimatsu, T. Ishikawa, S. Noji, and T. Nohno. 1999. Involvement of Wnt-5a in chondrogenic pattern formation in the chick limb bud. *Development, Growth, and Differentiation* 41:29–40.

Kawasaki K., T. Suzuki, and K. M. Weiss. 2004. Genetic basis for the evolution of vertebrate mineralized tissue. *Proceedings of the National Academy of Sciences, USA* 101:11356–11361.

Kearney, M. 2002. Appendicular skeleton in amphisbaenians (Reptilia: Squamata). *Copeia* 2002:719–738.

Kearney, M. 2003. Systematics of the Amphisbaenia (Lepidosauria: Squamata) based on morphological evidence from Recent and fossil forms. *Herpetological Monographs* 17:1–74.

Kelley, R. O. 1970. An electron microscopic study of mesenchyme during development of interdigital spaces in man. *Anatomical Record* 168:43–53.

Kelley, R. O., P. F. Goetinck, and J. A. MacCabe. 1982. *Limb development and regeneration, Part B.* New York: Liss.

Kellogg, R. 1932. Mexican tailless amphibians in the United States National Museum. *United States National Museum Bulletin* 160:1–224.

Kember, N. F., and J. K. Kirkwood. 1987. Cell kinetics and longitudinal bone growth in birds. *Cell and Tissue Kinetics* 20:625–629.

Kember, N. F., J. K. Kirkwood, P. J. Duignan, D. Godfrey, and D. J. Spratt. 1990. Comparative cell kinetics of avian growth plates. *Research Veterinary Science* 49:283–288.

Kemp, A. 2002. Unique dentition of lungfish. *Microscopy Research and Technique* 59:435–448.

Kemp, N. E. 1977. Banding pattern and fibrillogenesis of ceratotrichia in shark fins. *Journal of Morphology* 154:187–204.

Kemp, N. E., and J. H. Park. 1970. Regeneration of lepidotrichia and actinotrichia in the tail fin of the teleost *Tilapia mossambica. Developmental Biology* 22:321–342.

Kemp, T. S. 1982. *Mammal-like reptiles and the origin of mammals.* London: Academic Press.

Kenefick, J. H. 1954. Observations on egg laying of the tortoise *Gopherus polyphemus. Copeia* 1954:228–229.

Kengaku, M., J. Capdevila, C. Rodriguez-Esteban, J. De La Pena,

R. L. Johnson, J. C. Belmonte, and C. J. Tabin. 1998. Distinct WNT pathways regulating AER formation and dorsoventral polarity in the chick limb bud. *Science* 280:1274–1247.

Kerr, J. G. 1932. Archaic fishes—Lepidosiren, Protopterus, Polypterus—and their bearing upon the problems of vertebrate morphology. *Jenaisch Zeitschrift fur Naturwissenschaft* 67:419–433.

Kerr, J. F., B. Harmon, and J. Searle. 1972. Apoptosis: A basic biological phenomenon with wide-ranging implication in tissue kinetics. *British Journal of Cancer* 26:239–257.

Khokha, M. K., D. Hsu, L. J. Brunet, M. S. Dionne, and R. M. Harland. 2003. Gremlin is the BMP antagonist required for maintenance of Shh and Fgf signals during limb patterning. *Nature Genetics* 34:303–307.

Kida, Y., Y. Maeda, T. Shiraishi, T. Suzuki, and T. Ogura. 2004. Chick Dach1 interacts with the Smad complex and Sin3a to control AER formation and limb development along the proximodistal axis. *Development* 131:4179–4187.

Kielan-Jaworowska, Z. 1977. Evolution of the therian mammals in the Late Cretaceous of Asia. Part 2: Postcranial skeleton in Kennalestes and Asioryctes. *Acta Palaeontologica Polonica* 37:65–83.

Kielan-Jaworowska, Z., and P. P. Gambaryan. 1994. Postcranial anatomy and habits of Asian multituberculate mammals. *Fossils and Strata* 36:1–92.

Kielan-Jaworowska, Z., M. J. Novacek, B. A. Trofimov, and D. Dashzeveg. 2000. Mammals from the Mesozoic of Mongolia. In *The age of dinosaurs in Russia and Mongolia*, ed. M. J. Benton, M. A. Shishkin, D. M. Unwin, and E. N. Kurochkin, 573–626. Cambridge: Cambridge University Press.

Kieny, M., and A. Chevallier. 1979. Autonomy of tendon development in the embryonic chick wing. *Journal of Embryology and Experimental Morphology* 49:153–165.

Kierdorf, H., U. Kierdorf, T. Szuwart, U. Gath, and G. Clem. 1994. Light microscopic observations on the ossification process in the early developing pedicle of fallow deer (*Dama dama*). *Annals of Anatomy* 176:243–249.

Kim, H. T., W. M. Olson, and B. K. Hall. 2002. Effects of hind limb denervation on development of sesamoids in *Hymenochirus boettgeri. American Zoologist* 42:1256.

Kim, I. S., F. Otto, B. Zabel, and S. Mundlos. 1999. Regulation of chondrocyte differentiation by Cbfa1. *Mechanisms of Development* 80:159–170.

Kimmel, C. B., W. W. Ballard, S. R. Kimmel, B. Ulmann, and T. F. Schilling. 1995. Stages of embryonic development of the zebrafish. *Developmental Dynamics* 203:253–310.

Kimmel, R. A., D. H. Turnbull, V. Blanquet, W. Wurst, C. A. Loomis, and A. L. Joyner. 2000. Two lineage boundaries coordinate vertebrate apical ectodermal ridge formation. *Genes and Development* 14:1377–1389.

King, G. M. 1991. The aquatic *Lystrosaurus*: A palaeontological myth. *Historical Biology* 4:285–321.

King, J. A., P. C. Marker, K. J. Seung, and D. M. Kingsley. 1994. BMP5 and the molecular, skeletal, and soft-tissue alterations in short ear mice. *Developmental Biology* 166:112–122.

King, J. E. 1983. *Seals of the world.* 2nd ed. London: British Museum of Natural History.

King, M. W., T. Nguyen, J. Calley, M. W. Harty, M. C. Muzinich, A. L. Mescher, C. Chalfant, et al. 2003. Identification of genes expressed during *Xenopous laevis* limb regeneration using subtractive hybridization. *Developmental Dynamics* 226:398–409.

Kingdon, J. 1974. *East African mammals.* Vol. 2A. Chicago: University of Chicago Press.

Kingsley, D. M. 1994. What do BMPs do in mammals? Clues from the mouse short-ear mutation. *Trends in Genetics* 10:16–21.

Kingsley, D. M., A. E. Bland, J. M. Grubber, P. C. Marker, L. B. Russell, N. G. Copeland, and N. A. Jenkins. 1992. The mouse short ear skeletal morphogenesis locus is associated with defects in a bone morphogenetic member of the TGF beta superfamily. *Cell* 71:399–410.

Kingsley, J. S. 1892. The head of an embryo Amphiuma. *American Naturalist* 26:671–680.

Kirk, E. C., M. Cartmill, R. F. Kay, and P. Lemelin. 2003. Comment on "Grasping Primate Origins." *Science* 300:741.

Kirkwood, J. K., P. J. Duignan, N. F. Kember, P. M. Bennett, and D. J. Price. 1989a. The growth rate of the tarsometatarsus bone in birds. *Journal of Zoology, London* 217:403–416.

Kirkwood, J. K., D. M. Spratt, and P. J. Duignan, P. J. 1989b. Patterns of cell proliferation and growth rate in limb bones of the domestic fowl (*Gallus domesticus*). *Research in Veterinary Science* 47:139–147.

Kitagaki, J., M. Iwamoto, J. G. Liu, Y. Tamamura, M. Pacifci, and M. Enomoto-Iwamoto. 2003. Activation of beta-catenin-LEF/TCF signal pathway in chondrocytes stimulates ectopic endochondral ossification. *Osteoarthritis Cartilage* 11:36–43.

Kizirian, D. A., and R. W. McDiarmid. 1998. A new species of *Bachia* (Squamata: Gymnophthalmidae) with plesiomorphic limb morphology. *Herpetologica* 54:245–253.

Klembara, J., and I. Bartik. 2000. The postcranial skeleton of *Discosauriscus* Kuhn, a seymouriamorph tetrapod from the Lower Permian of the Boskovice Furrow (Czech Republic). *Transactions of the Royal Society of Edinburgh, Earth Sciences* 90:287–316.

Kline, S. C., R. N. Hotchkiss, M. A. Randolph, and A. J. Weiland. 1990. Study of growth kinetics and morphology in limbs transplanted between animals of different ages. *Plastic and Reconstructive Surgery* 85:273–280.

Kluge, A. G. 1962. Comparative osteology of the eublepharid lizard genus *Coleonyx* Gray. *Journal of Morphology* 110:299–332.

Kluge, A. G. 1976. Phylogenetic relationships in the lizard family Pygopodidae: An evaluation of theory, methods and data. *Miscellaneous Publications, Museum of Zoology, University of Michigan* 152:1–72.

Kluge, A. G., and R. A. Nussbaum. 1995. A review of the African-Madagascan gekkonid lizard phylogeny and biogeography (Squamata). *Miscellaneous Publications, Museum of Zoology, University of Michigan* 183:1–20.

Kmita, M., N. Fraudeau, Y. Herault, and D. Duboule. 2002. Serial deletions and duplications suggest a mechanism for the collinearity of Hoxd genes in limbs. *Nature* 420:145–150.

Knepton, J. C., Jr. 1954. A note on the burrowing habits of the salamander *Amphiuma means. Copeia* 1954:68.

Knüsel, P. L. 1944. *Beiträge zur Morphologie und Funktion der Crocodiliden-Extremitäten.* Sarnen, Switzerland: Louis Ehrli.

Koebernick, K., and T. Pieler. 2002. Gli-type zinc finger proteins as bipotential transducers of Hedgehog signaling. *Differentiation* 70:69–76.

Koedam, J. A., J. J. Smink, and S. C. van Buul-Offers. 2002. Glucocorticoids inhibit vascular endothelial growth factor expression in growth plate chondrocytes. *Molecular and Cellular Endocrinology* 197:35–44.

Kohn, L. P., P. Olson, and J. M. Cheverud. 1997. Age of epiphyseal closure in tamarins and marmosets. *American Journal of Primatology* 41:129–139.

Kohno, H., and Y. Taki. 1983. Comments on the development of fin-supports in fishes. *Japanese Journal of Ichthyology* 30:284–290.

Kollros, J. J. 1984. Limb regeneration in anuran tadpoles following repeated amputations. *Journal of Experimental Zoology* 232:217–229.

Komori, T., H. Yagi, S. Nomura, A. Yamaguchi, K. Sasaki, K. Deguchi, Y. Shimizu, R. T. Bronson, et al. 1997. Targeted disruption of Cbfa1 results in a complete lack of bone formation owing to maturational arrest of osteoblasts. *Cell* 89:755–764.

Koob, T. J., and A. P. Summers. 2002. Tendon: Bridging the gap. *Comparative Biochemistry and Physiology Part A* 133:905–909.

Koopman, K. F. 1984. Bats. In *Orders and families of recent mammals of the world*, ed. S. Anderson and J. K. Jones Jr., 145–186. New York: Wiley.

Koopman, K. F. 1993. Chiroptera. In *Mammalian species of the world*, ed. D. E. Wilson and D. M. Reeder, 137–241. Washington, DC: Smithsonian Press.

Kooyman, G. L. 1989. *Zoophysiology: Diverse divers, physiology and behavior.* Berlin: Springer-Verlag.

Korsmeyer, K. E., J. F. Steffensen, and J. Herskin. 2002. Energetics of median and paired fin swimming, body and caudal fin swimming, and gait transition in parrotfish (*Scarus schlegeli*) and triggerfish (*Rhinecanthus aculeatus*). *Journal of Experimental Biology* 205:1253–1263.

Koshiba, K., A. Kuroiwa, H. Yamamoto, K. Tamura, and H. Ide. 1998. Expression of Msx genes in regenerating and developing limbs of axolotl. *Journal of Experimental Zoology* 282:703–714.

Kostakopoulou, K., N. Vargesson, J. D. Clarke, P. M. Brickell, and C. Tickle. 1997. Local origin of cells in FGF-4-induced outgrowth of amputated chick wing bud stumps. *International Journal of Developmental Biology* 41:747–750.

Kostakopoulou, K., A. Vogel, P. Brickell, and C. Tickle. 1996. Regeneration of wing bud stumps of chick embryos and reactivation of *Msx-1* and *shh* expression in response to FGF-4 and ridge signals. *Mechanisms of Development* 55:119–131.

Kostenuik, P. J., J. Harris, B. P. Halloran, R. T. Turner, E. R. Morey-Holton, and D. D. Bikle. 1999. Skeletal unloading causes resistance of osteoprogenitor cells to parathyroid hormone and to insulin-like growth factor-I. *Journal of Bone and Mineral Research* 14:21–31.

Koster, R., R. Stick, F. Loosli, and J. Wittbrodt. 1997. Medaka spalt acts as a target gene of hedgehog signaling. *Development* 124:3147–3156.

Kotrschal, K., A. Goldschmid, H. Adam, and M. Whitear. 1985. The first dorsal fin of *Gaidropsarus mediterraneus* (Teleostei), a specialized chemosensory organ. *Progress in Zoology* 30:727–730.

Kottelat, M., and K. K. P. Lim. 1999. Mating behavior of *Zenarchopterus gilli* and *Zenarchopterus buffonis* and function of the modified dorsal and anal fin rays in some species of *Zenarchopterus* (Teleostei: Hemiramphidae). *Copeia* 1999:1097–1101.

Koumoundouros, G., P. Divanach, and M. Kentouri. 1999. Osteological development of the vertebral column and of the caudal complex in *Dentex dentex. Journal of Fish Biology* 54:424–436.

Koumoundouros, G., P. Divanach, and M. Kentouri. 2001. Osteological development of *Dentex dentex* (Osteichthyes: Sparidae). *Marine Biology* 138:399–406.

Kram, R., A. Domingo, and D. P. Ferris. 1997. Effect of reduced gravity on the preferred walk-run transition speed. *Journal of Experimental Biology* 200:821–826.

Kraus, P., D. Fraidenraich, and C. A. Loomis. 2001. Some distal limb structures develop in mice lacking Sonic hedgehog signaling. *Mechanisms of Development* 100:45–58.

Krause, D. W., and F. A. Jenkins Jr. 1983. The postcranial skeleton of North American multituberculates. *Bulletin of the Museum of Comparative Zoology, Harvard University* 150:199–246.

Krauss, S., J. P. Concordet, and P. W. Ingham. 1993. A functionally conserved homolog of the *Drosophila* segment polarity gene hh is expressed in tissues with polarizing activity in zebrafish embryos. *Cell* 75:1431–1444.

Kühlhorn, F. 1984. Grabanpassungen beim Burmeister-Gürtelmull, *Bermeisteria retusa* (Burmeister, 1863). *Säugetierkundliche Mitteilungen* 31:97–111.

Kuhn-Schnyder, E., and H. Rieber. 1986. *Handbook of paleozoology.* Baltimore: Johns Hopkins University Press.

Kumar, A., C. P. Velloso, Y. Imokawa, and J. P. Brockes. 2000. Plasticity of retrovirus labelled myotubes in the newt limb regeneration blastema. *Developmental Biology* 218:125–136.

Kume, K., K. Satomura, S. Nishisho, E. Kitaoka, K. Yamanouchi, S. Tobume, and M. Nagayama. 2002. Potential role of leptin during endochondral ossification. *Journal of Histochemistry and Cytochemistry* 50:1559–1569.

Kundrát, M., V. Seichert, A. P. Russell, and K. Smetana. 2002. Pentadactyl pattern of the avian wing autopodium and pyramid reduction hypothesis. *Journal of Experimental Zoology (Molecular and Developmental Evolution)* 294:152–159.

Kvellestad, A., S. Høie, K. Thorud, B. Tørud, and A. Lyngøy. 2000. Platyspondyly and shortness of vertebral column in farmed Atlantic salmon *Salmo salar* in Norway: Description and interpretation of pathologic changes. *Diseases of Aquatic Organisms* 39:97–108.

L

Lacasa Ruiz, A. 1989. Nuevo genero de ave fosil del Yacimiento Neocomiense del Montsec (Provincia de Lerida, España). *Estudios Geologicos* 45:417–425.

Lacey, D. L., E. Timms, H. L. Tan, M. J. Kelley, C. R. Dunstan, T. Burgess, R. Elliott, et al. 1998. Osteoprotegerin ligand is a cytokine that regulates osteoclast differentiation and activation. *Cell* 93:165–176.

Ladher, R. K., V. L. Church, S. Allen, L. Robson, A. Abdelfattah, N. A. Brown, G. Hattersley, et al. 2000. Cloning and expression of the Wnt antagonists Sfrp-2 and Frzb during chick development. *Developmental Biology* 218:183–198.

Laforest, L., C. W. Brown, G. Poleo, J. Géraudie, M. Tada, M. Ekker, and M. A. Akimenko. 1998. Involvement of the *Sonic Hedgehog, patched 1*, and *bmp2* genes in patterning of the zebrafish dermal fin rays. *Development* 125:4175–4184.

Lagler, K. F., J. E. Bardach, R. R. Miller, and D. R. M. Passino. 1977. *Ichthyology.* 2nd ed. New York: Wiley.

Lako, M., T. Strachan, P. Bullen, D. I. Wilson, S. C. Robson, and S. Lindsay. 1998. Isolation, characterisation and embryonic expression of WNT11, a gene which maps to 11q13.5 and has possible roles in the development of skeleton, kidney and lung. *Gene* 219:101–110.

Lanctot, C., B. Lamolet, and J. Drouin. 1997. The bicoid-related homeoprotein Ptx1 defines the most anterior domain of the embryo and differentiates posterior from anterior lateral mesoderm. *Development* 124:2807–2817.

Lande, R. 1978. Evolutionary mechanisms of limb loss in tetrapods. *Evolution* 32:73–92.

Landis, W. J., and F. H. Silver. 2002. The structure and function of normally mineralizing avian tendons. *Comparative Biochemistry and Physiology Part A* 133:1135–1157.

Landry, S. O., Jr. 1958. The function of the entepicondylar foramen in mammals. *American Midland Naturalist* 60:100–112.

Lang, J. W. 1987. Crocodilian behaviour: Implications for management. In *Wildlife management: Crocodiles and alligators,* ed. G. J. W. Webb, S. C. Manolis, and P. J. Whitehead, 273–294. Chipping Norton, Australia: Surrey Beatty and Sons.

Langston, W., and R. R. Reisz. 1981. *Aerosaurus wellesi*, new species, a varanopseid mammal-like reptile (Synapsida: Pelycosauria) from the Lower Permian of New Mexico. *Journal of Vertebrate Paleontology* 1:73–96.

Lanske, B., A. C. Karaplis, K. Lee, A. Luz, A. Vortkamp, A. Pirro, M. Karperien, et al. 1996. PTH/PTHrP receptor in early development and Indian hedgehog-regulated bone growth. *Science* 273:663–666.

Lanzing, W. J. R. 1976. The fine structure of fins and fin rays of *Tilapia mossambica* (Peters). *Calcified Tissue Research* 173:349–356.

Larson, A., and J. B. Losos. 1996. Phylogenetic systematics of adaptation. In *Adaptation,* ed. M. R. Rose and G. V. Lauder, 187–220. San Diego, CA: Academic Press.

Larsson, H. C. E., and G. P. Wagner. 2002. Pentadactyl ground state of the avian wing. *Journal of Experimental Zoology (Molecular and Developmental Evolution)* 294:146–151.

Lattanzi, G., A. Ognibene, P. Sabatelli, C. Capanni, D. Toniolo, M. Columbaro, S. Santi, et al. 2000. Emerin expression at the early stages of myogenic differentiation. *Differentiation* 66:208–217.

Lauder, G. V. 1995. Pectoral fin locomotion in fishes: Three dimensional kinematics and electromyography. *American Zoologist* 35:3A.

Lauder, G. V. 1996. The argument from design. In *Adaptation,* ed. M. R. Rose and G. V. Lauder, 55–91. San Diego, CA: Academic Press.

Lauder, G. V. 2000. Function of the caudal fin during locomotion in fishes: Kinematics, flow visualization, and evolutionary patterns. *American Zoologist* 40:101–122.

Laufer, E., R. Dahn, O. E. Orozco, C. Y. Yeo, J. Pisenti, D. Henrique, U. K. Abbott, J. F. Fallon, and C. Tabin. 1997. Expression of Radical fringe in limb-bud ectoderm regulates apical ectodermal ridge formation. *Nature* 386:366–373.

Laufer, E., C. E. Nelson, R. L. Johnson, B. A. Morgan, and C. Tabin. 1994. Sonic hedgehog and Fgf-4 act through a signaling cascade and feedback loop to integrate growth and patterning of the developing limb bud. *Cell* 79:993–1003.

Laurentus, J. N. 1768. *Specimen medicum exhibens synopsin reptilium emendatam cum experimentis circa venena et antidota reptilium Austriacorum.* Vienna: Trattnern.

Laurin, M. 1998a. The importance of global parsimony and historical bias in understanding tetrapod evolution. Part 1: Systematics, middle ear evolution, and jaw suspension. *Annals des Sciences Naturelles, Zoologie* 19:1–42.

Laurin, M. 1998b. The importance of global parsimony and historical bias in understanding tetrapod evolution. Part 2: Vertebral centrum, costal ventilation, and paedomorphosis. *Annals des Sciences Naturelles, Zoologie* 19:99–114.

Laurin, M., M. Girondot, and A. de Ricqlès. 2000. Early tetrapod evolution. *Trends in Ecology and Evolution* 15:118–123.

Laurin, M., and R. R. Reisz. 1995. A reevaluation of early amniote phylogeny. *Zoological Journal of the Linnean Society* 113:165–223.

Laurin, M., and R. R. Reisz. 1997. A new perspective on tetrapod phylogeny. In *Amniote origins: Completing the transition to land,* ed. S. S. Sumida and K. L. Martin, 9–59. London: Academic Press.

Laville, E. 1989. Etude cinématique du fouissage chez *Arvicola terrestris scherman* (Rodentia, Arvicolidae). *Mammalia* 53:177–189.

Laville, E., A. Casinos, J.-P. Gasc, S. Renous, and J. Bou. 1989. Les mécanismes du fouissage chez *Arvicola terrestris* et *Spalax ehrenbergi:* Étude fonctionelle et évolutive. *Anatomischer Anzeiger* 169:131–144.

Lawson, D. A. 1975. Pterosaur from the latest Cretaceous of West Texas: Discovery of the largest flying creature. *Science* 187:947–948.

Lawson, N. D., and B. M. Weinstein. 2002. In vivo imaging of embryonic vascular development using transgenic zebrafish. *Development* 248:307–318.

Le, A. X., T. Miclau, D. Hu, and J. A. Helms. 2001. Molecular aspects of healing in stabilized and non-stabilized fractures. *Journal of Orthopaedic Research* 19:78–84.

Leamy, L. 1974. Heritability of osteometric traits in a randombred population of mice. *Journal of Heredity* 65:109–120.

Leamy, L. 1975. Component analysis of osteometric traits in randombred house mice. *Systematic Zoology* 24:207–217.

Leamy, L. 1977. Genetic and environmental correlates of morphometric traits in randombred house mice. *Evolution* 31:357–369.

Leamy, L. 1981. Heritability of morphometric ratios in randombred house mice. In *Mammalian population genetics,* ed. M. H. Smith and J. Joule, 254–271. Athens: University of Georgia Press.

Leamy, L., and D. Bradley. 1982. Static and growth allometry of morphometric traits in randombred house mice. *Evolution* 36:1200–1212.

Leamy, L., and S. S. Sustarsic. 1978. A morphometric discriminant analysis of agouti genotypes in C57B/6 house mice. *Systematic Zoology* 27:49–60.

Lebedev, O. A. 1984. The first find of a Devonian tetrapod in the USSR. [In Russian.] *Dokladi Akademii Nauk SSSR* 278:1470–1473.

Lebedev, O. A., and M. I. Coates. 1995. The postcranial skeleton of the Devonian tetrapod *Tulerpeton curtum* Lebedev. *Zoological Journal of the Linnean Society* 114:307–348.

LeBoy, P. S., G. Grasso-Knight, M. D'Angelo, S. W. Volk, J. B. Lian, H. Drissi, G. S. Stein, and S. L. Adams. 2001. Smad-Runx interactions during chondrocyte maturation. *Journal of Bone and Joint Surgery* 83A (suppl., part 1): S1–15, S1–22.

Lee, B., K. Thirunavukkarasu, L. Zhou, L. Pastore, A. Baldini, J. Hecht, V. Geoffroy, P. Ducy, and G. Karsenty. 1997. Missense mutations abolishing DNA binding of the osteoblast-specific transcription factor OSF2/CBFA1 in cleidocranial dysplasia. *Nature: Genetics* 16:307–310.

Lee, B. S., S. Holliday, B. Ojikutu, I. Krits, and S. L. Gluck. 1996. Osteoclasts express the B2 isoform of vacuolar H+-ATPase intracellularly and on their plasma membranes. *American Journal of Physiology: Cell Physiology* 270:C382–C388.

Lee, K., J. D. Deeds, and G. V. Segre. 1995. Expression of parathyroid hormone-related peptide and its receptor messenger ribonucleic acids during fetal development of rats. *Endocrinology* 136:453–463.

Lee, K. K., and W. Y. Chan. 1991. A study on the regenerative potential of partially excised mouse embryonic fore-limb bud. *Anatomy and Embryolgy* 184:153–157.

Lee, K. K., A. K. Leung, M. K. Tang, D. Q. Cai, C. Schneider, C. Brancolini, and P. H. Chow. 2001. Functions of the growth arrest specific 1 gene in the development of the mouse embryo. *Developmental Biology* 234:188–203.

Lee, K. K., M. K. Tang, D. T. Yew, P. H. Chow, S. P. Yee, C. Schneider, and C. Brancolini. 1999. Gas2 is a multifunctional gene involved in the regulation of apoptosis and chondrogenesis in the developing mouse limb. *Developmental Biology* 207:14–25.

Lee, M. S. Y. 1997. The phylogeny of varanoid lizards and the affinities of snakes. *Philosophical Transactions of the Royal Society of London B* 352:53–91.

Lee, M. S. Y., and M. W. Caldwell. 1998. Anatomy and relationships of *Pachyrhachis problematicus,* a primitive snake with hindlimbs. *Philosophical Transactions of the Royal Society of London B* 353:1521–1552.

Lee, M. S. Y., and M. W. Caldwell. 2000. *Adriosaurus* and the affinities of mosasaurs, dolichosaurs and snakes. *Journal of Paleontology* 74:915–937.

Lefebvre, V., P. Li, and B. de Crombrugghe. 1998. A new long form of Sox5 (L-Sox5), Sox6 and Sox9 are coexpressed in chondrogenesis and cooperatively activate the type II collagen gene. *European Molecular Biology Organization Journal* 17:5718–5733.

Lehmann, K., P. Seemann, S. Stricker, M. Sammar, B. Meyer, K. Suring, F. Majewski, et al. 2003. Mutations in bone morphogenetic protein receptor 1B cause brachydactyly type A2. *Proceedings of the National Academy of Sciences, USA* 100:12277–12282.

Lehmann, W. H. 1963. The forelimb architecture of some fossorial rodents. *Journal of Morphology* 113:59–76.

Le Minor, J. M. 1987. Comparative anatomy and significance of the sesamoid bone of the peroneus longus muscle (*os perineum*). *Journal of Anatomy* 151:85–99.

Le Minor, J. M. 1994. The sesamoid bone of the Musculus abductor pollicis longus (*Os radiale externum* or Prepollex) in primates. *Acta Anatomica* 150:227–231.

Leptin, M. 1991. Twist and snail as positive and negative regulators during Drosophila mesoderm development. *Genes and Development* 5:1568–1576.

Lerner, A. L., and J. L. Kuhn. 1997. Characterization of regional and age-related variations in the growth of the rabbit distal femur. *Journal of Orthopaedic Research* 15:353–361.

Lerner, A. L., J. L. Kuhn, and S. J. Hollister. 1998. Are regional variations in bone growth related to mechanical stress and strain parameters? *Journal of Biomechanics* 31:327–335.

Le Roith, D., C. Bondy, S. Yakar, J.-L. Liu, and A. Butler. 2001.

The somatomedin hypothesis: 2001. *Endocrine Reviews* 22:43–74.

Lessa, E. P., and B. R. Stein. 1992. Morphological constraints in the digging apparatus of pocket gophers (Mammalia: Geomyidae). *Biological Journal of the Linnean Society* 47:439–453.

Lessa, E. P., and C. S. Thaeler Jr. 1989. A reassessment of morphological specializations for digging in pocket gophers. *Journal of Mammalogy* 70:689–700.

Lettice, L. A., T. Horikoshi, S. J. Heaney, M. J. van Baren, H. C. van der Linde, G. J. Breedveld, M. Joosse, et al. 2002. Disruption of a long-range cis-acting regulator for Shh causes preaxial polydactyly. *Proceedings of the National Academy of Sciences, USA* 99:7548–7553.

Lettice, L. A., S. J. Heaney, L. A. Purdie, L. Li, P. de Beer, B. A. Oostra, D. Goode, G. Elgar, R. E. Hill, and E. de Graaf. 2003. A long-range Shh enhancer regulates expression in the developing limb and fin and is associated with preaxial polydactyly. *Human Molecular Genetics* 12:1725–1735.

Lewis, J., and P. Martin. 1989. Vertebrate development: Limbs; A pattern emerges. *Nature* 342:734–735.

Lewis, K. E., G. Drossopoulou, I. R. Paton, D. R. Morrice, K. E. Robertson, D. W. Burt, P. W. Ingham, and C. Tickle. 1999. Expression of ptc and gli genes in talpid3 suggests bifurcation in Shh pathway. *Development* 126:2397–2407.

Lewis, O. J. 1958. The tubercle of the tibia. *Journal of Anatomy* 92:587–592.

Lewis, O. J. 1989. *Functional morphology of the evolving hand and foot*. Oxford: Clarendon Press.

Lewis, P. M., M. P, Dunn, J. A. McMahon, M. Logan, J. F. Martin, B. St-Jacques, and A. P. McMahon. 2001. Cholesterol modification of sonic hedgehog is required for long-range signaling activity and effective modulation of signaling by Ptc1. *Cell* 105:599–612.

Lheureux, E. 1983. The origin of tissues in the x-irradiated regenerating limb of the newt *Pleurodeles waltii*. In *Limb development and regeneration, part A*, ed. J. F. Fallon and A. I. Caplan, 455–465. New York: Liss.

Li, C., L. Chen, T. Iwata, M. Kitagawa, X. Y. Fu, and C. X. Deng. 1999. A Lys644Glu substitution in fibroblast growth factor receptor 3 (FGFR3) causes dwarfism in mice by activation of STATs and ink4 cell cycle inhibitors. *Human Molecular Genetics* 8:35–44.

Li, C., D. E. Clark, E. A. Lord, J. L. Stanton, and J. M. Suttie. 2002. Sampling technique to discriminate the different tissue layers of growing antler tips for gene discovery. *Anatomical Record* 268:125–130.

Li, C., and J. M. Suttie. 1994. Light microscopic studies of pedicle and early first antler development in red deer (*Cervus elaphus*). *Anatomical Record* 239:198–215.

Li, M., K. Chan, D. Cai, P. Leung, C. Cheng, K. Lee, and K. K. Lee. 2000. Identification and purification of an intrinsic human muscle myogenic factor that enhances muscle repair and regeneration. *Archives of Biochemistry and Biophysics* 384:263–268.

Li, S., and K. Muneoka. 1999. Cell migration and chick limb development: Chemotactic action of FGF-4 and the AER. *Developmental Biology* 211:335–347.

Libbin, R. M., and M. Weinstein. 1987. Sequence of development of innately regenerated growth-plate cartilage in the hindlimb of the neonatal rat. *American Journal of Anatomy* 180:255–265.

Libbin, R. M., I. J. Singh, A. Mirschman, and O. G. Mitchell. 1988. A prolonged cartilaginous phase in the newt forelimb skeletal regeneration. *Journal of Experimental Zoology* 248:238–242.

Liem, K., W. E. Bemis, L. Grande, and W. F. Walker. 2000. *Functional anatomy of the vertebrates: An evolutionary perspective*. Belmont, CA: Brooks/Cole.

Lighthill, M. J. 1960. Note on the swimming of slender fish. *Journal of Fluid Mechanics* 9:305–317.

Lighthill, M. J. 1969. Hydromechanics of aquatic animal propulsion. *Annual Review of Fluid Mechanics* 1:413–446.

Lighthill, M. J. 1970. Aquatic animal propulsion of high hydromechanical efficiency. *Journal of Fluid Mechanics* 44:265–301.

Lighthill, M. J. 1971. Large-amplitude elongated-body theory of fish locomotion. *Proceedings of the Royal Society of London B* 179:125–138.

Lillegraven, J. A. 1975. Biological considerations of the marsupial-placental dichotomy. *Evolution* 29:707–722.

Limpus, C. J., P. J. Couper, and M. A. Read. 1994. The loggerhead turtle, *Caretta caretta*, in Queensland: Population structure in a warm temperate feeding area. *Memoirs of the Queensland Museum* 37:195–204.

Lin, K., and O. Rieppel. 1998. Functional morphology and ontogeny of *Keichousaurus hui* (Reptilia, Sauropterygia). *Fieldiana: Geology* 39:1–35.

Linck, M. H., J. A. DePari, B. O. Butler, and T. E. Graham. 1989. Nesting behavior of the turtle, *Emydoidea blandingi*, in Massachusetts. *Journal of Herpetology* 23:442–444.

Lindsey, C. C. 1978. Form, function, and locomotory habits in fish. In *Fish physiology*, vol. 7, ed. W. S. Hoar and D. J. Randall, 1–100. New York: Academic Press.

Lingham-Soliar, T. 1991. Locomotion in mosasaurs. *Modern Geology* 16:229–248.

Lingham-Soliar, T. 1992. A new mode of locomotion in mosasaurs: Subaqueous flying in *Plioplatecarpus marshi*. *Journal of Vertebrate Paleontology* 12:405–421.

Lingham-Soliar, T. 1999. A reappraisal of subaqueous flight in mosasaurs. *Neues Jahrbuch für Geologie und Paläontologie, Abhandlungen* 213:145–167.

Lingham-Soliar, T. 2000. Plesiosaur locomotion: Is the four-wing problem real or merely an atheoretical exercise? *Neues Jahrbuch für Geologie und Paläontologie, Abhandlungen* 217:45–87.

Lingham-Soliar, T., and W.-E. Reif. 1998. Taphonomic evidence for fast tuna-like swimming in Jurassic and Cretaceous ichthyosaurs. *Neues Jahrbuch für Geologie und Paläontologie, Abhandlungen* 207:171–183.

Linnaeus, C. 1758. *Systema naturae*. Vol. 1. 10th ed. Stockholm: Laurentii Salvii.

List, J. C. 1955. External limb vestiges in *Leptotyphlops*. *Herpetologica* 11:15–16.

List, J. C. 1958. Notes on the skeleton of the blind snake, *Typhlops braminus*. *Spolia Zeylanica* 28:169–174.

List, J. C. 1966. Comparative osteology of the snake families Typhlopidae and Leptotyphlopidae. *Illinois Biological Monographs* 36:1–112.

Litingtung, Y., R. D. Dahn, Y. Li, J. F. Fallon, and C. Chiang. 2002. Shh and Gli3 are dispensable for limb skeleton formation but regulate digit number and identity. *Nature* 418:979–983.

Little, R. D., J. P. Carulli, R. G. Del Mastro, J. Dupuis, M. Osborne, C. Folz, S. P. Manning, et al. 2002. A mutation in the LDL receptor-related protein 5 gene results in the autosomal dominant high-bone-mass trait. *American Journal of Human Genetics* 70:11–19.

Liu, H., C. P. Ellington, K. Kawachi, C. Van den Berg, and A. Willmott. 1998. A computational fluid dynamic study of hawkmoth hovering. *Journal of Experimental Biology* 201:461–477.

Liu, H., R. J. Wassersug, and K. Kawachi. 1997. The three-dimensional hydrodynamics of tadpole locomotion. *Journal of Experimental Biology* 200:2807–2819.

Liu, Q., A. E. Kerstetter, E. Azodi, and J. A. Marrs. 2003. Cadherin-1, -2, and -11 expression and Cadherin-2 function in the pectoral limb bud and fin of the developing zebrafish. *Developmental Dynamics* 228:734–739.

Liu, Z., J. Xu, J. S. Colvin, and D. M. Ornitz. 2002. Coordination of chondrogenesis and osteogenesis by fibroblast growth factor 18. *Genes and Development* 16:859–865.

Livezey, B. C. 1995. Heterochrony and the evolution of avian flightlessness. In *Evolutionary change and heterochrony,* ed. K. J. McNamara, 169–193. New York: Wiley.

Livezey, B. C., and P. S. Humphrey. 1983. Mechanics of steaming in steamer-ducks. *Auk* 100:485–488.

Livezey, B. C., and R. L. Zusi. 2001. Higher-order phylogenetics of modern Aves based on comparative anatomy. *Netherlands Journal of Zoology* 51:179–205.

Lo, D.C., F. Allen, and J. P. Brockes. 1993. Reversal of muscle differentiation during urodele limb regeneration. *Proceedings of the National Academy of Sciences, USA* 90:7230–7234.

Lockwood, R., J. P. Swaddle, and J. M. V. Rayner. 1998. Avian wingtip shape reconsidered: Wingtip shape indices and morphological adaptations to migration. *Journal of Avian Biology* 29:273–292.

Loeffler, I. K., D. L. Stocum, J. F. Fallon, and C. U. Meteyer. 2001. Leaping lopsided: A review of the current hypotheses regarding etiologies of limb malformations in frogs. *Anatomical Record* 265:228–245.

Logan, C., A. Hornbruch, I. Campbell, and A. Lumsden. 1997. The role of Engrailed in establishing the dorsoventral axis of the chick limb. *Development* 124:2317–2324.

Logan, M., S. M. Pagan-Westphal, D. M. Smith, L. Paganessi, and C. J. Tabin. 1998. The transcription factor Pitx2 mediates situs-specific morphogenesis in response to left-right asymmetric signals. *Cell* 94:307–317.

Logan, M., and C. J. Tabin. 1999. Role of Pitx1 upstream of Tbx4 in specification of hindlimb identity. *Science* 283:1736–1739.

Lomaga, M. A., W. C. Yeh, I. Sarosi, G. S. Duncan, C. Furlonger, A. Ho, S. Morony, et al. 1999. TRAF6 deficiency results in osteopetrosis and defective interleukin-1, CD40, and LPS signaling. *Genes and Development* 13:1015–1024.

Lombard, R. E., and J. R. Bolt. 1995. A new primitive tetrapod *Whatcheeria deltae* from the Lower Carboniferous of Iowa. *Palaeontology* 38:471–494.

Lombard, R. E., and J. R. Bolt. 1999. A microsaur from the Mississippian of Illinois and a standard format for morphological characters. *Journal of Paleontology* 73:908–923.

Long, F., U. I. Chung, S. Ohba, J. McMahon, H. M. Kronenberg, and A. P. McMahon. 2004. Ihh signaling is directly required for the osteoblast lineage in the endochondral skeleton. *Development* 131:1309–1318.

Long, F., and T. F. Linsenmayer. 1998. Regulation of growth region cartilage proliferation and differentiation by perichondrium. *Development* 125:1067–1073.

Long, J. A. 1989. A new rhizodontiform fish from the Early Carboniferous of Victoria, Australia, with remarks on the phylogenetic position of the group. *Journal of Vertebrate Paleontology* 9:1–17.

Long, J. A. 2001. On the relationships of *Psarolepis* and the onychodontiform fishes. *Journal of Vertebrate Paleontology* 21:815–820.

Loomis, C. A., E. Harris, J. Michaud, W. Wurst, M. Hanks, and A. L. Joyner. 1996. The mouse Engrailed-1 gene and ventral limb patterning. *Nature* 382:360–363.

Loomis, C. A., R. A. Kimmel, C. X. Tong, J. Michaud, and A. L. Joyner. 1998. Analysis of the genetic pathway leading to formation of ectopic apical ectodermal ridges in mouse Engrailed-1 mutant limbs. *Development* 125:1137–1148.

Lorena, D., K. Uchio, A. M. Costa, and A. Desmouliere. 2002. Normal scarring: Importance of myofibroblasts. *Wound Repair and Regeneration* 10:86–92.

Louw, G. N., and E. Holm. 1972. Physiological, morphological and behavioural adaptations of the ultrapsammophilous, Namib Desert lizard *Aporosaura anchietae* (Bocage). *Madoqua,* series 2, 1:67–85.

Love, A. C. 2003. Evolutionary morphology, innovation and the synthesis of evolutionary and developmental biology. *Biology and Philosophy* 18:309–345.

Lovejoy, C. O., P. L. Reno, M. A. McCollum, M. W. Hamrick, R. S. Meindl, and M. J. Cohn. 2000. The evolution of primate hands: Growth scaling registers with posterior HOXD expression. *American Zoologist* 40:1109.

Loveridge, J. P. 1976. Strategies of water conservation in southern African frogs. *Zoologica Africana* 11:319–333.

Loveridge, N., and C. Farquharson. 1993. Studies on growth plate chondrocytes *in situ*: Cell proliferation and differentiation. *Acta Paediatrica* 391 (suppl.): 42–28.

Lu, H. C., J. P. Revelli, L. Goering, C. Thaller, and G. Eichele. 1997. Retinoid signaling is required for the establishment of a ZPA and for the expression of Hoxb-8, a mediator of ZPA formation. *Development* 124:1643–1651.

Lu, M.-F., H.-T. Cheng, A. R. Lacy, M. J. Kern, E. A. Argao, S. S. Potter, E. N. Olson, and J. F. Martin. 1999. Paired-related homeobox genes cooperate in handplate and hindlimb zeugopod morphogenesis. *Developmental Biology* 205:145–157.

Lubosche, W. 1910. *Bau und entstehung der wirbeltiergelenke.* Jena, Germany: Gustav Fischer.

Lucifora, L. O., and A. I. Vassallo. 2002. Walking in skates (Chondrichthyes, Rajidae): Anatomy, behaviour and analogies to tetrapod locomotion. *Biological Journal of the Linnean Society* 77:35–41.

Lui, V. C., L. J. Ng, J. Nicholls, P. P. Tam, and K. S. Cheah. 1995. Tissue-specific and differential expression of alternatively spliced alpha 1(II) collagen mRNAs in early human embryos. *Developmental Dynamics* 203:198–211.

Luiker, E. A., and E. D. Stevens. 1993. Effect of stimulus train duration and cycle frequency on the capacity to do work in pectoral fin muscle of the pumpkinseed sunfish, *Lepomis gibbosus. Canadian Journal of Zoology* 71:2185–2189.

Luke, C. 1986. Convergent evolution of lizard toe fringes. *Biological Journal of the Linnean Society* 27:1–16.

Lull, R. S. 1917. *Organic evolution.* New York: Macmillan.

Lull, R. S. 1933. A revision of the Ceratopsia or horned dinosaurs. *Peabody Museum of Natural History Bulletin* 3:1–175.

Lull, R. S., and N. E. Wright. 1942. Hadrosaurian dinosaurs of North America. *Geological Society of America, Special Papers* 40:1–242.

Luo, G., C. Hofmann, A. L. Bronckers, M. Sohocki, A. Bradley, and G. Karsenty. 1995. BMP-7 is an inducer of nephrogenesis, and is also required for eye development and skeletal patterning. *Genes and Development* 9:2808–2820.

Luo, Z. 1994. Sister-group relationships of mammals and transformations of diagnostic mammalian characters. In *The shadow of the dinosaurs: Early Mesozoic tetrapods*, ed. N. C. Fraser and H.-D. Sues, 98–130. Cambridge: Cambridge University Press.

Lussier, M., C. Canoun, C. Ma, A. Sank, and C. Shuler. 1993. Interdigital soft tissue separation induced by retinoic acid in mouse limbs cultured in vitro. *International Journal of Developmental Biology* 37:555–564.

Lupu, F., J. D. Terwiliger, K. Lee, G. V. Segre, and A. Efstratiadis. 2001. Roles of growth hormone and insulin-like growth factor 1 in mouse postnatal growth. *Developmental Biology* 229:141–162.

Lutfi, A. M. 1970. Study of cell multiplication in the cartilaginous upper end of the tibia of the domestic fowl by tritiated thymidine autoradiography. *Acta Anatomica* 76:454–463.

Lynch, V., J. Roth, K. Takahashi, C. W. Dunn, D. F. Nonaka, et al. 2004. Adaptive evolution of HoxA-11 and HoxA-13 at the origin of the uterus in mammals. *Proceedings of the Royal Society B* 271:2201–2207.

Lyons, K. M., B. L. Hogan, and E. J. Robertson. 1995. Colocalization of BMP 7 and BMP 2 RNAs suggests that these factors cooperatively mediate tissue interactions during murine development. *Mechanisms of Development* 50:71–83.

M

Mabee, P. M. 2000. Developmental data and phylogenetic systematics: Evolution of the vertebrate limb. *American Zoologist* 40:789–800.

Mabee, P. M., P. L. Crotwell, N. C. Bird, and A. C. Burke. 2002. Evolution of median fin modules in the axial skeleton of fishes. *Journal of Experimental Zoology* 294:77–90.

Mabee, P. M., and M. Noordsy. 2004. Development of the paired fins in the paddlefish, *Polyodon spathula*. *Journal of Morphology* 261:334–344.

Mabee, P. M., and T. A. Trendler. 1996. Development of the cranium and paired fins in *Betta splendens* (Teleostei: Percomorpha): Intraspecific variation and interspecific comparisons. *Journal of Morphology* 227:249–287.

MacCabe, J. A., J. Errick, and J. W. Saunders Jr. 1974. Ectodermal control of the dorsoventral axis in the leg bud of the chick embryo. *Developmental Biology* 39:69–82.

MacCracken, J. G., D. W. Uresk, and R. M. Hansen. 1985. Vegetation and soils of burrowing owl nest sites in Conata Basin, South Dakota. *Condor* 87:152–154.

Macias, D., Y. Gañan, T. K. Sampath, M. E. Piedra, M. A. Ros, and J. M. Hurlé. 1997. Role of BMP-2 and OP-1 (BMP-7) in programmed cell death and skeletogenesis during chick limb development. *Development* 124:1109–1117.

MacLeod, N., and K. D. Rose. 1993. Inferring locomotor behavior in paleogene mammals via eigenshape analysis. *American Journal of Science* 293A:300–355.

Maden, M. 1997. Retinoic acid and its receptors in limb regeneration. *Seminars in Cell and Developmental Biology* 8:445–453.

Maden, M., and N. Holder. 1984. Axial characteristics of nerve induced supernumerary limbs in the axolotl. *Wilhelm Roux's Archives of Developmental Biology* 193:394–401.

Magnan, A. 1929. Les caractéristiques géométriques et physiques des Poissons, avec contribution à l'étude de leur équilibre statique et dynamique. *Annales des Sciences Naturalles, Zoologie* 12:4–133.

Magnan, A. 1930. Les caractéristiques géométriques et physiques des Poissons, avec contribution à l'étude de leur équilibre statique et dynamique. *Annales des Sciences Naturalles, Zoologie* 13:355–489.

Mahendra, B. C. 1936. Contributions to the osteology of the Ophidia. I: The endoskeleton of the so-called "blind snake," *Typhlops braminus* Daud. *Proceedings of the Indian Academy of Sciences* 3:128–142.

Main, A. R. 1968. Ecology, systematics, and evolution of Australian frogs. In *Advances in ecological research,* vol. 5, ed. J. B. Cragg, 37–86. New York: Academic Press.

Main, A. R., M. J. Littlejohn, and A. K. Lee. 1959. Ecology of Australian frogs. In *Biogeography and ecology in Australia*, ed. A. Keast, R. L. Crocker, and C. S. Christian, 396–411. The Hague: Dr. W. Junk.

Maisano, J. A. 2001. A survey of the state of ossification in neonatal squamates. *Herpetological Monographs* 15:135–157.

Maisano, J. A. 2002a. Postnatal skeletal ontogeny in *Callisaurus draconoides* and *Uta stansburiana* (Iguania: Phrynosomatidae). *Journal of Morphology* 251:114–139.

Maisano, J. A. 2002b. The potential utility of postnatal skeletal developmental patterns in squamates phylogenetics. *Zoological Journal of the Linnean Society* 136:277–313.

Maisey, J. G. 1984. Chondrychthyan phylogeny: A look at the evidence. *Journal of Vertebrate Paleontology* 4:359–371.

Maisey, J. G. 1988. Phylogeny of early vertebrate skeletal induction and ossification patterns. *Evolutionary Biology* 22:1–36.

Mallo, M. 2001. Formation of the middle ear: Recent progress on the developmental and molecular mechanisms. *Developmental Biology* 231:410–419.

Mansuri, A. P. 1976. Effective muscle force: A contributory factor in determining functional ability of pectoral, pelvic and caudal fins of a fish, *Tilapia mossambica*. *Journal of Animal Morphology and Physiology* 23:32–39.

Manter, J. T. 1940. The mechanism of swimming in the alligator. *Journal of Experimental Zoology* 83:345–358.

Marcil, A., E. Dumontier, M. Chamberland, S. A. Camper, and J. Drouin. 2003. Pitx1 and Pitx2 are required for development of hindlimb buds. *Development* 130:45–55.

Marden, J. H. 1994. From damselflies to pterosaurs: How burst and sustainable flight performance scale with size. *American Journal of Physiology* 266:R1077–R1084.

Marey, E.-J. 1874. *Animal mechanism: A treatise on terrestrial and aerial locomotion.* 2nd ed. London: H. S. King.

Marey, E.-J. 1890. La locomotion aquatique étudiée par la chronophotographie. *Comptes Rendus de l'Académie des Sciences* 111:213–216.

Marey, E.-J. 1894. *Le mouvement.* Transl. E. Pritchard. New York: D. Appleton, 1895.

Margetic, S., C. Gazzola, C. G. Pegg, and R. A. Hill. 2002. Leptin: A review of its peripheral actions and interactions. *International Journal of Obesity* 26:1407–1433.

Mariani, F. V., and G. R. Martin. 2003. Deciphering skeletal patterning: Clues from the limb. *Nature* 423:319–325.

Marí-Beffa, M., M. C. Carmona, and J. Becerra. 1989. Elastoidin turn-over during tail fin regeneration in teleosts: A morphometric and radioautographic study. *Anatomy and Embryology* 180:465–470.

Marí-Beffa, M., P. Palmqvist, F. Marin-Giron, G. S. Montes, and J. Becerra. 1999. Morphometric study of the regeneration of individual rays in teleost tail fins. *Journal of Anatomy* 195:393–405.

Marí-Beffa, M., J. A. Santamaría, P. Fernandez-Llebrez, and J. Becerra. 1996. Histochemically defined cell states during tail fin regeneration in teleost fishes. *Differentiation* 60:139–149.

Marigo, V., R. L. Johnson, A. Vortkamp, and C. J. Tabin. 1996. Sonic hedgehog differentially regulates expression of GLI and GLI3 during limb development. *Developmental Biology* 180:273–283.

Marks, S. C., Jr. 1989. Osteoclast biology: Lessons from mammalian mutations. *American Journal of Medical Genetics* 34:43–54.

Marshak, D. R., D. Gottlieb, and R. L. Gardner. 2001. Introduction to *Stem cell biology*. New York: Cold Spring Harbor Laboratory Press.

Marshall, A. M. 1916. *The frog*. London: Macmillan.

Marshall, N. B. 1976. *The life of fishes*. New York: Universe Books.

Martill, D. M. 1993. Soupy substrates: A mechanism for the exceptional preservation of ichthyosaurs of the Posidonia Shale (Lower Jurassic) of Germany. *Kaupia—Darmstädter Beiträge zur Naturgeschichte* 2:77–97.

Martin, D. J. 1973. Selected aspects of burrowing owl ecology and behavior. *Condor* 75:446–456.

Martin, E. A., E. L. Ritman, and R. T. Turner. 2003. Time course of epiphyseal growth plate fusion in rat tibiae. *Bone* 32:261–267.

Martin, G. R. 1998. The roles of FGFs in the early development of vertebrate limbs. *Genes and Development* 12:1571–1586.

Martin, P. 1990. Tissue patterning in the developing mouse limb. *International Journal of Developmental Biology* 34:323–336.

Martin, P. 1997. Wound healing: Aiming for perfect skin regeneration. *Science* 276:75–81.

Maryanska, T. A. 1977. Ankylosauridae (Dinosauria) from Mongolia. *Palaeontologica Polonica* 37:85–151.

Massare, J. A. 1994. Swimming capabilities of Mesozoic marine reptiles: A review. In *Mechanics and physiology of animal swimming*, ed. L. Maddock, Q. Bone, and J. M. V. Rayner, 133–149, 205–229. Cambridge: Cambridge University Press.

Massare, J. A. 1997. Introduction to part VI: Faunas, behavior, and evolution. In *Ancient marine reptiles*, ed. J. M. Callaway and E. L. Nicholls, 401–421. San Diego, CA: Academic Press.

Massare, J. A., and J. M. Callaway. 1990. The affinities and ecology of Triassic ichthyosaurs. *Geological Society of America Bulletin* 102:409–416.

Masuya, H., T. Sagai, K. Moriwaki, and T. Shiroishi. 1997. Multigenic control of the localization of the zone of polarizing activity in limb morphogenesis in the mouse. *Developmental Biology* 182:42–51.

Mathur, J. K., and S. C. Goel. 1976. Patterns of chondrogenesis and calcification in the developing limb of the lizard, *Calotes versicolor. Journal of Morphology* 149:401–420.

Matthew, W. D. 1904. The arboreal ancestry of the Mammalia. *American Naturalist* 38:811–818.

Maunz, M., and R. Z. German. 1997. Ontogeny and limb bone scaling in two new world marsupials, *Monodelphis domestica* and *Didelphis virginiana. Journal of Morphology* 231:117–130.

Mauras, N. 2001. Growth hormone and sex steroids: Interactions at puberty. *Neuroendocrinology* 30:529–544.

Mayer, C. 1825. Über die hintere Extremität der Ophidier. *Novorum Actorum Academiae Caesareae Leopoldino-Carolinae Naturae Curiosorum* 12:819–842.

Maynard Smith, J., R. Burian, S. Kauffman, P. Alberch, J. Campbell, et al. 1985. Developmental constraints and evolution. *Quarterly Review of Biology* 60:265–287.

Maynard Smith, J., and R. J. G. Savage. 1956. Some locomotory adaptations in mammals. *Journal of the Linnean Society (Zoology)* 42:603–622.

Maynard Smith, J., and K. Sondhi. 1960. The genetics of a pattern. *Genetics* 45:1039–1050.

Mayr, E. 1983. How to carry out the adaptationist program? *American Naturalist* 121:324–334.

Mazin, J.-M. 2001. Mesozoic marine reptiles: An overview. In *Secondary adaptation of tetrapods to life in water: Proceedings of the International Meeting, Poitiers, 1996*, ed. J.-M. Mazin and V. de Buffrénil, 95–118. Munich: Pfeil.

Mazin, J.-M., and V. de Buffrénil, eds. 2001. *Secondary adaptation of tetrapods to life in water: Proceedings of the International Meeting, Poitiers, 1996*. Munich: Pfeil.

Mazin, J.-M., P. Hantzpergue, G. Lafaurie, and P. Vignaud. 1995. Des pistes de ptérosaures dans le Tithonien de Crayssac (Quercy, France). *Comptes Rendus de l'Académie des Sciences*, series 2A, 320:417–424.

Mazin, J.-M., and G. Pinna. 1993. Palaeoecology of the armoured placodonts. *Paleontologia Lombarda della Società Italiana di Scienze Naturali e del Museo Civico di Storia Naturale di Milano, Nuova serie* 2:83–91.

McClanahan, L., Jr. 1967. Adaptations of the spadefoot toad, *Scaphiopus couchi*, to desert environments. *Comparative Biochemistry and Physiology* 20:73–99.

McClure, J. 1983. The effect of diphosphonates on heterotopic ossification in regenerating Achilles tendon of the mouse. *Journal of Pathology* 139:419–430.

McDevitt, C. A., and R. J. Webber. 1990. The ultrastructure and biochemistry of meniscal cartilage. *Clinical Orthopaedics* 252:8–18.

McGann, C. J., S. J. Odelberg, and M. T. Keating. 2001. Mammalian myotube dedifferentiation induced by newt regeneration extract. *Proceedings of the National Academy of Sciences, USA* 98:13699–13704.

McGhee, G. R. 1999. *Theoretical morphology: The concept and its applications*. New York: Columbia University Press.

McGonnell, I. M. 2001. The evolution of the pectoral girdle. *Journal of Anatomy* 199:189–194.

McGowan, C. 1972. The distinction between latipinnate and longipinnate ichthyosaurs. *Life Sciences Occasional Papers, Royal Ontario Museum* 20:1–8.

McGowan, C. 1991. *Dinosaurs, spitfires, and sea dragons*. Cambridge, MA: Harvard University Press.

McGowan, C. 1992. The ichthyosaurian tail: Sharks do not provide an appropriate analogue. *Palaeontology* 35:555–570.

McHugh, K. P., K. Hodivala-Dilke, M. H. Zheng, N. Namba, J. Lam, D. Novack, X. Feng, F. P. Ross, R. O. Hynes, and S. L. Teitelbaum. 2000. Mice lacking beta3 integrins are osteosclerotic because of dysfunctional osteoclasts. *Journal of Clinical Investigations* 105:433–440.

McIlhenny, E. A. 1935. *The alligator's life history*. Boston: Christopher.

McIntosh, J. S. 1990. Sauropoda. In *The Dinosauria*, ed. D. B. Weishampel, P. Dodson, and H. Osmólska, 345–401. Berkeley: University of California Press.

McKenna, M. C., and S. K. Bell. 1997. *Classification of mammals above the species level*. New York: Columbia University Press.

McKibbin, B., and F. W. Holdsworth. 1967. The dual nature of epiphyseal cartilage. *Journal of Bone and Joint Surgery* 49B:351–361.

McLaughlin, R. L., and D. L. G. Noakes. 1998. Going against the flow: An examination of the propulsive movements made by young brook trout in streams. *Canadian Journal of Fisheries and Aquatic Sciences* 55:853–860.

McLean, W., and B. R. Olsen. 2001. Mouse models of abnormal skeletal development and homeostasis. *Trends in Genetics* 17:S38–S43.

McPherron, A. C., A. M. Lawler, and S. J. Lee. 1999. Regulation of anterior/posterior patterning of the axial skeleton by growth/differentiation factor 11. *Nature Genetics* 22:260–264.

Meckert, D. 1993. Der Schultergürtel des *Sclerocephalus haeuseri* Goldfuss, 1847 im vergleich mit Eryops Cope, 1877 (Eryopoida, Amphibia, Perm). *Palaeontographica A* 229:113–140.

Meers, M. B. 2002. Cross-sectional geometric properties of the crocodylian humerus: An exception to Wolff's Law? *Journal of Zoology, London* 258:405–418.

Meister, W. 1962. Histological structure of the long bones of penguins. *Anatomical Record* 143:377–388.

Menzies, J. I. 1958. Breeding behaviour of the chamaeleon (*Chamaeleo gracilis*) in Sierra Leone. *British Journal of Herpetology* 2:130–132.

Menzies, J. I., and M. J. Tyler. 1977. The systematics and adaptations of some Papuan microhylid frogs which live underground. *Journal of Zoology, London* 183:431–464.

Mercader, N., E. Leonardo, N. Azpiazu, A. Serrano, G. Morata, C. Martinez, and M. Torres. 1999. Conserved regulation of proximodistal limb axis development by Meis1/Hth. *Nature* 402:425–429.

Mercader, N., E. Leonardo, M. E. Piedra, A. C. Martinez, M. A. Ros, and M. Torres. 2000. Opposing RA and FGF signals control proximodistal vertebrate limb development through regulation of Meis genes. *Development* 127:3961–3970.

Mercanter, N., E. Leonardo, N. Azpiazu, A. Serrano, G. Morata, et al. 1999. Conserved regulation of proximodistal limb axis development by Meis1/Hth. *Nature* 402:425–429.

Mérida-Velasco, J. A., I. Sánchez-Montesinos, J. Espín-Ferra, J. F. Rodríguez-Vásquez, J. R. Mérida-Velasco, and J. Jiménez-Collado. 1997. Development of the human knee joint. *Anatomical Record* 248:269–278.

Merino, R., Y. Gañan, D. Macias, A. N. Economides, K. T. Sampath, and J. M. Hurlé. 1998. Morphogenesis of digits in the avian limb is controlled by FGFs, TGFbetas, and noggin through BMP signalling. *Developmental Biology* 200:35–45.

Merino, R., D. Macias, Y. Gañan, J. Rodriguez-Leon, A. N. Economides, C. Rodriguez-Esteban, J. C. Izpisúa Belmonte, and J. M. Hurlé. 1999a. Control of digit formation by active signalling. *Development* 126:2161–2170.

Merino, R., J. Rodriguez-Leon, D. Macias, Y. Gañan, A. N. Economides, and J. M. Hurlé. 1999b. The BMP antagonist Gremlin regulates outgrowth, chondrogenesis and programmed cell death in the developing limb. *Development* 126:5515–5522.

Mertens, R. 1960. *The world of amphibians and reptiles*. New York: McGraw-Hill.

Mescher, A. L. 1976. Effects on adult newt limb regeneration of partial and complete skin flaps over the amputation surface. *Journal of Experimental Zoology* 195:117–128.

Mescher, A. L. 1996. The cellular basis of limb regeneration in urodeles. *International Journal of Developmental Biology* 40:785–795.

Messel, H., and G. C. Vorlicek. 1989. Ecology of *Crocodylus porosus* in northern Australia. In *Crocodiles: Their ecology, management, and conservation*, 164–183. Gland, Switzerland: International Union for Conservation of Nature and Natural Resources.

Meteyer, C. U. 2000. *Field guide to malformations of frogs and toads with radiographic interpretations*. Madison, WI: USGS National Wildlife Center.

Meteyer, C. U., I. K. Loeffler, J. F. Fallon, K. A. Converse, E. Green, J. C. Helgen, S. Kersten, R. Levey, L. Eaton-Poole, and J. G. Burkhart. 2000. Hind limb malformations in free-living northern leopard frogs (*Rana pipiens*) from Maine, Minnesota, and Vermont suggest multiple etiologies. *Teratology* 62:151–171.

Methven, D. A. 1985. Identification and development of larval and juvenile *Urophycis chuss, U. tenuis* and *Phycis chesteri* (Pisces, Gadidae) from the Northwest Atlantic. *Journal of Northwest Atlantic Fishery Science* 6:9–20.

Metscher, B. D., and P. E. Ahlberg. 1999. Zebrafish in context: Uses of a laboratory model in comparative studies. *Developmental Biology* 210:1–4.

Metscher, B. D., K. Takahashi, K. D. Crow, C. Amemiya, D. F. Nonaka, and G. P. Wagner. 2005. Expression of Hoxa-11 and Hoxa-13 in the pectoral fin of a basal ray-finned fish, *Polyodon spathula*: Implications for the origin of tetrapod limbs. *Evolution and Development* 7:186–195.

Mettam, A. E. 1895. The rudimentary metacarpal and metatarsal bones of the domestic ruminants. *Journal of Anatomy and Physiology* 29:244–253.

Meunier, F. J. 1983. *Les tissus osseux des Ostéichthyens: Structure, genèse, croissance et évolution*. Paris: Archives et Documents, Micro-Edition, Institut d'Ethnologie, S. N. 82-600-328.

Meunier, F. J. 1989. The acellularisation process in Osteichthyan bone. *Progress in Zoology* 35:443–446.

Meunier, F. J., and A. Huysseune. 1992. The concept of bone tissue in Osteichthyes. *Netherlands Journal of Zoology* 42:445–458.

Michaud, J. L., F. Lapointe, and N. M. Le Douarin. 1997. The dorsoventral polarity of the presumptive limb is determined by signals produced by the somites and by the lateral somatopleure. *Development* 124:1453–1463.

Middleton, K. M., and S. M. Gatesy. 2000. Theropod forelimb design and evolution. *Zoological Journal of the Linnean Society* 128:149–187.

Mikić, B., T. L. Johnson, A. B. Chhabra, B. J. Schalet, M. Wong, and E. B. Hunziker. 2000a. Differential effects of embryonic immobilization on the development of fibrocartilaginous skeletal elements. *Journal of Rehabilitation Research and Development* 37:127–133.

Mikić, B., M. Wong, M. Chiquet, and E. B. Hunziker. 2000b. Mechanical modulation of tenascin-C and collagen-XII expression during avian synovial joint formation. *Journal of Orthopaedic Research* 18:406–415.

Milaire, J. 1967. Evolution des processus dégéneratifs dans la cape apicale au cours du dévelopment des members chez le rat et la souris. *Comptes Rendus de l'Academie des Sciences* (Paris) 265:137–140.

Milaire, J. 1976. Rudimentation digitale au cours du dévelopment normal de l'autopode chez les mammifères. In *Mechânismes de la rudimentation des organs chez les embryons des vertebraes,* ed. A. Raynaud, 221–233. Paris: Editions du C.N.R.S.

Milaire, J. 1974. Histochemical aspects of organogenesis in vertebrates. Part 1: The skeletal system, limb morphogenesis, the sense organs. In *Handbuch Der Histochemie,* vol. 8, pt. 3, ed. W. Grauman and K. Neumann, 1–135. Stuttgart: Gustav Fischer Verlag.

Milaire, J. 1977. Histochemical expression of morphogenetic gradients during limb morphogenesis. *Birth Defects: Original Articles Series* 13:37–67.

Milaire, J., and M. Roze. 1983. Hereditary and induced modifications of the normal necrotic patterns in the developing limb buds of the rat and mouse: Facts and hypothesis. *Archives in Biology* 94:459–490.

Miles, S. S. 1941. The shoulder anatomy of the armadillo. *Journal of Mammalogy* 22:157–169.

Millar, S. E. 2002. Molecular mechanisms regulating hair follicle development. *Journal of Investigative Dermatology* 118:216–225.

Miller, C. M. 1944. Ecologic relations and adaptations of the limbless lizards of the genus *Anniella. Ecological Monographs* 14:271–289.

Miller, G. S., Jr. 1907. The families and genera of bats. *U.S. National Museum Bulletin* 57:1–282.

Milligan, C. E., D. Prevette, H. Yaginuma, S. Homma, C. Cardwell, L. C. Fritz, K. J. Tomaselli, R. W. Oppenheim, and L. M. Schwartz. 1995. Peptide inhibitors of the ICE protease family arrest programmed cell death of motoneurons in vivo and in vitro. *Neuron* 15:385–393.

Millot, J., and J. Anthony, J. 1958. *Latimeria chalumnae* dernier des crossoptérygiens. In *Traité de zoologie: Agnathes et poissons—Anatomie, éthologie, systématique,* vol. 13, pt. 3, ed. P.-P. Grassé, 2553–2597. Paris: Masson.

Milner, A. C. 1980. A review of the Nectridea (Amphibia). In *The terrestrial environment and the origin of land vertebrates,* ed. A. L. Panchen, 377–405. London: Academic Press.

Milner, A. C., and W. Lindsay. 1998. Postcranial remains of *Baphetes* and their bearing on the relationships of the Baphetidae (= Loxommatidae). *Zoological Journal of the Linnean Society* 122:211–235.

Milner, A. R. 1993. The Paleozoic relatives of lissamphibians. *Herpetological Monographs* 7:8–27.

Milner, A. R. 2000. Mesozoic and tertiary Caudata and Albanerpetontidae. In *Amphibian biology,* vol. 4, *Palaeontology: The evolutionary history of amphibians,* ed. H. Heatwole and R. L. Carroll, 1412–1444. Chipping Norton, Australia: Surrey Beatty and Sons.

Milner, A. R., and S. E. K. Sequeira. 1994. The temnospondyl amphibians from the Viséan of East Kirkton, West Lothian, Scotland. *Transactions of the Royal Society of Edinburgh, Earth Sciences* 84:331–361.

Min, H., D. M. Danilenko, S. A. Scully, B. Bolon, B. D. Ring, J. E. Harpley, M. DeRose, and W. S. Simonet. 1998. Fgf-10 is required for both limb and lung development and exhibits striking functional similarity to Drosophila branchless. *Genes and Development* 12:3156–3161.

Mindell, D. P. 1997. *Avian molecular evolution and systematics.* San Diego, CA: Academic Press.

Miner, R. W. 1925. The pectoral limb of *Eryops* and other primitive tetrapods. *Bulletin of the American Museum of Natural History* 51:145–312.

Minetti, A. E. 1998. The biomechanics of skipping gaits: A third locomotion paradigm. *Proceedings of the Royal Society of London B* 265:1227–1235.

Minguillon, C., J. Del Buono, and M. P. Logan. 2005. Tbx5 and Tbx4 are not sufficient to determine limb-specific morphologies but have common roles in initiating limb outgrowth. *Developmental Cell* 8:75–84.

Minina, E., C. Kreschel, M. C. Naski, D. M. Ornitz, and A. Vortkamp. 2002. Interaction of FGF, Ihh/Pthlh, and BMP signaling integrates chondrocyte proliferation and hypertrophic differentiation. *Developmental Cell* 3:439–449.

Minina, E., H. M. Wenzel, C. Kreschel, S. Karp, W. Gaffield, A. P. McMahon, and A. Vortkamp. 2001. BMP and Ihh/PTHrP signaling interact to coordinate chondrocyte proliferation and differentiation. *Development* 128:4523–4534.

Mirkes, P. E., S. A. Little, and C. C. Umpierre. 2001. Co-localization of active caspase-3 and DNA fragmentation (TUNEL) in normal and hyperthermia-induced abnormal mouse development. *Teratology* 63:134–143.

Mishina, Y., A. Suzuki, N. Ueno, and R. R. Behringer. 1995. Bmpr encodes a type I bone morphogenetic protein receptor that is essential for gastrulation during mouse embryogenesis. *Genes and Development* 9:3027–3037.

Misof, B. Y., and G. P. Wagner. 1992. Regeneration in *Salaria pavo* (Blenniiddae, Teleostei): Histogenesis of the regenerating pectoral fin suggests different mechanisms for morphogenesis and structural maintenance. *Anatomy and Embryology* 186:153–165.

Mivart, St. G. 1879. On the fins of elasmobranchii. *Transactions of the Zoological Society of London* 10:439–484.

Miyake, T., A. M. Cameron, and B. K. Hall. 1997. Stage-specific expression patterns of alkaline phosphatase during development of the first arch skeleton in inbred C57BL/6 mouse embryos. *Journal of Anatomy* 190:239–260.

Miyake, T., and B. K. Hall. 1994. Development of in vitro organ culture techniques for differentiation and growth of cartilages and bones from teleost fish and comparisons with in vivo skeletal development. *Journal of Experimental Zoology* 268:22–43.

Miyazaki, K., K. Uchiyama, Y. Imokawa, and K. Yoshizato. 1996. Cloning and characterization of cDNAs for matrix metalloproteinases of regenerating newt limbs. *Proceedings of the National Academy of Sciences, USA* 93:6819–6824.

Mizell, M. 1968. Limb regeneration: Induction in the newborn opossum. *Science* 161:283–286.

Mizelle, J. D. 1935. Swimming of the muskrat. *Journal of Mammalogy* 16:22–25.

Mizuta, S., J. H. Hwang, and R. Yoshinaka. 2003. Molecular species of collagen in pectoral fin cartilage of skate (*Raja kenojei*). *Food Chemistry* 80:1–7.

Mlynarski, M., and L. Madej. 1961. The rudimentary limbs in Aniliidae (Serpentes). *British Journal of Herpetology* 3:1–6.

Modha, M. L. 1967. The ecology of the Nile crocodile (*Crocodylus niloticus* Laurenti) on Central Island, Lake Rudolf. *East African Wildlife Journal* 5:74–95.

Mohammed, M. B. H. 1991. A comparative survey on heterophalangeal elements of some geckos (Gekkonidae: Reptilia). *Journal of the Egyptian German Society of Zoology* 5:315–332.

Moll, D. 1979. Subterranean feeding by the Illinois mud turtle, *Kinosternon flavescens spooneri*. *Journal of Herpetology* 13:371–373.

Molnar, R. E., S. M. Kurzanov, and D. Zhiming. 1990. Carnosauria. In *The Dinosauria*, ed. D. B. Weishampel, P. Dodson, and H. Osmólska, 169–209. Berkeley: University of California Press.

Mommsen, T. P. 1998. Growth and metabolism. In *The physiology of fishes,* 2nd ed., ed. D. H. Evans, 65–99. New York: CRC Press.

Montero, J. A., Y. Gañan, D. Macias, J. Rodriguez-Leon, J. J. Sanz-Ezquerro, R. Merino, J. Chimal-Monroy, M. A. Nieto, and J. M. Hurlé. 2001. Role of FGFs in the control of programmed cell death during limb development. *Development* 128:2075–2084.

Montes, G. S., J. Becerra, O. M. S. Toledo, M. A. Gordilho, and L. C. U. Junqueira, L. C. U. 1982. Fine structure and histochemistry of the tail fin rays in teleosts. *Histochemistry* 75:363–376.

Moodie, G. E. E. 1978. Observations on the life history of the caecilian *Typhlonectes compressicaudus* (Dumeril and Bibron) in the Amazon basin. *Canadian Journal of Zoology* 56:1005–1008.

Mook, C. C. 1921. Notes on the postcranial skeleton in the Crocodilia. *Bulletin of the American Museum of Natural History* 44:67–100.

Moon, R. T., J. D. Brown, and M. Torres. 1997. WNTs modulate cell fate and behavior during vertebrate development. *Trends in Genetics* 13:157–162.

Moore, J. A. 1993. *Science as a way of knowing: The foundations of modern biology.* Cambridge, MA: Harvard University Press.

Morgan, T. H. 1898. Experimental studies of the regeneration of Planaria maculata. *Arch. Entw. Mech. Org* 7:364–397.

Morgan, T. H. 1900. Regeneration in teleosts. *Arch. Entw. Mech. Org* 10:120–131.

Morgan, T. H. 1902. Further experiments on the regeneration of the tail of fishes. *Arch. Entw. Mech. Org* 14:539–561.

Morgan, T. H. 1904. Notes on regeneration: The limitation of the regenerative power of *Dendrocoelum lacteum*. *Biological Bulletin* 6:159–163.

Morgan, T. H. 1906. The physiology of regeneration. *Journal of Experimental Zoology* 4:457–500.

Morris, C. 1892. The origin of lungs. *American Naturalist* 26:975–986.

Morse, E. S. 1872. On the tarsus and carpus of birds. *Annals of the Lyceum of Natural History of New York* 10:141–158.

Mortlock, D. P., C. Guenther, and D. M. Kingsley. 2003. A general approach for identifying distant regulatory elements applied to the Gdf6 gene. *Genome Research* 13:2069–2081.

Mortlock, D. P., L. C. Post, and J. W. Innis. 1996. The molecular basis of hypodactyly (Hd): A deletion in Hoxa 13 leads to arrest of digital arch formation. *Nature Genetics* 13:184–289.

Mosauer, W. 1932. Adaptive convergence in the sand reptiles of the Sahara and of California. *Copeia* 1932:72–78.

Moser, H. G. 1996. *The early stages of fishes in the California current region.* California Cooperative Oceanic Fisheries Investigations Atlas no. 33. La Jolla, CA: California Cooperative Oceanic Fisheries Investigations.

Moss, M. L. 1961a. Osteogenesis of acellular teleost fish bone. *American Journal of Anatomy* 108:99–109.

Moss, M. L. 1961b. Studies of the acellular bone of teleost fish. I: Morphological and systemic variations. *Acta Anatomica* 46:343–362.

Moss, M. L. 1962. Studies of the acellular bone of teleost fish. II: Response to fracture under normal and acalcemic conditions. *Acta Anatomica* 48:46–60.

Moss, M. L. 1963. The biology of acellular teleost bone. *Annals of the New York Academy of Sciences* 109:337–350.

Moss, M. L. 1977. Skeletal tissues in sharks. *American Zoologist* 17:335–342.

Moss, M. L., and L. Moss-Salentijn. 1983. Vertebrate cartilages. In *Cartilage: Structure, function, and biochemistry,* vol. 1, ed. B. K. Hall, 1–30. New York: Academic Press.

Motani, R. 1999. On the evolution and homologies of ichthyopterygian forefins. *Journal of Vertebrate Paleontology* 19:28–41.

Motani, R., H. You, and C. McGowan. 1996. Eel-like swimming in the earliest ichthyosaurs. *Nature* 382:347–348.

Moy-Thomas, J. A., and R. S. Miles. 1971. *Palaeozoic fishes.* London: Chapman and Hall.

Muir, H. 1995. The chondrocyte, architect of cartilage: Biomechanics, structure, function and molecular biology of cartilage matrix macromolecules. *BioEssays* 17:1039–1048.

Mukerji, D. D. 1931. Some observations on the burrowing toad, *Cacopus globulosus* Günther. *Journal and Proceedings, Asiatic Society of Bengal* 27:97–100.

Mulder, E. W. A. 2001. Co-ossified vertebrae of mosasaurs and cetaceans: Implications for the mode of locomotion of extinct marine reptiles. *Paleobiology* 27:724–734.

Mullen, L. M., S. V. Bryant, M. A. Torok, B. Blumberg, and D. M. Gardiner. 1996. Nerve dependency of regeneration: The role of Distal-less and FGF signaling in amphibian limb regeneration. *Development* 122:3487–3497.

Müller, G. B. 2003. Embryonic motility: Environmental influences and evolutionary innovation. *Evolution and Development* 5:46–60.

Müller, G. B., and P. Alberch. 1990. Ontogeny of the limb skeleton in Alligator mississippiensis: Developmental invariance and change in the evolution of Archosaur limbs. *Journal of Morphology* 203:151–164.

Müller, G. B., and G. P. Wagner. 1991. Novelty in evolution: Restructuring the concept. *Annual Review of Ecology and Systematics* 22:229–256.

Müller, J. 1831. Beiträge zur Anatomie und Naturgeschichte der Amphibien. *Zeitschrift für Physiologie* 4:190–275.

Müller, M. 1900. *Die Reduktion des Brustschultergürtels der Saurier bis zum völligen Verluste desselben.* Inaugural-Dissertation, Universität Leipzig. Leipzig: Hesse and Becker.

Müller, O. 1913. *Neue Untersuchungen über die Extremitätengürtel von* Anguis fragilis. Inaugural-Dissertation, Universität Bern. Bern: Max Drechsel.

Muller, T. L., V. Ngo-Muller, A. Reginelli, G. Taylor, R. Anderson, and K. Muneoka. 1999. Regeneration in higher vertebrates:

Limb buds and digit tips. *Seminars in Cell and Developmental Biology* 10:405–413.

Müller, U. K., B. L. E. Van den Heuvel, E. J. Stamhuis, and J. J. Videler. 1997. Fish foot prints: Morphology and energetics of the wake behind a continuously swimming mullet (*Chelon labrosus* Risso). *Journal of Experimental Biology* 200:2893–2906.

Mundlos, S. 1999. Cleidocranial dysplasia: Clinical and molecular genetics. *Journal of Medical Genetics* 36:177–182.

Mundlos, S., F. Otto, C. Mundlos, J. B. Mulliken, A. S. Aylsworth, S., Albright, D. Lindhout, et al. 1997. Mutations involving the transcription factor CBFA1 cause cleidocranial dysplasia. *Cell* 89:773–779.

Muneoka, K., and S. V. Bryant. 1982. Evidence that patterning mechanisms in developing and regenerating limbs are the same. *Nature* 298:369–371.

Muneoka, K., and S. V. Bryant. 1984. Cellular contribution to supernumerary limbs resulting from the interaction between developing and regenerating tissues in the axolotl. *Developmental Biology* 105:179–187.

Muneoka, K., S. V. Bryant, and D. M. Gardiner. 1989. Growth control in limb regeneration. In *Developmental biology of the axolotl*, ed. J. B. Armstrong and G. M. Malacinski, 143–156. New York: Oxford University Press.

Muneoka, K., W. Fox, and S. V. Bryant. 1986a. Cellular contribution from dermis and cartilage to the regenerating limb blastema in axolotls. *Developmental Biology* 16:256–260.

Muneoka, K., G. Holler-Dinsmore, and S. V. Bryant. 1986b. Intrinsic control of regenerative loss in *Xenopus laevis* limbs. *Journal of Experimental Zoology* 240:47–54.

Muneoka, K., and D. Sassoon. 1992. Limb development and regeneration. *Developmental Biology* 152:37–49.

Muneoka, K., N. Wanek, and S. V. Bryant. 1986c. Mouse embryos develop normally *ex utero. Journal of Experimental Zoology* 239:289–293.

Murciano, C., T. D. Fernandez, I. Duran, D. Maseda, J. Ruiz-Sanchez, J. Becerra, M.-A. Akimenko, and M. Marí-Beffa, M. 2002. Ray-interray interactions during fin regeneration of *Danio rerio. Developmental Biology* 252:214–224.

Murray, P. D. F. 1936. *Bones: A study of the development and structure of the vertebrate skeleton.* Cambridge: Cambridge University Press.

Murray, P. D. F., and J. S. Huxley. 1925. Self-differentiation in the grafted limb-bud of the chick. *Journal of Anatomy* 59:379–384.

Muruzábal, F. J., G. Frühbeck, J. Gómez-Ambrosi, M. Archanco, and M. A. Burrell. 2002. Immunocytochemical detection of leptin in non-mammalian vertebrate stomach. *General and Comparative Endocrinology* 128:149–152.

Mussi, M., A. P. Summers, and P. Domenici. 2002. Gait transition speed, pectoral fin-beat frequency and amplitude in *Cymatogaster aggregata, Embiotoca lateralis* and *Damalichthys vacca. Journal of Fish Biology* 61:1282–1293.

Muybridge, E. 1887. *Animal locomotion: An electro-photographic investigation of consecutive phases of animal movements, 1872–1885.* Philadelphia: University of Pennsylvania.

Muybridge, E. 1893. *Descriptive zoopraxography, or the science of animal locomotion made popular.* Philadelphia.

Muybridge, E. 1972. *Eadweard Muybridge: The Stanford Years, 1872–1882.* Stanford, CA: Stanford University Department of Art.

N

Nabrit, S. M. 1929. The role of the fin rays in the regeneration in the tail-fins of fishes (in *Fundulus* and goldfish). *Biology Bulletin* 56:235–266.

Nabrit, S. M. 1931. The role of the basal plate in regeneration in the tail-fins of fishes (*Fundulus* and *Carassius*). *Biology Bulletin* 60:60–63.

Naiche, L. A., and V. E. Papaioannou. 2003. Loss of Tbx4 blocks hindlimb development and affects vascularization and fusion of the allantois. *Development* 130:2681–2893.

Nakagawa, N., M. Kinosaki, K. Yamaguchi, N. Shima, H. Yasuda, K. Yano, T. Morinaga, and K. Higashio. 1998. RANK is the essential signaling receptor for osteoclast differentiation factor in osteoclastogenesis. *Biochemical and Biophysical Research Communications* 253:395–400.

Nakanishi, K., M. Maruyama, T. Shibata, and N. Morishima. 2001. Identification of a caspase-9 substrate and detection of its cleavage in programmed cell death during mouse development. *Journal of Biological Chemistry* 276:41237–41244.

Naples, V. L. 1999. Morphology, evolution and function of feeding in the giant anteater (*Myrmecophaga tridactyla*). *Journal of Zoology, London* 249:19–41.

Nardi, J. B., and D. L. Stocum. 1983. Surface properties of regenerating limb cells: Evidence for gradation along the proximo-distal axis. *Differentiation* 25:27–31.

Naruse, I., and Y. Kameyama. 1982. Morphogenesis of genetic preaxial polydactyly, Polydactyly Nagoya, Pdn, in mice. *Congenital Anomalalies* 22:137–144.

Naski, M. C., J. S. Colvin, J. D. Coffin, and D. M. Ornitz. 1998. Repression of hedgehog signaling and BMP4 expression in growth plate cartilage by fibroblast growth factor receptor 3. *Development* 125:4977–4988.

Neame, P. J., H. Tapp, and A. Azizan. 1999. Noncollagenous, non-proteoglycan macromolecules of cartilage. *Cellular and Molecular Life Sciences* 55:1327–1340.

Nechiporuk, A., and M. Keating. 2002. A proliferation gradient between proximal and msxb-expressing distal blastema directs zebrafish fin regeneration. *Development* 129:2607–2617.

Nechiporuk, A., K. D. Poss, S. L. Johnson, and M. T. Keating. 2003. Positional cloning of a temperature-sensitive mutant emmental reveals a role for sly1 during cell proliferation in zebrafish fin regeneration. *Developmental Biology* 258:291–306.

Neill, W. T. 1951. Notes on the role of crawfishes in the ecology of reptiles, amphibians, and fishes. *Ecology* 32:764–766.

Neill, W. T. 1971. *The last of the ruling reptiles: Alligators, crocodiles, and their kin.* New York: Columbia University Press.

Nelson, C. E., B. A. Morgan, A. C. Burke, E. Laufer, E. DiMambro, L. C. Murtaugh, E. Gonzales, L. Tessarollo, L. F. Parada, and C. Tabin. 1996. Analysis of Hox gene expression in the chick limb bud. *Development* 122:1449–1466.

Neumann, C. J., H. Grandel, W. Gaffield, F. Schulte-Merker, and C. Nüsslein-Volhard. 1999. Transient establishment of anterior-posterior polarity in the zebrafish pectoral fin bud in the absence of sonic hedgehog activity. *Development* 126:4817–4826.

Nevo, E. 1961. Observations on Israeli populations of the mole rat *Spalax e. ehrenbergi* Nehring 1898. *Mammalia* 25:127–144.

Nevo, E. 1979. Adaptive convergence and divergence of subterranean mammals. *Annual Review of Ecology and Systematics* 10:269–308.

Nevo, E. 1999. *Mosaic evolution of subterranean mammals: Regression, progression, and global convergence.* Oxford: Oxford University Press.

Newman, S. A. 1993. Why does a limb look like a limb? *Progress in Clinical and Biological Research* 383A:89–98.

Ng, J. K., Y. Kawakami, D. Buscher, A. Raya, T. Itoh, C. M. Koth, C. Rodriguez Esteban, et al. 2002. The limb identity gene Tbx5 promotes limb initiation by interacting with Wnt2b and Fgf10. *Development* 129:5161–5170.

Ng, L. J., S. Wheatley, G. E. Muscat, J. Conway-Campbell, J. Bowles, E. Wright, D. M. Bell, P. P. Tam, K. S. Cheah, and P. Koopman. 1997. SOX9 binds DNA, activates transcription, and coexpresses with type II collagen during chondrogenesis in the mouse. *Developmental Biology* 183:108–121.

Nicholls, E. L. 1997. Introduction to part III: Testudines. In *Ancient marine reptiles*, ed. J. M. Callaway and E. L. Nicholls, 219–223. San Diego, CA: Academic Press.

Nicholls, E. L., and S. J. Godfrey. 1994. Subaqueous flight in mosasaurs: A discussion. *Journal of Vertebrate Paleontology* 14:450–452.

Nicholls, E. L., and A. P. Russell. 1991. The plesiosaur pectoral girdle: The case for a sternum. *Neues Jahrbuch für Geologie und Paläontologie, Abhandlungen* 182:161–185.

Niederreither, K., J. Vermot, B. Schuhbaur, P. Chambon, and P. Dollé. 2002. Embryonic retinoic acid synthesis is required for forelimb growth and anteroposterior patterning in the mouse. *Development* 129:3563–3574.

Nieuwkoop, P. D., and J. Faber. 1967. *Normal table of Xenopus laevis (Daudin).* 2nd ed. Amsterdam: North-Holland.

Nigg, B. M., B. R. MacIntosh, and J. Mester. 2000. *Biomechanics and biology of movement.* Champaign, IL: Human Kinetics.

Nilsson, C., D. Swolin-Ede, C. Ohlsson, E. Eriksson, H.-P. Ho, P. Björnthorp, and A. Holmäng. 2003. Reductions in adipose tissue and skeletal growth in rat adult offspring after prenatal leptin exposure. *Journal of Endocrinology* 176:13–21.

Nishii, K., T. Tsuzuki, M. Kumai, N. Takeda, H. Koga, S. Aizawa, T. Nishimoto, and Y. Shibata. 1999. Abnormalities of developmental cell death in Dad1-deficient mice. *Gene Cells* 4:243–252.

Niswander, L., C. Tickle, A. Vogel, I. Booth, and G. R. Martin. 1993. FGF-4 replaces the apical ectodermal ridge and directs outgrowth and patterning of the limb. *Cell* 75:579–587.

Niswander, L., S. Jeffrey, G. R. Martin, and C. Tickle. 1994. A positive feedback loop coordinates growth and patterning in the vertebrate limb. *Nature* 371:609–612.

Nitecki, M. H., ed. 1990. *Evolutionary innovations.* Chicago: University of Chicago Press.

Niven, J. S. F. 1933. The development in vivo and in vitro of the avian patella. *Wilhelm Roux's Archives of Developmental Biology* 128:480–501.

Noble, G. K. 1931. *The biology of the amphibia.* New York: McGraw-Hill.

Noble, G. K., and M. E. Jaeckle. 1928. The digital pads of the tree frogs: A study of the phylogenesis of an adaptive structure. *Journal of Morphology and Physiology* 45:259–292.

Noji, S., E. Koyama, F. Myokai, T. Nohno, H. Ohuchi, K. Nishikawa, and S. Taniguchi. 1993. Differential expression of three chick FGF receptor genes, FGFR1, FGFR2 and FGFR3, in limb and feather development. *Progress in Clinical Biological Research* 383B:645–654.

Noonan, K. J., E. B. Hunziker, J. Nessler, and J. A. Buckwalter.

1998. Changes in cell, matrix compartment, and fibrillar collagen volumes between growth plate zones. *Journal of Orthopaedic Research* 16:500–508.

Noramly, S., J. Pisenti, U. Abbott, and B. Morgan. 1996. Gene expression in the limbless mutant: Polarized gene expression in the absence of Shh and an AER. *Developmental Biology* 179:339–346.

Norberg, U. M. 1972. Functional osteology and myology of the wing of the dog-faced bat *Rousettus aegypticus* (E. Geoffroy) (Pteropidae). *Zeitschrift für Morphologie der Tiere* 73:1–44.

Norberg, U. M. 1976. Kinematics, aerodynamics, and energetics of horizontal flight in the long-eared bat *Plecotus auritus*. *Journal of Experimental Biology* 65:179–212.

Norberg, U. M. 1981. Flight, morphology and the ecological niche in some bats and birds. *Symposium of the Zoological Society of London* 48:173–197.

Norberg, U. M. 1985. Flying, gliding, and soaring. In *Functional vertebrate morphology*, ed. M. Hildebrand, D. M. Bramble, K. F. Liem, and D. B. Wake, 129–158. Cambridge, MA: Belknap Press.

Norberg, U. M. 1990. *Vertebrate flight: Mechanics, physiology, morphology, ecology, and evolution.* Berlin: Springer-Verlag.

Norberg, U. M., A. P. Brooke, and W. J. Trewhella. 2000. Soaring and non-soaring bats of the family Pteropodidae (Flying Foxes, *Pteropus* spp.): Wing morphology and flight performance. *Journal of Experimental Biology* 203:651–664.

Norberg, U. M., and J. M. V. Rayner. 1987. Ecological morphology and flight in bats (Mammalia; Chiroptera): Wing adaptations, flight performance, foraging strategy, and echolocation. *Philosophical Transactions of the Royal Society of London B* 316:335–427.

Norell, M. A., and J. A. Clarke. 2001. Fossil that fills a critical gap in avian evolution. *Nature* 409:181–183.

Norell, M. A., J. M. Clark, and P. J. Makovicky. 2001. Phylogenetic relationships among coelurosaurian theropods. In *New perspectives on the origin and early evolution of birds: Proceedings of the International Symposium in Honor of John H. Ostrom*, ed. J. A. Gauthier and L. F. Gall, 49–67. New Haven, CT: Peabody Museum of Natural History, Yale University.

Norell, M. A., Q. Ji, K. Gao, C.-X. Yuan, Y. Zhao, and L. Wang. 2002. "Modern" feathers on a non-avian dinosaur. *Nature* 416:36–37.

Norman, D. B. 1985. *The illustrated encyclopedia of dinosaurs.* New York: Crescent Books.

Norman, D. B. 1990. Problematic theropoda: "Coelurosaurs." In *The Dinosauria*, ed. D. B. Weishampel, P. Dodson, and H. Osmólska, 280–305. Berkeley: University of California Press.

Norman, D. B., and D. B. Weishampel. 1990. Iguanodontidae and related ornithopods. In *The Dinosauria*, ed. D. B. Weishampel, P. Dodson, and H. Osmólska, 510–533. Berkeley: University of California Press.

Novack, D. V., and S. J. Korsmeyer. 1994. Bcl-2 protein expression during murine development. *American Journal of Pathology* 145:61–73.

Novacek, M. J. 1985. Evidence for echolocation in the oldest known bats. *Nature* 315:140–141.

Novacek, M. J. 1992. Mammalian phylogeny: Shaking the tree. *Nature* 356:121–125.

Nowak, R. M. 1991. *Walker's mammals of the world.* Baltimore: Johns Hopkins University Press.

Nowak, R. M. 1999. *Walker's mammals of the world.* Vol. 2. 5th ed. Baltimore: John Hopkins University Press.

Nudds, R. L., G. J. Dyke, and J. M. V. Rayner. 2004. Forelimb proportions and the evolutionary radiation of neornithes. *Proceedings of the Royal Society of London B* 271:S324–S327.

Nunley, J. A., P. V. Spiegl, R. D. Goldner, and J. R. Urbaniak. 1987. Longitudinal epiphyseal growth after replantation and transplantation in children. *Journal of Hand Surgery* 12A:274–279.

Nurminskaya, M., and T. F. Linsenmayer. 1996. Identification and characterization of up-regulated genes during chondrocyte hypertrophy. *Developmental Dynamics* 206:260–271.

Nussbaum, R. A. 1982. Heterotopic bones in the hindlimbs of frogs of the Families Pipidae, Ranidae and Sooglossidae. *Herpetologica* 38:312–320.

Nye, H. L., J. A. Cameron, E. A. Chernoff, and D. L. Stocum. 2003. Regeneration of the urodele limb: A review. *Developmental Dynamics* 226:280–294.

Nybelin, O. 1963. Zur Morphologie und Terminologie des Schwanzskelettes der Actinopterygier. *Arkiv för Zoologi* 15:485–516.

Nye, H. L. D., J. A. Cameron, E. A. G. Chernoff, and D. L. Stocum. 2003. Regeneration of the urodele limb: A review. *Developmental Dynamics* 226:280–294.

Nyhart, L. K. 2002. Learning from history: Morphology's challenges in Germany ca. 1900. *Journal of Morphology* 252:2–14.

O

Obst, B. S., W. M. Hamner, P. P. Hamner, E. Wolanski, M. Rubega, and B. Littlehales. 1996. Kinematics of phalarope spinning. *Nature* 384:121.

Odelberg, S. J., A. Kollhoff, and M. T. Keating. 2000. Dedifferentiation of mammalian myotubes induced by msx1. *Cell* 103:1099–1109.

Ogden, G. J., M. H. S. Conlogue, and A. G. J. Rhodin. 1981. Roentenographic indicators of skeletal maturity in marine mammals (*Cetacea*). *Skeletal Radiology* 7:119–123.

O'Gorman, F. 1963. Observations on terrestrial locomotion in Antarctic seals. *Proceedings of the Zoological Society of London* 141:837–850.

Ogren, L., and C. McVea Jr. 1982. Apparent hibernation by sea turtles in North American waters. In *Biology and conservation of sea turtles,* ed. K. A. Bjorndal, 127–132. Washington, DC: Smithsonian Institution Press.

Ohbayashi, N., M. Shibayama, Y. Kurotaki, M. Imanishi, T. Fujimori, N. Itoh, and S. Takada. 2002. FGF18 is required for normal cell proliferation and differentiation during osteogenesis and chondrogenesis. *Genes and Development* 16:870–879.

Ohlsson, C., B.-Å. Bengtsson, O. G. Isaksson, T. T. Andreassen, and M. C. Slootweg. 1998. Growth hormone and bone. *Endocrine Reviews* 19:45–79.

Ohsugi, K., D. M. Gardiner, and S. V. Bryant. 1997. Cell cycle length affects gene expression and pattern formation in limbs. *Developmental Biology* 189:13–21.

Ohuchi, H., T. Nakagawa, A. Yamamoto, A. Araga, T. Ohata, Y. Ishimaru, H. Yoshioka, et al. 1997. The mesenchymal factor, FGF10, initiates and maintains the outgrowth of the chick limb bud through interaction with FGF8, an apical ectodermal factor. *Development* 124:2235–2244.

Ohuchi, H., T. Nakagawa, M. Yamauchi, T. Ohata, H. Yoshioka, K. Kuwana, T. Mima, T. Mikawa, T. Nohno, and S. Noji. 1995. An additional limb can be induced from the flank of the chick embryo by FGF4. *Biochemical and Biophysical Research Communications* 209:809–816.

O'Keefe, F. R. 2001. Ecomorphology of plesiosaur flipper geometry. *Journal of Evolutionary Biology* 14:987–991.

Olsen, B. R., A. M. Reginato, and W. Wang. 2000. Bone development. *Annual Review of Cell and Developmental Biology* 16:191–220.

Olson, E. C. 1936. The ilio-sacral attachment of *Eryops. Journal of Paleontology* 10:648–651.

Olson, S. L. 1999. *Avian paleontology at the close of the 20th century: Proceedings of the 4th International Meeting of the Society of Avian Paleontology and Evolution, Washington, D.C., 4–7 June 1996.* Washington, DC: Smithsonian Institution Press.

Olson, S. L., and Y. Hasegawa. 1979. Fossil counterparts of giant penguins from the North Pacific. *Science* 206:688–689.

Olson, W. M. 1998. Evolutionary and developmental morphology of the dwarf African clawed frog, *Hymenochirus boettgeri* (Amphibia: Anura: Pipidae). PhD diss., University of California, Berkeley.

Olson, W. M. 2000. Phylogeny, ontogeny, and function: Extraskeletal bones in the tendons and joints of *Hymenochirus boettgeri* (Amphibia: Anura: Pipidae). *Zoology, Analysis of Complex Systems* 103:15–24.

O'Neill, W. C. 1999. Physiological significance of volume-regulatory transporters. *American Journal of Physiology* 276:C995–C1101.

O'Rahilly, R., D. J. Gray, and E. Gardner. 1957. Chondrification in the hands and feet of staged human embryos. *Carnegie Institute of Washington, Contributions to Embryology* 36:183–192.

O'Reilly, J. C., N. J. Kley, D. A. Ritter, A. P. Summers, and D. R. Carrier. 2001. Kinetics of burrowing in limbless tetrapods. *Journal of Morphology* 248:269.

O'Reilly, J. C., D. A. Ritter, and D. R. Carrier. 1997. Hydrostatic locomotion in a limbless tetrapod. *Nature* 386:269–272.

Ornitz, D. M., and P. J. Marie. 2002. FGF signaling pathways in endochondral and intramembranous bone development and human genetic disease. *Genetics and Development* 16:446–465.

Ornitz, D. M., J. Xu, J. S. Colvin, D. G. McEwen, C. A. MacArthur, F. Coulier, G. Gao, and M. Goldfarb. 1996. Receptor specificity of the fibroblast growth factor family. *Journal of Biological Chemistry* 271:15292–15297.

O'Rourke, M. P., K. Soo, R. R. Behringer, C. C. Hui, and P. P. Tam. 2002. Twist plays an essential role in FGF and SHH signal transduction during mouse limb development. *Developmental Biology* 248:143–156.

Orzack, S. H., and E. Sober, eds. 2001. *Adaptation and optimality.* Cambridge: Cambridge University Press.

Osborn, H. F. 1916. Skeletal adaptations of *Ornitholestes, Struthiomimus, Tyrannosaurus. Bulletin of the American Museum of Natural History* 35:733–771.

Osborn, H. F. 1917. *The origin and evolution of life: On the theory of the action, reaction, and interaction of energy.* New York: Charles Scribner's Sons.

Osborn, H. F. 1918. Equidae of the Oligocene, Miocene, and Pliocene of North America, iconographic type revision. *Memoirs of the American Museum of Natural History* 2:1–330.

Osburn, R. C. 1903. Adaptation to aquatic, arboreal, fossorial, and cursorial habits in mammals. I: Aquatic adaptations. *American Naturalist* 37:651–655.

Oshida, T., N. Hachiya, M. C. Yoshida, and N. Ohtaishi. 2000. Comparative anatomical note on the origin of the long accessory styliform cartilage of the Japanese giant flying squirrel, *Petaurista leucogenys*. *Mammal Study* 25:35–39.

Ostrom, J. H. 1969. Osteology of *Deinonychus antirrhopus*, an unusual theropod from the Lower Cretaceous of Montana. *Peabody Museum of Natural History Bulletin* 30:1–165.

Ostrom, J. H. 1973. The ancestry of birds. *Nature* 242:136.

Ostrom, J. H. 1974. *Archaeopteryx* and the origin of flight. *Quarterly Review of Biology* 49:27–47.

Ostrom, J. H. 1975. The origin of birds. *Annual Review of Earth and Planetary Science* 3:55–77.

Ostrom, J. H. 1976a. *Archaeopteryx* and the origin of birds. *Biological Journal of the Linnean Society* 8:91–182.

Ostrom, J. H. 1976b. On a new specimen of the Lower Cretaceous Theropod dinosaur *Deinonychus antirrhopus*. *Breviora* 439:1–21.

Ostrom, J. H. 1976c. Some hypothetical anatomical stages in the evolution of avian flight. *Smithsonian Contributions in Paleontology* 27:1–21.

Ostrom, J. H. 1978. The osteology of *Compsognathus longipes*. *Zitteliana* 4:73–118.

Otto, F., A. P. Thornell, T. Crompton, A. Denzel, K. C. Gilmour, I. R. Rosewell, G. W. Stamp, et al. 1997. Cbfa1, a candidate gene for cleidocranial dysplasia syndrome, is essential for osteoblast differentiation and bone development. *Cell* 89:765–771.

Owen, R. 1849. *The nature of limbs: A discourse delivered on Friday, February 8, at an evening meeting of the Royal Institution of Great Britain.* London: Jon Van Voorst.

Owen, R. 1849 [2007]. *The nature of limbs: A discourse delivered on Friday, February 8, at an evening meeting of the Royal Institution of Great Britain.* Repr. ed. by R. Amundson, with foreword by B. K. Hall. Chicago: University of Chicago Press, forthcoming.

Owre, O. T. 1967. Adaptations for locomotion and feeding in the anhinga and double-crested cormorant. *Ornithological Monographs* 6:1–138.

Oxnard, C. E. 1984. *The order of man: A biomathematical anatomy of the primates.* New Haven, CT: Yale University Press.

P

Pabst, D. A., S. A. Rommel, and W. A. McLellan. 1999. The functional morphology of marine mammals. In *Biology of marine mammals,* ed. J. E. Reynolds III and S. A. Rommel, 15–72. Washington, DC: Smithsonian Institution Press.

Pacifici, M., E. B. Golden, O. Oshima, I. M. Shapiro, P. S. LeBoy, and S. L. Adams. 1995. Hypertrophic chondrocytes: The terminal lineage of differentiation I the chondrogenic cell lineage. *Annals of the New York Academy of Sciences* 599:45–57.

Packard, D. S., E. J. Levinsohn, and D. R. Hootnick. 1993. Most human lower limb malformations appear to result from post-specification insults. In *Limb development and regeneration,* ed. J. F. Fallon, P. F. Goetinck, R. O. Kelley, and D. L. Stocum, 417–426. New York: Wiley-Liss.

Padian, K. 1983a. A functional analysis of flying and walking in pterosaurs. *Paleobiology* 9:218–239.

Padian, K. 1983b. Osteology and functional morphology of *Dimorphodon macronyx* (Buckland) (Pterosauria: Rhamphorhynchoidea) based on new material in the Yale Peabody Museum. *Postilla* 189:1–44.

Padian, K. 1985. The origins and aerodynamics of flight in extinct vertebrates. *Palaeontology* 28:413–433.

Padian, K. 1991. Pterosaurs: Were they functional birds or functional bats? In *Biomechanics in evolution,* vol. 36 of Seminar Series of the Society for Experimental Biology, ed. J. M. V. Rayner and R. J. Wootton, 146–160. Cambridge: Cambridge University Press.

Padian, K. 1997. *Encyclopedia of dinosaurs.* San Diego, CA: Academic Press.

Padian, K., and L. M. Chiappe. 1998. The origin and early evolution of birds. *Biological Reviews* 73:1–42.

Padian, K., A. J. De Ricqlès, and J. R. Horner. 2001. Dinosaurian growth rates and bird origins. *Nature* 412:405–408.

Padian, K., and P. E. Olsen. 1984a. Footprints of the Komodo monitor and the trackways of fossil reptiles. *Copeia* 1984:662–671.

Padian, K., and P. E. Olsen. 1984b. The fossil trackway *Pteraichnus*: Not pterosaurian, but crocodilian. *Journal of Paleontology* 58:178–184.

Padian, K., and J. M. V. Rayner. 1993a. Structural fibers of the pterosaur wing: Anatomy and aerodynamics. *Naturwissenschaften* 80:361–364.

Padian, K., and J. M. V. Rayner. 1993b. The wings of pterosaurs. *American Journal of Science* 293A:91–166.

Palmeirim, I., D. Henrique, D. Ish-Horowicz, and O. Pourquie. 1997. Avian hairy gene expression identifies a molecular clock linked to vertebrate segmentation and somitogenesis. *Cell* 91:639–648.

Palmer, M. 1960. Expanded ilio-sacral joint in the toad *Xenopus laevis*. *Nature* 187:797–798.

Panchen, A. L. 1975. A new genus and species of anthracosaur amphibian from the Lower Carboniferous of Scotland and the status of *Pholidogaster pisciformis* Huxley. *Philosophical Transactions of the Royal Society of London B* 269:581–640.

Panchen, A. L. 1985. On the amphibian *Crassigyrinus scoticus* Watson from the Carboniferous of Scotland. *Philosophical Transactions of the Royal Society of London B* 309:505–568.

Panchen, A. L., and T. R. Smithson. 1987. Character diagnosis, fossils, and the origin of tetrapods. *Biological Reviews* 62:341–438.

Panchen, A. L., and T. R. Smithson, T. R. 1988. The relationships of early tetrapods. In *The phylogeny and classification of the tetrapods,* vol. 1, *Amphibians, reptiles, birds,* ed. M. J. Benton, 1–32. Oxford: Clarendon Press.

Panchen, A. L., and T. R. Smithson. 1990. The pelvic girdle and hind limb of *Crassigyrinus scoticus* (Lydekker) from the Scottish Carboniferous and the origin of the tetrapod pelvic skeleton. *Transactions of the Royal Society of Edinburgh, Earth Sciences* 81:31–44.

Panman, L., and R. Zeller. 2003. Patterning the limb before and after SHH signalling. *Journal of Anatomy* 202:3–12.

Papadimitriou, H. M., S. M. Swartz, and T. H. Kunz. 1996. Ontogenetic and anatomic variation in mineralization of the wing skeleton of the Mexican free-tailed bat, *Tadarida brasiliensis*. *Journal of Zoology, London* 240:411–426.

Papageorgiou, S., and Y. Almirantis. 1996. Gradient model describes the spatial-temporal expression pattern of Hoxa genes in the developing vertebrate limb. *Developmental Dynamics* 207:461–469.

Papenfuss, T. J. 1982. The ecology and systematics of the amphisbaenian genus *Bipes*. *Occasional Papers of the California Academy of Sciences* 136:1–42.

Parchman, A. J., S. M. Reilly, and A. R. Biknevicius. 2003. Whole-body mechanics and gaits in the gray short-tailed opossum *Monodelphis domestica:* Integrating patterns of locomotion in a semi-erect mammal. *Journal of Experimental Biology* 206:1379–1388.

Parfitt, A. M. 2002. Misconceptions 1: Epiphyseal fusion causes cessation of growth. *Bone* 30:377–379.

Parker, H. W. 1931. Parallel modifications in the skeleton of the Amphibia Salientia. *Archivio Zoologico Italiano* 16:1239–1248.

Parker, W. K. 1868. *A monograph on the structure and development of the shoulder-girdle and sternum in the Vertebrata.* London: Ray Society.

Parmelee, J. R. 1990. Lack of burrowing ability in the blue-spotted salamander, *Ambystoma laterale*. *Bulletin of the Chicago Herpetological Society* 25:81–83.

Parr, B. A., and A. P. McMahon. 1995. Dorsalizing signal Wnt-7a required for normal polarity of D-V and A-P axes of mouse limb. *Nature* 374:350–353.

Parr, B. A., M. J. Shea, G. Vassileva, and A. P. McMahon. 1993. Mouse Wnt genes exhibit discrete domains of expression in the early embryonic CNS and limb buds. *Development* 119:247–261.

Parrish, J. M. 1987. The origin of crocodilian locomotion. *Paleobiology* 13:396–414.

Parsons, F. G. 1904. Observations on traction epiphyses. *Journal of Anatomy* 38:248–258.

Parsons, F. G. 1908. Further remarks on traction epiphyses. *Journal of Anatomy* 42:388–399.

Parsons, G. R., and J. L. Sylvester Jr. 1992. Swimming efficiency of the white crappie, *Pomoxis annularis*. *Copeia* 1992:1033–1038.

Pastoret, C., and T. A. Partridge. 1998. Muscle regeneration. In *Cellular and molecular basis of regeneration*, ed. P. Ferretti and J. Géraudie, 309–333. Chichester, UK: Wiley.

Pateder, D., R. N. Rosier, E. M. Schwarz, P. R. Reynolds, J. E. Puzas, M. D'Souza, and R. J. O'Keefe. 2000. PTHrP expression in chondrocytes, regulation by TGF-β, and interactions between epiphyseal and growth plate chondrocytes. *Experimental Cell Research* 256:555–562.

Pathi, S., J. B. Rutenberg, R. L. Johnson, and A. Vortkamp. 1999. Interaction of Ihh and BMP/Noggin signaling during cartilage differentiation. *Developmental Biology* 209:239–253.

Paton, R. L., T. R. Smithson, and J. A. Clack. 1999. An amniote-like skeleton from the Early Carboniferous of Scotland. *Nature* 398:508–513.

Patterson, C. 1977. Cartilage bones, dermal bones and membrane bones, or the exoskeleton versus the endoskeleton. In *Problems in vertebrate evolution*, ed. S. Andrews, R. Miles, and A. Walker, 77–121. London: Academic Press.

Patterson, C. 1980. Origin of tetrapods: historical introduction to the problem. In *The terrestrial environment and the origin of land vertebrates*, ed. A. L. Panchen, 159–175. London: Academic Press.

Patterson, C. 1993. Bird or dinosaur? *Nature* 365:21–22.

Paukstis, G. L., and L. E. Brown. 1987. Evolution of the intercalary cartilage in chorus frogs, genus *Pseudacris* (Salientia: Hylidae). *Brimleyana* 13:55–61.

Paukstis, G. L., and L. E. Brown. 1991. Evolutionary trends in the morphology of the intercalary phalanx of anuran amphibians. *Canadian Journal of Zoology* 69:1297–1301.

Pautou, M. P. 1975. Morphogenesis of the chick embryo foot. *Journal of Embryology and Experimental Morphology* 1975:511–529.

Pearce, J. J., G. Penny, and J. Rossant. 1999. A mouse cerberus/Dan-related gene family. *Developmental Biology* 209:98–110.

Pearse, R. V., II, K. J. Vogan, and C. J. Tabin. 2001. Ptc1 and Ptc2 transcripts provide distinct readouts of Hedgehog signaling activity during chick embryogenesis. *Developmental Biology* 239:15–29.

Pearson, D. M., and T. S. Westoll. 1979. The Devonian actinopterygian *Cheirolepis* Agassiz. *Transactions of the Royal Society of Edinburgh* 70:337–399.

Pearson, K., and A. G. Davin. 1921a. On the sesamoids of the knee-joint. Part 1: Man. *Biometrika* 13:133–175.

Pearson, K., and A. G. Davin. 1921b. On the sesamoids of the knee-joint. Part 2: Evolution of the sesamoids. *Biometrika* 13:350–400.

Pecorino, L. T., A. Entwistle, and J. P. Brockes. 1996. Activation of a single retinoic acid receptor isoform mediates proximodistal respecification. *Current Biology* 6:563–569.

Pedersen, H. E. 1949. The ossicles of the semilunar cartilages of rodents. *Anatomical Record* 105:1–9.

Peichel, C. L., K. Nereng, K. A. Ohgi, B. L. E. Cole, P. F. Colosimo, C. A. Buerkle, D. Schluter, and D. M. Kingsley. 2001. The genetic architecture of divergence between threespine stickleback species. *Nature* 414:901–905.

Peignoux-Deville, J., C. A. Baud, F. Lallier, and B. Vidal. 1985. Perichondral ossification of vertebral arches from dogfish to man. *Progress in Zoology* 30:65–68.

Peignoux-Deville, J., F. Lallier, and B. Vidal. 1982. Evidence for the presence of osseous tissue in dogfish vertebrae. *Cell and Tissue Research* 222:605–614.

Pennycuick, C. J. 1975. Mechanics of flight. In *Avian biology*, vol. 5, ed. D. S. Farmer and J. R. King, 1–75. London: Academic Press.

Pennycuick, C. J. 1986. Mechanical constraints of the evolution of flight. In *The origin of birds and the evolution of flight: Memoirs of the California Academy of Sciences*, vol. 8, ed. K. Padian, 83–98. San Francisco: California Academy of Sciences.

Pennycuick, C. J. 1987. Flight of seabirds. In *Seabirds: Feeding ecology and role in marine ecosystems*, ed. J. P. Croxall, 43–62. Cambridge: Cambridge University Press.

Pereira, F., and M. Gharib. 2002. Defocusing digital particle image velocimetry and the three-dimensional characterization of two-phase flows. *Measurement Science and Technology* 13:683–694.

Pereira, B. P., S. P. Cavanagh, and R. W. Pho. 1997. Longitudinal growth rate following slow physeal distraction: The proximal tibial growth plate studied in rabbits. *Acta Orthopaedica Scandanavica* 68:262–268.

Perle, A., L. M. Chiappe, R. Barsbold, J. M. Clark, and M. A. Norell. 1994. Skeletal morphology of *Mononykus olecranus* (Theropod: Avialae) from the Late Cretaceous of Mongolia. *American Museum Novitates* 3105:1–29.

Persson, P., K. Sundell, B. T. Björnsson, and H. Lundqvist. 1998. Calcium metabolism and osmoregulation during sexual maturation of river running Atlantic salmon. *Journal of Fish Biology* 52:334–349.

Peters, R. C., K. Kotrschal, and W. D. Krautgartner. 1991. Solitary chemoreceptor cells of *Ciliata mustela* (Gadidae, Teleostei) are tuned to mucoid stimuli. *Chemical Senses* 16:31–42.

Peters, W. C. H. 1882. *Zoologie III: Amphibien; Naturwissenschaftliche Reise nach Mossambique auf Befehl seiner Majestät des Königs Friedrich Wilhelm IV. In den Jahren 1842 bis 1848 ausgeführt von Wilhelm C. H. Peters.* Berlin: G. Reimer.

Peterson, R. L., T. Papenbrock, M. M. Davda, and A. Augulewitsch. 1994. The murine HoxC cluster contains five neighboring AbdB-related Hox genes that show unique spatially coordinated expression in posterior embryonic subregions. *Mechanisms of Development* 47:253–260.

Petranka, J. W. 1998. *Salamanders of the United States and Canada.* Washington, DC: Smithsonian Institution Press.

Pettigrew, J. B. 1874. *Animal locomotion.* New York: D. Appleton.

Pettigrew, J. D., B. G. M. Jamieson, S. K. Robson, L. S. Hall, K. I. McAnally, and H. M. Cooper. 1989. Phylogenetic relations between microbats, megabats and primates (Mammalia: Chiroptera and Primates). *Philosophical Transactions of the Royal Society of London B* 325:489–559.

Phillip, M., O. Moran, L. Lazar. 2002. Growth without growth hormone. *Journal of Pediatric Endocrinology and Metabolism* 15:1267–1272.

Piedra, M. E., J. M. Icardo, M. Albajar, J. C. Rodriguez-Rey, and M. A. Ros. 1998. Pitx2 participates in the late phase of the pathway controlling left-right asymmetry. *Cell* 94:319–324.

Pinna, G., and S. Nosotti. 1989. Anatomia, morfologia funzionale e paleoecologia del rettile placodonta *Psephoderma alpinum* Meyer, 1858. *Memorie della Società Italiana di Scienze Naturali e del Museo Civico di Storia Naturale di Milano* 25:17–50.

Pinto, J. P., M. C. P. Ohresser, and M. L. Cancela. 2001. Cloning of the bone Gla protein gene from the teleost fish *Sparus aurata:* Evidence for overall conservation in gene organization and bone-specific expression from fish to man. *Gene* 270:77–91.

Pispa, J., and I. Thesleff. 2003. Mechanisms of ectodermal organogenesis. *Developmental Biology* 262:195–205.

Pitsillides, A. A., C. W. Archer, P. Prehm, M. T. Bayliss, and J. C. W. Edwards. 1995. Alterations in hyaluronan synthesis during developing joint cavitation. *Journal of Histochemistry and Cytochemistry* 43:263–273.

Pizette, S., C. Abate-Shen, and L. Niswander. 2001. BMP controls proximodistal outgrowth, via induction of the apical ectodermal ridge, and dorsoventral patterning in the vertebrate limb. *Development* 128:4463–4474.

Pizette, S., and L. Niswander. 1999. BMPs negatively regulate structure and function of the limb apical ectodermal ridge. *Development* 126:883–894.

Pizette, S., and L. Niswander. 2000. BMPs are required at two steps of limb chondrogenesis: Formation of prechondrogenic condensations and their differentiation into chondrocytes. *Developmental Biology* 219:237–249.

Platt, J. F., and D. N. Nettleship. 1985. Diving depths of four alcids. *Auk* 102:293–297.

Platt, K. A., J. Michaud, and A. L. Joyner. 1997. Expression of the mouse Gli and Ptc genes is adjacent to embryonic sources of hedgehog signals suggesting a conservation of pathways between flies and mice. *Mechanisms of Development* 62:121–135.

Poleo, G., C. W. Brown, L. Laforest, and M.-A. Akimenko. 2001. Cell proliferation and movement during early fin regeneration in zebrafish. *Developmental Dynamics* 221:380–390.

Polinkovsky, A., N. H. Robin, J. T. Thomas, M. Irons, A. Lynn, F. R. Goodman, W. Reardon, et al. 1997. Mutations in CDMP1 cause autosomal dominant brachydactyly type C. *Nature: Genetics* 17:18–19.

Polk, J. D., B. Demes, W. L. Jungers, A. R. Biknevicius, R. E. Heinrich, and J. A. Runestad. 2000. A comparison of primate, carnivoran and rodent limb bone cross-sectional properties: Are primates really unique? *Journal of Human Evolution* 39:297–325.

Pollard, H. B. 1891. On the anatomy and phylogenetic position of Polypterus. *Anatomische Anzeiger* 6:338–344.

Pooley, A. C. 1969. The burrowing behaviour of crocodiles. *Lammergeyer* 10:60–63.

Pooley, A. C. 1977. Nest opening response of the Nile crocodile *Crocodylus niloticus. Journal of Zoology, London* 182:17–26.

Pooley, A. C. 1982. *Discoveries of a crocodile man.* London: William Collins Sons.

Pooley, A. C., and C. Gans. 1976. The Nile crocodile. *Scientific American* 234 (4): 114–119, 122–124.

Poore, S. O., A. Sánchez-Haiman, and G. E. Goslow Jr. 1997. Wing upstroke and the evolution of flapping flight. *Nature* 387:799–802.

Poss, K. D., M. T. Keating, and A. Nechiporuk. 2003. Tales of regeneration in zebrafish. *Developmental Dynamics* 226:202–210.

Poss, K. D., A. Nechiporuk, A. M. Hillam, S. L. Johnson, and M. T. Keating. 2002. Mps1 defines a proximal blastemal proliferative compartment essential for zebrafish fin regeneration. *Development* 129:5141–5149.

Poss, K. D., J. Shen, and M. T. Keating. 2000a. Induction of lef1 during zebrafish fin regeneration. *Developmental Dynamics* 219:282–286.

Poss, K. D., J. Shen, A. Nechiporuk, G. McMahon, B. Thisse, C. Thisse, and M. T. Keating. 2000b. Roles for Fgf signaling during zebrafish fin regeneration. *Developmental Biology* 222:347–358.

Post, L. C., and J. W. Innis. 1999. Altered Hox expression and increased cell death distinguish Hypodactyly from Hoxa13 null mice. *International Journal of Developmental Biology* 43:287–294.

Potthoff, T. 1975. Development and structure of the caudal complex, the vertebral column, and the pterygiophores in the black fin tuna (*Thunnus atlanticus*, Pisces, Scombridae). *Bulletin of Marine Science* 25:205–231.

Potthoff, T., and S. Kelley. 1982. Development of the vertebral column, fins and fin supports, branchiostegal rays, and squamation in the swordfish, *Xiphias gladius. Fishery Bulletin* 80:161–186.

Potthoff, T., S. Kelley, and L. A. Collins. 1988. Osteological development of the red snapper *Lutjanus campechanus* (Lutjanidae). *Bulletin of Marine Science* 43:1–40.

Pough, F. H. 1969. The morphology of undersand respiration in reptiles. *Herpetologica* 25:216–223.

Pough, F. H. 1970. The burrowing ecology of the sand lizard, *Uma notata. Copeia* 1970:145–157.

Pough, F. H., R. M. Andrews, J. E. Cadle, M. L. Crump, A. H.

Savitzky, K. D. Wells. 1998. *Herpetology*. Upper Saddle River, NJ: Prentice Hall.

Poulin, M. L., K. M. Patrie, M. J. Botelho, R. A. Tassava, and I. M. Chiu. 1993. Heterogeneity in the expression of fibroblast growth factor receptors during limb regeneration in newts (*Notophtalmus viridescens*). *Development* 119:353–360.

Poynton, J. C. 1964. The amphibia of Southern Africa: A faunal study. *Annals of the Natal Museum* 17:1–334.

Poynton, J. C., and S. Pritchard. 1976. Notes on the biology of *Breviceps* (Anura: Microhylidae). *Zoologica Africana* 11:313–318.

Prader, A., J. M. Tanner, and G. A. von Harnack. 1963. Catch-up growth following illness or starvation. *Journal of Pediatrics* 62:646–659.

Prasad, G. V. R., and M. Godinot. 1994. Eutherian tarsal bones from the Late Cretaceous of India. *Journal of Paleontology* 68:892–902.

Prentice, A. M., S. E. Moore, A. C. Collinson, and M. A. O'Connell. 2002. Leptin and undernutrion. *Nutrition Reviews* 60:S56–S67.

Presch, W. 1975. The evolution of limb reduction in the teiid lizard genus *Bachia*. *Bulletin of the Southern California Academy of Sciences* 74:113–121.

Presch, W. 1980. Evolutionary history of the South American microteiid lizards (Teiidae: Gymnophthalminae). *Copeia* 1980:36–56.

Pridmore, P. A. 1985. Terrestrial locomotion in monotremes (Mammalia: Monotremata). *Journal of Zoology, London A* 205:53–73.

Prochel, J., and M. R. Sánchez-Villagra. 2003. Carpal homology in *Monodelphis domestica* and *Caluromys philander* (Marsupialia). *Zoology* 106:73–84.

Proctor, J. B. 1928. On the remarkable gecko *Palmatogecko rangei* Andersson. *Proceedings of the Zoological Society of London* 1928:917–922.

Proctor, N. S., and P. J. Lynch. 1993. *Manual of ornithology: Avian structure and function*. New Haven, CT: Yale University Press.

Propper, C. R., R. E. Jones, M. S. Rand, and H. Austin. 1991. Nesting behavior of the lizard *Anolis carolinensis*. *Journal of Herpetology* 25:484–486.

Prum, R. O. 1999. Development and evolutionary origin of feathers. *Journal of Experimental Zoology (Molecular Development and Evolution)* 285:291–306.

Prum, R. O., and A. H. Brush. 2002. The evolutionary origin and diversification of feathers. *Quarterly Reviews of Biology* 77:261–295.

Pucek, Z. 1965. A case of polydactyly in Apodemus flavicollis (Melchior, 1834). *Acta Theriologica* 10:232–233.

Púgener, L. A., and A. M. Maglia. 1997. Osteology and skeletal development of *Discoglossus sardus* (Anura: Discoglossidae). *Journal of Morphology* 233:267–286.

Putta, S., J. J. Smith, J. Walker, M. Rondet, D. W. Weisrock, J. Monaghan, A. K. Samuels, et al. 2004. From biomedicine to natural history research: Expressed sequence tag resources for ambystomatid salamanders. *BMC Genomics* 5:54–70.

Puttick, G. M., and J. U. M. Jarvis. 1977. The functional anatomy of the neck and forelimbs of the Cape golden mole, *Chrysochloris asiatica* (Lipotyphla: Chrysochloridae). *Zoologica Africana* 12:445–458.

Q

Quint, E., A. Smith, F. Avaron, L. Laforest, J. Miles, W. Gaffield, and M. A. Akimenko. 2002. Bone patterning is altered in the regenerating zebrafish caudal fin after ectopic expression of *sonic hedgehog* and *bmp2b* or exposure to cyclopamine. *Proceedings of the National Academy of Sciences, USA* 99:8713–8718.

R

Raath, M. A. 1985. The theropod *Syntarsus* and its bearing on the origin of birds. In *The beginnings of birds,* ed. M. K. Hecht, J. H. Ostrom, G. Viohl, and P. Wellnhofer, 219–227. Eichstätt: Freunde des Jura-Museums.

Rackoff, J. S. 1980. The origin of the tetrapod limb and the ancestry of tetrapods. In *The terrestrial environment and the origin of land vertebrates,* ed. A. L. Panchen, 255–292. New York: Academic Press.

Radinsky, L. B. 1987. *The evolution of vertebrate design.* Chicago: University of Chicago Press.

Raff, R. A. 1996. *The shape of life.* Chicago: University of Chicago Press.

Rage, J.-C., and F. Escuillié. 2000. Un nouveau serpent bipède du Cénomanien (Crétacé). Implications phylétiques. *Comptes Rendus de l'Academie des Sciences, Paris, Sciences de la Terre et des Planètes* 330:513–520.

Rageh, M. A. E., L. Mendenhall, E. E. A. Moussad, S. E. Abbey, A. L. Mescher, and R. A. Tassava. 2002. Vasculature in preblastema and nerve-dependent blastema stages of regenerating forelimbs of the adult newt, *Notophthalmus viridescens*. *Journal of Experimental Zoology* 292:255–266.

Rahmani, T. M. Z. 1974. Le développment et la régression des bourgeons de membres antérieurs chez l'Ophisaure (*Ophisaurus apodus,* Pallas). *Annals of Embryology and Morphology* 7:159–170.

Raikow, R. J. 1985. Locomotor system. In *Form and function in birds,* ed. A. S. King and J. McLelland, 57–147. London: Academic Press.

Raikow, R. J., L. Bicanovsky, and A. H. Bledsoe. 1988. Forelimb joint mobility and the evolution of wing-propelled diving in birds. *Auk* 105:446–451.

Rainger, R. 1991. *An agenda for antiquity: Henry Fairfield Osborn and vertebrate paleontology at the American Museum of Natural History, 1890–1935.* Tuscaloosa: University of Alabama Press.

Rajendran, M. V. 1985. *Studies in Uropeltid snakes.* Madurai, India: Publications Division, Madurai Kamaraj University.

Rallis, C., B. G. Bruneau, J. Del Buono, C. E. Seidman, J. G. Seidman, S. Nissim, C. J. Tabin, and M. P. Logan. 2003. Tbx5 is required for forelimb bud formation and continued outgrowth. *Development* 130:2741–2751.

Rallis, C., J. Del Buono, and M. P. Logan. 2005. Tbx3 can alter limb position along the rostrocaudal axis of the developing embryo. *Development* 132:1961–1970.

Ralphs, J. R., M. Benjamin, and A. Thornett. 1991. Cell and matrix biology of the suprapatella in the rat: A structural and immunocytological study of fibrocartilage in a tendon subjected to compression. *Anatomical Record* 231:167–177.

Ralphs, J. R., R. N. S. Tyers, and M. Benjamin. 1992. Development of functionally distinct fibrocartilages at two sites in the quadriceps tendon of the rat: The suprapatella and the attachment to the patella. *Anatomy and Embryology* 185:181–187.

Rama, S., and G. Chandrakasan. 1984. Distribution of different molecular species of collagen in the vertebral cartilage of shark (*Carcharius acutus*). *Connective Tissue Research* 12:111–118.

Ramamurti, R., W. C. Sandberg, R. Lohner, J. A. Walker, and M. W. Westneat. 2002. Fluid dynamics of flapping aquatic flight in the bird wrasse: Three-dimensional unsteady computations with fin deformation. *Journal of Experimental Biology* 205:2997–3008.

Rancourt, D. E., T. Tsuzuki, and M. R. Capecchi. 1995. Genetic interaction between hoxb-5 and hoxb-6 is revealed by nonallelic noncomplementation. *Genes and Development* 9:108–122.

Rand, A. S. 1968. A nesting aggregation of iguanas. *Copeia* 1968:552–561.

Ranke, M. B. 2002. Catch-up growth: New lessons for the clinician. *Journal of Pediatric Endocrinology and Metabolism* 15:1257–1266.

Raven, H. C. 1950. *The anatomy of the gorilla.* New York: Columbia University Press.

Rawls, J. F., and S. L. Johnson. 2000. Zebrafish kit mutation reveals primary and secondary regulation of melanocyte development during fin stripe regeneration. *Development* 127:3715–3724.

Rawls, J. F., and S. L. Johnson. 2001. Requirements for the kit receptor tyrosine kinase during regeneration of zebrafish fin melanocytes. *Development* 128:1943–1949.

Rawls, J. F., E. M. Mellgren, and S. L. Johnson. 2001. How the zebrafish gets its stripes. *Developmental Biology* 240:301–314.

Raynaud, A. 1962. Étude histologique de la structure des ébauches des membres de l'embryon d'Orvet (*Anguis fragilis*, L.) au cours de leur développement et de leur régression. *Comptes Rendus des Séances de l'Academie des Sciences* 254:4505–4507.

Raynaud, A. 1963. La formation et la régression des ébauches des membres de l'embryon d'Overt (*Anguis fragilis*, L.): Observations effectuées sur les ébauches des membres postérieurs. *Bulletin de la Societe Zoologique de France* 88:299–324.

Raynaud, A. 1972. Étude embryologique de la formation des appendices postérieurs et de la ceinture pelvienne chez le Python réticulé (*Python reticulatus*). *Mémoires du Muséum National d'Histoire Naturelle, Paris,* series A, 76:1–31.

Raynaud, A. 1985. Development of limbs and embryonic limb reduction. In *Biology of the reptilia,* vol. 15, ed. C. Gans and F. Billett, 59–148. New York: Wiley.

Raynaud, A. 1986. Modifications précoces de l'ontogenèse des membres d'embryons de *Lacerta viridis* (Laur.) sous l'effect de la cytosine-arabinofuranoside: Comparaison avec lo'ontogenèse des membres de Reptiles serpentiformes. *Comptes Rendus des Séances de l'Academie des Sciences* 303:37–42.

Raynaud, A. 1987. Modalités ontogénétiques de la réduction digitale dans la patte des embryons de Seps tridactyle (*Chalcides chalcides,* L.). *Comptes Rendus de l'Académie des Sciences,* series 3, *Sciences de la Vie* 304:359–362.

Raynaud, A. 1989. Réduction expérimentale des membres chez les embryons de *Lacerta viridis* et formation des membres rudimentaires chez les Reptiles serpentiformes. *Geobios Memoire Spécial* 12:323–335.

Raynaud, A. 1990. Developmental mechanism involved in the embryonic reduction of limbs in reptiles. *International Journal of Developmental Biology* 34:233–243.

Raynaud, A. 1991. Modifications de la structure des mains et des pieds des embryons de lézard vert (*Lacerta viridis* Laur.) sous l'effet de la cytosine-arabinofuranoside. *Annales des Sciences Naturelles, Zoologie* 12:11–38.

Raynaud, A., and M. Clergue-Gazeau. 1986. Identification des doigts réduits ou manquants des les pattes des embryons de Lézard vert (Lacerta viridis) traités par la cystosine-arabinofuranoside: Comparison avec les réductions digitales naturelles des especes de reptiles serpentiformes. *Archives of Biology* (Brussels) 97:279–299.

Raynaud, A., J.-P. Gasc, and S. Renous-Lecuru. 1975. Les rudiments de membres et leur développment embryonnaire chez *Scelotes inornatus* (A. Smith), Scincidae, Sauria. *Bulletin du Muséum National d'Histoire Naturelle, Paris,* series 3, *Zoologie* 208:537–551.

Raynaud, A., and P. Kan. 1988. Données autoradiographiques, obtenues avec la trymidine tritiée, sur la réduction expérimentale et évolutive des membres, chez les Reptiles. *Comptes Rendus de l'Académie des Sciences,* series 3, *Sciences de la Vie* 303:37–42.

Rayner, J. M. V. 1979. A new approach to animal flight mechanics. *Journal of Experimental Biology* 80:17–54.

Rayner, J. M. V. 1981. Flight adaptations in vertebrates. *Symposium of the Zoological Society of London* 48:137–172.

Rayner, J. M. V. 1987. The mechanics of flapping flight in bats. In *Recent advances in the study of bats,* ed. M. B. Fenton, P. A. Racey, and J. M. V. Rayner, 23–42. Cambridge: Cambridge University Press.

Rayner, J. M. V. 1988. The evolution of vertebrate flight. *Biological Journal of the Linnean Society* 34:269–287.

Rayner, J. M. V. 2001a. Mathematical modelling of the avian flight power curve. *Mathematical Methods in the Applied Sciences* 24:1485–1514.

Rayner, J. M. V. 2001b. On the origin and evolution of flapping flight aerodynamics in birds. In *New perspectives on the origin and early evolution of birds: Proceedings of the International Symposium in Honor of John H. Ostrom,* ed. J. A. Gauthier and L. F. Gall, 363–385. New Haven, CT: Peabody Museum of Natural History, Yale University.

Rayner, J. M. V., and H. D. J. N. Aldridge. 1985. Three-dimensional reconstruction of animal flight paths and the turning flight of microchiropteran bats. *Journal of Experimental Biology* 118:247–265.

Rayner, J. M. V., and G. J. Dyke. 2003. Origins and evolution of diversity in the avian wing. In *Vertebrate biomechanics and evolution,* ed. V. L. Bels, J.-P. Gasc, and A. Casinos, 297–317. Oxford: BIOS Scientific.

Redford, K. H., and J. F. Eisenberg. 1992. *Mammals of the neotropics: The southern cone.* Vol. 2. Chicago: University of Chicago Press.

Reed, C. A. 1951. Locomotion and appendicular anatomy in three soricoid insectivores. *American Midland Naturalist* 45:513–671.

Reed, C. A. 1958. Observations on the burrowing rodent *Spalax* in Iraq. *Journal of Mammalogy* 39:386–389.

Reese, S., U. R. Pfuderer, K. Loeffler, and K.-D. Budras. 2001. Topography, structure and function of the patella and patelloid in marsupials. *Anatomia, Histologia, Embryologia* 30:289–294.

Reginelli, A. D., Y. Q. Wang, D. Sassoon, and K. Muneoka. 1995. Digit tip regeneration correlates with regions of Msx1 (Hox7) expression in fetal and newborn mice. *Development* 121:1065–1076.

Regnard, M. P. 1893. Dynamomètre permettant de mesurer la puissance musculaire de l'appareil caudal du Poisson. *Comptes Rendus Hebdomadaires des Séances et Mémoires de la Société de Biologie* 5:80–81.

Reichman, O. J., and S. C. Smith. 1990. Burrows and burrowing behavior by mammals. In *Current mammalogy*, vol. 2, ed. H. H. Genoways, 197–244. Lincoln: University of Nebraska State Museum.

Reid, R. E. H. 1984. The histology of dinosaurian bone, and its possible bearing on dinosaurian physiology. *Symposium Zoological Society of London* 52:629–663.

Reilly, S. M. 1998. Sprawling locomotion in the lizard *Sceloporus clarkii*: Speed modulation of motor patterns in a walking trot. *Brain Behavior and Evolution* 52:126–138.

Reilly, S. M., and J. A. Elias. 1998. Locomotion in *Alligator mississippiensis*: Kinematic effects of speed and posture and their relevance to the sprawling-to-erect paradigm. *Journal of Experimental Biology* 201:2559–2574.

Reimchen, T. E., and N. F. Temple. 2004. Hydrodynamic and phylogenetic aspects of the adipose fin in fishes. *Canadian Journal of Zoology* 82:910–916.

Reisz, R. R. 1972. Pelycosaurian reptiles from the Middle Pennsylvanian of North America. *Bulletin of the Museum of Comparative Zoology* 144:27–61.

Reisz, R. R. 1997. The origin and early evolutionary history of amniotes. *Trends in Ecology and Evolution* 12:218–222.

Reisz, R. R., D. S. Berman, and D. Scott. 1984. The anatomy and relationships of the Lower Permian reptile *Araeoscelis*. *Journal of Vertebrate Paleontology* 4:57–67.

Reiter, I., M. Tzukerman, and G. Maor. 2002. Spontaneous differentiating primary chondrocytic tissue culture: A model for endochondral ossification. *Bone* 31:333–339.

Renesto, S., and A. Tintori. 1995. Functional morphology and mode of life of the Late Triassic placodont *Psephoderma alpinum* Meyer from the Calcare di Zorzino (Lombardy, N. Italy). *Rivista Italiana di Paleontologia e Stratigrafia* 101:37–48.

Renous, S., and V. Bels. 1991. Étude cinématique de la palette natatoire antérieure de la tortue Luth, *Dermochelys coriacea* (Vandelli, 1761), au cours de sa locomotion terrestre. *Canadian Journal of Zoology* 69:495–503.

Renous, S., and V. Bels. 1993. Comparison between aquatic and terrestrial locomotions of the leatherback sea turtle (*Dermochelys coriacea*). *Journal of Zoology, London* 230:357–378.

Renous, S., V. Bels, and J. Davenport. 2000. Locomotion in marine Chelonia: Adaptation to the aquatic habitat. *Historical Biology* 14:1–13.

Renous, S., J. M. Exbrayat, and J. Estabel. 1997. Recherche d'indices de membres chez les Amphibiens Gymnophiones. *Annales des Sciences Naturelles, Zoologie, Paris*, series 13, 18:11–26.

Renous, S., and J.-P. Gasc. 1979. Etude des modalités de réduction des membres chez un Squamate serpentiforme: *Scelotes*, scincidé afro-malgache (1). *Annales des Sciences Naturelles, Zoologie, Paris*, series 13, 1:99–132.

Renous, S., and J.-P. Gasc. 1986. Le fouissage des Gymnophiones (Amphibia): Hypothèse morphofonctionelle fondée sur la comparaison avec d'autres Vertébrés apodes. *Zoologische Jahrbücher. Abteilung für Anatomie und Ontogenie der Tiere* 114:95–130.

Renous, S., J.-P. Gasc, and I. Ineich. 1993. Données préliminaires sur les capacités locomotrices des Amphibiens Gymnophiones. *Annales des Sciences Naturelles, Zoologie, Paris*, series 13, 14:59–79.

Renous, S., J.-P. Gasc, and A. Raynaud. 1991. Comments on the pelvic appendicular vestiges in an amphisbaenian: *Blanus cinereus* (Reptilia, Squamata). *Journal of Morphology* 209:23–38.

Renous, S., J. Lescure, J.-P. Gasc, and V. Bels. 1989. Intervention des membres dans la locomotion et le creusement du nid chez la tortue luth (*Dermochelys coriacea*) (Vandelli, 1961). *Amphibia-Reptilia* 10:355–369.

Renous, S., A. Raynaud, J.-P. Gasc, and C. Pieau. 1976. Caractères rudimentaires, anatomiques et embryologiques, de la ceinture pelvienne et des appendices postérieurs du Python réticulé (*Python reticulatus* Schneider, 1801). *Bulletin du Muséum National d'Histoire Naturelle, Paris*, series 3, Zoologie 267:547–583.

Renous-Lécuru, S. 1974. La musculature du membre antérieur de *Bipes canaliculatus* (Amphisbénidés, Reptiles). *Bulletin du Muséum National d'Histoire Naturelle, Paris*, series 3, Zoologie 172:1261–1282.

Reynolds, S., N. Holder, and M. Fernandes. 1983. The form and structure of supernumerary hindlimbs formed following skin grafting and nerve deviation in the newt *Triturus cristatus*. *Journal of Embryology and Experimental Morphology* 77:221–241.

Reznick, D., C. Ghalambor, and L. Nunney. 2002. The evolution of senescence in fish. *Mechanisms of Ageing and Development* 123:773–789.

Richards, W. J., R. V. Miller, and E. D. Houde. 1974. Egg and larval development of the Atlantic thread herring, *Opisthonema oglinum*. *Fishery Bulletin* 72:1123–1136.

Richardson, K. C., G. J. W. Webb, and S. C. Manolis. 2002. *Crocodiles: Inside out; A guide to the crocodilians and their functional morphology*. Chipping Norton, Australia: Surrey Beatty and Sons.

Richardson, M. K., and A. D. Chipman. Developmental constraints in a comparative framework: A test case using variations in phalanx number during amniote evolution. *Journal of Experimental Zoology (Molecular and Developmental Evolution)* 296B:8–22.

Richardson, M. K., and H. H. A. Oeschläger. 2002. Time, pattern and heterochrony: A study of hyperphalangy in the dolphin embryo flipper. *Evolutionary Development* 4:435–444.

Riddle, R. D., M. Ensini, C. Nelson, T. Tsuchida, T. M. Jessell, and C. Tabin. 1995. Induction of the LIM homeobox gene *Lmx1* by WNT7a establishes a dorsoventral pattern in the vertebrate limb. *Cell* 83:631–640.

Riddle, R. D., R. L. Johnson, E. Laufer, and C. Tabin. 1993. Sonic hedgehog mediates the polarizing activity of the ZPA. *Cell* 75:1401–1416.

Ridley, M. 1996. *Evolution*. 2nd ed. Malden, MA: Blackwell Science.

Riedl, R. 1978. *Order in living organisms*. New York: Wiley.

Rieppel, O. 1989. The hindlimb of *Macrocnemus bassanii* (Nopcsa) (Reptilia, Diapsida): Development and functional anatomy. *Journal of Vertebrate Paleontology* 9:373–387.

Rieppel, O. 1992a. Studies on skeleton formation in reptiles. I: The postembryonic development of the skeleton in *Cyrtodactylus pubisulcus* (Reptilia: Gekkonidae). *Journal of Zoology, London* 227:87–100.

Rieppel, O. 1992b. Studies on skeleton formation in reptiles. III: Patterns of ossification in the skeleton of *Lacerta vivipara* Jacquin. (Reptilia, Squamata). *Fieldiana: Zoology* 68:1–25.

Rieppel, O. 1993a. Studies on skeleton formation in reptiles. II: *Chamaeleo hoehnelii* (Squamata: Chamaeleoninae), with comments on the homology of carpal and tarsal bones. *Herpetologica* 49:66–78.

Rieppel, O. 1993b. Studies on skeletal formation in reptiles. V: Patterns of ossification in the skeleton of *Alligator mississippiensis* DAUDIN (Reptilia, Crocodylia). *Zoological Journal of the Linnean Society* 109:301–325.

Rieppel, O. 1993c. Studies on skeleton formation in reptiles: Patterns of ossification in the skeleton of *Chelydra serpentina* (Reptilia, Testudines). *Journal of Zoology, London* 231:487–509.

Rieppel, O. 1994. Osteology of *Simosaurus gaillardoti* and the relationships of stem-group Sauropterygia. *Fieldiana: Geology* 28:1–85.

Rieppel, O. 1995. The genus *Placodus*: Systematics, morphology, paleobiogeography, and paleobiology. *Fieldiana: Geology* 31:1–44.

Rieppel, O. 1997. Introduction to part II: Sauropterygia. In *Ancient marine reptiles*, ed. J. M. Callaway and E. L. Nicholls, 107–119. San Diego, CA: Academic Press.

Rieppel, O. 2003. Semaphoront, cladograms, and the roots of total evidence. *Biological Journal of the Linnean Society* 80:167–186.

Rieppel, O., and R. R. Reisz. 1999. The origin and early evolution of turtles. *Annual Review of Ecology and Systematics* 30:1–22.

Rieppel, O., H. Zaher, E. Tchernov, and M. J. Polcyn. 2003. The anatomy and relationships of *Haasiophis terrasanctus*, a fossil snake with well-developed hind limbs from the mid-Cretaceous of the Middle East. *Journal of Paleontology* 77:536–558.

Riess, J. 1986. Fortbewegunsweise, Schwimmbiophysik und Phylogenie der Ichthyosaurier. *Palaeontographica* A192:93–155.

Riess, J., and E. Frey. 1991. The evolution of underwater flight and the locomotion of plesiosaurs. In *Biomechanics in evolution*, Society for Experimental Biology Seminar series 36, ed. J. M. V. Rayner and R. J. Wootton, 131–144.

Roach, H. I., and N. M. P. Clarke. 2000. Physiological cell death of chondrocytes in vivo is not confined to apoptosis. *Journal of Bone and Joint Surgery* 82B:601–613.

Roach, H. I., J. Erenpreisa, and T. Aigner. 1995. Osteogenic differentiation of hypertrophic chondrocytes involves asymmetric cell divisions and apoptosis. *Journal of Cell Biology* 95:483–494.

Roach, H. I., G. Mehta, R. O. C. Oreffo, N. M. P. Clarke, and C. Cooper. 2003. Temporal analysis of rat growth plates: Cessation of growth with age despite presence of a physis. *Journal of Histochemistry and Cytochemistry* 51:373–383.

Robb, J. 1977. *The Tuatara*. Durham, UK: Meadowfield Press.

Roberts, T. R. 2002. The integrated function of muscles and tendons during locomotion. *Comparative Biochemistry and Physiology A (Molecular and Interactive Physiology)* 133:1087–1099.

Robertson, K. E., C. Tickle, and S. M. Darling. 1997. Shh, Fgf4, and Hoxd gene expression in the mouse limb mutant hypodactyly. *International Journal of Developmental Biology* 41:733–736.

Robinson, J. A. 1975. The locomotion of plesiosaurs. *Neues Jahrbuch für Geologie und Paläontologie, Abhandlungen* 149:286–332.

Robinson, J. A. 1977. Intracorporal force transmission in plesiosaurs. *Neues Jahrbuch für Geologie und Paläontologie, Abhandlungen* 153:86–128.

Robledo, R. F., L. Rajan, X. Li, and T. Lufkin. 2002. The Dlx5 and Dlx6 homeobox genes are essential for craniofacial, axial, and appendicular skeletal development. *Genes and Development* 16:1089–1101.

Roček, Z. 2000. Mesozoic anurans. In *Amphibian biology*, vol. 4, *Palaeontology: The evolutionary history of amphibians*, ed. H. Heatwole and R. L. Carroll, 1295–1331. Chipping Norton, Australia: Surrey Beatty and Sons.

Roček, Z., and J.-C. Rage. 2000. Proanuran stages (*Triadobatrachus, Czatkobatrachus*). In *Amphibian biology*, vol. 4, *Palaeontology: The evolutionary history of amphibians*, ed. H. Heatwole and R. L. Carroll, 1283–1294. Chipping Norton, Australia: Surrey Beatty and Sons.

Rodriguez-Esteban, C., J. W. Schwabe, J. De La Pena, B. Foys, B. Eshelman, and J. C. Belmonte. 1997. Radical fringe positions the apical ectodermal ridge at the dorsoventral boundary of the vertebrate limb. *Nature* 386:360–366.

Rodriguez-Esteban, C., J. W. Schwabe, J. De La Pena, D. E. Rincon-Limas, J. Magallon, J. Botas, and J. C. Belmonte. 1998. Lhx2, a vertebrate homologue of apterous, regulates vertebrate limb outgrowth. *Development* 125:3925–3934.

Rodriguez-Leon, J., R. Merino, D. Macias, Y. Gañan, E. Santesteban, and J. M. Hurlé. 1999. Retinoic acid regulates programmed cell death through BMP signalling. *Nature: Cell Biology* 1:125–126.

Roelink, H., and R. Nusse. 1991. Expression of two members of the Wnt family during mouse development: Restricted temporal and spatial patterns in the developing neural tube. *Genes and Development* 5:381–388.

Roff, D. A. 1997. *Evolutionary quantitative genetics*. New York: Chapman and Hall.

Rolfe, W. D. I., E. N. K. Clarkson, and A. L. Panchen, eds. 1994. Volcanism and early terrestrial biotas. *Transactions of the Royal Society of Edinburgh, Earth Sciences* 84:175–464.

Romankowowa, A. 1961. The sesamoid bones of the autopodia of bats. *Acta Theriologica* 10:125–140.

Romanoff, A. L. 1960. *The avian embryo: Structural and functional development*. New York: Macmillan.

Romer, A. S. 1922. The locomotor apparatus of certain primitive and mammal-like reptiles. *Bulletin of the American Museum of Natural History* 46:517–606.

Romer, A. S. 1923. Crocodilian pelvic muscles and their avian and reptilian homologues. *Bulletin of the American Museum of Natural History* 48:533–552.

Romer, A. S. 1933. *Vertebrate paleontology*. Chicago: University of Chicago Press.

Romer, A. S. 1956. *Osteology of the reptiles*. Chicago: University of Chicago Press.

Romer, A. S. 1957. The appendicular skeleton of the Permian embolomerous amphibian *Archeria*. *Contribution of the Museum of Geology, University of Michigan* 13:103–159.

Romer, A. S. 1966. *Vertebrate paleontology*. Chicago: University of Chicago Press.

Romer, A. S. 1974. Aquatic adaptations in reptiles: Primary or secondary? *Annals of the South African Museum* 64:221–230.

Romer, A. S., and T. S. Parsons. 1977. *The vertebrate body*. 5th ed. Philadelphia: W. B. Saunders.

Rondelet, G. 1554. *Liber de piscibus marinis, in quibus verae piscium effigies expressae sunt.* Lyons, France: Bonhomme.

Rooney, P. 1994. Intratendinous ossifications. In *Bone,* vol. 8, *Mechanisms of bone development and growth,* ed. B. K. Hall, 47–83. Boca Raton, FL: CRC Press.

Rooney, P., C. Archer, and L. Wolpert. 1984. Morphogenesis of cartilaginous long bone rudiments. In *The role of the extracellular matrix in development,* ed. R. L. Trelstad, 305–322. New York: Wiley-Liss.

Ros, M. A., R. D. Dahn, M. Fernandez-Teran, K. Rashka, N. C. Caruccio, S. M. Hasso, J. J. Bitgood, J. J. Lancman, and J. F. Fallon. 2003. The chick oligozeugodactyly (ozd) mutant lacks sonic hedgehog function in the limb. *Development* 130:527–537.

Ros, M. A., M. Sefton, and M. A. Nieto. 1997. Slug, a zinc finger gene previously implicated in the early patterning of the mesoderm and the neural crest, is also involved in chick limb development. *Development* 124:1821–1829.

Rosa-Molinar, E., S. E. Hendricks, J. F. Rodriguez-Sierra, and B. Fritzsch. 1994. Development of the anal fin appendicular support in the western mosquito fish, *Gambusia affinis affinis* (Baird and Girard, 1854): A reinvestigation and reinterpretation. *Acta Anatomica* 151:20–35.

Rosen, D. E., P. L. Forey, B. G. Gardiner, and C. Patterson. 1981. Lungfishes, tetrapods, paleontology, and plesiomorphy. *Bulletin of the American Museum of Natural History* 167:159–276.

Rosen, J. 1983. *A symmetry primer for scientists.* New York: Wiley.

Rosenberger, L. J. 2001. Pectoral fin locomotion in batoid fishes: Undulation versus oscillation. *Journal of Experimental Biology* 204:379–394.

Rosenberger, L. J., and M. W. Westneat. 1999. Functional morphology of undulatory pectoral fin locomotion in the stingray *Taeniura lymma* (Chondrichthyes: Dasyatidae). *Journal of Experimental Biology* 202:3523–3539.

Ross, C. A., ed. 1989. *Crocodiles and alligators.* New York: Facts on File.

Rosse, C., and P. Gaddum-Rosse. 1997. *Hollinshead's textbook of anatomy.* 5th ed. Philadelphia: Lippincott-Raven.

Rountree, R. B., M. Schoor, H. Chen, M. E. Marks, V. Harley, Y. Mishina, and D. M. Kingsley. 2004. BMP receptor signaling is required for postnatal maintenance of articular cartilage. *PLoS Biology* 2:1815–1827.

Rousseau, F., J. Bonaventure, L. Legeai-Mallet, A. Pelet, J. M. Rozet, P. Maroteaux, M. Le Merrer, and A. Munnich. 1994. Mutations in the gene encoding fibroblast growth factor receptor-3 in achondroplasia. *Nature* 371:252–254.

Rowe, T. B. 1988. Definition, diagnosis, and the origin of Mammalia. *Journal of Vertebrate Paleontology* 8:241–164.

Rowe, T. B. 1993. Phylogenetic systematics and the early history of mammals. In *Mammal phylogeny: Mesozoic differentiation, multituberculates, monotremes, early therians, and marsupials,* ed. F. S. Szalay, M. J. Novacek, and M. C. McKenna, 129–145. New York: Springer-Verlag.

Rowe, T. B., and J. A. Gauthier. 1990. Ceratosauria. In *The Dinosauria,* ed. D. B. Weishampel, P. Dodson, and H. Osmólska, 151–168. Berkeley: University of California Press.

Roy, S., D. M. Gardiner, and S. V. Bryant. 2000. Vaccinia as a tool for functional analysis in regenerating limbs: Ectopic expression of *Shh. Developmental Biology* 218:199–205.

Rubin, L., and J. W. Saunders Jr. 1972. Ectodermal-mesodermal interactions in the growth of buds in the chick embryo: Constancy and temporal limits of the ectodermal induction. *Developmental Biology* 28:94–112.

Rufai, A., M. Benjamin, and J. R. Ralphs. 1992. Development and ageing of phenotypically distinct fibrocartilages associated with the rat Achilles tendon. *Anatomy and Embryology* 186:611–618.

Ruff, C. B. 2000. Body size, body shape, and long bone strength in modern humans. *Journal of Human Evolution* 38:269–290.

Ruff, C. B., and J. A. Runestad. 1992. Primate limb bone structural adaptations. *Annual Review of Anthropology* 21:407–433.

Ruibal, R., L. Tevis Jr., and V. Roig. 1969. The terrestrial ecology of the spadefoot toad *Scaphiopus hammondii. Copeia* 1969:571–584.

Russell, A. P. 2002. Integrative functional morphology of the gekkotan adhesive system (Reptilia: Gekkota). *Integrative and Comparative Biology* 42:1154–1163.

Russell, A. P., and A. M. Bauer. 1988. Paraphalangeal elements of gekkonid lizards: A comparative survey. *Journal of Morphology* 197:221–240.

Russell, A. P., and A. M. Bauer. 1990. Substrate excavation in the Namibian web-footed gecko, *Palmatogecko rangei* Andersson 1908, and its ecological significance. *Tropical Zoology* 3:197–207.

Russell, D. A. 1967. Systematics and morphology of American mosasaurs (Reptilia, Sauria). *Peabody Museum of Natural History, Yale University, Bulletin* 23:1–237.

Ruta M., M. I. Coates, and D. L. J. Quicke. 2003. Early tetrapod relationships revisited. *Biological Reviews* 78:251–345.

Ruta, M., A. R. Milner, and M. I. Coates. 2002. The tetrapod *Caerorhachis bairdi* Holmes and Carroll from the Lower Carboniferous of Scotland. *Transactions of the Royal Society of Edinburgh, Earth Sciences* 92:229–261.

Ruther, U., D. Komitowski, F. R. Schubert, and E. F. Wagner. 1989. c-fos expression induces bone tumors in transgenic mice. *Oncogene* 4:861–865.

Rutledge, J. J., E. J. Eisen, and J. E. Legates. 1974. Correlated response in skeletal traits and replicate variation in selected lines of mice. *Theoretical and Applied Genetics* 45:26–31.

Ruvinsky, I., L. M. Silver, and J. J. Gibson-Brown. 2000. Phylogenetic analysis of T-Box genes demonstrates the importance of amphioxus for understanding evolution of the vertebrate genome. *Genetics* 156:1249–1257.

Ryan, A. K., B. Blumberg, C. Rodriguez-Esteban, S. Yonei-Tamura, K. Tamura, T. Tsukui, J. de la Pena, et al. 1998. Pitx2 determines left-right asymmetry of internal organs in vertebrates. *Nature* 394:545–551.

Ryan, M. C., and L. J. Sandell. 1990. Differential expression of a cysteine-rich domain in the amino-terminal propeptide of type II (cartilage) procollagen by alternative splicing of mRNA. *Journal of Biological Chemistry* 265:10334–10339.

Ryu, J. H., S. J. Kim, S. H. Kim, C. D. Oh, S. G. Hwang, C. H. Chun, S. H. Oh, J. K. Seong, T. L. Huh, and J. S. Chun. 2002. Regulation of the chondrocyte phenotype by beta-catenin. *Development* 129:5541–5550.

S

Sagai, T., M. Hosoya, Y. Mizushina, M. Tamura, and T. Shiroishi. 2005. Elimination of a long-range cis-regulatory module causes complete loss of limb-specific Shh expression and truncation of the mouse limb. *Development* 132:797–803.

Sagai, T., H. Masuya, M. Tamura, K. Shimizu, Y. Yada, S. Wakana, Y. Gondo, T. Noda, and T. Shiroishi. 2004. Phylogenetic conservation of a limb-specific, cis-acting regulator of Sonic hedgehog (*Shh*). *Mammalian Genome* 15:23–34.

Sahni, M., D.C. Ambrosetti, A. Mansukhani, R. Gertner, D. Levy, and C. Basilico. 1999. FGF signaling inhibits chondrocyte proliferation and regulates bone development through the STAT-1 pathway. *Genes and Development* 13:1361–1366.

Sahni, M., R. Raz, J. D. Coffin, D. Levy, and C. Basilico. 2001. STAT1 mediates the increased apoptosis and reduced chondrocyte proliferation in mice overexpressing FGF2. *Development* 128:2119–21129.

Salas-Vidal, E., C. Valencia, and L. Covarrubias. 2001. Differential tissue growth and patterns of cell death in mouse limb autopod morphogenesis. *Developmental Dynamics* 220:295–306.

Sánchez-Villagra, M. R. 2002. Comparative patterns of postcranial ontogeny in therian mammals: An analysis of relative timing of ossification events. *Journal of Experimental Zoology (Molecular and Developmental Evolution)* 294:264–273.

Sánchez-Villagra, M. R., and W. Maier. 2002. Ontogenetic data and the evolutionary origin of the mammalian scapula. *Naturwissenschaften* 89:459–461.

Sanchiz, B. 1998. Salentia. In *Encyclopedia of paleoherpetology*, vol. 4, ed. P. Wellnhofer, xii–275. Munich: Pfeil.

Sander, P. M. 2000. Ichthyosauria: Their diversity, distribution and phylogeny. *Paläontologische Zeitschrift* 74:1–35.

Sanders, J., and M. Davies. 1984. Burrowing behaviour and associated hindlimb myology in some Australian hylid and leptodactylid frogs. *Australian Zoologist* 21:123–142.

Santamaría, J. A., and J. Becerra. 1991. Tail fin regeneration in teleosts: Cell-extracellular matrix interaction in blastemal differentiation. *Journal of Anatomy* 176:9–21.

Santamaría, J. A., M. Marí-Beffa, and J. Becerra. 1992. Interactions of the lepidotrichial matrix components during tail fin regeneration in teleosts. *Differentiation* 49:143–150.

Santamaría, J. A., M. Marí-Beffa, L. Santos-Ruiz, and J. Becerra. 1996. Incorporation of bromodeoxyuridine in regenerating fin tissue of the goldfish *Carassius auratus*. *Journal of Experimental Zoology* 275:300–307.

Santos-Ruiz, L., J. A. Santamaría, J. Ruiz-Sanchez, and J. Becerra. 2002. Cell proliferation during blastema formation in the regenerating teleost fin. *Developmental Dynamics* 223:262–272.

Sanz, J. L., L. M. Chiappe, and A. D. Buscalioni. 1995. The osteology of *Concornis lacustris* (Aves: Enanthiornithes) from the Lower Cretaceous of Spain and a reexamination of its phylogenetic relationships. *American Museum Novitates* 3133:1–13.

Sanz-Ezquerro, J. J., and C. Tickle. 2000. Autoregulation of Shh expression and Shh induction of cell death suggest a mechanism for modulating polarising activity during chick limb development. *Development* 127:4811–4823.

Sanz-Ezquerro, J. J., and C. Tickle. 2003a. Digital development and morphogenesis. *Journal of Anatomy* 202:51–58.

Sanz-Ezquerro, J. J., and C. Tickle. 2003b. Fgf signalling controls the number of phalanges and tip formation in developing digits. *Current Biology* 13:1830–1836.

Sarin, V. K., and D. R. Carter. 2000. Mechanobiology and joint conformity regulate endochondral ossification of sesamoids. *Journal of Orthopaedic Research* 18:706–712.

Sarin, V. K., G. M. Erickson, N. J. Giori, A. B. Bergman, and D. R. Carter. 1999. Coincident development of sesamoid bones and clues to their evolution. *Anatomical Record (New Anatomist)* 257:174–180.

Sasayama, Y. 1999. Hormonal control of Ca homeostasis in lower vertebrates: Considering the evolution. *Zoological Science* 16:857–869.

Sato, A., Y. Matsumoto, U. Koide, Y. Kataoka, N. Yoshida, T. Yokota, M. Asashima, and R. Nishinakamura. 2003. Zinc finger protein sall2 is not essential for embryonic and kidney development. *Molecular Cell Biology* 23:62–69.

Satokata, I., L. Ma, H. Ohshima, M. Bei, I. Woo, K. Nishizawa, T. Maeda, et al. 2000. Msx2 deficiency in mice causes pleiotropic defects in bone growth and ectodermal organ formation. *Nature: Genetics* 24:391–395.

Saunders, J. W. 1948. The proximo-distal sequence of the origin of the parts of the chick wing and the role of the ectoderm. *Journal of Experimental Zoology* 108:363–404.

Saunders, J. W. 1977. The experimental analysis of chick limb bud development. In *Vertebrate limb and somite morphogenesis*, ed. D. A. Ede, J. R. Hinchliffe, and M. J. Balls, 1–24. Cambridge: Cambridge University Press.

Saunders, J. W., and J. F. Fallon. 1967. Cell death in morphogenesis. In *Major problems in developmental biology*, ed. M. Locke, 289–314. New York: Academic Press.

Saunders, J. W., and M. T. Gasseling. 1968. Ectodermal-mesenchymal interactions in the origin of limb symmetry. In *Epithelial-mesenchymal interactions*, ed. R. Fleischmeyer and R. E. Billingham, 78–97. Baltimore: Williams and Wilkins.

Savage, M. P., and J. F. Fallon. 1995. FGF-2 mRNA and its antisense message are expressed in a developmentally specific manner in the chick limb bud and mesonephros. *Developmental Dynamics* 202:343–53.

Savage, R. M. 1942. The burrowing and emergence of the spadefoot toad, *Pelobates fuscus fuscus* Wagler. *Proceedings of the Zoological Society of London A* 112:21–35.

Sawin, H. J. 1945. Amphibians from the Dockum Triassic of Howard County, Texas. University of Texas Publication no. 4401:361–399.

Scaal, M., F. Proels, E.-M. Fuchtbauer, K. Patel, C. Hornik, T. Koehler, B. Christ, and B. Brand-Saberi. 2002. BMPs induce dermal markers and ectopic feather tracts. *Mechanisms of Development* 110:41–60.

Scanlon, J. D., M. S. Y. Lee, M. W. Caldwell, and R. Shine. 1999. The palaeoecology of the primitive snake *Pachyrhachis*. *Historical Biology* 13:127–152.

Scapinelli, R., and K. Little. 1970. Observations on the mechanically induced differentiation of cartilage from fibrous connective tissue. *Journal of Pathology* 101:85–91.

Schaefer, S. L., K. A. Johnson, R. T. O'Brien. 1995. Compensatory tibial overgrowth following healing of closed femoral fractures in young dogs. *Veterinary and Comparative Orthopaedics and Traumatology* 8:159–162.

Schaeffer, B. 1941. The morphological and functional evolution of the tarsus in amphibians and reptiles. *Bulletin of the American Museum of Natural History* 78:395–472.

Schaeffer, B. 1965. The evolution of concepts related to the origin of the Amphibia. *Systematic Zoology* 14:115–118.

Schaeffer, B. 1977. The dermal skeleton in fishes. In *Problems in vertebrate evolution*, ed. S. M. Andrews, R. S. Miles, and A. D. Walker, 25–52. London: Academic Press.

Schaller, S. A., and K. Muneoka. 2001. Inhibition of polarizing activity in the anterior limb bud is regulated by extracellular factors. *Developmental Biology* 240:443–456.

Schew, W. A., and R. E. Ricklefs. 1998. Developmental plasticity. In *Avian growth and development: Evolution within the Altricial-Precocial spectrum*, ed. J. M. Starck and R. E. Ricklefs, 288–304. Oxford: Oxford University Press.

Schipani, E., B. Lanske, J. Hunzelman, A. Luz, C. S. Kovacs, K. Lee, A. Pirro, H. M. Kronenberg, and H. Juppner. 1997. Targeted expression of constitutively active receptors for parathyroid hormone and parathyroid hormone-related peptide delays endochondral bone formation and rescues mice that lack parathyroid hormone-related peptide. *Proceedings of the National Academy of Sciences, USA* 94:13689–13694.

Schmalhausen, I. 1916. On the functions of the fins of the fish. *Revue Zoologique Russe* (Moscow) 1:185–214.

Schmalhausen, I. I. 1968. *The origin of terrestrial vertebrates.* Transl. Leo Kelso. New York: Academic Press.

Schmidt, K. P., and R. F. Inger. 1957. *Living reptiles of the world.* Garden City, NY: Doubleday.

Schmidt-Nielsen, K. 1984. *Scaling: Why is animal size so important?* Cambridge: Cambridge University Press.

Schnell, G. D., and R. K. Selander. 1981. Environmental and morphological correlates of genetic variation in mammals. In *Mammalian population genetics*, ed. M. H. Smith and J. Joule. Athens: University of Georgia Press.

Schoch, R. R. 1992. Comparative ontogeny of Early Permian branchiosaurid amphibians from southwestern Germany. *Palaeontographica Abteilung A* 222:43–83.

Schoch, R. R., and R. L. Carroll. 2003. Ontogenetic evidence for the Paleozoic ancestry of salamanders. *Evolution and Development* 5:314–324.

Schrank, A. J., and P. W. Webb. 1998. Do body and fin form affect the abilities of fish to stabilize swimming during maneuvers through vertical and horizontal tubes? *Environmental Biology of Fishes* 53:365–371.

Schreiweis, D. O. 1982. A comparative study of the appendicular musculature of penguins (Aves: Sphenisciformes). *Smithsonian Contribution to Zoology* 341:1–46.

Schufeldt, R. W. 1882. Concerning some of the forms assumed by the patella in birds. *Proceedings of the United States National Museum* 3:324–331.

Schultz, A. H. 1936. Characters common to higher primates and characters specific for man (continued). *Quarterly Review of Biology* 11:425–455.

Schultze, H.-P. 1984. Juvenile specimens of *Eusthenopteron foordi* Whiteaves, 1881 (Osteolepiform Rhipidistian, Pisces) from the Late Devonian of Miguasha, Québec, Canada. *Journal of Vertebrate Paleontology* 4:1–16.

Schultze, H.-P. 1993. Patterns of diversity in the skulls of jawed fish. In *The skull*, vol. 2, *Patterns of structural and systematic diversity*, ed. J. Hanken and B. K. Hall, 189–254. Chicago: University of Chicago Press.

Schultze, H.-P., and C. R. Marshall. 1993. Contrasting the use of functional complexes and isolated characters in lungfish evolution. In *Palaeontological studies in honour of Ken Campbell*, ed. P. A. Jell, 211–224. Brisbane: Memoir 15 of the Association of Australasian Palaeontologists.

Schutt, W. A., Jr., J. S. Altenbach, Y.-H. Chang, D. M. Cullinane, J. W. Hermanson, F. Muradali, and J. E. A. Bertram. 1997. The dynamics of flight-initiating jumps in the common vampire bat *Desmodus rotundus. Journal of Experimental Biology* 200:3003–3012.

Schutt, W. A., Jr., and N. B. Simmons. 1998. Morphology and homology of the chiropteran calcar, with comments of the phylogenetic relationships of *Archaeopteropus. Journal of Mammalian Evolution* 5:1–32.

Schwartz, M. W., S. C. Woods, D. Porte, R. J. Seeley, and D. G. Baskin. 2000. Central nervous system control of food intake. *Nature* 404:661–671.

Schweitzer, R., K. J. Vogan, and C. J. Tabin. 2000. Similar expression and regulation of Gli2 and Gli3 in the chick limb bud. *Mechanisms of Development* 98:171–174.

Scott, W. B. 1940. Artiodactyla (the mammalian fauna of the White River Oligocene, part 4). *Transactions of the American Philosopical Society* 28:363–746.

Sedgwick, A. 1850. *A discourse on the studies of the University of Cambridge.* Cambridge.

Segev, O., I. Chumakov, Z. Nevo, D. Givol, L. Madar-Shapiro, Y. Sheinin, M. Weinreb, and A. Yayon. 2000. Restrained chondrocyte proliferation and maturation with abnormal growth plate vascularization and ossification in human FGFR-3(G380R) transgenic mice. *Human Molecular Genetics* 9:249–258.

Sekine, K., H. Ohuchi, F. Fujiwara, M. Yamasaki, T. Yoshizawa, T. Sato, N. Yagishita, et al. 1999. Fgf10 is essential for limb and lung formation. *Nature Genetics* 21:138–141.

Selever, J., W. Liu, M.-F. Lu, R. R. Behringer, and J. F. Martin. 2004. Bmp4 in limb bud mesoderm regulates digit pattern by controlling AER development. *Developmental Biology* 276:268–279.

Semlitsch, R. D. 1983. Burrowing ability and behavior of salamanders of the genus *Ambystoma. Canadian Journal of Zoology* 61:616–620.

Semon, R. 1893–1915. *Zoologische Forschungsreisen in Australien und dem Malayischen Archipel.* 5 vols. Jena, Germany: Gustav Fischer.

Semon, R. 1899. *In the Australian bush and on the coast of the Coral Sea: Being the experiences and observations of a naturalist in Australia, New Guinea, and the Molucas.* London: Macmillan.

Sereno, P. C. 1991. Basal archosaurs: Phylogenetic relationships and functional implications. *Journal of Vertebrate Paleontology* 11 (suppl.): 1–53.

Sereno, P. C. 1997. The origin and evolution of dinosaurs. *Annual Review of Earth and Planetary Sciences* 25:435–489.

Sereno, P. C., and A. B. Arcucci. 1993. Dinosaurian precursors from the Middle Triassic of Argentina: *Lagerpeton chanarensis. Journal of Vertebrate Paleontology* 13:385–399.

Sereno, P. C., and A. B. Arcucci. 1994. Dinosaurian precursors from the Middle Triassic of Argentina: *Marasuchus lilloensis,* gen. nov. *Journal of Vertebrate Paleontology* 14:53–73.

Sereno, P. C., and M. C. McKenna. 1995. Cretaceous multituberculate skeleton and the early evolution of the mammalian shoulder girdle. *Nature* 377:144–147.

Sessions, S. K., and S. V. Bryant. 1988. Evidence that regenerative ability is an intrinsic property of limb cells in *Xenopus. Journal of Experimental Zoology* 247:39–44.

Settle, S. H., Jr., R. B. Rountree, A. Sinha, A. Thacker, K. Higgins, and D. M. Kingsley. 2003. Multiple joint and skeletal patterning defects caused by single and double mutations in the mouse Gdf6 and Gdf5 genes. *Developmental Biology* 254:116–130.

Sewertzoff, A. N. 1931. Studien uber die Reduktion der Organe der Wirbeltiere. *Zoologische Jahrbücher. Abteilung für Anatomie und Ontogenie der Tiere* 53:611–700.

Seymour, R. S. 1982. Physiological adaptations to aquatic life. In *Biology of the reptilia,* vol. 13, ed. C. Gans and F. H. Pough, 1–51. New York: Academic Press.

Shadwick, R. E., H. S. Rapoport, and J. M. Fenger. 2002. Structure and function of tuna tail tendons. *Comparative Biochemistry and Physiology A (Molecular and Interactive Physiology)* 133:1109–1125.

Shaffer, H. B., J. M. Clark, and F. Kraus. 1991. When molecules and morphology clash: A phylogenetic analysis of the North American ambystomatid salamanders (Caudata: Ambystomatidae). *Systematic Zoology* 40:284–303.

Shapiro, M. D. 2002. Developmental morphology of limb reduction in *Hemiergis* (Squamata: Scincidae): Chondrogenesis, osteogenesis, and heterochrony. *Journal of Morphology* 254:211–231.

Shapiro, M. D., and T. F. Carl. 2001. Novel features of tetrapod limb development in two nontraditional model species: A skink and a direct-developing frog. In *Beyond heterochrony: The evolution of development,* ed. M. L. Zelditch, 337–361. New York: Wiley.

Shapiro, M. D., J. Hanken, and N. Rosenthal. 2003. Developmental basis of evolutionary digit loss in the Australian lizard *Hemiergis. Journal of Experimental Zoology (Molecular and Developmental Evolution)* 297B:48–56.

Shapiro, M. D., M. E. Marks, C. L. Peichel, B. K. Blackman, K. Nereng, B. Jónsson, D. Schluter, and D. M. Kingsley. 2004. Genetic and developmental basis of evolutionary pelvic reduction in threespine sticklebacks. *Nature* 428:717–723.

Shaw, C. E. 1960. Notes on the eggs, incubation and young of *Chamaeleo basiliscus. British Journal of Herpetology* 2:182–185.

Shea, B. T. 1993. Bone growth and primate evolution. In *Bone,* vol. 8, *Mechanisms of bone development and growth,* ed. B. K. Hall, 133–157. Boca Raton, FL: CRC Press.

Shea, K. G., S. S. Coleman, and D. A. Coleman. 1997. Growth of the proximal fibular physis and remodeling of the epiphysis after microvascular transfer. *Journal of Bone and Joint Surgery* 79A:583–586.

Sheil, C. A. 1999. Osteology and skeletal development of *Pyxicephalus adspersus* (Anura: Ranidae: Raninae). *Journal of Morphology* 240:49–75.

Sheil, C. A. 2003. Osteology and skeletal development of *Apalone spinifera* (Reptilia: Testudines: Trinychidae). *Journal of Morphology* 256:42–78.

Shiang, R., L. M. Thompson, Y. Z. Zhu, D. M. Church, T. J. Fielder, M. Bocian, S. T. Winokur, and J. J. Wasmuth. 1994. Mutations in the transmembrane domain of FGFR3 cause the most common genetic form of dwarfism, achondroplasia. *Cell* 78:335–342.

Shimer, H. W. 1903. Adaptation to aquatic, arboreal, fossorial, and cursorial habits in mammals. III: Fossorial adaptations. *American Naturalist* 37:819–825.

Shively, M. J. 1978. First metacarpal bone or proximal phalanx? *Journal of the American Veterinary Radiological Society* 19:40–52.

Shoshani, J., and M. C. McKenna. 1998. Higher taxonomic relationships among extant mammals based on morphology, with selected comparisons of results from molecular data. *Molecular Phylogenetics and Evolution* 9:572–584.

Shubin, N. H. 1995. The evolution of paired fins and the origin of tetrapod limbs. *Evolutionary Biology* 28:39–86.

Shubin, N. H. 2002. Origin of evolutionary novelty: Examples from limbs. *Journal of Morphology* 252:15–28.

Shubin, N. H., and P. Alberch. 1986. A morphogenetic approach to the origin and basic organization of the tetrapod limb. *Evolutionary Biology* 20:319–387.

Shubin, N. H., C. Tabin, and S. Carroll. 1997. Fossils, genes and the evolution of animal limbs. *Nature* 388:639–648.

Shubin, N. H., and D. B. Wake. 2003. Morphological variation, development, and evolution of the limb skeleton of salamanders. In *Amphibian biology,* vol. 5, *Osteology,* ed. H. Heatwole and M. Davies, 1782–1808. Chipping Norton, Australia: Surrey Beatty and Sons.

Shubin, N. H., D. B. Wake, and A. J. Crawford. 1995. Morphological variation in the limbs of *Taricha granulosa* (Caudata: Salamandridae): Evolutionary and phylogenetic implications. *Evolution* 49:874–884.

Sibley, C. G., and J. E. Ahlquist. 1990. *Phylogeny and classification of birds: A study in molecular evolution.* New Haven, CT: Yale University Press.

Sibonga, J. D., M. Zhang, G. L. Evans, K. C. Westerland, J. M. Cavolina, E. Morey-Holton, and R. T. Turner. 2000. Effects of spaceflight and simulated weightleness on longitudinal bone growth. *Bone* 27:535–540.

Siebler, T., H. Robson, S. M. Shalet, and G. R. Williams. 2001. Glucocorticoids, thyroid ormone and growth hormone interactions: Implications for the growth plate. *Hormone Resserch* 56 (suppl. 1): 7–12.

Simeone, A., D. Acampora, M. Pannese, M. D'Esposito, A. Stornaiuolo, M. Gulisano, A. Mallamaci, et al. 1994. Cloning and characterization of two members of the vertebrate Dlx gene family. *Proceedings of the National Academy of Sciences, USA* 91:2250–2254.

Simmons, N. B., and J. H. Geisler. 1998. Phylogenetic relationships of *Icaronycteris, Hassianycteris,* and *Palaeochiropteryx* to extant bat lineages, with comments on the evolution of echolocation and foraging strategies in Microchiroptera. *Bulletin of the American Museum of Natural History* 235:1–182.

Simon, H. G., C. Nelson, D. Goff, E. Laufer, B. A. Morgan, and C. J. Tabin. 1995. Differential expression of myogenic regulatory genes and Msx-1 during dedifferentiation and redifferentiation of regenerating amphibian limbs. *Developmental Dynamics* 202:1–12.

Simonet, W. S., D. L. Lacey, C. R. Dunstan, M. Kelley, M. S. Chang, R. Lüthy, H. Nguyen, et al. 1997. Osteoprotegerin: A novel secreted protein involved in the regulation of bone density. *Cell* 89:309–319.

Simpson, G. G. 1945. The principles of classification and a classification of mammals. *Bulletin of the American Museum of Natural History* 85:1–350.

Simpson, G. G. 1975. Fossil penguins. In *The biology of penguins,* ed. B. Stonehouse, 19–41. London: Macmillan.

Singer, M. 1952. The influence of the nerve in regeneration of the amphibian extremity. *Quarterly Reviews of Biology* 27:169–200.

Singer, M. 1978. On the nature of the neurotrophic phenomenon in urodele limb regeneration. *American Zoologist* 18:829–841.

Singer, M., and L. Craven. 1948. The growth and morphogenesis of the regenerating forelimb of adult *Triturus* following denervation at various stages of development. *Journal of Experimental Zoology* 108:279–308.

Sire, J. Y., and A. Huysseune. 2003. Formation of dermal skeletal and dental tissues in fish: A comparative and evolutionary approach. *Biological Reviews* 78:219–249.

Sironen, R. K., H. M. Karjalainen, M. A. Elo, K. Kaarniranta, K. Törrönen, M. Takigawa, H. Helminen, and M. Lammi. 2002. cDNA array reveals mechanosensitive genes in chondrocytic cells under hydrostatic pressure. *Biochimica et Biophysica Acta* 1591:45–54.

Sisson, S., and J. D. Grossman. 1938. *The anatomy of the domestic animals.* Philadelphia: W. B. Saunders.

Sivakumar, P., and G. Chandrakasan. 1998. Occurrence of a novel collagen with three distinct chains in the cranial cartilage of the squid *Sepia officinalis:* Comparison with shark cartilage collagen. *Biochimica et Biophysica Acta—General Subjects* 1381:161–169.

Skoczen, S. 1958. Tunnel digging by the mole (*Talpa europaea* Linne). *Acta Theriologica* 2:235–249.

Sledge, C. B. 1966. Some morphological and experimental aspects of limb development. *Clinical Orthopaedics* 44:241–264.

Slijper, E. J. 1946. *Comparative biologic-anatomical investigations on the vertebral column and spinal musculature of mammals.* Amsterdam: North-Holland.

Slonaker, J. R. 1920. Some morphological changes for adaptation in the mole. *Journal of Morphology* 34:335–373.

Smalian, C. 1885. Beiträge zur Anatomie der Amphisbaeniden. *Zeitschrift für Wissenschaftliche Zoologie* 42:126–202.

Small, K. M., and S. S. Potter. 1993. Homeotic transformations and limb defects in Hox A11 mutant mice. *Genes and Development* 7:2318–2328.

Smink, J. J. 2003. Glucocorticoid-induced effects on the growth plate and the IGF system. PhD diss., Utrecht University.

Smith, B. J., S. D. Holladay, and S. A. Smith. 1995. Patella of selected bats: Patterns of occurrence or absence and associated modifications of the quadriceps femoris tendon. *Anatomical Record* 242:575–580.

Smith, E. N. 1979. Behavioral and physiological thermoregulation of crocodilians. *American Zoologist* 19:239–247.

Smith, M. M., and B. K. Hall. 1990. Development and evolutionary origins of vertebrate skeletogenic and odontogenic tissues. *Biological Reviews* 65:277–373.

Smith, M. M., A. Hickman, D. Amanze, A. Lumsden, and P. Thorogood. 1994. Trunk neural crest origin of caudal fin mesenchyme in the zebrafish *Brachydanio rerio. Proceedings of the Royal Society of London B* 256:137–145.

Smith-Vaniz, W. F., L. S. Kaufman, and J. Glowacki. 1995. Species-specific patterns of hyperostosis in marine teleost fishes. *Marine Biology* 121:573–580.

Smithson, T. R. 1982. The cranial morphology of *Greererpeton burkemorani* Romer (Amphibia: Temnospondyli). *Zoological Journal of the Linnean Society* 76:29–90.

Smithson, T. R. 1985. The morphology and relationships of the Carboniferous amphibian *Eoherpeton watsoni* Panchen. *Zoological Journal of the Linnean Society* 85:317–410.

Smithson, T. R. 1989. The earliest known reptile. *Nature* 314:676–678.

Smithson, T. R. 1994. *Eldeceeon rolfei,* a new reptiliomorph from the Viséan of East Kirkton, West Lothian, Scotland. *Transactions of the Royal Society of Edinburgh, Earth Sciences* 84:377–382.

Smithson, T. R. 2000. Anthracosaurs. In *Amphibian biology,* vol. 4, *Palaeontology: The evolutionary history of amphibians,* ed. H. Heatwole and R. L. Carroll, 1053–1063. Chipping Norton, Australia: Surrey Beatty and Sons.

Smithson, T. R., R. L. Carroll, A. L. Panchen, and S. M. Andrews. 1994. *Westlothiana lizziae* from the Viséan of East Kirkton, West Lothian, Scotland and the amniote stem. *Transactions of the Royal Society of Edinburgh, Earth Sciences* 84:383–412.

Smits, P., P. Li, J. Mandel, Z. Zhang, J. M. Deng, R. R. Behringer, B. de Croumbrugghe, and V. Lefebvre. 2001. The transcription factors L-Sox5 and Sox6 are essential for cartilage formation. *Developmental Cell* 1:277–290.

Smuts, M. M. S., and A. J. Bezuidenhout. 1993. Osteology of the thoracic limb of the African elephant (*Loxodonta Africana*). *Onderstepoort Journal of Veterinary Research* 60:1–14.

Sokol, O. M. 1962. Feeding in the pipid frog *Hymenochirus boettgeri* (Tornier). *Herpetologica* 25:9–24.

Solloway, M. J., and E. J. Robertson. 1999. Early embryonic lethality in Bmp5: Bmp7 double mutant mice suggests functional redundancy within the 60A subgroup. *Development* 126:1753–68.

Song, K., Y. Wang, and D. Sassoon. 1992. Expression of Hox-7.1 in myoblasts inhibits terminal differentiation and induces cell transformation. *Nature* 360:477–481.

Sordino, P., and D. Duboule. 1996. A molecular approach to the evolution of vertebrate paired appendages. *Trends in Ecology and Evolution* 11:114–119.

Sordino, P., F. v. d. Hoeven, and D. Duboule. 1995. Hox gene expression in teleost fins and the origin of vertebrate digits. *Nature* 375:678–681.

Soriano, P., C. Montgomery, R. Geske, and A. Bradley. 1991. Targeted disruption of the c-src proto-oncogene leads to osteopetrosis in mice. *Cell* 64:693–702.

Sower, S. A., K. I. Reed, and K. J. Babbitt. 2000. Limb malformations and abnormal sex hormone concentrations in frogs. *Environmental Health Perspectives* 108:1085–1090.

Spedding, G. R., A. Hedenstrom, and M. Rosen. 2003. Quantitative studies of the wakes of freely flying birds in a low-turbulence wind tunnel. *Experiments in Fluids* 34:291–303.

Spedding, G. R., M. Rosén, and A. Hedenström. 2003. A family of vortex wakes generated by a thrush nightingale in free flight in a wind tunnel over its entire range of natural flight speeds. *Journal of Experimental Biology* 206:2313–2344.

Špinar, Z. V. 1952. Revision of some Moravian Discosauriscidae (Labyrinthodontia) [in Czech]. *Rozpravy Ústředního ústavu geologického* 15:1–129.

Spitz, F., F. Gonzalez, and D. Duboule. 2003. A global control region defines a chromosomal regulatory landscape containing the HoxD cluster. *Cell* 113:405–417.

Springer, M. S., E. C. Teeling, O. Madsen, M. J. Stanhope, and W. W. de Jong. 2001. Integrated fossil and molecular data reconstruct bat echolocation. *Proceedings of the National Academy of Sciences, USA* 98:6241–6246.

Stamhuis, E. J., and J. J. Videler. 1995. Quantitative flow analysis around aquatic animals using laser sheet particle image velocimetry. *Journal of Experimental Biology* 198:283–294.

Stanhope, M. J., V. G. Waddell, O. Madsen, W. W. de Jong, S. B. Hedges, G. C. Cleven, D. J. Kao, and M. S. Springer. 1998.

Molecular evidence for multiple origins of Insectivora and for a new order of endemic African insectivore mammals. *Proceedings of the National Academy of Sciences, USA* 95:9967–9972.

Starck, D. 1979. *Vergleichende Anatomie der Wirbeltiere auf evolutionsbiologischer Grundlage. Band. 2: Das Skeletsystem: Allgemeines, Skeletsubstanzen, Skelet der Wirbeltiere einschließlich Lokomotionstypen.* Berlin: Springer Verlag.

Starck, J. M. 1996. Comparative morphology and cytokinetics of skeletal growth in hatchlings of altricial and precocial birds. *Zoologischer Anzeiger* 235:43–75.

Starck, J. M. 1998. Structural variants and invariants in avian embryonic and postnatal development. In *Avian growth and development: Evolution within the altricial-precocial spectrum,* ed. J. M. Starck and R. E. Ricklefs, 59–88. Oxford: Oxford University Press.

Stearns, S. C. 1986. Natural selection and fitness, adaptation and constraint. In *Patterns and processes in the history of life,* ed. D. M. Raup and D. Jablonski, 23–44. Berlin: Springer Verlag.

Stearns, S. C. 1992. *The evolution of life histories.* Oxford: Oxford University Press.

Stebbins, R. C. 1943. Adaptations in the nasal passages for sand burrowing in the saurian genus *Uma. American Naturalist* 77:38–52.

Stebbins, R. C. 1944. Some aspects of the ecology of the iguanid genus *Uma. Ecological Monographs* 14:311–332.

Stein, B. R. 1981. Comparative limb mycology of two opossums, *Didelphis* and *Chironectes. Journal of Morphology* 169:113–140.

Stein, B. R. 2000. Morphology of subterranean rodents. In *Life underground: The biology of subterranean rodents,* ed. E. A. Lacey, J. L. Patton, and G. N. Cameron, 19–61. Chicago: University of Chicago Press.

Steiner, H. 1934. Über die embryonale Hand- und Fuss-Skelettanlage bei den Crocodiliern, sowie über ihre Beziehung zur Vogel-Flügelanlage und zur ursprünglichen Tetrapoden-Extremität. *Revue Suisse de Zoologie* 41:383–396.

Stéphan, P. 1900. Recherches histologiques sur la structure du tissu osseux des poissons. *Bulletin des Sciences de France et de la Belgique* 33:281–429.

Stephens, T. D., R. L. Beier, D.C. Bringhurst, S. R. Hiatt, M. Prestridge, D. E. Pugmire, and H. J. Willis. 1989. Limbness in the early chick embryo lateral plate. *Developmental Biology* 133:1–7.

Stephens, T. D., and T. R. McNulty. 1981. Evidence for a metameric pattern in the chick humerus. *Journal of Embryology and Experimental Morphology* 61:191–205.

Stephenson, N. G. 1960. The comparative osteology of Australian geckos and its bearing on their morphological states. *Journal of the Linnean Society* 44:278–299.

Stephenson, N. G. 1962. The comparative morphology of the head skeleton, girdles and hind limbs in the Pygopodidae. *Journal of the Linnean Society* 44:627–644.

Sterelny, K. 2000. Development, evolution and adaptation. *Philosophy of Science* 67:S369–S387.

Stevens, D. G., M. I. Boyer, and V. A. Bowen. 1999. Transplantation of epiphyseal plate allographs between animals of different ages. *Journal of Pediatric Orthopedics* 19:398–403.

Steyn, W. 1963. *Angolocaurus skoogi* (Andersson): A new record from South West Africa. *Cimbebasia* 6:8–11.

Steyn, W., and W. D. Haacke. 1966. A new webfooted gekko (*Kaokogecko vanzyli* gen. et sp. nov.) from the north-western South West Africa. *Cimbebasia* 18:1–23.

Stiassny, M. L. J. 2000. Skeletal system. In *The laboratory fish,* part 3, *Gross functional anatomy,* ed. G. K. Ostrander, 109–118. San Diego: Academic Press.

Stickel, L. F. 1950. Populations and home range relationships of the box turtle, *Terrapene c. carolina* (Linnaeus). *Ecological Monographs* 20:351–378.

Stickel, W. H., and L. F. Stickel. 1946. Sexual dimorphism in the pelvic spurs of *Enygrus. Copeia* 1946:10–12.

St-Jacques, B., M. Hammerschmidt, and A. P. McMahon. 1999. Indian hedgehog signaling regulates proliferation and differentiation of chondrocytes and is essential for bone formation. *Genes and Development* 13:2072–2086.

Stobutzki, I. C., and D. R. Bellwood. 1994. An analysis of the sustained swimming abilities of pre- and post-settlement coral reef fishes. *Journal of Experimental Marine Biology and Ecology* 175:275–286.

Stocum, D. L. 1984. The urodele limb regeneration blastema: Determination and organization of the morphogenetic field. *Differentiation* 27:13–28.

Stocum, D. L. 1991. Limb regeneration: A call to arms (and legs). *Cell* 67:4–8.

Stocum, D. L. 1996. A conceptual framework for analyzing axial patterning in regenerating urodele limbs. *International Journal of Developmental Biology* 40:773–783.

Stocum, D. L., and G. E. Dearlove. 1972. Epidermal-mesodermal interaction during morphogenesis of the limb regeneration blastemal in larval salamanders. *Journal of Experimental Zoology* 181:49–62.

Stokely, P. S. 1947a. Limblessness and correlated changes in the girdles of a comparative morphological series of lizards. *American Midland Naturalist* 38:725–754.

Stokely, P. S. 1947b. The post-cranial skeleton of *Aprasia repens. Copeia* 1947:22–28.

Stokes, I. A., P. L. Mente, J. C. Iatridis, C. E. Farnum, and D. D. Aronsson. 2002. Enlargement of growth plate chondrocytes modulated by sustained mechanical loading. *Journal of Bone and Joint Surgery* 84A:1842–1848.

Stopper, G. F., and G. P. Wagner. 2005. Of chicken wings and frog legs: A smorgasbord of evolutionary variation in mechanisms of tetrapod limb development. *Developmental Biology* 288:21–39.

Storer, R. W. 1960. Evolution in the diving birds. In *Proceedings of the 12th International Ornithological Congress, Helsinki, 1958,* 694–707.

Storm, E. E., T. V. Huynh, N. G. Copeland, N. A. Jenkins, D. M. Kingsley, and S.-J. Lee. 1994. Limb alterations in brachypodism mice due to mutations in a new member of the TGFβ superfamily. *Nature* 368:639–643.

Storm, E. E., and D. M. Kingsley. 1996. Joint patterning defects caused by single and double mutations in members of the bone morphogenetic protein (BMP) family. *Development* 122:3969–3979.

Storm, E. E., and D. M. Kingsley. 1999. GDF5 coordinates bone and joint formation during digit development. *Developmental Biology* 209:11–27.

Storrs, G. W. 1993. Function and phylogeny in sauropterygian (Diapsida) evolution. *American Journal of Science* 293A:63–90.

Stratford, T. H., C. Horton, and M. Maden. 1996. Retinoic acid is required for the initiation of outgrowth in the chick limb bud. *Current Biology* 6:1124–1133.

Stratford, T. H., K. Kostakopoulou, and M. Maden. 1997. Hoxb-8 has a role in establishing early anterior-posterior polarity in chick forelimb but not hindlimb. *Development* 124:4225–4234.

Strickler, T. L. 1978. *Functional osteology and myology of the shoulder in Chiroptera.* Basel, Switzerland: S. Karger.

Sues, H.-D. 1987. Postcranial skeleton of *Pistosaurus* and interrelationships of the Sauropterygia (Diapsida). *Zoological Journal of the Linnean Society* 90:109–131.

Sumida, S. S. 1989. The appendicular skeleton of the Early Permian genus *Labidosaurus* (Reptilia, Captorhinomorpha, Captorhinidae) and the hind limb musculature of captorhinid reptiles. *Journal of Vertebrate Paleontology* 9:295–313.

Sumida, S. S. 1997. Locomotor features of taxa spanning the origin of amniotes. In *Amniote origins: Completing the transition to land,* ed. S. S. Sumida and K. L. Martin, 353–398. San Diego, CA: Academic Press.

Sumida, S. S. 2003. Review of "Gaining Ground: The Origin and Evolution of Tetrapods." *Palaios* 18:193–194.

Sumida, S. S., D. S. Berman, and T. Martens. 1998. A trematopid amphibian from the Lower Permian of central Germany. *Palaeontology* 41:1–25.

Summerbell, D. 1974. A quantitative analysis of the effect of excision of the AER from the chick limb-bud. *Journal of Embryology and Experimental Morphology* 32:651–660.

Summerbell, D., J. H. Lewis, and L. Wolpert. 1973. Positional information in chick limb morphogenesis. *Nature* 244:492–496.

Summers, A. P., and T. J. Koob. 2002. The evolution of tendon: Morphology and material properties. *Comparative Biochemistry and Physiology Part A* 133:1159–1170.

Summers, A. P., and J. C. O'Reilly. 1997. A comparative study of locomotion in the caecilians *Dermophis mexicanus* and *Typhlonectes natans* (Amphibia: Gymnophiona). *Zoological Journal of the Linnean Society* 121:65–76.

Sun, X., F. V. Mariani, and G. R. Martin. 2002. Functions of FGF signalling from the apical ectodermal ridge in limb development. *Nature* 418:501–508.

Superti-Furga, A., L. Bonafé, and D. L. Rimoin. 2001. Molecular-pathogenetic classfication of genetic disorders of the skeleton. *American Journal of Medical Genetics* 106:282–293.

Suzuki, H. K. 1963. Studies on the osseous system of the slider turtle. *Annals of the New York Academy of Sciences* 109:351–410.

Suzuki, T., A. S. Srivastava, and T. Kurokawa. 2000. Experimental induction of jaw, gill and pectoral fin malformations in Japanese flounder, *Paralichthys olivaceus,* larvae. *Aquaculture* 185:175–187.

Suzuki, T., J. Takeuchi, K. Koshiba-Takeuchi, and T. Ogura. 2004. Tbx genes specify posterior digit identity through Shh and BMP signaling. *Developmental Cell* 6:43–53.

Swartz, S. M. 1997. Allometric patterning in the limb skeleton of bats: Implications for the mechanics and energetics of powered flight. *Journal of Morphology* 234:277–294.

Swartz, S. M. 1998. Skin and bones: Functional, architectural, and mechanical differentiation in the bat wing. In *Bat biology and conservation,* ed. T. H. Kunz and P. A. Racey, 109–126. Washington, DC: Smithsonian Institution Press.

Swartz, S. M., M. B. Bennett, and D. R. Carrier. 1992. Wing bone stresses in free flying bats and the evolution of skeletal design for flight. *Nature* 359:726–729.

Szalay, F. S. 1977. Phylogenetic relationships and a classification of the eutherian Mammalian. In *Major patterns in vertebrate evolution,* ed. M. K. Hecht, P. C. Goody, and B. M. Hecht, 315–374. New York: Plenum Press.

Szalay, F. S. 1981. Functional analysis and the practice of the phylogenetic method as reflected by some mammalian studies. *American Zoologist* 21:37–45.

Szalay, F. S. 1984. Arboreality: Is it homologous in metatherian and eutherian mammals? *Evolutionary Biology* 18:215–258.

Szalay, F. S. 1994. *Evolutionary history of the marsupials and an analysis of osteological characters.* Cambridge: Cambridge University Press.

T

Taber, F. W. 1945. Contribution to the life history and ecology of the nine-banded armadillo. *Journal of Mammalogy* 26:211–226.

Takada, S., K. L. Stark, M. J. Shea, G. Vassileva, J. A. McMahon, and A. P. McMahon. 1994. Wnt-3a regulates somite and tail-bud formation in the mouse embryo. *Genes and Development* 8:174–189.

Takahashi, I., G. H. Nuckolls, K. Takahashi, O. Tanaka, I. Semba, R. Dashner, L. Shum, and H. Slavkin. 1998. Compressive force promotes Sox9, type II collagen and aggrecan and inhibits IL-1β expression resulting in chondrogenesis in mouse embryonic limb bud mesenchymal cells. *Journal of Cell Science* 111:2067–2076.

Takahashi, M., K. Tamura, D. Buscher, H. Masuya, S. Yonei-Tamura, K. Matsumoto, M. Naitoh-Matsuo, et al. 1998. The role of Alx-4 in the establishment of anteroposterior polarity during vertebrate limb development. *Development* 125:4417–4425.

Takeda, K., O. Takeuchi, T. Tsujimura, S. Itami, O. Adachi, T. Kawai, H. Sanjo, K. Yoshikawa, N. Terada, and S. Akira. 1999. Limb and skin abnormalities in mice lacking IKKalpha. *Science* 284:313–316.

Takeda, S., J. P. Bonnamy, M. J., Owen, P. Ducy, and G. Karsenty. 2001. Continuous expression of Cbfa1 in nonhypertrophic chondrocytes uncovers its ability to induce hypertrophic chondrocyte differentiation and partially rescues Cbfa1-deficient mice. *Genes and Development* 15:467–481.

Takechi, M., and C. Itakura. 1995. Ultrastructural and histochemical studies of the epiphyseal plate in normal chicks. *Anatomical Record* 242:29–39.

Takenaka, A., A. Okada, K. Iwai, and Y. Ayaki. 2003. Separation of collagen from fish scales by treatment of dilute hydrochloric acid aqueous solution. *Journal of the Japanese Society for Food Science and Technology (Nippon Shokuhin Kagaku Kogaku Kaishi)* 50:67–71.

Takeuchi, J. K., K. Koshiba-Takeuchi, T. Suzuki, M. Kamimura, K. Ogura, and T. Ogura. 2003. Tbx5 and Tbx4 trigger limb initiation through activation of the Wnt/Fgf signaling cascade. *Development* 130:2729–2739.

Takeuchi, J. K., K. Koshiba-Takeuchi, K. Matsumoto, A. Vogel-Hopker, M. Naitoh-Matsuo, K. Ogura, N. Takahashi, K. Yasuda, and T. Ogura. 1999. Tbx5 and Tbx4 genes determine the wing/leg identity of limb buds. *Nature* 398:810–814.

Tamai, K., M. Semenov, Y. Kato, R. Spokony, C. Liu, Y. Katsuyama, F. Hess, J. P. Saint-Jeannet, and X. He. 2000. LDL-receptor-related proteins in Wnt signal transduction. *Nature* 407:530–535.

Tamura, K., R. Kuraishi, D. Saito, H. Masaki, H. Ide, and A. Yonei-Tamura. 2001. Evolutionary aspects of positioning and identification of vertebrate limbs. *Journal of Anatomy* 119:195–204.

Tamura, K., S. Yonei-Tamura, and J. C. Belmonte. 1999. Differential expression of Tbx4 and Tbx5 in zebrafish fin buds. *Mechanisms of Development* 87:181–184.

Tanaka, E. M., A. A. F. Gann, P. B. Gates, and J. P. Brockes. 1997. Newt myotubes reenter the cell cycle by phosphorylation of the retinoblastoma protein. *Journal of Cell Biology* 136:155–165.

Tanaka, M., M. J. Cohn, P. Ashby, M. Davey, P. Martin, and C. Tickle. 2000. Distribution of polarizing activity and potential for limb formation in mouse and chick embryos and possible relationships to polydactyly. *Development* 127:4011–4021.

Tanaka, M., L. A. Hale, A. Amores, Y. L. Yan, W. A. Cresko, T. Suzuki, and J. H. Postlethwait. 2005. Developmental genetic basis for the evolution of pelvic fin loss in the pufferfish *Takifugu rubripes*. *Developmental Biology* 281:227–239.

Tanaka, M., A. Münsterberg, W. G. Anderson, A. R. Prescott, N. Hazon, and C. Tickle. 2002. Fin development in a cartilaginous fish and the origin of vertebrate limbs. *Nature* 416:527–531.

Tanaka, M., K. Tamura, S. Noji, T. Nohno, and H. Ide. 1997. Induction of additional limb at the dorsal-ventral boundary of a chick embryo. *Developmental Biology* 182:191–203.

Tanaka, M., Y. Shigetani, S. Sugiyama, K. Tamura, H. Nakamura, and H. Ide. 1998. Apical ectodermal ridge induction by the transplantation of En-1-overexpressing ectoderm in chick limb bud. *Development Growth and Differentiation* 40:423–429.

Tang, M. K, A. K. Leung, W. H. Kwong, P. H. Chow, J. Y. Chan, V. Ngo-Muller, M. Li, and K. K. Lee. 2000. Bmp-4 requires the presence of the digits to initiate programmed cell death in limb interdigital tissues. *Developmental Biology* 218:89–98.

Tank, P. W., B. M. Carlson, and T. G. Connelly. 1976. A staging system for forelimb regeneration in the axolotl, *Ambystoma mexicanum*. *Journal of Morphology* 150:117–128.

Tank, P. W., and N. Holder. 1979. The distribution of cells in the upper forelimbs of the axolotl. *Journal of Experimental Zoology* 209:435–442.

Tare, R. S., R. O. C. Oreffo, K. Sato, H. Rauvala, N. M. P. Clarke, and H. I. Roach. 2002. Effects of targeted overexpression of pleiotrophin on postnatal bone development. *Biochemical and Biophysical Research Communications* 298:324–332.

Tarasoff, F. J., A. Bisaillon, J. Piérard, and A. P. Whitt. 1972. Locomotor patterns and external morphology of the river otter, sea otter, and harp seal (Mammalia). *Canadian Journal of Zoology* 20:915–929.

Tassava, R. A., and R. J. Goss. 1966. Regeneration rate and amputation level in fish fins and lizard tails. *Growth* 30:9–21.

Tavares, A. T., J. C. Izpisúa Belmonte, and J. Rodriguez-Leon. 2001. Developmental expression of chick twist and its regulation during limb patterning. *International Journal of Developmental Biology* 45:707–713.

Tavares, A. T., T. Tsukui, and J. C. Izpisúa Belmonte. 2000. Evidence that members of the Cut/Cux/CDP family may be involved in AER positioning and polarizing activity during chick limb development. *Development* 127:5133–5144.

Tavormina, P. L., R. Shiang, L. M. Thompson, Y. Z. Zhu, D. J. Wilkin, R. S. Lachman, W. R. Wilcox, D. L. Rimoin, D. H. Cohn, and J. J. Wasmuth. 1995. Thanatophoric dysplasia (types I and II) caused by distinct mutations in fibroblast growth factor receptor 3. *Nature: Genetics* 9:321–328.

Tawk, M., D. Tuil, Y. Torrente, S. Vriz, and D. Paulin. 2002. High-efficiency gene transfer into adult fish: A new tool to study fin regeneration. *Genesis* 32:27–31.

Taylor, C. R., N. C. Heglund, and G. M. O. Maloiy. 1982. Energetics and mechanics of terrestrial locomotion. I. Metabolic energy consuption as a function of speed and body size in birds and mammals. *Journal of Experimental Biology* 97:1–21.

Taylor, C. R., N. C. Heglund, T. A. McMahon, and T. R. Looney. 1980. Energetic cost of generating muscular force during running: A comparison of small and large mammals. *Journal of Experimental Biology* 86:9–18.

Taylor, B. K. 1978. The anatomy of the forelimb in the anteater (*Tamandua*) and its functional implications. *Journal of Morphology* 157:347–368.

Taylor, C. W. 1985. Calcium regulation in vertebrates: An overview. *Comparative Biochemistry and Physiology* 82A:249–255.

Taylor, E. H. 1968. *The Caecilians of the world: A taxonomic review*. Lawrence: University of Kansas Press.

Taylor, E. H. 1969. Skulls of Gymnophiona and their significance in the taxonomy of the group. *University of Kansas Science Bulletin* 48:585–687.

Taylor, G. P., R. Anderson, A. D. Reginelli, and K. Muneoka. 1994. FGF-2 induces regeneration of the chick limb bud. *Developmental Biology* 163:282–284.

Taylor, M. A. 1993. Stomach stones for feeding or buoyancy? The occurrence and function of gastroliths in marine tetrapods. *Philosophical Transactions of the Royal Society of London B* 341:163–175.

Taylor, M. A. 1994. Stone, bone or blubber? Buoyancy control strategies in aquatic tetrapods. In *Mechanics and physiology of animal swimming*, ed. L. Maddock, Q. Bone, and J. M. V. Rayner, 151–161, 205–229. Cambridge: Cambridge University Press.

Taylor, M. A. 2000. Functional significance of bone ballast in the evolution of buoyancy control strategies by aquatic tetrapods. *Historical Biology* 14:15–31.

Taylor, M. A. 2001. Locomotion in Mesozoic marine reptiles. In *Palaeobiology II,* ed. D. E. G. Briggs and P. R. Crowther, 404–407. Oxford: Blackwell Scientific Press.

Taylor, M. A. 2002. Origins. In *Encyclopedia of marine mammals*, ed. W. F. Perrin, B. Würsig, and J. G. M. Thewissen, 833–837. San Diego, CA: Academic Press.

Taylor, M. E. 1974. The functional anatomy of the forelimb of some African Viverridae (Carnivora). *Journal of Morphology* 143:307–336.

Tchernov, E., O. Rieppel, H. Zaher, M. J. Polcyn, and L. L. Jacobs. 2000. A fossil snake with limbs. *Science* 287:2010–2012.

Teeling, E. C., O. Madsen, R. A. Van Den Bussche, W. W. de Jong, M. J. Stanhope, and M. S. Springer. 2002. Microbat paraphyly and the convergent evolution of a key innovation in Old World rhinolophid microbats. *Proceedings of the National Academy of Sciences, USA* 99:1431–1436.

Teeling, E. C., M. Scally, D. J. Kao, M. L. Romagnoli, M. S. Springer, and M. J. Stanhope. 2000. Molecular evidence regarding the origin of echolocation and flight in bats. *Nature* 403:188–192.

Teitelbaum, S. L. 2000. Bone resorption by osteoclasts. *Science* 289:1504–1508.

Teodecki, E. E., E. D. Brodie Jr., D. R. Formanowicz Jr., and R. A. Nussbaum. 1998. Head dimorphism and burrowing speed in the African caecilian *Schistometopum thomense* (Amphibia: Gymnophiona). *Herpetologica* 54:154–160.

Teugels, G. G. 1983. La structure de la nageoire adipeuse dans les genres *Dinotopterus, Heterobranchus* et *Clarias* (Pisces, Siluriformes, Clariidae). *Cybium* 7:11–14.

te Welscher, P., M. Fernandez-Teran, M. A. Ros, and R. Zeller. 2002a. Mutual genetic antagonism involving GLI3 and dHAND prepatterns the vertebrate limb bud mesenchyme prior to SHH signaling. *Genes and Development* 16:421–426.

te Welscher, P., A. Zuniga, S. Kuijper, T. Drenth, H. J. Goedemans, F. Meijlink, and R. Zeller. 2002b. Progression of vertebrate limb development through SHH-mediated counteraction of GLI3. *Science* 298:827–830.

Thacher, J. K. 1877. Median and paired fins, a contribution to the history of vertebrate limbs. *Transactions of the Conneticut Academy* 3:281–310.

Thewissen, J. G. M., and D. M. Badoux. 1986. The descriptive and functional myology of the fore-limb of the aardvark (*Orycteropus afer*, Pallas 1766). *Anatomischer Anzeiger* 162:109–123.

Thewissen, J. G. M., and F. E. Fish. 1997. Locomotor evolution in the earliest cetaceans: Functional model, modern analogues, and paleontological evidence. *Paleobiology* 23:482–490.

Thewissen, J. G. M., S. T. Hussain, and M. Arif. 1994. Fossil evidence for the origin of aquatic locomotion in Archaeocete whales. *Science* 263:210–212.

Thewissen, J. G. M., E. M. Williams, L. J. Roe, and S. T. Hussain. 2001. Skeletons of terrestrial cetaceans and the relationship of whales to artiodactyls. *Nature* 412:277–281.

Thissen, J.-P., J.-M. Keterslegers, and L. E. Underwood. 1994. Nutritional regulation of the insulin-like growth factors. *Endocrine Reviews* 15:80–101.

Thompson, D'A. W. 1884. On the nature and action of certain ligaments. *Journal of Anatomy and Physiology* 18:406–410.

Thompson, D'A. W. 1890. On the systematic position of *Hesperornis*. Vol. 1. Studies from the Museum of Zoology in University College, Dundee. Berlin: R. Friedländer.

Thompson, D'A. W. 1917. *On growth and form*. London: Cambridge University Press.

Thompson, E. E. 1896. *Anatomy of animals: Studies in the forms of mammals and birds*. London: Macmillan.

Thompson, S. D. 1985. Bipedal hopping and seed-dispersion selection by heteromyid rodents: The role of locomotion energetics. *Ecology* 66:220–229.

Thomson, K. S. 1968. A critical review of the diphyletic theory of rhipidistian-amphibian relationships. In *Current problems in lower vertebrate phylogeny*, ed. T. Ørvig, 285–305. Stockholm: Almqvist and Wiksell.

Thomson, K. S. 1991. *Living fossil: The story of the coelacanth*. New York: Norton.

Thomson, K. S., and K. V. Hahn. 1968. Growth and form in fossil rhipidistian fishes (Crossopterygii). *Journal of Zoology, London* 156:199–223.

Thorbjarnarson, J. B. 1989. Ecology of the American crocodile, *Crocodylus acutus*. In *Crocodiles: Their ecology, management, and conservation*, 228–259. Gland, Switzerland: International Union for Conservation of Nature and Natural Resources.

Thorington, R. W., Jr., K. Darrow, and C. G. Anderson. 1998. Wing tip anatomy and aerodynamics in flying squirrels. *Journal of Mammalogy* 79:245–250.

Thorington, R. W., Jr., and K. Darrow. 2000. Anatomy of the squirrel wrist: Bones, ligaments and muscles. *Journal of Morphology* 246:85–102.

Thorington, R. W., Jr., and B. J. Stafford. 2001. Homologies of the carpal bones in flying squirrels (Pteromynidae): A review. *Mammal Study* 26:61–68.

Thorngren, K.-G., and L. I. Hansson. 1981. Cell production in different growth plates in the rabbit. *Acta Anatomica* 110:121–127.

Thornton, C. S. 1968. Amphibian limb regeneration. *Advances in Morphogenesis* 7:205–249.

Thorp, B. H. 1988. Relationship between the rate of longitudinal bone growth and physeal thickness in growing fowl. *Research in Veterinary Science* 45:83–85.

Thorpe, R. S. 1981. The morphometrics of the mouse: A review. In *Biology of the house mouse*, ed. R. J. Berry, 85–125. *Symposium of the Zoological Society of London* 47. London: Zoological Society of London.

Tickle, C. 1995. Vertebrate limb development. *Current Opinion in Genetics and Development* 5:478–484.

Tickle, C. 2002. Vertebrate limb development and possible clues to diversity in limb form. *Journal of Morphology* 252:29–37.

Tickle, C., B. Alberts, L. Wolpert, and J. Lee. 1982. Local application of retinoic acid to the limb bond mimics the action of the polarizing region. *Nature* 296:564–566.

Tickle, C. 2003. Patterning systems: From one end of the limb to the other. *Developmental Cell* 4:449–458.

Tickle, C., and A. Münsterberg. 2001. Vertebrate limb development: The early stages in chick and mouse. *Current Opinions in Genetics and Development* 11:476–481.

Tiedemann, F. 1976. Vergleichend anatomische Untersuchungen an Muskeln und Knochen des Beckengürtels von *Ophisaurus harti* BLGR., *Ophisaurus apodus* PALL. und *Ophisaurus koellikeri* GTHR. *Annalen des Naturhistorischen Museums in Wien* 80:325–335.

Tiedemann, M.-A., and F. Tiedemann. 1975. Vergleichend anatomische Untersuchungen an Schulter- und Beckengürtel verschiedener südafrikanischer Skinkarten mit besonderer Berücksichtigung von Reduktionserscheinungen. *Zoologica (Stuttgart)* 43:1–80.

Tihen, J. A. 1945. Notes on the osteology of typhlopid snakes. *Copeia* 1945:204–210.

Tihen, J. A. 1958. Comments on the osteology and phylogeny of ambystomatid salamanders. *Bulletin of the Florida State Museum, Biological Sciences* 3:1–50.

Timmons, M., J. Wallin, P. W. Rigby, and R. Balling. 1994. The role of *Pax-1* in the development of the cervico-occipital transitional zone. *Anatomical Embryology* 192:221–227.

Tobalske, B. W., and K. P. Dial. 1996. Flight kinematics of black-billed magpies and pigeons over a wide range of speeds. *Journal of Experimental Biology* 199:263–280.

Tobalske, B. W., T. L. Hedrick, K. P. Dial, and A. A. Biewener. 2003. Comparative power curves in bird flight. *Nature* 421:363–366.

Todt, W. L., and J. F. Fallon. 1986. Development of the apical ectodermal ridge in the chick leg bud and a comparison with the wing bud. *Anatomical Record* 215:288–304.

Tondravi, M. M., S. R. McKercher, K. Anderson, J. M. Erdmann, M. Quiroz, R. Maki, and S. L. Teitelbaum. 1997. Osteopetrosis in mice lacking haematopoietic transcription factor PU.1. *Nature* 386:81–84.

Tone, S., S. Tanaka, and Y. Kato. 1983. The inhibitory effect of 5-bromodeoxyuridine on the programmed cell death in the chick limb. *Development Growth and Differentiation* 25:381–391.

Toole, B. P. 1991. Glycosaminoglycans in morphogenesis. In *Cell biology of the extracellular matrix*, ed. E. Hay, 259–294. New York: Plenum Press.

Topol, L., X. Jiang, H. Choi, L. Garrett-Beal, P. J. Carolan, and

Y. Yang. 2003. Wnt-5a inhibits the canonical Wnt pathway by promoting GSK-3-independent beta-catenin degradation. *Journal of Cell Biology* 162:899–908.

Tracy, C. R., and K. A. Christian. 1985. Are marine iguana tails flattened? *British Journal of Herpetology* 6:434–435.

Trenham, P. C. 2001. Terrestrial habitat use by adult California tiger salamanders. *Journal of Herpetology* 35:343–346.

Tria, A. J., Jr., and J. A. Alicea. 1995. Embryology and anatomy of the patella. In *The patella,* ed. G. R. Scuderi, 11–23. New York: Springer-Verlag.

Tricker, R. A. R., and B. J. K. Tricker. 1967. *The science of movement.* New York: American Elsevier.

Trivett, M. K., R. A. Officer, J. G. Clement, T. I. Walker, J. M. Joss, P. M. Ingleton, T. J. Martin, and J. A. Danks. 1999. Parathyroid hormone-related protein (PTHrP) in cartilaginous and bony fish tissues. *Journal of Experimental Zoology* 284:541–548.

Trueb, L. 1996. Historical constraints and morphological novelties in the evolution of the skeletal system of pipid frogs (Anura: Pipidae). In *The biology of Xenopus,* ed. R. C. Tinsley and H. R. Kobel, 349–377. Oxford: Clarendon Press.

Trueb, L., L. Analía Púgener, and A. M. Maglia. 2000. An osteological description of *Pipa pipa* (Anura: Pipidae), with an account of skeletal development in the species. *Journal of Morphology* 243:75–104.

Trueb, L., and D.C. Cannatella. 1982. The cranial osteology and hyolaryngeal apparatus of *Rhinophrynus dorsalis* (Anura: Rhinophrynidae) with comparisons to recent pipid frogs. *Journal of Morphology* 171:11–40.

Trueb, L., and R. Cloutier. 1991. Toward an understanding of the amphibians: Two centuries of systematic history. In *Origins of the higher groups of tetrapods,* ed. H.-P. Schultze and L. Trueb, 175–193. Ithaca, NY: Cornell University Press.

Trueb, L., and C. Gans. 1983. Feeding specializations of the Mexican burrowing toad, *Rhinophrynus dorsalis* (Anura: Rhinophrynidae). *Journal of Morphology* 199:189–208.

Trueb, L., and J. Hanken. 1992. Skeletal development in *Xenopus laevis* (Anura: Pipidae). *Journal of Morphology* 214:1–41.

Trueb, L., L. A. Púgener, and A. M. Maglia. 2000. Ontogeny of the bizarre: An osteological description of *Pipa pipa* (Anura: Pipidae), with an account of skeletal development in the species. *Journal of Morphology* 243:75–104.

Trueta, J., and V. P. Amato. 1960. The vascular contribution to osteogenesis. III: Changes in the growth cartilage caused by experimentally induced ischaemia. *Journal of Bone and Joint Surgery* 42B:571–587.

Tschumi, P. A. 1957. The growth of the hind limb of Xenopus laevis and its dependence on the epidermis. *Journal of Anatomy* 9:149–173.

Tsvetkov, V. I. 1992. Toward an assessment of locomotor function of the pectoral girdle and fins of actinopterygian fishes. *Journal of Ichthyology* 32:79–91.

Tumpel, S., J. J. Sanz-Ezquerro, A. Isaac, M. C. Eblaghie, J. Dobson, and C. Tickle. 2002. Regulation of Tbx3 expression by anteroposterior signalling in vertebrate limb development. *Developmental Biology* 250:251–262.

Tyler, M. J., J. D. Roberts, and M. Davies. 1980. Field observations on *Arenophryne rotunda* Tyler, a leptodactylid frog of coastal sandhills. *Australian Wildlife Research* 7:295–304.

Tyler, S. J., and S. J. Ormerod. 1994. *The dippers.* London: Poyser.

Tyrcha, J. 2001. Age-dependent cell cycle models. *Journal of Theoretical Biology* 213:89–101.

U

Udagawa, N., N. Takahashi, T. Akatsu, H. Tanaka, T. Sasaki, T. Nishihara, T. Koga, T. J. Martin, and T. Suda. 1990. Origin of osteoclasts: Mature monocytes and macrophages are capable of differentiating into osteoclasts under a suitable microenvironment prepared by bone marrow-derived stromal cells. *Proceedings of the National Academy of Sciences, USA* 87:7260–7264.

Ueta, C., M. Iwamoto, N. Kanatani, C. Yoshida, Y. Liu, M. Enomoto-Iwamoto, T. Ohmori, et al. 2001. Skeletal malformations caused by overexpression of Cbfa1 or its dominant negative form in chondrocytes. *Journal of Cell Biology* 153:87–100.

Uhen, M. D. 1998. Middle to late Eocene basilosaurines and dorudontines. In *The emergence of whales: Evolutonary patterns in the origin of Cetacea,* ed. J. G. M. Thewissen, 29–61. New York: Plenum Press.

Underwood, G. 1957. On the lizards of the family Pygopodidae: A contribution to the morphology and phylogeny of the Squamata. *Journal of Morphology* 100:207–268.

Unger, S. 2002. A genetic approach to the diagnosis of skeletal dysplasia. *Clinical Orthopaedics and Related Research* 401:2–38.

Unwin, D. M. 1987. Pterosaur locomotion: Joggers or waddlers? *Nature* 327:13–14.

Unwin, D. M. 1988a. New pterosaurs from Brazil. *Nature* 332:398–399.

Unwin, D. M. 1988b. New remains of the pterosaur *Dimorphodon* (Pterosauria: Rhamphorhynchoidea) and the terrestrial ability of early pterosaurs. *Modern Geology* 13:57–68.

Unwin, D. M. 1993. Aves. In *The fossil record 2,* ed. M. J. Benton, 717–738. London: Chapman and Hall.

Unwin, D. M. 1995. Preliminary results of a phylogenetic analysis of the Pterosauria (Diapsida: Archosauria). In *Sixth Symposium on Mesozoic Terrestrial Ecosystems and Biota, short papers,* ed. A.-I. Sun and Y. Wang, 69–72. Beijing: China Ocean Press.

Unwin, D. M. 1997. Pterosaur tracks and the terrestrial ability of pterosaurs. *Lethaia* 29:373–386.

Unwin, D. M. 1999. Pterosaurs: Back to the traditional model? *Trends in Ecology and Evolution* 14:263–268.

Unwin, D. M., and N. N. Bakhurina. 1994. *Sordes pilosus* and the nature of the pterosaur flight apparatus. *Nature* 371:62–64.

Unwin, D. M., E. Frey, D. M. Martill, J. B. Clarke, and J. Rieß. 1996. On the nature of the pteroid in pterosaurs. *Proceedings of the Royal Society of London B* 263:45–52.

Unwin, D. M., and D. M. Henderson. 2002. On the trail of the totally integrated pterosaur. *Trends in Ecology and Evolution* 17:58–59.

Urist, M. R. 1980. Heterotopic bone formation. In *Fundamental and clinical bone physiology,* ed. M. R. Urist, 369–393. Philadelphia: J. B. Lippencott.

V

Vainio, S., I. Karavanova, A. Jowett, and I. Thesleff. 1993. Identification of BMP-4 as the signal mediating secondary induction between epithelial and mesenchymal tissues during early tooth development. *Cell* 75:45–58.

Vanden Berge, J. C., and R. W. Storer. 1995. Intratendinous ossifications in birds: A review. *Journal of Morphology* 226:47–77.

Van der Eerden, B. C. J., J. van de Ven, C. W. G. M. Lowik,

J. M. Wit, and M. Karperien. 2002. Sex steroid metabolism in the tibial growth plate of the rat. *Endocrinology* 143:4048–4055.

van Dijk, D. E. 2001. Osteology of the ranoid burrowing African anurans *Breviceps* and *Hemisus*. *African Zoology* 36:137–141.

van Eeden, F. J. M., M. Granato, U. Schach, M. Brand, M. Furutani-Seiki, P. Haffter, M. Hammerschmidt, et al. 1996. Genetic analysis of fin formation in the zebrafish, *Danio rerio*. *Development* 123:255–262.

Vanky, P., U. Brockstedt, A. Hjerpe, and B. Wikström. 1998. Kinetic studies on epiphyseal growth cartilage in the normal mouse. *Bone* 22:331–339.

Van Hoepen, E. C. N. 1915. Stegocephalia of Senekal. O.F.S. *Annals Transvaal Museum* 5:124–149.

van Tuinen, M., C. G. Sibley, and S. B. Hedges. 2000. The early history of modern birds inferred from DNA sequences of nuclear and mitochondrial ribosomal genes. *Molecular Biology and Evolution* 17:451–457.

Van Valkenburgh, B. 1985. Locomotor diversity within past and present guilds of large predatory mammals. *Paleobiology* 11:406–428.

Van Valkenburgh, B. 1987. Skeletal indicators of locomotor behavior in living and extinct carnivores. *Journal of Vertebrate Paleontology* 7:162–182.

Vargesson, N., J. D. Clarke, K. Vincent, C. Coles, L. Wolpert, et al. 1997. Cell fate in the chick limb bud and relationship to gene expression. *Development* 124:1909–1918.

Vargesson, N., K. Kostakopoulou, G. Drossopoulou, S. Papageorgiou, and C. Tickel. 2001. Characterization of Hoxa gene expression in the chick limb bud in response to FGF. *Developmental Dynamics* 220:87–90.

Vassallo, A. I. 1998. Functional morphology, comparative behaviour, and adaptation in two sympatric subterranean rodents genus *Ctenomys* (Caviomorpha: Octodontidae). *Journal of Zoology, London* 244:415–427.

Vasse, J., J.-P. Gasc, and S. Renous-Lecuru. 1974. Les membres rudimentaires chez l'adulte et chez l'embryon de *Scelotes brevipes* (Hewitt) (Scincidae, Sauria). *Annales d'Embryologie et de Morphogenèse* 7:417–424.

Vaughn, T. A. 1959. Functional morphology of three bats: *Eumops, Myotis, Macrotus*. *University of Kansas Publications: Museum of Natural History* 12:1–153.

Vazquez, R. J. 1994. The automating skeletal and muscular mechanisms of the avian wing (Aves). *Zoomorphology* 114:59–71.

Velloso, C. P., A. Kumar, E. M. Tanaka, and J. P. Brockes. 2000. Generation of mononucleate cells from post-mitotic myotubes proceeds in the absence of cell cycle progression. *Differentiation* 66:239–246.

Verreijdt, L., E. Vandervennet, J. Y. Sire, and A. Huysseune. 2002. Developmental differences between cranial bones in the zebrafish (*Danio rerio*): Some preliminary light and TEM observations. *Connective Tissue Research* 43:109–112.

Vial, J. L., ed. 1973. *Evolutionary biology of the anurans: Contemporary research on major problems*. Columbia: University of Missouri Press.

Vialleton, L. M. 1924. *Membres et ceintures des Vertébrés tétrapodes*. Paris: Doin.

Videler, J. J. 1981. Swimming movements, body structure and propulsion in cod, *Gadus morhua*. *Symposia of the Zoological Society of London* 48:1–27.

Videler, J. J. 1993. *Fish swimming*. London: Chapman and Hall.

Videler, J. J. 2003. *Comparative biomechanics: Life's physical world*. Princeton, NJ: Princeton University Press.

Videler, J. J., and J. T. Jorna. 1985. Functions of the sliding pelvis in *Xenopus laevis*. *Copeia* 1985:251–254.

Villwock, W. 1984. Schuppen- und Ventralflossenreduktion, Phänomene regressiver Evolution am Beispiel altweltlicher Zahnkarpfen der Tribus Aphaniini (Pisces: Cyprinodontidae). In *Regressive Evolution und Phylogenese: Eine Zusammenschau unter genetischen, öklogischen und ethologischen Gesichtspunkten*, ed. C. Schemmel, J. Parzefall, N. Peters, G. Peters, H. Wilkens, W. Villwock, C. D. Zander, and M. Dzwillo, 72–100. Hamburg: Paul Parey.

Vizcaíno, S. F., R. A. Fariña, and G. V. Mazzetta. 1999. Ulnar dimensions and fossoriality in armadillos. *Acta Theriologica* 44:309–320.

Vizcaíno, S. F., and N. Milne. 2002. Structure and function in armadillo limbs (Mammalia: Xenarthra: Dasypodidae). *Journal of Zoology, London* 257:117–127.

Vlaskalin, T., C. J. Wong, and C. Tsilfidis. 2004. Growth and apoptosis during larval forelimb development and adult forelimb regeneration in the newt (*Notophthalmus viridescens*). *Development Genes and Evolution* 214:423–431.

Vogel, A., C. Rodriguez, and J.-C. Izpisúa Belmonte. 1996. Involvement of FGF-8 in initiation, outgrowth and patterning of the vertebrate limb. *Development* 122:1737–1750.

Vogel, A., C. Rodriguez, W. Warnken, J. C. Izpisúa Belmonte. 1995. Dorsal cell fate specified by chick Lmx1 during vertebrate limb development. *Nature* 378:716–720.

Vogel, K. G., and T. J. Koob. 1989. Structural specialization in tendons under compression. *International Review of Cytology* 115:237–293.

Vogel, S. 2001. *Prime mover: A natural history of muscle*. New York: Norton.

Vogel, S. 2003. *Comparative biomechanics: Life's physical world*. Princeton, NJ: Princeton University Press.

von Meyer, H. 1861a. *Archaeopteryx lithographica* und *Pterodactylus* von Solnhofen. *Neues Jahrbuch für Mineralogie, Geologie, und Paläontologie* 1861:678–679.

von Meyer, H. 1861b. Vögel-Federn und *Palpipes priscus* von Solnhofen. *Neues Jahrbuch für Mineralogie, Geologie, und Paläontologie* 1861:5–61.

von Meyer, H. 1862. On the *Archaeopteryx lithographica* from the lithographic slate of Solenhofen. *Annals and Magazine of Natural History* 9 (3): 366–370.

Vorobyeva, E. 1992. *The problem of the terrestrial vertebrate origin*. Moscow: Nauka.

Vorobyeva, E. 2000. Morphology of the humerus in the rhipidistian Crossopterygii and the origin of tetrapods. *Paleontological Journal* 34:632–641.

Vorobyeva, E., and A. Kuznetsov. 1992. The locomotor apparatus of *Panderichthys rhombolepis* (Gross), a supplement to the problem of fish-tetrapod transition. In *Fossil fishes as living animals*, ed. E. Mark-Kurik, 131–140. Tallinn: Academia 1.

Vorobyeva, E., and H.-P. Schultze. 1991. Description and systematics of panderichthyid fishes with comments on their relationships to tetrapods. In *Origins of the higher groups of tetrapods: Controversy and consensus*, ed. H.-P. Schultze and L. Trueb, 68–109. Ithaca, NY: Cornell University Press.

Vortkamp, A., K. Lee, B. Lanske, G. V. Segre, H. M. Kronenberg, and C. J. Tabin. 1996. Regulation of rate of cartilage differentiation by Indian hedgehog and PTH-related protein. *Science* 273:613–622.

Vu, T. H., J. M. Shipley, G. Bergers, J. E. Berger, J. A. Helms, D. Hanahan, S. D. Shapiro, R. M. Senior, and Z. Werb. 1998. MMP-9/gelatinase B is a key regulator of growth plate angiogenesis and apoptosis of hypertrophic chondrocytes. *Cell* 93:411–422.

W

Wada, N., Y. Kawakami, R. Ladher, P. H. Francis-West, and T. Nohno. 1999. Involvement of Frzb-1 in mesenchymal condensation and cartilage differentiation in the chick limb bud. *International Journal of Developmental Biology* 43:495–500.

Wada, N., I. Kimura, H. Tanaka, H. Ide, and T. Nohno. 1998. Glycosylphosphatidylinositol-anchored cell surface proteins regulate position-specific cell affinity in the limb bud. *Developmental Biology* 202:244–252.

Wagner, G. P., and C.-H. Chiu. 2001. The tetrapod limb: A hypothesis on its origin. *Journal of Experimental Zoology (Molecular Development and Evolution)* 291:226–240.

Wagner, G. P., C.-H. Chiu, and M. Laubichler. 2000. Developmental evolution as a mechanistic science: The inference from developmental mechanisms to evolutionary processes. *American Zoologist* 40:819–831.

Wagner, G. P., and J. A. Gauthier. 1999. 1, 2, 3 = 2, 3, 4: A solution to the problem of the homology of the digits in the avian hand. *Proceedings of the National Academy of Sciences, USA* 96:5111–5116.

Wagner, G. P., P. A. Khan, M. J. Blanco, B. Y. Misof, and R. A. Liversage. 1999. Evolution of Hoxa-11 expression in amphibians: Is the urodele autopodium an innovation? *American Zoologist* 39:686–694.

Wagner, G. P., and V. J. Lynch. 2005. Molecular evolution of evolutionary novelties: The vagina and uterus of placental mammals. *Journal of Experimental Zoology B (Molecular and Developmental Evolution)* 304B:580–592.

Wagner, G. P., and B. Y. Misof. 1992. Evolutionary modification of regenerative capability in vertebrates: A comparative study on teleost pectoral fin regeneration. *Journal of Experimental Zoology* 261:62–78.

Wagner, G. P., and G. B. Müller. 2002. Evolutionary innovations overcome ancestral constraints: A re-examination of character evolution in male sepsid flies (Diptera: Sepsidae). *Evolution and Development* 4:1–6.

Wagner, T., J. Wirth, J. Meyer, B. Zabel, M. Held, J. Zimmer, J. Pasantes, et al. 1994. Autosomal sex reversal and campomelic dysplasia are caused by mutations in and around the SRY-related gene SOX9. *Cell* 79:1111–1120.

Wainwright, P. C., D. R. Bellwood, and M. W. Westneat. 2002. Ecomorphology of locomotion in labrid fishes. *Environmental Biology of Fishes* 65:47–62.

Wainwright, W. P. 1880. *Radical-mechanics of animal locomotion, with remarks on the setting-up of soldiers, horse and foot, and on the supplying of cavalry horses.* New York: D. Van Nostrand.

Waitkuwait, W. E. 1985. Investigations of the breeding biology of the West-African slender-snouted crocodile *Crocodylus cataphractus* Cuvier, 1824. *Amphibia-Reptilia* 6:387–399.

Waitkuwait, W. E. 1989. Present knowledge on the west African

slender-snouted crocodile, *Crocodylus cataphractus* Cuvier 1824 and the west African dwarf crocodile *Osteolaemus tetraspis*, Cope 1861. In *Crocodiles: Their ecology, management, and conservation*, 260–275. Gland, Switzerland: International Union for Conservation of Nature and Natural Resources.

Wake, D. B., and N. H. Shubin. 1994. Urodele limb development in relation to phylogeny and life history. *Journal of Morphology* 220:407–408.

Wake, M. H. 1993a. Non-traditional characters in the assessment of caecilian phylogenetic relationships. *Herpetological Monographs* 7:42–55.

Wake, M. H. 1993b. The skull as a locomotor organ. In *The skull*, vol. 3, *Functional and evolutionary mechanisms*, ed. J. Hanken and B. K. Hall, 197–240. Chicago: University of Chicago Press.

Wake, M. H. 2003. The osteology of caecilians. In *Amphibian biology*, vol. 5, *Osteology*, ed. H. Heatwole and M. Davies, 1809–1876. Chipping Norton, Australia: Surrey Beatty and Sons.

Wake, M. H., and J. Hanken. 1982. Development of the skull of *Dermophis mexicanus* (Amphibia: Gymnophiona), with comments on skull kinesis and amphibian relationships. *Journal of Morphology* 173:203–223.

Walker, C. F. 1938. The structure and systematic relationships of the genus *Rhinophrynus*. *Occasional Papers of the Museum of Zoology, University of Michigan* 372:1–11.

Walker, J. A., and M. W. Westneat. 1997. Labriform propulsion in fishes: Kinematics of flapping aquatic flight in the bird wrasse *Gomphosus varius* (Labridae). *Journal of Experimental Biology* 200:1549–1569.

Walker, J. A., and M. W. Westneat. 2000. Mechanical performance of aquatic rowing and flying. *Proceedings of the Royal Society of London B* 267:1875–1881.

Walker, J. A., and M. W. Westneat. 2002. Performance limits of labriform propulsion and correlates with fin shape and motion. *Journal of Experimental Biology* 205:177–187.

Walker, J. M., S. E. Trauth, J. M. Britton, and J. E. Cordes. 1986. Burrows of the parthenogenetic whiptail lizard *Cnemidophorus laredoensis* (Teiidae) in Webb Co., Texas. *Southwestern Naturalist* 31:408–410.

Walker, M., D. Phalan, J. Jensen, J. Johnson, M. Drew, V. Samii, G. Henry, and J. McCauley. 2002. Meniscal ossicles in large non-domestic cats. *Veterinary Radiology and Ultrasound.* 43:249–254.

Walker, R. 1985. *A guide to the post-cranial bones of East African mammals.* Norwich, UK: Hylochoerus Press.

Walker, W. F., Jr. 1971. Swimming in sea turtles of the family Cheloniidae. *Copeia* 1971:229–233.

Walker, W. F., Jr. 1973. The locomotor apparatus of Testudines. In *Biology of the reptilia*, vol. 4, ed. C. Gans and T. S. Parsons, 1–100. New York: Academic Press.

Wall, W. P. 1983. The correlation between high limb-bone density and aquatic habits in recent mammals. *Journal of Paleontology* 57:197–207.

Wallace, H. 1981. *Vertebrate limb regeneration.* Chichester, UK: Wiley.

Walton, D. W., and G. M. Walton. 1970. Post-cranial osteology of bats. In *About bats: A chiropteran biology symposium*, ed. B. H. Slaughter and D. W. Walton, 93–126. Dallas: Southern Methodist University Press.

Wanek, N., K. Muneoka, and S. V. Bryant. 1989. Evidence for regulation following amputation and tissue grafting in the developing mouse limb. *Journal of Experimental Zoology* 249:45–61.

Wang, B., J. F. Fallon, and P. A. Beachy 2000. Hedgehog-regulated processing of Gli3 produces an anterior/posterior repressor gradient in the developing vertebrate limb. *Cell* 100:423–434.

Wang, C. K., M. Omi, D. Ferrari, H. C. Cheng, G. Lizarraga, H. J. Chin, W. B. Upholt, C. N. Dealy, R. A. Kosher. 2004. Function of Bmps in the apical ectoderm of the developing mouse limb. *Developmental Biology* 269:109–122.

Wang, Q., R. P. Green, G. Zhao, and D. M. Ornitz. 2001. Differential regulation of endochondral bone growth and joint development by FGFR1 and FGFR3 tyrosine kinase domains. *Development* 128:3867–3876.

Wang, X., and J. Lü. 2001. Discovery of a pterodactylid pterosaur from the Yixian Formation of western Liaoning, China. *Chinese Science Bulletin* 46:1112–1117.

Wang, X., Z. Zhou, F. Zhang, and X. Xu. 2002. A nearly completely articulated rhamphorhynchoid pterosaur with exceptionally well-preserved wing membranes and "hairs" from Inner Mongolia, northeast China. *Chinese Science Bulletin* 47:226–232.

Wang, Y. 2000. A new salamander (Amphibia: Caudata) from the Early Cretaceous Jehol Biota. *Vertebrata PalAsiatica* 38:100–103.

Wang, Y., and D. Sassoon. 1995. Ectoderm-mesenchyme and mesenchyme mesenchyme interactions regulate Msx-1 expression and cellular differentiation in the murine limb bud. *Developmental Biology* 168:374–382.

Warburton, D., M. Schwarz, D. Tefft, G. Flores-Delgado, K. D. Anderson, and W. V. Cardoso. 2000. The molecular basis of lung morphogenesis. *Mechanical Development* 92:45–81.

Warner, R. 1946. Pectoral girdles vs. hyobranchia in the snake genera *Liotyphlops* and *Anomalepis*. *Science* 103:720–722.

Warren, A. A. 2000. Secondarily aquatic temnospondyls of the Upper Permian and Mesozoic. In *Amphibian biology,* vol. 4, *Palaeontology: The evolutionary history of amphibians,* ed. H. Heatwole and R. L. Carroll, 1121–1149. Chipping Norton, Australia: Surrey Beatty and Sons.

Warren, A. A., and M. N. Hutchinson. 1983. The last Labyrinthodont? A new Brachyopoid (Amphibia, Temnospondyli) from the early Jurassic Evergreen formation of Queensland, Australia. *Philosophical Transactions of the Royal Society of London B* 303:1–62.

Warren, A. A., and N. Snell. 1991. The postcranial skeleton of Mesozoic temnospondyl amphibians: A review. *Alcheringa* 15:43–64.

Waskiewicz, A. J., H. A. Rikhof, R. E. Hernandez, and C. B. Moens. 2001. Zebrafish Meis functions to stabilize Pbx proteins and regulate hindbrain patterning. *Development* 128:4139–4151.

Watson, A. G., L. E. Stein, C. Marshall, and G. A. Henry. 1994. Polydactyly in a bottlenose dolphin, *Tursiops truncatus*. *Marine Mammal Science* 10:93–100.

Watson, D. M. S. 1913. On the primitive tetrapod limb. *Anatomische Anzeiger* 46:24–27.

Watson, D. M. S. 1919. Structure, evolution and origin of the Amphibia: The "orders" Rachitomi and Sterospondyli. *Philosophical Transactions of the Royal Society of London B* 209:1–74.

Watson, D. M. S. 1926. Croonian lecture: The evolution and origin of the Amphibia. *Philosophical Transactions of the Royal Society of London B* 214:189–257.

Watson, D. M. S. 1957. On *Millerosaurus* and the early history of the sauropsid reptiles. *Philosophical Transactions of the Royal Society of London B* 240:325–400.

Watson, D. M. S. 1958. A new labyrinthodon (Paracyclotosaurus) from the Upper Trias of New South Wales. *Bulletin of the British Museum of Natural History: Geology* 3:239–263.

Watson, D. M. S., and E. L. Gill. 1923. The structure of certain palaeozoic dipnoi. *Journal of the Linnean Society,* no. 435:163.

Wayne, R. K., and C. B. Ruff. 1993. Domestication in bone growth. In *Bone,* vol. 7, *Bone growth—B,* ed. B. K. Hall, 105–131. Boca Raton, FL: CRC Press.

Webb, G. J. W., and C. Gans. 1982. Galloping in *Crocodylus johnstoni*: A reflection of terrestrial activity? *Records of the Australian Museum* 34:607–618.

Webb, J. F. 1999. Larvae in fish: Development and evolution. In *The origin and evolution of larval forms,* ed. B. K. Hall and M. H. Wake, 109–158. San Diego, CA: Academic Press.

Webb, P. W. 1973. Kinematics of pectoral fin propulsion in *Cymatogaster aggregata*. *Journal of Experimental Biology* 59:697–710.

Webb, P. W. 1975. Hydrodynamics and energetics of fish propulsion. *Bulletin of the Fisheries Research Board of Canada* 190:1–159.

Webb, P. W. 1978. Hydrodynamics: nonscombroid fish. In *Fish physiology,* vol. 7, ed. W. S. Hoar and D. J. Randall, 189–237. New York: Academic Press.

Webb, P. W. 1994. The biology of fish swimming. In *Mechanics and physiology of animal swimming,* ed. L. Maddock, Q. Bone, and J. M. V. Rayner, 45–62. Cambridge: Cambridge University Press.

Webb, P. W., and R. W. Blake. 1985. Swimming. In *Functional vertebrate morphology,* ed. M. Hildebrand, D. M. Bramble, K. F. Liem, and D. B. Wake, 110–128. Cambridge, MA: Harvard University Press.

Webb, P. W., and V. de Buffrénil. 1990. Locomotion in the biology of large aquatic vertebrates. *Transactions of the American Fisheries Society* 119:629–641.

Webb, P. W., and A. G. Fairchild. 2001. Performance and maneuverability of three species of teleostean fishes. *Canadian Journal of Zoology* 79:1866–1877.

Webb, P. W., G. D. LaLiberte, and A. J. Schrank. 1996. Does body and fin form affect the maneuverability of fish traversing vertical and horizontal slits? *Environmental Biology of Fishes* 46:7–14.

Webb, R. G. 1969. Survival adaptations of tiger salamanders (*Ambystoma tigrinum*) in the Chihuahuan Desert. In *Physiological systems in semiarid environments,* ed. C. C. Hoff and L. Riedesel, 143–147. Albuquerque: University of New Mexico Press.

Weil, A. 2002. Upwards and onwards. *Nature* 416:798–799.

Weinstein, R. B., and Full, R. J. 1999. Intermittent locomotion increases endurance in a gecko. *Physiological and Biochemical Zoology* 72:732–739.

Weir, E. C., W. M. Philbrick, M. Amling, L. A. Neff, R. Baron, and A. E. Broadus. 1996. Targeted overexpression of parathyroid hormone-related peptide in chondrocytes causes chondrodysplasia and delayed endochondral bone formation. *Proceedings of the National Academy of Sciences, USA* 93:10240–10245.

Weisel, G. F. 1967. Early ossification in the skeleton of the sucker (*Catostomus macrocheilus*) and the guppy (*Poecilia reticulata*). *Journal of Morphology* 121:1–18.

Weishampel, D. B., and J. R. Horner. 1990. Hadrosauridae. In *The Dinosauria*, ed. D. B. Weishampel, P. Dodson, and H. Osmólska, 534–561. Berkeley: University of California Press.

Weiss, R. E., and N. Watabe. 1979. Studies on the biology of fish bone. III: Ultrastructure of osteogenesis and resorption in osteocytic (cellular) and anosteocytic (acellular) bones. *Calcified Tissue International* 28:43–56.

Weiss Sachdev, S., U. H. Dietz, Y. Oshima, M. R. Lang, E. W. Knapik, Y. Hiraki, and C. Shukunami. 2001. Sequence analysis of zebrafish *chondromodulin-l* and expression profile in the notochord and chondrogenic regions during cartilage morphogenesis. *Mechanisms of Development* 105:157–162.

Wellborn, V. 1933/1997. *Comparative osteological examinations of geckonids, eublepharids and uroplatids.* Herpetologica Translations no. 1. Transl. A. P. Russell, A. M. Bauer, and A. Deufel, with an introduction by A. M. Bauer and A. P. Russell. Utah: Logan, Herpbooks.

Wellik, D. M., and M. R. Capecchi. 2003. Hox10 and Hox11 genes are required to globally pattern the mammalian skeleton. *Science* 301:363–367.

Wellnhofer, P. 1970. Die Pterodactyloidea (Pterosauria) der Oberjura-Plattenkalke Süddeutschlands. *Abhandlungen der Bayerischen Akademie der Wissenschaften zu München, Mathematisch-Naturwissenschaftlichen Klasse* 141:1–133.

Wellnhofer, P. 1974a. *Campylognathoides liasicus* (Quenstedt), an Upper Liassic pterosaur from Holzmaden: The Pittsburgh specimen. *Annals of Carnegie Museum* 45:5–34.

Wellnhofer, P. 1974b. Das fünfte Skelettexemplar von *Archaeopteryx. Palaeontographica Abteilung A* 147:169–216.

Wellnhofer, P. 1975a. Die Rhamphorhynchoidea (Pterosauria) der Oberjura-Plattenkalke Süddeutschlands. I: Allgemeine Skelettmorphologie. *Palaeontographica Abteilung A* 148:1–33.

Wellnhofer, P. 1975b. Die Rhamphorhynchoidea (Pterosauria) der Oberjura-Plattenkalke Süddeutschlands. II: Systematische beschreibung. *Palaeontographica Abteilung A* 148:132–186.

Wellnhofer, P. 1975c. Die Rhamphorhynchoidea (Pterosauria) der Oberjura-Plattenkalke Süddeutschlands. III: Palökologie und Stammesgeschichte. *Palaeontographica Abteilung A* 149:1–30.

Wellnhofer, P. 1978. *Pterosauria.* Stuttgart: Gustav Fischer Verlag.

Wellnhofer, P. 1988. Terrestrial locomotion in pterosaurs. *Historical Biology* 1:3–16.

Wellnhofer, P. 1991. *The illustrated encyclopedia of pterosaurs.* New York: Crescent.

Wellnhofer, P. 1993. Das siebte Exemplar von *Archaeopteryx* aus den Solnhofener Schichten. *Archaeopteryx* 11:1–48.

Wellstead, C. F. 1991. Taxonomic revision of the Lysorophia, Permo-Carboniferous lepospondyl amphibians. *Bulletin of the American Museum of Natural History* 209:1–90.

Wermel, J. 1934. Untersuchungen über die Kinetogenese und ihre Bedeutung in der onto- und phylogenetischen Entwicklung (Experimente und Vergleichungen an Wirbeltierextremitäten). I: Allgemeine Einleitung, Veränderungen der Länge der Knochen. *Gegenbaurs Morphologisches Jahrbuch* 74:143–169.

Werneberg, R., and J. S. Steyer. 2002. Redescription of the type species *Cheliderpeton vranyi* Fritsch 1877 (Amphibia, Temnospondyli) from the Lower Permian of Czech Republic (Bohemia). *Paläontologische Zeitschrift* 76:149–162.

Westneat, M. W. 1996. Functional morphology of aquatic flight in fishes: Kinematics, electromyography, and mechanical modeling of labriform locomotion. *American Zoologist* 36:582–598.

Westneat, M. W., and J. A. Walker. 1997. Motor patterns of labriform locomotion: Kinematic and electromyographic analysis of pectoral fin swimming in the labrid fish *Gomphosus varius. Journal of Experimental Biology* 200:1881–1893.

Westoll, T. S. 1943. The origin of the primitive tetrapod limb. *Proceedings of the Royal Society of London B* 131:373–393.

Whidden, H. P. 2000. Comparative myology of moles and the phylogeny of the Talpidae (Mammalia, Lipotyphla). *American Museum Novitates* 3294:1–53.

Whitaker, R., and Z. Whitaker. 1984. Reproductive biology of the mugger (*Crocodylus palustris*). *Journal of the Bombay Natural History Society* 81:297–317.

Whitaker, R., and Z. Whitaker. 1989. Ecology of the mugger crocodile. In *Crocodiles: Their ecology, management, and conservation*, 276–296. Gland, Switzerland: International Union for Conservation of Nature and Natural Resources.

White, A., and G. Wallis. 2001. Endochondral ossification: A delicate balance between growth and mineralization. *Current Biology* 11:R589–R591.

White, J. A., M. B. Boffa, B. Jones, and M. Petkovich. 1994. A zebrafish retinoic acid receptor expressed in the regenerating caudal fin. *Development* 120:1861–1872.

White, T. D. 1989. An analysis of epipubic bone function in mammals using scaling theory. *Journal of Theoretical Biology* 139:342–357.

White, T. D. 1991. *Human osteology.* New York: Academic Press.

White, T. E. 1939. Osteology of *Seymouria baylorensis* Broili. *Harvard University Museum of Comparative Zoology Bulletin* 85:326–409.

Whiting, A. S., A. M. Bauer, and J. W. Sites Jr. 2003. Phylogenetic relationships and limb loss in sub-Saharan African scincine lizards (Squamata: Scincidae). *Molecular Phylogenetics and Evolution* 29:582–598.

Whitten, J. M. 1969. Cell death during early morphogenesis: Parallels between insect limb and vertebrate limb development. *Science* 163:1456–1457.

Whoriskey, F. G., and R. J. Wootton. 1987. Swimming endurance of threespine sticklebacks, *Gasterosteus aculeatus* L., from the Afon Rheidol, Wales. *Journal of Fish Biology* 30:335–340.

Whyte, M. A., and M. Romano. 2002. A dinosaur ichnocoenosis from the Middle Jurassic of Yorkshire, UK. *Ichnos* 8:223–234.

Wiedersheim, R. 1879. *Die Anatomie der Gymnophionen.* Jena, Germany: Gustav Fischer.

Wiens, J. J., and J. L. Slingluff. 2001. How lizards turn into snakes: A phylogenetic analysis of body-form evolution in anguid lizards. *Evolution* 55:2303–2318.

Wild, R. 1978. Die Flugsaurier (Reptilia, Pterosauria) aus der Oberen Trias von Cene bei Bergamo, Italien. *Bollettino della Società Paleontologica Italiana* 17:176–256.

Wild, R. 1984a. Flugsaurier aus der Obertrias von Italien. *Naturwissenschaften* 71:1–11.

Wild, R. 1984b. A new pterosaur (Reptilia: Pterosauria) from the Upper Triassic (Norian) of Friuli, Italy. *Gortania Atti di Musei Friuli di Storia Naturali* 5:45–62.

Wilga, C. D., and G. V. Lauder. 1999. Locomotion in sturgeon: Function of the pectoral fins. *Journal of Experimental Biology* 202:2413–2432.

Wilga, C. D., and G. V. Lauder. 2000. Three-dimensional kinematics and wake structure of the pectoral fins during locomotion in leopard sharks *Triakis semifasciata*. *Journal of Experimental Biology* 203:2261–2278.

Wilga, C. D., and G. V. Lauder. 2001. Functional morphology of the pectoral fins in bamboo sharks, *Chiloscyllium plagiosum*: Benthic vs. pelagic station-holding. *Journal of Morphology* 249:195–209.

Wilkie, A. O. 2003. Why study human limb malformations? *Journal of Anatomy* 202:27–35.

Wilkie, A. O., S. J. Patey, S. H. Kan, A. M. van den Ouweland, and B. C. Hamel. 2002. FGFs, their receptors, and human limb malformations: Clinical and molecular correlations. *American Journal of Medical Genetics* 112:266–278.

Wilkins, A. S. 2002. *The evolution of developmental pathways.* Sunderland, MA: Sinauer Association.

Wilkins, R. J., J. A. Browning, and J. C. Ellory. 2000. Surviving in a matrix: Membrane transport in articular chondrocytes. *Journal of Membrane Biology* 177:95–108.

Wilkinson, M., and R. A. Nussbaum. 1999. Evolutionary relationships of the lungless caecilian *Atretochoana eiselti* (Amphibia: Gymnophiona: Typhlonectidae). *Zoological Journal of the Linnean Society* 126:191–223.

Willert, C. E., and M. Gharib. 1991. Digital particle image velocimetry. *Experiments in Fluids* 10:181–193.

Williams, E. M. 1999. Synopsis of the earliest cetaceans. In *The emergence of whales: Evolutionary patterns in the origin of Cetacea,* ed. J. G. M. Thewissen, 1–28. New York: Plenum Press.

Williams, P. L., and R. Warwick. 1980. *Gray's anatomy.* 36th ed. Edinburgh: Churchill Livingstone.

Williams, T. M. 1983. Locomotion in the North American mink, a semi-aquatic mammal. *Journal of Experimental Biology* 103:155–168.

Williams, T. M. 1989. Swimming by sea otters: Adaptations for low energetic cost locomotion. *Journal of Comparative Physiology A* 164:815–824.

Williams, T. M., and G. A. J. Worthy. 2002. Anatomy and physiology: The challenge of aquatic living. In *Marine mammal biology,* ed. A. R. Hoelzel, 73–141. Oxford: Blackwell.

Williams, T. M., W. A. Friedl, M. L. Fong, R. M. Yamada, P. Sedivy, and J. E. Huan. 1992. Travel at low energetic cost by swimming and wave-riding bottlenose dolphins. *Nature* 355:821–823.

Williamson, H. C. 1893. On the anatomy of the pectoral arch of the grey gurnard (*Trigla gurnardus*), with special reference to its innervation. *Annual Report of the Fishery Board for Scotland* 12:322–332.

Williamson, M. R., K. P. Dial, and A. A. Biewener. 2001. Pectoralis muscle performance during ascending and slow level flight in mallards (*Anas platyrhynchos*). *Journal of Experimental Biology* 204:595–507.

Williston, S. W. 1910. *Cacops, Desmospondylus:* New genera of Permian vertebrates. *Bulletin of the Geological Society of America* 21:249–284.

Williston, S. W. 1911. *American Permian vertebrates.* Chicago: University of Chicago Press.

Williston, S. W. 1912. Restoration of *Limnoscelis,* a cotylosaur reptile from New Mexico. *American Journal of Sciences* 34:457–468.

Williston, S. W. 1915. *Trimerorhachis,* a Permian temnospondyl amphibian. *Journal of Geology* 23:246–255.

Williston, S. W. 1925. *Osteology of the reptiles.* Ed. W. K. Gregory. Cambridge, MA: Harvard University Press.

Wilsman, N. J., C. E. Farnum, E. M. Green, E. M. Leiferman, and M. K. Clayton. 1996a. Cell cycle analysis of proliferative zone chondrocytes in growth plates elongating at different rates. *Journal of Orthopaedic Research* 14:562–572.

Wilsman, N. J., C. E. Farnum, E. M. Leiferman, M. Fry, and C. Barreto. 1996b. Differential growth by growth plates as a function of multiple parameters of chondrocytic kinetics. *Journal of Orthopaedic Research* 11:927–936.

Winnier, G., M. Blessing, P. A. Labosky, and B. L. Hogan. 1995. Bone morphogenetic protein-4 is required for mesoderm formation and patterning in the mouse. *Genes and Development* 9:2105–2116.

Wit, J.-M., and B. Boersma. 2002. Catch-up growth: Definition, mechanisms, and models. *Journal of Pediatric Endocrinology and Metabolism* 15:1229–1241.

Witmer, L. M. 1991. Perspectives on avian origins. In *Origins of the higher groups of tetrapods: Controversy and consensus,* ed. H.-P. Schultze and L. Trueb, 427–465. Ithaca, NY: Comstock.

Witmer, L. M. 2002. The debate on avian ancestry: Phylogeny, function, and fossils. In *Mesozoic birds: Above the heads of dinosaurs,* ed. L. M. Chiappe and L. M. Witmer, 3–30. Berkeley: University of California Press.

Witten, P. E. 1992. Beitrag zur Kenntnis der Entwicklung des Flossenskelettes bei *Cyprinodon* (Lacepede 1803) (Pisces, Cyprinodontidae). Diploma thesis, Zoological Institute and Zoological Museum, University of Hamburg.

Witten, P. E. 1995. Auf- und Abbau des azellulären Knochens bei Teleosteern. Histologische und enzymhistochemische Untersuchungen an *Oreochromis niloticus* (L. 1757) (Teleostei: Cichlidae). PhD diss., University of Hamburg.

Witten, P. E. 1997. Enzyme histochemical characteristics of osteoblasts and mononucleated osteoclasts in a teleost fish with acellular bone (*Oreochromis niloticus,* Cichlidae). *Cell and Tissue Research* 287:591–599.

Witten, P. E., and B. K. Hall. 2002. Differentiation and growth of kype skeletal tissues in anadromous male Atlantic salmon (*Salmo salar*). *International Journal of Developmental Biology* 46:719–730.

Witten, P. E., and B. K. Hall. 2003. Seasonal changes in the lower jaw skeleton in male Atlantic salmon (*Salmo salar* L.): Remodeling and regression of the kype after spawning. *Journal of Anatomy* 203:435–450.

Witten, P. E., A. Hansen, and B. K. Hall. 2001. Features of mono- and multinucleated bone resorbing cells of the zebrafish *Danio rerio* and their contribution to skeletal development, remodeling, and growth. *Journal of Morphology* 250:197–207.

Witten, P. E., L. S. Holliday, G. Delling, and B. K. Hall. 1999. Immunohistochemical identification of a vacuolar proton pump (V-ATPase) in bone-resorbing cells of an advanced teleost species (*Oreochromis niloticus*). *Journal of Fish Biology* 55:1258–1272.

Witten, P. E., and W. Villwock. 1997. Growth requires bone resorption at particular skeletal elements in a teleost fish with acellular bone (*Oreochromis niloticus,* Teleostei: Cichlidae). *Journal of Applied Ichthyology* 13:149–158.

Witten, P. E., W. Villwock, N. Peters, and B. K. Hall. 2000. Bone resorption and bone remodelling in juvenile carp, *Cyprinus carpio* L. *Journal of Applied Ichthyology* 16:254–261.

Witz, B. W., D. S. Wilson, and M. D. Palmer. 1991. Distribution of *Gopherus polyphemus* and its vertebrate symbionts in three burrow categories. *American Midland Naturalist* 126:152–158.

Wolfe, A. D., H. L. D. Nye, and J. A. Cameron. 2000. Extent of ossification at the amputation plane is correlated with the decline of blastema formation and regeneration in *Xenopus laevis* hindlimbs. *Developmental Dynamics* 218:681–697.

Wolfman, N. M., G. Hattersley, K. Cox, A. J. Celeste, R. Nelson, N. Yamaji, J. L. Dube, et al. 1997. Ectopic induction of tendon and ligament in rats by growth and differentiation factors 5, 6, and 7, members of the TGF-β gene family. *Journal of Clinical Investigation* 100:321–330.

Woloshin, P., K. Song, C. Degnin, A. M. Killary, D. J. Goldhamer, D. Sassoon, and M. J. Thayer. 1995. MSX1 inhibits myoD expression in fibroblast X 10T1/2 cell hybrids. *Cell* 82:611–620.

Wolpert, L. 1969. Positional information and the spatial pattern of cellular differentiation. *Journal of Theoretical Biology* 25:1–47.

Wolpert, L. 1989. Positional information revisited. *Development* 107 (suppl.): 3–12.

Wong, M., and D. R. Carter. 1990. A theoretical model of endochondral ossification and bone architectural construction in long bone ontogeny. *Anatomy and Embryology* 181:523–532.

Wood, A. 1982. Early pectoral fin development and morphogenesis of the apical ectodermal ridge in the killifish, *Aphyosemion scheeli. Anatomical Record* 204:349–356.

Wood, A., and P. Thorogood. 1984. An analysis of *in vivo* cell migration during teleost fin morphogenesis. *Journal of Cell Science* 66:205–222.

Wood, W., M. Turmaine, R. Weber, V. Camp, R. A. Maki, S. R. McKercher, and P. Martin. 2000. Mesenchymal cells engulf and clear apoptotic footplate cells in macrophage-less PU.1 null mouse embryos. *Development* 127:5245–5252.

Woodbury, A. M., and R. Hardy. 1948. Studies of the desert tortoise, *Gopherus agassizii. Ecological Monographs* 18:145–200.

Woodward, A. S. 1907. On a new dinosaurian reptile (*Scleromochlus taylori* gen. et sp. nov.) from the Trias of Lossiemouth, Elgin. *Quarterly Journal of the Geological Society of London* 63:140–144.

Wright, A. H., and A. A. Wright. 1949. *Handbook of frogs and toads of the United States and Canada.* 3rd ed. Ithaca, NY: Cornell University Press.

Wright, A. H., and A. A. Wright. 1957a. *Handbook of snakes of the United States and Canada.* Vol. 1. Ithaca, NY: Cornell University Press.

Wright, A. H., and A. A. Wright. 1957b. *Handbook of snakes of the United States and Canada.* Vol. 2. Ithaca, NY: Cornell University Press.

Wright, B. 2000. Form and function in aquatic flapping propulsion: Morphology, kinematics, hydrodynamics, and performance of the triggerfishes (Tetraodontiformes: Balistidae). PhD diss., University of Chicago.

Wright, E., M. R. Hargrave, J. Christiansen, L. Cooper, J. Kun, T. Evans, U. Gangadharan, A. Greenfield, and P. Koopman. 1995. The Sry-related gene Sox9 is expressed during chondrogenesis in mouse embryos. *Nature: Genetics* 9:15–20.

Wright, J. L., D. M. Unwin, M. G. Lockley, and E. C. Rainforth. 1997. Pterosaur tracks from the Purbeck Limestone Formation of Dorset, England. *Proceedings of the Geologists' Association* 108:39–48.

Wu, Q., Y. Zhang, and Q. Chen. 2001. Indian hedgehog is an essential component of mechanotransduction complex to stimulate chondrocyte proliferation. *Journal of Biological Chemistry* 276:35290–35296.

Wu, T. Y.-T. 1971a. Hydromechanics of swimming of fishes and cetaceans. *Advances in Applied Mechanics* 11:1–63.

Wu, T. Y.-T. 1971b. Hydromechanics of swimming propulsion. Part 1: Swimming of a two-dimensional flexible plate at variable forward speeds in an inviscid fluid. *Journal of Fluid Mechanics* 46:337–355.

Wu, T. Y.-T. 1971c. Hydromechanics of swimming propulsion. Part 2: Some optimum shape problems. *Journal of Fluid Mechanics* 46:521–544.

Wu, T. Y.-T. 1971d. Hydromechanics of swimming propulsion. Part 3: Swimming and optimum movements of slender fish with side fins. *Journal of Fluid Mechanics* 46:545–568.

X

Xu, X., Z.-L. Tang, and X. Wang. 1999a. A therizinosauroid dinosaur with integumentary structures from China. *Nature* 399:350–354.

Xu, X., X. Wang, and X.-C. Wu. 1999b. A dromaeosaurid dinosaur with a filamentous integument from the Yixian Formation of China. *Nature* 401:262–266.

Xu, X., M. Weinstein, C. Li, M. Naski, R. I. Cohen, D. M. Ornitz, P. Leder, and C. Deng. 1998. Fibroblast growth factor receptor 2 (FGFR2)–mediated reciprocal regulation loop between FGF8 and FGF10 is essential for limb induction. *Development* 125:753–765.

Xu, X., Z. Zhou, and R. O. Prum. 2001. Branched integumental structures in *Sinornithosaurus* and the origin of feathers. *Nature* 410:200–204.

Xu, X., Z. Zhou, and X. Wang. 2000. The smallest known non-avian theropod dinosaur. *Nature* 408:705–708.

Xu, X., Z. Zhou, X. Wang, X. Kuang, F. Zhang, and X. Du. 2003. Four-winged dinosaurs from China. *Nature* 421:335–340.

Y

Yablokov, A. V. 1974. *Variability of mammals.* New Delhi: Amerind.

Yada, Y., S. Makino, S. Chigusa-Ishiwa, and T. Shiroishi. 2002. The mouse polydactylous mutation, luxate (lx), causes anterior shift of the anteroposterior border in the developing hindlimb bud. *International Journal of Developmental Biology* 46:972–982.

Yalden, D. W. 1966. The anatomy of mole locomotion. *Journal of Zoology, London* 149:55–64.

Yamaguchi, T. P., A. Bradley, A. P. McMahon, and S. Jones. 1999. A Wnt5a pathway underlies outgrowth of multiple structures the vertebrate embryos. *Development* 126:1211–1223.

Yamaji, N., A. J. Celeste, R. S. Thies, J. S. Song, S. M. Bernier, D. Goltzman, K. M. Lyons, J. Nove, V. Rosen, and J. M. Wozney. 1994. A mammalian serine/threonine kinase receptor specifically binds BMP-2 and BMP-4. *Biochemical and Biophysical Research Communications* 205:1944–1951.

Yan, Y.-L., C. T. Miller, R. M. Nissen, A. Singer, D. Liu, A. Kirn, B. Draper, et al. 2002. A zebrafish *sox9* gene required for cartilage morphogenesis. *Development* 129:5065–5079.

Yang, E. V., D. M. Gardiner, M. R. J. Carlson, C. A. Nugas, and

S. V. Bryant. 1999. Expression of *Mmp-9* and related matrix metalloproteinase genes during axolotl limb regeneration. *Developmental Dynamics* 216:2–9.

Yang, Y., and L. Niswander. 1995. Interaction between the signaling molecules WNT7a and SHH during vertebrate limb development: Dorsal signals regulate anteroposterior patterning. *Cell* 80:939–47.

Yang, Y., G. Drossopoulou, P. T. Chuang, D. Duprez, E. Marti, D. Bumcrot, N. Vargesson, et al. 1997. Relationship between dose, distance and time in Sonic Hedgehog–mediated regulation of anteroposterior polarity in the chick limb. *Development* 124:4393–4404.

Yang, Y., L. Topol, H. Lee, and J. Wu. 2003. Wnt5a and Wnt5b exhibit distinct activities in coordinating chondrocyte proliferation and differentiation. *Development* 130:1003–1015.

Yasuda, H., N. Shima, N. Nakagawa, K. Yamaguchi, M. Kinosaki, S. Mochizuki, A. Tomoyasu, et al. 1998. Osteoclast differentiation factor is a ligand for osteoprotegerin/osteoclastogenesis-inhibitory factor and is identical to TRANCE/RANKL. *Proceedings of the National Academy of Sciences, USA* 95:3597–3602.

Yelon, D., B. Ticho, M. E. Halpern, L. Ruvinsky, R. K. Ho, L. M. Silver, and D. Y. Stainier. 2000. The bHLH transcription factor hand2 plays parallel roles in zebrafish heart and pectoral fin development. *Development* 127:2573–2582.

Yi, S. E., A. Daluiski, R. Pederson, V. Rosen, and K. M. Lyons. 2000. The type I BMP receptor BMPRIB is required for chondrogenesis in the mouse limb. *Development* 127:621–630.

Yokoi, H., T. Kobayashi, M. Tanaka, Y. Nagahama, Y. Wakamatsu, H. Takeda, K. Araki, K. I. Morohashi, and K. Ozato. 2002. sox9 in a teleost fish, medaka (*Oryzias latipes*): Evidence for diversified function of Sox9 in gonad differentiation. *Molecular Reproduction and Development* 63:5–16.

Yokouchi, Y., S. Nakazato, M. Yamamoto, Y. Goto, T. Kameda, et al. 1995. Misexpression of Hoxa-13 induces cartilage homeotic transformation and changes cell adhesiveness in chick limb buds. *Genes and Development* 9:2509–2522.

Yokouchi, Y., J. Sakiyama, T. Kameda, H. Iba, A. Suzuki, N. Ueno, and A. Kuroiwa. 1996. BMP-2/-4 mediates programmed cell death in chicken limb buds. *Development* 122:3725–3734.

Yokouchi, Y., H. Sasaki, and A. Kuroiwa. 1991. Homeobox gene expression correlated with the bifurcation process of limb cartilage development. *Nature* 353:443–445.

Yokoyama, H., H. Ide, and K. Tamura. 2001. FGF-10 stimulates limb regeneration ability in *Xenopus laevis*. *Developmental Biology* 233:72–79.

Yokoyama, H., S. Yonei-Tamura, T. Endo, J. C. Izpisúa Belmonte, K. Tamura, and H. Ide. 2000. Mesenchyme with *fgf-10* expression is responsible for regenerative capacity in *Xenopus* limb buds. *Developmental Biology* 219:18–29.

Yonei-Tamura, S., T. Endo, H. Yajima, H. Ohuchi, H. Ide, and K. Tamura. 1999a. FGF7 and FGF10 directly induce the apical ectodermal ridge in chick embryos. *Developmental Biology* 211:133–143.

Yonei-Tamura, S., K. Tamura, T. Tsukui, and J. C. Izpisúa Belmonte. 1999b. Spatially and temporally-restricted expression of two T-box genes during zebrafish embryogenesis. *Mechanisms of Development* 80:219–221.

Yoshioka, H., C. Meno, K. Koshiba, M. Sugihara, H. Itoh, Y. Ishimaru, T. Inoue, et al. 1998. Pitx2, a bicoid-type homeobox gene, is involved in a lefty-signaling pathway in determination of left-right asymmetry. *Cell* 94:299–305.

Young, J. Z. 1981. *The life of vertebrates.* 3rd ed. Oxford: Clarendon Press.

Yu, M., P. Wu, R. B. Widelitz, and C.-M. Chuong. 2002. BMP and SHH in the branching morphogenesis of feathers and the evolution of feather forms. *Nature* 420:308–312.

Yu, X. 1998. A new porolepiform-like fish, *Psarolepis romeri* gen. et sp. nov. (Sarcopterygii, Osteichthyes) from the Lower Devonian of Yunnan, China. *Journal of Vertebrate Paleontology* 18:261–274.

Z

Zákány, J., C. Fromental-Ramain, X. Warot, and D. Duboule. 1997. Regulation of number and size of digits by posterior Hox genes: A dose-dependent mechanism with potential evolutionary implications. *Proceedings of the National Academy of Sciences, USA* 94:13695–13700.

Zakeri, Z. F., and H. S. Ahuja. 1994. Apoptotic cell death in the limb and its relationship to pattern formation. *Biochemistry and Cell Biology* 72:603–613.

Zangerl, R. 1944. Contributions to the osteology of the skull of the Amphisbaenidae. *American Midland Naturalist* 31:417–454.

Zangerl, R. 1945. Contributions to the osteology of the postcranial skeleton of the Amphisbaenidae. *American Midland Naturalist* 33:764–780.

Zappalorti, R. T., E. W. Johnson, and Z. Leszczynski. 1983. The ecology of the northern pine snake, *Pituophis melanoleucus melanoleucus* (Daudin) (Reptilia, Serpentes, Colubridae), in southern New Jersey, with special notes on habitat and nesting behavior. *Bulletin of the Chicago Herpetological Society* 18:47–72.

Zardoya, R., and A. Meyer. 2001. Vertebrate phylogeny: Limits of inference of mitochondrial genome and nuclear rDNA sequence data due to an adverse signal/noise ratio. In *Major events in early vertebrate evolution,* Systematics Association Special Volume series no. 61, ed. P. E. Ahlberg, 135–155. London: Taylor and Francis.

Zauner, H., G. Begemann, M. Marí-Beffa, and A. Meyer. 2003. Differential regulation of msx genes in the development of the gonopodium, an intromittant organ, and of the "sword," a sexually selected trait of swordtail fishes (*Xiphophorus*). *Evolution and Development* 5:466–477.

Zelzer, E., D. J. Glotzer, C. Hartmann, D. Thomas, N. Fukai, S. Soker, and B. R. Olsen. 2001. Tissue specific regulation of VEGF expression during bone development requires Cbfa1/Runx2. *Mechanics of Development* 106:97–106.

Zeng, X., J. A. Goetz, L. M. Suber, W. J. Scott Jr., C. M. Schreiner, and D. J. Robbins. 2001. A freely diffusible form of Sonic hedgehog mediates long-range signalling. *Nature* 411:716–720.

Zenmyo, M., S. Komiya, R. Kawabata, Y. Sasaguri, A. Inoue, and M. Morimattsu. 1996. Morphological and biochemical evidence for apoptosis in the terminal hypertrophic chondrocytes of the growth plate. *Journal of Pathology* 180:430–433.

Zhang, H., and A. Bradley. 1996. Mice deficient for BMP2 are nonviable and have defects in amnion/chorion and cardiac development. *Development* 122:2977–2986.

Zhang, Z., X. Yu, Y. Zhang, B. Geronimo, A. Lovlie, S. H.

Fromm, and Y. Chen. 2000. Targeted misexpression of constitutively active BMP receptor-IB causes bifurcation, duplication, and posterior transformation of digit in mouse limb. *Developmental Biology* 220:154–167.

Zhao, Q., H. Eberspaecher, V. Lefebvre, and B. De Crombrugghe. 1997. Parallel expression of Sox9 and Col2a1 in cells undergoing chondrogenesis. *Development Dynamics* 209:377–386.

Zhou, Z. 1992. Preliminary report on a Mesozoic bird from Liaoning, China. *Chinese Science Bulletin* 37:1365–1368.

Zhou, Z., and X. Wang. 2000. A new species of *Caudipteryx* from the Yixian Formation of Liaoning, northeast China. *Vertebrata PalAsiatica* 38:111–127.

Zhou, Z., X. Wang, F. Zhang, and X. Xu. 2000. Important features of *Caudipteryx*: Evidence from two nearly complete specimens. *Vertebrata PalAsiatica* 38:241–254.

Zhou, Z., and F. Zhang. 2001. Two new ornithurine birds from the Early Cretaceous of western Liaoning, China. *Chinese Science Bulletin* 46:1258–1264.

Zhou, Z., and F. Zhang. 2002a. Largest bird from the Early Cretaceous and its implications for the earliest avian ecological diversification. *Naturwissenschaften* 89:34–38.

Zhou, Z., and F. Zhang. 2002b. A long-tailed, seed-eating bird from the Early Cretaceous of China. *Nature* 418:405–409.

Zhou, Z., and F. Zhang. 2002c. *Proceedings of the 5th symposium of the Society of Avian Paleontology and Evolution, Beijing, 1–4 June 2000*. Beijing: Science Press.

Zhu, M., and H.-P. Schultze. 2001. Interrelationships of basal osteichthyans. In *Major events in early vertebrate evolution*, ed. P. E. Ahlberg, 289–332. London: Taylor and Francis.

Zhu, M., X. Yu, and P. E. Ahlberg. 2001. A primitive sarcopterygian fish with an eyestalk. *Nature* 410:81–84.

Zhu, M., X. Yu, and P. Janvier. 1999. A primitive fossil fish sheds light on the origin of bony fishes. *Nature* 397:607–610.

Zimmer, C., and C. Buell. 1998. *At the water's edge*. New York: Simon and Schuster.

Zou, H., and L. Niswander. 1996. Requirement for BMP signalling in interdigital apoptosis and scale formation. *Science* 272:738–741.

Zou, H., R. Wieser, J. Massague, and L. Niswander. 1997. Distinct roles of type I bone morphogenetic protein receptors in the formation and differentiation of cartilage. *Genes and Development* 11:2191–2203.

Zug, G. R. 1971. Buoyancy, locomotion, morphology of the pelvic girdle and hindlimb, and systematics of cryptodiran turtles. *Miscellaneous Publications, Museum of Zoology, University of Michigan* 142:1–98.

Zug, G. R. 1972. Anuran locomotion: Structure and function. I: Preliminary observations on relation between jumping and osteometrics of appendicular and postaxial skeleton. *Copeia* 1972:613–624.

Zug, G. R. 1974. Crocodilian galloping: An unique gait for reptiles. *Copeia* 1974:550–552.

Zug, G. R. 1978. Anuran locomotion-Structure and function. 2: Jumping performance of semiaquatic, terrestrial, and arboreal frogs. *Smithsonian Contributions to Zoology* 276:1–31.

Zug, G. R., L. J. Vitt, and J. P. Caldwell. 2001. *Herpetology: An introductory biology of amphibians and reptiles*. San Diego, CA: Academic Press.

Zuñiga, A., A.-P. G. Haramis, A. P. McMahon, and R. Zeller. 1999. Signal relay by BMP antagonism controls the SHH/FGF4 feedback loop in vertebrate limb buds. *Nature* 401:598–602.

Zuñiga, A., R. Quillet, F. Perrin-Schmitt, and R. Zeller R. 2002. Mouse Twist is required for fibroblast growth factor-mediated epithelial-mesenchymal signalling and cell survival during limb morphogenesis. *Mechanisms of Development* 114:51–59.

Zuzarte-Luís, V., M. T. Berciano, M. Lafarga, and J. M. Hurlé. 2006. Caspase redundancy and release of mitochondrial apoptotic factors characterize interdigital apoptosis. *Apoptosis* 11(3).

Zuzarte-Luís, V., and J. M. Hurlé. 2002. Programmed cell death in the developing limb. *International Journal of Developmental Biology* 46:871–876.

Zuzarte-Luís, V., J. A. Montero-Simon, J. Rodriguez-Leon, R. Merino, J. C. Rodriguez-Rey, and J. M. Hurlé. 2004. Involvement of BMP5 in the control of interdigital apoptosis in the developing limb. *Developmental Biology* 272:39–52.

Zwilling, E., and L. Hansborough. 1956. Interaction between limb bud ectoderm and mesoderm in the chick embryo. III: Experiments with polydactylous limbs. *Journal of Experimental Zoology* 132:219–239.

Contributors

Marie-Andrée Akimenko
Ottawa Health Research Institute
725 Parkdale Avenue
Ottawa, Ontario K1Y 4E9
Canada
makimenko@ohri.ca

Charles W. Archer
Connective Tissue Biology Laboratories
School of Biosciences
Cardiff University
Museum Avenue
Cardiff, Wales CF10 3US
United Kingdom
Archer@Cardiff.ac.uk

Peter J. Bowler
School of Anthropological Studies
The Queen's University
Belfast, N. Ireland BT7 1NN
United Kingdom
pbowler@clio.arts.qub.ac.uk

Susan V. Bryant
University of California
School of Biological Sciences
100 BSA, MC 1450
Irvine, CA 92697
svbryant@uci.edu

Robert L. Carroll
Redpath Musuem
McGill University
859 Sherbrooke Street West
Montreal, Quebec H3A 2K6
Canada
Bob@Bio1.lan.Mcgill.ca

Michael I. Coates
Organismal Biology and Anatomy
University of Chicago
1027 E. 57th Street
Chicago, IL 60637
mcoates@midway.uchicago.edu

Jason P. Downs
Department of Geology and Geophysics
Yale University
New Haven, CT 06511
jason.downs@yale.edu

Gary P. Dowthwaite
School of Biosciences and Cardiff Institute of Tissue
Engineering and Repair
Cardiff University
Museum Avenue
Cardiff, Wales CF10 3US
United Kingdom
dowthwaite@cardiff.ac.uk

Eliot G. Drucker
Museum of Comparative Zoology
Harvard University
26 Oxford Street
Cambridge, MA 02138
eliot@washingtontrout.org

Cornelia E. Farnum
Department of Biomedical Sciences
College of Veterinary Medicine
Cornell University
Ithaca, NY 14853
cef2@cornell.edu

Philippa H. Francis-West
Department of Craniofacial Development
KCL, Guy's Tower, Floor 28
Guy's Hospital
London, England SE1 9RT
United Kingdom
pfrancis@hgmp.mrc.ac.uk

David M. Gardiner
Department of Developmental and Cell Biology and
Developmental Biology Center
University of California, Irvine
Irvine, CA 92697
dmgardin@uci.edu

Stephen M. Gatesy
Department of Ecology and Evolutionary Biology
Brown University
Box G-B209
Providence, RI 02912
Stephen_Gatesy@brown.edu

Brian K. Hall
Department of Biology
Dalhousie University
Halifax, Nova Scotia B3H 4J1
Canada
bkh@dal.ca

Robert B. Holmes
Paleobiology
Canadian Museum of Nature
PO Box 3443 Station D
Ottawa, Ontario K1P 6P4
Canada
RHolmes@mus-nature.ca

Juan M. Hurlé
Departamento de Anatomia y Biologia Celular
Facultad de Medicina
Universidad de Cantabria
C/ Cardenal Herrera Oria s/n
Santander 39011
Spain
hurlej@unican.es

Ann Huysseune
Vakgroep Biologie
Universiteit Gent
Ledeganckstraat 35
Gent B-9000
Belgium
Ann.Huysseune@UGent.be

Maureen Kearney
Department of Zoology
Field Museum of Natural History
Chicago, IL 60605
mkearney@fieldmuseum.org

Nathan J. Kley
Department of Anatomical Sciences
Health Sciences Center, T8 (069)
Stony Brook University
Stony Brook, NY 11794-8081
Nathan.Kley@stonybrook.edu

Hans C. E. Larsson
Redpath Museum
McGill University
859 Sherbrooke Street West
Montreal, Quebec H3A 2K6
Canada
hans.ce.larsson@mcgill.ca

Kevin M. Middleton
Department of Ecology and Evolutionary Biology
Brown University
Box G-B204
69 Brown Street
Providence, RI 02912
Kevin_Middleton@brown.edu

Lee Niswander
Howard Hughes Medical Institute
University of Colorado
Health Sciences Center
Developmental Biology
Building RD-1 North, Room P-18 4402E
12800 E. 18th Avenue
Aurora, CO 80010-7174
Lee.Niswander@uchsc.edu
and
Section of Developmental Biology
Department of Pediatrics
University of Colorado Health Sciences Center
12800 E. 19th Avenue
Aurora, CO 80045

Wendy M. Olson
Department of Biology
University of Northern Iowa
Cedar Falls, IA 50614
wendy.olson@uni.edu

P. David Polly
Department of Geological Sciences
Indiana University
1001 E. 10th Street
Bloomington, IN 47405-1405
pdpolly@indiana.edu

Marcello Ruta
Department of Earth Sciences
Wills Memorial Building
University of Bristol
Queen's Road, Bristol BS8 1RJ, UK
M.Ruta@bristol.ac.uk

Michael D. Shapiro
Department of Biology
University of Utah
257 South 1400 East
Salt Lake City, UT 84112
shapiro@biology.utah.edu

Neil H. Shubin
Department of Organismal Biology and Anatomy
University of Chicago
Chicago, IL 60637
nshubin@uchicago.edu

Amanda Smith
Ottawa Health Research Institute
725 Parkdale Avenue
Ottawa, Ontario K1Y 4E9
Canada

Adam P. Summers
Ecology and Evolutionary Biology
321 Steinhaus Hall
University of California
Irvine, CA 92697-2525
asummers@uci.edu

Mikiko Tanaka
Graduate School of Bioscience and Biotechnology
Tokyo Institute of Technology B-17
4259 Nagatsuta-cho, Midori-ku
Yokohama 226-8501
Japan

Michael A. Taylor
Department of Natural Sciences
National Museums of Scotland
Chambers Street
Edinburgh, Scotland EH1 1JF
United Kingdom
M.Taylor@nms.ac.uk

J. G. M. Thewissen
Department of Anatomy
Northeastern Ohio Universities College of Medicine
Box 95
4209 State Route 44
Rootstown, OH 44272
thewisse@neoucom.edu

Cheryll Tickle
Division of Cell and Developmental Biology
School of Life Sciences
University of Dundee
Dundee, Scotland DD1 5EH
United Kingdom
catickle@dundee.ac.uk

Matthew K. Vickaryous
Department of Biology
Dalhousie University
Life Sciences Center
1355 Oxford Street
Halifax, Nova Scotia B3H 4J1
Canada
mvickary@dal.ca

Gunter P. Wagner
Department of Ecology and Evolutionary Biology
Yale University
New Haven, CT 06511
gunter.wagner@yale.edu

Scott D. Weatherbee
Memorial Sloan Kettering Cancer Center
Developmental Biology Program
New York, NY 10021
s-weatherbee@ski.mskcc.org

P. Eckhard Witten
AKVAFORSK (The Institute of Aquaculture Research)
N-6600 Sunndalsøra, Norway
and
Department of Biology
Dalhousie University
Halifax, Nova Scotia B3H 4J1
Canada
eckhard.witten@akvaforsk.no

Vanessa Zuzarte-Luís
Departamento de Anatomia y Biologia Celular
Facultad de Medicina
Universidad de Cantabria
C/ Cardenal Herrera Oria s/n
Santander 39011
Spain
zuzartev@unican.es

Index